MAGNETOTAIL PHYSICS

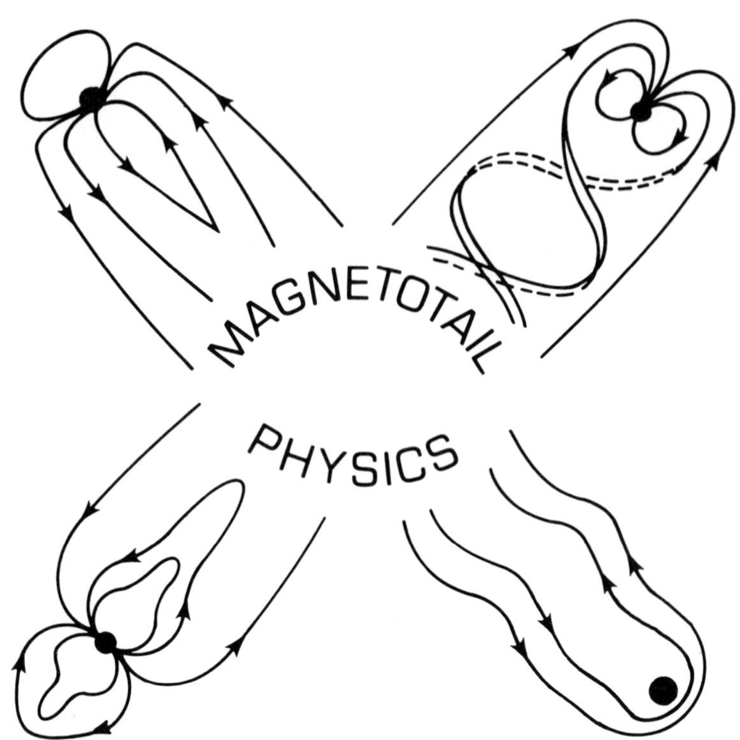

MAGNETOTAIL
PHYSICS

Magnetotail Physics

Edited by

Anthony T. Y. Lui

The Johns Hopkins University
Applied Physics Laboratory
Laurel, Maryland

THE JOHNS HOPKINS UNIVERSITY PRESS
BALTIMORE and LONDON

1987

The Johns Hopkins University Press, 701 West 40th Street,
Baltimore, Maryland 21211
The Johns Hopkins Press Ltd., London

The paper used in this publication meets the minimum requirements of American National Standard for Information Sciences—Permanence of Paper for Printed Library Materials, ANSI Z39.48-1984.

Library of Congress Cataloging-in-Publication Data

Magnetotail Physics.

   (Johns Hopkins studies in Earth and planetary sciences)
   Papers from a conference held at The Johns Hopkins University Applied Physics Laboratory, Oct. 28–31, 1985, which was organized by the American Geophysical Union.
   1. Magnetotails—Congresses. I. Lui, Anthony T. Y.
II. American Geophysical Union. III. Series.
QC809.M35M335 1987   538′.7   86-27614
ISBN 0-8018-3496-1 (alk. paper)

PREFACE

Magnetotails seem to be found universally in association with celestial objects. The most popularly known magnetotails are the tails of comets. Akin to comets, all planets that we have explored so far possess magnetotails whose volumes are generally many thousands to millions of times greater than those of the planets themselves. Magnetotails have also been considered as nearby space laboratories, accessible to *in situ* measurements and bearing prototype physical processes important in a large variety of astrophysical situations.

Naturally, the most extensively surveyed of all magnetotails is the earth's, for which the first direct measurements began more than two decades ago. Magnetotail research has advanced over this time from mere theoretical speculation and space exploration to a phase of in-depth theoretical analysis, numerical simulations, and active experimentation.

A conference on the subject was held at The Johns Hopkins University Applied Physics Laboratory on October 28-31, 1985. It was organized through the American Geophysical Union. The Program Committee consisted of

S.-I. Akasofu	Geophysical Institute, University of Alaska
D. N. Baker	Los Alamos National Laboratory
D. H. Fairfield	Goddard Space Flight Center/NASA
L. A. Frank	The University of Iowa
A. T. Y. Lui	Applied Physics Laboratory, The Johns Hopkins University
L. R. Lyons	The Aerospace Corporation
G. K. Parks	Washington University
T. Sato	Hiroshima University
S. D. Shawhan	NASA Headquarters
B. T. Tsurutani	Jet Propulsion Laboratory
D. J. Williams	Applied Physics Laboratory, The Johns Hopkins University

The conference was entitled the Chapman Conference on Magnetotail Physics to honor the late Sydney Chapman (1888 1970) who pioneered in many scientific disciplines. It brought together one hundred seventy scientists from the United States, Canada, Japan, and several European nations.

This monograph is a compendium based on "a meeting of minds" of the conferees. It is assembled to reflect the main conference goals, which are:

1. To consolidate our understanding of the earth's magnetotail, obtained from the many spacecraft missions since the advent of artificial satellites in space exploration about three decades ago,
2. To promote the utilization of computer simulation as a tool for advancing our comprehension of complex magnetotail systems,
3. To enhance the interchange of knowledge resulting from research on the earth's magnetotail and similar systems in space and in the laboratory, and
4. To encourage young scientists to participate and identify future research in this specialized discipline.

The content, primarily based on the presentations at the conference, is organized into eight chapters. Chapter One provides a brief introduction to various magnetotail domains and a historical perspective of the subject. The formation of the magnetotail and the basic processes operating within it are dealt with in Chapter Two. Chapter Three covers the fluid-like aspects of magnetotail dynamics. Chapter Four addresses the behavior of individual particles and how their collective effects modify the system. The results of active experiments in the magnetotail and of the creation of artificial comet tails are highlighted in Chapter Five. Chapter Six presents investigations of planetary and cometary magnetotail systems. A dialog in the form of a panel debate on three controversial topics constitutes Chapter Seven. The final chapter summarizes the achievements of the conference and suggests problems to be tackled in the immediate future. Tutorial articles, which are derived from kick-off talks given during the meeting (with one exception in Chapter Three), are included on each of the topics. This monograph as a whole is designed to be suitable as a reference book for graduate courses related to magnetospheric, space plasma, and astrophysics research.

The specialty of magnetotail research is still a relatively young science. Our present perception of magnetotail physics is achieved in the majority of cases through consistency arguments, mostly because of our inability to conduct controlled investigations in space. It is prudent to maintain a certain degree of skepticism in reading scientific articles, including those from leading authorities in the field. There is an old Chinese proverb that says "Accepting what you read with complete faith is worse than not reading at all." Even the most widely held view on a scientific issue is occasionally not the correct one.

Many individuals have contributed significantly to the success of the conference and to the completion of the monograph. I am indebted to the technical and organizational support provided by the American Geophysical Union staff (B. Weaver, C. L. Bravo, and A. F. Spilhaus, Jr.). The conference staff, consisting of B. A. Northrop, K. L. Blankenship, V. J. Franke, B. L. Laub, J. F. Patarini, G. W. Snyder, and L. C. Sussman, admirably handled a multitude of conference arrangements, last-minute changes, and numerous nearly impossible tasks. The projectionists, G. A. Bennett and W. Hill, demonstrated their superb professionalism in providing audiovisual support during the meeting. I am particularly thankful for the outstanding and dedicated secretarial assistance of K. L. Blankenship who typed and sent out hundreds of letters during the preparations for both the conference and the monograph. The Word Processing Center staff (B. A. Northrop, V. J. Franke, B. L. Laub, L. C. Sussman, and S. L. Testerman), the staff of the Technical Publications Group (M. W. Burgan, J. W. Kaufman, A. L. Machurek, J. Elbaz, B. L. Bankert, N. L. Zepp, and P. H. Zurvalec), and the editorial staff at The Johns Hopkins University Press (A. Richter and J. S. Johnston) did a marvelous and accurate transformation of manuscripts of heterogeneous type styles into a uniform and attractive monograph. I deeply appreciate the efforts and prompt responses of all the referees who helped in improving the quality of the manuscripts published here. In particular, the members of the Editorial Board, S.-I. Akasofu, D. N. Baker, L. R. Lyons, J. A. Slavin, G. L. Siscoe, and D. J. Williams, have kindly volunteered their valuable time to provide me with guidance and advice during the review process. I have also benefitted from many useful exchanges with my APL colleagues, T. A. Potemra, S. M. Krimigis, K. B. Baker, B. H. Mauk, D. G. Sibeck, and L. J. Zanetti. I am grateful to the Air Force Geophysics Laboratory, the University of California Institute of Geophysics and Planetary Physics, the National Aeronautics and Space Administration, and the National Science Foundation for financial support for the publication of this monograph.

August 1986

Anthony T. Y. Lui
Applied Physics Laboratory,
The Johns Hopkins University

MAGNETOTAIL PHYSICS

Table of Contents

III. Fluid Aspects of Magnetotail Dynamics

IV. Kinetic Aspects of Magnetotail Dynamics

Color Plates

I. INTRODUCTION

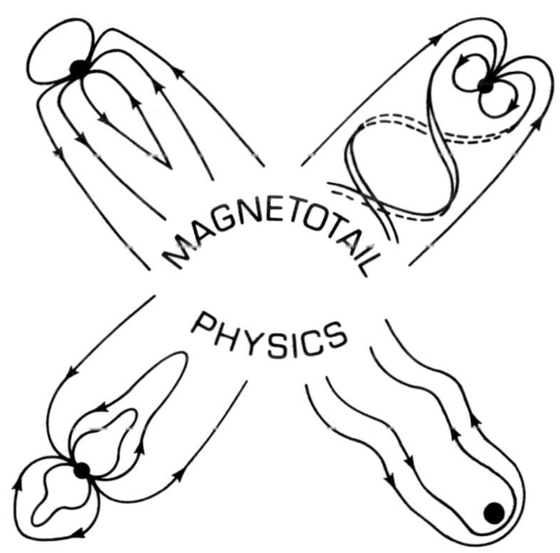

ROAD MAP TO MAGNETOTAIL DOMAINS

A. T. Y. Lui*

A definition is introduced that is applicable to the two generic classes of magnetotails that are now recognized. The earth's magnetotail is the most extensively surveyed of all magnetotails and is used as a prototype to illustrate the various magnetotail plasma domains. Typical values and variations of magnetotail parameters are discussed briefly.

WHAT IS A MAGNETOTAIL?

A comet presents the most visible and popularly known magnetotail, having a length that is usually many hundreds of times larger than the associated comet coma. The presence of an extended tail region attached to a celestial body is not unique to comets but seems in fact to be universal to large celestial objects embedded in a medium of flowing magnetized plasma. In general, a magnetotail may be defined as a region in space created by the interaction between the intrinsic magnetic field or ionized atmosphere of a large-scale object and a flowing magnetized plasma. It is in the downwind region of the object and contains a magnetic field configuration in which the field is oriented predominantly along the direction of flow of the incident magnetized medium.

TWO CLASSES OF MAGNETOTAILS

The most extensively surveyed of all magnetotails is that of the earth. It is formed by the outflowing plasma from the sun—the solar wind—interacting with the earth's intrinsic magnetic field. The interaction results in an electric current system that distorts the geomagnetic field and creates a magnetic field configuration in which the field points predominantly parallel or antiparallel to the incident solar wind flow, a configuration commonly referred to as tail-like. The earth's magnetotail may be considered to be an extension of the atmosphere, elongated in the downwind direction, not only because the magnetic field in the magnetotail can often be traced back to the earth but also because a significant portion of the particle population in the magnetotail originates from the earth. The earth's magnetotail represents the class of magnetotails that are associated with celestial objects possessing a substantial intrinsic magnetic field. Comets, on the other hand, represent the other class of magnetotails—the class in which a tail-like magnetic field configuration is formed by impeding the flow of magnetized medium in the vicinity of the object. In the comet class, the ionized atmosphere of the object rather than the intrinsic magnetic field (as in the case of the earth) represents the obstacle to the flowing medium.

*The Johns Hopkins University Applied Physics Laboratory, Laurel, Maryland 20707.

Since the earth's magnetotail has been surveyed in more detail than all other magnetotails, we shall use it as a prototype to describe the plasma domains commonly found within a magnetotail. The magnetotail structures in the near-earth portion, at less than about 60 R_e (R_e = earth radius) from earth, are described first, followed by discussions on major differences that are notable in the more distant magnetotail.

MAGNETOSPHERE

Plate I-1 illustrates the various plasma domains within the earth's magnetotail. The solar-wind flow is obstructed and becomes deflected by the earth's magnetic field, resulting in the formation of a cavity, known as the earth's magnetosphere, carved out in the solar-wind stream. The magnetotail is the cylindrically-shaped portion of the nightside magnetosphere. Because of the high speed of the solar wind, exceeding the local sound or Alfvén speed, a shock front called the bow shock is formed ahead of the magnetosphere. Behind the bow shock is a region known as the magnetosheath where the solar-wind flow is reduced and the particle population of the solar wind is thermalized. Adjacent to the magnetosheath is the magnetopause which defines the outer surface of the magnetosphere. As illustrated in Plate I-1, on the dayside at high latitudes, there is an abrupt transition between magnetic field lines that permeate only the dayside magnetosphere and adjacent field lines at higher latitudes that extend downstream toward the magnetotail. This demarcation gives rise to an indentation region on the dayside magnetopause called the cleft or the polar cusp.

MAGNETOPAUSE BOUNDARY LAYER

The magnetopause is an imperfect shield that allows a small fraction ($\leq 1\%$) of magnetosheath plasma to cross the magnetopause and enter into the magnetosphere. The intruded plasma forms a layer called the magnetopause boundary layer. One entry location is adjacent to the cleft region and, accordingly, the magnetopause boundary layer at that point is called the entry layer. Magnetosheath plasma enters rather freely across the magnetopause in the entry layer and penetrates deep into the low-altitude region of the magnetosphere before retreating to higher altitudes into the magnetotail.

A further distinction is often made between the magnetopause boundary layer at high latitudes in the magnetotail, termed the plasma mantle, and that at lower latitudes at the flanks of the magnetopause, termed the low-latitude boundary layer. An important difference between the latter two regions of the magnetopause boundary layer is that field lines in the plasma mantle are open, i.e., field lines when traced in the downstream direction are finally linked with the interplanetary magnetic field in the solar wind, whereas field lines in the low-latitude boundary layer are closed, i.e., field lines when traced in the downstream direction eventually return to the earth in the other half of the magnetotail.

TAIL LOBE, PLASMA SHEET, AND NEUTRAL SHEET

A large portion of the magnetotail consists of two low-density regions known as the tail lobes, one in the northern half of the magnetotail and the other in the southern half. The field lines threading through the tail lobes are also open. Particles populating this region include ions from the cleft region as mentioned previously, ions from the polar region at low altitudes, and electrons from the solar wind entered into the tail lobe on open field lines. The energy density in the tail lobe is still dominated by the magnetic field.

Bordering the tail lobe at its lower latitude interface is the plasma-sheet boundary layer. This region is often the most dynamic plasma domain of the magnetotail, where ion beams coming from the earth and from further downstream are often found. It is also where a lot of plasma wave activities in the magnetotail are detected. Magnetic field-aligned currents, flowing toward or away from the earth, are often observed in this region. These activities gradually diminish as one approaches the center plane of the magnetotail where the central plasma sheet resides. The plasma sheet as a whole (i.e., the central plasma sheet and the plasma-sheet boundary layer) is reported to be thinnest near the midnight region and is about twice as thick near the flanks of the magnetotail. The plasma energy density is comparable to the magnetic field energy density in the plasma-sheet region.

In the middle of this reservoir of particles in the plasma sheet lies the neutral sheet where the magnetic field is weak (a few nanoteslas). The field reverses direction from pointing sunward to pointing tailward or vice versa as the neutral sheet is crossed. The region of the neutral sheet is considered to be an ideal site for a process called magnetic reconnection, where oppositely directed magnetic field lines are brought together and the associated magnetic-field energy is dissipated by accelerating charged particles. Magnetic reconnection is a prime candidate to account for many impulsive energy-release phenomena observed by space and astrophysics investigations.

Complex field structures, such as flux ropes and magnetic islands, are occasionally observed. Flux ropes are associated with strong currents along magnetic field lines, giving rise to the field lines spiralling like the fibers of a rope. Magnetic islands, in which the entrapped magnetic field lines form closed loops, can be considered as a degenerate case of magnetic flux ropes.

DISTANT MAGNETOTAIL

In general, all the plasma regimes and some complex magnetic structures found in the near-earth region can be identified, by and large, in the distant magnetotail (~ 100–200 R_e) also. For instance, inference on the presence of magnetic islands with lengths 50–100 R_e, also known as plasmoids, have been reported in the distant magnetotail. In addition, influence and penetration of the solar wind are more vividly seen in the distant magnetotail than in the near-earth portion. Large lateral motion of the distant magnetotail can result from slight changes in the incident flow direction of the solar wind, analogous to the motions of a windsock. The entry location of magnetosheath plasma in the tail lobe and the north-south dimension of the distant magnetotail are found to be dependent on the dawn-dusk component of the interplanetary magnetic field in the solar wind. Flux ropes having their axes aligned nearly in the direction of the magnetotail axis have also been reported. The clear correspondence between the near-earth magnetotail and the distant magnetotail appears to be lost at the very distant region. Beyond about 500 R_e downstream, the magnetotail is barely detectable because of the special orientation of the magnetotail magnetic field, i.e., sunward or antisunward.

SPATIAL VARIATIONS OF MAGNETOTAIL PARAMETERS

More quantitative variations of the magnetic field strength, number density, flow speed, and ion temperature across the different regions of the magnetotail are illustrated in Fig. 1. The profiles are given in terms of the distance from the neutral sheet so that the approximate thickness of the various magnetotail domains can be determined by the scale provided. The values given are appropriate for the magnetotail in the downstream distances of about 20–30 R_e. Although the exact values of these quantities vary somewhat with the downstream distance, the trends shown remain valid throughout the magnetotail. The dotted portions of the curves denote a general lack of measurement and thus a certain amount of uncertainty.

The magnetic field profile shown in Fig. 1 indicates that the field strength is highest in the tail lobe and lowest in the neutral sheet. The number density has the opposite trend. It is lowest in the tail lobe and highest in the neutral sheet, although its value there is still less than that in the magnetosheath. The number density in the plasma mantle is found to decrease with increasing separation from the magnetopause. High plasma flow speeds are detected typically in the plasma-sheet boundary layer with greatly reduced speeds seen in the central plasma sheet. The plasma mantle shows a gradual transition from the high flow speeds in the magnetosheath to vanishingly low flow speeds toward the tail lobe, much like

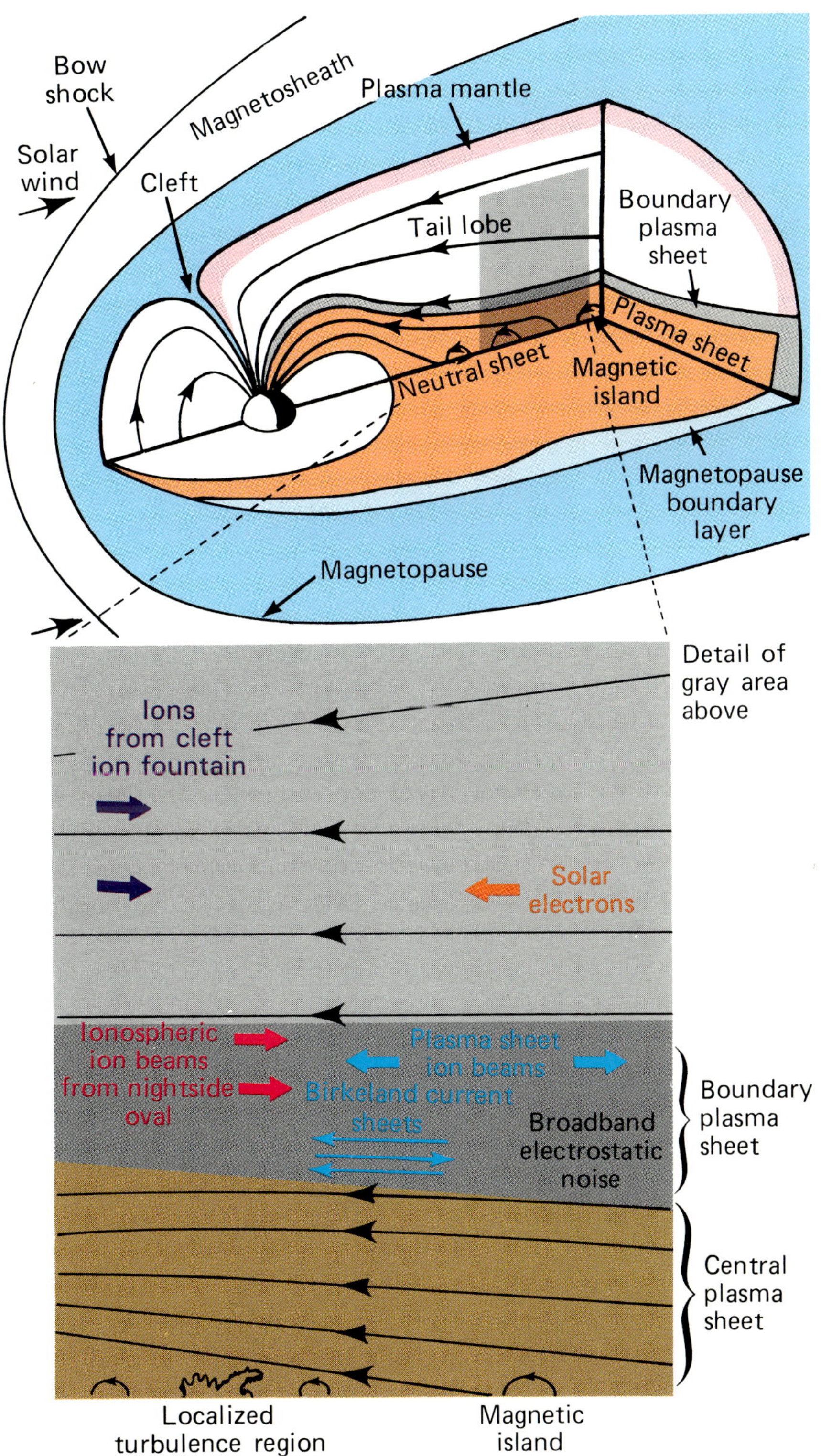

Plate I-1—A three-dimensional drawing of the magnetotail, exhibiting the plasma domains within.

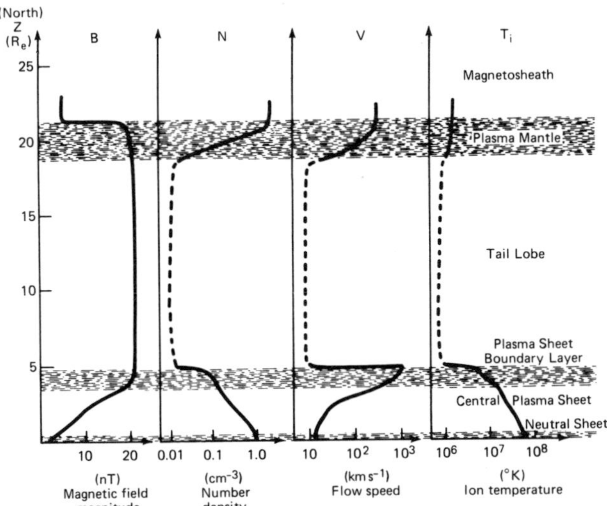

Figure 1—North-south profiles of magnetic field (*B*), number density (*N*), plasma bulk flow speed (*V*), and ion temperature (T_i) in the midnight region of the magnetotail.

the trend seen in the number density variation in the plasma mantle. The highest ion temperature occurs in the central plasma sheet but takes up lower values at the plasma-sheet boundary layer, which is still above the temperatures sensed in the plasma mantle region.

The electron temperature, although not shown, is generally lower than the ion temperature, but its variation across different magnetotail domains is very similar to that of the ion temperature. It is to be cautioned that the general structures of the magnetotail illustrated in Plate I-1 and the profiles given in Fig. 1 portray only a general picture of the magnetotail. Notable deviations can occassionally be found. In particular, there are observations that suggest that the plasma-sheet boundary layer may intrude into the midnight section of the tail lobe and bisect it when the interplanetary magnetic field in the solar wind has a strong northward component.

ELECTRICAL CURRENT SYSTEM

The tail-like geometry of the magnetotail is maintained by a large electric current system. The currents are driven by the dynamo action derived from plasma flowing across a magnetic field. Such a condition is almost always realized in the magnetopause, plasma mantle, and the low-latitude boundary layer. One important part of the current circuit is the current flowing from dusk to dawn on both the north and south halves of the magnetotail magnetopause. This current closes by flowing within the plasma sheet and perhaps also in the polar region, linked by magnetic field-aligned currents. Similarly, another part of the current circuit involves the low-latitude boundary layer as the dynamo region driving a dawn-to-dusk current in the midplane of the magnetotail and its closure is in the low-altitude polar region via magnetic field-aligned currents also. These currents are shown schematically in Fig. 2.

PLASMA CONVECTION

The interaction of the geomagnetic field with the solar wind also results in a dynamo that drives the plasma within the magnetotail into a large-scale circulation pat-

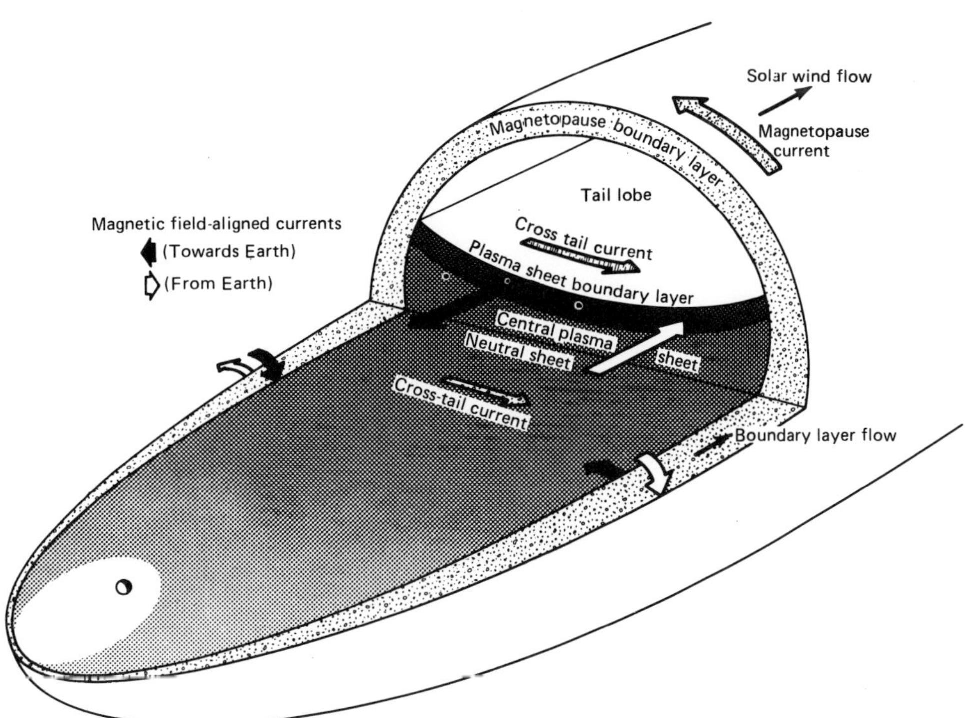

Figure 2—Schematic representation of the electrical current system flowing within the magnetotail.

Table 1—Typical values of some parameters in the earth's magnetotail.

		Magnetosheath	Plasma mantle	Tail lobe	Plasma-sheet boundary layer	Central plasma sheet	Neutral sheet
Characteristics							
number density	n (cm^3)	5	1	10^{-2}	10^{-1}	5×10^{-1}	1
ion temperature	T_i (K)	10^6	10^6	10^6	10^7	5×10^7	5×10^7
electron temperature	T_e (K)	5×10^5	5×10^5	5×10^5	5×10^6	10^7	10^7
magnetic field	B (nT)	5	25	25	20	10	2
electric field	E (mV/m)	1	1	10^{-1}	10^{-1}	10^{-2}	10^{-2}
Lengths							
Debye length	$\lambda_D \propto (T/n)^{1/2}$ (m)	1.8×10^1	4.0×10^1	4.0×10^2	4.0×10^2	2.8×10^2	2.0×10^2
plasma skin depth	$c/\omega_{pe} \propto n^{-1/2}$ (m)	2.4×10^3	5.3×10^3	5.3×10^4	1.7×10^4	7.5×10^3	5.3×10^3
electron gyroradius	$\rho_e \propto T_e^{1/2} B^{-1}$ (R$_e$)	4.9×10^{-4}	9.8×10^{-5}	9.8×10^{-5}	3.9×10^{-4}	1.1×10^{-3}	5.5×10^{-3}
ion gyroradius	$\rho_i \propto T_i^{1/2} B^{-1}$ (R$_e$)	3.0×10^{-2}	5.9×10^{-3}	5.9×10^{-3}	2.3×10^{-2}	1.0×10^{-1}	5.2×10^{-1}
Frequencies							
electron gyrofrequency	$f_{ce} \propto B$ (Hz)	1.4×10^2	7.0×10^2	7.0×10^2	5.6×10^2	2.8×10^2	5.6×10^1
ion gyrofrequency	$f_{ci} \propto B$ (Hz)	7.6×10^{-2}	3.8×10^{-1}	3.8×10^{-1}	3.0×10^{-1}	1.5×10^{-1}	3.0×10^{-2}
electron plasma frequency	$f_{pe} \propto n^{1/2}$ (Hz)	2.0×10^4	9.0×10^3	9.0×10^2	2.8×10^3	6.3×10^3	9.0×10^3
ion plasma frequency	$f_{pi} \propto n^{1/2}$ (Hz)	4.7×10^2	2.1×10^2	2.1×10^1	6.6×10^1	1.5×10^2	2.1×10^2
electron plasma/gryofrequency ratio	$f_{pe}/f_{ce} \propto n^{1/2} B^{-1}$	1.4×10^2	1.3×10^1	1.3×10^0	5.0×10^0	4.5×10^1	1.6×10^2
Velocities							
ion sound speed	$c_s \propto T_e^{1/2}$ (km s^{-1})	8.3×10^1	8.3×10^1	8.3×10^1	2.6×10^2	3.7×10^2	3.7×10^2
Alfvén speed	$v_A \propto Bn^{-1/2}$ (km s^{-1})	4.9×10^1	5.5×10^2	5.5×10^3	1.4×10^3	3.0×10^2	4.4×10^1
electron thermal speed	$vT_e \propto T_e^{1/2}$ (km s^{-1})	2.8×10^3	2.8×10^3	2.8×10^3	8.7×10^3	1.2×10^4	1.2×10^4
ion thermal speed	$vT_i \propto T_i^{1/2}$ (km s^{-1})	9.1×10^1	9.1×10^1	9.1×10^1	2.9×10^2	6.4×10^2	6.4×10^2
convection speed	$v_E \propto EB^{-1}$ (km s^{-1})	2.0×10^2	4.0×10^0	4.0×10^0	5.0×10^0	1.0×10^0	5.0×10^0
Miscellaneous							
thermal/magnetic energy ratio	$\beta \propto nT_i B^{-2}$	6.9×10^0	5.6×10^{-2}	5.6×10^{-4}	8.7×10^{-2}	8.8×10^0	4.3×10^2
magnetic energy density	$W_B \propto B^2$ (J m^{-3})	9.9×10^{-18}	2.5×10^{-16}	2.5×10^{-16}	1.6×10^{-16}	4.0×10^{-17}	1.6×10^{-18}

tern. The large-scale plasma convection perpendicular to the magnetic field is intimately associated with a large-scale electric field pattern.

Dominant plasma convection in the midplane of the magnetotail is illustrated in Fig. 3a. At distances close to the earth, the plasma is flowing predominantly sunward, with the exception of the magnetopause boundary layers where the flow is antisunward. Between ~60 and 180 R_e, the plasma convection in the central region of the magnetotail is often mixed between sunward and tailward. Beyond ~180 R_e, the convection direction is mainly downstream. Figure 3b is a schematic diagram of the plasma convection on the plane normal to the magnetotail axis within ~60 R_e downstream. At high latitudes, the convection is toward the central plane, whereas in the central and boundary plasma-sheet regions, the convection component is toward the flanks of the magnetotail. These general patterns can be profoundly altered when activities in the magnetotail are enhanced or when the interplanetary magnetic field in the solar wind is strongly northward. Furthermore, there is mounting evidence that plasma convection in the near-earth region (<15 R_e downstream) is probably not a steady-state process.

PLASMA WAVES

Several types of plasma waves are detected in the magnetotail. They are a means for energy dissipation in the collisionless plasma of the magnetotail and can provide important clues to the plasma processes operative in the magnetotail. Below is a brief account of some waves observed across the magnetotail.

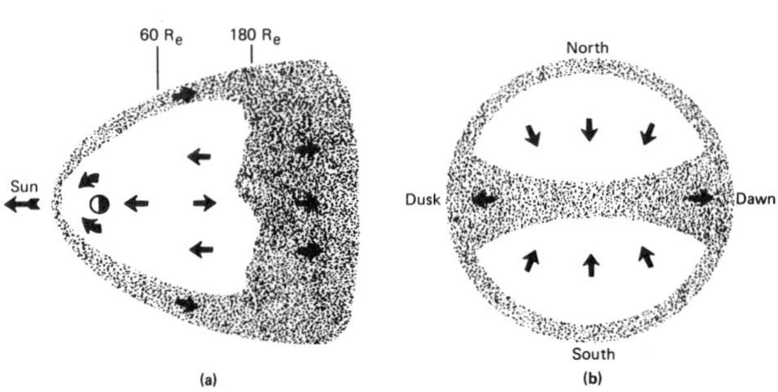

Figure 3—(a) Plasma convection within the equatorial plane of the magnetotail. (b) Component of plasma convection on a cross-sectional plane of the magnetotail.

In the magnetosheath, signatures of nonoscillatory drift mirror waves have been reported. Inside the magnetopause, an electromagnetic continuum radiation with frequency above the local electron plasma frequency is observed. In the plasma-sheet boundary layer, broadband electrostatic emissions, whistler magnetic bursts, and magnetosonic waves are detected. The more common broadband electrostatic emission occurs in the frequency range from about 10 Hz (near the lower hybrid frequency) to a few kilohertz (close to the local electron gyrofrequency). Whistler magnetic bursts consist of nearly monochromatic tones lasting from a few seconds to a few tens of seconds. The frequency varies rapidly with time, but it remains below the local electron gyrofrequency. Magnetosonic waves are detected in the distant plasma-sheet boundary layer where signatures of slow mode shocks have also been reported. Within the central plasma sheet, the wave emission, which tends to be associated with hot plasma and energetic processes, is the electrostatic electron cyclotron waves that occur near the electron gyrofrequency and its harmonics. These waves are thought to be responsible for dissipation well above the classic resistivity expected for the collisionless plasma in the magnetotail. The neutral-sheet region is usually quiescent in terms of wave activity.

VALUES OF MAGNETOTAIL PARAMETERS

Table 1 provides values for some important physical parameters relevant to the study of the various magnetotail domains. These include several fundamental quantities for lengths, frequencies, velocities, and some miscellaneous parameters. The basic characteristics in each magnetotail region assumed in deriving the values quoted are listed at the top of the table under each magnetotail domain. Protons are assumed to be the ion species. Naturally, the exact values measured in a particular location of the magnetotail at a particular time may vary considerably dependent on a large number of factors, but nevertheless, the tabulated values are useful as a guide. Furthermore, the dependencies of these values on the characteristic parameters are shown in the table so that more appropriate values can be obtained readily.

FINAL REMARKS

In closing, it is appropriate to mention that the structure and dynamics observed in the earth's mangetotail generally transpire to other magnetotails as well, including those of comets. The similarities and differences between various magnetotail systems are still being actively investigated. Nonetheless, there is little doubt that the earth's magnetotail serves as a good starting and reference point in our pursuit to understand all magnetotails in the universe. It is a window for the earthling to gain some perspective of physical processes in the cosmos.

ACKNOWLEDGMENT—The author is grateful to B. H. Mauk for valuable comments on the manuscript. This work is supported by the Atmospheric Sciences Section of the National Science Foundation, grant ATM-8611354 to The Johns Hopkins University.

MAGNETOTAIL RESEARCH: THE EARLY YEARS

N. F. Ness*

INTRODUCTION

The dictionary definition of "early" is: (a) near the beginning of a period of time and (b) in a distant past time. In preparing this talk, I felt that both forms of this definition apply equally well to my topic. Indeed, it is now two decades since the first publication of results from the IMP-1 spacecraft entitled "The Earth's Magnetic Tail"[1] appeared. This marked the beginning of a concerted effort to study a most unique feature of our terrestrial space environment formed by the solar wind interaction with Earth's magnetic field: the geomagnetotail. Since then, spacecraft of the USA have discovered and explored the magnetic tails of other magnetized planets in the solar system: Mercury, Jupiter, and Saturn. The discovery and study of the geomagnetic tail have been principally the result of USA launched spacecraft, and the USSR did not play any significant role.

In reviewing the subject area, its publication record, and significant conceptual developments, I have concluded that the decade from approximately 1961-71 represents that period of time in geomagnetotail research that is well described by the phrase "The Early Years." An excellent contemporary overview of plasmas in the Earth's magnetotail has recently appeared.[2]

It will not be possible to include a detailed description of the many individual research efforts, both theoretical and observational, that were conducted throughout this period. Nor will it be possible to refer explicitly to all those spacecraft whose scientific data have contributed to research on Earth's magnetic tail. And, finally, it is difficult to discuss in perfect chronological sequence these developments, while also maintaining a coherent and brief overview. I ask the reader to be forbearing in his criticisms on this last issue.

Figure 1 summarizes the history of space exploration launch activities from 1958-69 and indicates the large number of launches and hence the vigorous nature of the USA and USSR space programs at that time. Notice should be made of the very short intervals of operation of most of the spacecraft launched in the first half of that period. Subsequent technological improvements in spacecraft systems' reliability extended operational lifetimes notably.

In the overview that is presented, much of the research was conducted with my colleagues, K. W. Behannon and D. H. Fairfield, who contributed very significantly in many ways to our present-day understanding of the geomagnetic tail.

---

*Goddard Space Flight Center, Greenbelt, Maryland 20771.

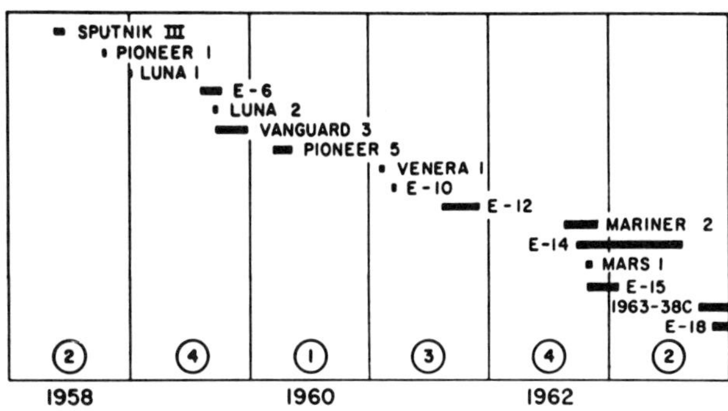

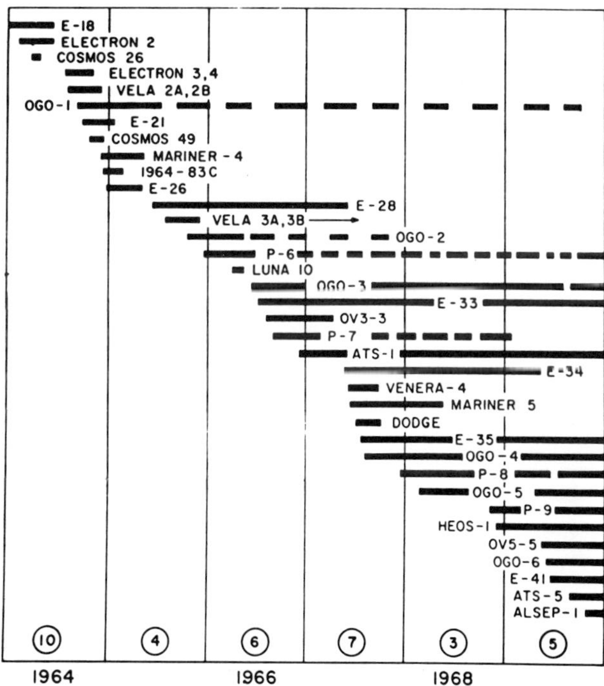

Figure 1—1958-69 launch history and operational periods of spacecraft carrying magnetometers and plasma and energetic particle detectors.[3] E represents Explorer, P = Pioneer, and OGO = Orbiting Geophysical Observatory. The number following indicates the sequential ordering of the spacecraft in that series.

PRE-IMP 1 (EXPLORER 18) PERIOD: 1958-63

The geomagnetic tail is formed as a direct result of the interaction of Earth's magnetic field with the continuous flux of low-energy solar coronal plasmas known as the solar wind. The existence of the solar wind was established in 1961-62 with the plasma measurements of Explorer 10[4] and Mariner 2.[5] (See Ref. 6 for a sum-

"

mary of the early studies, mainly theoretical, on the solar wind.)

Whether Earth's magnetic field in the surrounding space would be confined to a "closed" region or extended to an "open" region as a result of its interaction with the solar wind was among the early topics considered by theoreticians.[7-9] Figure 2 illustrates the nature of the dichotomy between the closed and open magnetosphere topologies. The upper portion is similar to that of the teardrop model of the magnetosphere proposed by Johnson.[8] The lower portion of the figure portrays the interconnection between an interplanetary magnetic field and the geomagnetic field as postulated by Dungey.[7] Figure 3 shows the development proposed by Piddington,[9] in which the transient formation of a magnetic tail was invoked to explain the origin of the main phase of a geomagnetic storm.

Paralleling the appearance of these qualitative sketches of concepts for the distortion of the geomagnetic field by the solar wind, theoretical computations to determine quantitatively the distorted magnetic field in the geomagnetic "cavity" thus formed and the position of the cavity boundary were under way (see review by Beard[10]). Figure 4 shows the basic model of the solar wind interaction with the geomagnetic field: specular reflection of the solar wind from the boundary of the geomagnetic cavity and no detached bow shock wave. This was the physical model assumed for the determination of the position of the cavity boundary and the interior magnetic field in the mathematically posed "free" boundary val-

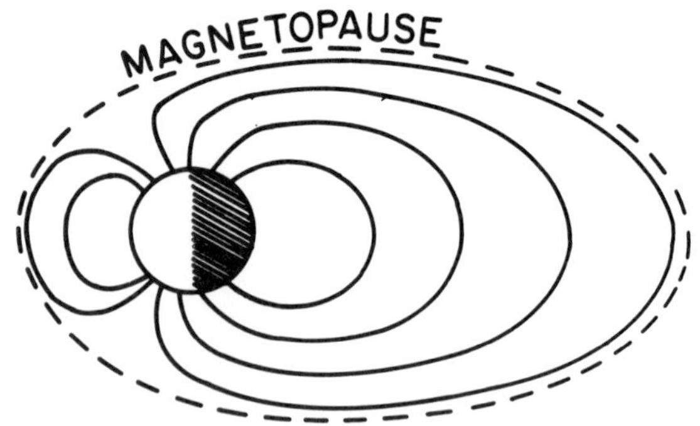

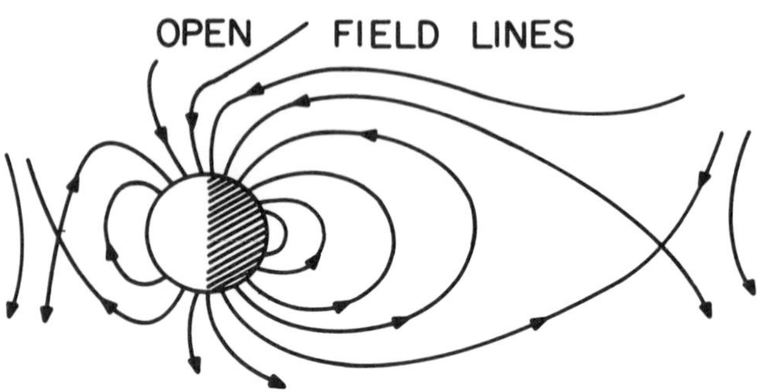

Figure 2—Schematic models, in noon-midnight meridian plane, of closed (upper-Johnson[8]) and open (lower-Dungey[7]) magnetospheres showing field line geometry.

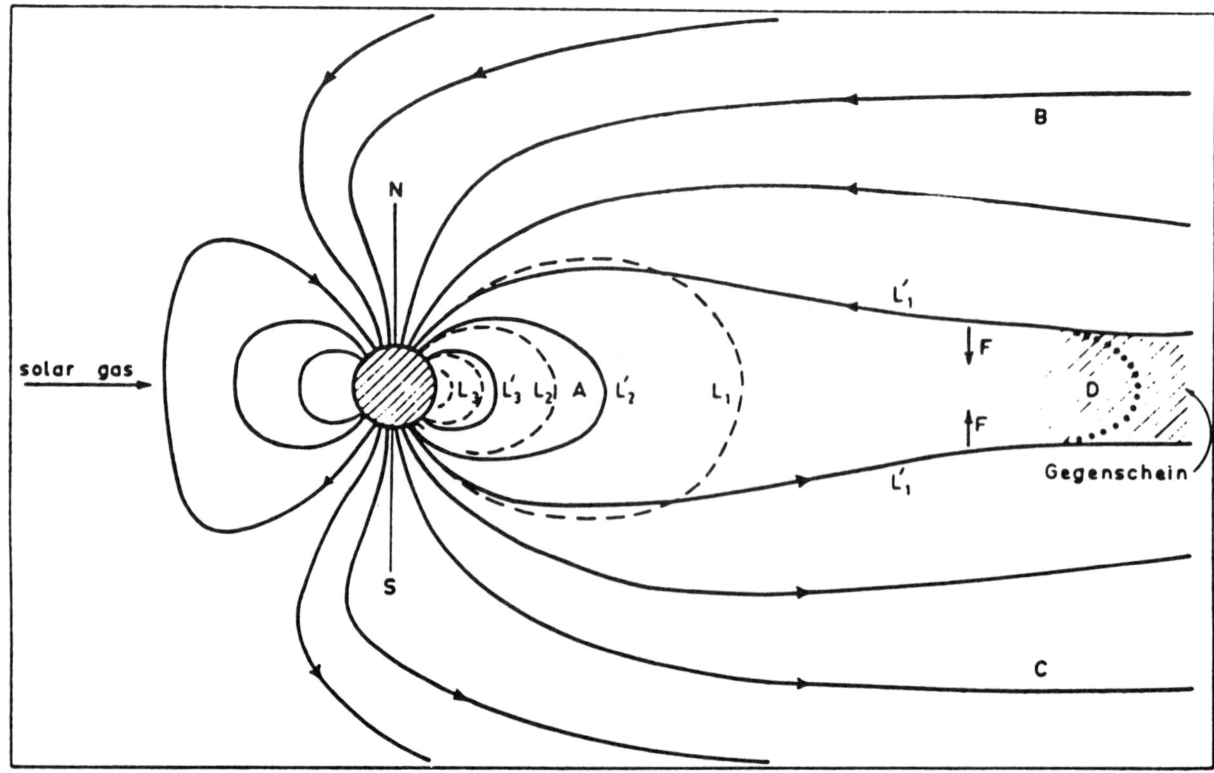

Figure 3—Piddington's[9] model of formation of magnetic tail during main phase of magnetic storm.

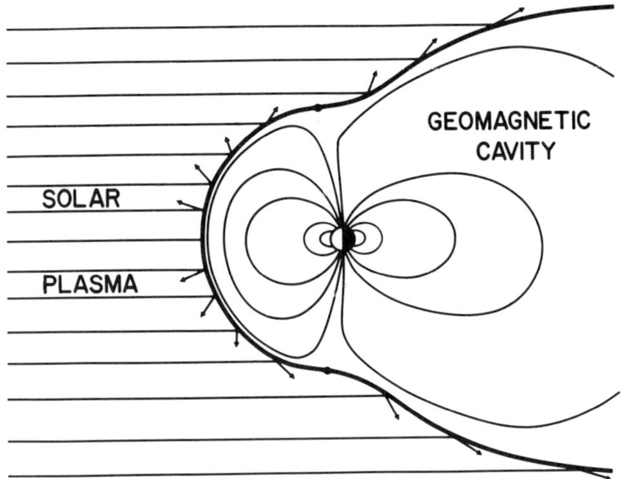

Figure 4—Early model of solar wind interaction with geomagnetic field forming cavity or magnetosphere and used as basis for early quantitative models of geomagnetosphere field.[10]

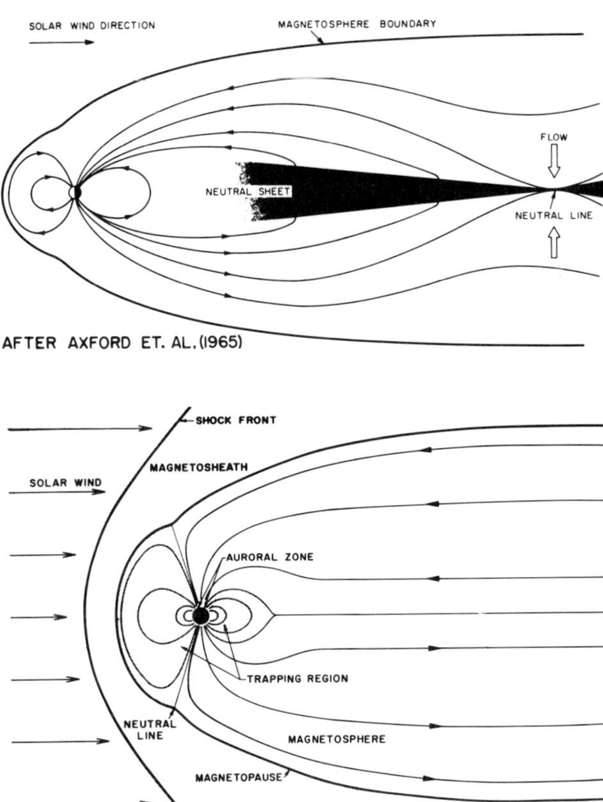

Figure 5—Two conceptual models illustrating extreme cases of tail field structure with the position of the neutral line emphasizing reconnection of lobe field lines across the neutral sheet. The Dessler and Juday[16] model is nearly identical with an earlier version of Dessler.[14]

ue problem. Beard and his collaborator, Gilbert Mead, contributed significantly to the development of such quantitative models.

The first experimental study of the nightside boundary of the geomagnetic cavity and the magnetic tail region of Earth was by Explorer 10, the first all NASA-Goddard Space Flight Center spacecraft, launched in March 1961. This spacecraft was battery powered (hence the short lifetime of several days, out to apogee at 43 R_e) and equipped with only a plasma probe[4] and a magnetometer instrumentation system.[11] It penetrated the duskside boundary of the geomagnetic cavity at a distance of 21 R_e and at a local time of 2100-2200. In the absence of repeated traversals through the nightside magnetosphere region, it was not possible to deduce the full nature of the magnetic field geometry at that time from the limited data available.

Later spacecraft contributed to the then rapidly evolving evidence for a strong day-night asymmetry in the terrestrial outer radiation belts and plasma environment. Results from Explorer 12[12] and Explorer 14[13] clearly demonstrated the effects of a strong perturbation of the geomagnetic field by the solar wind and an extension, in the antisolar direction, of the observed energetic particle and plasma populations. This experimental evidence combined with earlier ideas such as those of Johnson and Piddington led to the construction of qualitative theoretical models containing a full-fledged permanent geomagnetic tail by Dessler[14] and Axford et al.,[15] illustrated in Fig. 5.

EXPLORER 18 TO EXPLORER 35 PERIOD: 1963-68

Following the 1961 launches of Explorers 10 and 12, the NASA/GSFC proposed a series of spin stabilized Interplanetary Monitoring Platforms to continue the study of the radiation environment of Earth and cis-lunar space, preparatory to manned space flight. The first in this outstandingly successful series of 10 spacecraft (1963-73) was the IMP 1 (or Explorer 18), launched in late November 1963. Figure 6 shows the initial results obtained by this spacecraft as it surveyed and mapped the positions of the boundary of the geomagnetic cavity, the magnetopause, and the position from the subsolar region around through the dawnside sector of the detached bow shock wave, associated with the supersonic and super Alfvénic solar wind interaction with the geomagnetic field.

Overlain on this diagram, in which cylindrical symmetry about the Earth-Sun line is assumed for the boundary positions, is the trajectory of Explorer 10 during the time interval when it observed the duskside boundary of the geomagnetic cavity. As can be seen, the Explorer 10 and IMP 1 positions agreed quite well and showed that the diameter of the magnetosphere tail region is approximately 40 R_e at a distance of 20 R_e behind Earth.

Data from an orbit of IMP 1, while in Earth's magnetic tail, are shown in Fig. 7. The dashed curves represent the expected parameters using only an internal geomagnetic field model. These vector data are presented as latitude (θ) and longitude (ϕ) in what was then a newly

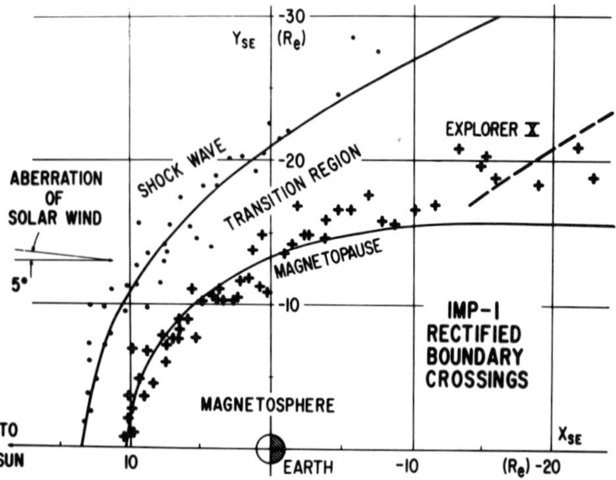

Figure 6—Magnetopause and bow shock boundaries observed by IMP 1 and Explorer 10.[1]

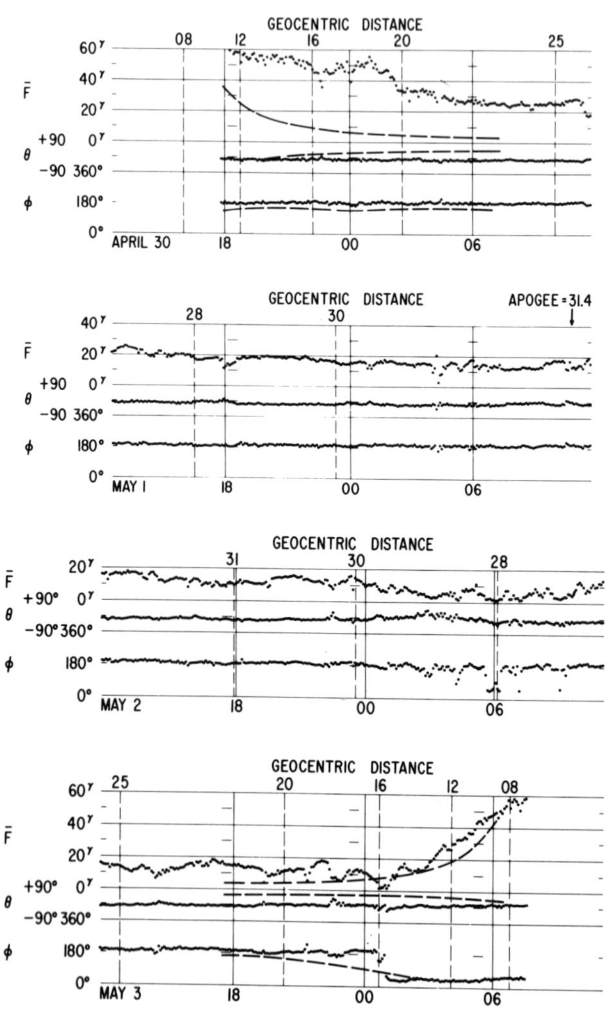

Figure 7—Representative magnetic field data (F = field magnitude in nanoteslas; θ, φ = latitude, longitude in solar magnetospheric coordinates) from IMP 1 when in geomagnetic tail in early 1964.[1]

defined geocentric coordinate system, solar magnetospheric, *SM*. The noninertial coordinate system was introduced to take into account the permanent existence of the solar wind (X_{SM} axis from Earth to Sun) and the diurnal and annual "wobbling" of Earth's magnetic dipole axis (the plane $X_{SM} - Z_{SM}$ includes the instantaneous position of the geomagnetic dipole axis) associated with both the rotation of Earth and its obliquity. In this *SM* coordinate system, the magnetic field vector at large distances from Earth ($R > 12$ R_e) is seen to be directed preferentially away ($\theta = 0°$, $\phi = 180°$) or toward ($\theta \sim 0°$, $\phi \sim 0°$) the Sun. (Solar ecliptic (*SE* coordinates) were defined earlier in the study of the Explorer 10 observations.[11])

A summary of the IMP-1 vector magnetic field observations in the tail is shown in Fig. 8 and clearly indicates the spatial distortion of the geomagnetic field, whereby a "bending" of the lines of force toward the nightside of Earth occurs, forming the magnetic tail region. In the data set shown, the field is directed away from the Sun. Although not shown, measurements in the complementary "upper" portion of the tail region indicate clearly a sunward directed magnetic field. This forms the basic field geometry of the geomagnetic tail: two regions, or lobes, of magnetic flux bundles, oppositely directed, and separated by a weak field region, the neutral sheet, where the field rapidly reverses direction. This geometry is described approximately as a theta configuration in cross section.

An equally important result of the early observational studies of the geomagnetic tail was the detection and identification of the plasma sheet region associated with the magnetic field reversal or neutral sheet. Figure 9 illustrates the variation in plasma density and energy that would be expected across an ideal magnetic neutral sheet. It is the presence of this plasma, surrounding the neutral sheet, that leads to the generation of an electrical

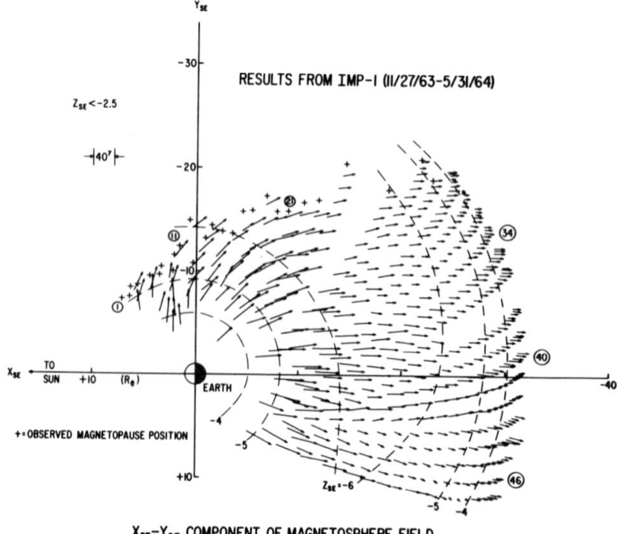

Figure 8—Summary of magnetic vector observations in the tail's southern lobe by IMP 1.[1]

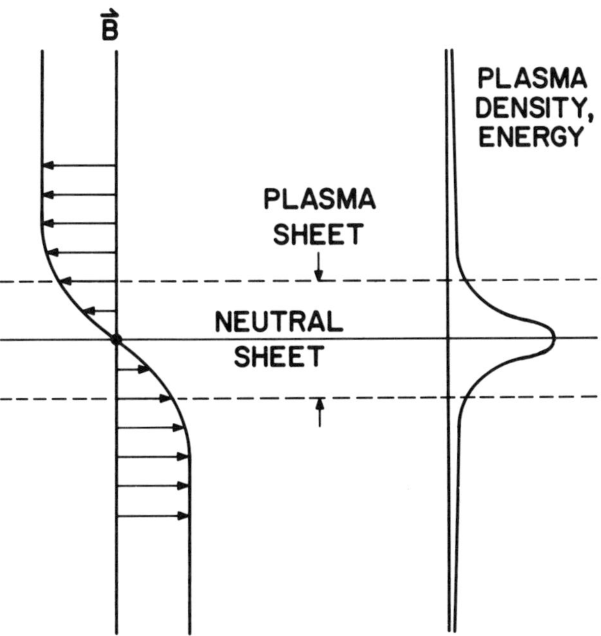

Figure 9—Variation of magnetic field and plasma parameters across ideal 2-dimensional neutral sheet.

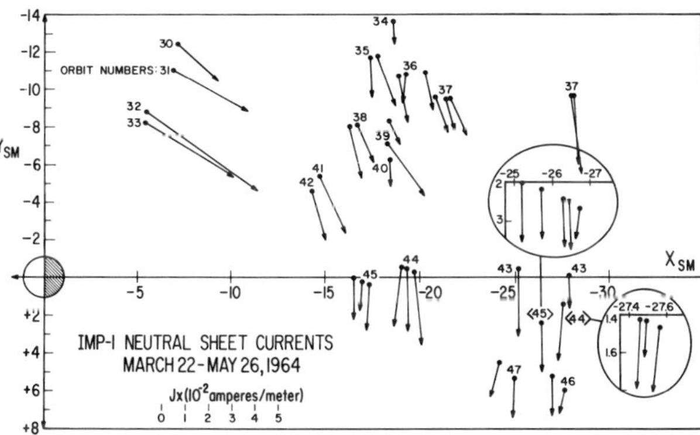

Figure 10—Deduced plasma sheet current vectors derived from IMP 1 data[17] plotted in X_{SM}-Z_{SM} plane.

current flow across the tail. This can also be thought of as generating the magnetic flux observable in the tail. A determination of the current vectors associated with the plasma sheet were derived from IMP 1 data by Speiser and Ness[17] and some of these results are shown in Fig. 10. Observations of the current intensities and directions are shown as well as an indication that multiple crossings occurred in relatively short periods of time. This suggested that the geomagnetic tail neutral sheet was often in rapid motion.

One puzzling feature of some of the distant nightside observations of energetic electrons on IMP 1 was their frequent impulsive behavior. This led to the short-lived concept of confined regions or electron "islands" in the tail.[18] After these observations were first reported, it was soon recognized that it was most likely that the IMP 1 electrons ($E_e > 40$ keV) were only the high-energy

tail of the plasma sheet electron distribution.[19] Correlation of the magnetic effects of these energetic electrons with the magnetic field on IMP 1 were studied by Anderson and Ness.[20]

The first definitive measurements of the spectra of the electrons in the plasma sheet and its size and location were obtained by the Vela 2[19] and the Vela 3A spacecraft.[21] These spacecraft, in distant circular Earth orbit near 17 R_e, had been launched in support of a U.S. National Security nuclear test detection system. They were instrumented to provide information on the geophysical space environment as well. As a result, the variable and low-energy characteristics of the electrons populating the plasma sheet were immediately noted and studied.

At this point, I should state that both theoretical models and in-situ observations of the magnetosphere were advancing very rapidly and nearly simultaneously in this time period. In a seminal paper, Levy et al.[22] gave an overview of the entire solar wind interaction with the geomagnetic field, anticipating many fundamental features including: the detached bow shock wave, field line reconnection at the dayside magnetopause, and multiple neutral lines in the fully developed geomagnetic tail (see Plate I-2).

Temporal variations of the geomagnetic tail field intensity were observed by the IMP 1 spacecraft and found to be positively correlated with global geomagnetic activity, as measured by the planetary magnetic activity index K_p.[23] Figure 11 summarizes the statistics of the tail field magnitude for low and high K_p periods. The average intensity of the magnetic field throughout the IMP 1 data set was found to be 16 nT.

Assuming conservation of magnetic flux from the polar regions of Earth connecting to the magnetospheric tail, the corresponding colatitude of the auroral polar cap region could be computed (see Fig. 12). This showed that for a tail radius = 20 R_e, and tail field intensity = 16 nT, the colatitude of the polar cap $\approx$ 13°. This agreed well with observations of the northern auroral zone and was consistent with the concept of a permanent magnetic tail whose field lines directly connect to the polar caps. A further consequence of this configuration is that aurora occur on field lines at and near the boundary separating tail field lines from lower latitude field lines.

Williams and Mead[24] showed theoretically and observationally that a correlation should exist between temporal changes of the latitudinal boundary of the trapping region of Earth's radiation belts with the intensity of the magnetic field in the geomagnetic tail. This result was obtained with a quantitative model of the terrestrial magnetosphere using a 3-dimensional dipole with a 2-dimensional tail current sheet. Computation of the configuration of the nightside magnetosphere was made in order to determine the quantitative relationship between field intensity and trapping boundary. Observational studies by Ness and Williams[25] correlated IMP 1 tail field data with radiation belt high latitude boundaries determined by the spacecraft 1963-38c.

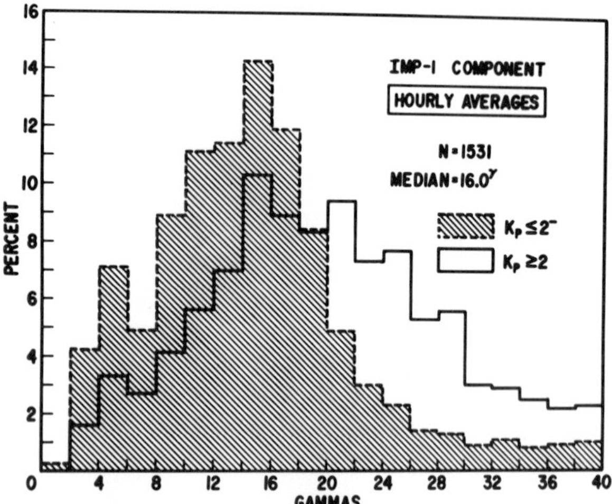

Figure 11—Variation of intensity of tail field (nanoteslas) with planetary magnetic activity index, K_p.[23]

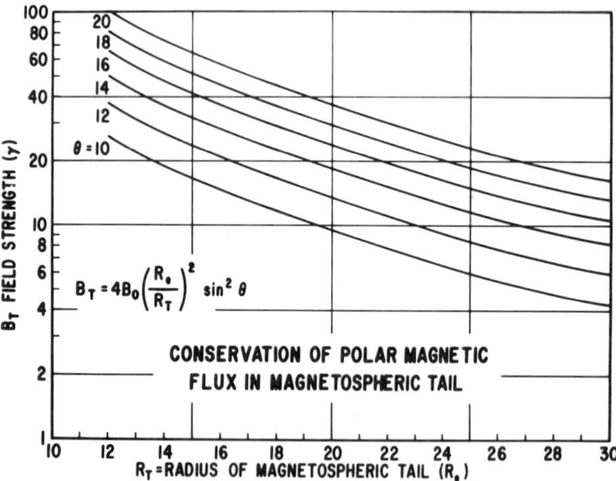

Figure 12—Variation of colatitude of polar cap region θ_{pc} (degrees) connecting total magnetic flux to bipolar magnetic tail with radius R_T (units of Earth radii, R_e = 6378 km) and intensity B_T (nanoteslas).

Theorists were routinely using magnetospheric configurations that took into account the permanent presence of both the solar wind and the geomagnetic tail. Features of the magnetic tail that challenged theorists and observers then (and now) were the nature of the magnetic field configuration in the neutral sheet region, i.e., any possible connection across it between the two lobes, the location of any neutral line or lines, and the length of the geomagnetic tail. Figure 5 illustrates two extreme models of the geomagnetic tail that were proposed to illustrate where a neutral line in the geomagnetic tail would occur: close to Earth[15] or extremely distant.[14] Data from the early series of IMP spacecraft (1, 2, 3, and 4) showed that the magnetic field did connect across the neutral sheet but that the tail neutral line,

if one existed, was beyond spacecraft apogee (31 to 35 R_e).

On July 1, 1966, the Explorer 33 (Anchored IMP D) spacecraft was launched. It extended the study of the geomagnetic tail to a distance of 80 R_e. Explorer 35 (Anchored IMP E) was launched on July 19, 1967 and also contributed to the study of the tail. The primary objective of the Anchored IMP D and E mission was to place an IMP type spacecraft in a captured, stable orbit about the Moon. IMP E was successfully injected into lunar orbit and although IMP D did not achieve a lunar orbit, it was successful in studying and surveying the geomagnetic tail: confirming its existence and well-defined two-lobe structure out to 80 R_e.[26-28]

Earlier, and much more limited, tail region data were obtained by the USSR lunar orbiting spacecraft, Luna 10, launched in April 1966. Dolginov et al.,[29] in a study of magnetic field data, had reported the absence of any sensible lunar magnetic field or any steady geomagnetic tail-like field near the time of full moon. However, plasma observations on Luna 10 showed the existence of a sheet-like plasma.[30]

Ness[31] proposed that the inconsistent and conflicting Luna 10 results could be reconciled with the existence of a geomagnetic tail by taking into account the fact that the Luna 10 observations probably occurred within the plasma sheet in the magnetic tail. This was substantiated by simultaneous tail field observations by IMP 3 in the distant tail during the intervals of Luna 10 data, which showed the existence of a stable and well-developed tail structure.

Thus, the 1963-68 period of study of the geomagnetic tail established the many basic features that we now accept as common knowledge.

LAST OF THE EARLY YEARS: 1968-72

Pioneers 7 and 8, injected into heliocentric orbits in 1966 and 1968, crossed the distant geomagnetic tail at distances of 1000 R_e[32] and 500 R_e.[33] Since these Pioneer spacecraft were instrumented primarily to study the solar wind, the plasma detectors onboard were not optimized for detection of the plasma sheet. The correlated observations in the distant tail between magnetic field and plasma data were thus characterized by solar wind "dropouts" associated with weak fields separating oppositely directed interplanetary magnetic fields.

At the Solar Terrestrial Physics Symposium held in Washington, D.C. in 1968, the following major issues were posed concerning the geomagnetic tail:[34]

1. What mechanism forms the tail? Is it due to reconnection with the interplanetary field or a viscous interaction between the solar wind and the magnetopause? Or a combination of these two?
2. To what distance does a coherent, theta configuration tail extend? And where is the neutral line (or lines)?
3. Does merging occur continuously or impulsively across the neutral sheet?
4. Are precipitating auroral particles accelerated in the tail?

Plate I·2—Figure used by Avco Everett Research Labs in their 1963 Christmas card, based on research reported in Levy et al.[22] (courtesy of Harry E. Petschek).

5. What and where is the source of the tail plasma, especially the transient events?

6. Is this tail region the source of long period micropulsations observed on Earth's surface?

Studies of the Earth's magnetic tail continued with the last several IMP spacecraft (Explorers 41, 43, 47, and 50 or IMPs 5, 6, 7, and 8). Fairfield and Ness[35] studied the response of the geomagnetic tail to substorms. In the presubstorm phase, the geomagnetic tail was found to be compressed and the field intensity enhanced as more flux was added to the tail and thus more energy stored by the solar wind. Upon subsequent release of this energy, creating the substorm phenomena, the geomagnetic tail field relaxed to a more dipolar-like and a less tail-like configuration. These two states of the magnetic tail are illustrated in Figs. 13 and 14. The first suggestions of plasma sheet contraction and expansion in association with magnetic bays in the nighttime auroral zone were made on the basis of Vela 2A and 2B spacecraft.[36]

Detailed studies of the magnetic field topology in the plasma sheet region were complicated by temporal variations of the plasma sheet-neutral sheet regime. This is illustrated in Fig. 15, where detailed vector magnetic field measurements from the Explorer 34 spacecraft are shown, typical of the neutral sheet crossings observed with high resolution experiments. It was difficult to ascertain the orientation of the neutral sheet and hence the normal component of the magnetic field crossing the neutral sheet because of the obvious temporal variations of the field structure and/or its position relative to the spacecraft.

In an attempt to compare observations with theory, a statistical approach was thought appropriate. Subsets of magnetic field observations when the X_{SM} component of the field reversed (and the field was weak) were used as indicators of proximity or immersion within the plasma sheet. These data were then statistically summarized and compared with theoretical models of multiple X and O neutral sheet structures derived from plasma equilibrium theory. Schindler and Ness[37] showed the consistency of the observations with such neutral sheet structures. The geometry of the field lines in the region of such multiple neutral line sheet structures are shown in Fig. 16.

Other research studies utilizing the IMP data by Fairfield elaborated more fully on the nature and temporal variations of the geomagnetic tail and its physical properties. This set the stage for subsequent investigations by the more recent spacecraft ISEE 1, 2, and 3 with much higher telemetry rates and, most importantly, much more sophisticated second and third generation plasma detectors capable of observing the low density, low energy, particle fluxes resident within the geomagnetic tail and important in its dynamics.

In summary, the results of the studies in this first decade of exploration of the geomagnetic tail are summarized in Table 1.

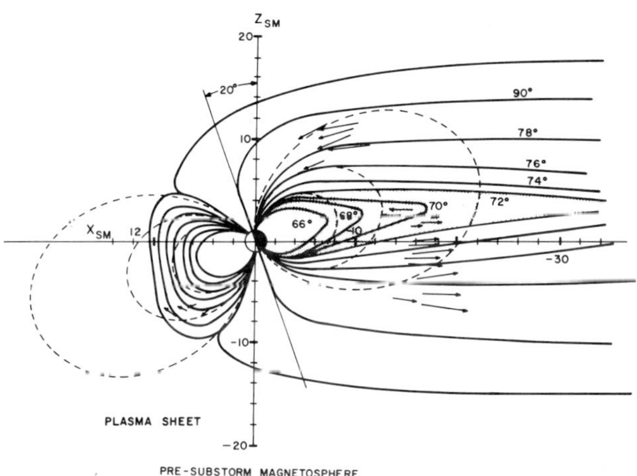

Figure 13—Configuration of magnetic tail field and plasma sheet (presubstorm).[35]

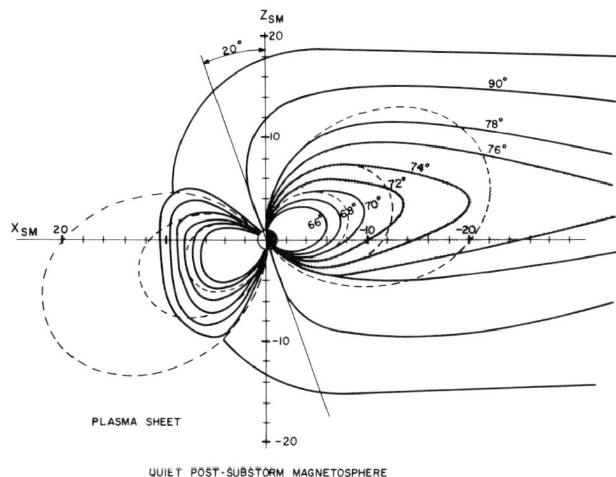

Figure 14—Configuration of magnetic tail field and plasma sheet (postsubstorm).[35]

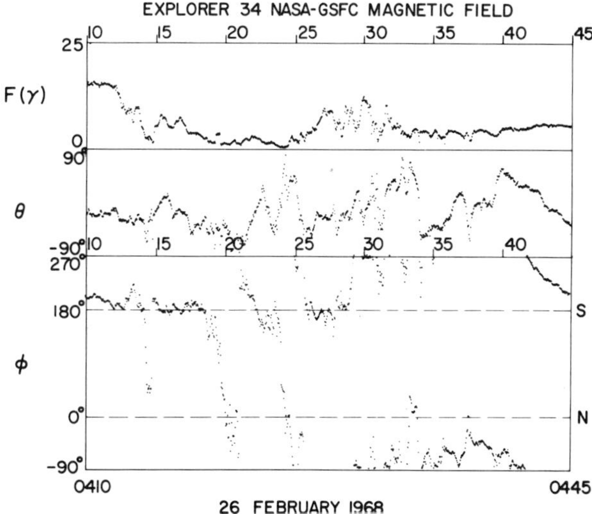

Figure 15—Detailed IMP 4 observations of vector magnetic field (F = magnitude in nanoteslas; θ, ϕ = latitude, longitude in *SM* coordinates) during crossing of neutral sheet.[37]

19

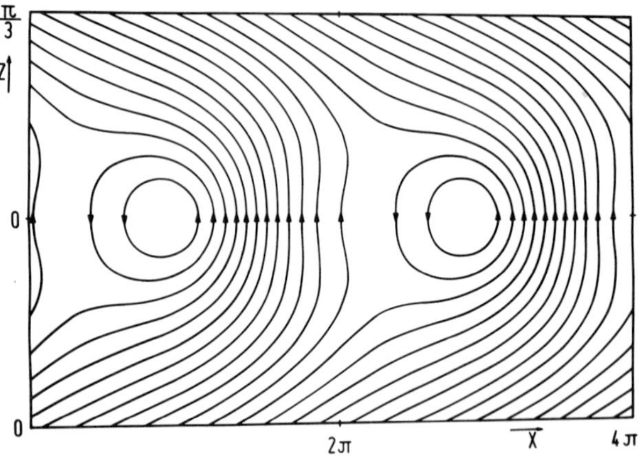

Figure 16—Structure of X and O line neutral sheet.[37] The two scales (X,Z) are dimensionless parameters used in the generalized analysis.

Table 1—Main Results in First Decade (1961-71).

1. Permanent geomagnetic tail and imbedded plasma sheet-neutral sheet discovered.

2. Average configuration: Boundaries and their locations.
   Gradients of plasmas and fields.
   Neutral sheet and plasma sheet locations.

3. Dynamical changes: Energy storage and loss.
   Configuration changes and correlations.
   Plasma sheet thinning.
   Reconnection across neutral sheet.

4. Direct but asymmetrical access of interplanetary charged particles to polar caps that implied interconnection with the interplanetary magnetic field and a dependence on sector structure.

5. Length of tail: Ordered "theta"   >   80 R_e
   Observable   > 1000 R_e

6. "Magnetic Comet" or "Earth Flare" metaphor used to describe tail formation and position by comparison with Type II ion tails or dynamical changes.

ACKNOWLEDGMENT—In closing, I wish to acknowledge the contributions of those colleagues with whom I have shared some of this exciting work, only briefly summarized here. I look forward to hearing of the new results of subsequent studies to be reported at this meeting.

REFERENCES

[1] N. F. Ness, "The Earth's Magnetic Tail," *J. Geophys. Res.* **70**, 2989 (1965).

[2] L. A. Frank, "Plasmas in the Earth's Magnetotail," *Space Sci. Rev.* **42**, 211 (1985).

[3] N. F. Ness, "Magnetometers for Space Research," *Space Sci. Rev.* **11**, 111 (1970).

[4] A. Bonetti, H. S. Bridge, A. J. Lazarus, B. Rossi, and F. Scherb, "Explorer 10 Plasma Measurements," *J. Geophys. Res.* **68**, 4017 (1963).

[5] M. Neugebauer and C. Snyder, "The Mission of Mariner 2: Preliminary Observations, Solar Plasma Experiments," *Science* **138**, 1095 (1962).

[6] E. N. Parker, *Interplanetary Dynamical Processes*, Interscience Publishers, New York (1963).

[7] J. W. Dungey, "Interplanetary Magnetic Field and the Auroral Zones," *Phys. Rev. Lett.* **6**, 47 (1961).

[8] F. Johnson, "Gross Character of the Geomagnetic Field in the Solar Wind," *J. Geophys. Res.* **65**, 3049 (1961).

[9] J. H. Piddington, "Geomagnetic Storm Theory," *J. Geophys. Res.* **65**, 93 (1960).

[10] D. B. Beard, "The Solar Wind Geomagnetic Field Boundary," *Rev. Geophys.* **2**, 335 (1964).

[11] J. P. Heppner, N. F. Ness, T. L. Skillman, and C. S. Scearce, "Explorer 10 Magnetic Field Measurements," *J. Geophys. Res.* **68**, 1 (1963).

[12] J. W. Freeman, "Electron Distribution in the Outer Radiation Zone," *J. Geophys. Res.* **69**, 1691 (1964).

[13] L. A. Frank, "A Survey of Electrons Beyond 5 R_E with Explorer 14," *J. Geophys. Res.* **70**, 1593 (1965).

[14] A. J. Dessler, "Length of the Magnetosphere Tail," *J. Geophys. Res.* **69**, 3913 (1964).

[15] W. I. Axford, H. E. Petschek, and G. L. Siscoe, "The Tail of the Magnetosphere, *J. Geophys. Res.* **70**, 1433 (1965).

[16] A. J. Dessler and R. D. Juday, "Configuration of Auroral Radiation in Space," *Planet. Space Sci.* **13**, 63 (1965).

[17] T. W. Speiser and N. F. Ness, "The Neutral Sheet in the Geomagnetic Tail: Its Motion, Equivalent Currents and Field Line Connection Through It," *J. Geophys. Res.* **72**, 131 (1967).

[18] K. A. Anderson, H. K. Harris, and R. J. Paoli, "Energetic Electron Fluxes In and Beyond the Earth's Outer Magnetosphere," *J. Geophys. Res.* **70**, 1039 (1965).

[19] S. J. Bame, J. R. Asbridge, H. E. Felthauser, R. A. Olson, and I. B. Strong, "Electrons in the Plasmas Sheet of the Earth's Magnetic Tail," *Phys. Rev. Lett.* **16**, 138 (1966).

[20] K. A. Anderson and N. F. Ness, "Correlation of Magnetic Fields and Energetic Electrons on the IMP-1 Satellite," *J. Geophys. Res.* **71**, 3705 (1966).

[21] S. J. Bame, J. R. Asbridge, H. E. Felthauser, E. W. Hones, and I. B. Strong, "Characteristics of the Plasma Sheet in the Earth's Magnetotail," *J. Geophys. Res.* **72**, 113 (1967).

[22] R. H. Levy, H. E. Petschek, and G. L. Siscoe, "Aerodynamic Aspects of Magnetosphere Flow," *AIAA J.* **2**, 2065 (1964).

[23] K. W. Behannon, and N. F. Ness, "Magnetic Storms in the Earth's Magnetic Tail," *J. Geophys. Res.* **71**, 2327 (1966).

[24] D. J. Williams and G. D. Mead, "Nightside Magnetosphere Configurations as Obtained from Trapped Electrons at 1100 km," *J. Geophys. Res.* **70**, 3017 (1965).

[25] N. F. Ness and D. J. Williams, "Correlated Magnetic Tail and Radiation Belt Observations," *J. Geophys. Res.* **71**, 322 (1966).

[26] K. W. Behannon, "Mapping of the Earth's Bow Shock and Magnetic Tail by Explorer 33," *J. Geophys. Res.* **73**, 907 (1968).

[27] K. W. Behannon, "Geometry of the Geomagnetic Tail," *J. Geophys. Res.* **75**, 743 (1970).

[28] J. D. Mihalov, D. S. Colburn, and C. P. Sonett, "Configuration and Reconnection of the Geomagnetic Tail," *J. Geophys. Res.* **73**, 943 (1968).

[29] Sh. Sh. Dolginov et al., "Possible Interpretation of the Results of Measurements on the Lunar Orbiter Luna 10," *Dokl. Akad. Nauk. S.S.S.R.* **170**, 570 (1966).

[30] K. I. Gringauz et al., "Experimental Results Concerning the Lunar Ionosphere Detected on the First Artificial Lunar Satellite," *Dokl. Akad. Nauk. S.S.S.R.* **170**, 1306 (1966).

[31] N. F. Ness, "Remarks on the Interpretation of Lunik 10 Magnetometer Results," *Geomagn. Aeron.* **3**, 431 (1967).

[32] N. F. Ness, C. S. Scearce, and S. C. Cantarano, "Probable Observations of the Geomagnetic Tail at 10^3 R_E," *J. Geophys. Res.* **72**, 3769 (1967).

[33] F. Mariani and N. F. Ness, "Observations of the Geomagnetic Tail at 500 R_E by Pioneer 8, *J. Geophys. Res.* **74**, 5633 (1969).

[34] N. F. Ness, "The Geomagnetic Tail," *Rev. Geophys.* **7**, 97 (1968).

[35] D. H. Fairfield and N. F. Ness, "Configuration of the Geomagnetic Tail During Substorms," *J. Geophys. Res.* **75**, 7032 (1970).

[36] E. W. Hones, Jr., J. R. Asbridge, S. J. Bame, and I. B. Strong, "Outward Flow of Plasma in the Magnetotail Following Geomagnetic Bays," *J. Geophys. Res.* **72**, 5879 (1967).

[37] K. Schindler and N. F. Ness, "Neutral Sheet Fine Structure," *J. Geophys. Res.* **77**, 91 (1972).

II. MAGNETOTAIL CONFIGURATION

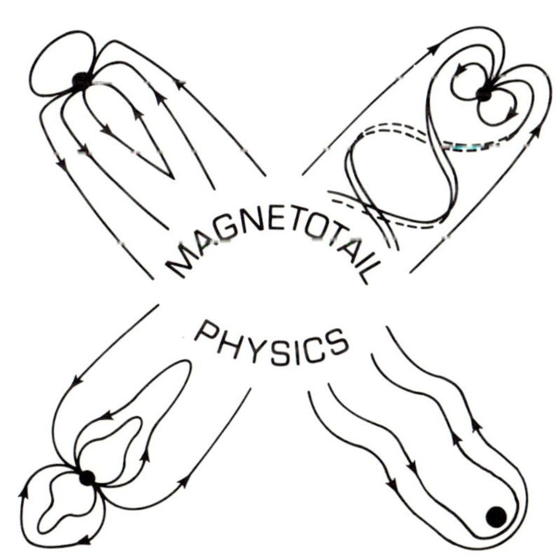

STRUCTURE OF THE GEOMAGNETIC TAIL

D. H. Fairfield*

Solar wind impinging on the geomagnetic dipole draws magnetic field lines out behind the earth forming a geomagnetic tail that extends to at least hundreds of earth radii. The tail lobes consist of tenuous, low-beta plasmas that exist on field lines that emanate from the northern and southern polar caps. Evidence for interconnection of tail lobe and interplanetary field lines is obtained from anisotropic lobe electron distributions that result from solar electrons freely entering the lobe that is directly connected to the sun. The lobes are separated by a current or "neutral" sheet that is embedded within a thicker high-beta, central plasma sheet with a typical north-south dimension of several R_e. A plasma sheet boundary layer lies between the low density lobes and the central plasma sheet and is characterized by time-varying streaming ions traveling in various directions and with various energies. Inside the tail magnetopause are magnetosheath-like plasmas of the high latitude plasma mantle and the low latitude boundary layer. Magnetotail pressure 30 R_e behind the earth is typically 1.8×10^{-9} dynes/cm^2, both in the tail lobes where the average field is 20 nT and in the higher beta plasma sheet where the average density is 0.25/cc and proton and electron average energies are near 5 and 1 keV, respectively. In the distant tail, regions similar to those seen nearer the earth can be identified, although the boundary plasma gradually fills a larger portion of the tail lobe at more distant locations. The highly time-variable nature of the magnetotail is partially explained by theoretical work indicating that a realistic, steady-state magnetotail cannot be achieved even with constant boundary conditions.

INTRODUCTION

More than two decades of magnetospheric research have demonstrated that the geomagnetic tail is one of the most interesting and important regions of the earth's magnetosphere. It is the site of high velocity flows, large electric currents, highly tenuous plasmas, and large spatial gradients that lead to a plethora of plasma physical phenomena including waves and instabilities that transfer energy and accelerate particles. This paper will concentrate on describing the properties of various magnetotail regions where these phenomena take place (e.g., Fig. 1). Discussions of dynamical effects can be found in Refs. 2-4.

Most investigators seem to agree that the primary mechanism involved in forming the geomagnetic tail is the dayside reconnection of geomagnetic and interplanetary field lines and their subsequent transport to the nightside hemisphere. These field lines form two antiparallel bundles of magnetic flux, often called the magnetotail lobes (see Fig. 1). The lobes contain only very tenuous plasmas and hence are field-dominated, low-β regions ($\beta = 8\pi nk(T_p + T_e)/B^2$). Anisotropic, field aligned electrons with energies of the order of 100 eV have recently been detected in the lobes and identified as solar electrons that freely enter the tail along open field lines and precipitate over the polar cap as "polar rain".[5] This process occurs in one hemisphere or the other depending on whether the IMF (interplanetary magnetic field) is directed toward or away from the sun. This preferential entry helps confirm that one lobe is connected to interplanetary field lines leading back to the sun while the other is connected to fields that lead further out into the solar system.

*Laboratory for Extraterrestrial Physics, NASA Goddard Space Flight Center, Greenbelt, Maryland 20771.

TRANSPORT REGIONS OF THE MAGNETOSPHERE

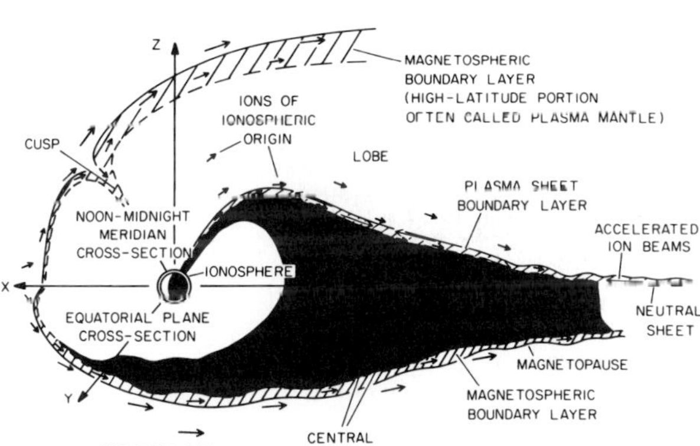

Figure 1—Schematic representation of different regions in the earth's magnetosphere in the *X-Z* meridian plane (top) and *X-Y* equatorial plane (bottom) (Eastman et al.[1]).

Solar-wind protons readily enter a limited region of the dayside high-latitude magnetosphere known as the polar cusp (see Fig. 1) (e.g., Ref. 6). Most of these protons mirror at lower altitudes and return to the outer magnetosphere where they form a boundary-layer region adjacent to the magnetopause called the plasma mantle.[7] Plasmas of this region are similar to the exterior magnetosheath, except mantle densities and velocities are slightly lower and the field lines are magnetospheric (i.e., they have an orientation characteristic of the magnetosphere and a relatively low level of fluctuations).

Additional plasma penetrates the dayside magnetopause via diffusive or other processes and transports closed field lines to the tail (e.g., Ref. 8), but this process appears to be less important than reconnection, except under northward IMF conditions when reconnection is of minimal importance (e.g., Ref. 9). This region, the low-latitude portion of which is sometimes called the low-latitude boundary layer,[10] has plasma properties that are similar to the plasma mantle as befits their common solar wind origin. However, the existence of the low-latitude boundary layer on closed field lines allows the intermingling of trapped energetic magnetospheric particles. Eastman et al.[1] give the name magnetospheric boundary layer to the collection of all these regions that are adjacent to the magnetosheath.

Separating the anti-parallel bundles of tail magnetic flux is a current sheet whose presence is dictated by the $\nabla \times B$ of this field configuration. Alternatively, one can view this current as the source of the tail lobe field. This current sheet is commonly called a neutral sheet because of the small, and perhaps sometimes zero, field at its center. On the average, however, there is a small northward B_z field at the center of the neutral sheet indicating that some field lines still close from the southern to the northern hemisphere (e.g., Ref. 11).

The remaining tail regions in the interior of the magnetotail are characterized primarily by their plasma properties. Isotropic plasma surrounding the neutral sheet forms the central plasma sheet whose plasma β usually, but not always, assumes a high value near or above unity in the outer magnetosphere. This region approaches the earth along high-latitude field lines (see Fig. 1). In the equatorial plane, the earthward edge of the plasma sheet typically lies between 5 and 10 R_e.[12] Between the lobes and the central plasma sheet lies a region of anisotropic plasmas characterized by rapidly flowing ion beams that is termed the plasma-sheet boundary layer. The majority of workers seem to believe that the central plasma sheet projected along magnetic field lines to the earth would correspond to a region of diffuse subvisual aurora that occur equatorward of the brighter, structured aurora of the conventional auroral oval as defined by Feldstein and Starkov.[13] The structured aurora would then map to the plasma-sheet boundary layer (e.g., Ref. 14). Feldstein and Galperin,[15] however, have recently argued that the structured aurora map to the central plasma sheet and the plasma-sheet boundary layer maps to a diffuse region poleward of the auroral oval. The limited accuracy of existing static magnetic field models along with the variability of the real magnetosphere precludes distinguishing definitively between these alternatives.

INTERPLANETARY CONTROL OF THE TAIL

The variability of the magnetotail is illustrated in Fig. 2, which shows locations of the outer boundary of the tail as seen by the Explorer 33 spacecraft over a 3-year period.[16] The extremes of positions of the boundary are seen to be comparable with the dimensions of the tail, especially beyond 40 R_e. We now review a number of reasons for these boundary motions, which are all due, either directly or indirectly, to changes in the solar wind and its embedded magnetic field.

The simplest way to view the magnetic tail is as a huge wind sock aligned with the solar wind. If the direction of the solar wind changes, and it does vary by up to 5°

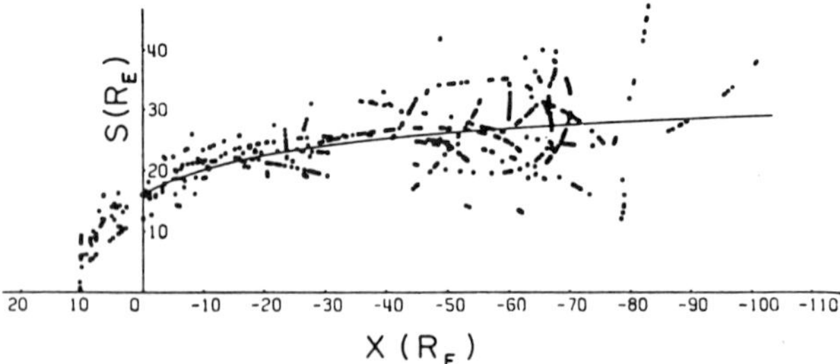

Figure 2—Locations of tail boundary crossings detected by the Explorer 33 spacecraft. The abcissa is distance down the tail and the ordinate is distance from the tail axis (Howe and Binsack[16]).

to 10°,[17] the tail will realign itself with the new direction and the boundary will move. These boundary motions will be greatest at great distances; every 1° solar wind change will produce a 3.5 R_e boundary motion at 200 R_e.

Other causes of boundary motions can be understood by considering how solar-wind pressure affects the magnetosphere. Assuming pressure balance at the magnetopause, we equate the external pressure of the solar wind to the internal pressure in the magnetosphere:

$$Kn_{sw}m_p v^2 \sin^2\Theta + n_{sw}kT_{sw} + B_{sw}^2/8\pi$$

$$= nkT + B^2/8\pi , \qquad (1)$$

where K is a constant that is a measure of how efficiently solar wind particles transfer their momentum to the magnetosphere ($K=1$ for inelastic collisions, which we assume; $k=2$ for elastic collisions, etc. (e.g., Ref. 18)), T_{sw} is the sum of the solar wind electron and proton temperatures, n_{sw}, B_{sw}, and v are solar-wind density, field strength, and velocity, m_p is the proton mass (heavy ions are neglected), Θ is the angle between the solar-wind velocity and the magnetopause surface, B is the magnetosphere field strength, n is the magnetosphere plasma density, T is the sum of the magnetosphere electron and proton temperatures, and k is the Boltzmann constant. On the dayside of the earth, Θ is large and the first term in Eq. 1 is the dominant term on the left-hand side. On the nightside, the flaring angle Θ becomes smaller at larger distances down the tail and eventually the other two terms on the left-hand side become of comparable magnitude. The tail pressure on the right-hand side decreases down the tail as Θ decreases on the left-hand side. At some point, the tail will cease flaring, Θ will become zero, the tail radius will remain constant, and B will be constant and balanced by only the external thermal and magnetic pressure. ISEE 3 measurements have shown that, on the average, this situation occurs at about $X_{sm} \cong 120$ R_e with a tail radius of 30 R_e[19] or less.[20] If sin Θ is not zero and solar-wind density or velocity increases, the tail will undergo a compression to increase B and T. This will move the boundaries inward. If sin Θ is zero it will take an increase in $n_{sw}T_{sw}$ and/or B_{sw} to produce compression. If flux is added to the tail, say by increased dayside reconnection due to increased southward IMF, Θ will increase in the near-earth tail and again compression will result. In this case, the field strength in the distant tail will remain unchanged, but the radius will expand to accommodate the increased flux. Significant changes in polar cap dimensions are thought to reflect these flux changes.[21] Fairfield[20] argues that the open polar cap flux may vary by as much as a factor of 10 between very quiet times and magnetic storms, a condition that would produce quite different tail cross sections at different times.

Still another way the IMF may influence boundary location is illustrated by Fig. 3.[22] The view is a *Y-Z* plane cross section of the distant tail with interplanetary field lines shown draped over the top and bottom of the tail. The exterior field will exert a maximum pressure on the top and bottom of the tail and a minimum pressure on the sides of the tail causing an elongation of the tail boundary in the plane of the IMF and the solar wind. As the *Y-Z* component of the IMF changes direction, the major axis of the elliptical cross section will follow its direction and produce local boundary motions. Sibeck et al.[22] have presented evidence that this effect may be important in the distant tail, but it is not apt to be significant closer than to the earth where the first term in Eq. 1 is dominant. These authors have sketched in magnetic field lines entering the tail in a limited region or window.[23,24] There is indeed evidence that plasma is seen to enter the tail preferentially in certain quadrants that depend on the IMF direction (Ref. 25 and references therein).

MAGNETIC FIELD STRUCTURE

Numerous studies have helped define the typical magnetic field structure of the magnetotail. Figure 4, from Ref. 19, shows the average lobe field strength in the tail as a function of distance down the tail as measured by the ISEE 3 spacecraft. This spacecraft covered the distance range out to 238 R_e in 1982-83. The earlier data of Behannon[26] has been superposed for comparison. The difference between the two data sets is related

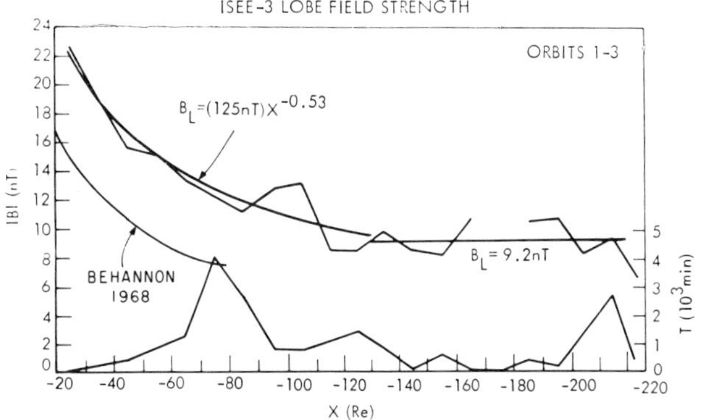

Figure 4—Tail lobe field strength as a function of distance down the tail. The solar maximum data of Behannon[27] for $K_p < 2$ have been superposed on the ISEE 3 data of Slavin et al.[19]

primarily to the fact that Behannon's data[26] is for $K_p \leq 2$ conditions whereas the data of Slavin et al.[19] are for all conditions; disturbed conditions are well known to be associated with stronger magnetotail fields (e.g., Ref. 27). Also Behannon's data were taken near solar maximum when the average solar-wind pressure was unusually low (e.g., Ref. 28), whereas the data of Slavin et al. are from the decreasing portion of the sunspot cycle when pressure was higher. Both data sets show a decrease in field strength with distance down the tail, associated with the decreasing flaring angle in Eq. 1. The approximately constant ISEE 3 data beyond 120 R_e indicate that tail flaring has terminated and the sin Θ term in Eq. 1 has gone to zero.

The flaring of the lobe field away from the tail axis inside lunar orbit is a readily detectable feature of the tail magnetic field. Figure 5[11] shows the B_z component of the field plotted as a function of distance from the

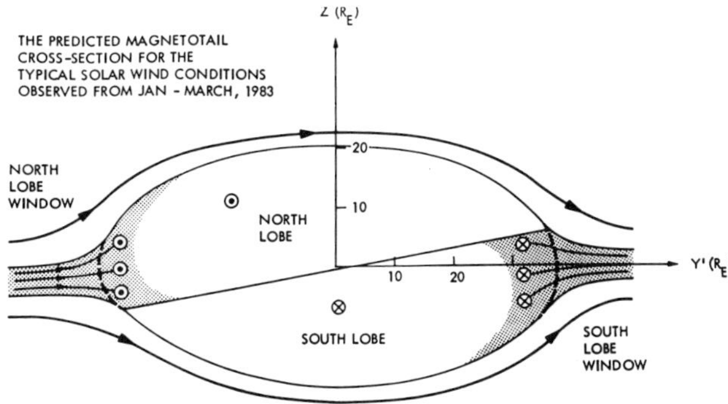

Figure 3—A view of the distant tail cross section illustrating how increased pressure of the IMF draped over the tail might produce an elliptical shape. Additional IMF lines are shown connected to the tail lobes in restricted regions of the northern and southern hemispheres (Sibeck et al.[22]).

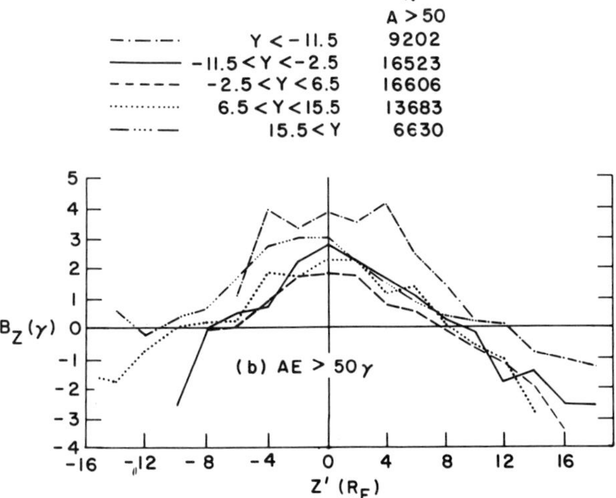

Figure 5—The average north-south field component in the tail, plotted as a function of distance from the equatorial plane. Closed field lines produce a positive B_z near the equatorial plane, whereas diverging open field lines produce a negative B_z at high latitudes (Fairfield[11]).

neutral sheet for five longitudinal sectors of the tail. Data are from the IMP 6 spacecraft and the distance range $-33 < X_{sm} < -20$ R_e. The average negative values for $|Z| > 8$ R_e are a manifestation of tail flaring. Comparable values of B_y (not shown) also confirm flaring in the Y direction.

Another significant feature in Fig. 5 is the positive B_z's near the equatorial plane. The average values are all positive, but it is significant that the minimum B_z occurs near midnight and the maximum B_z's occur in the dusk and dawn sectors. This longitudinal variation shows that field lines are most tail-like near midnight and more dipole-like near dawn and dusk (see also Ref. 29).

Slavin et al.[19] also find positive B_z components at larger distances. From 60 to 100 R_e the B_z average is slightly greater than 1 nT and from 100–210 R_e, B_z is often negative but the average is still slightly positive. Beyond 210 R_e the average is $B_z = -0.2$ nT. More significantly, when Slavin et al. investigate B_z as a function of Y in the different distance ranges, they find that between 100 and 180 R_e, B_z is negative near midnight and larger on the flanks as was the case near earth. They suggest that on the average, a neutral line exists at roughly 130 R_e near midnight ($Y = 0$) but it curves outward to greater distances for Y values away from midnight as suggested by Russell.[30] They note that simulations also predict neutral line formation near midnight.[31]

PLASMA-SHEET BOUNDARY LAYER

The magnetotail region between the central plasma sheet and the tail lobes known as the plasma-sheet boundary layer has undergone intense investigation in recent years. The primary distinguishing feature of this region is the presence of anisotropic distributions of energetic particles with flow velocities greater than thermal velocities.[14] The University of Iowa three-dimensional plasma experiment clearly reveals the presence of field-aligned ion beams flowing in either one or both directions along the field with energies from tens of electron volts up to the experimental limit of 45 keV.[1] These streaming ions are frequently seen at even higher energies[32] particularly during the late phases of substorms.[33] These ions are thought to be accelerated in the deep tail[32,34] such that they flow toward the earth where they mirror in the strong lower altitude fields and return tailward to form the counterstreaming beams. A cross-tail electric field along with the different particle velocities spatially disperses ions of different energies that helps create the complicated, structured distributions seen by spacecraft. Since spacecraft moving from the lobe to the central plasma sheet typically detect earthward beams followed by counterstreaming beams followed by the isotropic plasma-sheet distributions, Eastman et al.[1] argue that the plasma-sheet boundary layer is the source of energetic plasma in the central plasma sheet.

Eastman et al.[1] also argue that the plasma-sheet boundary is a continually present feature of the magnetotail, although it may be detected more frequently at the time of substorms as plasma-sheet contractions

and expansions move it past an observing spacecraft. Figure 6 illustrates the regions where the ISEE 1 spacecraft detected the plasma-sheet boundary layer (heavy lines) in relation to the central plasma sheet (dotted lines) and tail lobe (dashed lines). From this plot Eastman et al. estimate that the thickness of the plasma-sheet boundary layer is less than 0.5 to 3 R_e.

A number of other observations reveal other features of the plasma-sheet boundary layer that tend to occur on shorter time scales that presumably correspond to smaller distance scales. Parks et al.[35] were the first to detect features on the edge of the plasma sheet. Their instrument had high time resolution rather than the ability to detect anisotropies and they noted the frequent presence of a layer of few keV electrons and ions adjacent to the more intense plasma sheet but with intensities intermediate between the intense plasma sheet and the tenuous lobes. They estimated the thickness of this layer at 200–10,000 km. A likely example of this phenomenon from the work of Cattell et al.[36] is presented in Fig. 7. Although not as clear as Parks' original examples, the somewhat reduced and variable fluxes in 4.8–6.6 keV electrons can be seen in panel d for some 30 s after and before the high intensity changes at 0740 55s and 0742 30s. Associated with these intermediate fluxes are variable electric fields (panel b) as measured by the ISEE 1 double probe electric field experiment. These burst-like electric fields typically occur for intervals of the order of a minute at the boundary of the region of more intense plasma fluxes. Peak amplitudes within these bursts typically are 30 mV/m. The field varies on time scales of a fraction of a second and the large spikes often occur within a few seconds of large particle gradients. These large electric field bursts are apparently seen on less than half the crossings. Cattell et al.[36] estimate the width of individual spikes as 1–100 km.

A study of how these electric-field type events in the boundary layer vary as a function of distance from the earth reveals a decrease in amplitude with increasing distance that is consistent with the electrostatic mapping

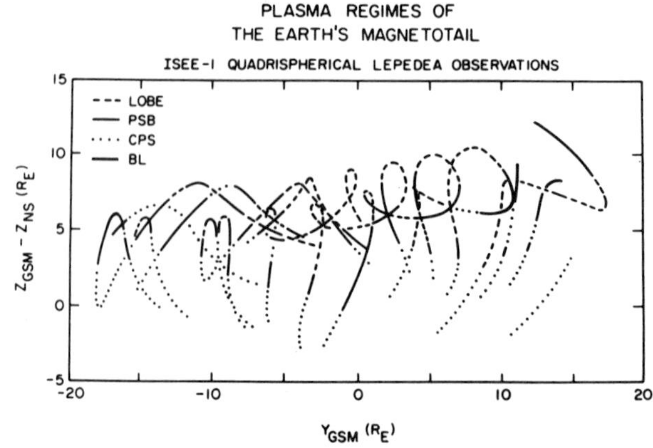

Figure 6—Illustration of the locations of various magnetotail regions in a cross-sectional projection of numerous ISEE 1 orbits (Eastman et al.[1]).

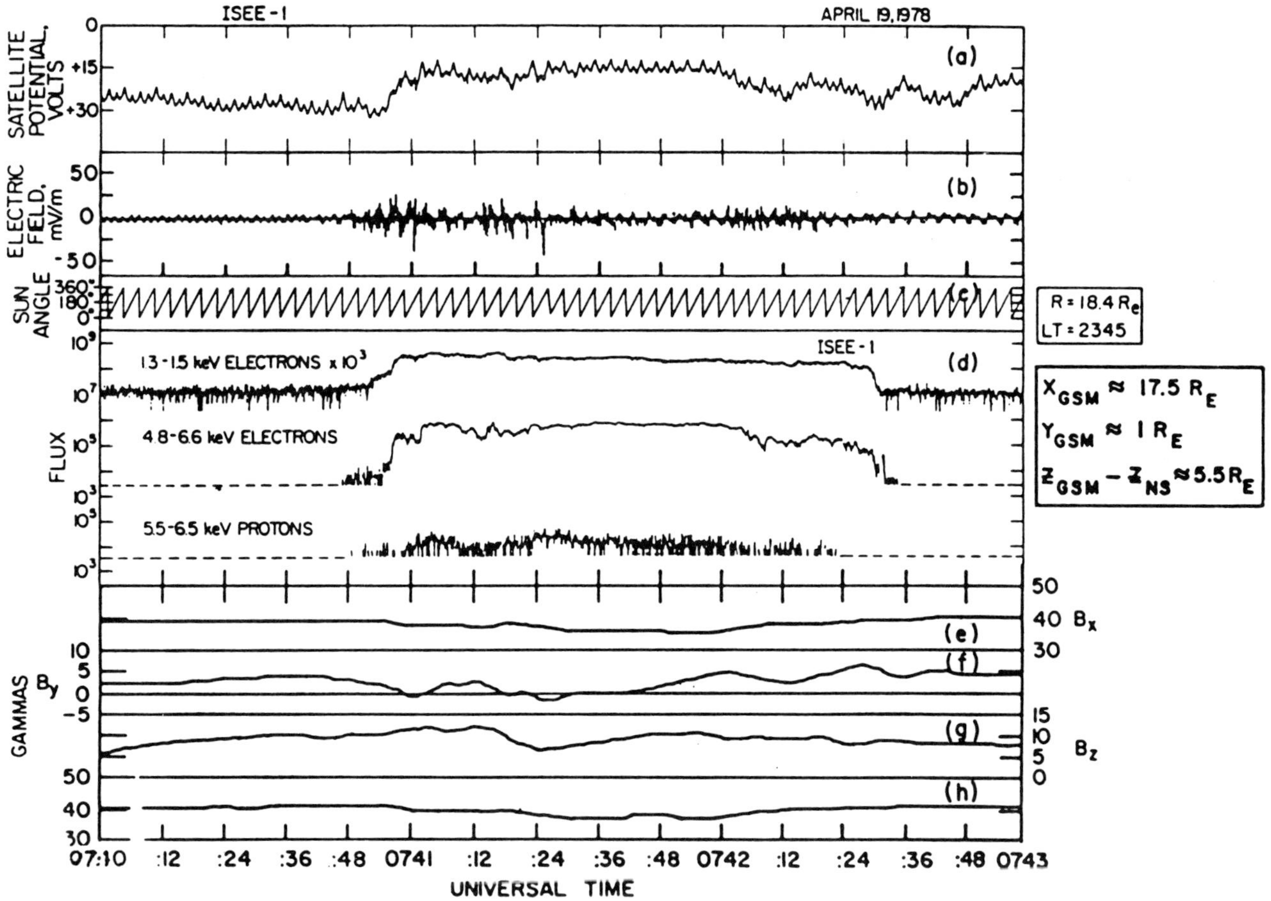

Figure 7—Illustration of regions of intense variable electric fields (panel b) associated with boundaries of the plasma sheet as seen in energetic particles (panel d) and the magnetic field (panels e–h) (Cattell et al.[36]).

of low-altitude electric fields along equipotential magnetic field lines to the magnetotail.[37] Since the low-altitude electric fields are thought to be producing auroral arcs, this result suggests that auroral arcs are linked to the plasma-sheet boundary layer. This same correspondence has been suggested independently by Lyons and Evans.[38] Using low-altitude spacecraft data, these authors noted that auroral arcs, as indicated by a large precipitating electron flux, tended to occur where precipitating energetic ions exhibited a full loss cone near their high-latitude boundary. Furthermore the low-altitude ion spectra were similar to those measured in the plasma-sheet boundary layer and also to those predicted by Lyons and Speiser[34] to result from acceleration in the distant tail current sheet. In their picture, ions accelerated in the distant tail current sheet form the plasma-sheet boundary layer and precipitate in the high-latitude portion of the auroral oval on the same field lines where auroral arcs occur. The auroral electrons are accelerated at low altitudes, but, except for the spatial correspondence, their relation to the ions is not clear. As mentioned earlier, Feldstein and Galperin[15] argue that spatial association of the standard auroral oval and the plasma-sheet boundary layer is incorrect.

Another characteristic of the plasma-sheet boundary layer is broad-band electrostatic noise,[39,40] which is much more common in the plasma-sheet boundary layer than in the adjacent lobe or central plasma-sheet regions.[1,41] Since this noise is closely associated with anisotropic ion distributions,[1] which often have a positive slope in the reduced distribution function,[1] it is attractive to envisage these beams as the free energy source for the noise.[42]

MAGNETOTAIL PLASMA PARAMETERS

Although plasmas in the earth's magnetotail have been intensively studied and typical values of densities and average energies in various regions are often cited, occurrence frequency distributions of such parameters have not heretofore been available from the most frequently studied region within 40 R_e. In Figs. 8–10 such distributions are shown for density and for proton and electron average energies.[43] In these figures plasma β has been used to define different tail regions and a value of 0.25 has been used to separate the low β tail lobe from two higher β regions in the plasma sheet. This value of 0.25 is somewhat arbitrary, but the substantially different character of the distributions of βs above and be-

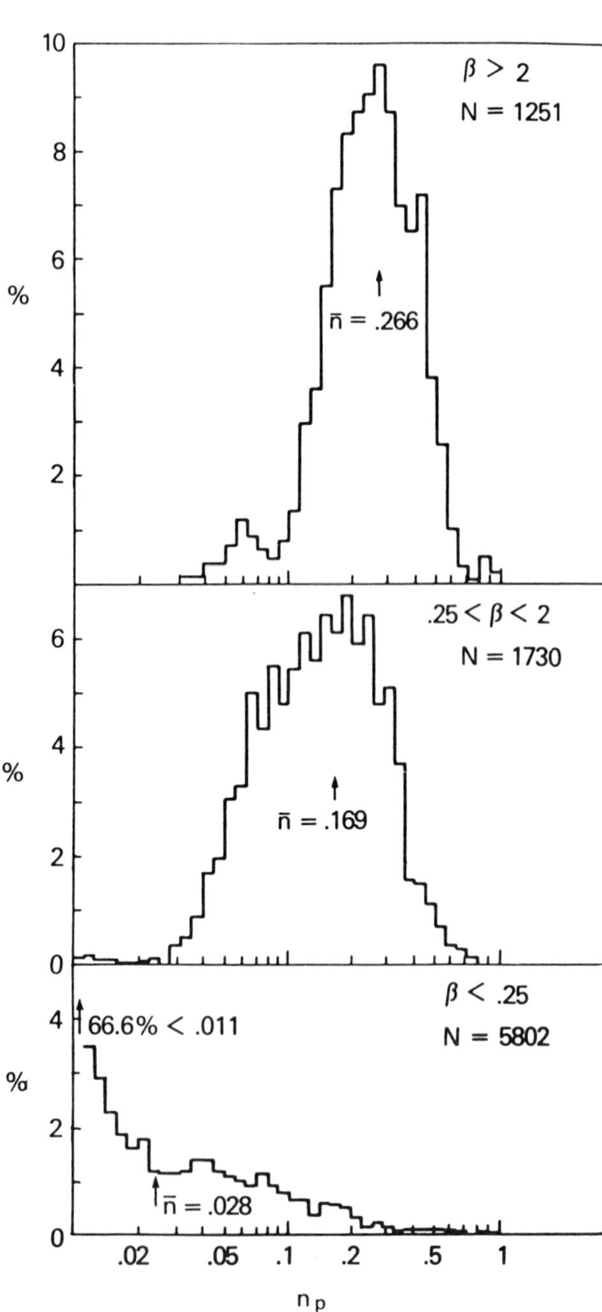

Figure 8—Occurrence distributions for magnetotail plasma density in three different plasma β regimes. $\beta < 0.25$ corresponds to the lobe and the two higher β distributions encompass the plasma sheet (Fairfield and Frank[43]).

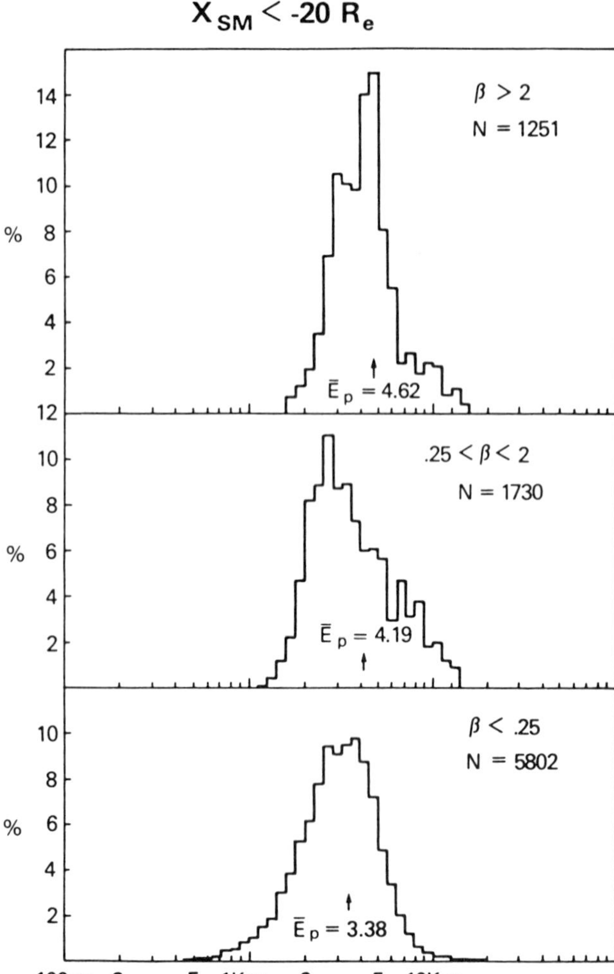

Figure 9—Occurrence distributions of proton average energies for the three β regimes of Fig. 8. (Fairfield and Frank[43]).

low this value suggests that this is a reasonable and natural choice. The $\beta > 2$ values undoubtedly come from a region nearest the equatorial plane that corresponds to the central plasma sheet. The $0.25 < \beta < 2$ values tend to come from the region between the $\beta < 0.25$ lobe and the central plasma sheet. This region contains the plasma-sheet boundary layer, but it prob-

ably also contains some central plasma sheet as well, since the number of measurements seems to be disproportionately large compared to the expected thickness of this boundary region.

If data for the two $\beta > 0.25$ distributions in Figs. 8–10 are combined, the average plasma sheet proton density and proton and electron average energies are 0.21/cc, 4.4 keV, and 1.0 keV, respectively. Average values for the various distributions are shown on the figures. If the average values for the two higher β regions are combined with the average field strengths of 7.3 nT and 15.3 nT, plasma pressures of 1.8×10^{-9} dynes/cm^2 are obtained for both regions. In the lobe region where plasma pressure is negligible and the field is 19.7 nT, total pressure is virtually the same at 1.7×10^{-9} dynes/cm^2. This result supports an earlier study based on individual traversals of the plasma-sheet/lobe boundary that showed that total pressure tends to be the same in the plasma sheet and in the tail lobes.[44] It should be noted, however, that both the earlier work supporting pres-

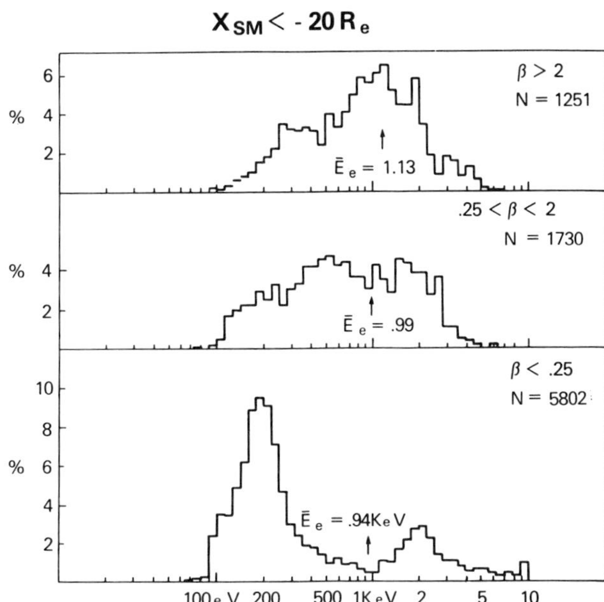

Figure 10—Same as Fig. 9, only for electrons (Fairfield and Frank[43]).

sure balance[44] and the more recent work[43] were accomplished only after a recalibration that reduced the measured plasma densities by 35%. ISEE 1 measurements of Lennartsson and Shelley[45] also appear to yield higher densities than those shown here (but comparable average energies) that would also lead to excessive plasma pressures. Whether the 35% reduction is warranted or whether there actually is a surprising pressure, nonequilibrium is not known.

Although conventional plasma experiments span the range of energies that make important contributions to the magnetotail energy density, they cannot always measure the very lowest energy particles that can influence the number density. Recently Etcheto and Saint-Marc[46] have circumvented this problem by obtaining particle density by measuring the plasma frequency with a relaxation sounder. They report the occurrence of cool particle densities up to 5/cc in narrow regions near the plasma-sheet boundary. It is interesting to speculate that these particles may be related to field-aligned currents that Atkinson[47] has argued will redistribute plasma in the magnetotail and produce high density regions.

Velocities in the plasma sheet range up to many hundreds of km/s but the average is near 100 km/s, with large velocities tending to occur at larger IMP distances of 20–30 R_e.[43,48] Flows in the earthward quadrant are slightly more frequent than the other directions but by less than a factor of two.[43,48] Flows in the $0.25 < \beta < 2$ region are slightly larger and most often in the earthward or tailward direction.

Plasma measurements in the more distant tail on ISEE 3 are limited to electrons. These measurements along with the magnetic field measurements generally reveal the same magnetotail regions as are present near earth.[49] Plasma-sheet parameters are quite compatible with those nearer earth with average densities of 0.2 – 0.3 and average temperatures of the order of 2×10^6 K (E_e = 170 eV)[19] that are consistent with lower tail-lobe field strengths at these distances. Flow velocities, however, are quite different from those nearer earth. Earthward of 120 R_e, the flows are preferentially along the earth-sun line with tailward flows being predominant. Beyond 120 R_e the flows are still larger in magnitude and almost exclusively tailward.[50]

COMPOSITION

One of the most significant magnetospheric discoveries of recent years is the realization that O^+ ions are often present in the magnetotail in significant numbers.[45,51-53] The source of these ions must be the ionosphere and hence their presence in the magnetotail identifies this low-altitude region as a significant source that supplements the solar wind in supplying plasma to the tail. He^{++} ions, on the other hand, originate in the solar wind and provide a tracer for this source.

Figure 11 (Sharp et al.,[53] as reproduced by Balsiger[54]) shows the relative occurrence frequencies of number densities of He^{++} and O^+ relative to H^+ in the energy range 100 eV–16 keV for quiet ($AE < 100$ nT) and disturbed ($AE > 500$ nT) geomagnetic conditions. The most striking result of Fig. 11 is the enhancement of O^+/H^+ during active conditions; on the average the O^+ density is 39% of the H^+ density during disturbed conditions but only 2% during quiet conditions. Although less dramatic, the He^{++} ions from the solar wind vary in the opposite manner. He^{++} is always at least 0.5% during quiet conditions with an average value of 2.0% whereas values are frequently less than 0.5% during disturbed conditions with an average of 1.1%. Ionospheric He^+ ions also increase with increasing activity but by a much smaller factor than O^+; this difference suggests that the polar wind is not the source of ionospheric ions but rather an ionospheric acceleration process must be involved.[55]

Evidence that the ionosphere is a source of magnetotail protons as well as O^+ ions can be deduced from the observation that field-aligned beams of H^+ and O^+ are frequently present in the plasma sheet, tail lobe, and plasma-sheet boundary layer.[1,55] An example of streaming O^+ ions in the plasma sheet is shown in Fig. 12. The bottom trace shows the counting rate for O^+ ions plotted versus time. The energy being measured changed from 630 eV to 1.6 keV at 0216 13.3s. The top trace indicates the pitch angle at various times during the 3 s spacecraft spin cycle. The measured ions at each energy are aligned near the field direction, with the 630 eV ions moving generally tailward and the 1.6 keV ions moving earthward. The fact that the ions are not moving exactly along the field is due to their drift motion in the existing magnetospheric E and B fields. Candidi et al.[56] and Orsini et al.[57] have conducted further studies of this nonalignment and even used it to obtain the north-south component of the magnetospheric electric field.

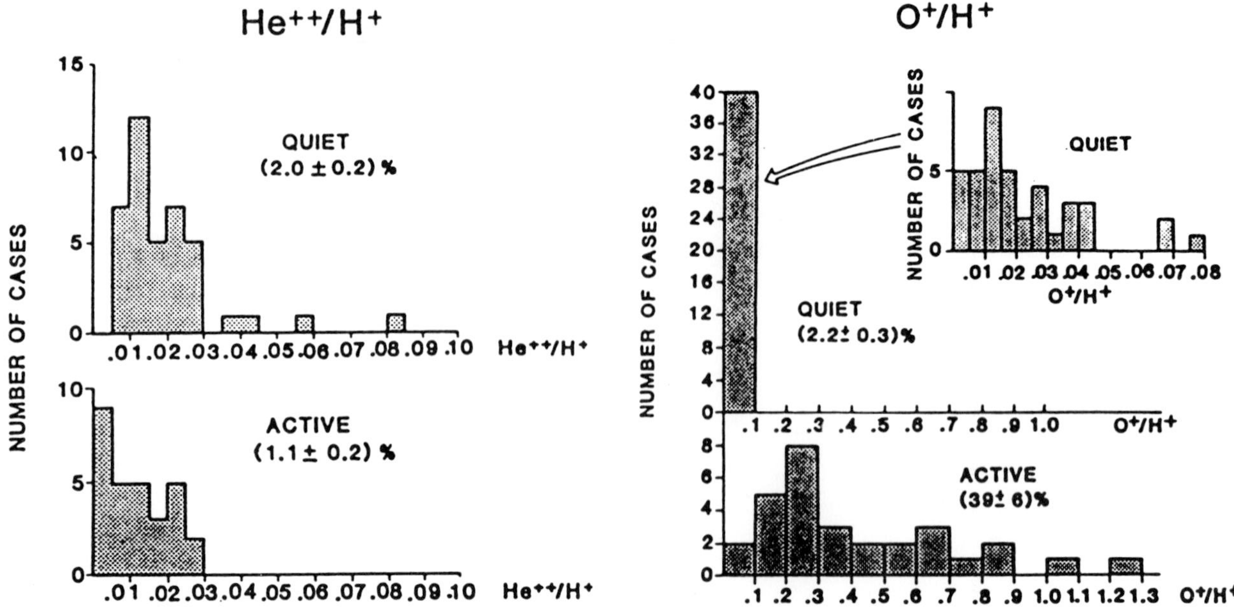

Figure 11—Distributions of plasma sheet O⁺ and He⁺ densities relative to proton densities for geomagneticly active and quiet conditions. Much more O⁺ is present during active times. The energy per charge range is 0.1–16 keV (Sharp et al.[53]).

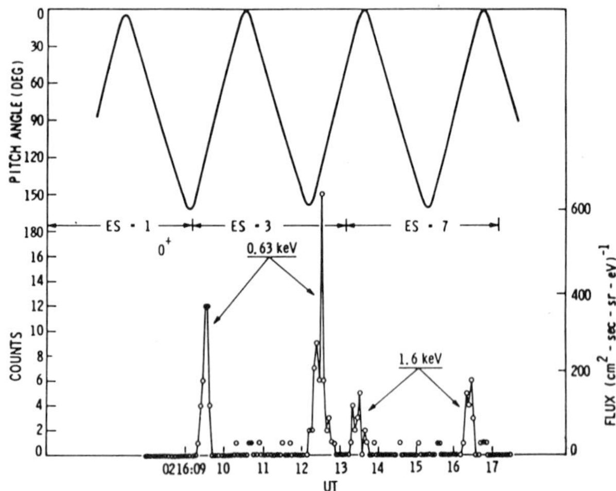

Figure 12—High time resolution ion mass spectrometer measurements of streaming O⁺ ions in the plasma sheet. The ion peaks occur near but not at the field direction (see text) (Sharp et al.[55]).

The H⁺ and O⁺ beams in the magnetotail are seen 10 to 30% of the time with O⁺ being slightly more prevalent than H⁺.[53] Sharp et al.[53] find that these beams typically have energies of hundreds of electron volts and the tail-lobe energy distributions of H⁺ and O⁺ are very similar, as if they were accelerated through the same electric potential. Frank et al.[51] and Candidi et al.,[56] on the other hand, find that the H⁺ and O⁺ usually have equal velocities rather than equal energies. Plasma-sheet oxygen beams tend to be more energetic and are seen more often when K_p is high.[53] The tail beams are always flowing tailward in the lobes, but they can be moving in either direction in the plasma sheet, as in Fig. 12, where field lines are thought to be closed. The field-aligned nature of the beams suggests that the ions have been accelerated at low altitudes and their pitch angles decreased as they adiabatically move to the weaker fields of the tail. A higher perpendicular temperature for H⁺, however, suggests additional heating of this component in the distant magnetotail.[55] These ion beams are clearly distinguished from beams in the magnetospheric boundary layer that are usually more intense and are thought to be due to a solar-wind source because they exhibit little O⁺ and have a solar-wind-like proportion of He⁺⁺ moving at the same velocity as H⁺.

The general characteristics of these lobe plasma-sheet beams seem to identify them as the same outflowing beams observed at low altitudes (Yau et al.,[58] and references therein). It should be noted that both the upflowing ions and plasma-sheet composition exhibit large solar cycle dependences.[45,59] Fortunately the "proton" statistics in Figs. 8–10 were taken relatively near solar minimum when O⁺ contamination would not be too large.

LACK OF A STEADY-STATE MAGNETOTAIL

The recent Fairfield and Frank study[43] confirms that average plasma pressure in the plasma sheet between X_{sm} of -10 and -20 R_e is not more than twice as large as that between -20 and -33 R_e. This result is not new or surprising, but it is the basis for an argument that a steady-state magnetotail cannot exist. One of the clearest, most-widely-accepted experimental results in magnetospheric physics is the existence of a global dawn-dusk electric field across the polar cap, at least during periods when the IMF is southward. It is also widely believed that this electric field is mapped to the magnetosphere along equipotential field lines. Associated with such a uniform cross-tail field one would expect earthward convection of plasma in the plasma sheet. However, when Erickson and Wolf[60] tried to model such a simple, realistic magnetotail quantitatively, they found that a steady-state solution did not exist. Schindler and Birn[61] essentially confirmed this result; they carried out a somewhat more rigorous calculation that demonstrated the existence of steady-state solutions, but only for unrealistic magnetotails that conflict with observations. The fundamental problem that Erickson and Wolf discovered is that when closed plasma-sheet flux tubes undergo earthward adiabatic convection from the deep magnetotail in the cross tail E field, they undergo a compression such that the near-earth plasma-sheet pressure is much too large. The problem can only be avoided by starting with an unrealistic tail configuration or by postulating an unrealistically large loss or cooling of plasma at the earthward end of the tail.[61] Birn and Schindler[62] confirmed this result in a three-dimensional calculation, and Birn and Schindler[63] justified the use of a quasistatic approximation that had been used in carrying out the earlier time-dependent calculations.

What the time-dependent, quasistatic models show is that the magnetotail avoids the above problem by assuming an increasingly tail-like field configuration with a thinner plasma sheet, increased energy storage, and a smaller equatorial field strength. Erickson[64] shows how a minimum in the equatorial ($B = B_z$) field strength develops near 10 R_e just outside of the region where the dipole field begins to increase very rapidly. All the above authors note that this slow evolution in the presence of steady-state boundary conditions is in agreement with the observed characteristics of the substorm growth phase. The small equatorial B_z also is just the condition that will trigger collisionless or resistive instabilities thought to be associated with neutral line formation (e.g., Schindler and Birn[61]).

There is also recent additional statistical evidence supporting these small equatorial fields and the resulting neutral line formation between 10 and 20 R_e. Figure 13 (adapted from Fairfield[29]) shows distributions of B_z within 3 R_e of the expected location of the current sheet for distance ranges of $-10\ R_e > X_{sm} > -20\ R_e$ and $-20\ R_e > X_{sm} > -34\ R_e$. At more distant locations one finds an average B_z of 1.7 nT and 15.5% of the B_z's are negative (southward). Nearer the earth, where one might expect the northward dipole field to become more important, one instead finds 29.0% of the fields are negative. This result supports the theoretical expectations of Erickson[64] and experimental results (e.g., Hones[65]) that indicate that neutral lines are indeed apt to form in the near-earth region.

FUTURE STUDIES

In this section some significant problems are outlined whose solution promises to advance significantly our understanding of the magnetotail. In spite of the empha-

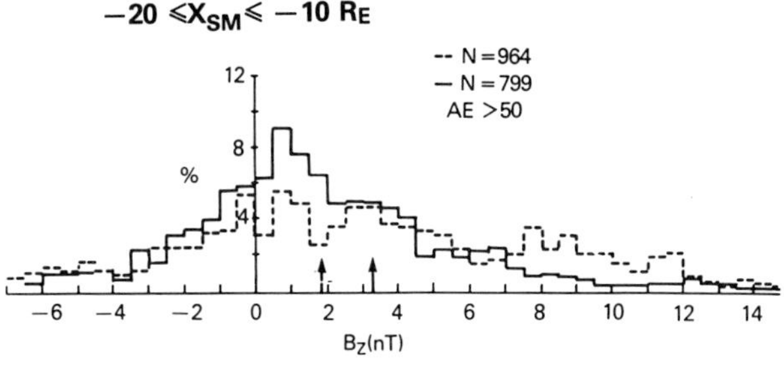

Figure 13—Equatorial distributions of B_z in the midnight sector for distance ranges on either side of 20 R_e. The fact that southward fields are seen almost twice as often nearer the earth is supportive of the idea that neutral lines form preferentially inside of 20 R_e (Fairfield[29]).

sis on the steady or average magnetotail in this paper, it will be necessary to include some discussion of time variations.

One of the most serious deficiencies in magnetotail knowledge is the lack of a quantitative time-varying magnetic field model. Such a model would permit quantitative study of the relation between high- and low-altitude processes such as auroral electron acceleration and magnetotail current-sheet acceleration. Although various useful quantitative models exist, at best they are valid for a designated average level of geomagnetic disturbance, whereas in reality the field configuration frequently exhibits significant changes on a scale of tens of minutes. In future studies of the time-varying magnetotail, it will be important to separate the relatively well-understood influences of the solar-wind velocity and density in order to obtain better information on the effects of IMF direction and related internal dynamic processes. Additional information is needed on the B_z component of the magnetotail field and its temporal variations at different locations. Also the locations and amounts of flux crossing the magnetotail boundary are of interest. The configuration and dynamics of the magnetotail under northward-IMF, geomagnetically quiet conditions is also a significant problem.

Magnetotail electric-field information is very important and not well known, due to the difficulties of making direct in situ measurements in tenuous plasmas. Although a simple global cross-tail field is presumed to exist on the basis of low-altitude measurements, the actual field is undoubtedly much more complex.

The complexities of the plasma-sheet boundary layer continue to pose many problems. Future studies in this region should involve data from various experiments, since at present it is often impossible to relate the results of different studies whose individual experiments do not even necessarily identify the same region as the boundary layer. The field-aligned currents and high-density/low-energy regions are particularly intriguing topics related to the boundary layer.

Further information on ion composition will continue to be useful in distinguishing between ionospheric and solar-wind plasma sources. Higher time-resolution data are particularly necessary in order to understand the exact role of substorms in low-altitude ion acceleration. Detailed study of the magnetotail boundary may shed further light on the solar-wind entry process.

The distant magnetotail remains a vast and variable region whose thermal ion characteristics are almost completely unknown. Continued study of this region and its relation to near-earth phenomena promises further insights into magnetotail behavior.

REFERENCES

[1] T. E. Eastman, L. A. Frank, W. K. Peterson, and W. Lennartsson, "The Plasma Sheet Boundary Layer," *J. Geophys. Res.* **89**, 1553-1572 (1984).

[2] E. W. Hones, Jr., ed., *Magnetic Reconnection in Space and Laboratory Plasma*, Geophysical Monograph No. 30, American Geophysical Union, Washington, D.C. (1984).

[3] D. M. Butler and K. Papadopoulou, eds., *Solar Terrestrial Physics: Present and Future*, Reference Publication 1120, National Aeronautics and Space Administration, Washington, D.C., Chapter 8 (1984).

[4] G. Rostoker and T. E. Eastman, "A Boundary Layer Model for a Magnetospheric Substorm, *J. Geophys. Res.* (in press, 1987).

[5] D. H. Fairfield and J. D. Scudder, "Polar Rain: Solar Coronal Electrons in the Earth's Magnetosphere," *J. Geophys. Res.* **90**, 4055-4068 (1985).

[6] G. Haerendel and G. Paschmann, "Interaction of the Solar Wind with the Dayside Magnetosphere," in *Magnetosphere Plasma Physics*, A. Nishida, ed., D. Reidel, Dordrecht, Holland, pp. 49-142 (1982).

[7] H. Rosenbauer, H. Grunwaldt, M. D. Montgomery, G. Paschmann, and N. Sckopke, "HEOS 2 Plasma Observations in the Distant Polar Magnetosphere: The Plasma Mantle," *J. Geophys. Res.* **80**, 2723-2737 (1975).

[8] T. E. Eastman, B. Popielawska, and L. A. Frank, "Three-Dimensional Plasma Observations Near the Outer Magnetosphere Boundary," *J. Geophys. Res.* **90**, 9519-9539 (1985).

[9] S. W. H. Cowley, "The Causes of Convection in the Earth's Magnetosphere: A Review of Developments During the IMS," *Rev. Geophys. Space Phys.* **20**, 531-565 (1982).

[10] N. Sckopke, G. Paschmann, G. Haerendel, B. U. O. Sonnerup, S. J. Bame, T. G. Forbes, E. W. Hones, Jr., and C. T. Russell, "Structure of the Low-Latitude Boundary Layer," *J. Geophys. Res.* **86**, 2099-2110 (1981).

[11] D. H. Fairfield, "On the Average Configuration of the Geomagnetic Tail," *J. Geophys. Res.* **84**, 1950-1958 (1979).

[12] D. H. Fairfield and A. F. Viñas, "The Inner Edge of the Plasma Sheet and the Diffuse Aurora," *J. Geophys. Res.* **89**, 841-854 (1984).

[13] Y. I. Feldstein and G. V. Starkov, "Dynamics of Auroral Belt and Polar Geomagnetic Disturbances," *Planet, Space Sci.* **15**, 209-222 (1967).

[14] T. E. Eastman, L. A. Frank, and C. Y. Huang, "The Boundary Layers as the Primary Transport Regions of the Earth's Magnetotail," *J. Geophys. Res.* **90**, 9541-9560 (1985).

[15] Y. I. Feldstein, and Y. I. Galperin, "The Auroral Luminosity Structure in the High-Latitude Upper Atmosphere: Its Dynamic and Relationship to the Large-Scale Structure of the Earth's Magnetosphere, *Rev. Geophys.* **23**, 217-275 (1985).

[16] H. C. Howe, Jr., and J. H. Binsack, "Explorer 33 and 35 Plasma Observations of Magnetosheath Flow," *J. Geophys. Res.* **77**, 3334-3344 (1972).

[17] J. H. Wolfe, "The Large-Scale Structure of the Solar Wind," in *Solar Wind*, C. P. Sonnett et al., eds., NASA SP-308, U.S. Government Printing Office, Washington, D.C., pp. 179-196 (1972).

[18] M. A. Schield, "Pressure Balance Between Solar Wind and Magnetosphere," *J. Geophys. Res.* **74**, 1275-1286 (1969).

[19] J. A. Slavin, E. J. Smith, D. G. Sibeck, D. N. Baker, R. D. Zwickl, and S.-I. Akasofu, "An ISEE-3 Study of Average and Substorm Conditions in the Distant Magnetotail," *J. Geophys. Res.* **90**, 10875-10895 (1985).

[20] D. H. Fairfield, "Time Variations of the Distant Geomagnetic Tail," *Geophys. Res. Lett* **13**, 80-83 (1986).

[21] K. Makita, C.-I. Meng, and S.-I. Akasofu, "Temporal and Spatial Variations of the Polar Cap Dimension Inferred from the Precipitation Boundaries," *J. Geophys. Res.* **90**, 2744-2752 (1985).

[22] D. G. Sibeck, G. L. Siscoe, J. A. Slavin, E. J. Smith, B. T. Tsurutani, and S. J. Bame, "Magnetic Field Properties of the Distant Magnetotail Magnetopause and Boundary Layer," *J. Geophys. Res.* **90**, 9561-9575 (1985).

[23] D. P. Stern, "A Study of the Electric Field in an Open Magnetospheric Model," *J. Geophys. Res.* **78**, 7292-7305 (1973).

[24] S.-I. Akasofu and M. Roederer, "Dependence of the Polar Cap Geometry on the IMF," *Planet. Space Sci.* **32**, 111-118 (1984).

[25] J. T. Gosling, D. N. Baker, S. J. Bame, W. C. Feldman, R. D. Zwickl, and E. J. Smith, "North-South and Dawn-Dusk Plasma Asymmetries in the Distant Tail Lobes: ISEE 3," *J. Geophys. Res.* **90**, 6354-6360 (1985).

[26] K. W. Behannon, "Mapping of the Earth's Bow Shock and Magnetic Tail by Explorer 33," *J. Geophys. Res.* **73**, 907-930 (1968).

[27] K. W. Behannon and N. F. Ness, "Magnetic Storms in the Earth's Magnetic Tail," *J. Geophys. Res.* **71**, 2327-2351 (1966).

[28] D. H. Fairfield, "Global Aspects of the Earth's Magnetopause," in *Magnetospheric Boundary Layers*, B. Battrick, ed., ESA Scientific and Technical Publications Branch, ESTEC, Noordwijk, The Netherlands, pp. 5-13 (1979).

[29] D. H. Fairfield, "The Magnetic Field of the Equatorial Magnetotail from 10 to 40 R_e," *J. Geophys. Res.* **91**, 4238-4244 (1986).

[30] C. T. Russell, "Some Comments on the Topology of the Geomagnetic Tail," *J. Geophys. Res.* **82**, 1625-1627 (1977).

[31] J. Birn and E. W. Hones, Jr., "Three-Dimensional Computer Modeling of Dynamic Reconnection in the Geomagnetic Tail," *J. Geophys. Res.* **86**, 6802-6808 (1981).

[32] D. J. Williams, "Energetic Ion Beams at the Edge of the Plasma Sheet: ISEE 1 Observations Plus a Simple Explanatory Model," *J. Geophys. Res.* **86**, 5507-5518 (1981).

[33] A. T. Lui, T. E. Eastman, D. J. Williams, and L. A. Frank, "Observations of Ion Streaming During Substorms," *J. Geophys. Res.* **88**, 7753-7764 (1983).

[34] L. R. Lyons and T. W. Speiser, "Evidence for Current Sheet Acceleration in the Geomagnetic Tail," *J. Geophys. Res.* **87**, 2276-2286 (1982).

[35] G. K. Parks, C. S. Lin, K. A. Anderson, R. P. Lin, and H. Reme, "ISEE 1 and 2 Particle Observatories of Outer Plasma Sheet Boundary," *J. Geophys. Res.* **84**, 6471-6476 (1979).

[36] C. Cattell, M. Kim, R. P. Lin, and F. Mozer, "Observations of Large Electric Fields Near the Plasma Sheet Boundary by ISEE 1," *Geophys. Res. Lett.* **9**, 539-542 (1982).

[37] S. Levin, K. Whitley, and F. S. Mozer, "A Statistical Survey of Large Electric Field Events in the Earth's Magnetotail," *J. Geophys. Res.* **88**, 7765-7768 (1983).

[38] L. R. Lyons and D. S. Evans, "An Association Between Discrete Aurora and Energetic Particle Boundaries," *J. Geophys. Res.* **89**, 2395-2400 (1984).

[39] D. A. Gurnett, L. A. Frank, and R. P. Lepping, "Plasma Waves in the Distant Magnetotail," *J. Geophys. Res.* **81**, 6059-6071 (1976).

[40] D. A. Gurnett and L. A. Frank, "A Region of Intense Plasma Wave Turbulence on Auroral Field Lines," *J. Geophys. Res.* **82**, 1031-1050 (1977).

[41] G. M. Parks, M. McCarthy, R. J. Fitzenreiter, J. Etcheto, K. A. Anderson, R. A. Anderson, T. E. Eastman, L. A. Frank, D. A. Gurnett, C. Huang, R. P. Lin, A. T. Y. Lui, K. W. Ogilvie, A. Peterson, H. Reme, and D. J. Williams, "Particle Field Characteristics of the High-Latitude Plasma Sheet Boundary Layer," *J. Geophys. Res.* **89**, 8885-8906 (1984).

[42] C. L. Grabbe and T. E. Eastman, "Generation of Broadband Electrostatic Noise by Ion Beam Instabilities in the Magnetotail," *J. Geophys. Res.* **89**, 3865-3872 (1984).

[43] D. H. Fairfield and L. A. Frank, "IMP-6 Plasma Statistics in the Geomagnetic Tail" (to be submitted, 1986).

[44] D. H. Fairfield, R. P. Lepping, E. W. Hones, Jr., S. J. Bame, and J. R. Asbridge, "Simultaneous Measurements of Magnetotail Dynamics by IMP Spacecraft," *J. Geophys. Res.* **86**, 1396-1414 (1981).

[45] W. Lennartsson and E. G. Shelley, "Survey of 0.1-16 keV/e Plasmasheet Ion Composition," *J. Geophys. Res.* **91**, 3061-3076 (1986).

[46] J. Etcheto and A. Saint-Marc, "Anomalously High Plasma Densities in the Plasma Sheet Boundary Layer," *J. Geophys. Res.* **90**, 5338-5344 (1985).

[47] G. Atkinson, "Field-Aligned Currents as a Diagnostic Tool: Result, a Renovated Model of the Magnetosphere," *J. Geophys. Res.* **89**, 217-226 (1984).

[48] L. A. Frank and K. L. Ackerson, "Several Recent Findings Concerning the Dynamics of the Earth's Magnetotail," *Space Sci. Rev.* **23**, 375-392 (1979).

[49] S. J. Bame, R. C. Anderson, J. R. Asbridge, D. N. Baker, W. C. Feldman, J. T. Gosling, E. W. Hones, Jr., D. J. McComas, and R. D. Zwickl, "Plasma Regimes in the Deep Geomagnetic Tail: ISEE 3," *Geophys. Res. Lett.* **10**, 912-915 (1983).

[50] R. D. Zwickl, D. N. Baker, S. J. Bame, W. C. Feldman, J. T. Gosling, E. W. Hones, Jr., D. J. McComas, B. T. Tsurutani, and J. A. Slavin, "Evolution of the Earth's Distant Magnetotail: ISEE 3 Electron Plasma Results," *J. Geophys. Res.* **89**, 11007-11012 (1984).

[51] L. A. Frank, K. L. Ackerson, and D. M. Yeager, "Observations of Atomic Oxygen (O^+) in the Earth's Magnetotail," *J. Geophys. Res.* **82**, 129-134 (1977).

[52] W. K. Peterson, R. D. Sharp, E. G. Shelley, R. G. Johnson, and H. Balsiger, "Energetic Ion Composition of the Plasma Sheet," *J. Geophys. Res.* **86**, 761-767 (1981).

[53] R. D. Sharp, W. Lennartsson, W. K. Peterson, and E. G. Shelley, "The Origins of the Plasma in the Distant Plasma Sheet," *J. Geophys. Res.* **87**, 10420-10424 (1982).

[54] H. Balsiger, "On the Composition of the Ring Current and the Plasma Sheet and What It Tells About the Sources of These Hot Plasmas," in *High Latitude Space Plasma Physics*, B. Hultqvist and T. Hagfors, eds., Plenum, New York, pp. 313-333 (1983).

[55] R. D. Sharp, D. L. Carr, W. K. Peterson, and E. G. Shelley, "Ion Streams in the Magnetotail," *J. Geophys. Res.* **86**, 4639-4648 (1981).

[56] M. Candidi, S. Orsini, and V. Formisano, "The Properties of Ionospheric O^+ Ions as Observed in the Magnetotail Boundary Layer and Northern Plasma Lobe," *J. Geophys. Res.* **87**, 9097-9106 (1982).

[57] S. Orsini, M. Candidi, H. Balsiger, and A. B. Ghielmetti, "Ionosphere Ions in the Near Earth Geomagnetic Tail Plasma Lobes," *Geophys. Res. Lett.* **9**, 163-166 (1982).

[58] A. W. Yau, E. G. Shelley, W. K. Peterson, and L. Lenchyshyn, "Energetic Auroral and Polar Ion Outflow at DE 1 Altitudes: Magnitude, Composition, Magnetic Activity Dependence, and Long-Term Variations," *J. Geophys. Res.* **90**, 8417-8432 (1985).

[59] A. W. Yau, P. H. Beckwith, W. K. Peterson, and E. G. Shelley, "Long Term (Solar Cycle) and Seasonal Variations of Upflowing Ionospheric Ion Events at DE 1 Altitudes," *J. Geophys. Res.* **90**, 6395-6407 (1985).

[60] G. M. Erickson and R. A. Wolf, "Is Steady Convection Possible in the Earth's Magnetotail?" *Geophys. Res. Lett.* **7**, 897-900 (1980).

[61] K. Schindler and J. Birn, "Self-Consistent Theory of Time-Dependent Convection in the Earth's Magnetotail," *J. Geophys. Res.* **87**, 2263-2275 (1982).

[62] J. Birn and K. Schindler, "Self-Consistent Theory of Three-Dimensional Convection in the Geomagnetic Tail," *J. Geophys. Res.* **88**, 6969 (1983).

[63] J. Birn and K. Schindler, "Computer Modeling of Magnetotail Convection," *J. Geophys. Res.* **90**, 3441-3447 (1985).

[64] G. Erickson, "On the Cause of X-Line Formation in the Near-Earth Plasma Sheet: Results of Adiabatic Convection of Plasma-Sheet Plasma," in *Magnetic Reconnection in Space and Laboratory Plasma*, E. W. Hones, Jr., ed., American Geophysical Union, Washington, D.C., pp. 296-302 (1984).

[65] E. W. Hones, Jr., "Transient Phenomena in the Magnetotail and Their Relation to Substorms," *Space Sci. Rev.* **23**, 393-410 (1979).

DISCUSSION

R. A. Hoffman: What is the cause of the often sharp delineation between the central plasma sheet and the plasma sheet boundary layer?

D. H. Fairfield: My tentative suggestion would be that the more isotropic, higher β central plasma sheet is on closed field lines where thermal ions undergo bounce motion along field lines conserving their first adiabatic invariant. The lower-density, lower-β plasma sheet boundary layer would be on field lines closing through the more distant plasma sheet with weak equatorial B_z where ions are unable to undergo bounce motion but rather where the beams are produced in the current sheet via the Lyons-Speiser mechanism. This suggestion is based mainly on general knowledge of the tail configuration, but could perhaps be further investigated by observing particle behavior such as the energy dependence of the transition (i.e., higher energies with larger gyroradii unable to bounce on higher latitude field lines). Equatorial drift in a dawn-dusk electric field would complicate this picture, however.

DYNAMICAL FEATURES OF THE PLASMA-SHEET ION COMPOSITION, DENSITY, AND ENERGY

W. Lennartsson*

A statistical study of more than 1500 hours of data from the energetic ion mass spectrometer on ISEE 1 has been undertaken in an effort to determine the average properties of the major plasma sheet ions at different levels of geomagnetic and solar activity. The data have been obtained between 10 and 23 R_e and cover energies between 0.1 and 16 keV/e. Four kinds of ions, the H^+, He^{++}, He^+, and O^+, make up the vast majority at all times, but the relative abundance of these varies strongly with varying activity. Two of the most prominent statistical trends are a decrease in the density of H^+ and He^{++} ions and an increase in the density of O^+ ions with rising levels of substorm activity, as measured by the hourly AE index. The H^+ ion usually constitutes well over 90% of the ions when $AE < 100$ γ but is sometimes matched or even surpassed by the O^+ when $AE \gtrsim 1000$ γ, at least locally. These trends represent a decline of the solar ion content in the plasma sheet with increasing substorm activity, and a partial replacement by ions of terrestrial origin. The peak densities of H^+ and He^{++} ions ($\gtrsim 1$ cm^{-3} total) are found in the flanks of the plasma sheet (GSM $Y \sim \pm 10$ R_e), whereas the peak O^+ densities ($\gtrsim 0.1$ cm^{-3}) are found in the middle section of the plasma sheet (GSM$|Y| < 10$ R_e), that is, in a region where cross tail currents are believed to be diverted through the ionosphere during substorms. The terrestrial ions also include the He^+, but this component is generally much less numerous than the O^+ and also less variable. The terrestrial component of the H^+ is strongly obscured by the solar component and is barely discernible only at the highest levels of activity.

Terrestrial and solar ions differ markedly in regard to energy as well. The mean energies of O^+ and He^+ ions remain about 3–5 keV at all levels of activity, but those of the H^+ and He^{++} ions increase monotonically with increasing AE index, ranging from about 1 keV/nucleon when AE approaches zero to well over 5 keV/nucleon when $AE \gtrsim 1000$ γ. The peak energies of the solar ions are found in the middle of the plasma sheet rather than the flanks, however. The energies of the H^+ and He^{++} ions, when combined with the densities, may suggest that solar ions have continual access to the plasma sheet without a significant increase in the bulk energy, but are temporarily heated during substorms, and reduced in numbers by injections into the inner magnetosphere, for instance.

Only the terrestrial ions are found to vary with the solar cycle, and the only clearly significant effect is in the overall densities of O^+ and He^+ ions. The O^+ is found to increase by about a factor of 3 over the time span of the measurements, this factor being the same at all levels of the AE index, whereas the He^+ only increases by a factor of 1.5. Since the measurement period coincides with a rising phase of the solar cycle, the effect may be due to an increase in the solar EUV flux, as discussed elsewhere in the literature in conjunction with similar observations at geostationary altitude.

INTRODUCTION

Among the several spacecraft that have carried energetic-ion mass spectrometers to date, the ISEE 1 has the most favorable orbit for surveying the plasma sheet, a near-equatorial orbit with apogee at 23 R_e.[1-3] During the course of several months each year (winter and spring) this orbit provides a fairly complete coverage of the plasma sheet inside of 23 R_e, making it possible to map the spatial distribution of different ions under varying geomagnetic conditions. This opportunity has been exploited in a recently completed statistical survey by Lennartsson and Shelley, a study that will be referred to here as Ref. 4. The present paper summarizes some of the results obtained in that study and focuses on the dynamical features that are associated with substorm activity.

The statistical material consists of approximately 900 plasma samples, each of which is based on a 1–3 hour long time-average of the differential flux of the four major ions, H^+, He^{++}, He^+, and O^+. The flux has been separately averaged in 30 contiguous energy channels, covering the energy-per-charge range,

$$0.1 - 16 \text{ keV/e} , \tag{1}$$

and in five contiguous pitch-angle ranges, and subsequently integrated to yield various moments of the ion velocity. The duration of each sampling has been chosen so as to correspond to roughly 1 R_e of spacecraft travel distance. The spatial distribution of these samples is illustrated in Fig. 1, which is taken from Ref. 4, as are the remaining figures.

These samples have been selected from a larger set of magnetotail samples as representing the plasma sheet proper. The selection has been based on a minimum total density of each sample (0.1 cm^{-3}), a minimum average ion energy (1 keV averaged over all ions), and a maximum tailward streaming velocity of H^+ and He^{++} ions (50 km/s). In addition to these criteria, which contain 1–3 hour averages of the ion fluxes, the selection has also been based on data having a much higher time resolution ($\sim$ 1 minute) obtained by the conventional electrostatic analyzer section of the instrument. These data have been used to exclude all samples obtained while the spacecraft made temporary excursions into the tail lobes or the magnetosheath. The sampling has been limited to $GSM\ R \geq 10$ R_e and $GSM\ X < 0$. The unequal samplings in 1978 and 1979 are due to

*Lockheed Palo Alto Research Laboratory, Palo Alto, California 94304.

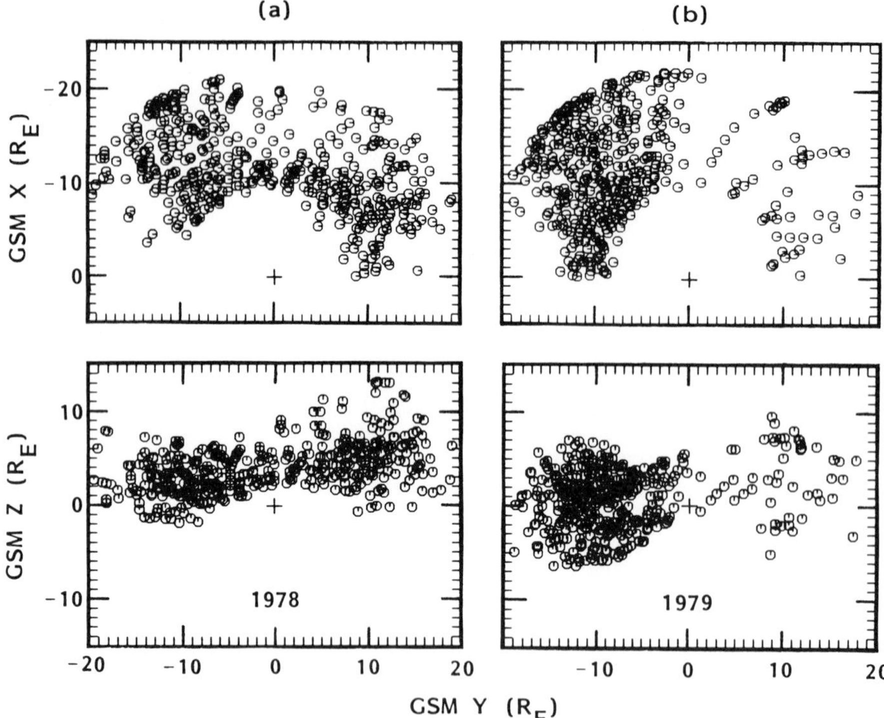

Figure 1—Distribution of plasma sheet samples in GSM coordinates during 1978 (left-hand panels) and 1979 (right-hand panels). The position of the earth is indicated by a cross in each panel.

changes in the spacecraft orbit and in the instrument operation procedures (not always covering the full range in Eq. 1). For a technical description of the instrument and its modes of operation, see Ref. 5.

PRINCIPAL STATISTICAL RESULTS

The bulk parameters of the four major ions show a large variance, especially the number densities. A part of this is related to the different locations of different samples, another part to variations in the geomagnetic activity, and a third part may be related to variations in the solar EUV flux.[6]

To determine some gross features of the spatial variance, the samples have been sorted into bins of 5 R$_e$ length in each of the three GSM coordinates. For geomagnetic correlations, the samples have been sorted according to the maximum hourly AE index during the time period consisting of the sampling itself (1–3 hours) and the hour immediately preceding (see for example Ref. 7). Long-term effects of changing solar EUV flux have been investigated by comparing data from 1978 and 1979.

Ion Densities

Of the four major ions the O^+ varies the most, by as much as three orders of magnitude in density, and is apparently strongly influenced both by the geomagnetic conditions and by the intensity of solar EUV radiation. This is illustrated in Fig. 2, which contains averaged data from approximately the same region of space, obtained one year apart. Note by comparison with Fig. 1 that the two sets of data have similar sampling in terms of GSM X and Y, as well as the absolute value

of Z. The later data period (1979) is close to the maximum of the solar cycle and has a higher average flux of solar EUV. Although there is no convenient index available for the EUV radiation itself, it is commonly assumed to follow the trends of the solar 10.7 cm radio

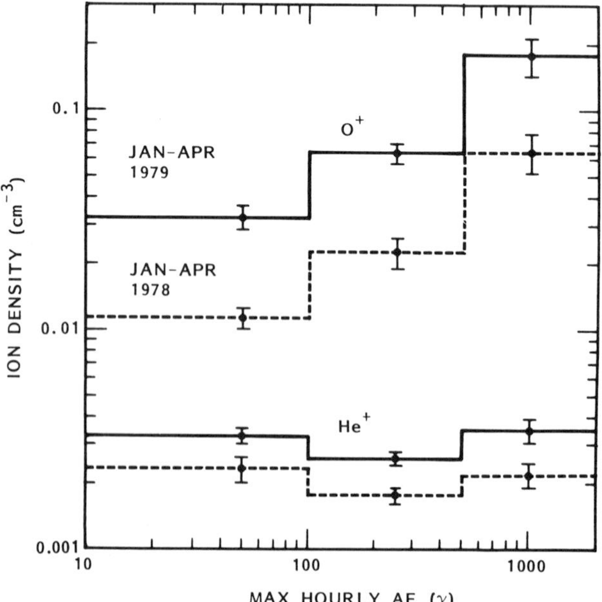

Figure 2—A comparison of the O^+ and He^+ densities during the early parts of 1978 (dashed lines) and 1979 (solid lines). Data from both years have been limited to the dawnside (GSM $Y <$ 0 and GSM $X < -5$ R$_e$). All error bars in this and following figures refer to the statistical uncertainty of the average itself ($\pm 1 \sigma$).

emission for which there is such an index. The later data period corresponds to an average daily 10.7 cm flux of 191.7 (in units of 10^{-22} Wm^{-2} Hz^{-1}) to be compared with an average flux of only 132.8 during the earlier data period. The daily flux is generally increasing during the two periods, but undergoes substantial fluctuations as well. The He$^+$ shows a lesser dependence on the solar cycle, as well as a lesser dependence on AE. Both ions are presumably of terrestrial origin.

All four ions are represented in Fig. 3, which shows the average density of all samples from 1978 and 1979 at each of 18 different levels of the AE index (defined by the tick marks on the abscissa). The He^{++} is presumably almost entirely of solar origin, whereas the H$^+$ may be dominated by solar protons, at least at the lower levels of activity. Both ions appear to decline in numbers with increasing activity (on average), but the decline is somewhat slower for the H$^+$, especially at $AE > 500\ \gamma$ and may reflect a growing proportion of terrestrial H$^+$ ions. Despite the strong increase in the O$^+$, the total ion density is significantly reduced during active conditions (dotted line).

The distribution of the total density in GSM X, Y, and Z is illustrated in Fig. 4a, b, and c, along with the distribution in Y of the O$^+$ density (Fig. 4d). The data have been grouped into only two ranges of the hourly AE index and have also been limited in some coordinate (Fig. 4b, c, and d) or by year (Fig. 4b and d). The distributions of the H$^+$ and He^{++} densities are similar to that of the total density. It may be noted, by comparing panels (Fig. 4a, b, and c) that the higher total density (and higher H$^+$ and He^{++} densities) during quiet conditions is primarily concentrated to the flanks of the plasma sheet. Conversely, the higher O$^+$ density during active times is concentrated around $Y = 0$ (Fig. 4d).

It should perhaps be mentioned that the two peaks in the dashed curve in Fig. 4b are present also when data samplings are limited to $X < -10\ R_e$, although the peaks are slightly lower then (about 10%).

Ion Energies

The distribution functions of the H$^+$ and the He^{++} are both fairly isotropic in these samples and are often similar to a Maxwell-Boltzmann distribution. The O$^+$ and the He$^+$ also have wide angular distributions, in general, but tend to be somewhat field-aligned (see Ref. 4 for details).

The average energies of all four ions including drift as well as "thermal" motion, are illustrated in Fig. 5 and are sorted by the hourly AE index in the same fashion as the densities in Fig. 3. The thermal part of these energies is generally strongly dominant, especially for the H$^+$ and He^{++} ions. Only the H$^+$ and He^{++} energies show a systematic dependence on the activity level. The He^{++} is seen to have a lesser energy per nucleon than the H$^+$ (on average), but a crude correction for the finite energy range of the data (Eq. 1) can be seen to bring the two ions closer together in terms of average energy per nucleon, as indicated by the dashed lines. The downturn of the H$^+$ energy at the highest activity levels may again reflect the increased admixture of terrestrial H$^+$.

As a last example of the different characteristics of solar and terrestrial ions, the distribution of the He^{++} and O$^+$ energies in GSM Y is shown in Fig. 6, which is constructed in analogy with Fig. 4b. Both energies vary somewhat with spatial coordinate (in opposite directions), but only the He^{++} energy varies significantly with the activity level. The spatial distribution of the He^{++} energy is similar to that of the H$^+$ energy, and the distribution of the O$^+$ energy resembles that of the He$^+$ energy.

COMMENTS

Possibly the most striking feature of the plasma sheet ion composition is the apparent mixing of solar and terrestrial ions. The traditional notion that the plasma sheet is populated by solar ions is supported by these data, in the sense that H$^+$ and He^{++} ions are found to carry most of the energy density most of the time (Figs. 3 and 5 combined), but this notion fails to address the whole story. Energetic O$^+$ ions are often a significant component as well, especially during disturbed conditions, and these ions are clearly of terrestrial rather than solar origin.[8] The frequent occurrence of O$^+$ ions with keV energy in the plasma sheet could hardly have been foreseen prior to the employment of space-borne mass spectrometers but may come as no surprise at this time. In fact, the O$^+$ in the plasma sheet has many features in common with the O$^+$ already observed at lower altitudes.

The apparent solar-cycle dependence in Fig. 2 is fairly consistent with results obtained for the O$^+$ at geosynchronous altitude by Young et al.[6] Their monthly averaged densities of O$^+$ (limited to $K_p < 2$) show an increase by a factor of 3–5 between the early parts of

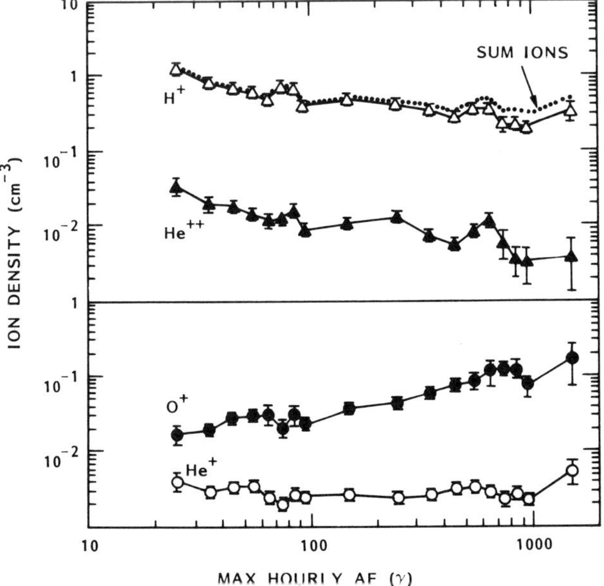

Figure 3—Average ion densities at different levels of magnetic activity. The dotted line represents the sum of the four densities shown by solid lines.

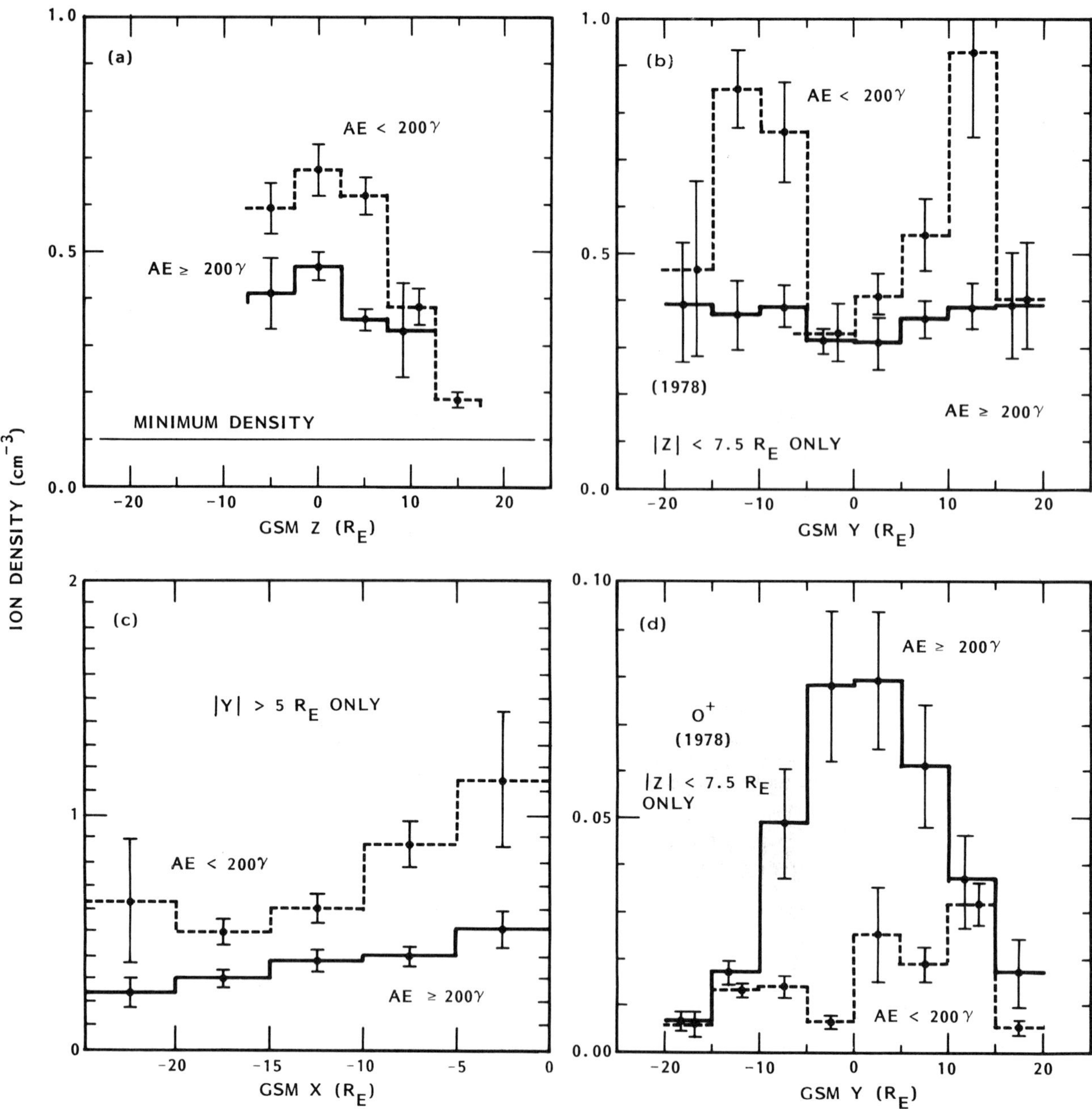

Figure 4—·Spatial distributions of the total ion density (a, b, and c) and the O$^+$ density (d) in different GSM coordinates. Note the different scales on the ordinates. The solid lines represent averages of data from "disturbed" times, when the maximum hourly $AE \geq 200$ γ. Data from quieter times are represented by the dashed lines.

1978 and 1979. Figure 2 is also qualitatively consistent with the explanation proposed by these authors. The basis for their explanation is an increased ionization and heating of the upper atmosphere, associated with increased solar EUV radiation. Of particular importance is the resulting increase in the ion scale heights, which brings a larger portion of the ions to altitudes of several thousand km, where auroral acceleration processes, in turn, raise the ion energies into the keV range and beyond (e.g., Refs. 9 and 10). The effect is expected to be the strongest for ions with the smallest initial scale height[6] and thus stronger for the O$^+$ than for the He$^+$ in agreement with Fig. 2.

The much briefer but stronger variations of the O$^+$ density that are seen during geomagnetic substorms, as measured by the AE index here (Figs. 2, 3, and 4d), are again consistent with results from lower altitudes (e.g., Refs. 6, 9-11). These variations may also be due, in part at least, to variations in the scale height of the ionospheric O$^+$ source, the heating in this case being supplied by electric currents.[6] It is conceivable, however, that changes in the spatial location and extent of the auroral acceleration processes themselves may be important as well. In any case, the substorm-related effects that are seen in the energetic (~keV) O$^+$ population in the plasma sheet are mainly in the number density. The aver-

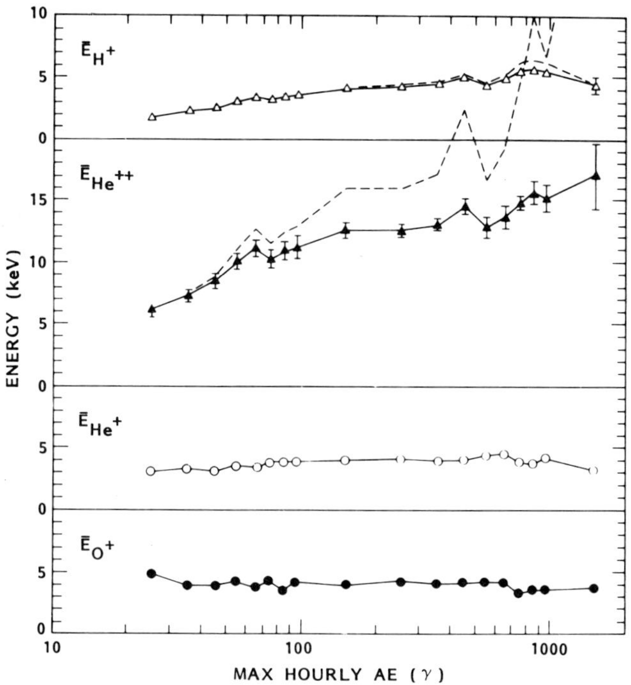

Figure 5—Average energies at different levels of magnetic activity. Error bars are shown only when larger than the data symbol. The dashed lines in the top two panels show the true average energy ($3kT/2$) of a Maxwell-Boltzmann distribution that would appear to have the same average energy as the solid lines, when only measured within the 0.1 – 16 keV/e energy window.

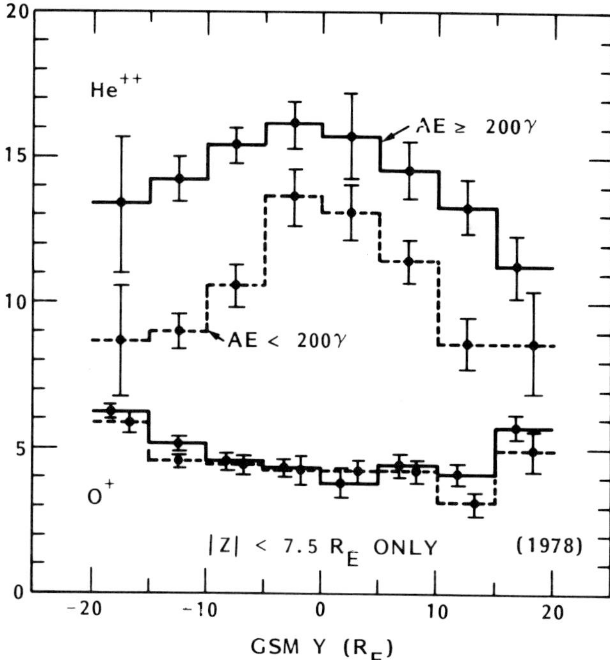

Figure 6—Distribution of the average energies of He^{++} and O$^+$ ions with respect to the GSM Y coordinate. No correction has been made here for the finite energy range of the samples.

age energy bestowed on these ions by the time they reach the plasma sheet is virtually constant (Figs. 5 and 6). The fact that the highest O$^+$ densities are found during substorms and in the neighborhood of local midnight (Figs. 3 and 4d) may suggest that the acceleration and injection of O$^+$ ions are physically related to the diversion of cross-tail currents through the ionosphere (cf. Ref. 12, Chapter 7 and Ref. 13).

In contrast, the H$^+$ and the He^{++} in the plasma sheet show many features that are not obvious in composition data from lower altitude. Earlier studies of the plasma sheet itself, using data from the ISEE 1 ion mass spectrometer[1-3,14], did uncover some of the differences between magnetically quiet and disturbed conditions, but the statistical material in those studies was too limited to show the nearly continuous functional relationships in Figs. 3 and 5. Neither did those studies have sufficient material to yield the spatial distributions in Figs. 4 and 6.

What is perhaps the most surprising feature in Fig. 3 is that the spatially averaged densities of H$^+$ and He^{++} ions approach a maximum when the hourly AE index approaches zero. Actually, the lowest values of the hourly AE index tend to occur during extended periods (up to 6 hours or more) of magnetic quiet, as shown in Ref. 4. This seems to imply that solar ions can enter the plasma sheet (beyond 10 R$_e$) without causing substorm activity. The latter is rather associated with a loss of H$^+$ and He^{++} ions from the plasma sheet, as indicated by the lower average densities at higher values of the AE index. A reasonable interpretation, it seems, would be that the solar ions have access to the plasma sheet at all times and that the loss rate of these ions is temporarily enhanced during substorms, either by injection into the inner magnetosphere or expulsion tailward, or both.[15]

A comparison of Fig. 3 with Fig. 5 shows that the maximum number of H$^+$ and He^{++} ions in the plasma sheet corresponds to a minimum in the energies of these ions. In fact, the spatially averaged energies approach values that are comparable to solar wind bulk energies (~ 1 keV/nucleon), when the AE index approaches zero. The one remaining difference is that the velocities are isotropized and "thermalized" in the plasma sheet (refer back to Ion Energies). This implies that solar ions can enter the plasma sheet without being substantially energized by the large-scale dawn-to-dusk electric field that is commonly assigned to an open magnetosphere.[16,17] A plausible interpretation might be that the plasma sheet is "open" to solar ions even with a very weak electric field component in the dawn-dusk direction. On the other hand, the reduction in the H$^+$ and He^{++} densities that is seen at higher levels of the AE index in Fig. 3 is associated with energization of these ions, according to Fig. 5, and may be the result of an enhanced dawn-dusk electric field. That would be consistent with the interpretation in the previous paragraph that the solar ions are injected into the inner magnetosphere (earthward of a "neutral line") or expelled tailward (tailward of a neutral line.[15])

ACKNOWLEDGMENT—This work was supported by NASA under contract NAS5-28702.

REFERENCES

[1] W. K. Peterson, R. D. Sharp, E. G. Shelley, and R. G. Johnson, "Energetic Ion Composition of the Plasma Sheet," *J. Geophys. Res.* **86**, 761 (1981).

[2] R. D. Sharp, D. L. Carr, W. K. Peterson, and E. G. Shelley, "Ion Streams in the Magnetotail," *J. Geophys. Res.* **86**, 4639 (1981).

[3] R. D. Sharp, W. Lennartsson, W. K. Peterson, and E. G. Shelley, "The Origins of the Plasma in the Distant Plasma Sheet," *J. Geophys. Res.* **87**, 10420 (1982).

[4] W. Lennartsson and E. G. Shelley, "Survey of 0.1 − 16 keV/e Plasma Sheet Ion Composition," *J. Geophys. Res.* **91**, 3061 (1986).

[5] E. G. Shelley, R. D. Sharp, R. G. Johnson, J. Geiss, P. Eberhardt, H. Balsiger, G. Haerendel, and H. Rosenbauer, "Plasma Composition Experiment on ISEE-A," *IEEE Trans. Geosci. Electr.* **GE-16**, 266 (1978).

[6] D. T. Young, H. Balsiger, and J. Geiss, "Correlation of Magnetospheric Ion Composition with Geomagnetic and Solar Activity," *J. Geophys. Res.* **87**, 9077 (1982).

[7] T. Kamei and H. Maeda, "Auroral Electrojet Indices (AE) for January-June 1979," in *Data Book 5*, World Data Center C2 for Geomagn., Kyoto Univ., Kyoto, Japan (Apr 1982).

[8] S. J. Bame, J. R. Asbridge, W. C. Feldman, M. D. Montgomery, and P. D. Kearney, "Solar Wind Heavy Ion Abundances," *Solar Phys.* **43**, 463 (1975).

[9] A. G. Ghielmetti, R. G. Johnson, R. D. Sharp, and E. G. Shelley, "The Latitudinal, Diurnal, and Altitudinal Distributions of Upward Flowing Energetic Ions of Ionospheric Origin," *Geophys. Res. Lett.* **5**, 59 (1978).

[10] H. Balsiger, P. Eberhardt, J. Geiss, and D. T. Young, "Magnetic Storm Injection of 0.9–16 keV/e Solar and Terrestrial Ions into the High Altitude Magnetosphere," *J. Geophys. Res.* **85**, 1645 (1980).

[11] A. W. Yau, E. G. Shelley, W. K. Peterson, and L. Lenchyshyn, "Energetic Auroral and Polar Ion Outflow at DE-1 Altitudes: Magnitude, Composition, Magnetic Activity Dependence, and Long-Term Variations," *J. Geophys. Res.* **90**, 8417 (1985).

[12] S.-I. Akasofu, *Physics of Magnetospheric Substorms*, D. Reidel, Hingham, Mass. (1977).

[13] D. N. Baker, T. A. Fritz, W. Lennartsson, B. Wilken, and H. W. Kroehl, "The Role of Heavy Ionospheric Ions in the Localization of Substorm Disturbances on 22 March 1979: CDAW 6," *J. Geophys. Res.* **90**, 1273 (1985).

[14] W. Lennartsson, R. D. Sharp, and R. D. Zwickl, "Substorm Effects on the Plasma Sheet Ion Composition on March 22, 1979 (CDAW 6)," *J. Geophys. Res.* **90**, 1243 (1985).

[15] E. W. Hones, Jr., in "Plasma Flow in the Magnetotail and Its Implications for Substorm Theories," *Dynamics of the Magnetosphere*, S.-I. Akasofu, ed., D. Reidel, Hingham, Mass., p. 545 (1979).

[16] T. W. Speiser, "Particle Trajectories in Model Current Sheets: 1. Analytical Solutions," *J. Geophys. Res.* **70**, 4219 (1965).

[17] S. W. H. Cowley, "Plasma Populations in a Simple Open Model Magnetosphere," *Space Sci. Rev.* **26**, 217 (1980).

DISCUSSION

J. Birn: Suppose there was an isolated substorm, followed by an extended quiet period. Then the high oxygen content during the substorm would eventually be followed by a more solar wind like composition after the quiet period. This could suggest that plasma that is lost from the plasma sheet during the substorm would eventually be replaced by solar wind plasma. Can this be observed on a shorter time scale at substorm recovery?

W. Lennartson: A separate study of that problem is under way, using 15–20 min. averages of the ion fluxes, but I am not yet prepared to answer your question. There are several practical problems associated with that kind of study. For example, the truly isolated substorm is rare and the spacecraft may not happen to be in a good position to make the necessary observations. Furthermore, the spacecraft is moving, and the plasma sheet may also be moving, making it difficult to separate spatial and temporal effects in the handful of events that may fit the definition of an "isolated" substorm.

R. Elphic: The equal energy per nucleon result that you discuss for the H^+ and He^{++} suggests an acceleration mechanism giving equal velocities to both species. Reconnection would do this. Are the energies you are showing the beam energies or the thermal energies?

W. Lennartson: The energies of the H^+ and He^{++} are almost entirely "thermal" in these plasma sheet samples. I should add that the H^+ and the He^{++} often have velocity distributions that are significantly different from each other, sometimes giving the same energy/charge. The velocity distributions are also often rather different from the Maxwell-Boltzmann shape that I have used in my crude "correction procedure." However, in a majority of the samples the H^+ and He^{++} appear to have more nearly equal energy/nucleon than equal energy/charge.

AURORAL ZONE FIELD-ALIGNED CURRENTS OBSERVED IN THE MAGNETOTAIL AND AT INTERMEDIATE ALTITUDES: AN ISEE PERSPECTIVE

R. C. Elphic*

T. J. Kelly[†]

H. E. Spence, R. J. Walker, and C. T. Russell*

M. Sugiura[‡]

Statistical studies of field-aligned currents observed at low altitudes in the auroral zone have revealed the now familiar Region 1 and 2 large-scale current systems. One should expect to observe these currents mapping to higher altitudes, presumably to the boundary of the plasma sheet and regions interior to it. What is observed by the dual ISEE spacecraft at intermediate altitudes (2–7 R_e) in the near-midnight local time sector is often (54%) of the cases) multiple current structures and ambiguous magnetic signatures. In only 27% of the cases is a Region 1 and 2 sense observed, and these observations are closely related to the onset of the substorm expansion phase. At the boundary of the plasma sheet between 10 and 20 R_e down the tail, the highest latitude currents are observed to be predominantly earthward-flowing in the near-midnight region. Most cases of tailward-flowing currents are observed dawnward of midnight, the overall sense thus opposing the expected Region 1 polarity. These results differ from earlier results, and may be due to the restricted local time sampling. Comparisons between DE and ISEE during times of magnetic conjunction also suggest that the highest latitude currents tend to be earthward near midnight. We discuss these results in light of near-earth and distant neutral lines and the resultant expected plasma convection.

INTRODUCTION

The communication of stresses and energy between the magnetotail and the terrestrial ionosphere is accomplished by field-aligned or Birkeland currents, readily observed by low-altitude spacecraft. Since the studies of Zmuda and Armstrong,[1] Sugiura,[2] and Iijima and Potemra[3] magnetospheric researchers have grown comfortable with the idea of large-scale Region 1 and 2 current systems arising from the stresses associated with magnetospheric dynamics. These currents have been well-observed at low altitudes by a variety of spacecraft; the Region 1 and 2 currents have been less thoroughly studied at high altitudes. Fairfield,[4] using IMP 4 and 5 data out to 17 R_e, found an overall agreement with the Region 1 sense in the 32 nightside cases he examined.

In this paper we examine the signatures of field-aligned currents in the near-midnight auroral zone at intermediate altitudes (2–7 R_e) and at high altitudes (10–20 R_e) using the dual ISEE spacecraft. For the cases at intermediate altitudes, we take advantage of occasions when the dipole is highly inclined with respect to the ecliptic, and when the ISEE orbit inclination is also large. At these times it is possible for the two spacecraft to pass rapidly across auroral zone field lines at relatively low altitudes. For the cases down the tail, we depend on plasma sheet expansion and thinning to take the spacecraft across field lines mapping to different auroral zone latitudes.

Ideally it is possible to use the spacecraft separation to determine the spatial structure of the currents. In practice this is more feasible for the intermediate altitudes than for the cases downtail. We shall show here a number of cases of currents observed near the earth in the midnight auroral zone, and then move on to some survey results for currents observed at the plasma sheet boundary layer.

BIRKELAND CURRENTS IN THE INNER MAGNETOSPHERE

Kelly et al.[5] have examined ISEE data from high-latitude, near-perigee passes for signatures of Birkeland currents. Figure 1 shows an outbound pass of ISEE on January 31, 1978, when the spacecraft traveled through the auroral zone field lines between about 3 and 6 R_e from the earth between 0100 and 0200 UT. The plane of the figure is the meridional plane containing the Z GSM axis and the instantaneous spacecraft position. The field lines shown are those of the model of Tsyganenko and Usmanov[6] in the 2400 LT plane. Shaded areas correspond to regions of field-aligned currents observed on ISEE.

The time series of the magnetic field data for this pass are shown in Fig. 2. Here the data from ISEE 1 (thick traces) and ISEE 2 (thin traces) are shown in detrended field-aligned coordinates, with the Z component along the model background field, the Y component azimuthal (eastward), and the X component completing the system (orthogonal to both the background field and the

*Institute of Geophysics and Planetary Physics, University of California, Los Angeles, California 90024.
[†]Applied Research Corporation, 8201 Corporate Drive, Landover, Maryland 20785.
[‡]Geophysical Institute, Kyoto University, Kyoto 606 Japan.

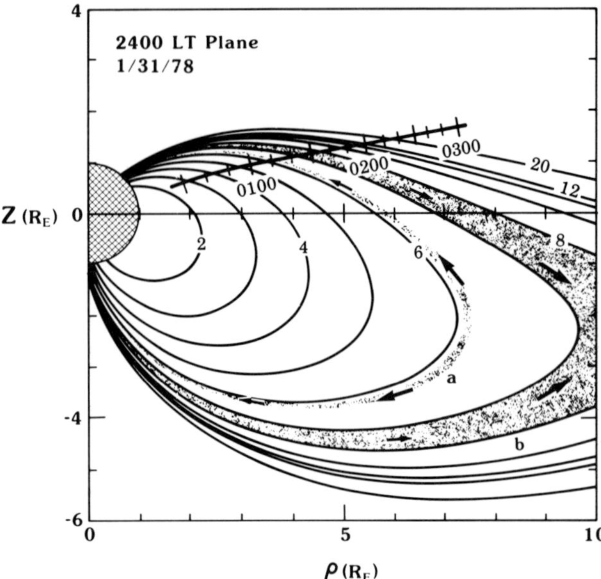

Figure 1—Trajectory of ISEE 1 and 2 on January 31, 1978 between 0100 and 0200. Plane of the figure is the meridional plane containing the Z GSM axis and the instantaneous spacecraft position. Also shown are model field lines in the 2400 LT plane.

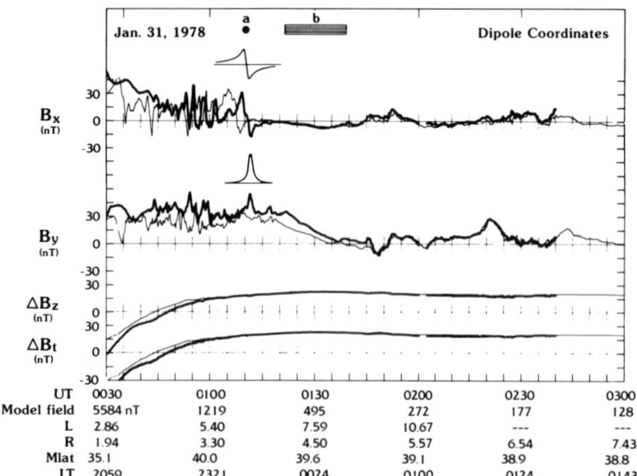

UT	0030	0100	0130	0200	0230	0300
Model field	5584 nT	1219	495	272	177	128
L	2.86	5.40	7.59	10.67	---	---
R	1.94	3.30	4.50	5.57	6.54	7.43
Mlat	35.1	40.0	39.6	39.1	38.9	38.8
LT	2059	2321	0024	0100	0124	0143

Figure 2—Time series of ISEE 1 (thick traces) and ISEE 2 (thin traces) for January 31, 1978. The coordinate system is a field-aligned system with Z along the model field and Y azimuthal and eastward.

azimuthal direction). The data, once rotated into this local field coordinate system, are "detrended" in the sense that the model field has been removed (Olson and Pfitzer [unpublished manuscript, 1977]). Field-aligned currents lying in a sheet at a constant L value will produce perturbations in only the B_y component as the spacecraft fly through that L shell.

Figure 2 shows considerable structure in the X and Y components around 0100, suggesting that the currents flowing here are highly filamented. Indeed, there is one clear signature of what appears to be a line current at about 0115. This signature consists of a bipolar excur-

sion in X and a single peak in Y; insets show the signature expected from an earthward line current traveling azimuthally at a location equatorward of the spacecraft. Later, beginning at 0120, there is a long, slow decrease in the B_y component, corresponding to the outward passage of the spacecraft across a sheet of current flowing out of the ionosphere. The separation of the two spacecraft and the lag of the ISEE 1 signature behind that of ISEE 2 suggest a thickness of about 4000 km, with current densities of 20 or 30 nA/m^2.

The upward flowing current at about 0130 could be interpreted as part of the high-latitude Region 1 system, but note that variations in the field continued for at least another hour. These variations may be due to currents flowing in the plasma sheet boundary layer. Unfortunately, the lack of separation between the two spacecraft traces at this time precludes a calculation of current structure. Nevertheless it is apparent that no clear large-scale Region 1 and 2 signatures are seen on this pass. Instead, at these altitudes, the currents tend to be highly structured.

To help bring these observations into perspective, one should note that between 0100 and 0200 substorm recovery was observed by ground stations in Iceland and Norway. AE averaged 89 nT for this hour. The highly structured currents discussed above may be a recovery feature.

Figure 3 shows another outbound pass of ISEE, this time on February 2, 1978. The field lines shown here correspond to the 2330 LT meridian. Regions of field-aligned currents are shown by the shaded areas. In Fig. 4 we see the time series for this day. ISEE 2 is again leading and observes the $+Y$ excursion at about 1030, a perturbation corresponding to a current flowing into the ionosphere. Following this, the Y component returns to zero, corresponding to passage through a current

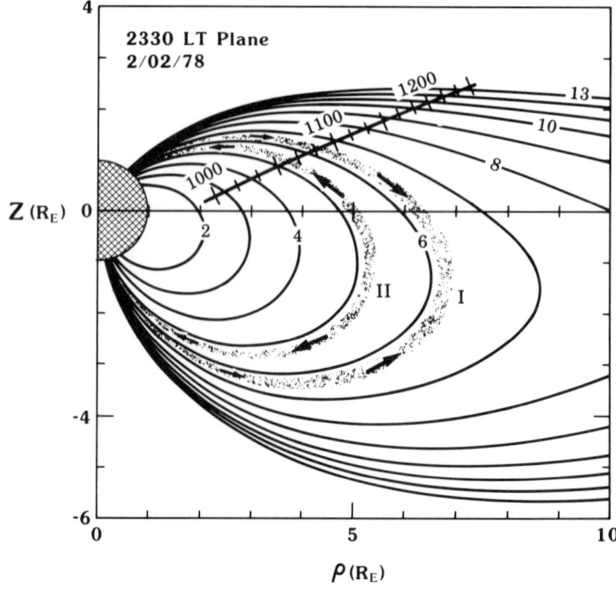

Figure 3—Same as Fig. 1 but for a pass on February 2, 1978.

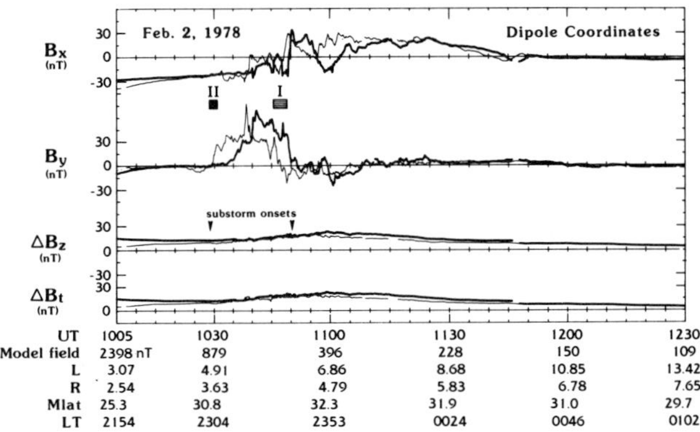

Figure 4—Same as Fig. 2 but for the pass on February 2, 1978.

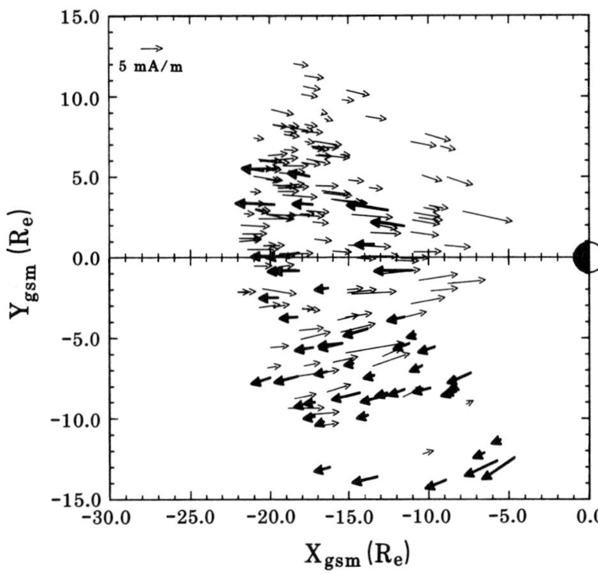

Figure 5—Magnitude and direction of field-aligned currents observed at the outermost edge of the plasma sheet, projected onto the X-Y GSM plane. From Elphic et al.[7]

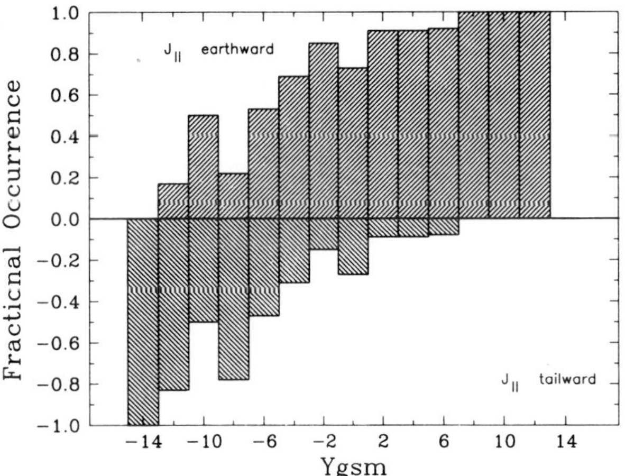

Figure 6—Fractional occurrence of earthward and tailward field-aligned currents as a function of position in Y GSM. Note that tailward currents tend to be found toward dawn.

flowing out of the ionosphere. Current densities for each current layer are about 20 nA/m². One further feature to note is the extreme structure in the X component, which one would not expect if the currents are sheets lying in the Y-Z plane. Finally, perturbations continue to be seen beyond the Region 1 current until almost 1130.

Two substorm onsets were observed in ground data at this time, one at about 1030, and the other at about 1050. *AE* averaged around 300 nT between 1000 and 1200. At College, Alaska, near the spacecraft's conjugate point, substorm expansion was observed beginning at 1041, with recovery extending from 1059 to about 1200. It is frequently the case that the currents observed at ISEE are associated with substorm activity.

Of the 180 orbits surveyed by Kelly et al.,[5] only 104 had sufficient data coverage to show at least one large-scale Birkeland current if such a current were present. Of these, 28 actually showed such currents. Only 11 of these were within 3 hours of midnight, and of these 11, four contained what could be regarded as large-scale Region 1 and/or 2 current systems. The remaining seven passes were quiet or contained small-scale field-aligned current signatures. One does not gain the impression that distinct Region 1 and 2 signatures are at all evident in the near-midnight auroral zone sampled at intermediate altitudes by ISEE.

FIELD-ALIGNED CURRENTS OBSERVED DOWNTAIL

Elphic et al.[7] recently published a survey of field-aligned currents observed by ISEE at the plasma-sheet boundary layer between 10 and 20 R_e down the tail. They found that most currents at the outermost boundary of the plasma sheet (presumably the highest latitude currents) flow earthward throughout the near-midnight region. Tailward flowing currents, while in the minority, tend to be found toward dawn. A plot of the location, strength, and polarity of these currents can be seen in Fig. 5. The fractional occurrence of the two polarities are shown in Fig. 6, which clearly reveals the quasi-Region 2 polarity.

One should bear in mind that these currents are found on the outermost boundary of the plasma sheet, and therefore should map to the highest latitudes of the auroral region. There may be other currents deeper within the plasma-sheet boundary layer. How these results differ from those of Fairfield[4] and Sugiura[2] is not immediately clear, but may be due to the selection or only the outermost currents, not necessarily the total current in the boundary. Moreover, we have confined our attention to near-midnight local times, where a clear Region 1 sense may not exist.

It is interesting to note that these currents vary in magnitude, depending on magnetospheric conditions. Figure 7 shows sheet current amplitudes as a function of the three-hour K_p index for the nearly 200 cases in our survey. The upper panel shows the behavior of the earthward-flowing currents, and the lower panel that of the tailward currents. The asterisks correspond to cases

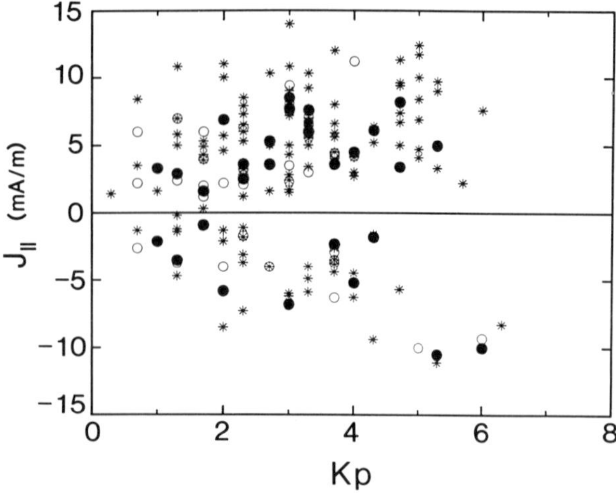

Figure 7—Sheet current amplitudes from the ISEE survey as a function of K_p. Current strength tends to increase with higher levels of activity. Asterisks correspond to cases of plasma-sheet expansion, filled circles to cases of brief fluctuations of the boundary across the spacecraft, and open circles to plasma-sheet thinnings.

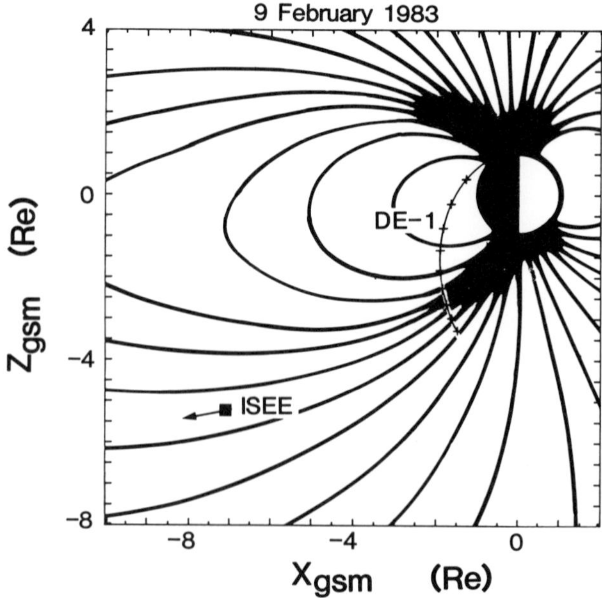

Figure 8—Trajectory of DE 1 and the ISEE 1 and 2 spacecraft for a period of magnetic conjunction on February 9, 1983. Plane of the figure is the *X-Z* GSM plane. Field direction at ISEE is denoted by the arrow.

of plasma-sheet expansion over the spacecraft, filled circles to cases where the plasma-sheet boundary fluctuated briefly over the spacecraft, and open circles to cases of plasma-sheet thinning. There is a clear tendency for the currents to intensify with higher K_p, just as magnetospheric currents at low altitudes do, but there is no clear dependence of current magnitude on plasma-sheet thinning or expansion.

SIMULTANEOUS DE AND ISEE OBSERVATIONS

Additional insight into the mapping of near-midnight auroral zone Birkeland currents may be gained by comparing ISEE and DE (Dynamics Explorer) observations when the spacecraft are in magnetic conjunction. This may lead to a better understanding of the details of magnetotail-ionosphere coupling. As a first step, we can check that the current magnitudes and polarities seen at low altitudes on DE and at high altitudes on ISEE are in agreement. One should be cautious in this comparison since the magnetic conjunction is based purely on model field mapping. At DE, we can usually rely on the relatively high spacecraft velocity to provide a profile of auroral zone currents; for ISEE, however, it is usually plasma-sheet expansion and thinning that allows the spacecraft to sound the currents in the boundary layer. We can remove a little of the resulting positional ambiguity by narrowing our attention to the highest latitude currents at DE, and the outermost boundary layer currents at ISEE.

Figure 8 shows the positions of the ISEE and DE 1 spacecraft in the *X-Z* GSM plane for a period on February 9, 1983. Also shown are field lines based on the Usmanov and Tsyganenko model. The field direction at ISEE is shown by the arrow. Figure 9a shows the DE 1 magnetic field time series as the spacecraft passed equatorward through the auroral zone. The coordinate system has Z along the model field, Y positive in the eastern azimuthal direction, and X completes the system. The first evidence of field-aligned currents is seen at about 1738 when DE 1 passes through a downward current, as indicated by the net $+Y$ field excursion.

Data from ISEE 2 are shown in Fig. 9b in the same coordinate system. Note that the baseline for the *BZ* and *BT* traces is 80 nT. At 1738, ISEE is in the plasma sheet boundary layer based on Berkeley particle data, and the field there has a positive Y component. Later, upon emergence from the plasma sheet near 1750 and again at 1806, the Y component returns to 0, suggesting the presence of an earthward-flowing current at the boundary of the plasma sheet. Assuming a simple mapping of the DE currents to ISEE altitudes, the expected perturbation in the Y component would be that indicated by the error bars. This appears to be in agreement with the perturbation observed at ISEE.

Other comparisons of this sort suggest that the highest latitude current observed by both ISEE and DE tend to be downward currents, in keeping with the statistical results described in the last section. Presumably these currents are carried by ionospheric electrons flowing out of the ionosphere poleward of the precipitating auroral electrons.

SUMMARY AND DISCUSSION

We have reviewed some of the observations of field-aligned currents in the near-midnight auroral zone from ISEE when the spacecraft were at intermediate and high altitudes, and for times of ISEE and DE magnetic conjunction. The ISEE intermediate altitude results suggest the absence near midnight of any large-scale, organized

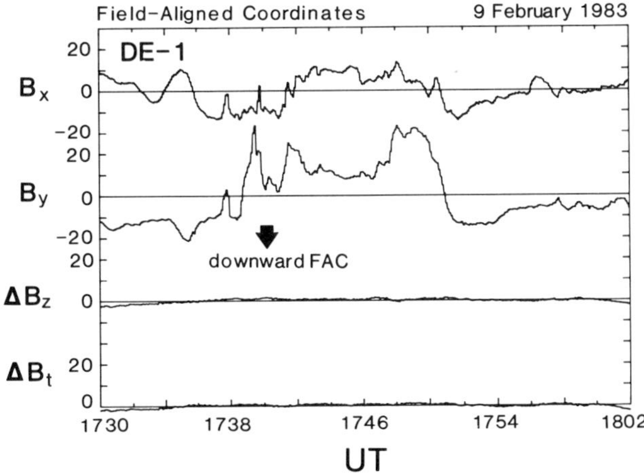

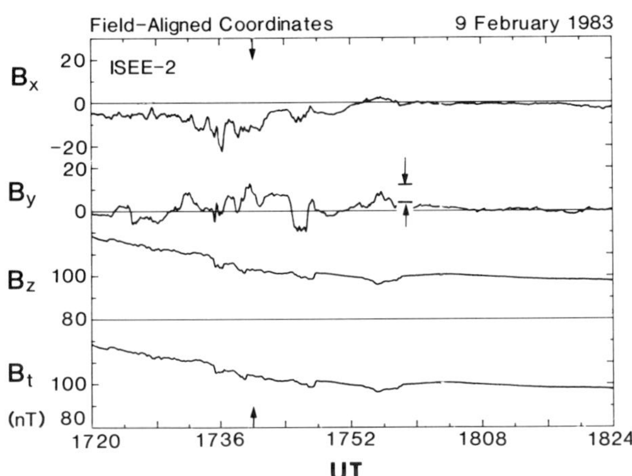

Figure 9—(a) DE 1 field data in field-aligned coordinates (see text). Deflection of *Y* component at about 1740 corresponds to a net downward current. (b) ISEE 2 field data from February 9, 1983 in field-aligned coordinates. The spacecraft is in the plasma-sheet boundary layer at about 1740, but by 1600 is in the lobe. Error bar denotes the magnitude of the field deflection corresponding to the DE 1 currents observed at lower altitudes.

Region 1 or 2 structure; instead, the currents are dominated by small-scale structures. At distances between 10 and 20 R_e downtail, the outer edge of the boundary of the plasma sheet is often observed to carry field-aligned currents, with earthward-flowing currents dominating throughout, but with tailward currents occurring toward dawn. This is opposite to the large-scale Region 1 sense. Finally, we looked at low-altitude DE observations as the spacecraft passed through the midnight auroral zone while ISEE was near the boundary of the plasma sheet at higher altitudes. This and other cases suggest that the currents map reasonably well between the two locations, that the poleward-most currents tend to flow into the ionosphere, and that upward-going ionospheric electrons probably carry the current.

Some perspective on these observations may be gained by considering the global current flow in the magnetosphere as deduced from model fields, in this case those of Tsyganenko and Usmanov.[6] This is accomplished by taking the curl of the model field near the earth, aver-

aging the field-aligned component over several R_e in altitude, and then mapping these currents to the equatorial plane. The model and the currents vary with activity level. The results are shown in Fig. 10, for $K_p > 3+$. The usual Region 1 system can be seen mapping out to the near-magnetopause region on the dayside and through dusk and dawn. A similar Region 1 system can be seen on the nightside near the earth. However, at higher latitudes on the nightside (at greater radial distances in the equatorial plane), there is a nightside system with a Region 2 sense, somewhat like the statistical results from ISEE downtail.

One rather simplistic explanation may be that these current systems define regions of similar plasma vorticity. Such vortical convection communicates stress to the ionosphere via field-aligned currents. For the dayside Region 1 system, the plasma circulation is nightward at the magnetopause with a return flow at lower latitudes. A similar circulation governs the Region 1 system on the nightside near the earth, but the Region 2 system at larger distances is governed by circulation associated with the existence of near-earth and distant neutral lines. In this picture the boundary between the two Region 1 systems and the distant Region 2-like system is demarcated by the neutral lines. It is worth noting, however, that for lower values of K_p the distant Region 1-like system diminishes, and more of the nightside is dominated by the Region 2-like system. Pressure gradients may be responsible for these currents.

One note of caution: while the currents are inferred from the near-earth model field, mapping these currents to the equator in the plasma sheet may lead to unrealistic equatorial locations because the field in the plasma sheet is much more uncertain than that outside. Consequently the inferred convection boundaries and neutral

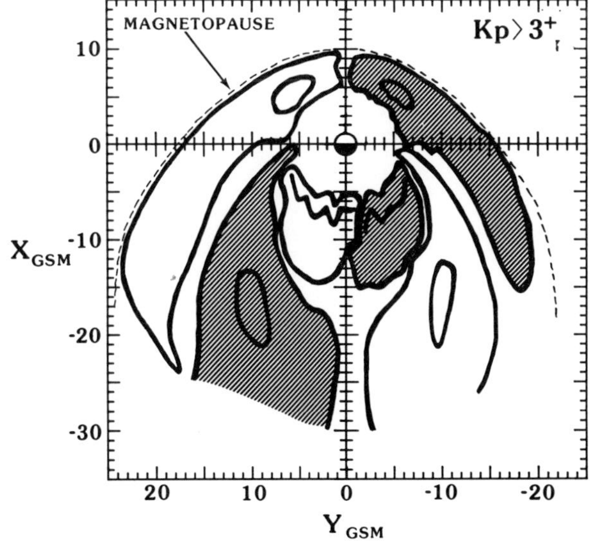

Figure 10—Regions of field-aligned current flowing into the ionosphere (hatched regions) and out of the ionosphere (unhatched regions) based on currents inferred from the magnetospheric model of Tsyganenko and Usmanov.[6] As shown here, the currents have been mapped to the equatorial plane.

lines may not be in their true physical locations. And there is another difficulty: at low values of K_p the near-earth Region 1-like system diminishes and more of the nightside is dominated by the Region 2-like system; this is not expected for a single distant neutral line.

ACKNOWLEDGMENT—We wish to thank D. Fairfield for helpful discussions, and gratefully acknowledge G. Parks for the Berkeley particle data. This work was supported by NASA grants NAS5-28448 and NAG5-474.

REFERENCES

[1] A. J. Zmuda and J. C. Armstrong, "The Diurnal Flow Pattern of Field-Aligned Currents, *J. Geophys. Res.* **79**, 4611 (1974).

[2] M. Sugiura, "Identifications of the Polar Cap Boundary and the Auroral Belt in the High-Latitude Magnetosphere: A model for Field-Aligned Currents," *J. Geophys. Res.* **80**, 2057 (1975).

[3] T. Iijima and T. A. Potemra, "Large-Scale Characteristics of Field-Aligned Currents Associated with Substorms," *J. Geophys. Res.* **83**, 599-615 (1975).

[4] D. H. Fairfield, "Magnetic Field Signatures of Substorm on High-Latitude Field Lines in the Nighttime Magnetosphere," *J. Geophys. Res.* **78**, 1553-1562 (1973).

[5] T. J. Kelly, C. T. Russell, R. J. Walker, G. K. Parks, and J. T. Gosling, "ISEE 1 and 2 Observations of Birkeland Currents in the Earth's Inner Magnetosphere," *J. Geophys. Res.* **91** (in press, 1986).

[6] N. S. Tsyganenko and A. V. Usmanov, "Determination of the Magnetospheric Current System Parameters and Development with Experimental Geomagnetic Field Models Based on Data from IMP and HEOS Satellites," *Planet. Space Sci.* **30**, 985 (1982).

[7] R. C. Elphic, P. A. Mutch, and C. T. Russell, "Observations of Field-Aligned Currents at the Plasma Sheet Boundary: An ISEE 1 and 2 Survey," *Geophys. Res. Lett.* **12**, 631-634 (1985).

DISCUSSION

G. Rostoker: Concerning your magnetic field (B_y) signatures – What criteria did you apply in terms of magnitude (ΔB_y) and duration of the observed signature? To clarify this further for a given ΔB_y satisfying your magnitude criterion, what length of duration of the perturbation would you use as a threshold above which you would eliminate the event from your sample?

R. C. Elphic: The boundary between the lobe and plasma sheet is often seen in the field data by the diamagnetic effect in field strength, the shear in the field (due to field-aligned currents), and an increase in variability of the field in the plasma sheet. We looked for unambiguous entries into or exits from the plasma sheet, and analyzed the current responsible for the changes. Our effective limit on ΔB_y was 1 nT, the background fields being between 20 and 40 nT. We sought rapid crossings of the boundary current, in effect crossings not requiring more than about 3 min. total.

V. M. Vasyliunas: You restricted your attention to small-scale structures, for obvious reasons (among others, a large-scale variation of B_y may be due to causes other than Birkeland currents). However, the Region 1 and 2 currents are fairly large-scale structures. Is it possible you may have missed them for this reason?

R. C. Elphic: For the downtail results, yes, we could have missed them. But that's less likely for the ISEE 1,2 near-earth auroral zone current observations, which indicate a lack of Region 1 and 2 currents about 70% of the time overall. Moreover, the downtail cases are crossings of the lobe/plasma sheet boundary layer interface, presumably the highest latitudes of the auroral zone. It seems unlikely that the Region 1 system extends further north than this. I still regard it as an interesting and challenging result that these currents, even if embedded in the high latitude edge of the larger Region 1 system, still exhibit a Region 2 polarity. We should reexamine the very lowest altitude results to look for evidence of this "subsystem".

BIDIRECTIONAL ELECTRON ANISOTROPIES IN THE DISTANT TAIL: ISEE 3 OBSERVATIONS OF POLAR RAIN

D. N. Baker, S. J. Bame, W. C. Feldman, J. T. Gosling, R. D. Zwickl,[*]
J. A. Slavin, and E. J. Smith[†]

A detailed observational treatment of bidirectional electrons (50–500 eV) in the distant magnetotail ($r \gtrsim 100\ R_e$) is presented. It is found that electrons in this energy range commonly exhibit strong, field-aligned anisotropies in the tail lobes. Because of large tail motions, the ISEE 3 data provide extensive sampling of both the north and south lobes in rapid succession, demonstrating directly the strong asymmetries that exist between the north and south lobes at any one time. The bidirectional fluxes are found to occur predominantly in the lobe directly connected to the sunward IMF in the open magnetosphere model (north lobe for away sectors and south lobe for toward sectors). Electron anisotropy and magnetic field data are presented, which show the transition from unidirectional (sheath) electron populations to bidirectional (lobe) populations. Taken together, the present evidence suggests that the bidirectional electrons that we observe in the distant tail are closely related to the polar rain electrons observed previously at lower altitudes. Furthermore, these data provide strong evidence that the distant tail is comprised largely of open magnetic field lines in contradistinction to some recently advanced models.

INTRODUCTION

A prevalent feature of the ISEE 3 plasma electron data is the occurrence of strong bidirectional (field-aligned) anisotropies above ~50 eV.[1] Early analysis (which regarded the boundary layer as distinct from the lobes) suggested that bidirectional distributions were present everywhere except in the tail lobes. However, later analysis built on a synthesis of all preceding work has shown that bidirectional electron distributions are, in fact, primarily a feature of the boundary plasmas found within the lobes.[2,3]

Our purpose in this paper is to demonstrate the common occurrence of bidirectional electron anisotropies throughout the distant tail. We will show the dependence of locations of such bidirectionality on solar wind and interplanetary magnetic field conditions. We will also show evidence that the source of the electrons exhibiting bidirectional behavior is the distant magnetosheath and that large asymmetries generally exist between the north and south deep tail lobes at any one time, depending upon IMF orientation. These results suggest that bidirectional tail lobe electrons are a consequence of open magnetic field lines and support the idea that the source of this population is the electron heat flux associated with the solar wind and the earth's bow shock. Further, we will demonstrate that the $\gtrsim 50$ eV electrons that exhibit bidirectionality are quite likely the same electrons that form the polar rain at lower altitudes.[4,5]

OBSERVATIONS

The plasma electron data have been examined using plots of the various fluid parameters, as well as 3D plots of the 2D distribution function in terms of energy, angle, and time.[6] These representations were supplemented by color-coded energy-time (E-t) and angle-time spectrograms, which were particularly useful for identifying the occurrence of bidirectional anisotropies. As shown in Plate II-1, the top four panels of each spectrogram display electron energy spectra averaged over the viewing quadrants centered, respectively, upon the spacecraft noon, dusk, midnight, and dawn meridians. The bottom four panels display electron angular distributions in different energy passbands. Color coding of the energy spectra and angular distributions is related to the log of the counts measured by the detector as indicated by the vertical calibration bar on the left. Numbers at the right-hand edges of the panels refer to either the energy in keV or the look angle of the measurement, as appropriate.

ISEE 3 was on the dawn side of the aberrated magnetotail at ~195 R_e geocentric distance on January 16. An episode of strong bidirectional electron fluxes is clearly evident in the second and third panels from the bottom (137–362 eV and 37–99 eV, respectively) between ~1900 UT and ~2030 UT. A few other sporadic instances of bidirectional fluxes are seen (e.g., ~2110 UT and ~2350 UT), but for most of the remaining time there were either no significant anisotropies above ~100 eV (e.g., ~1500–1700 UT) or else there were strong unidirectional anisotropies (~1200–1500 UT and ~2115–2200 UT).

Figure 1 gives the ISEE 3 electron data in the form of plasma moments (top four panels) plus the ISEE 3 magnetic field data (bottom three panels). Using the color spectrograms, the moments, and the magnetic field together, we can define the plasma regimes sampled by ISEE 3 in nearly all instances.[7] We have made such identifications for January 16 and these are shown at the top of Fig. 1.

From 1200 to ~1450 UT ISEE 3 was in the magnetosheath wherein densities were relatively high (~4 cm^{-3}), temperatures were low (2×10^5 K), plasma flows were strong (600 km/s) and steadily tailward ($\phi \sim 0°$), and the magnetic field strength was moderately high (~8 nT) with an azimuth near 210° (in GSM coordinates). Between ~1450 UT and ~2035 UT, ISEE 3

---

[*]Los Alamos National Laboratory, Los Alamos, New Mexico 87545.
[†]Jet Propulsion Laboratory, Pasadena, California 91109.

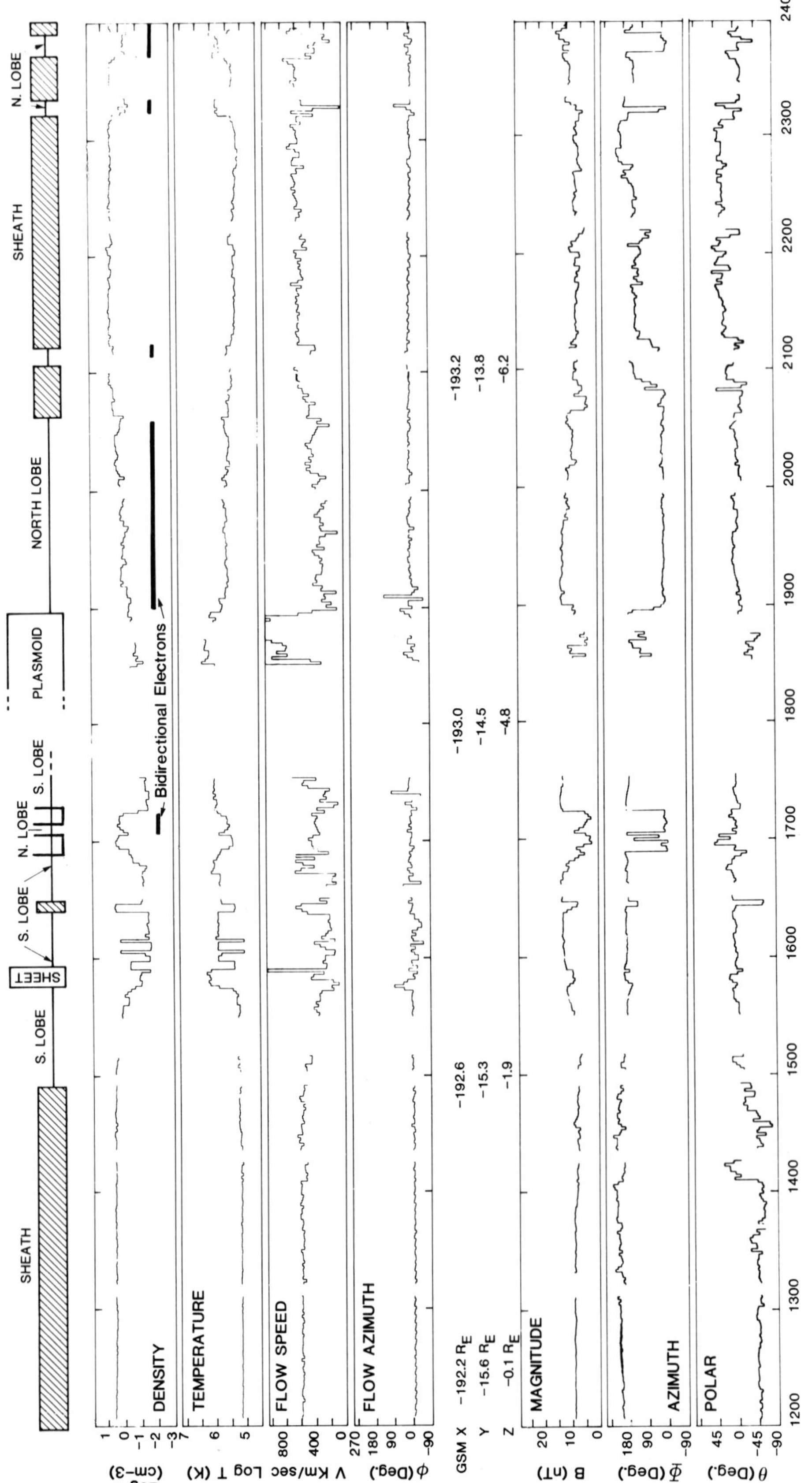

Figure 1—Plasma electron moments (upper four panels) and magnetic field data (lower three panels) for the period 1200–2400 UT on Jan 16, 1983. Identified plasma regions in which the spacecraft was located are shown along the top of the figure.

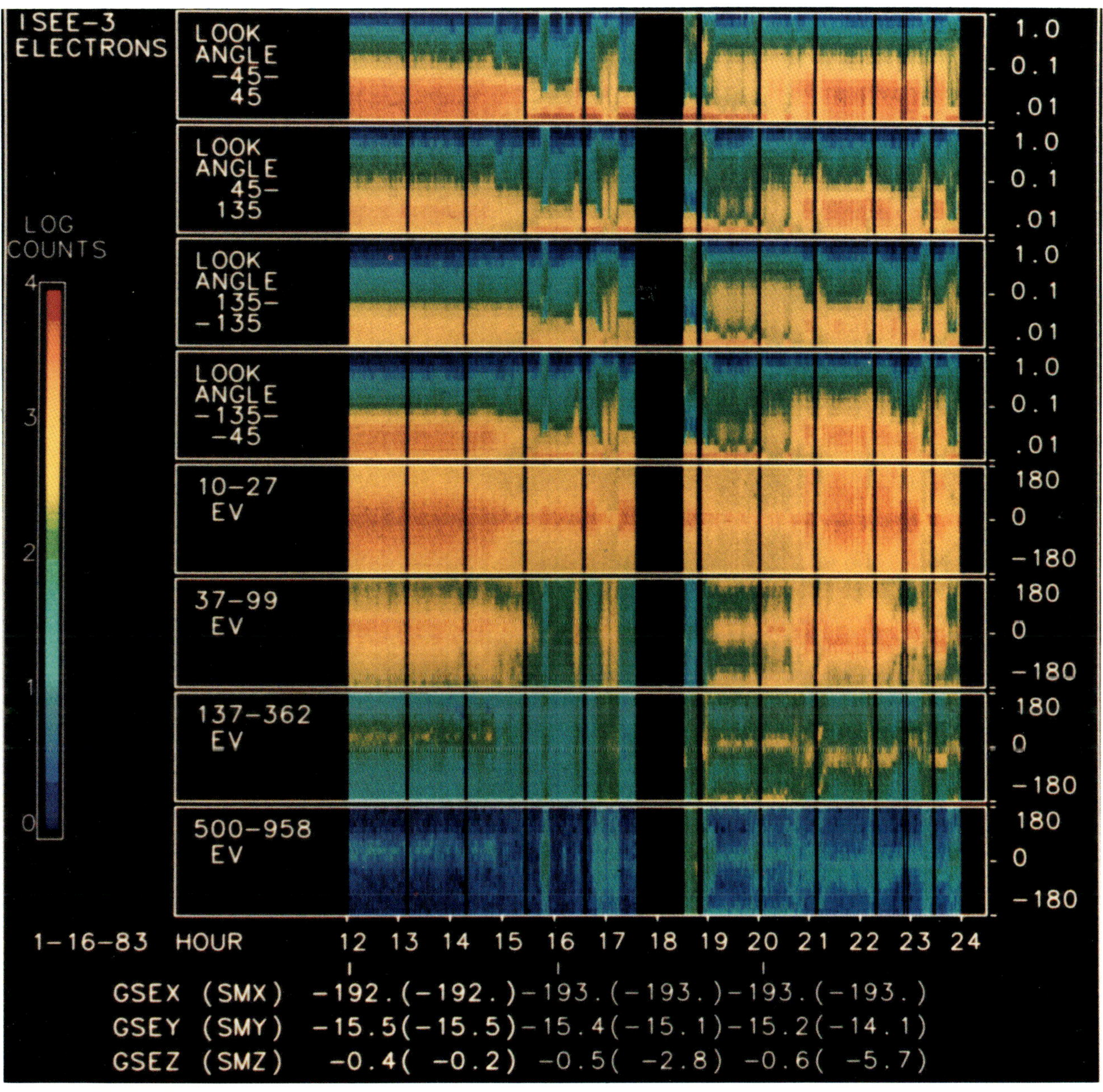

Plate II-1—A color spectrogram representation (as discussed in text) of ISEE 3 electron data for the period 1200-2400 UT on January 16, 1983. The spacecraft location (both in geocentric solar magnetospheric (GSM) and geocentric solar ecliptic (GSE) coordinates) is shown at the bottom of the figure.

was mostly in a tail lobe environment characterized by lower densities, higher temperatures, lower (variable) flow speeds, and strong, regular magnetic fields (at azimuths of either 0° or 180°). Some brief encounters with the current (plasma) sheet and the magnetosheath during this broad interval are identified at the top of Fig. 1. Finally, from 2035 UT to 2400 UT, ISEE 3 was again mostly in the magnetosheath.

From the magnetic field azimuth (ϕ) in Fig. 1, it is seen that between ~1450 UT and ~1900 UT, ϕ was ~180°. This is indicative[8] of the south tail lobe (**B** antisunward). From ~1900 UT to ~2050 UT, ϕ was ~0°, indicative of the north tail lobe. Other brief intervals of ϕ ~ 0° were seen around 1700 UT, ~2315 UT, and 2350 UT, thus implying that ISEE 3 was in the north lobe also during these times. The period around 1700 UT is complex in that plasma density was often high and the field strength was often low. Only from ~1705–1710 UT was ISEE really in a north lobe environment—for the remainder of the period 1650–1720 UT ISEE was probably in the northern portion of the distant plasma sheet. The distinction between north lobe and south lobe has been drawn in the region identifier band at the top of Fig. 1.

Comparison with Plate II-1 shows that bidirectional anisotropies occurred only in the north lobe part of the distant tail during this 12-hr interval. The times of bidirectional fluxes (from Plate II-1) are shown by the heavy band in the upper panel of Fig. 1. Conversely, in the south lobe regions ISEE 3 observed relatively little plasma at high energies (Plate II-1), and the plasma was nearly isotropic. Note from examination of Plate II-1 and Fig. 1 that a good case can be made to suggest that ISEE 3 crossed the tail completely from the south sheath all the way to the north sheath. The data suggest that the bidirectional electrons totally fill the region between the plasma sheet and the magnetopause in the north lobe.

IMP 8 was in the upstream solar wind during this day and, although large data gaps existed, substantial interplanetary magnetic field (IMF) data were obtained. The available data for the 12-hr interval of Fig. 1 reveal that for most of the time in which measurements were made, the IMF had an azimuth of ϕ_B ~135° in GSM coordinates. This corresponds to a $-B_x$ and $+B_y$ configuration, which is a classical "away" sector with the IMF at the expected garden-hose angle. Under such circumstances previous workers (e.g., Fennell et al.,[9] Yeager and Frank,[10] Meng and Kroehl[11]) have found enhanced polar rain electrons primarily in the northern polar cap and northern tail lobe near the earth. This result has been interpreted in terms of an open magnetosphere model where the solar-connected end of interplanetary magnetic field lines link directly to northern lobe field lines for away sectors and the outer-heliosphere end connects directly to the south lobe.[12]

Since the magnetic field lines being sampled by ISEE 3 in the distant tail lobes probably map directly to the earth's polar caps, it is quite likely that the field-aligned population that we observe would, at least in part, travel down to low-altitude, high-latitude regions. Hence, the

present observations are consistent with the possibility that tail lobe electrons are the source of polar rain since bidirectional electrons are a striking feature of the north lobe region yet no bidirectional electrons are evident in the southern lobe during the time shown in Fig. 1.

Careful inspection of plasma-electron angular-distribution data near the magnetopause often shows transitions from nearly unidirectional to bidirectional behavior with a smooth, gradual evolution in character. This evolution in the intense, anisotropic component suggests that one is observing a largely unidirectional source population develop into the bidirectional population.[3]

An example is shown in Plate II-2, which is a color spectrogram for the period 0000–0300 UT on January 3, 1983. Two slight penetrations of ISEE 3 from the sheath toward the tail lobes occurred at ~0050 UT and again at ~0150 UT. In these instances the unidirectional sheath electrons (137–362 eV) evolved into bidirectional electrons. The spacecraft did not penetrate deeply into the tail in either case (Fig. 2) since no clear lobelike magnetic fields were seen. After the longer bidirectional electron episode ending at ~0225 UT, ISEE 3 again went well back into the sheath where unidirectional electrons were seen (Plate II-2).

Inspection of the bottom two panels of Plate II-2 between 0140 and 0240 UT shows that the angular peak in the magnetosheath electron distribution centered on the spacecraft roll angle (azimuth) of $-40°$, corresponding to the antisunward heat flux, maps directly into the angular peak in the lobe distribution centered near $-160°$. This peak in the lobe distribution therefore represents field-aligned electron fluxes streaming toward the earth. Overlays of angular distributions of electron count rates for energies in the range spanning 71 eV to 363 eV for a sample magnetosheath and lobe spectrum on either side of the 0145 UT magnetopause crossing are shown in Fig. 3a. A quantitative comparison demonstrating the phase space density of the tailward-directed sheath population and the earthward-directed angular peak as a function of electron energy is shown in Fig. 3b. The straight lines are segmented linear least-squares fits using the lowest six energy points of each spectrum. They can be brought into exact coincidence by sliding the curve for the magnetosheath spectrum to the left by ~10 eV. In other words, the lobe and magnetosheath electron velocity distributions are equal for electrons in the range 71 eV $\leq$ E $\leq$ 361 eV if each electron loses 10 eV upon crossing the magnetopause from the magnetosheath to the lobe.[5] Such small magnetopause potentials suggest that these potentials are controlled by the bulk of the thermal (≤ 50 eV) electron distribution.[3] Hence the bidirectional population that we have focussed on here is something of a high-energy ($\gtrsim 50$ eV) test particle population.

A further quantitative comparison showing the close similarity of magnetosheath and lobe angular distributions on either side of the magnetopause crossing at 0145 UT on January 3 comes from the average FWHM angular widths of each of the count rate plots in Fig. 5a. Whereas the widths of the magnetosheath angular distributions average ~65° for 71 $\leq$ E $\leq$ 692 eV, the

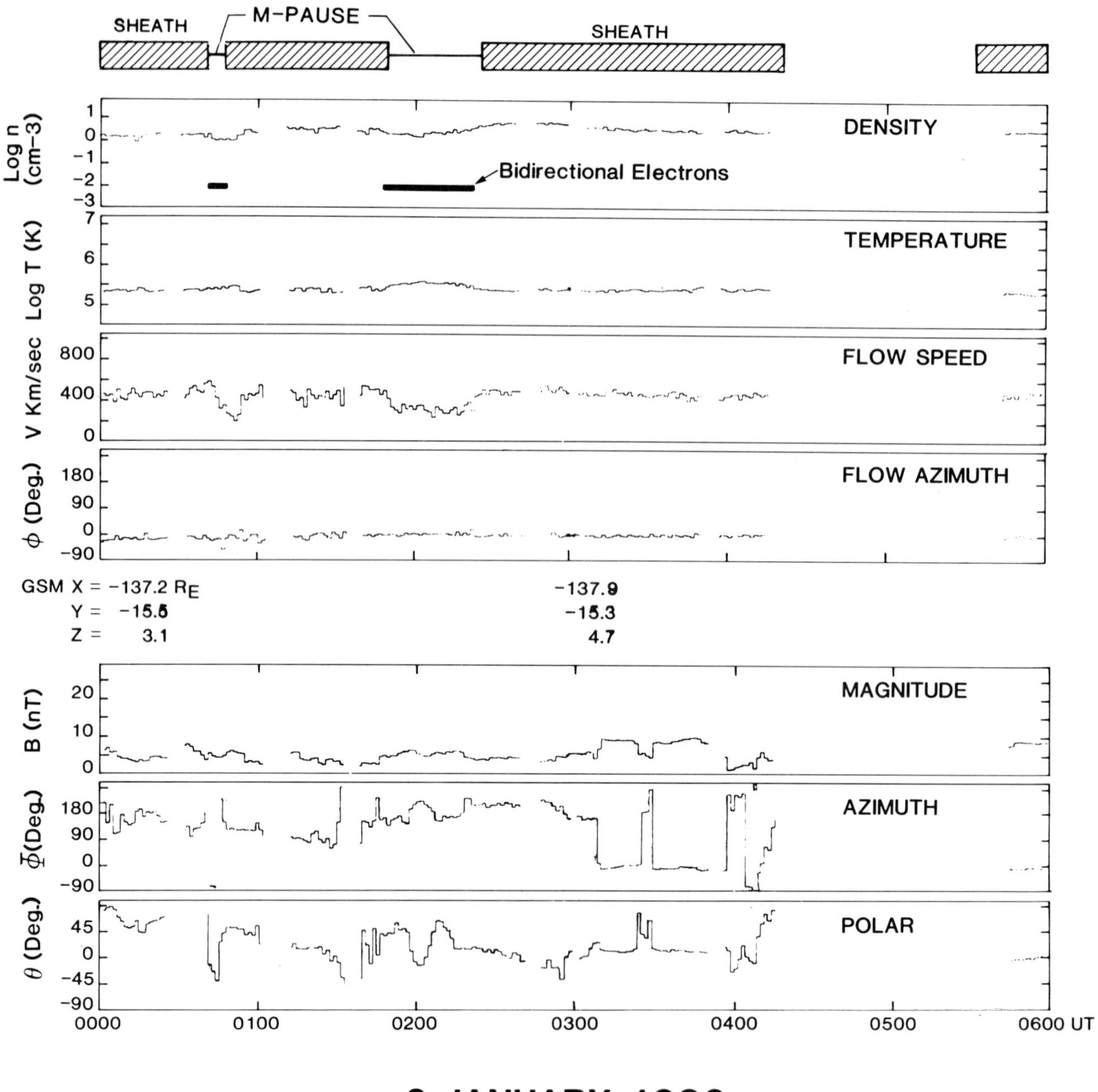

3 JANUARY 1983

Figure 2—Similar to Fig. 1 showing data for 0000–0600 UT on Jan 3, 1983.

widths of the lobe distributions average $\sim 75°$. Assuming adiabatic motion of the electrons across the magnetopause, the angular distribution widths should scale as $\sim \sin^2\alpha/B$. Using the ratio of the lobe field strength to sheath field strength as observed ($B_L/B_S \sim 7/6 \sim 1.16$), and using $\langle\alpha\rangle_s \sim 65°$, we have transformed lobe average angular width $\langle\alpha\rangle_l \sim 78°$. The agreement of this transformation with the observations further supports the suggestion that electrons are adiabatically crossing the magnetopause from the magnetosheath to the lobe at this time.

We also conclude from magnetic field minimum variance analyses that for many cases of the kind shown in Fig. 2, the magnetopause is a rotational discontinuity with a substantial normal component.[3]

SUMMARY AND DISCUSSION

In summary, ISEE 3 instrumentation has observed the common occurrence of strong bidirectional plasma (50 to 500 eV) electron anisotropies in all regions of the magnetospheric tail. These episodes are associated primari-

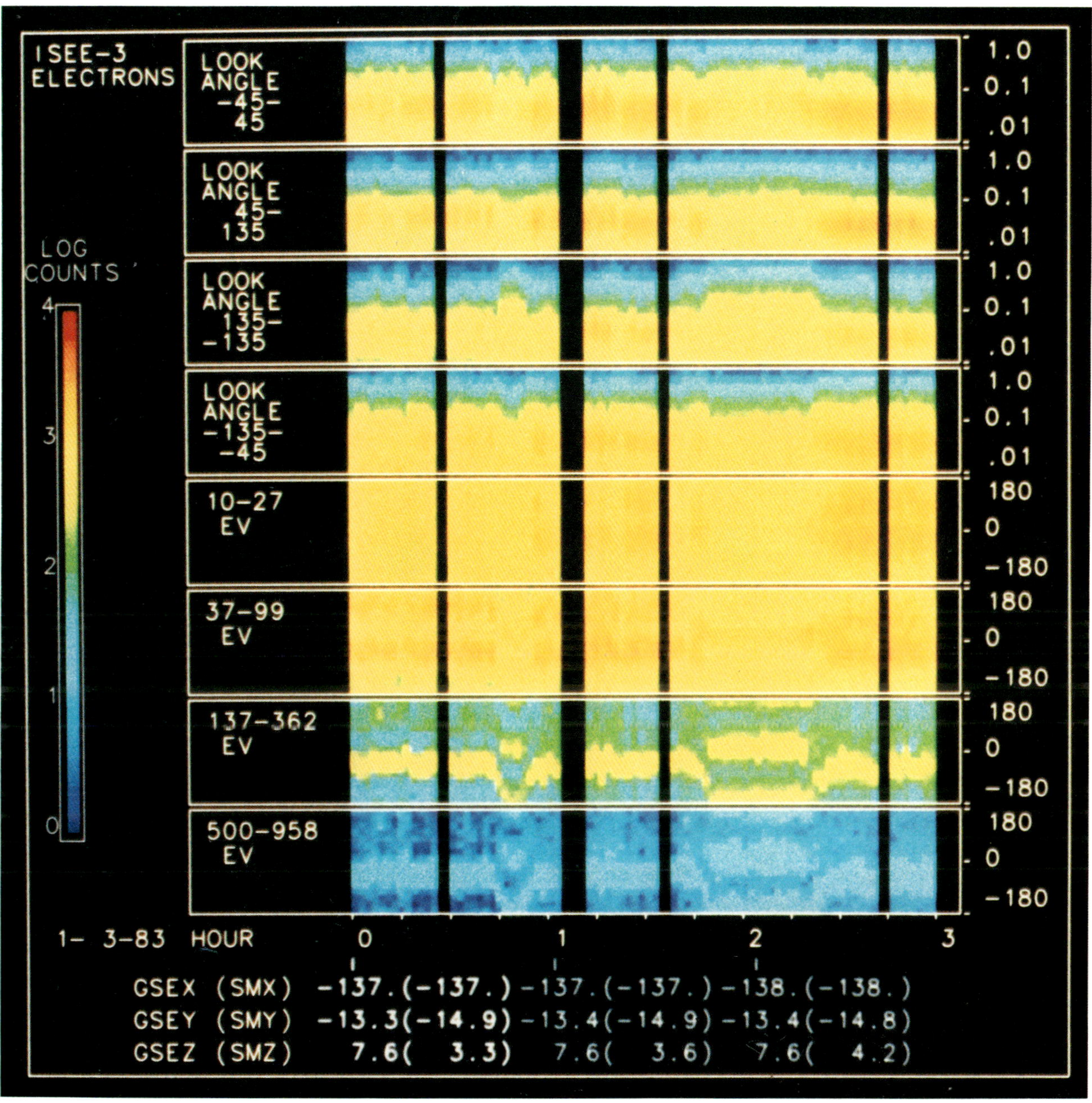

Plate II-2—Similar to Plate II-1, but showing data for the period 0000–0300 UT on January 3, 1983.

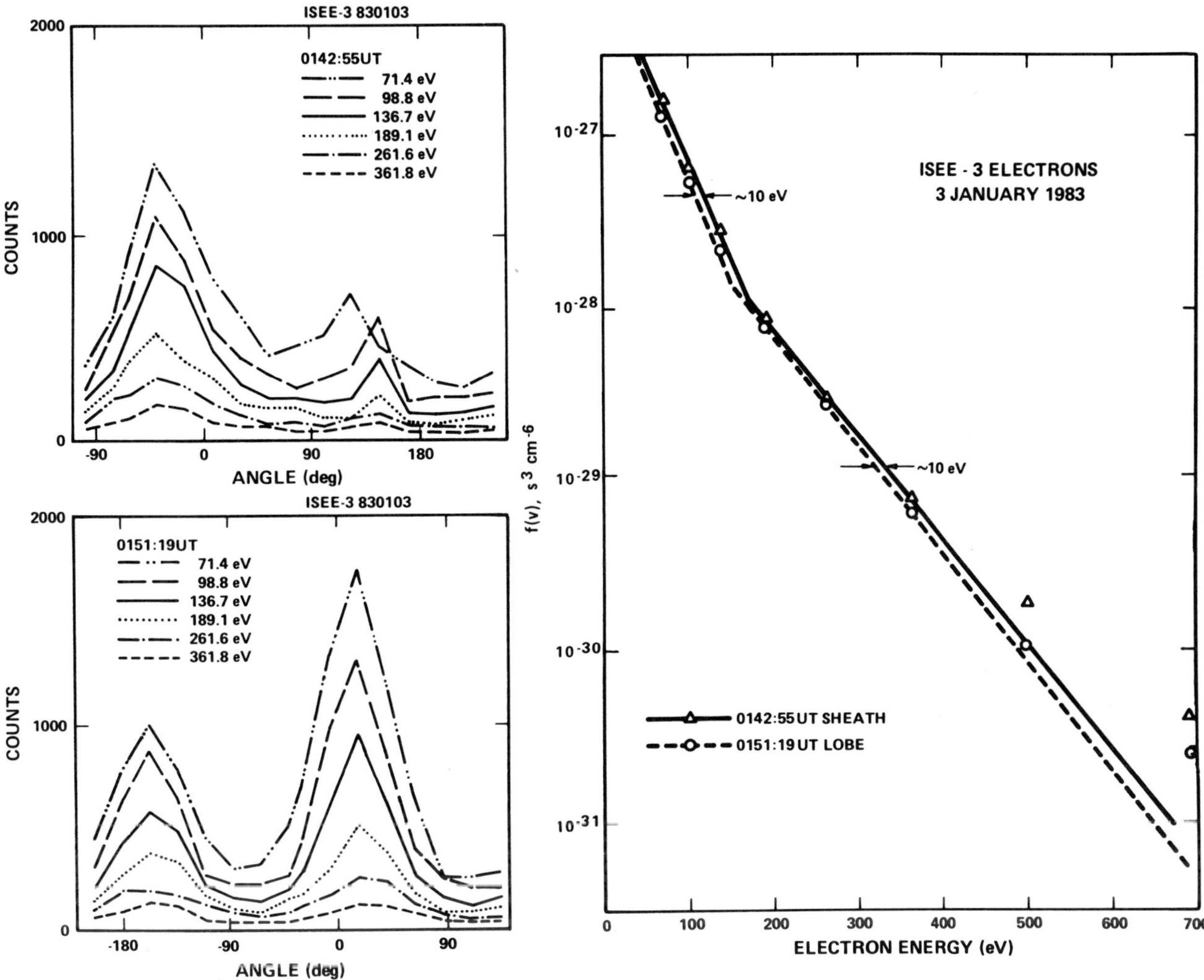

Figure 3—(a) Angular distributions in the magnetosheath and lobe as a function of electron energy near the time of a magnetopause crossing at 0145 UT on Jan 3, 1983. (b) Spectral analysis for the January 3 case showing a small inferred electrostatic potential at the magnetopause at the time of the ISEE 3 boundary crossing.

ly with the tail lobes, but occasionally the bidirectionality extends into the boundary of the current (plasma) sheet and, in many instances, out into the magnetosheath. Although the absolute intensity of the bidirectional electrons varies substantially, one has the clear impression from these data that this phenomenon is quite pervasive in that the tail lobe topologically is connected to the sun through the solar wind magnetic field in the open magnetosphere model.[12]

The present ISEE 3 observations bear directly on the magnetic topology of the distant tail and therefore on the source of polar cap precipitating particles. Most of the observations suggest that the bidirectional electrons are closely related to the polar rain electron population previously studied at low altitudes near the earth.

The total number densities of electrons during bidirectional events typically range from 10^{-1} to 5×10^{-1} cm^{-3}, but similar events can be discerned at density levels as low as 10^{-2} cm^{-3}. Directional energy fluxes in the bidirectional population at ISEE 3 normally range

from 10^{-3} ergs/cm^2-s-sr to 10^{-1} ergs/cm^2-s-sr. This flux compares favorably with energy fluxes observed for polar rain at lower altitudes.[4,9] We find, contrary to earlier results presented at lower altitudes, that higher energy plasma electrons in the distant tail lobes are, in fact, highly bidirectional rather than isotropic.

Figure 4a shows a modification of the model diagram of Fairfield and Scudder.[5] The local distribution function plots that we show encased in dashed boxes represent the measurements we have made directly with ISEE 3.

Note that the exaggerated field-aligned "loss cone" feature in Fig. 4a is in reality only ~1–2° wide and has not been observed directly. This shaded region of the various distributions is meant to represent those narrowly field-aligned electron populations that would be largely lost by precipitation at the earth. In our current data sets we have directly observed the broad unidirectional sheath populations, we have seen the bidirectional distributions (both in the sheath and in the lobes) that constitute the mirrored solar wind source, and we have seen

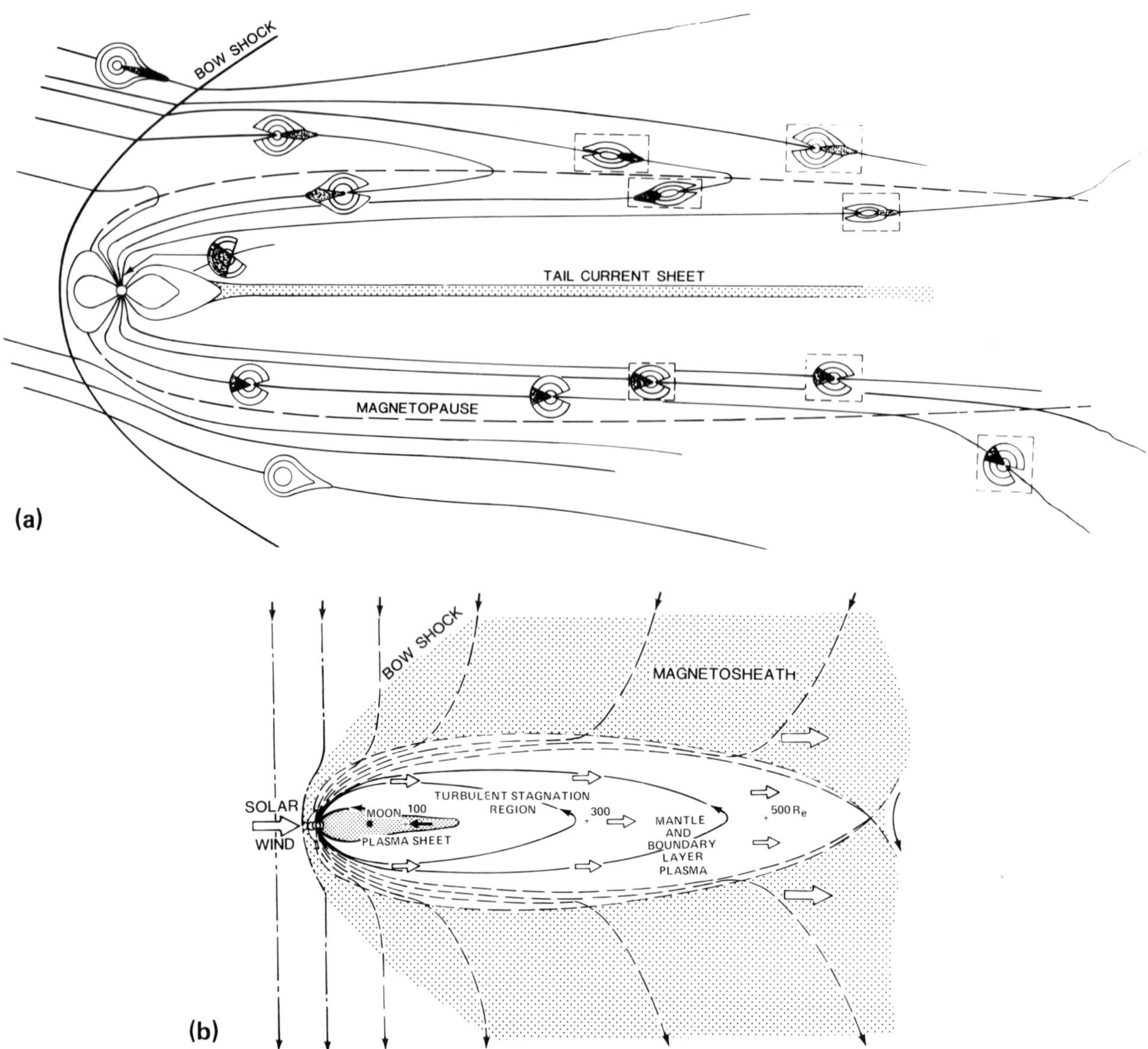

Figure 4—(a) A schematic showing the types of electron distributions that exist in the upstream solar wind, in the magnetosheath, and inside the magnetosphere. The distributions inside dashed boxes are those observed directly by ISEE 3. (Adapted from Fairfield and Scudder.[5]) (b) A noon-midnight cross section of the model of Heikkila[15] as discussed in the text. The model suggests that the distant tail is filled with closed magnetic field lines due to an expanded low-latitude boundary layer.

the low-density, isotropic lobe population in the "empty" lobe. Notably, we have seen all of these distributions, in both the connected and unconnected lobes, in very rapid succession—thus establishing the model shown in Fig. 4a.

We would therefore support the conclusion of Fairfield and Scudder[5] that a highly directional solar wind and/or bow shock population is very likely the source of the unidirectional sheath electron population observed by ISEE 3 and, also, the bidirectional lobe population. This, in turn, would then be the dominant source for the polar rain. Our results have little direct bearing on those of Gussenhoven et al.[13] regarding the strong and variably enhanced dayside polar cap precipitation that

most probably is due to polar cusp plasma entry. This contribution (see Fig. 7 in Ref. 13) would be complementary to the distant tail entry population that would dominate over most of the polar cap.

In light of these results, we wish to comment on the implications of our present observations for the magnetic topology of the distant tail. As noted by Bame et al.,[1] the occurrence of bidirectional electron anisotropies is often invoked as evidence of "closed" magnetic field lines (e.g., Baker and Stone[14]). Such an interpretation has been placed upon the bidirectional electron anisotropies reported here (W. Heikkila, private communication, 1984). In fact, recent models[15,16] have suggested that the low-latitude boundary layer—consisting of closed magnetic field lines—has expanded to fill the entire distant tail beyond $\sim 150\ R_e$.

Figure 4b shows the model of Heikkila[15] in a noon-midnight cross section. This model suggests that the near-earth plasma sheet is a small cavity that is completely engulfed in the distant tail by a "turbulent stagnation region" and a greatly expanded region of boundary layer plasmas that are threaded by closed field lines. Open field lines are found (in this model) only as a small layer near the magnetopause. All of these features are pictured as having developed rather fully by $\sim 150\ R_e$. A similar picture seems to be envisaged by Rostoker.[16]

Our present results from ISEE 3 suggest a substantially different picture than that of Fig. 4b. We find strong evidence of large regions of magnetic lobes in the distant tail. In these lobes there are bidirectional electrons, but these are observed principally in one lobe at a given time. The opposite lobe often contains very low density plasma and shows no bidirectional electron behavior. As described above, this is strong evidence that the distant lobe field lines are open to the solar wind throughout the lobes. Obviously, if distant tail field lines were closed, there would be similar plasmas (and similar electron anisotropies) in both lobes at any one time. In further disagreement with the model of Fig. 4b, we find that an identifiable current (plasma) sheet is detectable all the way out to 240 R_e. It is spatially thin (see also Ref. 2) except when plasmoids are present. Thus, ISEE 3 results provide no evidence that the low-latitude boundary layer has expanded to subsume the near-earth plasma (current) sheet nor that it has merged to fill most of the north-south extent of the tail. On the contrary, most of the distant tail is filled with open field lines upon which plasma densities either are high due to local plasma entry (given a proper IMF orientation) or else are low due to improper IMF connectivity.

ACKNOWLEDGMENT—This work was done under the auspices of the U.S. Department of Energy with support from NASA. The authors thank J. F. Fennell, D. H. Fairfield, and M. S. Gussenhoven for helpful discussions. R. C. Anderson, K. Sofaly, S. Kedge, and J. Tubb provided data analysis and graphics support.

REFERENCES

[1] S. J. Bame, R. C. Anderson, J. R. Asbridge, D. N. Baker, W. C. Feldman, J. T. Gosling, E. W. Hones, Jr., D. J. McComas, and R. D. Zwickl, "Plasma Regimes in the Deep Geomagnetic Tail: ISEE-3," _Geophys. Res. Lett._ **10**, 912 (1983).

[2] 2J. T. Gosling, D. N. Baker, S. J. Bame, W. C. Feldman, R. D. Zwickl, and E. J. Smith, "North-South and Dawn-Dusk Plasma Asymmetries in the Distant Tail Lobes: ISEE-3," _J. Geophys. Res._ **90**, 6354 (1985).

[3] D. N. Baker, S. J. Bame, W. C. Feldman, J. T. Gosling, R. D. Zwickl, J. A. Slavin, and E. J. Smith, "Strong Electron Bidirectional Anisotropies in the Distant Tail: ISEE-3 Observations of Polar Rain," _J. Geophys. Res._ **91** (in press, 1986).

[4] J. D. Winningham and W. J. Heikkila, "Polar Cap Auroral Electron Fluxes Observed with Isis 1," _J. Geophys. Res._ **79**, 949 (1974).

[5] D. H. Fairfield and J. D. Scudder, "Polar Rain: Solar Coronal Electrons in the Earth's Magnetosphere," _J. Geophys. Res._ **90**, 4055 (1985).

[6] W. D. Feldman, S. J. Schwartz, S. J. Bame, D. N. Baker, J. T. Gosling, E. W. Hones, Jr., D. J. McComas, R. D. Zwickl, and E. J. Smith, "Evidence for Slow Mode Shocks in the Deep Geomagnetic Tail," _Geophys. Res. Lett._ **11**, 599 (1984).

[7] R. D. Zwickl, D. N. Baker, S. J. Bame, W. C. Feldman, J. T. Gosling, E. W. Hones, Jr., D. J. McComas, and E. J. Smith, "Evolution of the Earth's Distant Magnetotail: ISEE-3 Electron Plasma Results," _J. Geophys. Res._ **89**, 11007 (1984).

[8] J. A. Slavin, B. T. Tsurutani, E. J. Smith, D. E. Jones, and D. G. Sibeck, "Average Configuration of the Distant ($<220\ R_e$) Magnetotail: Initial ISEE-3 Magnetic Field Results," _Geophys. Res. Lett._ **10**, 973 (1983).

[9] J. F. Fennell, P. F. Mizera, and D. R. Croley, Jr., "Low Energy Polar Cap Electrons During Quiet Times," in _Proc. Int. Conf. Cosmic Rays_ **14**, MG8-3, 1267 (1975).

[10] D. M. Yeager and L. A. Frank, "Low Energy Electron Intensities at Large Distances Over the Earth's Polar Cap," _J. Geophys. Res._ **81**, 3966 (1976).

[11] C.-I. Meng and H. W. Kroehl, "Intense Uniform Precipitation of Low-Energy Electrons Over the Polar Cap," _J. Geophys. Res._ **82**, 2305 (1977).

[12] G. Paulikas, "Tracing of High-Latitude Magnetic Field Lines by Solar Particles," _Rev. Geophys. Space Phys._ **12**, 117 (1974).

[13] M. S. Gussenhoven, D. A. Hardy, N. Heinemann, and R. K. Burkhardt, "Morphology of the Polar Rain," _J. Geophys. Res._ **89**, 9785 (1984).

[14] D. N. Baker and E. C. Stone, "Energetic Electron Anisotropies in the Magnetotail: Identification of Open and Closed Field Lines," _Geophys. Res. Lett._ **3**, 557 (1976).

[15] W. J. Heikkila, "Magnetospheric Topology of Fields and Currents," in _Magnetospheric Currents_ (Geophys. Monog. 28), Amer. Geophys. Un., 208 (1983).

[16] G. Rostoker, "Definition of a Substorm, Physical Processes in a Substorm and Sources of Discomfort," in _Magnetic Reconnection in Space and Laboratory Plasmas_ (Geophys. Monog. 30), Amer. Geophys. Un., 380 (1984).

DISCUSSION

R. C. Elphic: You performed a minimum variance analysis of the magnetic field for several magnetopause crossings in order to determine whether the boundary is open (a rotational discontinuity) or closed. Such an analysis is notoriously ambiguous. Could you do an analysis of tangential momentum balance such as has been done for the dayside magnetopause for reconnection events?

D. N. Baker: Our analysis of the magnetopause field changes by minimum variance methods was done in support of the other ISEE 3 results suggestive of a locally "open" magnetopause. Although the minimum variance axis was only moderately well determined in many of our cases, nonetheless the minimum variance technique provided good support for our interpretation of a rotational discontinuity in these instances. When combined with the striking signatures in the plasma electron pitch angle distributions, the overall interpretation of interconnection from the sheath to the tail lobes is very firmly established. Of course, we would like to perform a complete momentum balance analysis for our distant tail magnetopause crossings. This is made difficult since we do not have plasma ion data available from the spacecraft. We will, however, be doing all possible analyses in order to substantiate further the nature of the distant sheath-tail interface region.

MAGNETIC CONFIGURATION OF THE DISTANT PLASMA SHEET: ISEE 3 OBSERVATIONS

J. A. Slavin,* P. W. Daly,† E. J. Smith,* T. R. Sanderson,‡ K.-P. Wenzel,‡ R. P. Lepping,§ and H. W. Kroehl‖

Observations from the JPL vector helium magnetometer and the EPAS energetic ion spectrometer on ISEE 3 are used to investigate the magnetic configuration of the central plasma sheet as a function of distance down the tail, the magnitude and orientation of the IMF, and substorm activity. The ISEE 3 magnetotail measurements are supplemented with IMP 8 GSFC interplanetary magnetic field observations and University of Kyoto/NOAA AE indices. Southward B_z in the plasma sheet is observed to be correlated with high-speed antisolar bulk flows as predicted by the Dungey model of the magnetosphere. It is also found that southward plasma-sheet magnetic fields are correlated with southward IMF upstream of the magnetosphere. An effective dayside reconnection efficiency of 25 ±4% is inferred, in good agreement with theory and near-earth studies of magnetic flux transfer. This unique data set has also allowed an examination of the variation in plasma-sheet B_z and V_x with distance down the tail for both substorm and quiet conditions. During substorms, $|AL| > 100$ nT, negative magnetic fields, and high-speed antisolar flows are observed in the central plasma sheet at $|X| > 80$–100 R_e. Earthward of that distance positive B_z and solar-directed V_x are measured. At nearly all distances, $\langle V_x B_z \rangle$ is positive (i.e., dawn-to-dusk electric field) under substorm conditions indicating that magnetic field energy is being converted to plasma kinetic energy as required in the open model of the magnetosphere. During quiet conditions, $|AL| < 100$ nT, the magnetic field in the plasma sheet is positive out to the most distant ISEE 3 apogee, $|X| = 240$ R_e, despite the presence of tailward flow in the plasma sheet beyond $|X| = 80$ R_e. This result suggests that the quiet time central plasma sheet at $|X| > 80$–100 R_e is threaded by closed field lines that are being carried slowly tailward. The energy required to do work against these closed field lines must be supplied by the kinetic energy in the mantle and low latitude boundary layer flows as described in the quasiviscous models of the solar wind interaction with the geomagnetic field. These quiet-time results stand out in contrast with the ISEE 3 observations during substorms, which show very good agreement with the original merging models by Dungey.

INTRODUCTION

ISEE 3 spent most of 1983 in the earth's magnetotail executing two "figure-eight" orbits with apogees of 220 and 240 R_e plus several smaller orbits. This first extensive set of observations in the translunar tail (Pioneer 7 and 8 made rapid crossings at distances of 1000 and 500 R_e in 1966 and 1968; see Ref. 1) has returned much new information on a variety of magnetospheric phenomena (see *Geophys. Res. Lett.* **11**, no. 10 (1984)), but few comprehensive multiinstrument investigations have been conducted. In this study, the magnetic field observations from the JPL vector helium magnetometer[2] and bulk flows derived from the >35 keV EPAS ion anisotropy measurements[3] have been merged for the magnetotail phase of the ISEE 3 mission (i.e., DOY 356, 1982 to 293, 1983). The main objective of the study is to ascertain the magnetic configuration of the central plasma sheet as a function of distance down the tail, the B_z component of the interplanetary magnetic field (IMF), and the substorm activity. For this purpose IMF observations from the GSFC fluxgate magnetometer on IMP 8 and AE indices prepared by the University of Kyoto/NOAA for 1982-3 have also been utilized. The results lend strong support to the

hypothesis that neutral lines form in the tail following intervals of southward IMF B_z and in association with substorm activity.

ISEE 3 OBSERVATIONS OF B_z AND V_x IN THE PLASMA SHEET

Early in the development of magnetospheric physics, two mechanisms were proposed through which the solar wind might drive magnetospheric convection. Dungey[4] suggested that "reconnection" between interplanetary and geomagnetic fields would provide a means for the solar wind to pull field lines back into the magnetotail lobes as illustrated in the top panel of Fig. 1. This process can be viewed either as a transfer of magnetic flux to the nightside magnetosphere or a Poynting flux of energy into the tail. Reconnection at the neutral line in the near-tail sporadically releases stress in the form of substorms.[5] The other mechanism was proposed by Axford and Hines[6] who suggested that the solar wind exerts a quasiviscous drag on the geomagnetic field and pulls closed field lines tailward to produce a teardrop shaped magnetospheric cavity. The "drag" force may be associated with either true MHD viscosity or mass diffusion across the magnetopause.[7] The stress built up by the antisolar flow containing closed field lines at the edges of the magnetosphere is relieved by an earthward flow in the central region of the cavity as shown in the lower panel of Fig. 1.

The magnetospheric observations compiled over the last 25 years have left little doubt that quasiviscous processes operate at the magnetopause,[8] but magnetic

*Jet Propulsion Laboratory, California Institute of Technology, Pasadena, California 91109.
†Max-Planck Institut für Aeronomie, D-3411 Katlenburg-Lindau 3, Federal Republic of Germany.
‡ESA Space Science Department, ESTEC, 2200 AG, The Netherlands.
§NASA/Goddard Space Flight Center, Greenbelt, Maryland 20771.
‖NOAA/National Geophysical Data Center, Boulder, Colorado, 80303.

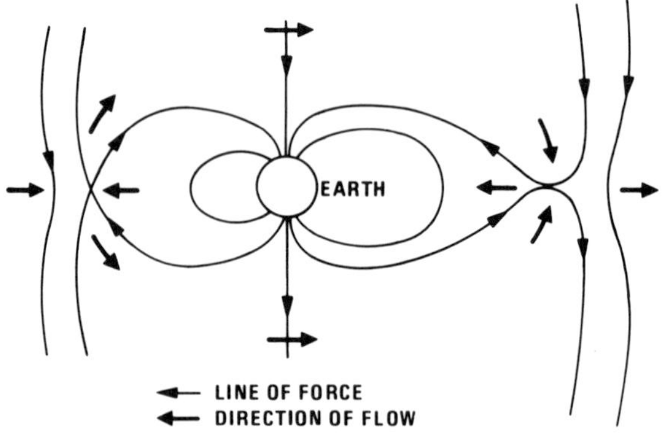

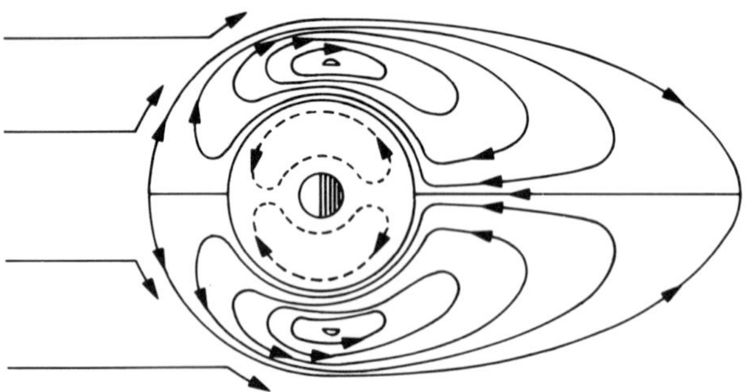

Figure 1—The original reconnection[4] and viscosity[5] models for driving magnetospheric convection are displayed in the top and bottom panels, respectively.

reconnection has been shown to be responsible for the vast majority of the energy input to the magnetosphere (e.g., see review by Cowley[9]). However, the in situ observation of the flows and magnetic field orientations predicted in Fig. 1 have been limited to the cis-lunar tail where the results have been inconclusive.[10,11] The ISEE 3 observations have the advantage of being made in a region that is expected to be nearly always tailward of the substorm neutral lines, which are believed to form at $|X| = 10$–30 R_e. Under these circumstances a positive correlation between B_z and V_x is expected with negative (antisolar) V_x accompanying negative (southward) B_z in the plasma sheet.

For the purpose of this study all intervals during which ISEE 3 was immersed in the plasma sheet for greater than ~ 10 min have been identified in the same manner as in the study by Slavin et al.[2] The selection criteria, described in that study, are intended to exclude the plasma-sheet boundary layer (i.e., the plasma sheet-lobe interface region) and include only the central plasma sheet. It is in this region that bulk flows transport magnetic flux in the Dungey-type models of the magnetosphere.[9] Since reconnection appears to occur preferentially toward the east-west center of the plasma sheet,[2,12] this investigation has limited itself to the region $-15 < Y' < 15$ R_e. The prime designation indicates that the spacecraft location has been aberrated by $4°$ to take into account the orbital motion of the earth.

Figure 2 displays a plot of plasma sheet V_x versus B_z (GSM coordinates) observed by ISEE 3 at $|X| = 20$ to 240 R_e. The total number of 10-min averaged points is 1419 corresponding to 236.5 hours, or about 10 days in the plasma sheet. Overall B_z and V_x do appear to be positively correlated as predicted by the Dungey model with a correlation coefficient of 0.5. The slopes and intercepts for the two linear regression fits are 1.37 (± 0.07) $\times 10^2$ and -513 (± 10) km/s for V_x on B_z and 1.52 (± 0.08) $\times 10^{-3}$ and 1.11 (± 0.01) nT for B_z on V_x. The unexpected aspect of the correlation is the offset toward large tailward flow speeds (i.e., $B_z = 0$ corresponds to $V_x \sim -500$ km/s). Similar results have been obtained by Slavin et al.[2] using a smaller data set with flow speeds determined from the plasma electron measurements.[13] The retreat of the near-earth neutral line(s) during substorm recovery phase[14] should bias the observations somewhat toward tailward flows, but the magnitude of the offset is larger than might have been predicted. The cause appears to be a predominance of tailward flow during nonsubstorm intervals, when the magnetic field is northward, as will be discussed later.

EFFECT OF INTERPLANETARY B_z

In the open model of the magnetosphere, southward IMF results in dayside reconnection which transfers magnetic flux to the lobes of the magnetotail.[4,5,11,15-17] The flux added to the magnetotail must eventually be balanced by a return of field lines to the dayside magnetosphere (and the interplanetary medium) to limit the stress exerted on the magnetosphere by the solar wind. Accordingly, the prediction of the Dungey model in Fig. 1 is that southward interplanetary magnetic fields up-

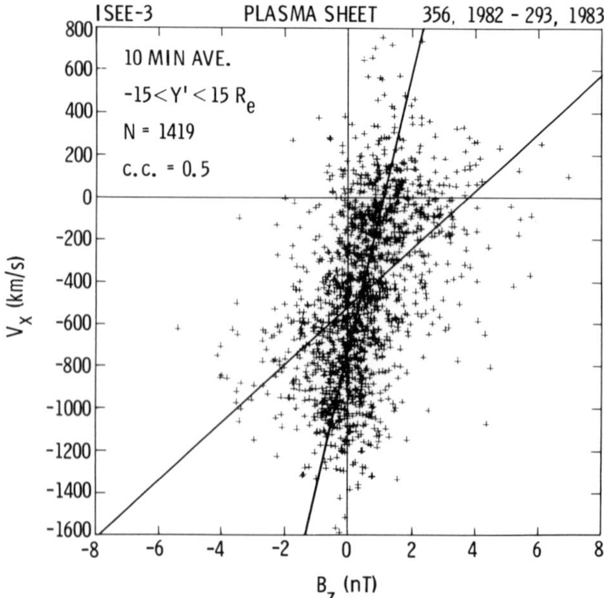

Figure 2—Ten-minute averages of the GSM X component of the flow velocity and the GSM Z component of the magnetic field measured by ISEE 3 in the plasma sheet are plotted against each other. All downstream distances, $|X| = 20$ to 240 R_e are displayed, but the observations are limited to intervals when ISEE 3 was at $-15 < Y' < 15$ R_e.

stream of the magnetosphere should correlate with southward magnetic fields in the central plasma sheet tailward of the neutral line.

In Fig. 3, plasma sheet B_z observed at ISEE 3 is plotted against IMF B_z observed at IMP 8 for three different downstream regions. As described earlier, the data set is composed of 10-min averages in GSM coordinates. A time lag is necessary to allow for the initiation of reconnection in the tail and for the reconnected flux to be transported to ISEE 3. A constant 30 min was assumed for the first lag, based on the results of previous magnetic flux transfer studies (e.g., Ref. 16). The second lag depends upon the distance between ISEE 3

and the location of the neutral line. For this purpose it was assumed that reconnection commences at a distance of 20 R_e behind the earth,[14] and the reconnected flux tubes move down the plasma sheet with a mean flow speed of 800 km/s.[2,17] The expression for the total time lag, in minutes, is therefore

$$\tau = 30 + (|X| - 20)/(800 \times 60/6400), \quad (1)$$

which yields time delays of approximately 1 hour near ISEE 3 apogee at 220 to 240 R_e. Examination of different time lags and flow speeds indicated that the values in Eq. 1 are near optimum for the 10-min averaged data set used in this study. As in the previous section, the analysis was limited to the region between $Y' = \pm 15$ R_e where reconnection appears to take place most frequently.[2]

Figure 3 shows that southward magnetic fields in the plasma sheet are clearly correlated with southward IMF. For each of the three distance regimes, the observations have been fit with a least square line of the form

$$B_z \text{ (ISEE 3)} = a\, B_z \text{ (IMP 8)} + b \quad (2)$$

with the slope and intercept displayed in each panel. The modest 0.3–0.4 correlation coefficients indicate that there is a large amount of unexplained variance, but for these numbers of points the existence of a correlation between the ISEE 3 and IMP 8 B_z is certain at the >99% confidence level (e.g., Bevington[18]). The scatter in the distributions may be due to tail motion associated with changing solar wind conditions, normal mode oscillations, MHD waves, plasmoids, and turbulence in the plasma sheet.

The slopes of the lines in Fig. 3 provide a measure of the efficiency of the dayside merging process. If the effective widths of the dayside and nightside neutral lines are assumed comparable (e.g., both ~30 R_e), and the flow speed in the central plasma sheet is taken to be approximately twice that of the solar wind as suggested in Eq. 1, then the effective dayside reconnection efficiency is simply twice the slope of the regression lines. Weighting the values in Fig. 3 by the number of points in each fit and combining them yields an effective dayside reconnection efficiency of 25 ±4% in excellent agreement with the results of previous near-earth studies of magnetic flux transfer (e.g., Holzer and Slavin[16]).

EFFECT OF SUBSTORM ACTIVITY

In Figs. 4 and 5 the ISEE 3 observations of plasma sheet B_z and V_x have been averaged into 10 R_e bins along X for disturbed ($|AL| > 100$ nT and quiescent, $|AL| > 100$ nT) conditions with respect to substorm activity. Also displayed are $\langle V_x B_z \rangle$, the local magnetic flux transfer rate or motional electric field, and the number of minutes spent by ISEE 3 in each bin. The interpretation of $\langle V_x B_z \rangle$ in terms of flux transfer or a motional electric field tacitly assumes that the flow direction and magnetic field vector are not highly aligned as is the case for the plasma sheet boundary layer.[19] However, this assumption appears reasonable for an aver-

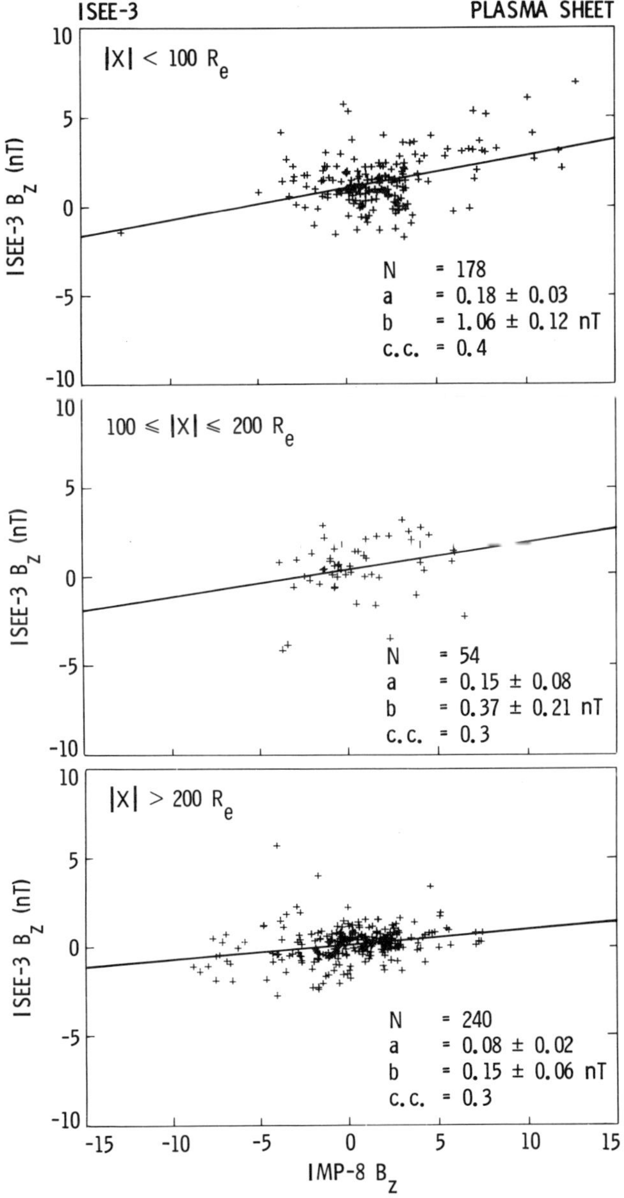

Figure 3—The B_z component of the plasma-sheet magnetic field measured by ISEE 3 at three different downstream distances is plotted against IMF B_z observed by IMP 8. The time lag used is described in the text and only intervals when ISEE 3 was located at $-15 < Y' < 15$ R_e are included.

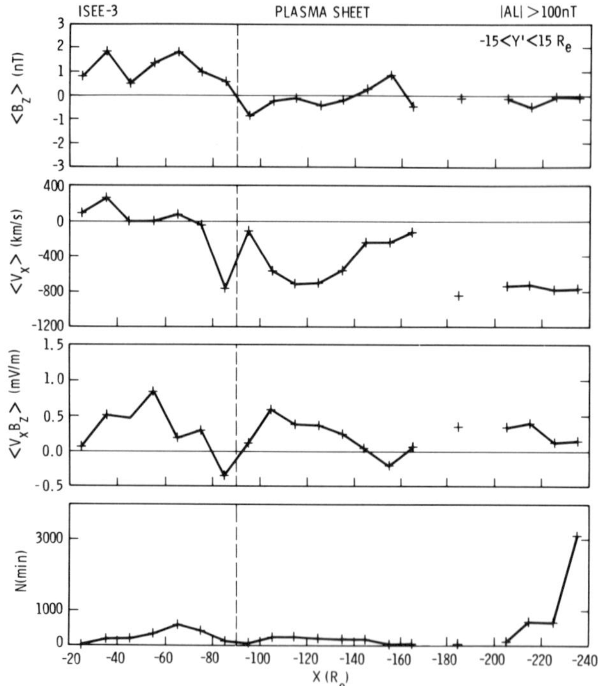

Figure 4—Plasma sheet B_z, V_x, magnetic flux transfer rate, and $V_x B_z$, observed by ISEE 3 for $|AL| > 100$ nT have been averaged into 10 R_e bins and plotted against X. The number of 10-min-averaged data points for each bin is displayed in the bottom panel and the time lag assumed is described in the text. The analysis is limited to intervals when ISEE 3 was in the region $-15 < Y' < 15$ R_e.

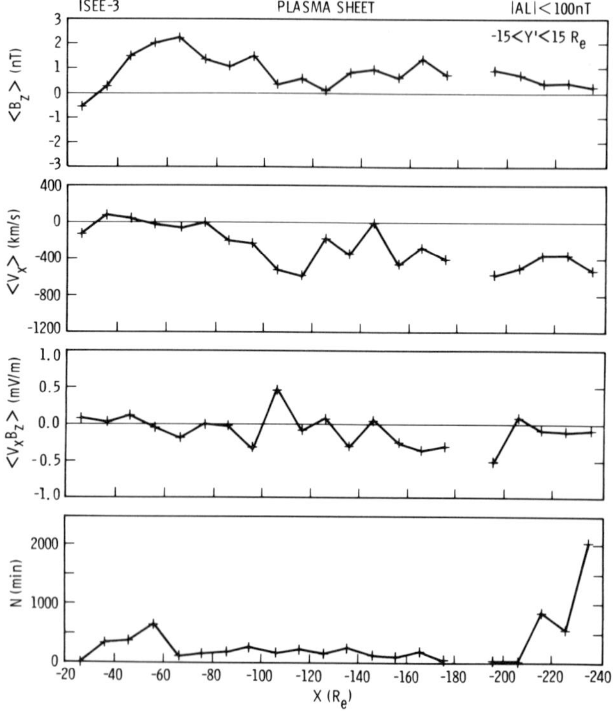

Figure 5—ISEE 3 plasma sheet parameters are displayed as in Fig. 4, but only intervals with little substorm activity, $|AL| < 100$ nT, are included.

age over all of the central plasma sheet region including the neutral sheet (e.g., Cattell et al.[20]).

During substorm conditions, ISEE 3 observed northward magnetic fields and earthward flow when it was inside $|X| = 90$ R_e. Beyond that distance the magnetic fields were predominantly southward, $B_z \sim -0.1$ to -0.3 nT, and the flow tailward, $V_x \sim -200$ to -800 km/s. These results are in excellent agreement with the predictions of the Dungey model in Fig. 1 with an average neutral line location of $|X| = 90$ R_e. Given that neutral lines are expected to form close to the earth, but later move down the tail, the mean neutral line position in Fig. 4 appears to be not inconsistent with existing phenomenological substorm models.[14]

Figure 5 displays the corresponding analysis for low substorm activity levels. In agreement with the reconnection based models, the plasma sheet magnetic field is northward, $B_z \sim 0.2$ to 0.8 nT, everywhere except for a bin at 25 R_e with few counts. However, under these conditions the Dungey model would predict a slow earthward convection of plasma at all distances where $B_z > 0$. Figure 5 shows instead a net tailward flow of plasma beyond $|X| \sim 80$ R_e. The flow speed is only about half of that during substorms, but it suggests that closed field lines are continuously being carried slowly downstream during nonsubstorm quiet intervals. This implies that the plasma is doing work on the field as indicated by the largely negative motional electric field in the third panel. In this respect alone, the ISEE 3 observations appear to resemble the viscous interaction depicted at the bottom of Fig. 1, but with the low-latitude boundary layer extending across the full width of the quiet-time plasma sheet as predicted by some modern quasiviscous models.[21]

SUMMARY

In this study the magnetic configuration of the central plasma sheet from $|X| = 20$ to 240 R_e has been examined using ISEE 3 measurements. Following intervals of southward IMF B_z and substorm activity, the ISEE 3 observations show high-speed antisolar bulk flows threaded by southward magnetic fields in the distant plasma sheet. These findings are in excellent agreement with the classical Dungey model of the magnetosphere. Under these conditions much of the plasma sheet lies on field lines that have been disconnected from the earth. An effective dayside reconnection efficiency of 25 ±4% is inferred from the positive correlation between ISEE 3 B_z in the plasma sheet and IMP 8 B_z in the interplanetary medium. This value is in good agreement with the 20 ± 10% estimates for dayside reconnection efficiency which have been derived from near-earth observational studies (e.g., Holzer and Slavin[16]) and theory (e.g., Vasyliunas[22]). However, during quiet intervals with little substorm activity, the magnetic field in the plasma sheet is northward out to ISEE 3 apogee, $|X| = 240$ R_e, despite the presence of slow tailward flow beyond $|X| > 80$–100 R_e. This result does not agree with the models in which the ground state of the open magnetosphere possesses a stationary distant neutral line. Rather, the quiet-time distant plasma sheet appears to be on closed field lines. The momentum of the plasma-sheet particles being supplied by the mantle and low latitude boundary layers is apparently great enough to carry these geomagnetic field lines down the tail as predicted by some theoretical models (e.g., Heikkila[21] and Hill and Reiff[23]). The magnetic stress accumulated as a result of this process is presumably released during the next substorm when near-earth reconnection disconnects these highly distended field lines.

ACKNOWLEDGMENT—The authors are pleased to acknowledge support by J. Wolf in the computational aspects of this study. The research contained in this report was carried out at the Jet Propulsion Laboratory, California Institute of Technology under contract to the National Aeronautics and Space Administration.

REFERENCES

[1] U. Villante, "An Overview of Pioneer Observations of the Distant Geomagnetic Tail," *Space Sci. Rev.* **20**, 123 (1977).

[2] J. A. Slavin, E. J. Smith, D. G. Sibeck, D. N. Baker, R. D. Zwickl, and S.-I. Akasofu, "An ISEE 3 Study of Average and Substorm Conditions in the Distant Magnetotail," *J. Geophys. Res.* **90**, 10875 (1985).

[3] P. W. Daly, T. R. Sanderson, and K.-P. Wenzel, "Survey of Energetic (>35 keV) Ion Anisotropies in the Deep Geomagnetic Tail," *J. Geophys. Res.* **89**, 10733 (1984).

[4] J. W. Dungey, "Interplanetary Magnetic Field and the Auroral Zones," *Phys. Rev. Lett.* **6**, 47 (1961).

[5] C. T. Russell and R. L. McPherron, "The Magnetotail and Substorms," *Space Sci. Rev.* **15**, 205 (1973).

[6] W. I. Axford and C. O. Hines, "A Unifying Theory of High Latitude Geophysical Phenomena and Geomagnetic Storms," *Can. J. Phys.* **3**, 1433 (1961).

[7] B. U. O. Sonnerup, "Theory of the Low-Latitude Boundary Layer," *J. Geophys. Res.* **85**, 2017 (1980).

[8] T. E. Eastman, E. W. Hones, Jr., S. J. Bame, and J. R. Asbridge, "The Magnetospheric Boundary Layer: Site of Plasma, Momentum, and Energy Transfer into the Magnetosphere," *Geophys. Res. Lett.* **3**, 685 (1976).

[9] S. W. H. Cowley, "The Causes of Convection in the Earth's Magnetosphere: A Review of Developments During the IMS," *Rev. Geophys. Space Phys.* **20**, 531 (1982).

[10] A. T. Y. Lui, C.-I. Meng, and S.-I. Akasofu, "Search for the Magnetic Neutral Line in the Near-Earth Plasma Sheet, 3. An Extensive Study of Magnetic Field Observations at Lunar Distances," *J. Geophys. Res.* **82**, 3603 (1977).

[11] M. N. Caan, R. L. McPherron, and C. T. Russell, "Solar Wind and Substorm Related Changes in the Lobes of the Geomagnetic Tail," *J. Geophys. Res.* **84**, 1971 (1979).

[12] H. Hayakawa, A. Nishida, E. W. Hones, Jr., and S. J. Bame, "Statistical Characteristics of Plasma Flow in the Magnetotail," *J. Geophys. Res.* **87**, 277 (1982).

[13] R. D. Zwickl, D. N. Baker, S. J. Bame, W. C. Feldman, J. T. Gosling, E. W. Hones, Jr., D. J. McComas, B. T. Tsurutani, and J. A. Slavin, "Evolution of the Earth's Distant Magnetotail: ISEE 3 Electron Plasma Results," *J. Geophys. Res.* **89**, 11007 (1984).

[14] R. L. McPherron, "Magnetospheric Substorms," *Rev. Geophys. Space Phys.* **17**, 657 (1979).

[15] D. H. Fairfield and N. F. Ness, "Configuration of the Geomagnetic Tail During Substorms," *J. Geophys. Res.* **75**, 7032 (1970).

[16] R. E. Holzer and J. A. Slavin, "A Correlative Study of Magnetic Flux Transfer in the Magnetosphere," *J. Geophys. Res.* **84**, 10894 (1979).

[17] D. N. Baker, S. J. Bame, R. D. Belian, W. C. Feldman, J. T. Gosling, P. R. Higbie, E. W. Hones, Jr., D. J. McComas, and R. D. Zwickl, "Correlated Dynamical Changes in the Near-Earth and Distant Magnetotail Regions: ISEE 3," *J. Geophys. Res.* **89**, 3855 (1984).

[18] P. R. Bevington, *Data Reduction and Error Analysis for the Physical Sciences,* McGraw-Hill, New York, pp. 310-312 (1969).

[19] A. T. Y. Lui, L. A. Frank, K. L. Ackerson, C.-I. Meng, and S.-I. Akasofu, "Plasma Flows and Magnetic Field Vectors in the Plasma Sheet During Substorms," *J. Geophys. Res.* **83**, 3849 (1978).

[20] C. A. Cattell, F. S. Mozer, E. W. Hones, Jr., R. R. Anderson, and R. D. Sharp, "ISEE Observations of the Plasma Sheet Boundary, Plasma Sheet, and Neutral Sheet, 1. Electric Field, Magnetic Field, Plasma, and Ion Composition," *J. Geophys. Res.* **91**, 5663 (1986).

[21] W. J. Heikkila, "Magnetospheric Topology of Fields and Currents," in *Magnetospheric Currents,* T. A. Potemera, ed., *Amer. Geophys. Union,* Washington, D.C. pp. 208-222 (1984).

[22] V. M. Vasyliunas, "Theoretical Models of Magnetic Field Line Merging," *Rev. Geophys. Space Phys.* **13**, 303 (1975).

[23] T. W. Hill and P. H. Reiff, "On the Cause of Plasma-Sheet Thinning During Substorms," *Geophys. Res. Lett.* **7**, 1977 (1980).

DISCUSSION

R. Zwickl: In your detailed statistical study of the ISEE 3 plasma sheet data, you did not mention how the data were selected, an important point since any small systematic effect would have a large impact on the results.

J. A. Slavin: Our intent was to limit our study to the central plasma sheet and exclude the boundary layers. The procedure was to identify the intervals of high β plasma based upon the strong diamagnetic decreases they produce in the magnetic field (see, e.g., Slavin et al., J. Geophys. Res., 1985). The intervals were then examined a second time using the ISEE 3 LANL plasma/JPL magnetic field microfiche to confirm the plasma sheet identifications. Only intervals greater in duration than 10–15 minutes were used in the study. The shorter plasma sheet encounters were excluded in order to minimize aliasing of our 10 minute averaged data set by the boundary layer as ISEE 3 enters and leaves the plasma sheet.

V. M. Vasyliunas: While I do not necessarily disagree with your conclusion that the reconnection model works quite well. I do feel the evidence you present is rather weak; there is enormous scatter in your plots of dependence on the IMF z component, and correlation coefficients of at most 0.3–0.4 are insignificant (they indicated that at most 10% of the variability is accounted for by the claimed dependence).

J. A Slavin: A correlation coefficient does indeed specify what fraction of the independent quantity (i.e., $\gamma^2 \sim 10$–20% in these cases). However, it is the correlation coefficient and the number of events in the sample space that determine the "confidence level" for the linear trend produced by the regression. Given the numbers of points in our studies, there is no doubt that B_z and V_x both become negative in the distant plasma sheet following southward IMF and substorm activity. Reconnection theory requires that this relationship exists, but it does not insist that nothing else be disturbing the magnetic field. Solar wind $\vec{V}$ changes, tail "flapping", plasma waves, and plasmoids (which produce large $\pm B_z$, but transport no net flux out of the tail) all contribute to the large variances measured by ISEE 3. Against this background we can clearly observe the effects of reconnection, but they are not always as dominant as one might expect.

NEUTRAL-SHEET CROSSINGS IN THE DISTANT MAGNETOTAIL

W. J. Heikkila*

We have analyzed the magnetic field data from ISEE 3 in the distant magnetotail for 18 crossings of the cross-tail current sheet (or so-called neutral sheet) to determine the direction of the normal component B_z. The crossings occurred near the middle of the aberrated magnetotail ($0 < y < 30$ R$_e$, $-10 < z < 5$) in GSM coordinates, at a distance of about 220 R$_e$, January 28 to February 12, 1983, tailward of the curved X-line deduced by previous workers. In two cases we found B_z negative (southward), as would be required with a magnetic neutral line (reconnection line) earthward of the spacecraft. In 10 acceptable cases B_z was clearly northward ($B_z > 0.4$ nT), consistent with closed field lines connected to the earth; in two cases B_z was very close to zero. In several instances there was structure in B_y, suggesting localized currents with x or z directions. One may have been a magnetopause crossing. Consequently, we have not used the data for four cases. The average B_z for the other 14 cases was 0.7 ±0.3 nT. In addition, we have inspected the magnetic record, and have found that B_z increased to positive values whenever the field magnitude became small (< 2 nT); the common assumption that B_z is independent of z is not accurate. The strong preponderance of northward B_z favors a model of the magnetotail that is dominated by boundary layer plasma, flowing tailward on closed magnetic field lines, which requires the existence of an electric field in the sense from dusk to dawn. Since the observed flow was usually less than the magnetosheath flow speed, these observations are consistent with the idea that magnetospheric processes are powered by a boundary layer dynamo; this is a form of viscous interaction first proposed by Axford and Hines. The steady state reconnection model of Dungey is not supported by these observations.

INTRODUCTION

The presence of an extended magnetotail at all times has had considerable influence on our concepts of the plasma physics of the magnetosphere. Many workers (e.g., the reviews by Cowley[1,2]) believe that models based on the reconnection process[3] are supported by the observations, especially now that ISEE 3 has accomplished its geotail mission (see the special issue of *Geophys. Res. Lett.* **11** (Oct 1984)). In particular, the z-component (in GSM coordinates) of the magnetic field beyond 200 R$_e$ was reported to be slightly negative (pointing southward), implying that the steady state reconnection line, or X-line, was inside of that distance.[4,5] Perhaps the most convincing evidence has been the observation[6] that the plasma velocity was always tailward at the greatest distances (more than 97% of the time when the spacecraft was in the magnetotail at $x < -180$ R$_e$). With the usually assumed dawn-dusk electric field across the magnetosphere, consistent with the reconnection process, the magnetic field would have a southward component so that the plasma velocity $V = E \times B/B^2$ would be tailward. Zwickl et al.[6] concluded that the X-line was ". . . usually within 120 R$_e$ of earth, and rarely beyond 180 R$_e$." Slavin et al.,[7] using 5-minute average values of the data in boxes 10 R$_e$ in the x and y directions, concluded that the reconnection line was curved (see Fig. 20 in Ref. 7 and our Fig. 1). They found a weak southward B_z from y(GSM) $= 0$ to 30 R$_e$, at $-180 < x < -225$ R$_e$, in the central part of the aberrated magnetotail (aberration of approximately 4.5° because of the earth's orbital velocity).

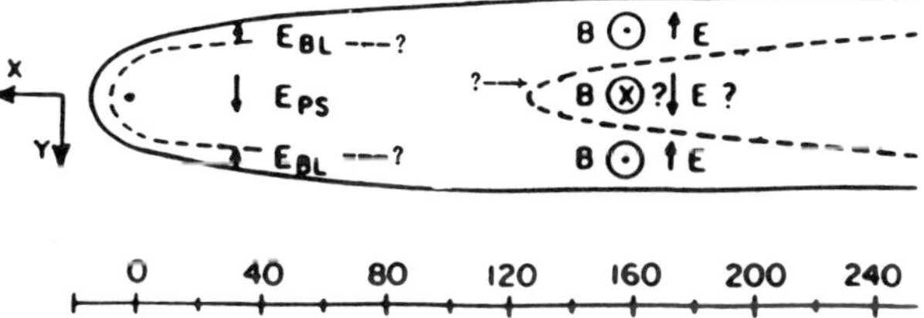

Figure 1—Equatorial view of the magnetotail, showing the curved X-line deduced by Slavin et al.[7]. The low-latitude boundary layer near the earth has been added, as well as the sense of the magnetic and electric field observations from Slavin et al. Our event locations were chosen for a test of the curved X-line; B_z should be negative at the neutral sheet beyond the X-line.

However, it may be premature to accept this one model of the magnetotail. In the dawn and dusk flanks Slavin et al.[7] found northward B_z (see Fig. 12 in Ref. 7 and our Fig. 1). Since the plasma velocity was tailward, this finding implies that the convection electric field was directed from dusk to dawn. This sense is opposite to the dawn-dusk electric field in the plasma sheet closer to earth for earthward convection, or the region of the deduced interplanetary magnetic field (IMF) lines in the distant tail for tailward convection (see Fig. 1). It is in the same sense[8,9] as the electric field in the low-latitude magnetospheric boundary layer (LLBL). Consequently, it is tempting to conclude that the broad regions (≈ 15 R$_e$) on the flanks of the distant magnetotail are a continuation and enlargement of the LLBL closer to the earth, a feature that was ignored in the other articles referred to above. If the flanks are indeed the LLBL on closed field lines, an immediate question then concerns the reality of the curved X-line (where does it begin and end?).

---

*University of Texas at Dallas, Richardson, Texas, 75083-0688.

These early statistical studies of the ISEE 3 observations are also difficult to reconcile with Villante's[10] analysis of Pioneer 7 data obtained in September 1966 at the still greater distance of 1000 R_e. In a detailed study of 14 magnetic field reversals (neutral sheet crossings), Villante found an average northward B_z of 1.5 nT. Cowley[2] has dismissed this result, saying that it is "considerably at odds with expectations." However, Cowley apparently had only the reconnection model in mind in reaching his negative conclusion; Villante's result is in excellent agreement with a model in which the LLBL is assumed to dominate the physics of the magnetotail.[11] In this model it is a reversal of the electric field (rather than the magnetic field) that is associated with tailward flow in the distant magnetotail.[12]

It is true that a number of difficulties and uncertainties can indeed occur in the interpretation of the data; some of these have been discussed by Lui[13] and by Heikkila et al.[14] What really matters for the reconnection models is the sense in which the magnetic field crosses through the neutral sheet, and not so much its orientation within the lobes or plasma sheet, which can be affected by a flapping motion of the magnetotail. The spacecraft was at its apogee tailward of the deduced curved X-line only for about 2 weeks (January 28 to February 12, 1983), and unusual solar wind conditions may have affected the result. In addition, a lumpy structure, perhaps caused by a filamentation of the cross-tail current, might show negative B_z with inadequate sampling. The level of geomagnetic activity may have an effect, as negative values do occur when plasmoids (magnetic islands) are present.[4] Such difficulties and complications can be avoided by concentrating attention on the magnetic field exactly at the field reversal (or the so-called neutral sheet) region.

Here a complication could arise since B_y may also have a characteristic change at the same time. If, however, the field changes all occur in a constant (inclined) plane, then it may be possible to find a new coordinate system in which one component (essentially the cross-tail B_y component) is negligible, and to follow changes in the other two when the primary field reverses; thus if $B_y' \approx 0$, and $B_x' = 0$ at the reversal, then the field has only the one component, an adjusted B_z', where the primes denote values in this slightly inclined coordinate system. This can be accomplished by doing a minimum variance analysis and is what Villante[10] did with the Pioneer 7 data.

It is the purpose of the present paper to analyze the ISEE 3 data with similar methods, using high resolution data, in order to resolve this controversy. The total amount of data produced by ISEE 3 in the distant magnetotail is enormous, and in this first analysis we will focus our attention on the central part of the tail when the spacecraft was tailward of the curved X-line, as deduced by Slavin et al.[7] The actual event locations that were analyzed are shown in Table 1 and are a random sampling of smooth crossings of the neutral sheet in this time interval.

ISEE 3 MEASUREMENTS

We have used data from the Jet Propulsion Laboratory vector helium magnetometer on ISEE 3,[15] as well as the Los Alamos National Laboratory plasma analyzer.[16] The magnetometer makes six measurements per second, with negligible drift; these can be averaged for the kind of temporal resolution required, commonly 1 second, 3 second, 1 minute, 5 minute, or 1 hour. The plasma instrument performs a two-dimensional electron measurement with 16 energy sweeps in 3 seconds, repeated every 84 seconds. A detailed discussion of the methods used to derive plasma bulk parameters from the electron measurements and their limitations is found in Ref. 6. These parameters are the bulk density, temperature, flow speed, and direction.

As has been pointed out before,[6,7] we do need to look at both kinds of data in order to identify correctly the various regions; this we have done for the complete time interval (about 2 weeks) analyzed in this paper. A sample of representative data was shown in Ref. 14 for a one-hour period on February 5, 1983. Within this region there was a crossing from the southern lobe into the northern lobe at 15:55:50 UT, that appears to be at a uniform speed, judging by the smooth B_x record. This is the kind of crossing we want to analyze, wishing to avoid transient phenomena such as plasmoids.

It was noted in that paper that at the precise moment of the field reversal when $B_x = 0$, and the magnitude $|B|$ was very small, the component B_z has a sharp positive peak. Many models have assumed that B_z is constant across the neutral sheet, but this crossing shows this assumption may not be accurate.

There was also a characteristic change in B_y at the same time. A minimum variance analysis of the magnetometer data has been done for this field reversal region (see number 15 in Table 1 for information such as the eigenvectors and Fig. 3 of Ref. 14 for the hodogram). We used the method of Siscoe et al.[17] in which one component is assumed to be negligible, so the change occurs through rotation in a constant plane. (We have also used Sonnerup and Cahill's method,[18] which does allow a finite normal component, with essentially the same results.) In this case B3 was the minimum variance direction, and we identified B3 with B_y' since the rotation was only about 30°. The maximum variance direction was B1 (i.e., the earthward component). The normal component B2 (a modified B_z') was indeed positive, as shown by the hodogram track remaining above zero in the hodogram as B1 reverses, being about 3 nT at the reversal of B1 (Fig. 3 of Heikkila et al.[14]). Therefore, the data for this neutral sheet crossing are consistent with closed field lines, earthward of a steady-state x-line.

Figure 2 presents a hodogram for another crossing, that at 01:46:27 UT on January 28. The main difference between the two is that Fig. 2 shows enhanced low-frequency wave activity; the plot of B3 vs. B2 indicates a wave of circular polarization (with a period of about one minute, as deduced from the recorded data). It is

Table 1—Listing of plasma and magnetic field data for 18 crossings of the field reversal region by ISEE 3. The value of B_z at each crossing is shown, as well as a corrected (by eye) B_z' to remove wave activity. The eigenvectors and λ's for a minimum variance analysis yield a normal component (B2 or B3).

# Date Time	X	Y	Z (Re)	K_p	log N (m^{-3})	log T	V (km/s)	φ (deg)	$\|B\|_{min}$	B_z (nT)	(B_z')	e_1	e_2	e_3	λ_1	λ_2	λ_3	λ_1/λ_2	λ_1/λ_3	λ_3/λ_2	B_n (nT)	BN	(B_n')	Sign
1) 28/83 01:28	−215	4.2	−6.1	1−	−0.9	6.0	400	280	1.4	1.0	(1.5)	0.845	0.520	−0.126	3,970	519	135	7.6	29.4	0.26	0.9 ± 1.7	B2	(1.5)	+
												−0.433	0.525	−0.733										
												−0.315	0.674	0.699										
2) 28/83 01:46	−215	4.1	−6.2	1−	−0.8	6.0	50	270	0.7	0	(0.6)	0.944	0.102	0.050	8,000	417	109	19.2	73.4	0.26	−0.16 ± 0.4	B2	(+0.6)	+
												0.063	−0.126	−0.990										
												−0.905	0.987	−0.132										
3) 28/83 02:10	−215	4.1	−6.2	1−	−0.7	5.7	200	0	0.4	0.4	(0.4)	0.947	0.301	−0.112	1,143	21.1	3.3	54.0	346	0.16	0.4 ± 0.2	B2	(0.4)	+
												−0.139	0.067	−0.988										
												−0.290	0.951	0.105										
4) 29/83 08:28	−217	4.0	−7.7	1+	−0.2	6.0	300	30	5.0	4.5	(3.0)	0.956	−0.277	−0.099	10,900	2,970	168	3.7	66	0.53	4.8 ± 0.9?	B2	(2.4)	+ ?
												−0.260	−0.637	−0.726										
												0.138	−0.719	−0.681										
5) 29/83 10:59	−217	4.0	−7.9	1+	−0.6	6.2	200	140	1.5	−0.8	(0)	0.918	0.222	−0.330	10,700	308	48	35	222	0.16	0.1 ± 1	B3		0
												0.340	0.869	−0.360										
												0.206	0.442	0.873										
6) 29/83 22:54	−217	7.8	−5.3	3−	0.4	5.5	400	0	6	3	(1.5)	0.883	0.396	−0.252	12,020	3,880	401	3.1	30	0.1	1.2 ± 1.5	B3	(1.5)	+
												−0.468	0.781	−0.413										
												0.034	0.483	0.875										
7) 30/83 09:52	−218	5.8	−8.2	3+	−0.8	6.4	200	−30	3	−2.0	(−2.5)	0.878	−0.191	−0.438	3,070	664	71	4.6	43	0.11	−2.5 ± 1.7	B2	(−2.5)	−
												−0.478	−0.314	−0.820										
												0.019	0.930	−0.367										
8) 30/83 11:19	−218	5.8	−8.3	5−	−1.0	6.3	300	140	5.5	5.5	(3.0)	0.825	−0.148	−0.545	13,670	4,560	85	3.0	160	0.19	5.5 ± 3.7?	B2	(3.0)	+ ?
												−0.556	−0.377	−0.740										
												−0.096	0.914	−0.394										
9) 30/83 12:08	−218	6.0	−8.2	4	−1.0	6.2	500	0	1.5	0.5	(+)	0.816	0.538	−0.213	8,016	582	95	13.7	84	0.16	0.1 ± 0.9	B3	(+)	0
												−0.544	0.588	−0.599										
												−0.197	0.605	0.772										
10) 33/83 16:12	−220	13.4	−7.1	1−	0.0	5.8	100	0	1.0	1.5	(0.9)	0.974	0.220	0.052	29,800	4,230	2,070	7.0	14.4	0.5	2.0 ± 1.3	B3	(1.5)	+
												−0.226	0.954	0.199										
												−0.057	−0.206	0.979										
11) 33/83 16:32	−220	13.5	−6.8	1−	−0.1	5.8	50	150	0.6	0.6	(0.7)	0.896	−0.435	−0.083	7,130	1,290	75.6	5.5	94	0.06	1.9 ± 0.9	B2	(1.5)	+
												0.271	0.687	−0.674										
												0.351	0.582	0.734										
12) 33/83 22:23	−220	15.2	−3.5	3−	−0.8	5.8	100	0	0.7	−0.5	(−1.0)	0.980	0.030	0.196	6,770	681	349	9.95	19.4	0.5	0.7 ± 1.0	B2	(−1.0)	−
												0.160	−0.704	−0.692										
												0.118	0.710	−0.695										
13) 33/83 22:39	−220	15.2	−3.4	3−	0.0	5.7	200	0	6.0	0.4	(0.4)	0.960	−0.135	−0.260	2,820	2,030	101	1.39	28	0.5	−0.4 ± 2.3	B3	(?)	?
												0.113	0.989	−0.098										
												0.271	0.064	0.961										
14) 35/83 10:08	−220	15.4	−9.8	2+	−0.5	5.8	0	—	0.4	0.4	(0.4)	0.942	0.272	0.195	617	91	4.7	6.8	132	0.5	0.4 ± 0.7	B2	(0.4)	+
												0.001	0.580	−0.815										
												−0.334	0.768	0.546										
15) 36/83 15:56	−220	19.3	−7.0	6	−1.0	6.2	400	90	3.6	3.0	(1.5)	0.988	−0.048	−0.148	66,100	4,140	207	16	319	0.05	3.0 ± 2.4	B2	(1.5)	+
												−0.150	−0.560	−0.815										
												−0.044	0.827	−0.561										
16) 41/83 21:20	−219	30.0	2.3	3−	−0.2	5.6	100	0	2	1.1	(1.1)	0.976	0.216	−0.026	1,456	232	20.2	6.3	72	0.09	0.8 ± 0.5	B2	(0.8)	+
												−0.126	0.466	−0.876										
												−0.177	0.858	0.482										
17) Magnetopause crossing																								
18) 43/83 04:41	−219	32.2	−2.0	3	0.3	5.8	200	0	2.5	1.5	(1.5)	0.891	0.091	0.444	6,160	334	19.8	18.5	310	0.06	0.5 ± 1.3	B3	(1.0)	+
												−0.346	0.770	0.536										
												−0.294	−0.632	0.718										

likely that the observed value of the B_z (or B2) component at the crossing of 0(−0.16) nT should be increased to $B_z' = +0.6$ nT for a measure of the true normal component, as an allowance for the wave activity. By the same token, B_z' for the crossing on February 5 should be reduced to $B_z' = 1.5$ nT. These qualitative corrections are listed under B_z' (B_n') in Table 1, and also in Table 2. The conclusion from Table 2 is that this correction would not influence the main result to any great extent.

The very next crossing at 02:09:40 (not shown) indicates that the amplitude of the waves can be small. How-ever, most of the crossings that we have analyzed showed appreciable wave activity.

Table 1 also shows the λ's and their ratios for a test of the significance of the minimum variance analysis.[19] All cases pass this test. Two cases (4 and 8) are suspect because B_z was extraordinarily large. Number 13 had a large B_y component, indicating tilt of 90°. These (and number 17) were consequently not used in the averaging process (indicated by ? in the last column of Table 1).

The average values of B_z (B_n) are listed in Table 2, under varying assumptions as to which events to choose. The cases with B2 > 0 identified as the normal (1, 2,

ISEE-3 MAGNETOMETER

83,28 01:45:00

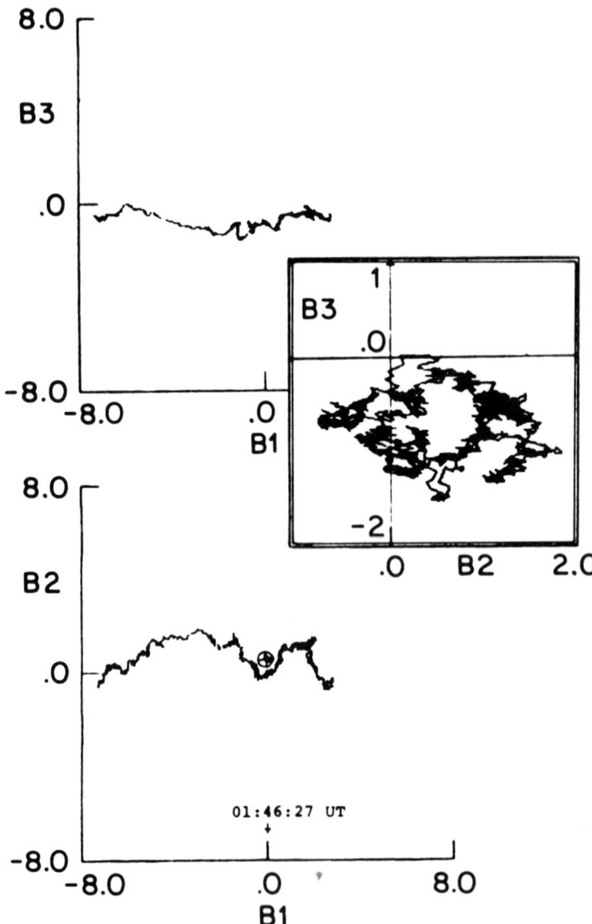

Figure 2—A hodogram for a crossing of the field reversal region. The plot of B3 vs. B2 indicates that a wave of circular polarization was propagating along the field. This may require a correction to B2, from −0.16 nT to +0.6 nT, for the component pointing northward across the field reversal region.

Table 2—Several types of averages, depending on which cases are used from Table 1. The result is consistent with $\overline{B_z} = 0.7 \pm 0.3$ nT.

	$\overline{B_z}$	$\overline{B_z'}$	$\overline{B_n}$	$\overline{B_n'}$
1. All 14 cases	0.7	0.5	0.5	0.5
2. Only $B_z > 0$, 12 cases	1.0	0.8	0.8	0.9
3. B2 only, 9 cases	0.4	0.3	0.5	0.4
4. B2 > 0, 7 cases	0.9	0.9	1.0	1.0

3, 7, 11, 12, 14, 15 and 16) are the easiest to interpret, with tilts of < 45° (see the eigenvectors). It may be possible to identify B3 with the normal component in five cases; this suggestion is verified by the similarity in the structure of B3 with B_z.

There are a few cases when B_z was negative at the neutral sheet crossing when apparently plasmoids were not present. Of the 18 cases we studied in this time interval, we have found only two such cases. One was the crossing on January 30 at 9:52:14 UT (Fig. 3), but note that there are several cases when $B_z > 0$ when $|B| < 2$ nT in this record. The hodogram (not included) shows a clearly negative B_z' ($\approx B_z$), with some small wave activity present. The plasma shows heating 20 s after the neutral sheet crossing (when the magnitude was 3 nT). In view of the high activity during this period ($K_p = 3+$), we suspect that a small plasmoid may be present. Nevertheless, this value has been included in the averaging process in Table 2. The averages listed in Table 2 range from 0.3 to 1.0 nT. Our result is consistent with $B_z = 0.7 \pm 0.3$ nT.

Actually, we have found that the record of the magnetic field is often sufficient to recover the sign of B_z in the neutral sheet. Figure 3 shows many features where BMAG is small (< 2 nT) and THTA < 90° (northward B_z). For another example, on February 10 at 21:20 UT (Fig. 4) the spacecraft dipped into the neutral sheet for about 5 minutes, and again at 22:35, 23:30, and 23:45 UT, with actual crossings at 23:42 and 23:50 UT. Each time the B_z component increased when the magnitude $|B|$ decreased to low values. This fact un-

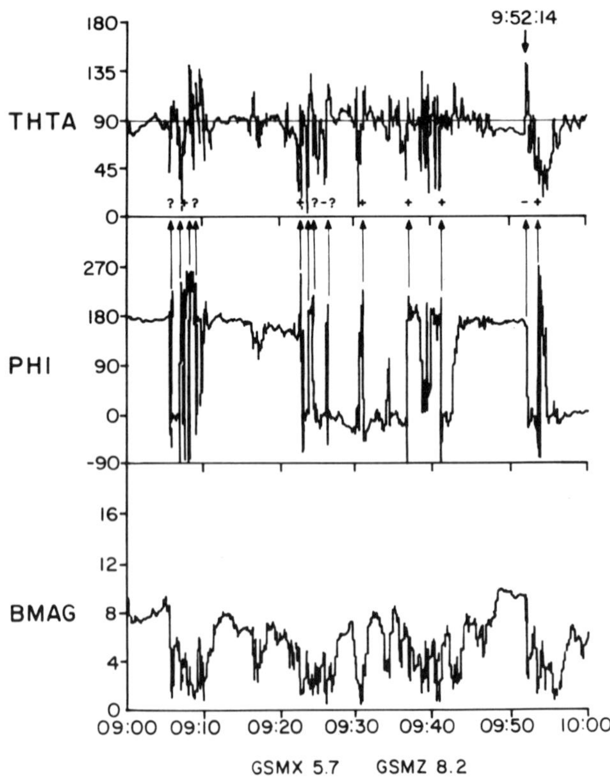

Figure 3—A one-hour record showing THTA < 90° (where THTA is a co-latitude angle) indicating northward B_z when $|B| < 2$ nT except for one crossing at 09:52:14 UT.

derscores the weakness in trying to estimate B_z from its value away from the field reversal region.

CONCLUSIONS

There were several hundred partial crossings of the field reversal region (as shown in Figs. 3 and 4) during this two-week interval, but only a few tens of uniform complete crossings. Eighteen cases were chosen for minimum variance analysis covering some 30 minutes of high-resolution data, for at least 400 vectors in each case. Our findings suggest the following conclusions:

1. The component B_z normal to the neutral sheet (or field reversal region) points decisively northward in 10 out of 14 cases (70%).
2. This component was largest at the actual crossing, and became smaller on either side. The common assumption that B_z is independent of z is erroneous.
3. B_z was negative in two crossings.
4. B_z was very close to zero in two cases.
5. Average $B_z = 0.7 \pm 0.3$ nT.
6. Use of minimum variance coordinates apparently had little effect on the magnitude, and none on the sign of B_n.
7. Wave activity was usually present, suggesting waves traveling along the magnetic field with circular polarization.
8. High resolution data (3-second averaging, or less) are most important, since B_z depends on z.

In addition, the type of record shown in Figs. 3 and 4 clearly indicate that B_z increased to positive values whenever the spacecraft was in the field reversal region.

DISCUSSION

It is generally a good idea (in some cases a necessity) to hypothesize a model based on fundamental physical principles in approaching a new problem, especially if the total amount of data available is very limited in either quantity or quality. Such is the case for the magnetosphere, where measurements are usually made on only a relatively few moving spacecraft. It goes without saying that any model must be falsifiable on the basis of observational evidence, should it prove to be contrary.

One such model is provided by the idea, or process, of magnetic reconnection pioneered by Dungey.[3] A key feature is a magnetospheric electric field that always points from dawn to dusk, and a magnetic X-line somewhere in the magnetotail. The convection velocity, $V_E = E \times B/B^2$, depends on this topology of the magnetic and electric fields.

However, the LLBL just inside the magnetopause seems to have been overlooked in the analysis carried out thus far. This LLBL provides a form of viscous interaction in a totally different model of the magnetosphere first proposed by Axford and Hines.[20] Figure 5 is based on their ideas, but with the addition of a magnetospheric boundary layer (which was discovered a decade later).

Two important recent findings now enter the picture. One is the conclusion that the LLBL closer to the earth is polarized to tens of kilovolts.[9,11,21-24] The same is true of the distant tail,[7] where the flow was tailward in the broad regions on the flanks where B_z was directed northward. This requires that the electric field was directed from dusk to dawn; we identify this region as the boundary layer.

The second new observation reported by Slavin at this meeting is that at low levels of geomagnetic activity ($|AL| < 100$ nT), the direction of B_z was northward across the entire magnetotail at 200 R_e. Since the flow was still tailward, they have concluded that "the low-

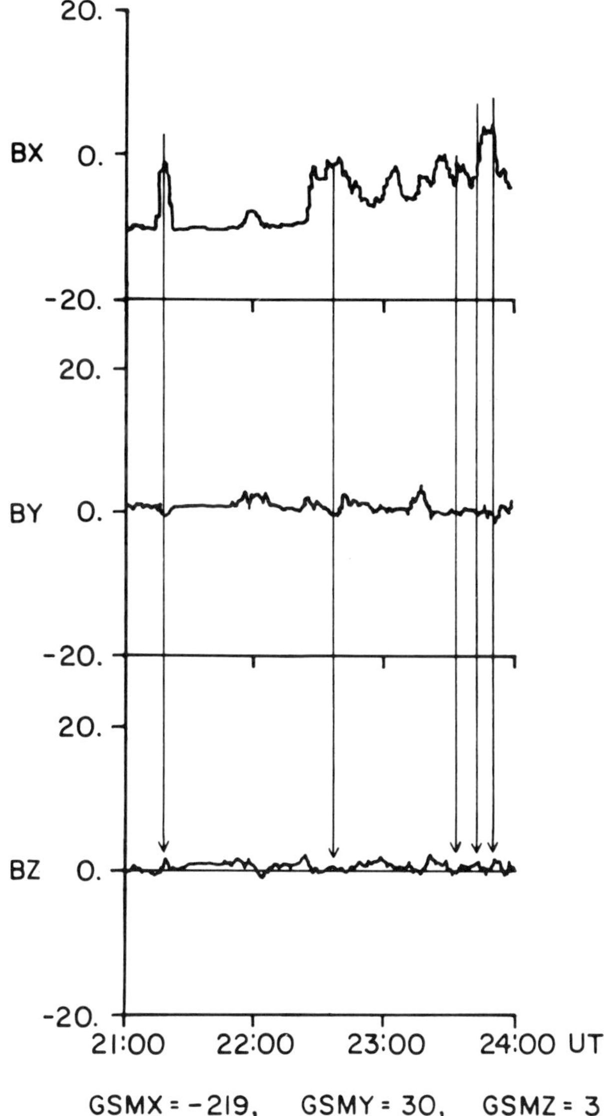

Figure 4—A three-hour record within the magnetotail showing that each time the magnitude of the magnetic field became low ($|B| < 2$ nT) the normal component B_z increased to larger positive values.

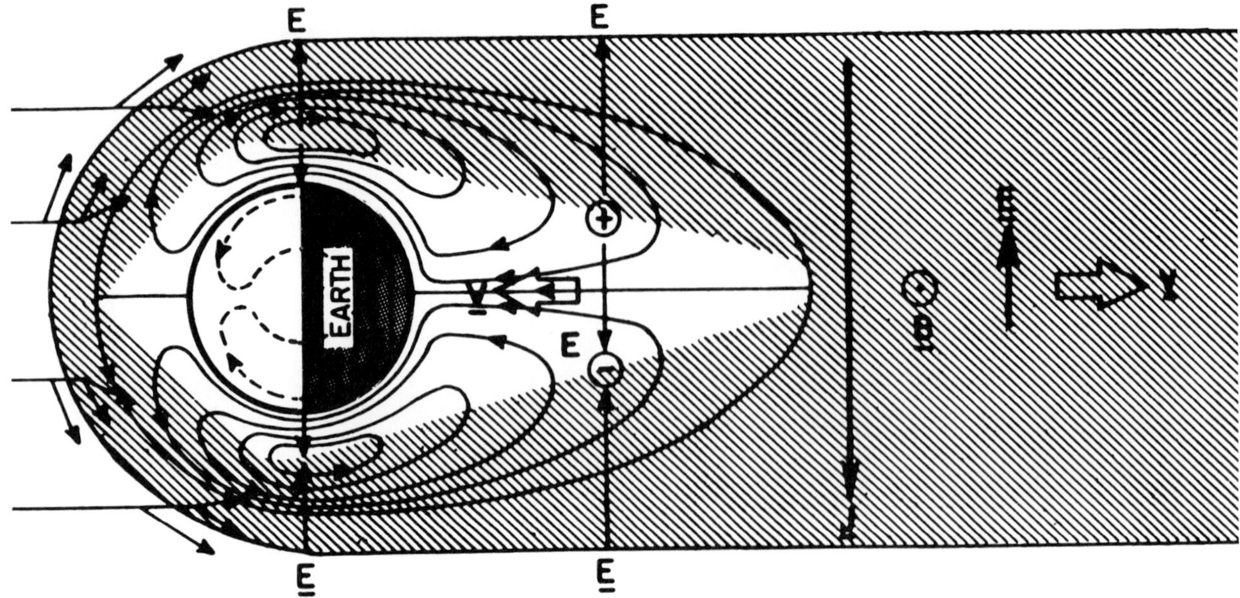

Figure 5—The Axford and Hines[20] model of the magnetotail, but with the addition of the low-latitude boundary layer. Since $E \cdot J < 0$, the plasma loses energy and momentum, showing that it behaves as a viscous medium.

latitude boundary layer extended across the full width of the quiet-time plasma sheet.'' This feature of the magnetotail was predicted by Heikkila.[25,26] According to this model, the plasma sheet closer to the earth is a small cavity filled with low-density plasma with earthward convection and energization, engulfed by boundary-layer plasma flowing tailward (see Fig. 5).

Our present study has concentrated on B_z during the two weeks when ISEE 3 was tailward of the curved X-line deduced by Slavin et al.,[7] a period when considerable geomagnetic activity occurred, but also with some quiet periods. We have found that B_z was clearly northward most of the time, agreeing with Villante.[10] This finding casts suspicion on the presumed existence of the curved X-line at all times. Instead, it supports the notion that tailward boundary layer flow is the primary agent for plasma sheet processes, including auroral phenomena. This tailward flow beyond 200 R_e found by Zwickl et al.[6] and the northward sense of the magnetic field (shown here and also by Slavin et al. at this meeting) is, in our view, compelling evidence for the reality of this effect.

A crucial question now is the process of entry of solar-wind plasma through the magnetopause and its exit at large distances downtail. Most of this plasma cannot return (there is too much to be returned via the plasma sheet, and its momentum is too great); the circulation that produces viscous interaction is not cyclic (as proposed by Cowley[1]), but rather open-ended. It follows that frozen field convection of magnetic flux in the magnetotail must somehow be violated. This point has been addressed by Heikkila,[12] who concludes that MHD is violated at the magnetopause; it is meaningless to apply the frozen field condition within the thin magnetopause.

What really matters is what the plasma particles do, and apparently they are able to cross the magnetopause into the LLBL (on closed field lines) quite easily, to produce a new MHD plasma inside. The magnetopause should be regarded as a sink of magnetic flux on one side and a source on the other. In this scheme, there is no requirement for a return convection of magnetic flux in the magnetotail, any more than a return flow of all the plasma in the LLBL.

Perhaps it is possible that both processes (reconnection as well as viscous interaction) can operate at certain times or certain locations (as has been suggested before, e.g., Cowley[1]); however, such a combination has yet to be elucidated (what happens to the X-line in the boundary layer?). Whatever the true answer is, it must be in agreement with our present result, namely that we have found a persistent northward B_z where previous interpretations have suggested that only southward B_z should be present.

ACKNOWLEDGMENT—I appreciate the use of the plasma data provided by Dan Baker, Sam Bame, and Ron Zwickl. Also, I wish to thank E. J. Smith and J. A. Slavin for providing access to the ISEE 3 field data and L. D. Wigglesworth for assistance in the computational aspects of the study.

This work was supported by the American Geophysical Union, the National Aeronautics and Space Administration, and the Danish Board of Space Activities.

REFERENCES

[1] S. W. H. Cowley, "Interpretation of Observed Relations Between Solar Wind Characteristics and Effects at Ionospheric Altitudes," in *High-Latitude Space Plasma Physics,* Hultqvist and Hagfors, p. 225 (1983).

[2] S. W. H. Cowley, "The Distant Geomagnetic Tail in Theory and Observation," in *Magnetic Reconnection in Space and Laboratory Plasmas,* Geophysical Monograph 30, E. W. Hones, Jr., ed., American Geophysical Union, Washington, D.C. p. 228 (1984).

[3] J. W. Dungey, "Interplanetary Field and the Auroral Zones," *Phys. Rev. Lett.* **6**, 47 (1961).

[4] G. L. Siscoe, D. E. Jones, G. D. Sibeck, J. A. Slavin, E. J. Smith, and B. T. Tsurutani, "ISEE 3 Magnetic Field Observations in the Magnetotail: Implications for Reconnection," in *Magnetic Reconnection in Space and Laboratory Plasmas,* E. W. Hones, Jr., ed., Geophysical Monograph 30, American Geophysical Union, Washington, D.C. (1984).

[5] B. T. Tsurutani, D. E. Jones, J. A. Slavin, D. G. Sibeck, and J. E. Smith, "Plasmasheet Magnetic Fields in the Distant Tail," *Geophys. Res. Lett.* **11**, 1062 (1984).

[6] R. D. Zwickl, D. N. Baker, S. J. Bame, W. C. Feldman, J. T. Gosling, E. W. Hones, Jr., D. J. McComas, B. T. Tsurutani, and J. A. Slavin, "Evolution of the Earth's Distant Magnetotail: ISEE 3 Electron Plasma Results," *J. Geophys. Res.* **89**, 11,007 (1984).

[7] J. A. Slavin, E. J. Smith, D. G. Sibeck, D. N. Baker, R. D. Zwickl, and S. I. Akasofu, "An ISEE 3 Study of Average and Substorm Conditions in the Distant Magnetotail," *J. Geophys. Res.* **90**, 10,875 (1985).

[8] T. E. Eastman, E. W. Hones, Jr., S. J. Bame, and J. R. Asbridge, "The Magnetospheric Boundary Layer: Site of Plasma, Momentum and Energy Transfer from the Magnetosheath into the Magnetosphere," *Geophys. Res. Lett.* **11**, 685 (1976).

[9] R. Lundin and D. S. Evans, "Boundary Layer Plasmas as a Source for High-Latitude, Early Afternoon, Auroral Arcs," *Planet. Space Sci.* **32**, 1389 (1985).

[10] U. Villante, "Neutral Sheet Observations at 1000 R_e," *J. Geophys. Res.* **81**, 212 (1976).

[11] W. J. Heikkila, "Transport of Plasma Across the Magnetopause," in *Solar Wind-Magnetosphere Coupling,* Y. Kamide and J. A. Slavin, eds., p. 337 (1986).

[12] W. J. Heikkila, "Transport of Plasma Across the Magnetopause," to be published in the proceedings of the AGU Chapman Conference on Solar Wind-Magnetosphere Interactions, held at Jet Propulsion Laboratory, 11-15 Feb (1986).

[13] A. T. Y. Lui, "Characteristics of the Cross-Tail Current in the Earth's Magnetotail," in *Magnetospheric Currents,* T. A. Potemra, ed., Geophysical Monograph 28, American Geophysical Union, Washington, D.C., p. 158 (1984).

[14] W. J. Heikkila, J. A. Slavin, E. J. Smith, D. N. Baker, and R. D. Zwickl, "Neutral Sheet Crossings by ISEE-3 in the Distant Magnetotail," in *Comparative Study of Magnetospheric Systems* R. Pellat, ed., p. 315 (1986).

[15] A. M. A. Frandsen, B. V. Connor, J. V. Amersfoort, and E. J. Smith, "The ISEE-C Vector Helium Magnetometer," *IEEE Trans. Geosci, Electr.* **GE-16**, 195-198 (1978).

[16] S. J. Bame, J. R. Asbridge, H. E. Felthauser, J. P. Glore, H. L. Hawk, and J. Chavez, "ISEE-C Solar Wind Plasma Experiment," *IEEE Trans. Geosci. Electr* **GE-16**, 160 (1978).

[17] G. L. Siscoe, L. Davis, Jr., P. J. Coleman, Jr., E. J. Smith, and D. E. Jones, "Power Spectra and Discontinuities of the Interplanetary Magnetic Field: Mariner 4," *J. Geophys. Res.* **73**, 61 (1968).

[18] B. U. O. Sonnerup and L. J. Cahill, "Magnetopause Structure and Attitude from Explorer 12 Observations," *J. Geophys. Res.* **72**, 171 (1967).

[19] G. L. Siscoe and R. W. Suey, "Significance Criteria for Variance Matrix Applications," *J. Geophys. Res.* **77**, 1321 (1972).

[20] W. I. Axford and C. O. Hines, "A Unifying Theory of High-Latitude Geophysical Phenomena and Geomagnetic Storms," *Can. J. Phys.* **39**, 1433 (1961).

[21] J. C. Foster, "Ionospheric Signatures of Magnetospheric Convection," *J. Geophys. Res.* **89**, 855 (1984).

[22] T. E. Eastman, B. Popielawska, and L. A. Frank, "Three-Dimensional Plasma Observations Near the Outer Magnetospheric Boundary," *J. Geophys. Res.* **90**, 9519 (1985).

[23] R. Lundin and E. M. Dubinin, "Solar Wind Energy Transfer Regions Inside the Dayside Magnetopause: Accelerated Heavy Ions as Tracers for MHD-Processes in the Dayside Boundary Layer," *Planet Space Sci.* **33**, 891 (1985).

[24] F. S. Mozer, "Electric Field Evidence on the Viscous Interaction at the Magnetopause," *Geophys. Res. Lett.* **11** (1984).

[25] W. J. Heikkila, "Magnetospheric Topology of Fields and Currents," in *Magnetospheric Currents*, Geophysical Monograph 28, T. A. Potemra, ed., American Geophysical Union, Washington, D. C., p. 208 (1984).

[26] W. J. Heikkila, "The Electromagnetic Field for an Open Magnetosphere," in *Magnetic Reconnection in Space and Laboratory Plasmas,* E. W. Hones, Jr., ed., p. 39 (1984).

ISEE 3 MAGNETOPAUSE CROSSINGS: EVIDENCE FOR THE KELVIN-HELMHOLTZ INSTABILITY

D. G. Sibeck*

J. A. Slavin and E. J. Smith[†]

The daily number of magnetopause crossings that ISEE 3 observed during its first distant magnetotail transit is moderately well correlated with the upstream solar wind/magnetosheath velocity observed by IMP 8, suggesting that some boundary crossings are caused by waves driven Kelvin-Helmholtz unstable. A statistical study of parameters attending the magnetopause crossings shows that the magnetopause is almost never unstable to purely longitudinal waves, but in over half the crossings the instability criterion for waves with a circumferential component was satisfied. Growth rates for waves driven Kelvin-Helmholtz unstable are probably sufficient for the waves to explain shorter-period (boundary interarrival times ≤ 20 minutes) magnetopause motion.

INTRODUCTION

There are many possible causes for distant magnetotail magnetopause motion. External causes include a variable solar-wind pressure, flow direction, or aberration angle.[1] Boundary waves may be driven Kelvin-Helmholtz unstable at the magnetopause.[2] Passing flux tubes of interconnected magnetosphere-magnetosheath field lines may also mimic magnetopause crossings.[3] Flux may be added to, or removed from, the magnetotail, causing its radius to increase or decrease.[4] The magnetotail may oscillate if its eigenmodes are excited by changes in solar-wind parameters.[5] Finally, the passage of a tailward-moving plasmoid may cause the magnetotail radius to increase briefly and then decrease.[6]

Some evidence exists for each mechanism; here we single out the contribution of Kelvin-Helmholtz boundary waves to distant magnetotail magnetopause motion. As shown below, this is possible because the frequency of magnetopause crossings produced by the other mechanisms is not proportional to the solar-wind or magnetosheath velocity.

Many studies have treated the Kelvin-Helmholtz instability at the magnetopause, showing that it is favored for weak magnetic fields or fields transverse to the flow, denser plasmas, and higher velocities. Thus Southwood[7] and Lee et al.[8] have predicted the instability to be more likely at and tailward of the dawn-dusk meridian than near the subsolar point. Observationally, Boller and Stolov[9-11] have verified that conditions on the near-earth magnetotail flanks satisfy the local instability criteria.

McKenzie[12] and Ershkovich et al.[5] considered whole-body near-earth magnetotail oscillations in response to the Kelvin-Helmholtz instability. McKenzie treated tailward propagating waves with a planar geometry and showed that waves of increasingly longer wavelength go unstable as the solar-wind velocity rises. Ershkovich et al. considered both azimuthal and longitudinal waves in a cylindrical geometry. They found that as the solar-wind velocity increases, the instability threshold for the azimuthal waves is crossed prior to that for the longitudinal waves.

Sizable boundary waves will not simply occur because the magnetopause is unstable; their growth rates must be suitably large. For a first approximation, it is reasonable to assume that the growth rates increase with solar-wind velocity.[9] Pu and Kivelson[13] solve for the growth rates and find that they are finite for a band of solar-wind velocities near the magnetotail Alfvén velocity, but increase with the solar-wind velocity when that velocity exceeds the magnetotail Alfvén velocity by a factor of 1.25.

One expects boundary waves to move tailward with the magnetosheath flow, and such waves are frequently inferred from magnetopause boundary normal oscillations.[14,15] There is also evidence for significant azimuthal normal components.[16,17] It is interesting to note that a dense boundary-layer plasma, which would favor the instability, lay just inside the magnetopause during the crossings reported by Sibeck et al.

The characteristics of the Kelvin-Helmholtz instability allow us to distinguish its contribution to magnetopause motion from those of the other mechanisms. In particular, the Kelvin-Helmholtz instability makes no contribution to magnetopause motion when the solar-wind velocity fails to exceed some instability threshold. Its contribution should increase as the solar-wind velocity rises, and may be the dominant cause of magnetopause crossings when the solar-wind velocity is large.

In contrast, the frequency of magnetopause crossings resulting from magnetopause motion driven by the other mechanisms does not depend so explicitly on the solar-wind velocity. Magnetopause motion or magnetotail eigenmode oscillations driven by solar-wind pressure and velocity fluctuations do not depend on the magnitude of the velocity, but rather on the magnitude of its variations. Although the solar-wind variability increases during intervals of high solar-wind velocity, it is still greater (by about a factor of 2) on the leading edge of fast solar-wind streams,[18] preventing a direct solar-wind velocity-magnetopause crossing correlation. Flux tube production, and hence their observation frequency, does not depend on the solar-wind velocity.[19] Flux addition to,

*The Johns Hopkins University Applied Physics Laboratory, Laurel, Maryland 20707.

[†]Jet Propulsion Laboratory, California Institute of Technology, 4800 Oak Grove Drive, Pasadena, California 91103.

and removal from, the magnetotail depends on the direction of the solar-wind magnetic field. Plasmoid production is irregular and associated with substorms as measured by the geomagnetic *AE* index.

In this paper we relate the frequency of distant magnetotail magnetopause crossings observed by ISEE 3 to the upstream solar-wind and magnetosheath velocities observed by IMP 8. The observations provide evidence for the Kelvin-Helmholtz instability.

DATA

ISEE 3 observed over 700 tangential discontinuity magnetopause crossings during its first distant ($|x|$ ~ 200 R_e) magnetotail pass from January to March 1983. During this dawn-to-dusk pass, the spacecraft moved from YGSM = -20 to 45 R_e, passing about 5 R_e below the nominal tail center. The method by which these crossings were identified and their characteristics have been described by Sibeck et al.[15,20-22] All identified magnetopause crossings separated by 2.5 or more minutes are included in this study, and their locations were displayed and used to infer a mean tail diameter by Slavin et al.[23] During ISEE 3's magnetotail pass, solar-wind and magnetosheath observations were intermittently available from the MIT plasma detector on IMP 8. For this paper, we average 1 hour average velocities once again to produce daily values. Because of the sporadic coverage, particularly in the magnetosheath, these IMP 8 values should be treated with caution. The magnetosheath velocity outside the distant magnetotail should be less than, but near, these solar-wind values.[24]

RESULTS

Figure 1 shows the number of magnetotail magnetopause crossings observed by ISEE 3 each day (continuous curve) and the upstream solar-wind/magnetosheath velocity (dot-dash curve) observed by IMP 8. Magnetopause crossings were recorded nearly every day except during the period from days 79 to 83 when ISEE 3 remained within the magnetotail. The crossings occur in bursts and are not spread evenly throughout each day. The daily number of magnetopause crossings seems to rise and fall in response to the solar-wind velocity, encouraging us to correlate the two variables.

Figure 2 shows a plot of the solar-wind/magnetosheath velocity each day versus the number of ISEE 3 magnetopause crossings that day. The two are positively correlated with a coefficient of 0.50. There is a large amount of scatter especially at low velocities and at low number of crossings. The best fit line intercepts the vertical (velocity) axis at 420 km/s, which can be taken as an estimate of the instability threshold. We were unable to find a clear correlation between the frequency of magnetopause crossings and the solar-wind velocity at shorter time scales (e.g., ~ 1 hour). This was due in part to periods of high solar-wind velocity with few magnetopause crossings (ISEE 3 remaining either inside or outside the magnetotail), but also to the sparse IMP

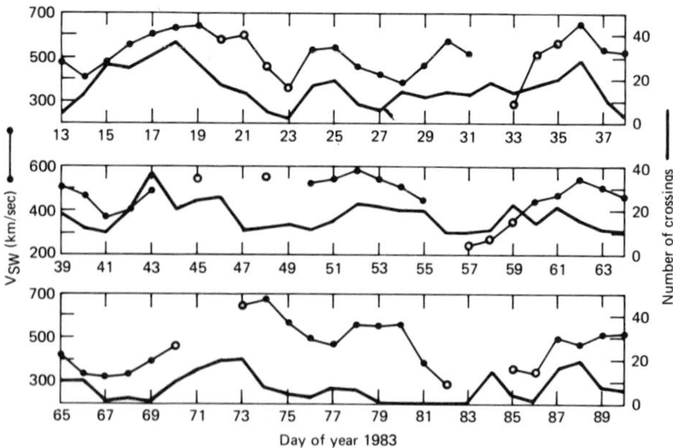

Figure 1 — The dot-dash curve shows the average daily solar-wind (solid circles) or magnetosheath (open circles) velocities observed by IMP 8, and the solid curve shows the number of magnetotail magnetopause crossings observed by ISEE 3 each day. Breaks in the dot-dash curve (solar-wind/magnetosheath velocity) occur when IMP 8 passes through the magnetotail. Both curves exhibit maxima near days 18, 25, 36, 43, 52, 73, and 88 and minima near days 23, 27, 41, 50, 57, 67, 77, 82, and 86. All the velocities must be treated with particular caution, and this is particularly true for those in the magnetosheath, which are based on only a few measurements.

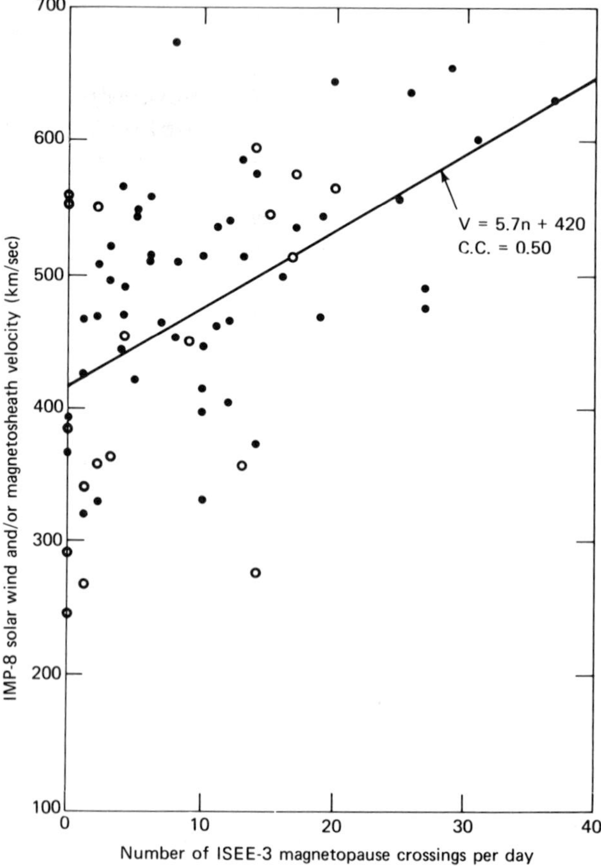

Figure 2 — A plot of the IMP 8 solar-wind (solid circles) or magnetosheath (open circles) velocity versus the number of ISEE 3 magnetopause crossings shows them to be positively correlated with a large amount of scatter. The best-fit line intercepts the ordinate at a velocity of 420 km/s.

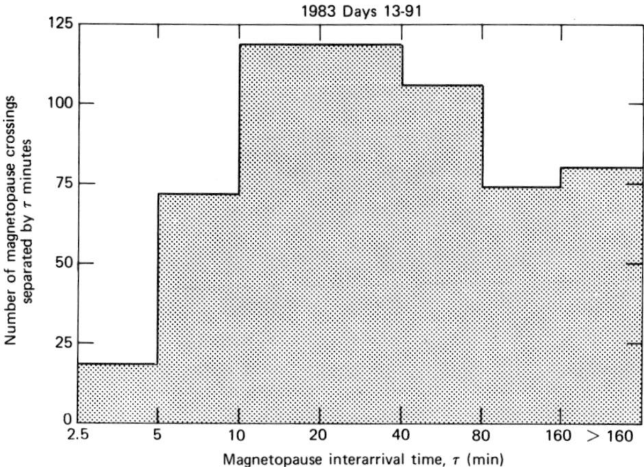

Figure 3 — A histogram of the times between successive ISEE 3 magnetopause crossings.

8 solar-wind/magnetosheath coverage, which precluded a sizable overlap with ISEE 3 crossings.

Figure 3 shows a histogram of boundary interarrival times, i.e., the time between two successive ISEE 3 magnetopause crossings. This is generally about 10–40 minutes, corresponding to wavelengths of 100–400 R_e if the crossings are caused by waves moving tailward with a solar-wind velocity of 500 km/s. If the waves are propagating tailward more slowly or obliquely (i.e., have azimuthal components), they could have shorter wavelengths.

INTERPRETATION

We interpret the above observations as evidence for the Kelvin-Helmholtz instability. It is easy to show that the distant magnetotail magnetopause is almost always unstable. According to the criteria of Ershkovich et al., we must show that

$$V_{SW} \geq A_1 - A_2, \qquad (1)$$

for waves with azimuthal components, and that

$$V_{SW} \geq A_1 + A_2, \qquad (2)$$

for solely longitudinal waves, where V_{SW} is the solar wind velocity and A_1 and A_2 are the magnetotail and magnetosheath Alfvén velocities. For a typical boundary-layer density just inside the magnetopause of 0.4 cm^{-3},[20] a magnetotail field strength of 9.1 nT or less in the boundary layers,[23] a magnetosheath density of 5 cm^{-3}, and a field strength of 5 nT,[20] the solar-wind and magnetotail Alfvén velocities are 50 and 310 km s^{-1}, respectively. Thus the magnetopause should be unstable to azimuthal waves when the solar-wind velocity exceeds 260 km s^{-1}, and unstable to purely longitudinal waves when that velocity exceeds 360 km s^{-1}. Since solar-wind velocities typically exceed these values (Fig. 1) the distant magnetotail magnetopause should be generally unstable. The 420 km s^{-1} instability threshold inferred from Fig. 2 is reasonably near the predicted 360 km s^{-1} threshold. However, we have used 5 minute averages of the ISEE 3 magnetic field and plasma parameters on either side of the magnetotail magnetopause to test the instability criteria at each crossing. Because not all the magnetopause crossings, particularly those with short boundary interarrival times, can be identified in these averages, only 466 crossings were studied. Of these, 236 satisfied Eq. 1, but only 24 satisfied Eq. 2. This indicates that magnetotail waves driven Kelvin-Helmholtz unstable must have azimuthal (circumferential) components, resulting in a fluted tail of sorts. Any circumferential wave motion is probably small compared to tailward propagation, as only the latter has been reported.

The growth rates provided by Pu and Kivelson[13] determine the wavelengths of waves that grow to sizable amplitudes during their tailward propagation from earth to ISEE 3 distances. We will use their results for the magnetotail configuration in which the fields, flow, and wave number vector are aligned and the ratios of magnetosphere to magnetosheath magnetic field strengths and densities are 1.5 and 0.1, respectively. Given the magnetosheath velocity and the magnetotail Alfvén velocity, we can read ϵ (a function of the wavelength and growth rate) from their plots. With ϵ we try various wavelengths to find growth rates greater than or equal to the inverse of the period during which the solar wind advects from earth to ISEE 3. We find that slow surface waves with boundary interarrival times of 2.1 minutes or less grow rapidly enough ($\epsilon = 0.06$) when the ratio of the solar-wind velocity to magnetotail Alfvén velocity is near one, and that incompressible waves with interarrival times of 3.4 minutes or less grow rapidly enough ($\epsilon = 0.4$) when the velocity ratio is 2. Thus we might only explain the shortest boundary interarrival times in Fig. 3.

However, the actual conditions at the magnetotail magnetopause are more favorable for instability than those in the model of Pu and Kivelson.[13] The ratio of the boundary layer to magnetosheath densities can exceed 0.1,[25] the ratio of boundary layer to magnetosheath field strengths is typically less than 1.5,[15] and because boundary-layer field lines bend in the direction of the magnetosheath magnetic field, they have components transverse to the flow. Furthermore, both the magnetosheath magnetic field and wave number vectors can have components transverse to the flow. Pu and Kivelson[13] find that simply orienting the magnetosphere magnetic field transverse to the flow causes the growth rate to increase by factors of as much as 5. Although the magnetotail magnetic field lies more nearly parallel than transverse to the flow, we will adopt this factor to include all the above favorable modifications. Then we might explain boundary interarrival times of 10.5 to 17.0 minutes or less, a significant fraction of the distribution in Fig. 3. The balance, and the low correlation coefficient in Fig. 2, must be explained by the other causes of boundary motion mentioned in the introduction. Efforts to determine the cause of long period motion are under way.

Now consider the amplitude of the waves. Slavin et al.[23] have noted that magnetopause crossings were seen frequently, even near the nominal magnetotail center. One possibility is that a strong and highly variable solar wind buffeted the tail and brought the magnetopause of a circular magnetotail to the nominal tail center. However, boundary normals for these crossings persistently point in the $-z$ direction,[21] and it is unlikely that the solar wind could manage to bring only the highest southern latitudes to the nominal tail axis so often. Another possibility is that the boundary waves have 25 R_e amplitudes on a tail of radius 30 R_e. This also seems unlikely. A simpler possibility is that the magnetotail is flattened in the north-south direction, but elongated in the east-west direction.[26] Sibeck et al.[20,21] described such a tail, whose total east-west dimension is 76 R_e and whose north-south dimension is 37 R_e at $x = -200$ R_e. Then the dawn-dusk ISEE 3 trajectory, which reached $Z_{GSM} = -10$ R_e, paralleled and remained near the southern magnetopause boundary. Waves of ~ 8 R_e amplitude could have caused the magnetopause crossings, and the normals at these crossings would have pointed nearly southward, as observed.

SUMMARY

IMP 8 solar wind/magnetosheath velocities and the number of ISEE 3 magnetotail magnetopause crossings on each day of its first distant tail pass are moderately well correlated, indicating that the Kelvin-Helmholtz instability drives substantial magnetopause motion. The observed plasma and magnetic field measurements at each magnetopause crossing seldom satisfy the instability criterion for purely longitudinal waves, but over 50% satisfy the criterion for waves with an azimuthal component. Waves driven Kelvin-Helmholtz unstable might reasonably account for magnetotail boundary interarrival times of up to 20 minutes, but observed longer-period motion probably has other causes. Daily magnetopause crossings along the ISEE 3 trajectory indicate bulk magnetotail motion, significant wave amplitudes, and perhaps most importantly, the spacecraft's close proximity to the magnetopause of a flattened magnetotail.

ACKNOWLEDGMENT — We wish to thank B. T. Tsurutani for helpful suggestions and A. Lazarus of the Massachusetts Institute of Technology for kindly supplying the IMP 8 plasma observations. Referee comments significantly improved the paper. Portions of this work were performed at the Jet Propulsion Laboratory, California Institute of Technology under contract with NASA and were carried out at the California State University at Los Angeles with support by a grant from NSF (ATM 81-20455).

REFERENCES

[1] R. L. Kaufmann and A. Konradi, "Explorer 12 Magnetopause Observations: Large-Scale Nonuniform Motion," *J. Geophys. Res.* **74**, 3609-3627 (1969).

[2] J. W. Dungey, "Electrodynamics of the Outer Atmosphere," in *The Physics of the Ionosphere,* 1954 Cambridge Conference, The Physical Society, London, pp. 229-236 (1955).

[3] D. G. Sibeck and G. L. Siscoe, "Downstream Properties of Magnetic Flux Transfer Events," *J. Geophys. Res.* **89**, 10709-10715 (1984).

[4] K. Maezawa, "Magnetotail Boundary Motion Associated with Geomagnetic Substorms," *J. Geophys. Res.* **80**, 3543-3548 (1975).

[5] A. I. Ershkovich and A. A. Nusinov, "Geomagnetic Tail Oscillations," *Cosmic Electro.* **2**, 471-480 (1972).

[6] J. A. Slavin, E. J. Smith, B. T. Tsurutani, D. G. Sibeck, H. J. Singer, D. N. Baker, J. T. Gosling, E. W. Hones, and F. L. Scarf, "Substorm Associated Traveling Compression Regions in the Distant Tail: ISEE 3 Geotail Observations," *Geophys. Res. Lett.* **11**, 657-660 (1984).

[7] D. J. Southwood, "The Hydromagnetic Stability of the Magnetospheric Boundary," *Planet. Space Sci.* **16**, 587-605 (1968).

[8] L.-C. Lee, R. K. Albano, and J. R. Kan, "Kelvin-Helmholtz Instability in the Magnetopause-Boundary Layer Region," *J. Geophys. Res.* **86**, 54-58 (1981).

[9] B. R. Boller and H. L. Stolov, "Kelvin-Helmholtz Instability and the Semiannual Variation of Geomagnetic Activity," *J. Geophys. Res.* **75**, 6073-6084 (1970).

[10] B. R. Boller and H. L. Stolov, "Explorer 18 Study of the Stability of the Magnetopause Using a Kelvin-Helmholtz Instability Criterion," *J. Geophys. Res.* **78**, 8078-8086 (1973).

[11] B. R. Boller and H. L. Stolov, "Investigation of the Association of Magnetopause Instability with Interplanetary Sector Structure," *J. Geophys. Res.* **79**, 673 (1974).

[12] J. F. McKenzie, "Hydromagnetic Wave Coupling Between the Solar Wind and the Plasma Sheet," *J. Geophys. Res.* **76**, 2958-2966 (1971).

[13] Z.-U. Pu and M. G. Kivelson, "Kelvin-Helmholtz Instability at the Magnetopause: Solution for Compressible Plasmas," *J. Geophys. Res.* **88**, 841-852 (1983).

[14] D. H. Fairfield, "Structure of the Magnetopause: Observations and Implications for Reconnection," *Space Sci. Rev.* **23**, 427-448 (1979).

[15] D. G. Sibeck, G. L. Siscoe, J. A. Slavin, E. J. Smith, B. T. Tsurutani, and R. P. Lepping, "The Distant Magnetotail's Response to a Strong IMF B_y: Twisting, Flattening, and Field Line Bending," *J. Geophys. Res.* **90**, 4011-4019 (1985).

[16] J. D. Mihalov, D. S. Colburn, and C. P. Sonnett, "Observations of Magnetopause Geometry and Waves at the Lunar Distance," *Planet. Space Sci.* **18**, 239-258 (1970).

[17] D. G. Sibeck, "An Explanation of Multiple Boundary Crossings Near the Magnetotail Magnetopause," M.S. Thesis, University of California, Los Angeles (1982).

[18] J. W. Belcher and L. Davis, Jr., "Large-Amplitude Alfvén Waves in the Interplanetary Medium 2," *J. Geophys. Res.* **76**, 3534-3563 (1971).

[19] R. P. Rijnbeek, "Flux Transfer Events: Impulsive Reconnection Signatures at the Earth's Magnetopause," Ph.D. Thesis, Imperial College of Science and Technology, London (1984).

[20] D. G. Sibeck, G. L. Siscoe, J. A. Slavin, E. J. Smith, B. T. Tsurutani, and S. J. Bame, "Magnetic Field Properties of the Distant Magnetotail Magnetopause and Boundary Layer," *J. Geophys. Res.* **90**, 9561-9575 (1985).

[21] D. G. Sibeck, G. L. Siscoe, J. A. Slavin, and R. P. Lepping, "Major Flattening of the Distant Geomagnetic Tail," *J. Geophys. Res.* **91**, 4223-4237 (1986).

[22] D. G. Sibeck, J. A. Slavin, E. J. Smith, and B. T. Tsurutani, "Twisting of the Geomagnetic Tail," in *Proceedings of the Chapman Conference on Solar Wind-Magnetosphere Coupling,* Y. Kamide and J. A. Slavin, eds., Terra/Reidel Publishing Co., Tokyo/Dordrecht (in press, 1986).

[23] J. A. Slavin, E. J. Smith, D. G. Sibeck, D. N. Baker, R. D. Zwickl, and S.-I. Akasofu, "An ISEE-3 Study of Average and Substorm Conditions in the Distant Magnetotail," *J. Geophys. Res.* **90**, 10875-10895 (1985).

[24] G. L. Siscoe, F. L. Scarf, D. S. Intrilligator, J. H. Wolfe, J. H. Binsack, H. S. Bridge, and V. M. Vasyliunas, "Evidence for a Geomagnetic Wake at 500 Earth Radii," *J. Geophys. Res.* **75**, 5319-5330 (1970).

[25] S. J. Bame, R. C. Anderson, J. R. Asbridge, D. N. Baker, W. C. Feldman, J. T. Gosling, E. W. Hones, Jr., D. J. McComas, and R. D. Zwickl, "Plasma Regimes in the Deep Geomagnetic Tail: ISEE 3," *Geophys. Res. Lett.* **10**, 912-915 (1983).

[26] F. C. Michel and A. J. Dessler, "Diffuse Entry of Solar-Flare Particles into Geomagnetic Tail," *J. Geophys. Res.* **75**, 6061-6072 (1970).

A THEORETICAL MODEL OF CONVECTION IN THE NEAR-EARTH PLASMA SHEET

G. Atkinson*

Equations describing convection in the plasma sheet and auroral zone are developed. The assumptions include hot protons, cold electrons, and Birkeland currents to a uniformly conducting ionosphere. Unlike previous models, a gradient of the plasma content of flux tubes is assumed. In addition, asymmetric nightside boundary conditions are studied. The equations are simplified to the point where analytic solutions are possible. The resulting solutions show both dayside and nightside Harang discontinuities and a cross-tail component to the convection. Birkeland currents (distributed and sheet) are calculated. The best agreement to observations occurs for enhanced sunward flow on the duskside of the tail and a weak gradient of the plasma content of flux tubes along the oval.

INTRODUCTION

This paper is a summary of recent work on a theoretical model of convection that includes the effects of a radial gradient of the plasma content, N, of flux tubes. The study will be published in more detail elsewhere.

The basic physics of hot protons and cold electrons undergoing electric field drift and magnetic field gradient drifts in a magnetosphere with an ionosphere was spelled out two decades ago by Fejer[1] and further developed by Vasyliunas.[2] It has been expanded to include other effects (e.g., ionospheric conductivity variations) in the Rice University numerical modeling approach (e.g., Harel et al.[3]) and is currently used to predict individual events (e.g., Harel and Wolf[4]). Wave propagation effects and temporal variations have also been included in numerical modeling.[5] Various analytic solutions have been developed (e.g., Southwood[6] and Crooker and Siscoe[7]). Recent computer solutions include Senior and Blanc.[8]

In a recent series of papers[9-12] the author used an empirical approach based on observational data and a relationship between Birkeland currents, convection, and plasma distribution in the outer magnetosphere to develop a model for the convection. Features arising from this empirical model included a dusk-to-dawn cross-tail flow, the Harang discontinuity, and sunward flow in the tail confined primarily to the duskside. These effects fall out of the theoretical model developed here. There are also some differences from the above work. These result from the attempt, in the above papers, to interpret all the observed Birkeland currents as closure of plasma pressure-gradient associated currents. In this paper we shall find that all the region 2 and part of the region 1 current are of this nature, but that the remainder of the region 1 current closes to the magnetopause.

DEVELOPMENT OF THE THEORY

Some basic assumptions that go into the model are as follows:

1. The magnetospheric tail is in pressure equilibrium (i.e., inertial effects are negligible).
2. Protons (ions) are hot and electrons are cold.
3. The height-integrated ionospheric conductivity is uniform within the auroral oval, and magnetic field lines are approximated as vertical at ionospheric height.
4. Magnetic field lines are electric equipotentials.
5. A steady-state solution is a valid approximation to the convection, at least between substorm expansions.

The approach used closely parallels that of Vasyliunas.[2] All quantities are mapped along magnetic field lines to the ionosphere and their values taken there. Further, we shall neglect corotation velocities. Equating the Birkeland currents consistent with the plasma drifts to those consistent with ionospheric currents (Eqs. 4 and 7 in Ref. 2) we get

$$\nabla \cdot e N \boldsymbol{v}_B = \nabla \cdot \boldsymbol{\Sigma} \cdot \boldsymbol{E} \tag{1}$$

where Σ is the height-integrated conductivity tensor, N is the number of electron-ion pairs on a flux tube of unit area at ionospheric altitude, and $\boldsymbol{v}_B = \boldsymbol{E} \times \boldsymbol{B}/B^2$ is the convection velocity.

Using the above relation for $\boldsymbol{v}_B$ and the assumptions at the beginning of this section, Eq. 1 can be reduced to

$$\Sigma_p \left(\frac{\partial^2 \phi}{\partial x^2} + \frac{\partial^2 \phi}{\partial y^2} \right)$$

$$+ \frac{e}{B_i} \left(\frac{\partial \phi}{\partial y} \frac{\partial N}{\partial x} - \frac{\partial \phi}{\partial x} \frac{\partial N}{\partial y} \right) = 0 \tag{2}$$

where x and y is a Cartesian system at ionospheric heights with y parallel to the oval (positive eastward) and x at right angles positive southward. In doing this we neglect the curvature of the oval.

We also make the assumption that $\partial N/\partial x$ is constant and negative corresponding to N increasing with higher L value in order to study the effects of a radial gradient.

---

*Herzberg Institute of Astrophysics, National Research Council of Canada, Ottawa, Ontario K1A 0R6, Canada.

For the initial part of the paper we also assume $\partial N/\partial y = 0$, allowing Eq. 2 to be reduced to the form

$$\frac{\partial^2 \phi}{\partial x^2} + \frac{\partial^2 \phi}{\partial y^2} - C \frac{\partial \phi}{\partial y} = 0 \qquad (3)$$

where

$$C = - \frac{e\partial N/\partial x}{B_i \, \Sigma_p} . \qquad (4)$$

Since $\partial N/\partial x$ is negative, C is a positive constant.

It is of interest that the quantity eN/B_i is the equivalent Hall conductivity, Σ^*, identified by Vasyliunas.[2] (In this paper we are assuming a constant gradient of Σ^* rather than a constant value.) In fact, the procedure and its justification can be compared quite closely with that used by Vasyliunas. The constant-gradient case can be viewed as the superposition of an infinite number of constant N cases, each with its own Alfvén shielding layer (e.g., Ref. 3). Thus, specifying a value of $\partial N/\partial x$ is consistent with defining the plasma distribution function for a given magnetic field model. In addition, each energy may be thought of as having a basically circular shielding layer (in a dipole field) that is perturbed by the convection. This last point is the justification for the $\partial N/\partial x = $ constant and $\partial N/\partial y = 0$ approximations.

THE GENERAL SOLUTION

For a steady state, the boundary conditions repeat every 24 hours in local time and hence a general solution periodic in y is appropriate:

$$\phi = \sum_{n=1}^{\infty} e^{-\alpha_n x} \left\{ A_n \cos(k_n y - \beta_n x) \right.$$

$$\left. + B_n \sin(k_n y - \beta_n x) \right\} . \qquad (5)$$

If L is the east-west length of the oval then

$$K_n = 2\pi n/L . \qquad (6)$$

α_n and β_n are greater than zero and are determined from K_n by

$$\alpha_n^2 - \beta_n^2 = k_n^2 \qquad (7a)$$

$$\alpha_n \beta_n = Ck_n/2 \qquad (7b)$$

A specific solution is obtained by equating A_n and B_n to the Fourier coefficients of the boundary condition, $\phi(0,y)$.

With inflow primarily on the nightside and outflow on the dayside, the fundamental mode, $n = 1$, must dominate. From this we introduce a degree of self-consistency by equating the scale length of the gradient, $\partial N/\partial x$, to the decay scale length of the fundamental mode of the solution, $1/\alpha_1$. That is

$$\partial N/\partial x = -N_0 \alpha_1 . \qquad (8)$$

where N_0 is the number of electron-ion pairs per unit-area flux tube at the northern boundary of the oval, and it is assumed that N goes to zero at the southern edge. The introduction of Eq. 8 specifies the plasma distribution function as discussed earlier. A comparison of the distribution function with observations indicates the degree of success of this approach to self-consistency. This is in fact done in the more-detailed version of this work and it is found that the distribution agrees satisfactorily with the observed temperature in the plasma sheet in the important energy range (several keV).

There is a useful approximation for the lower modes (n small), which corresponds to neglecting $\partial^2 \phi/\partial y^2$ in Eq. 3. The approximation is

$$\alpha_n = \beta_n = (Ck_n/2)^{1/2} . \qquad (9)$$

Using this with Eqs. 4, 6, and 8 gives a scale length $1/\alpha_1 = LB_i\Sigma_p/\pi e N_0$ for the width of the oval. For the values $B_i = 5 \times 10^{-5} T$, $N_0/B_i = 4 \times 10^{20}$ (Atkinson[12]), $\Sigma_p = 6$ ohm (the sum of both hemispheres), and $L = 10^7$ m, the scale length is 3×10^5 m in agreement with the observed width of the oval.

SPECIFIC SOLUTIONS

The first three figures show solutions in which the convection into the nightside oval from the polar cap exhibits a maximum at midnight (Fig. 1), a maximum before midnight (Fig. 2), and a maximum after midnight (Fig. 3). Dayside maxima are at noon in all three figures. The actual boundary conditions are:

$$\phi_1 (0,y) = \phi_0 \sin k_1 y \qquad (10)$$

$$\phi_2 (0,y) = \phi_0 [\sin (k_1 y + 0.42)$$

$$+ 0.25 \cos(2k_1 y + 0.84)] \qquad (11)$$

$$\phi_3 (0,y) = \phi_0 [\sin (k_1 y - 0.42)$$

$$- 0.25 \cos(2k_1 y - 0.84)] \qquad (12)$$

Figures 1a, 2a, and 3a show equipotentials. The plots are in polar form since that is more familiar for high-

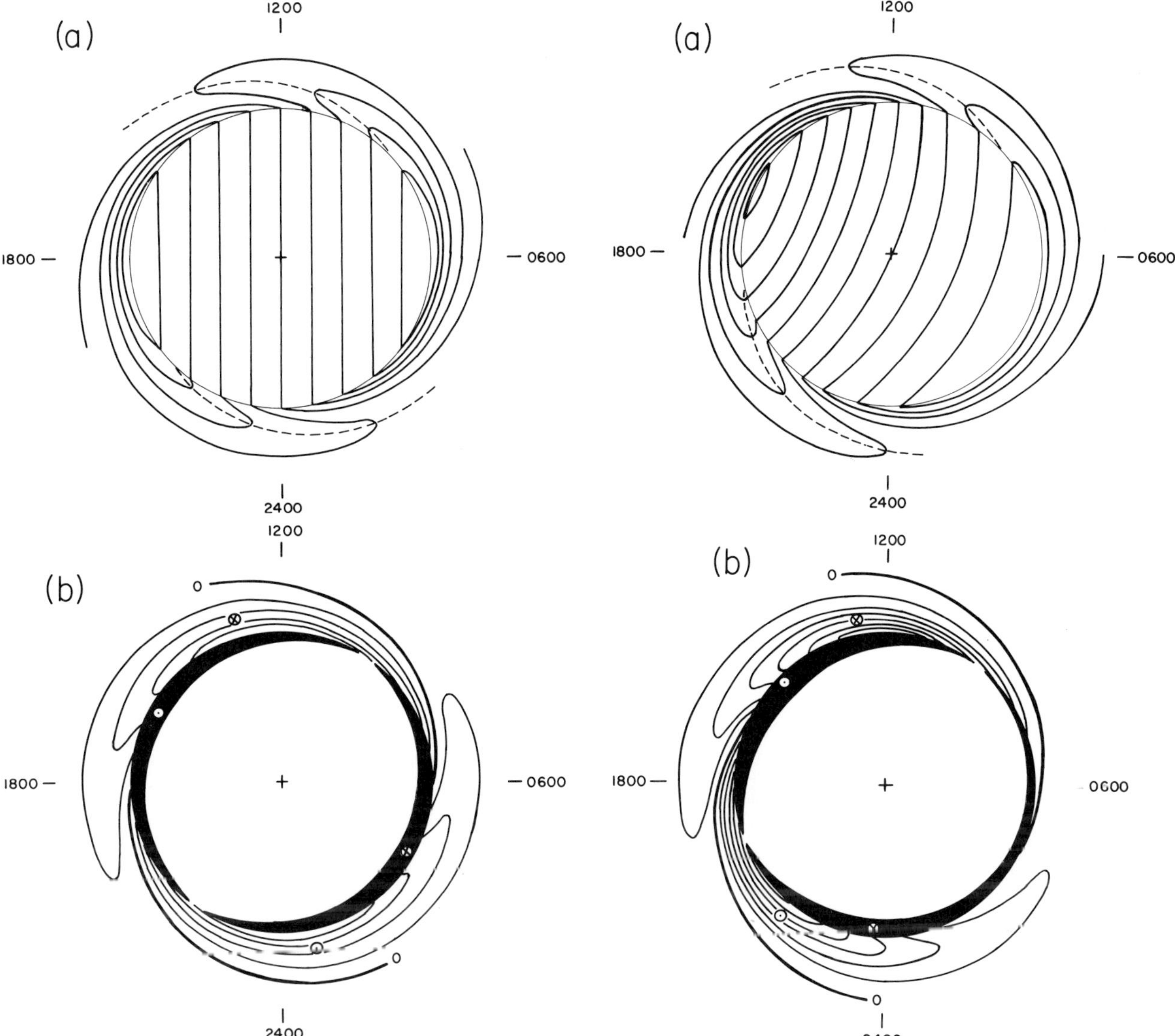

Figure 1—Solution for the symmetric boundary condition $\phi = \phi_0 \sin k_1 y$: (a) Equipotentials of the convection separated by 1/10 of the cross-polar-cap potential. The Harang discontinuities are shown by the dashed lines. (b) Birkeland currents: contours of distributed current density are shown within the oval separated by 6×10^{-5} A/m^2 (for 100 keV and other values as in text). The zero contours (darker lines) divide regions of upward (dots in circles) and downward (crosses in circles) current. The sheet current at the polar-cap boundary is proportional to the thickness of the black strip (1/5 of the polar-cap radius = 1 A/m^2).

Figure 2—Same as Fig. 1 but with maximum convection into the oval before midnight.

latitude plots; however, the reader should remember that the curvature was neglected in calculating the pattern. The equipotentials are closed in the polar cap as solutions of the Laplace equation, and are spaced at one tenth of the cross-polar-cap potential.

The solution in Fig. 1a shows symmetric dayside and nightside Harang discontinuities. Thus the Harang discontinuity is the result of the gradient of N since it does not occur in constant N solutions. It can be seen from Figs. 2a and 3a that the nightside Harang is stronger if

the inflow maximum is displaced toward dusk. Maximum inflow toward dusk also gives the best fit inside the polar cap to the models of Heppner[13] and radar observations (e.g., Ref. 14). These include a substantial dawn-to-dusk component on the nightside.

The value of the distributed Birkeland current is given by

$$ J_\parallel = -\frac{e}{B_i}\frac{\partial N}{\partial x}\frac{\partial \phi}{\partial y}. \tag{13} $$

It is shown as a contour plot of the current intensity in Figs. 1b, 2b, and 3b. Upward and downward currents are indicated by dots and crosses in circles and the heavier zero contour separates regions of upward and downward currents. There is a maximum downward current

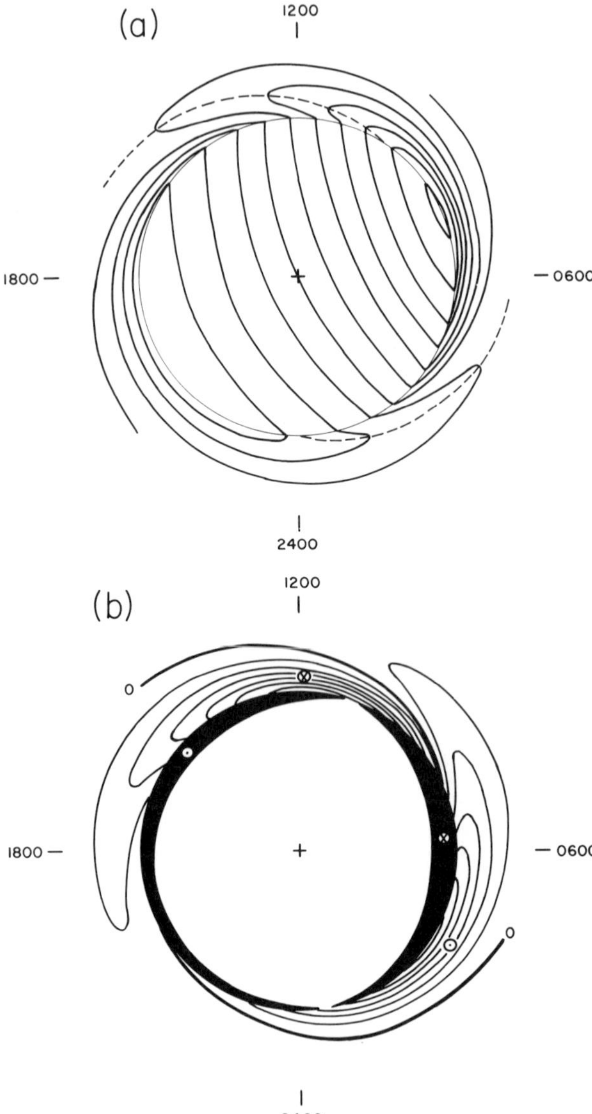

Figure 3—Same as Fig. 1 but with maximum convection into the oval after midnight.

near noon just south of the polar cap boundary in all cases and a maximum upward current displaced in the same direction as the maximum in inflow.

In addition to the distributed currents there are sheet currents at the boundary of the polar cap. We assume that the conductivity in the polar cap is zero and hence the northern boundary condition on the oval is

$$I_{\parallel} = \Sigma_p \frac{\partial \phi}{\partial x} + \Sigma_H \frac{\partial \phi}{\partial y} \qquad (14)$$

where $I_{\parallel}$ is positive upward. The sheet current is also shown in Figs. 1b, 2b, and 3b where its intensity is proportional to the width of the black strip inside the polar cap boundary. Its direction is again shown by dots (upward) and crosses (downward) in circles. In this case,

we used $\Sigma_H / \Sigma_p = 2$. Increasing the ratio to 4 rotates the sheet currents anticlockwise about 20° in local time, but does not affect the distributed currents. The sheet current is presumably the "driving" current in this model and connects to the plasma sheet and eventually to the magnetopause. It is clear that one cannot look at empirically determined region 1 currents and associate them with driving currents since at some local times the direction of the Birkeland currents is the same for the sheet current and the adjacent distributed current.

Comparison with observations of the convection, the location of the Harang, the dawn-dusk asymmetry in the sunward convection (e.g., Foster[15,16]; Baumjohann and Haerendel[17]; Heppner[13]), the location and intensity of the Birkeland currents[18,19]) gives the overall impression that a maximum inflow in the premidnight sector (Fig. 2) gives the best fit to the observed convection and Birkeland current systems. However, the theoretical models fail to reproduce some features of the observed dayside Birkeland currents. One way to improve the model is to introduce $\partial N / \partial y \neq 0$.

CONTOURS OF N NOT PARALLEL TO THE OVAL

Up to this point, we have considered the y axis to be parallel to contours of N and applied the boundary conditions along a constant N contour. It seems likely in reality that some of the contours terminate at the polar cap boundary rather than forming closed loops around the polar cap.

To this end we assume $\partial N / \partial y = (2N_0 / L) \sin k_1 y$. Ignoring the term in $\partial^2 \phi / \partial y^2$ (valid for lower order modes), Eq. 3 was solved numerically for the boundary condition with maximum inflow before midnight (Eq. 11) and the above value of $\partial N / \partial y$. Contours of N that exist in the northernmost 190 km of the midnight oval terminate at the polar cap boundary before reaching noon in this model (assuming $1/\alpha = 300$ km). The results are shown in Fig. 4.

A comparison of Fig. 4 with Fig. 2 shows the following effects of introducing the above value for $\partial N / \partial y$:

1. The dayside Harang discontinuity is suppressed whereas the nightside Harang extends from earlier local times to further around toward dawn.
2. The region of convection is narrowed on the dawnside and widened at dusk, although the potential across it stays the same.
3. The depth of penetration of the convection is increased on the nightside and decreased on the dayside.
4. Both sheet and distributed currents are enhanced at dawn and decreased at dusk.

The above effects were even more evident in a calculation using three times the above value for $\partial N / \partial y$. This result is shown in Fig. 5.

In conclusion to this section, it is apparent that introduction of a weak $\partial N / \partial y$ makes some features of the solution agree better with observation. This clearly is one parameter to be adjusted in future studies of this type.

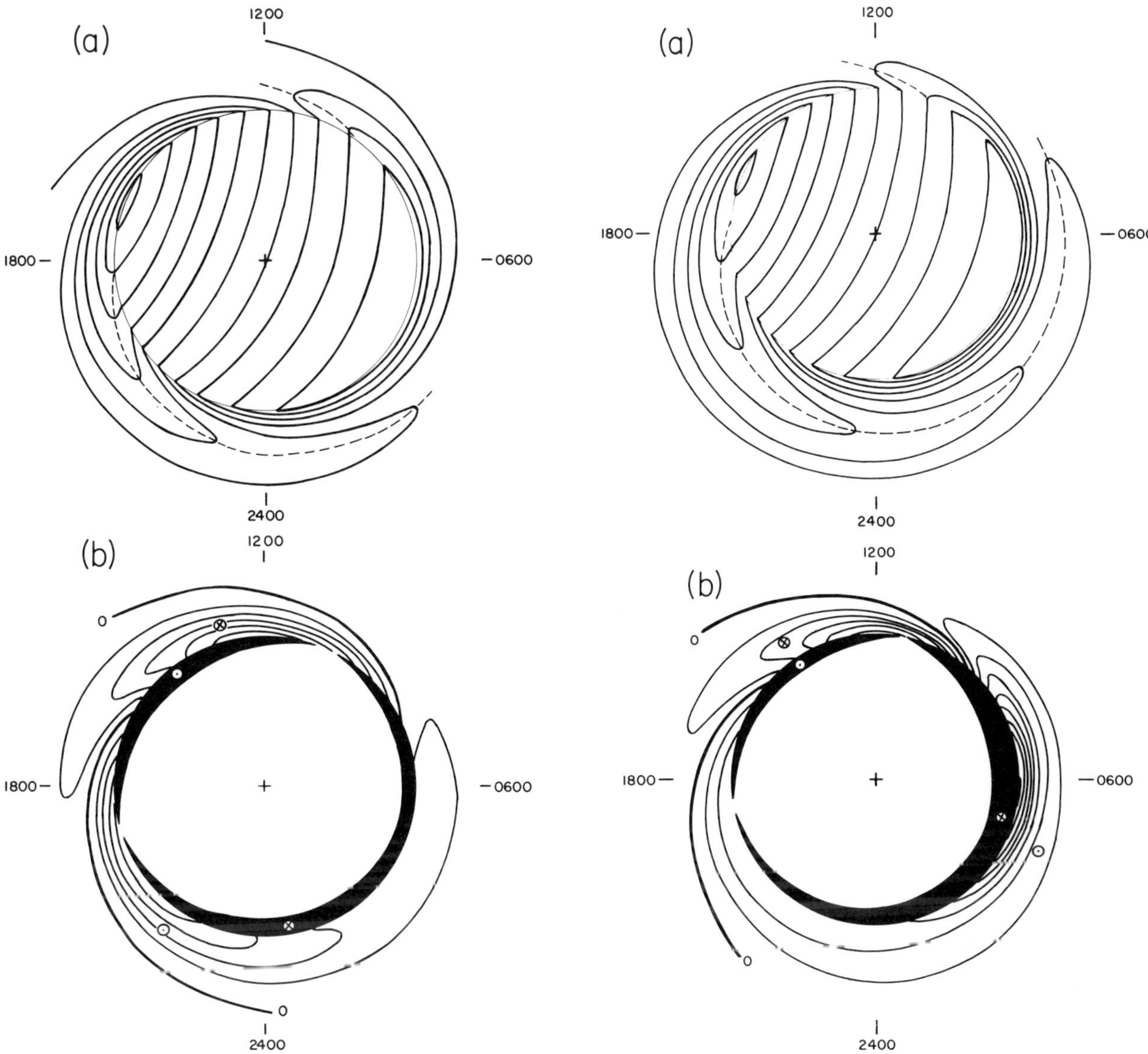

Figure 4—Same as Fig. 2 but with a weak $\partial N/\partial y$.

Figure 5—Same as Fig. 3 but with a stronger $\partial N/\partial y$.

REFERENCES

[1] J. A. Fejer, "Theory of the Geomagnetic Daily Disturbance Variations," *J. Geophys. Res.* **69**, 123-135 (1964).

[2] V. M. Vasyliunas, "The Interrelationship of Magnetospheric Processes," in *Earth's Magnetospheric Processes,* B. M. McCormac, ed., D. Reidel, pp. 29-38 (1972).

[3] M. Harel, R. A. Wolf, P. H. Reiff, R. W. Spiro, W. J. Burke, and F. J. Rich, "Quantitative Simulation of a Magnetospheric Substorm 1. Model Logic and Overview," *J. Geophys. Res.* **86**, 2217-2241 (1981).

[4] M. Harel and R. A. Wolf, "Convection," in *Physics of Solar-Planetary Environments,* Vol. II. D. J. Williams, ed., American Geophysical Union, Washington, D.C. (1976).

[5] R. J. Kan, and W. Sun, "Simulation of the Westward Travelling Surge and Pi2 Pulsations During Substorms," *J. Geophys. Res.* **90**, 10911-10922 (1985).

[6] D. J. Southwood, "The Role of Hot Plasma in Magnetospheric Convection," *J. Geophys. Res.* **82**, 5512-5520 (1977).

[7] N. U. Crooker and G. L. Siscoe, "Birkeland Currents as the Cause of the Low-Latitude Asymmetric Disturbance Field," *J. Geophys. Res.* **86**, 11201-11210 (1981).

[8] C. Senior and M. Blanc, "On the Control of Magnetospheric Convection by the Spatial Distribution of Ionospheric Conductivities," *J. Geophys. Res.* **89**, 261-284 (1984).

[9] G. Atkinson, "The Role of Currents in Plasma Redistribution," in *Magnetospheric Currents,* T. Potemra, ed., Geophysical Monograph 28, American Geophysical Union, Washington, D.C. (1983).

[10] G. Atkinson, "Field-Aligned Currents as a Diagnostic Tool. Result: A Renovated Model of the Magnetosphere," *J. Geophys. Res.* **89**, 217-226 (1984).

[11] G. Atkinson, "Thick Current Sheets in the Renovated Model of the Magnetosphere," *J. Geophys. Res.* **89**, 8949-8955 (1984).

[12] G. Atkinson, "Time-Dependent Flows in the Renovated Model of the Magnetosphere," *J. Geophys. Res.* **90**, 10843-10850 (1985).

[13] J. P. Heppner, "Empirical Models of High-Latitude Electric Fields," *J. Geophys. Res.* **82**, 1115-1125 (1977).

[14] J. V. Evans, J. M. Holt, W. L. Oliver, and R. H. Wand, "Millstone Hill Incoherent Scatter Observations of Auroral Convection over $60° \leq \Lambda \leq 75°$ 2. Initial Results," *J. Geophys. Res.* **85**, 41-54 (1980).

[15] J. C. Foster, "An Empirical Electric Field Model Derived from Chatanika Radar Data," *J. Geophys. Res.* **88**, 981-987 (1983).

[16] J. C. Foster, "Ionospheric Signatures of Magnetospheric Convection," *J. Geophys. Res.* **89**, 855-865 (1984).

[17] W. Baumjohann and G. Haerendel, "Magnetospheric Convection Observed Between 0600 and 2100 LT: Solar Wind and IMF Dependence," *J. Geophys. Res.* **90**, 6370-6378 (1985).

[18] T. Iijima and T. A. Potemra, "The Amplitude Distribution of Field-Aligned Currents at Northern High Latitudes Observed by Triad," *J. Geophys. Res.* **81**, 2165-2174 (1976).

[19] T. Iijima and T. A. Potemra, "Large-Scale Characteristics of Field-Aligned Currents Associated with Substorms," *J. Geophys. Res.* **83**, 599-615 (1978).

INTERPRETATION OF THE NORTHWARD B_z (NBZ) BIRKELAND CURRENT SYSTEM AND POLAR CAP CONVECTION PATTERNS IN TERMS OF THE IMPULSIVE PENETRATION MODEL

J. Lemaire*

According to the impulsive penetration (IP) theory, solar-wind plasmoids penetrate predominantly in the two lobes of the magnetotail when the interplanetary magnetic field (IMF) has a northward B_z (*Planet. Space Sci.* **27**, 47-57 (1979)). Momentum density of the penetrating solar-wind plasma element is transferred to the surrounding plasma and to the central polar-cap ionospheric regions, which, as a matter of consequence, is dragged over the poles in the sunward direction. This transient magnetotail plasma convection and the associated nonstationary polar-cap convection can be added and compared to steady-state convection patterns considered in earlier stationary interaction models.

In the northern hemisphere, the region of preferred penetration is shifted toward dusk for a positive B_y, and toward dawn for $B_y < 0$. The direction of these shifts is reversed in the southern hemisphere. The polar-cap convection pattern associated with impulsive penetration of solar-wind plasma irregularities is therefore also shifted toward dusk or dawn, according to the value of B_y, as mentioned above.

Furthermore, the NBZ Birkeland currents driven at the interfaces between the penetrating plasmoids and the ambient geomagnetic field are downward (upward) in the dusk (dawn) side for both hemispheres. The amplitudes of NBZ currents are expected to be larger when B_z is larger (and positive). The region of reversal of the NBZ currents is also shifted toward dawn or dusk depending on the sign of B_y as described above. These transient field-aligned currents, driven upward and downward, are associated with the nonzero field-aligned component of curl $\boldsymbol{B}$ (i.e., with magnetic shears) in the vicinity of the penetrating diamagnetic plasmoids.

SMALL-SCALE SOLAR-WIND PLASMA IREGULARITIES

From measurements of the interplanetary magnetic field (IMF) with high time resolution (i.e., 10 $\boldsymbol{B}$-vectors every second of time), it can be observed that the solar-wind magnetic field distribution and, consequently, the solar-wind plasma distribution itself are almost never uniform or stationary. Small-amplitude fluctuations in the IMF magnitude or/and direction are observed almost all the time. It is sometimes rare to find high-resolution magnetograms with no changes in B_x, B_y, or B_z over periods of time larger than 30 seconds. Large-amplitude changes in the IMF (e.g., current sheaths, rotational discontinuities, tangential discontinuities, magnetic holes, as illustrated in Fig. 1, or shocks) are, of course, less frequently observed than small-amplitude and small-scale fluctuations.[1,2] But the presence even of only 5% variation of $\boldsymbol{B}$ in the solar wind indicates that the interplanetary field and plasma are both patchy, nonstationary, and formed out of small-scale "plasma magnetic field entities" – i.e., helicoidal plasmoids, poloidal plasmoids, and toroidal plasmoids.[3]

Solar-wind plasma measurements are usually sampled with a rather low time resolution (e.g., $t \geqslant 10$ seconds) and are therefore inappropriate to identify plasma small-scale irregularities. However, these small-scale plasma elements have slightly different densities, different temperatures, perhaps different ionic abundances, different magnetizations, different vorticities, and different momentum densities. As a consequence, a steady-state and almost-uniform radial expansion of the solar at-

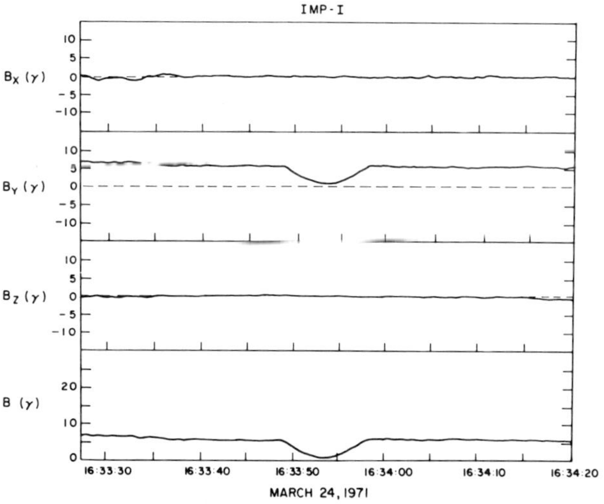

Figure 1—High-resolution magnetogram of the interplanetary magnetic field components (in nT) measured with the IMP I (Explorer 43). The sampling rate of the magnetometer is 12.5/s. The characteristic variation in the B_y component and in the total field B have been identified as "magnetic holes" by Turner et al.[2] The total duration of this event is less than 10 seconds, i.e., less than 20 average ion Larmor radii in extent. The magnetic field variation has been interpreted as a traversal of a field-aligned solar-wind density irregularity of less than 4000 km in radial extent.

mosphere is a rather oversimplified mathematical representation of physical reality. The supersonic nature of this radial expansion and the strong nonstationarity of phenomena at the sun's surface and in the solar corona lead to the expectation that the interplanetary medium at 1 AU cannot be anything else but nonstationary, high-

*Institut d'Aéronomie Spatiale de Belgique, 3 Ave Circulaire, B-1180 Brussels, Belgium.

ly sheared, and nonuniform over a wide range of scales, ranging from 1 AU down to a few ion Larmor gyroradii (i.e., 1 AU $>$ L $>$ 2000 km).

IMPULSIVE PENETRATION AND ADIABATIC DECELERATION

The solar-wind plasma irregularities or eddies with largest momentum penetrate deeper into the geomagnetic field.[4,5] This results from conservation of the total energy. It has been shown in laboratory experiments[6,7] as well as from kinetic plasma theory[8] that the kinetic energy ($\frac{1}{2} mv^2$) of a plasmoid is converted adiabatically into thermal energy ($kT_\perp^+$ and $kT_\perp^-$) when it penetrates into a region of higher magnetic field intensity, $B(r)$. Indeed, as a consequence of adiabatic conservation of μ, the magnetic moment of ions and electrons, the ratio ($kT_\perp^+ + kT_\perp^-$)/$B(r)$ is constant when the plasma cloud enters into the geomagnetic field. Since the magnetic field intensity, $B(r)$, increases as r^{-3} when the geocentric distance r decreases, there is always a position r_1 where the incident kinetic energy ($\frac{1}{2} mv_0^2$) of the plasmoid particles is fully converted into gyro-motion (i.e., into perpendicular thermal energy). The radial component of the bulk velocity of the plasmoid vanishes at r_1 where the geomagnetic field intensity is equal to

$$B(r_1) = \frac{\frac{1}{2} mv_0^2 + (kT_\perp^+)_0 + (kT_\perp^-)_0}{\bar{\mu}^+ + \bar{\mu}^-} \quad (1)$$

where $\bar{\mu}^+$ and $\bar{\mu}^-$ are the (conserved) average adiabatic moments of the impinging solar-wind ions and electrons forming the intruding plasma irregularity (see Ref. 9 for generalization and application of Schmidt's theory in the case of a sheared magnetic field distribution).

Equation 1 indicates that magnetosheath plasma clouds with larger incident velocities (v_0) are able to penetrate deeper into the geomagnetic field than average solar-wind plasma elements with a smaller momentum density. Solar-wind plasmoids with the largest momentum densities in the solar wind are therefore stopped or deflected sideward closer to the earth, i.e., where $B(r_1)$ is larger.

Experimental evidence for solar-wind plasma intrusions or plasmoids into the magnetosphere can be identified in a wide range of observations near the magnetopause and more specifically in the measurements reported by Lundin *et al.*,[10] Lundin and Aparicio,[11] Lundin and Dubinin,[12,13] and Eastman et al.[14] (See also Appendix 2.)

NONADIABATIC DECELERATION OF INTRUDING PLASMA ELEMENTS

In addition to the adiabatic deceleration discussed above, a plasma element injected impulsively across geomagnetic field lines is also decelerated nonadiabatically. Indeed, geomagnetic field lines are linked into the conducting dayside cusp (cleft) ionosphere. The integrat-

ed Pedersen conductivity along these magnetic field lines is not infinitely large nor is it equal to zero. Its value ranges between 1 and 10 Siemens as a result of the large transverse conductivity in the ionospheric E-region. As a consequence, the penetrating eddies are slowed down nonadiabatically as emphasized by Lemaire.[4,15] The nonadiabatic deceleration of collisionless plasma streams across magnetic field lines anchored in conducting "walls," like the earth's ionosphere, has clearly been demonstrated in laboratory experiments by Baker and Hamel.[16,17] Part of the incident kinetic energy ($\frac{1}{2} mv_0^2$) of the penetrating solar-wind plasma cloud is therefore also dissipated by Joule heating in the ionosphere at altitudes of the E-region and above.[18]

An additional consequence of the finiteness of the integrated Pedersen conductivity is that the ambient geomagnetic flux "diffuses" irreversibly into the engulfed solar-wind plasma elements. The B-field inside and at the surface of the plasma intrusion rotates to become gradually parallel to the ambient geomagnetic field.[19] B-field hodograms observed during magnetopause crossings support rotation of B better than a reversal of the magnetic field direction along hypothetical neutral lines.[20,21]

THE INFLUENCE OF IMF B_z ON IMPULSIVE PENETRATION

Plasmoids with an excess momentum that are moving with the background solar-wind bulk velocity have an excess mass density. Assuming nearly equal perpendicular plasma temperatures inside and outside the plasma irregularity, it can be inferred that the perpendicular pressure ($nkT_\perp^+ + nkT_\perp^-$) is larger inside than outside the element of plasma. As a matter of consequence, the magnetic energy density ($B^2/2\mu_0$) must be smaller inside than outside in order to satisfy pressure-balance equilibrium. The magnetization and the magnetic dipole moment (M) of the diamagnetic currents circulating in the plasmoid as well as its surface are then both pointing in a direction opposite to the ambient interplanetary magnetic field B.[22]

When the IMF has a northward component (i.e., B_z $>$ 0), the magnetic dipole moment M of a diamagnetic plasmoid with an excess density has a southward component (i.e., M_z $<$ 0). It can be shown that the magnetic force, $\nabla (M \cdot B_E)$ exerted on a southward oriented magnetic dipole moment M by the geomagnetic field, B_E, which has a southward oriented dipole component, M_E, is directed away from the earth. Indeed, when such a plasmoid is at low latitudes near the front side magnetopause, M being there nearly parallel to M_E, both dipoles then repel each other. In other words, the diamagnetic current loops responsible for the magnetic field depression inside the plasmoid are then pushed away by the southward oriented earth's dipole M_E. The dipole-dipole interaction acts then to reject the intruding small-scale plasma current system out of the inhomogeneous geomagnetic field distribution.[19]

Above the northern and southern magnetotail lobes the IMF field lines are draped along the magnetopause surface. When IMF $B_z > 0$ in front of the bow shock, the directions of magnetic field lines in magnetosheath are tilted in the antisunward (sunward) direction above the northern (southern) magnetotail surface. A plasmoid with an excess momentum density and an excess thermal pressure has necessarily a magnetic moment pointing in a direction opposite to background magnetic field in the magnetosheath, i.e., $M_x > 0$ above the northern magnetopause where $B_x < 0$; $M_x < 0$ above the southern magnetopause surface where $B_x > 0$. The magnetic force, $\nabla(M \cdot B_E)$, acting on the dipole moment M is directed toward the interior of the magnetotail over an extended area of the magnetopause beyond the magnetospheric neutral points in the northern and southern hemispheres. In other words, solar-wind irregularities with an excess momentum are attracted toward the inside of the magnetotail when the IMF is northward. On the contrary, for a southward IMF, the dipole-dipole interaction between plasmoids and the geomagnetic field favors impulsive penetration in the front side magnetosphere, and not in the northern nor in the southern magnetotail lobes.

The same conclusions had already been reached in a previous article by Lemaire et al.[22] Unfortunately, the captions of Figs. 6 and 8 in Ref. 22 have been mixed up. Furthermore, it contains incorrect statements concerning repulsive and attractive current systems on pages 50 and 51. However, the conclusions in this article remain essentially valid when the two-dimensional planar current sheaths are replaced by three-dimensional diamagnetic current loops, which have a finite magnetic dipole moment M.

THE INFLUENCE OF IMF B_y ON IMPULSIVE PENETRATION

The dipole-dipole force acting on a magnetosheath diamagnetic plasmoid is maximum in the vicinity of the polar cusps where the spatial derivatives of $(B_E)_x$, $(B_E)_y$, and $(B_E)_z$ are largest. For any IMF direction and any orientation of M, there is always a place in the vicinity of the neutral points where the magnetic field direction in the magnetosheath is antiparallel to the magnetospheric field. This is where the magnetic force $\nabla(M \cdot B_E)$ is maximum and directed toward the interior of the magnetosphere. When IMF $B_y > 0$, this place is shifted toward dusk (dawn) with respect to location of the northern (southern) polar cusp when IMF $B_y = 0$. As a consequence, the region of preferred impulsive penetration of solar-wind plasmoids is then shifted toward dusk (dawn) in the northern (southern) hemisphere as illustrated in Fig. 2a and c. The directions of these shifts is reversed in both hemispheres when IMF $B_y < 0$.

MAGNETOSPHERIC AND IONOSPHERIC CONVECTION PATTERNS RESULTING FROM IMPULSIVE PENETRATION

When a solar-wind plasma density irregularity is injected in the magnetotail as illustrated in Figs. 2b and 3a, the ambient magnetospheric plasma is pushed aside

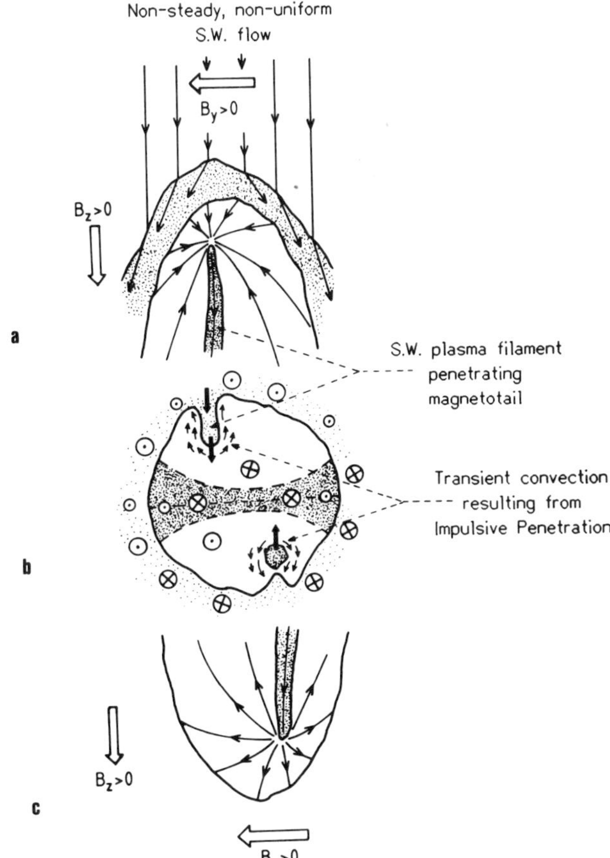

Figure 2—Cartoons illustrating the bumpy shape of the magnetosphere in a nonsteady and nonuniform solar-wind flow. The top panel (a) shows an elongated solar-wind plasma filament penetrating across the surface of the northern magnetotail lobe when the IMF has a northward B_z (NBZ). The neutral point of the northern polar cusp is shifted toward dusk when the IMF B_y component is positive. The bottom panel (c) correspond to the penetration of a solar-wind plasmoid in the southern tail lobe for the same IMF condition. Note that the southern polar cusp is shifted toward dawn. The central panel (b) represents a cross section of the magnetospheric tail lobes and of the plasma sheet (dotted area). The transient flow of magnetospheric plasma around impulsively injected solar-wind plasmoids is illustrated by small arrows. In the central panel, the observer is facing the sun.

and flows along the flanks of the intruding plasma body. Figure 3a represents a cross section through the northern magnetotail. The reader is looking toward the direction of the sun.

The direction of the bulk velocity vectors in the surrounding magnetospheric plasma is opposite to the impact velocity for the solar-wind plasma irregularity. This necessarily leads to a transient flow pattern of magnetospheric plasma in the tail lobes. This transient flow pattern is illustrated in Figs. 2b and 3a by small arrows directed away from the center of the magnetotail.

The convection electric field, E, associated with this transient flow of magnetospheric plasma across geomagnetic field lines is indicated by open arrows in Fig. 3a. Since the magnetospheric B-field is pointing toward the

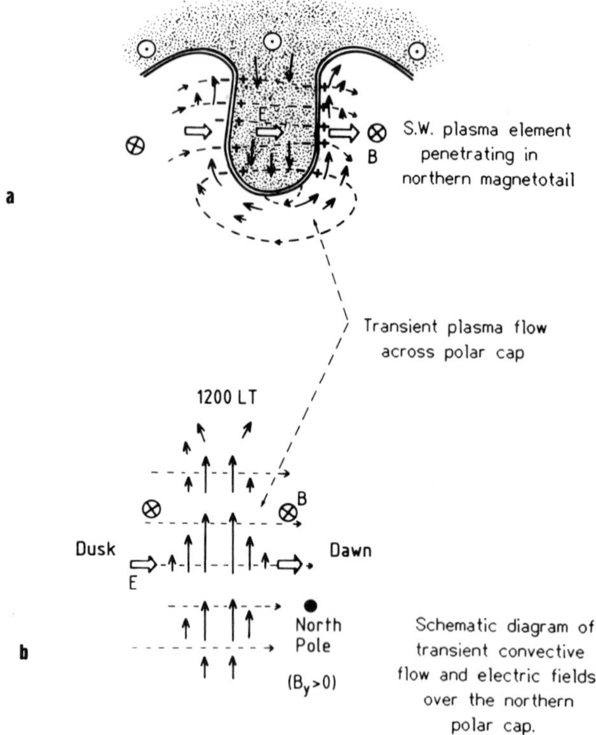

Figure 3—Transient flow pattern around a solar-wind plasma element penetrating impulsively in the northern magnetotail lobe. The observer is facing the sun in the top panel (a). The dusk-to-dawn convection electric field is indicated as well as electric field lines (dashed lines). The bottom panel (b) illustrates the convection electric field and sunward plasma flow pattern at ionospheric heights over the northern polar cap when IMF $B_z > 0$ and IMF $B_y > 0$.

sun in the northern tail lobe, the electric field, $E = -V \times B$, outside the plasma element is oriented from dusk to dawn. This convection electric field maps down into the polar-cap ionosphere as illustrated in Fig. 3b. Note that the convection electric field inside the intruding plasmoid does not necessarily map into the ionosphere since the magnetic field lines traversing this plasma element may not yet be connected to the polar cap. The dusk-to-dawn E-field drags ionospheric plasma over the polar cap in the direction of the sun as indicated in Fig. 3b by the arrows pointing toward 1200 LT.

Sunward flow of ionospheric plasma over the northern and southern polar caps has indeed been observed when the IMF has a northern B_z component. Sunward flow in the polar-cap ionosphere was first presented by Maezawa[23] and substantiated by data by Burke et al.[24] and by Zanetti et al.[25] The locations of these observations were shifted toward the dawn or dusk side of the polar caps depending on the sign of IMF B_y. The directions of these shifts correspond precisely to those of the preferred region of penetration of solar-wind plasma irregularities into the magnetotail lobes, when the direction of IMF B_y changes.

Note that the transient flow patterns illustrated in Fig. 3a and b should not necessarily be considered as part of a steady-state flow pattern usually illustrated and modeled in terms of closed equipotential contours and closed stream lines. Closed equipotential contours are appropriate to represent magnetospheric or ionospheric convection inferred from steady-state interaction models, but in the case of transient flow patterns like those expected in the nonsteady interaction model discussed here, it is not relevant to draw closed equipotential contours. Indeed, like in nonstationary hydrodynamic fluid motion (e.g., eddies or whirls), stream lines are not closed; drawing stationary equipotential patterns is then misleading.

The nonstationary flow patterns shown in Fig. 3a and b can, of course, be superimposed on the stationary convection flow patterns inferred from steady-state interaction models like those proposed by Crooker,[26] Reiff,[27] Reiff and Burch,[28] Lyons et al.,[29] Lyons,[30] or Kan and Burke.[31] Note, however, that a large number of small-scale solar-wind plasma elements penetrating continuously through a wide area of the tail lobes can drive a large-scale quasistationary sunward convection flow pattern over the poles, quite like those described in the steady-state antiparallel merging models mentioned above. Indeed, like it is for the large number of droplets forming a rain shower and pouring into surface water, the large number of plasma-density irregularities forming a disturbed solar-wind flow can penetrate in the magnetotail and change the convection in the plasma mantle as well as in the coupled ionosphere over a much wider volume or area than just one single small-scale plasmoid. Each individual plasma "droplet" contributes locally to the overall stream, but the duration of time as well as the extent in latitude of the plasma boundary layer or of the polar-cap ionosphere influenced by impulsive penetration of magnetosheath small-scale plasma irregularities does not depend so much on the size of these individual irregularities as it does on the width and length of the solar-wind volume where the plasma is turbulent and patchy. If the solar wind is nonuniform and patchy over heliocentric radial distances greater than 35,000,000 km, the shower of plasmoids penetrating in the magnetosphere will last longer than one day. In these circumstances a quasistationary convection flow pattern can eventually build up in the magnetosphere and in the ionosphere, but a true stationary regime can be established only when the small-scale plasma irregularities are evenly distributed in that solar-wind volume.

NBZ BIRKELAND CURRENT SYSTEM

In the laboratory experiments of Demidenko et al.[6,7] and Baker and Hamel[17] reported above, the injected plasmoids were characterized by low-β values. The solar-wind plasmoids are characterized by β values of the order of unity, and the kinetic energy density is then of the order of the magnetic energy density. A larger value of β does not impede penetration of plasmoids when their dipole magnetic moment has the right orientation. The entry of high-β diamagnetic solar-wind plasma elements in the magnetotails, however, perturbs the geomagnetic field distribution as illustrated in Fig. 4a and b.

The superimposed geomagnetic-field and plasmoid magnetic-field distribution are not curl free. When the direction of the magnetic field inside the plasmoid is not strictly parallel to the ambient geomagnetic field, curl B has a nonzero component in the direction parallel to B; i.e., parallel to the magnetic field lines. The parallel component of curl B is nonzero when the magnetic-field distribution is sheared, i.e., when magnetic lines are not parallel to each other. These magnetic shears are produced by Birkeland currents whose intensity is equal to

$$ J_\parallel \ = \ \frac{1}{\mu_0} \ (\text{curl } B)_\parallel \qquad (2) $$

Birkeland-current sheaths can be generated in collisionless plasma layers like those studied by Lemaire and Burlaga.[32] The intensity of the field-aligned currents in a diamagnetic plasma sheath is determined by the velocity distribution of the electrons and ions in the transition layer between the plasma of magnetospheric origin and the plasma of magnetosheath origin. The kinetic models of plasma slabs studied by Lemaire and Burlaga[32] and Roth[33] indicate that not only the magnetic field intensity but also the direction of B vary smoothly across plasma sheaths when the plasma densities, temperatures, ionic compositions, and magnetizations are different on both sides of the transition region. These plasma sheaths are at least a few ion Larmor gyroradii thick. The hodograms of calculated magnetic field vectors are very much like those measured when a spacecraft traverses a magnetopause surface or, equivalently, when it penetrates through the surface of one of the many intruding plasma elements, as illustrated in Fig. 4b.

Figure 4a and b, tries to illustrate how magnetic shears produced by the solar wind diamagnetic elements in the geomagnetic field are associated with Birkeland currents flowing downward (upward) in the afternoon (morning) side of the polar-cap region. These Birkeland currents can extend down into the conducting ionosphere where they produce small-amplitude magnetic field perturbations that indeed have been observed in the polar cap with the MAGSAT satellite. These Birkeland currents are detected in the polar caps only when the IMF has a northward B_z component. This is why they have been called NBZ Birkeland currents by Iijima et al.,[34] Potemra et al.,[35] Zanetti et al.[25] and Baumjohann and Friis-Christensen.[36]

In the northern (southern) polar cap, the location where NBZ currents shift toward dusk (dawn) have been observed when the IMF B_y becomes positive. These shifts correspond precisely to those expected for the preferred impulsive penetration region of solar-wind plasmoids into the magnetotail lobes when IMF B_y is positive.

The observations indicate also that the NBZ field-aligned current intensities are generally larger when the IMF is larger. This comes from the fact that the magnetic shears at the surface of a diamagnetic plasma element are enhanced when the IMF intensity is increased. As a consequence, curl B is enhanced as well as $J_\parallel$ according to Eq. 2 and to the model calculations by Lemaire and Burlaga.[32]

The observations indicate also that the NBZ field-aligned current intensities are generally larger when the

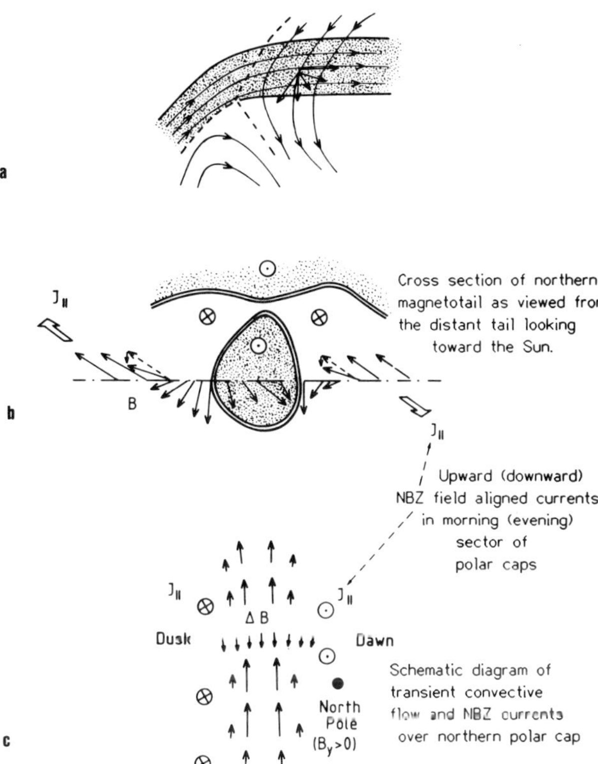

Figure 4—The top panel (a) shows a meridional cross section of a solar-wind plasma element bent into the magnetotail lobe behind the neutral point in the northern polar cusp. The magnetic field-line direction changes from sunward (outside the plasmoid) to antisunward in the middle of the plasmoid. The central panel (b) shows a cross section of the northern magnetotail lobe perpendicular to the earth-sun direction (x-axis). The observer is facing the sun. The solid arrows indicate the direction of the magnetic field along an S/C trajectory (dashed-dotted line) traversing an engulfed plasmoid. It is a perspective representation of a B-field hodogram. The magnetic shears at the surface of the plasma element (where curl B has a component parallel to B) drive sunward (downward) field-aligned currents on the duskside and antisunward (upward) field-aligned currents on the dawnside of the plasmoid. When the polar-cap ionosphere is a good conductor (i.e., when it is illuminated by UV photons from the sun) these NBZ Birkeland currents can more easily close into the ionosphere. In the bottom panel (c), the arrows pointing toward 1200 LT (i.e., toward the top of the figure) represent the sunward plasma convection over the polar cap for NBZ condition (also shown in Fig. 3b). The array of smaller arrows pointing in the 0000 LT direction represent the magnetic field perturbation produced by the NBZ Birkeland current system flowing down into the ionosphere on the duskside and out of the ionosphere on the dawnside of the patch where sunward convection is observed. The magnetic field signature of Hall and Pedersen currents is not illustrated.

polar angle of the IMF (measured from the $+z$ axis) is small, i.e., for IMF $B_z > 0$ but $B_y \sim 0$.[37,38] These experimental results can be interpreted in a similar manner when the polar angle of the IMF is small, i.e., when the angle between the IMF field lines in the magnetosheath near the magnetotail surface is almost antiparallel to the geomagnetic field as shown in Fig. 2a and c. The magnetic shear is then maximum; consequently, curl $\boldsymbol{B}_\parallel$ and the NBZ current densities, $J_\parallel$, are also maximum.

It is not yet clear why the IMF B_z must be larger than 5 nT for significant NBZ Birkeland current and convection flow patterns to occur in the polar cap.[34] A quantitative and time-dependent simulation of impulsive penetration of high-β plasmoids will probably be needed to answer this specific question. To our knowledge, there is no explanation for such a threshold at $B_z = 5$ nT in the steady-state antiparallel merging theories mentioned above.

PEDERSEN CURRENTS

In a quasisteady state, the field-aligned currents generated at the interface between an engulfed diamagnetic solar-wind plasma element and the surrounding geomagnetic field must necessarily close somewhere via electric currents transverse to the magnetic field lines either (a) at high altitude in the vicinity of the intruding plasmoid via complex curvature drift currents, as discussed in Appendix 1, or (b) in the ionospheric E-region via Hall and Pedersen currents where the electric conductivity has a finite (nonzero) value.

The amount of field-aligned currents that are diverted toward the ionosphere and that close there via horizontal currents depends on the value of the ionospheric Hall and Pedersen conductivity. The intensity of the transverse currents as well as the density of the associated field-aligned currents are therefore larger in the illuminated dayside polar caps than in the nightside region beyond the terminator where the integrated Pedersen conductivity, Σ_p, is drastically reduced. Observations by Bythrow et al.[39] confirm that NBZ Birkeland currents are indeed larger in the dayside than over the nightside polar caps where the conductivity of the ionosphere is low because of the lack of photoionization processes.

The most intense Pedersen current should occur at the interface between upward- and downward-flowing field-aligned current regions. Since the ratio of the Hall to the Pedersen height integrated conductivities is generally greater than unity, the Hall currents are expected to be greater than the Pedersen currents. This is possibly why the magnetic signature of Hall currents is generally prevailing in MAGSAT magnetograms. Furthermore, the intensity of Hall currents should also peak at the limit between Birkeland currents of opposite signs, i.e., where the dusk-to-dawn convection electric field, E, is maximum. This is confirmed by MAGSAT data.[25]

An additional reason for the magnetic signature of transverse Pedersen currents to be small and almost unnoticed in magnetograms is that the length of the Pedersen current sheaths is rather short, i.e., $0.3 - 3$ degrees in latitude, corresponding to the distance between footprints of geomagnetic-field lines linked to the volume of magnetospheric plasma regions perturbed by intruding plasmoids. Therefore, the magnetic-field perturbations produced by such short currents decrease rapidly with distance—more rapidly than those produced by the wider and more elongated Hall current sheaths.

APPENDIX 1: FIELD ALIGNED AND CURVATURE DRIFT CURRENTS

In the one-dimensional planar plasma sheaths or tangential discontinuities modeled by Lemaire and Burlaga[32] or Roth[33,40] the parallel (field-aligned) current as well as the perpendicular component of the diamagnetic current layer are both flowing parallel to the plane of the discontinuity. They "close" at infinity or in nonconducting walls wherein the straight magnetic field lines are supposed to be anchored. Furthermore, in these one-dimensional tangential discontinuity models there are no currents due to curvature of magnetic field lines.

However, intruding solar-wind plasmoids are never really flat slabs extending to infinity in two directions like in Lemaire and Burlaga's one-dimensional model; they are plasma clouds of finite extent in all directions and the field-aligned currents created at their surfaces are not uniform planar current sheaths but are nonuniformly distributed in a volume of finite extent. These field-aligned currents are more likely to be filamentary than ideally flat current layers. But, as a result of localized and patchy parallel currents, the geomagnetic field lines are deformed—they acquire helicity, i.e., additional curvature. The additional curvature drifts of the electrons and ions can give rise to complex additional electric currents that are perpendicular to the radius of curvature of the magnetic field lines. In three-dimensional models these additional curvature drift currents can be closure currents for the field-aligned currents, $J_\parallel$, generated elsewhere.

But three-dimensional plasma current distributions and magnetic field distributions produced by high-β plasma clouds in an external background magnetic field have not yet been worked out in general.

Although simulations of the geomagnetic field perturbations by simple cylindrical current systems have been illustrated by Lemaire,[41] a three-dimensional generalization of Lemaire and Burlaga's one-dimensional plasma slab model for tangential discontinuities is presently beyond our grasp.

In conclusion, although complete three-dimensional models of high-β plasmoids engulfed in an external magnetic field are not yet available, it can be considered that field-aligned currents generated in the vicinity of such plasmoid can possibly be closed by perpendicular drift currents resulting from enhanced curvature (helicity) of the magnetic field lines in the vicinity of the diamagnetic plasma cloud.

APPENDIX 2: IMPULSIVE PENETRATION OF SOLAR-WIND PLASMOIDS IN THE FRONT-SIDE MAGNETOSPHERE

Plasmoid penetration in the front-side magnetosphere is expected when the IMF has a southward B_z. Indeed, the magnetic moment of plasma density enhancements has then a component antiparallel to the magnetic moment of the earth.

Plasmoids penetrating in the front-side plasma boundary layers produce a wide spectrum of diamagnetic field signatures among which flux transfer events (FTE) are especially characteristic ones.[21,42-45] Impulsive penetration events (IPE) or plasma transfer events (PTE) most likely correspond to the same "events" as well as plasma inclusions or intrusions observed by Sckopke et al.[46] near the magnetopause. The clouds or blobs of boundary layer plasmas considered by Lundin and Evans[47] to be a source for high-altitude, early afternoon auroral arcs are also other words for the same physical plasma field entities, i.e., plasmoids as pointed out by Lemaire[4] and Heikkila.[48]

The ionospheric signature of plasmoids penetrating in the front-side magnetosphere for southward B_z conditions has been observed in the region of the high-latitude troughs or polar clefts. Impulsive magnetosheath particle precipitation in the polar cleft was then reported by Carlson and Torbert.[49] Furthermore, short-lived small-scale irregularities moving northward have been observed in dayside auroral events by Sandholt et al.[50] Similar northward motions have been detected in the high-latitude trough ionosphere, using the STARE radar system.[51] The observed northward motion of these iono-spheric plasma irregularities corresponds to a transient and localized flow of magnetospheric plasma toward the magnetopause boundary as illustrated in Fig. 3a. The magnetospheric plasma flows around the intruding plasmoid in a direction opposite to the velocity of the solar-wind plasmoid itself. Therefore, these observations by Sandholt et al.[50] and Goertz et al.,[51] fully support the IPE theory, quite contrary to conclusions proposed in these papers.

ACKNOWLEDGMENT—I wish to thank Larry Lyons and Rickard Lundin for constructive comments, as well as Walter Heikkila and Michel Roth for interesting discussions.

REFERENCES

[1] L. F. Burlaga, J. Lemaire, and J. M. Turner, "Interplanetary Current Sheets at 1 AU," *J. Geophys. Res.* **82**, 3191-3200 (1977).

[2] M. J. Turner, L. F. Burlaga, N. F. Ness, and J. Lemaire, "Magnetic Holes in the Solar Wind," *J. Geophys. Res.* **82**, 1921-1924 (1977).

[3] W. H. Bostick, "Experimental Study of Ionized Matter Projected Across a Magnetic Field," *Phys. Rev.* **104**, 292-299 (1956).

[4] J. Lemaire, "Impulsive Penetration of Filamentary Plasma Elements into the Magnetospheres of the Earth and Jupiter," *Planet. Space Sci.* **25**, 887-890 (1977).

[5] J. Lemaire and M. Roth, "Penetration of Solar Wind Plasma Elements into the Magnetosphere," *J. Atm. Terr. Phys.* **40**, 331-335 (1978).

[6] I. I. Demidenko, N. S. Lomino, V. G. Padalka, B. N. Rutkevich, and K. D. Sinel'nikov, "Motion of a Plasmoid in a Nonuniform Transverse Magnetic Field," *Sov. Phys.-Techn. Phys.* **11**, 1354-1358 (1967).

[7] I. I. Demidenko, N. S. Lomino, V. G. Padalka, B. N. Rutkevich, and K. D. Sinel'nikov, "Plasma Stream in an Inhomogeneous Transverse Magnetic Field," *Sov. Phys.-Techn. Phys.* **14**, 16-22 (1969).

[8] G. Schmidt, "Plasma Motion across Magnetic Fields," *Phys. Fluids,* **3**, 961-965 (1960).

[9] J. Lemaire, "Plasmoid Motion Across a Tangential Discontinuity (with Application to the Magnetopause)," *J. Plasma Phys.* **33-3**, 425-436 (1985).

[10] R. Lundin, B. Hultqvist, N. Pisarenko, and A. Zakharov, "Composition of the Hot Magnetospheric Plasma as Observed with the PROGNOZ-7 Satellite," in *Energetic Ion Composition in the Earth's Magnetosphere,* pp. 307-351 (1983).

[11] R. Lundin and B. Aparicio, "Observations of Penetrated Solar Wind Plasma Elements in the Plasma Mantle," *Planet. Space Sci.* **30**, 81-91 (1982).

[12] R. Lundin and E. M. Dubinin, "Solar Wind Energy Transfer Regions Inside the Dayside Magnetopause – I. Evidence for Magnetosheath Plasma Penetration," *Planet. Space Sci.* **32**, 745-755 (1984).

[13] R. Lundin and E. M. Dubinin, "Solar Wind Energy Transfer Regions Inside the Dayside Magnetopause: Accelerated Heavy Ions as Tracers for MHD-Processes in the Dayside Boundary Layer," *Planet. Space Sci.* **33**, 891-907 (1985).

[14] T. J. Eastman, B. Popielawska, and L. A. Frank, "Three-Dimensional Plasma Observations near the Outer Magnetospheric Boundary," *J. Geophys. Res.* **90**, 9519-9539 (1985).

[15] J. Lemaire, "The Magnetosphere Boundary Layer: A Stopper Region for a Gusty Solar Wind," in *Quantitative Modeling of the Magnetospheric Processes,* Geophysical Monograph 21, W. P. Olson, ed., American Geophysical Union, Washington, D.C. (1979).

[16] D. A. Baker and J. E. Hammel, "Demonstration of Classical Plasma Behavior in a Transverse Magnetic Field," *Phys. Rev. Lett.* **8**, 157-158 (1962).

[17] D. A. Baker and J. E. Hammel, "Experimental Studies of the Penetration of a Plasma Stream into a Transverse Magnetic Field," *Phys. Fluids.* **8**, 713-722 (1965).

[18] J. Lemaire, "Impulsive Penetration of Solar Wind Plasma and Its Effects on the Upper Atmosphere," in *Proc. Magnetospheric Boundary Layers Conference, Alpbach, 11-15 Jun 1979,* ESA-SP148, pp. 365-373 (1979).

[19] J. Lemaire (in preparation, 1986).

[20] B. U. O. Sonnerup and L. J. Cahill, Jr., "Explorer 12 Observations of the Magnetopause Current Layer," *J. Geophys. Res.* **73**, 1757-1770 (1968).

[21] J. Berchem and C. T. Russell, "Flux Transfer Events on the Magnetopause: Spatial Distribution and Controlling Factors," *J. Geophys. Res.* **89**, 6689-6703 (1984).

[22] J. Lemaire, M. J. Rycroft, and M. Roth, "Control of Impulsive Penetration of Solar Wind Irregularities into the Magnetosphere by the Interplanetary Magnetic Field Direction," *Planet. Space Sci.* **27**, 47-57 (1979).

[23] K. Maezawa, "Magnetospheric Convection Induced by the Positive and Negative Z Components of the Interplanetary Magnetic Field: Quantitative Analysis Using Polar Cap Magnetic Records," *J. Geophys. Res.* **81**, 2289-2303 (1976).

[24] W. J. Burke, M. C. Kelley, R. C. Sagalyn, M. Smiddy, and S. T. Lai, "Polar Cap Electric Field Structures with a Northward Interplanetary Magnetic Field," *Geophys. Res. Lett.* **6**, 21-24 (1979).

[25] L. J. Zanetti, T. A. Potemra, T. Iijima, W. Baumjohann, and P. F. Bythrow, "Ionospheric and Birkeland Current Distributions for Northward Interplanetary Magnetic Field: Inferred Polar Convection," *J. Geophys. Res.* **89**, 7453-7458 (1984).

[26] N. U. Crocker, "Dayside Merging and Cusp Geometry," *J. Geophys. Res.* **84**, 951-959 (1979).

[27] P. H. Reiff, "Sunward Convection in Both Polar Caps," *J. Geophys. Res.* **87**, 5976-5980 (1982).

[28] P. H. Reiff and J. L. Burch, "IMF B_y-Dependent Plasma Flow and Birkeland Currents in the Dayside Magnetosphere. 2. A Global Model for Northward and Southward IMF," *J. Geophys. Res.* **90**, 1595-1609 (1985).

[29] L. R. Lyons, T. L. Killeen, and R. L. Walterscheid, "The Neutral Wind 'Flywheel' as a Source of Quiet-Time, Polar-Cap Currents," *Geophys. Res. Lett.* **12**, 101-104 (1985).

[30] L. R. Lyons, "A Simple Model for Polar Cap Convection Patterns and Generation of Auroras," *J. Geophys. Res.* **90**, 1561-1567 (1985).

[31] J. R. Kan and W. J. Burke, "A Theoretical Model of Polar Cap Auroral Red Arcs," *J. Geophys. Res.* **90**, 4171-4177 (1985).

[32] J. Lemaire and L. F. Burlaga, "Diamagnetic Boundary Layers: A Kinetic Theory," *Astrophys. Space Sci.* **45**, 303-325 (1976).

[33] M. Roth, "Structure of Tangential Discontinuities at the Magnetopause: The Nose of the Magnetopause," *J. Atm. Terr. Phys.* **40**, 323-329 (1978).

[34] T. Iijima, T. A. Potemra, L. J. Zanetti, and P. F. Bythrow, "Large-Scale Birkeland Currents in the Dayside Polar Region during Strongly Northward IMF: A New Birkeland Current System," *J. Geophys. Res.* **89**, 7441-7452 (1984).

[35] T. A. Potemra, L. J. Zanetti, P. F. Bythrow, A. T. Y. Lui, and T. Iijima, "B_y-Dependent Convection Patterns During Northward Interplanetary Magnetic Field," *J. Geophys. Res.* **89**, 9753-9760 (1984).

[36] W. Baumjohann and E. Friis-Christensen, "Dayside High-Latitude Ionospheric Current Systems," in *The Polar Cusp,* J. A. Holtet and A. Egeland eds., D. Reidel, pp. 223-234 (1985).

[37] T. Iijima and T. A. Potemra, "The Relationship Between Interplanetary Quantities and Birkeland Current Densities," *Geophys. Res. Lett.* **9**, 442-445 (1982).

[38] L. J. Zanetti and T. A. Potemra, "The Relationship of Birkeland and Ionospheric Current Systems to the Interplanetary Magnetic Field," in *Proc. Chapman Conf. on Solar Wind-Magnetosphere Coupling,* Pasadena, California (1985).

[39] P. F. Bythrow, W. J. Burke, T. A. Potemra, L. J. Zanetti, and A. T. Y. Lui, "Ionospheric Evidence for Irregular Reconnection and Turbulent Plasma Flows in the Magnetotail during Periods of Northward Interplanetary Magnetic Field," *J. Geophys. Res.* **90**, 5319-5325 (1985).

[40] M. Roth, "La Structure Interne de la Magnétosphère," *Acad. Roy. Belg, Memoire de la Classe des Sciences,* Collection in 8, 2e série, T.XLIV, Fascicule 7 (1984).

[41] J. Lemaire, "Simulation of the Interconnection Between Geomagnetic and Interplanetary Magnetic Field Lines," video-cassette, IASB, Brussels (1982).

[42] C. T. Russell and R. C. Elphic, "ISEE Observations of Flux Transfer Events at the Dayside Magnetopause," *Geophys. Res. Lett.* **6**, 33 (1979).

[43] R. C. Elphic and C. T. Russell, "ISEE-1 and -2 Magnetometer Observations of the Magnetopause," *Eur. Space Agency Publ.,* SP-148 (1979).

[44] R. P. Rijnbeek, S. W. H. Cowley, D. J. Southwood, and C. T. Russell, "A Survey of Dayside Flux Transfer Events Observed by ISEE 1 and 2 Magnetometers," *J. Geophys. Res.* **89**, 786 (1984).

[45] M. A. Saunders, C. T. Russell, and N. Sckopke, "A Dual-Satellite Study of the Spatial Properties of FTEs," in *Magnetic Reconnection in Space and Laboratory Plasmas,* Geophysical Monograph No. 30, E. W. Hones, Jr., ed., American Geophysical Union, Washington, D.C., pp. 145-152 (1984).

[46] N. Sckopke, G. Paschmann, G. Haerendel, B. U. O. Sonnerup, S. J. Bame, T. G. Forbes, E. W. Hones, Jr., and C. T. Russell, "Structure of the Low-Latitude Boundary Layer," *J. Geophys. Res.* **86**, 2099 (1981).

[47] R. Lundin and D. S. Evans, "Boundary Layer Plasmas as a Source for Height-Latitude, Early Afternoon, Auroral Arcs," *Planet. Space Sci,* **33**, 1389-1406 (1985).

[48] W.-I. Heikkila, "Impulsive Plasma Transport Through the Magnetopause," *Geophys. Res. Lett.,* **9**, 159-162 (1982).

[49] C. W. Carlson and R. B. Torbert, "Solar Wind Injection in the Morning Auroral Oval," *J. Geophys. Res.* **85**, 2903-2908 (1980).

[50] P. E. Sandholt, C. S. Deehr, A. Ageland, and B. Lybekk, "Signatures in the Dayside Aurora of Plasma Transfer from the Magnetosheath," University of Oslo report 86-04 (1986).

[51] C. K. Goertz, E. Nielsen, A. Korth, K. H. Glassmeier, C. Haldoupis, P. Hoeg, and D. Hayward, "Observations of a Possible Ground Signature of Flux Transfer Events," *J. Geophys. Res.* **90**, 4069-4078 (1985).

THE INFLUENCE OF THE IMF B_y COMPONENT ON THE EARTH'S MAGNETO-HYDROSTATIC MAGNETOTAIL

G.-H Voigt and R. V. Hilmer*

Statistical data analyses have revealed a positive correlation between the IMF B_y and the magnetotail B_y components. The observations were interpreted in terms of partial penetration of the IMF. It was suggested that the IMF is shielded less in the tail plasma sheet than in the tail lobes. However, the MHD equilibrium theory offers a different explanation.

Linear solutions to the Grad-Shafranov equation for two-dimensional tail equilibria show that an azimuthal tilting of closed magnetotail field lines results in a shear of the magnetic field that gives rise to an enhancement of the B_y component in the plasma sheet, where MHD equilibrium must be maintained. The total B_y in the tail consists of a constant background field plus a field that exists only in the plasma sheet where it depends on the strength of the thermal plasma pressure and on the magnetic B_z component. B_y is strongest at the center of the plasma sheet, decreases monotonically toward the plasma-sheet boundary, and reaches the minimum value of the background part of B_y in the tail lobes.

The existence of B_y in the plasma sheet is associated with field-aligned currents flowing along plasma-sheet field lines. For positive B_y the field-aligned currents flow toward the earth in the northern plasma sheet and away from the earth in the southern plasma sheet. If B_y has the opposite polarity, the field-aligned currents flow in the opposite sense.

For a given B_y in the equatorial plane, the angle of the azimuthal field line tilt, $\phi = \arctan (B_y/B_z)$, increases with increasing thermal plasma-sheet pressure. The field-line tilt is only partially caused by direct IMF penetration into the tail. More important is the tilt caused by polar-cap convection flows that rotate closed plasma-sheet field lines under the influence of the IMF B_y in opposite directions in the two ionospheric polar caps.

INTRODUCTION

Statistical analyses of magnetic field data from earth's magnetotail ($X_{\mathrm{GSM}} > -30\,\mathrm{R_e}$) have revealed a positive correlation between the IMF B_y and the magnetotail B_y components. Fairfield[1] analyzed IMP 6 magnetic field data and found that—averaged over the whole tail—the magnetotail B_y amounts to 13% of the IMF B_y. Lui[2] used the same data set (IMP 6 tail crossings), but he restricted the analysis to data obtained near the plasma sheet and found a 50% correlation. Thus, he concluded that the IMF is shielded less in the plasma sheet than in the tail lobes.

It is the purpose of this study to offer a different interpretation. On the basis of the MHD equilibrium theory we argue that the enhancement of B_y in the plasma sheet cannot be explained by partial penetration of the IMF. In the next section of this paper we develop a two-dimensional magneto-hydrostatic tail model and show that the total B_y in the tail consists of an IMF-related vacuum background field and an additional field that exists only in the plasma sheet where MHD equilibrium must be maintained. The resulting additional field-line shear is most likely caused by ionospheric polar-cap convection, a conjecture that has been supported by Cowley[3] and by Moses et al.[4]

In the third section of this paper we demonstrate that the magnetic field and current components in the tail depend sensitively on the strength of the thermal-plasma pressure measured in the plasma sheet. We show in particular that, for a given B_y in the equatorial plane, the angle of the azimuthal field-line inclination, $\phi = \arctan$

(B_y/B_z), increases with increasing plasma-sheet pressure. This result agrees well with observations. The concluding section is devoted to the discussion of the physical mechanisms that lead to the enhancement of B_y in various regions of the plasma sheet.

THE TWO-DIMENSIONAL MAGNETOTAIL

The magnetotail model is sketched in Fig. 1. MHD equilibrium is maintained in the plasma sheet, the thickness of which is $2z_0$. The labels $z = +z_0$ and $z = -z_0$ mark the northern and southern positions of the plasma-sheet boundaries, which are assumed to be straight lines to allow for analytic equilibrium solutions. The magnetic field in the tail lobes, above $z = z_0$ and below $z = -z_0$, is described in terms of a vacuum configuration that is characterized by $P = 0$ and $\nabla \times \boldsymbol{B} = 0$.

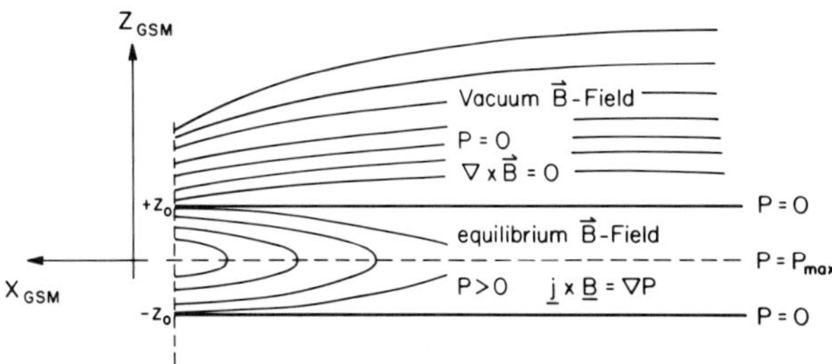

Figure 1—Sketch of the 2D magnetotail. The magnetic field in the plasma sheet satisfies the requirement of magneto-hydrostatic force balance. The plasma-sheet boundaries are located at $z = \pm z_0$. The boundary conditions at the plasma-sheet boundaries are $P = 0$, $B_z = 0$, and $\nabla \times \boldsymbol{B} = 0$. The symmetry conditions at the center of the plasma sheet are $B_x = 0$, and $P = P_{max}$. The magnetic field in the tail lobes is described in terms of a vacuum magnetic field.

*Department of Space Physics and Astronomy, Rice University, Houston, Texas 77251.

For isotropic thermal plasma, the magnetic-field configuration in the plasma sheet must meet the equilibrium conditions

$$(\nabla \times \mathbf{B}) \times \mathbf{B} = \mu_0 \nabla P \qquad (1)$$

$$\nabla \cdot \mathbf{B} = 0 \qquad (2)$$

Note that Eq. 1 represents the MHD equation of motion in the slow-flow approximation, i.e., we have ignored the inertia term $\rho(dv/dt)$. This approximation is applicable in the magnetosphere whenever plasma convection velocities are small compared to the Alfvén velocity and time scales are long compared to MHD wave travel times. In two dimensions ($\partial/\partial y = 0$) Eqs. 1 and 2 are equivalent to the Grad-Shafranov equation (e.g., Voigt[5]),

$$\nabla^2 A + \frac{d}{dA} \left[\mu_0 P(A) + \tfrac{1}{2} B_y(A)^2 \right] = 0 \quad (3)$$

where the scalar flux function $A(x,z)$ is the y-component of the magnetic vector potential A which satisfies the Coulomb gauge $\nabla \cdot A = 0$. The magnetic B_x and B_z components are calculated from $\mathbf{B} = \nabla \times A$ and read

$$B_x = -\frac{\partial A}{\partial z} \qquad B_z = +\frac{\partial A}{\partial x} \qquad (4)$$

The MHD equilibrium system contains two free physical functions, namely, the thermal plasma pressure P and the magnetic B_y component,

$$P = P(A) \qquad (5)$$

$$B_y = \pm [I(A)^2 + B_0^2]^{1/2} \qquad (6)$$

Note that $A(x,z)$ is constant along magnetic field lines. As long as thermodynamic conditions are not specified, the two functions $P(A)$ and $I(A)$ can be chosen arbitrarily. Thus, we assume, for the sake of mathematical convenience,

$$P(A) = \frac{1}{2\mu_0} k^2 A^2 \qquad (7)$$

$$I(A) = \alpha A \qquad (8)$$

The parameter k measures the thermal plasma pressure, according to Eq. 7. B_0 is the constant background B_y component that results from partial penetration of the constant IMF B_y. The background field is present in the whole tail, i.e., in the plasma sheet and in the tail lobes. The term αA describes the additional B_y component that exists owing to magnetic shear forces in regions that are plasma dominated; therefore, this term does not exist in the tail lobes. The sign of α corresponds

to the sign of the IMF B_y because the orientation of the IMF determines the azimuthal inclination of closed magnetotail field lines. It will become obvious below that the parameter α also measures the strength of the field-aligned currents that are associated with the field-line shear.

With Eqs. 7 and 8 (assumptions) the Grad-Shafranov equation (Eq. 3) becomes linear, namely

$$\nabla^2 A + (k^2 + \alpha^2) A = 0 \qquad (9)$$

The plasma currents follow from Eqs. 4, 6, and 9 via $\mu_0 j = \nabla \times B$. For convenience we introduce the function

$$Q(A) = \frac{I(A)}{B_y} = \frac{1}{\alpha}\frac{dB_y}{dA} = \left[1 + \left(\frac{B_0}{\alpha A}\right)^2\right]^{-1/2} \qquad (10)$$

and obtain the current components

$$\mu_0\, j_x = \alpha \cdot Q(A)\, B_x \qquad (11)$$

$$\mu_0\, j_y = \alpha \cdot Q(A)\, B_y + k^2 A \qquad (12)$$

$$\mu_0\, j_z = \alpha \cdot Q(A)\, B_z \qquad (13)$$

By introducing the plasma β parameter

$$\beta(x,z) = \frac{2\mu_0 P}{B^2} = \frac{k^2 A^2}{B^2} \qquad (14)$$

the field aligned currents

$$\mu_0\, j_\parallel = \left[\frac{dB_y}{dA} + \mu_0 \frac{B_y}{B^2} \cdot \frac{dP}{dA}\right] \cdot B \qquad (15a)$$

assume the convenient form

$$\mu_0\, j_\parallel = \alpha \cdot Q(A) \left[1 + \beta \cdot \left(\frac{1}{Q(A)}\right)^2\right] \cdot B \qquad (15b)$$

Note that there are no field-aligned currents if B_y in Eq. 6 reduces to zero. Note also that the strength of the field-aligned currents depends on the plasma-sheet pressure via $\beta(x,z)$ in Eq. 14. In order to obtain the magnetic field components, we solve Eq. 9 by separation of variables (see, for example, Ref. 6). This leads to a complete set of eigenfunctions

$$A_n(x,z) = \frac{B_n}{\eta_n} \cos(\eta_n z) \cdot \exp(-\lambda_n |x|) \qquad (16)$$

The variables x and z correspond to magnetospheric GSM coordinates. The physical boundary conditions at the plasma-sheet boundaries $z = +z_0$ and $z = -z_0$ are $B_z = 0$, $P = 0$, and $\nabla \times \mathbf{B} = 0$. At the center of the plasma sheet, we assume the symmetry conditions $B_x = 0$ and $P = P_{max}$. Thus, the eigenvalues η_n become

$$\eta_n = \frac{2n - 1}{z_0} \cdot \frac{\pi}{2} \tag{17}$$

The two physical parameters α and k are related to the eigenvalues through

$$\eta_n^2 = \lambda_n^2 + \alpha^2 + k^2 \tag{18}$$

The two parameters can be chosen arbitrarily in the range $0 \leq (\alpha^2 + k^2) \leq \eta_1^2$. Given B_0, k, and α in Eqs. 6-8, the three magnetic field components, Eqs. 4 and 6, follow from Eq. 16. A more detailed mathematical discussion of the eigenvalue problem is given by Voigt.[5] Without losing generality, we assume henceforth that the magnetotail configuration be determined mainly by the first leading eigenfunction ($n = 1$) of the vector potential Eq. 16. The higher eigenfunctions ($n > 1$) decrease rapidly down the tail according to Eq. 16. We define the x component of the magnetic field in the tail lobes,

$$B_{LX} = B_1 \cdot \exp(-\lambda_1 |x|) \tag{19}$$

and from Eq. 14 at the neutral sheet $z = 0$, we find

$$\beta_0 = \beta(x,z=0) = \frac{k^2}{\lambda_1^2 + (\alpha/Q_0)^2} \tag{20}$$

where $Q_0 = Q(A[x,z=0])$ according to Eq. 10. The two parameters k and α in Eqs. 7 and 8 can be expressed in terms of the measurable quantity β_0,

$$k^2 = \left(\frac{\pi}{2z_0}\right)^2 \cdot \frac{\beta_0}{(1 + \beta_0)} \left[1 + \left(\frac{B_0}{B_{LX}}\right)^2\right] \tag{21}$$

$$\alpha^2 = \left(\frac{\pi}{2z_0}\right)^2 \cdot \frac{1}{(1 + \beta_0)} \left[1 - \beta_0 \left(\frac{B_0}{B_{LX}}\right)^2\right] - \lambda_1^2 \tag{22}$$

The parameter λ_1 in Eq. 22 is determined by the magnitude of the magnetic B_z component, according to Eqs. 4 and 16. $B_z(x,z=0)$ and β_0 can be obtained directly from measurements when the spacecraft traverses the neutral sheet ($z = 0$). The angle of the azimuthal field-line inclination (see Fig. 2) is defined as

$$\phi = \text{arc tan } [B_y/B_z]$$

$$= \text{arc tan } [\alpha/(\lambda_1 Q)] \tag{23}$$

Thus, the field-line inclination can be determined directly by measuring the magnetic B_y and B_z components. The additional piece of information, the plasma β parameter β_0 in Eq. 22, can be used to test the validity of the MHD equilibrium concept. We shall see in the following section how β_0 influences the plasma-sheet configuration and the shear of closed plasma-sheet field lines.

MODEL RESULTS AND OBSERVATIONS

Within the limits of the linear theory, a realistic physical situation can be calculated by finding an appropriate set of the three parameters, B_0 in Eq. 6, k in Eq. 7, and α in Eq. 8, that brings the model in agreement with measurements. In particular, the distribution of the B_y component in the tail depends on the strength of the background field B_0 and on the parameter α. In this study, we demonstrate the importance of the plasma-related part of B_y in the plasma sheet. Thus, we ignore the vacuum background field and assume $B_0 = 0$ in Eq. 6, which leads to $Q(A) = 1$ in Eq. 10. Due to this simplification, the angle of the field-line tilt ϕ in Eq. 23 becomes constant throughout the plasma sheet, as sketched in Fig. 2.

First, we illustrate the influence of the plasma parameter $\beta_0 = \beta(x,z=0)$ on the plasma-sheet configuration. Figure 3 depicts plasma-sheet field lines for the special case $B_y = 0$. The labels $z = +z_0$ and $z = -z_0$ mark the positions of the northern and southern plasma-sheet boundaries where the thermal plasma pressure vanishes. The three field-line configurations in Fig. 3 are calculated for $\beta_0 = 50$ (top), 200 (middle), and 1000 (bottom). We see that the magnetic-tail field lines become stretched as β_0 increases; the magnetic B_z component decreases accordingly. The decrease of B_z is

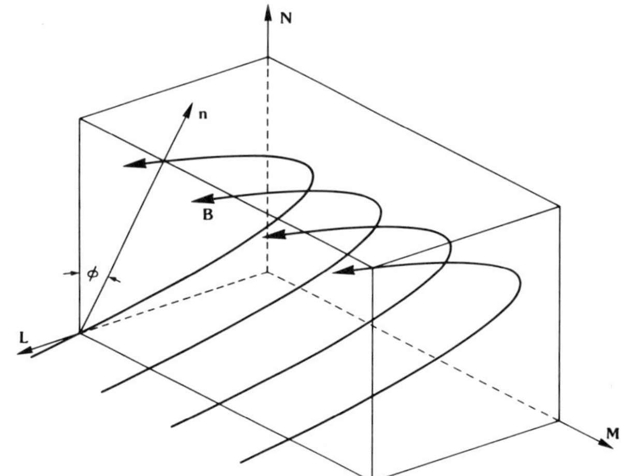

Figure 2—Magnetic field-line inclination in the plasma sheet (reproduced from figure in Ref. 7). The (L, M, N) coordinate system is aligned with the cross-tail current sheet in order to eliminate time changes of the neutral sheet orientation. The planes of magnetic field lines are inclined with respect to the neutral-sheet normal by the angle $\phi = $ arc tan (B_M/B_N). For the purpose of this study the (L, M, N) axes are assumed to be parallel to the GSM (X, Y, Z) axes, so that $B_y = B_M$ and $B_z = B_N$.

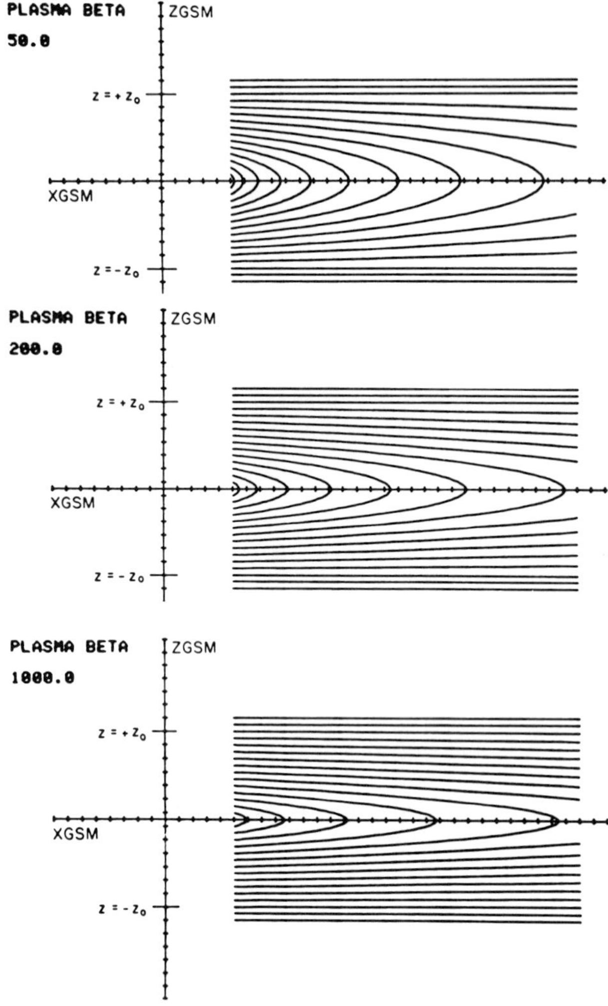

Figure 3—Plasma-sheet magnetic field lines according to linear solutions of the Grad-Shafranov Eq. 3. The B_y component is zero in this computation. The three panels depict configurations for $\beta_0 = 50$ (top), 200 (middle), and 1000 (bottom). The plasma parameter $\beta_0 = \beta(x, z = 0)$ is explained in Eqs. 14 and 20. Note that the field lines become stretched as β_0 increases.

always associated with an enhancement of the cross-tail current density at the center of the plasma sheet.

The decrease of the B_z component with increasing β_0 is not restricted to the special case $B_y = 0$. The effect of field-line stretching can be observed for $|B_y| > 0$ as well. As an example, the top panel of Fig. 4 shows B_z plotted versus β_0 for a dawn-dusk field $B_y = 1$ nT.

The middle panel in Fig. 4 emphasizes a different point of view: here we have assumed a normal component $B_z = 3$ nT at the center of the neutral sheet. The figure shows the magnitude of B_y one would expect for different values of the thermal plasma pressure confined on closed plasma-sheet field lines. Note that B_y decreases with increasing β_0, until B_y approaches zero for the maximum upper limit $\beta_{0,max} = 43.44$. The quantity $\beta_{0,max}$ becomes larger for smaller neutral sheet B_z values. For a given $\beta_0 < \beta_{0,max}$, B_y increases with decreasing B_z. This behavior can be understood by investigating Eqs. 21 and 22 and the eigenvalue Eq. 18.

Thus for a given plasma β at the center of the neutral sheet, the magnetic dawn-dusk component is more prominent in situations with the smaller B_z component.

The presence of a magnetic dawn-dusk component leads to an inclination of closed plasma-sheet field lines. McComas et al.[7] conducted three case studies of ISEE 1 and 2 tail current sheet crossings ($X_{GSM} \geq -22\ R_e$) and found indeed a field-line inclination depending on B_y. For their analysis, they first eliminated time variations of the neutral sheet orientation relative to the GSM coordinate system by defining an (L, M, N) system that is always aligned with the cross-tail current sheet. With respect to this coordinate system, they defined the angle $\phi = \text{arc tan}\ (B_M/B_N)$, as depicted in Fig. 2. In this study, we assume that the (L, M, N) axes are parallel to the GSM (X, Y, Z) axes; thus, we use the notations $B_y = B_M$ and $B_z = B_N$ in Eq. 23.

McComas and co-authors found that the thinner, higher-current-density neutral-sheet crossings were associated with the greater field-line inclinations. This observational result is in agreement with the MHD equilibrium tail model. In the bottom panel of Fig. 4, the angle of the azimuthal field-line tilt, Eq. 23, is plotted versus β_0. For this plot we have assumed an observed $B_y = 1$ nT at the center of the plasma sheet. Note that the inclination angle ϕ becomes larger for current sheets with the higher plasma β, which implies stronger cross-tail current densities.

In the special case of vanishing background field (i.e., $B_0 = 0$), the functional form of the dawn-dusk component in the plasma sheet is given by $B_y \propto \cos{(\eta_1 z)}$, according to Eqs. 6, 8, and 16. Thus, B_y is strongest at the center of the plasma sheet ($z = 0$) and decreases monotonically toward the plasma sheet boundary ($z = z_0$). The existence of B_y in the plasma sheet causes field-aligned currents to flow along plasma-sheet field lines. For positive B_y the field-aligned currents (Eq. 15b) flow toward the earth in the northern plasma sheet and away from the earth in the southern plasma sheet. If B_y has the opposite polarity, the field-aligned currents flow in the opposite sense. This relation between B_y and the field-aligned currents has been found in the ISEE 3 far tail data as well.[8]

CONCLUSIONS

The observational fact that, for a given IMF B_y, the magnetotail B_y is stronger in the plasma sheet than in the tail lobes can be understood in the framework of the MHD equilibrium theory. We have demonstrated that the magnetotail B_y (Eq. 6) consists of two parts. The first, $B_{y1} = B_0$, represents a certain fraction of the IMF B_y that is present in the tail lobes and in the plasma sheet. Its presence can in fact be explained in terms of "partial penetration" of the IMF into the magnetotail. B_{y1} is a vacuum field that does not lead to an enhancement of B_y in the plasma sheet.

Superimposed is the second field, $B_{y2} = I(A)$, which exists in the plasma sheet only. One can show that B_{y2} vanishes when the magnetic field in the plasma sheet ap-

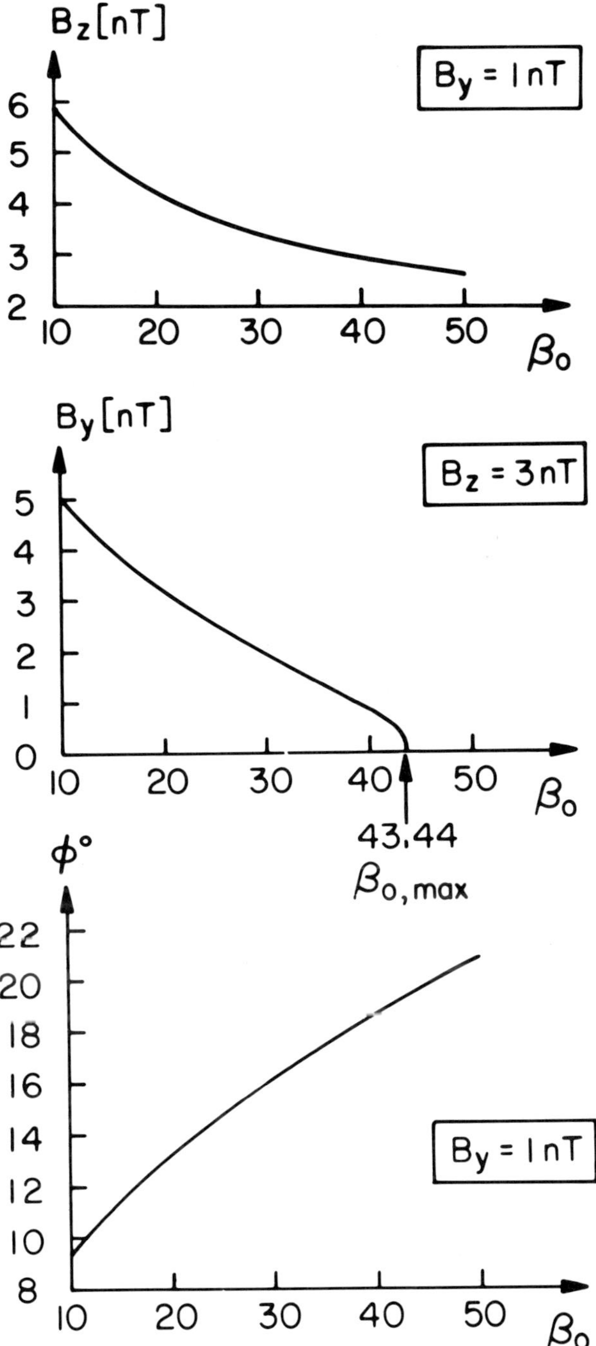

Figure 4—Magnetic field components at the center of the plasma sheet ($z = 0$). The plasma parameter is $\beta_0 = \beta(x,z=0)$, according to Eqs. 14 and 20. Plotted versus β_0 are B_z (top), B_y (middle), and $\phi = $ arc tan (B_y/B_z) (bottom). The quantities in the boxes are held fixed at the values indicated. Note that for a given B_y the inclination angle ϕ of plasma-sheet field lines (see Fig. 2) increases with increasing strength of the cross-tail current density.

proaches the unrealistic (but mathematically possible) vacuum configuration. Thus, B_{y2} exists only in regions that are plasma dominated. B_{y2} vanishes at the plasma-sheet boundaries where the thermal plasma pressure is zero.

Of course, the analytic results obtained in this study are model dependent through the very special assumptions (Eqs. 7 and 8) that specify the two free functions, Eqs. 5 and 6, in the Grad-Shafranov equation (Eq. 3). The two assumptions lead to linear solutions of the equilibrium problem. The magnetic field derived from these solutions is parameterized by the two physical constants k in Eq. 7 and α in Eq. 8. k measures the strength of the thermal plasma pressure, and α measures the plasma-sheet field B_{y2} and the density of the field-aligned currents. The two constants can be expressed in terms of the plasma β parameter, $\beta_0 = \beta(x,z=0)$, which is measured at the center of the plasma sheet.

Irrespective of our constraint imposing linear equilibrium solutions, the characteristics of the B_y component in the magnetotail are determined by three physical conditions: first, the thermal plasma pressure and the cross-tail current density reach their maximum values at the center of the plasma sheet ($z = 0$); second, the plasma-sheet boundaries are defined as zero-pressure magnetic surfaces; third, the tail lobes are described in terms of vacuum magnetic fields. Based on these three conditions, we find that the total B_y in the tail is strongest at the center of the plasma sheet, decreases monotonically toward the plasma-sheet boundaries, and reaches the minimum value of the vacuum background field, $B_{y1} = B_0$, in the tail lobes. We emphasize again that partial penetration of the IMF B_y (i.e., a simple vacuum superposition of IMF and magnetotail fields) cannot explain the observed enhancement of B_y in the plasma sheet.

The presence of a magnetospheric B_y component in response to the external IMF B_y has been reported extensively. The IMF B_y affects the convection patterns in the polar cap region (e.g., Potemra et al.[9] and Moses et al.[4]); it is correlated with the B_y component at geosynchronous orbit,[10] with B_y in the near-earth tail region,[1,2,7] and with B_y in the distant tail.[8,11] However, very different physical mechanisms explain the existence of B_y in different magnetospheric regions.

The IMF B_y affects the polar-cap-convection flow patterns through interconnection with open polar-cap field lines. According to this mechanism, the IMF acts primarily on open field lines, but the resulting azimuthal convection flow extends into the closed field-line region as well and generates an azimuthal shift of closed field lines moving with the plasma flow.[3,4] The plasma flow in the southern hemisphere is assumed to convect in the opposite sense, so that closed field lines that connect the northern and southern polar caps should be tilted to produce a B_y component in the same direction as the IMF.

Without considering the details of polar-cap convection, one can easily illustrate how the IMF directly affects the closed field-line region. As an example, Fig. 5 shows a perspective view of closed dipolar vacuum field lines distorted by a superimposed homogeneous B_y component. In the top panel of Fig. 5, B_y is zero. Computed are earth's dipole field lines that originate at $70°$ latitude from $-30°$ to $+30°$ longitude (22:00 to 2:00

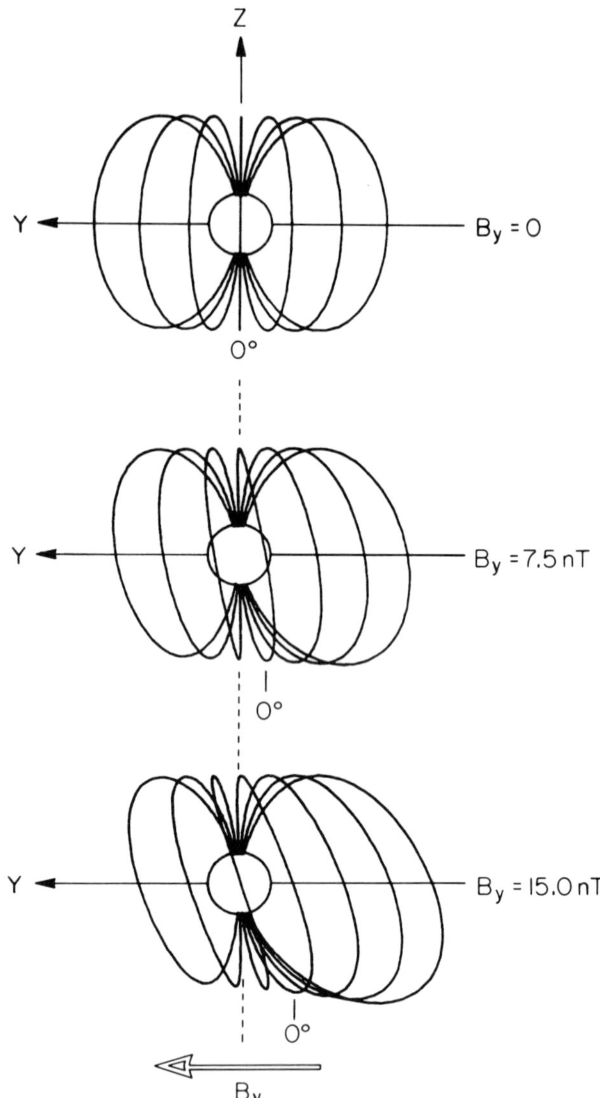

Figure 5—Perspective view of closed dipole field lines. Computed are field lines that originate at 70° geomagnetic latitude, from −30° to +30° longitude in steps of 10°. The three panels show the increasing influence of a superimposed vacuum dawn-dusk magnetic field. It is $B_y = 0$ (top), 7.5 nT (middle), and 15 nT (bottom). Note that the field lines become rotated and tilted under the influence of the B_y component. An inclination of closed field lines will ultimately lead to an enhancement of B_y in the plasma sheet, where MHD equilibrium must be maintained.

LT) in steps of 10°. Since the dipole field is not distorted for $B_y = 0$, the 0° longitude (24:00 LT) field line is projected into a vertical line in the perspective view. In the middle and bottom panels of Fig. 5, a homogeneous B_y of 7.5 nT and 15 nT, respectively, is superimposed. Note how the closed magnetic-field lines are rotated and tilted under the influence of the additional B_y. In the bottom panel, the −10° field line is rotated such that it crosses the midnight meridian. Of course, the two values of the superimposed B_y component are unrealistically high and have been used in Fig. 5 for computational convenience. Much smaller values of B_y

would generate the same tilt of dipole field lines that originate at geomagnetic latitudes higher than 70°.

The B_y induced rotation and tilt of closed field lines affect the equilibrium magnetic field configuration in the near-earth tail plasma sheet, a conjecture that has been supported by Moses et al.[4] Voigt[12] pointed out that the requirement of magneto-hydrostatic force balance causes the near-earth tail plasma sheet to adjust its shape and position to the orientation of earth's dipolar field lines. Thus, it is not surprising that the azimuthal tilt of closed polar-cap field lines leads to a shear of plasma-sheet field lines and, therefore, to an enhancement of B_y in the plasma sheet.

Note that the above mechanism is substantially different from the mechanism that generates the shear of magnetic field lines in the distant magnetotail. Tsurutani et al.[8] and especially Sibeck et al.[11] explain the existence of a prominent B_y component at $X_{GSM} = -180$ R_e in terms of a mechanism originally proposed by Cowley,[3] according to which the IMF exerts an external ($j \times B$) torque on the open magnetotail, which results in a twist of both the plasma sheet and the tail lobes.

In summary, we conclude that the enhancement of the B_y component in the near-earth tail plasma sheet is caused by the IMF, not directly by superposition or partial penetration, but rather indirectly via ionospheric polar-cap convection.

ACKNOWLEDGMENT—G.-H. Voigt benefited from many helpful conversations with David McComas and David Sibeck. We are grateful to Frank Toffoletto for providing us with his perspective field-line plotting program that generated the plots in Fig. 5.

This work was supported by AFGL under contract F-19628-83-K-0016, and by the National Science Foundation under grant ATM 84-17580.

REFERENCES

[1] D. H. Fairfield, "On the Average Configuration of the Geomagnetic Tail," *J. Geophys. Res.* **84**, 1950-1958 (1979).

[2] A. T. Y. Lui, "Characteristics of the Cross-Tail Current in the Earth's Magnetotail," in *Magnetospheric Currents*, Geophys. Monograph., **28**, American Geophysical Union, Washington, D.C., pp. 158-170 (1983).

[3] S. W. H. Cowley, "Magnetospheric Asymmetries Associated with the y-Component of the IMF," *Planet. Space Sci.* **29**, 79-96 (1981).

[4] J. J. Moses, N. U. Crooker, D. J. Gorney, and G. L. Siscoe, "High-Latitude Convection on Open and Closed Field Lines for Large IMF B_y," *J. Geophys. Res.* **90**, 11,078-11,082 (1985).

[5] G.-H. Voigt, "Magnetospheric Equilibrium Configurations and Slow Adiabatic Convection," in *Solar Wind Magnetosphere Coupling*, Y. Kamide and J. Slavin, eds., Terra/Reidel, Tokyo, pp. 233-273 (1986).

[6] G.-H. Voigt and R. A. Wolf, "On the Configuration of the Polar Cusps in Earth's Magnetosphere," *J. Geophys. Res.* **90**, 4046-4054 (1985).

[7] D. J. McComas, C. T. Russell, R. C. Elphic, and S. J. Bame, "The Near-Earth Cross-Tail Current Sheet: Detailed ISEE-1 and -2 Case Studies," *J. Geophys. Res.* **91**, 4287-4301 (1986).

[8] B. T. Tsurutani, J. A. Slavin, E. J. Smith, R. Okida, and D. E. Jones, "Magnetic Structure of the Distant Geotail from -60 to -220 R_e: ISEE-3," *Geophys. Res. Lett.* **11**, 1-4 (1984).

[9] T. A. Potemra, L. J. Zanetti, P. F. Bythrow, and A. T. Y. Lui, "B_y-Dependent Convection Patterns During Northward Interplanetary Magnetic Field," *J. Geophys. Res.* **89**, 9753-9760 (1984).

[10] S. W. H. Cowley and W. J. Hughes, "Observation of an IMF Sector Effect in the y Magnetic Field Component at Geostationary Orbit," *Planet. Space Sci.* **31**, 73-90 (1983).

[11] D. G. Sibeck, G. L. Siscoe, J. A. Slavin, E. J. Smith, B. T. Tsurutani, and R. P. Lepping, "The Distant Magnetotail's Response to a Strong Interplanetary Magnetic Field B_y: Twisting, Flattening, and Field-Line Bending," *J. Geophys. Res.* **90**, 4011-4019 (1985).

[12] G.-H. Voigt, "The Shape and Position of the Plasma Sheet in Earth's Magnetotail," *J. Geophys. Res.* **89**, 2169-2179 (1984).

DISCUSSION

J. Birn: Doesn't the dependence of B_y on the plasma β depend on the assumed dependence of B_y on the flux function A?

H. Voigt: Yes, of course! B_y does depend on the flux function A via $B_y = [I(A)^2 + B_0^2]^{1/2}$. Therefore, B_y depends on the plasma β-parameter via the functional form that we assume for $I(A)$. This function cannot be specified from the equilibrium theory, unless one includes an additional thermodynamic condition, for example, quasistatic convection.

III. FLUID ASPECTS OF MAGNETOTAIL DYNAMICS

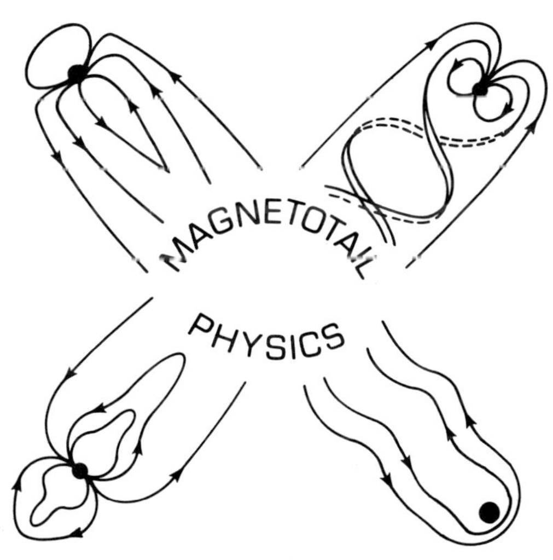

OBSERVATIONS ON THE FLUID ASPECTS OF
MAGNETOTAIL DYNAMICS

A. T. Y. Lui*

Although the earth's magnetotail consists of very rarefied plasmas, many global features of this region in space can be well described by fluid parameters rather than statistical quantities. Here we give a brief overview of our present knowledge of a host of substorm phenomena as well as of nonsubstorm related structures and dynamics in the magnetotail. Models of a substorm for the magnetotail are described and briefly commented on. Examples that indicate the limitations of fluid treatment of magnetotail dynamics are also discussed.

INTRODUCTION

The most abundant (over 99%) state of matter in the universe is in the plasma state. The magnetotail particle population is in this state since it exhibits the collective behavior characteristic of a plasma. The Debye length λ_D in the magnetotail is typically less than a fraction of a kilometer (see, for example, Table 1 in Ref. 1). This is much less than the scale-size L within the magnetotail, which is usually at least of the order of 10^{-2} to 10^{-1} R_e. Therefore, the quasineutrality condition holds rather well within the magnetotail region. The number of particles within a Debye sphere ($\sim n\lambda_D^3$) is of the order of 10^9, making the plasma parameter $(n\lambda_D^3)^{-1}$ a very small quantity. Under these conditions, collective effects are important as is expected for a plasma.

Plasmas in the earth's magnetotail are extremely tenuous. For comparison, a reasonably good vacuum in the laboroatory has a pressure of 10^{-8} torr, which is about eight orders of magnitude more dense than the most compact regime of the magnetotail. Consequently, collisions between particles in the magnetotail are extremely infrequent and magnetotail plasmas are described as collisionless. The classical mean-free-path of particles in the magnetotail plasma sheet is on the order of 10^{13} km, or about 10^9 R_e, which is many orders of magnitude larger than the magnetotail dawn-dusk or north-south dimension.

In spite of the long mean-free-path, plasmas in the magnetotail can often be treated rather adequately as fluids, especially for phenomena of very-low-frequency and very-long-spatial scales. The resolution of this dilemma lies in the consideration of magnetic-field and collective processes present in magnetotail plasmas. The presence of a magnetic field imposes a constraint to particle motions, causing particles to gyrate around the field direction, much like collisions restricting particles from traveling in undistorted linear trajectories. In addition, there is a large number of collective processes operative within plasmas enabling individual particles to communicate forces they experience, portraying the essential role of collisions in a collisional medium. However, the lack

of collisions is indicated by the departure of the velocity distribution of magnetotail particles from a Maxwellian distribution. Often, only the low-energy portion of the population is Maxwellian-like, but the high-energy portion of the distribution has a power-law[2] or exponential[3] dependence on the energy. The former shape of the velocity distribution, known as a κ-distribution, is found in many planetary plasmas and has been suggested to be a consequence of a plasma in a superthermal radiation field.[4]

The intent of this paper is to give a brief overview of observed fluid characteristics of the magnetotail plasmas. Most studies on magnetotail dynamics are focused on a transient process called a magnetospheric substorm. Before embarking on a discussion of substorm phenomena in the magnetotail, however, we shall describe the dynamic features believed to be unrelated to the substorm process. Some of these features are semipermanent magnetotail structures probably associated with the formation of the magnetotail. The concept of a magnetospheric substorm is then introduced, followed by a review of the substorm morphology in the magnetotail. To a great extent, interpretations of the observed magnetotail substorm phenomena are given separately, along with substorm models and some controversial issues. Limitations on fluid treatment of magnetotail dynamics are briefly discussed before the section on final remarks.

NONSUBSTORM RELATED MAGNETOTAIL STRUCTURES

It is increasingly evident that the earth's magnetotail does not attain a steady state, especially in plasma convection. Erickson and Wolf[5] have discovered that steady-state convection in the magnetotail leads to unrealistically high compression of plasma in the near-earth part of the plasma sheet. The attempt by Schindler and Birn[6] to derive a self-consistent theory of magnetotail convection leads to a similar conclusion. They have shown that a self-consistent model of the magnetotail can in principle be constructed theoretically, but the resulting tail configuration departs considerably from the observed tail characteristics. In examining the consequences of steady-state convection theoretically, Lui and

*The Johns Hopkins University Applied Physics Laboratory, Laurel, Maryland 20707.

Hasegawa[7] have found that plasma convection is intrinsically time-dependent when the solar-wind plasma enters through the flanks of the magnetotail as observed. These studies therefore indicate that the magnetotail is always evolving and dynamic.

Nevertheless, one may distinguish magnetotail phenomena in two categories. There are transient magnetotail features that occur in association with a process known as a magnetospheric substorm during which the magnetotail tends to readjust itself to a more stable configuration. There are also other magnetotail features that appear to be independent of substorm occurrence. Some of these may be regarded as semipermanent features in the magnetotail formation process and are discussed below.

Vortices

The large-scale circulation pattern of the plasma sheet has been inferred at the flanks of the magnetotail. Rotation of the direction of plasma flow, primarily in the equatorial plane, with a period of ~5 to 15 minutes, has been seen occasionally by ISEE 1 and 2.[8-11] This kind of flow rotation is sometimes observed up to five cycles and has been interpreted as plasma vortices. The preferred rotation sense is clockwise in the morning sector and anticlockwise in the evening sector when viewed from north of the equatorial plane. Signatures of a large vortex have been observed in the distant magnetotail also.[12]

Waves in the local magnetic field accompanying flow rotations are seen, although phase velocities and phase lags determined from plasma flow and magnetic field often differ. From phase lags of the flow rotations between the two spacecraft, ISEE 1 and 2, the wavelengths of the vortices have been deduced to be about one to many tens of earth radii, but these values appeared to be larger than those deduced from energetic particle data.

The vortices have been suggested as hydromagnetic waves driven by Kelvin-Helmholtz instability at the boundary between the magnetotail and the magnetosheath. Recently, Sibeck et al.[13] have found evidence in support of this contention. They have shown that the daily number of magnetopause crossings at the distant tail encountered by the ISEE 3 is linearly correlated with the solar-wind or magnetosheath speed. Furthermore, they have verified that the parameters at the interface satisfy the instability criterion for waves driven by Kelvin-Helmholtz instability for transverse propagation in over half of the sample magnetopause crossings.

Neutral Sheet Structures

The neutral sheet is usually envisaged as a plane current sheet separating regions of magnetic field with opposite directions.[14] Although this picture is useful as an average description of the field geometry, the actual observation of the neutral sheet often reveals more complicated field structures. From detailed dual-spacecraft studies, McComas et al.[15] have obtained current density profiles that show the main cross-tail current sheet to be many thermal ion gyroradii thick and to contain a lot of fine structures in current density.

A magnetic field with either a northward or southward component is detected in the neutral sheet region.[16,17] The propagation of a wave along the neutral-sheet surface, tilting the sheet normal away from its average orientation, can cause reversing sign in the north-south field component. This wave motion would also tilt the field adjacent to the neutral sheet to give rise to a northward or a southward component.

Another cause for a fluctuating north-south field component is the presence of magnetic islands (or magnetic bubbles) within which magnetic field lines form closed loops.[18-20] Magnetic islands are detected at all levels of activity in the magnetotail.[21] The occurrence of magnetic islands reflects that the neutral-sheet current is seldom uniform throughout. The cross-tail current density fluctuates continuously at various locations in the neutral sheet, occasionally resulting in the formation of magnetic islands. Thus magnetic islands should be considered as a persistent feature of the neutral sheet, although each individual magnetic island may be a rather transient local phenomenon. The formation of magnetic islands within the neutral sheet involves magnetic reconfiguration or reconnection. However, the fact that energetic phenomenon in particles is not often observed in association with these events[21] suggests that the magnetic reconnection process[22-24] leading to the magnetic island formation does not necessarily involve a substantial conversion of energy from magnetic field to particles.

Slow-Mode Shock

Evidence of slow-mode shocks thought to be associated with reconnection on the nightside magnetotail has been reported using the ISEE 3 measurements in the distant magnetotail.[25-28] The presence of these shocks was predicted about two decades ago by some theories on reconnection[23] that suggest that slow-mode shocks should be formed at the plasma-sheet boundaries. Figure 1 is a schematic diagram of the slow-mode shocks together with some key features that have been reported. The locations of the shock are generally found in the downstream distances between $100\ R_e$ and $200\ R_e$. The shock is detected in quiet and magnetically disturbed intervals,[29] indicating that it may be a semipermanent feature of the distant magnetotail.

The Rankine-Hugoniot jump conditions for the slow-mode shock transition have been tested and there is a good agreement between predicted and observed values, although a more definitive identification is precluded by the lack of plasma ion measurements from the ISEE 3 spacecraft for these studies. The average characteristics of the shock have been determined. The shock thickness is approximately 10^3 km.[27,28] In the upstream region (tail lobe), strong electron plasma oscillations and magnetosonic waves are detected in association with a heat flux along the magnetic field line directed usually down the tail and is carried by the high energy tail portion of the electron population. Within the shock layer, intense low-frequency broadband electrostatic waves are

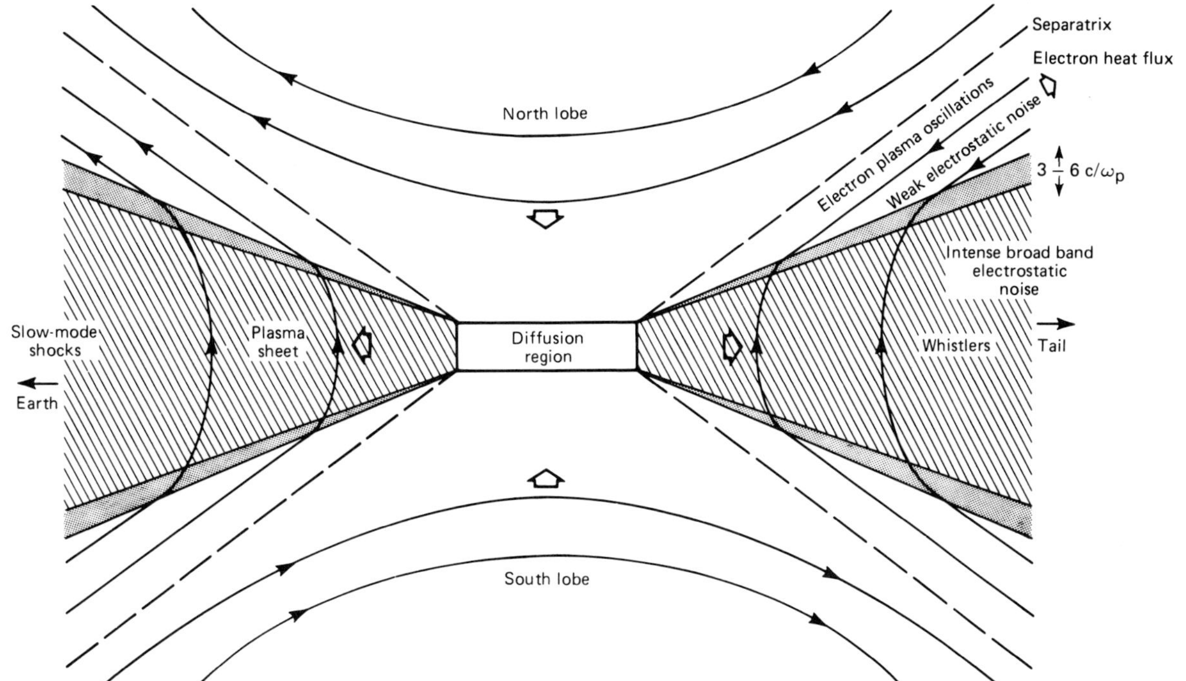

Figure 1—Illustration of the key features of the slow-mode shock observed by the ISEE 3 spacecraft. The open magnetic field lines in the north and south lobes are convected into the diffusion region for magnetic reconnection. A separatrix and a slow-mode shock are formed between the transition from the lobe to the plasma sheet. Between the separatrix and the shock, strong electron plasma oscillations, weak electrostatic noise and an electron heat flux are seen. The shock thickness is estimated to be 3 to 6 times the ion inertia length (c/ω_{pi}). Intense broadband electrostatic waves are detected at the shock. Whistler emissions are seen in the plasma-sheet region (after Feldman et al.[25] and Scarf et al.[28]).

observed.[28] The associated cross-tail electric field is found to be ~ 1–2 mV/m with an upstream Alfvén Mach number of ~ 0.1–0.3 and a decrease in the Poynting vector across the shock as $\sim (2$–$17) \times 10^{-3}$ erg/cm^2/s. Feldman et al.[26] have calculated that if the shock is assumed to extend over a surface area of 7 R$_e$ by 100 R$_e$ in each half of the magnetotail, tailward of the reconnection diffusion region, then the energy dissipation at the shock transition is about 5×10^{18}ergs/s —a value matching the energy dissipation of 3×10^{18} ergs/s for a moderate size substorm.[30] Suggestions have been made that the turbulence in the shock layer can provide dissipation through anomalous resistivity although the exact plasma instability and wave modes have yet to be identified.[28]

NONSUBSTORM-RELATED CONFIGURATIONAL CHANGES

Tail Lobe Bifurcation

During periods of strong and sustained northward interplanetary magnetic field (IMF), investigations of electric field pattern, Birkeland currents, and global auroral distributions in the polar region have suggested that the magnetotail may undergo rather drastic configurational changes between southward and northward IMF conditions.[31-35] Figure 2 illustrates schematically a proposed tail bifurcation topology.[32,36] The shaded region, which

includes the plasma sheet and its boundary layer, is depicted to extend through the usual tail lobe region to the magnetopause. The asymmetry between the northern and southern halves of the magnetotail in the loca-

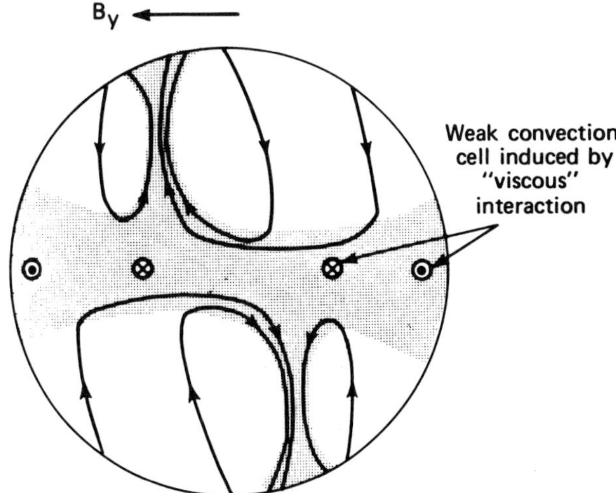

Figure 2—A dawn-dusk cross section of the magnetotail to illustrate the suggestion of bifurcation of the tail lobes and the associated convection during strongly northward IMF conditions. The asymmetry introduced by the IMF $B_y > 0$ situation is depicted here (after Potemra et al.[32]).

tion of intrusion is suggested to be related to the IMF B_y (the dawn-dusk component). When the IMF B_y is pointed duskward as shown, the magnetopause region, which gives field orientation antiparallel to the IMF, is located duskward of the cusp region.[33] The resulting plasma convection within the tail is denoted by the solid lines. In addition, there are weak convection cells induced by viscous-like interaction in the equatorial region of the plasma sheet. Further evidence of this topological change has been reported by Huang et al.[37] who presented observations indicating filamentary plasma structures intruding into the tail lobe. The thin plasma filaments may provide the source for discrete auroral particles in polar cap arcs.

It should be noted that the suggestion of tail lobe bifurcation is not universally accepted.[38-40] An alternative interpretation of the reported phenomena is that auroral activities that are usually seen along the auroral oval latitudes migrated poleward as a result of substorm activity and the polar cap is not bisected.

Tail Twisting and Flattening

Observations generally indicate an approximately circular tail cross section with a slight elongation in the north-south direction.[41,42] In addition, it is generally thought that the neutral sheet lies parallel to the ecliptic plane. However, departure from these two general characteristics are sometimes suggested by observations in the distant magnetotail. Evidence of tail twisting has been presented by Sibeck et al.[43] They have shown from ISEE 3 observations that during an interval of strong IMF B_y, the north lobe plasma mantle occurs below the ecliptic plane and the magnetopause boundary normal deviates markedly from the radial direction of the nominal tail axis. These observations are consistent with a twisted neutral sheet and the north-south dimension of the tail cross section flattened by the strong IMF B_y during that time. The sense of twisting is in agreement with the expected torque exerted on the magnetopause.[44]

SUBSTORM PHENOMENA

Magnetospheric Substorm

A magnetospheric substorm[45,46] is a transient process initiated on the nightside of the earth in which a significant amount of energy derived from the solar-wind/magnetosphere interaction is deposited in the auroral ionosphere and in the magnetosphere.[47] Manifestations of a substorm can be found on the ground in the geomagnetic field, in the ionosphere as well as in the magnetosphere. The most visible exhibition of a substorm is in the evolution of a global auroral display. The time sequence of auroral development can be regarded as a movie of the substorm disturbance in the magnetosphere projected along a magnetic field line onto the ionosphere.

A schematic illustration of a substorm from the global auroral development is given in Fig. 3. Prior to the onset of a magnetospheric substorm, auroras are distributed in an oval-shaped region known as the auroral oval[46,48]

around the magnetic pole. It is displaced about 3° toward the midnight. On a global scale, two types of auroras, discrete and diffuse, constitute the instantaneous auroral oval.[39,49] When there is very little auroral activity, the auroral oval is typically small in size. It is now recognized that this state of the auroral oval is associated usually with prolonged periods of northward IMF.

Three phases of this disturbance have been proposed, namely, growth, expansion, and recovery phases.[45,50] When the IMF turns southward, the growth phase of a substorm is considered to start. The size of the visible auroral oval starts to expand to lower latitudes. Several studies indicate that a good correlation exists between the IMF B_z (the north-south component) and the size of the auroral oval.[51,52]

The onset of the expansion phase of a substorm is marked by a dramatic change in the auroral luminosity in the night sector. A discrete auroral arc, typically the most equatorward one, suddenly brightens locally (Fig. 3a). The brightening spreads rapidly both westward toward the evening sector and eastward toward the morning sector. This is followed by discrete auroras moving rapidly poleward, creating a bulge-shaped auroral disturbance region known as the auroral bulge (Fig. 3b).

The expansion of the auroral bulge into the evening sector is led by a surge or a fold in the discrete aurora commonly referred to as a westward traveling surge, although several surges may be seen at the poleward boundary of the auroral bulge. An actual image of the auroral bulge is shown in Plate III-1, which is obtained by the ultraviolet auroral imager[53] on board the Viking spacecraft. The picture was taken at 1149:49 UT on October 15, 1986 during a substorm.[54] The auroral bulge occupies a substantial local time region—about 4 hours from about 19 to about 23 MLT (magnetic local time). There are several localized auroral intensifications that appear as bright spots on the global picture. Three bright spots occurred in the afternoon sector at 13.2, 14.5, and 16.1 MLT that are about 1.5 MLT apart. Six bright spots are found at the poleward boundary of the nightside auroral bulge at about 19.3, 19.9, 20.4, 20.9, 21.2, and 21.6 MLT that are separated by about 0.5 MLT.

The evening expansion of the surge activity is often very nonuniform,[55,56] staying practically still and then taking sudden shifts in the westward and poleward directions. Each of these shifts appears as an intensification in magnetic disturbances for a ground magnetic station located underneath the surge form.[57,58] Extension of substorm activity in the morning sector is seen by the appearance of auroral patches drifting eastward (Fig. 3c).

The poleward expansion of the auroras eventually halts and the recovery phase of a substorm begins (Fig. 3d). The auroras gradually diminish in intensity and may move equatorward. The auroral oval distribution without the deformation of the auroral bulge is slowly restored. The actual temporal sequence of global auroral substorm development has recently been obtained by snapshots every 12 min by the DE imager[56] and at every 40 s by the Viking imager.[59]

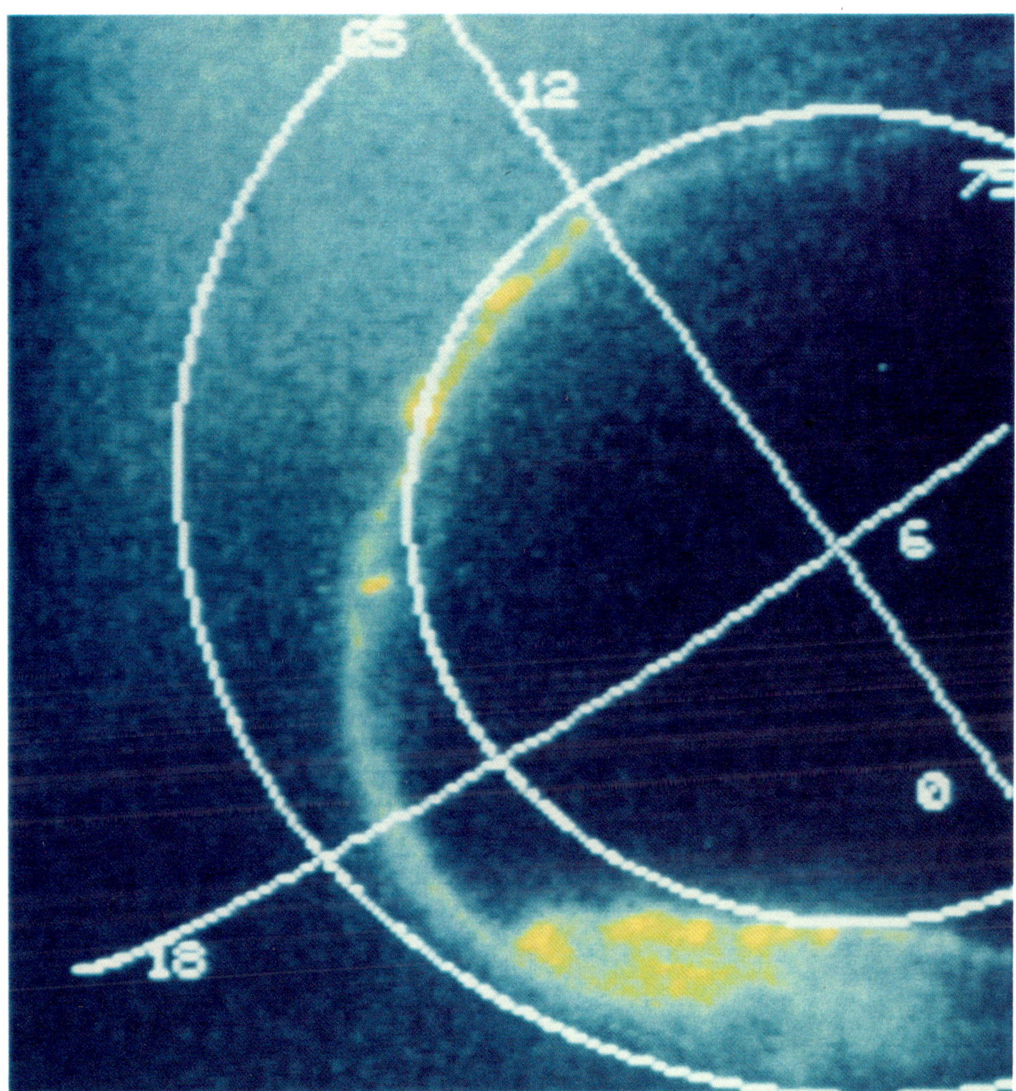

Plate III-1—Auroral distributions in the northern polar region recorded by the Viking ultraviolet auroral imager during orbit 1296 on October 15, 1986 at 1149:49 UT. The colors in the order of blue, green, yellow, and orange represent progressively higher luminosity in N_2 Lyman-Birge-Hopfield (LBH) band emission at $1400 - 1600$ Å. Magnetic meridians of 00, 06, 12, 18 MLT and magnetic latitudes of 65° and 75° are overlaid on the image. This picture shows an auroral bulge extending from ~19 to ~23 MLT with six bright spots at its poleward boundary.

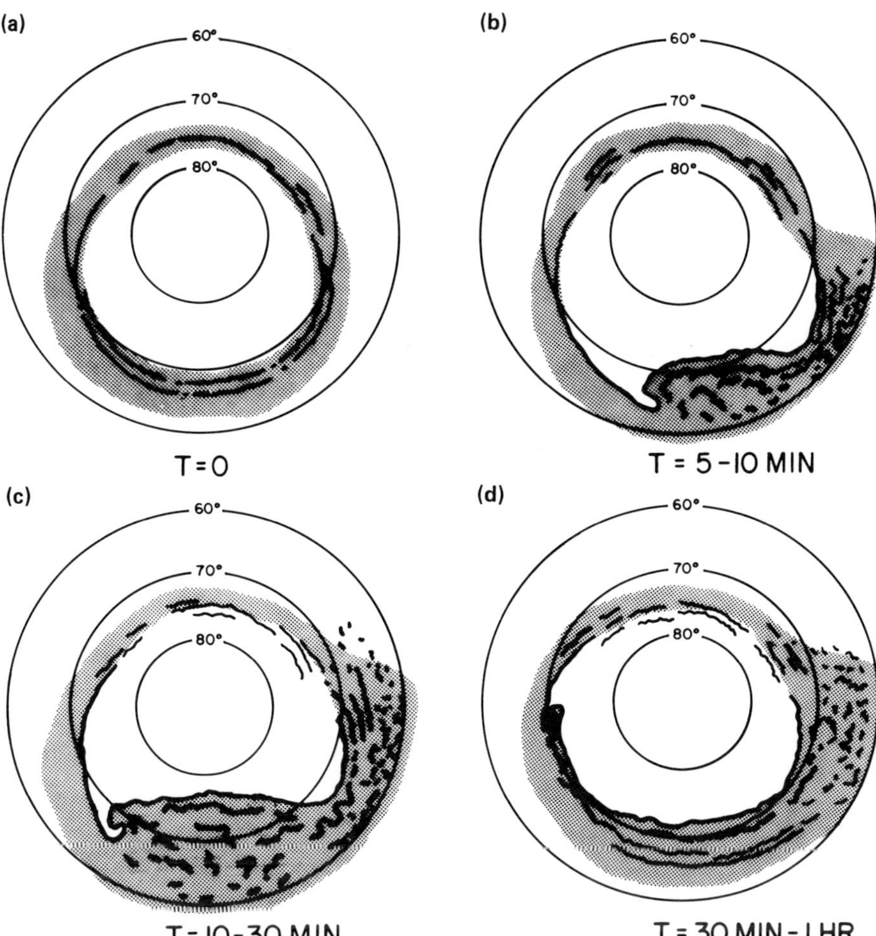

(a) T = 0

(b) T = 5 – 10 MIN

(c) T = 10 – 30 MIN

(d) T = 30 MIN – 1 HR

Figure 3—A schematic diagram to illustrate the substorm development of auroras. The time *T* is referenced to the onset of the substorm expansion phase (after Akasofu[45,46]).

Near-Earth Tail

It is convenient to organize substorm phenomena in the magnetotail by differentiating the near-earth portion (≤ 15 R_e), the midtail section (~ 15–60 R_e), and the distant region (beyond ~ 60 R_e).

Several changes are reported to take place in the near-earth magnetotail during the growth phase of the substorm.[50] The plasma sheet is reported to become thin gradually, the magnetic field becomes more stretched out into a tail-like configuration, and the energetic electrons develop an angular distribution known as cigar-shaped, i.e., the particle intensity is highest along the magnetic field line and is lowest perpendicular to it.[60] At the onset of the substorm expansion phase the near-earth portion of the plasma sheet becomes thicker. Particles are energized and/or transported rapidly to regions earthward of the presubstorm plasma sheet. Thereafter, these particles undergo adiabatic drifts expected under the magnetic and electric field configuration within the magnetosphere.[61-67] In one case study, Arnoldy[68] noted a gradual enhancement of particle flux in one energy channel prior to a nearly simultaneous flux increase in other channels. This feature is suggested as a signature of injection, existing initially over a spatially-limited local time sector.

The injection of particles is accompanied by a relaxation of the stretched tail-like magnetic field to a more dipolar-like configuration. It is also associated with enhanced broadband electrostatic waves and an induced dawn-dusk electric field of ~ 10 mV/m.[69-71] Occassionally, there is indication from auroral images that surface waves may be formed at the earthward edge of the plasma sheet.[72,73]

Mid-Tail

In the mid-tail region, an increase in the tail cross section[74] and the magnetic field magnitude are seen[75] prior to the onset of the substorm expansion phase. At the onset, plasma-sheet thinning is often observed.[76-78] The reported thinning speed is shown in Table 1. However, there are also observations during substorms in which plasma-sheet thinning is not detected, suggesting that the thinning phenomenon may be rather localized both in the radial and dawn-dusk dimensions.[78] Flapping of the plasma sheet is sometimes detected during substorm activity.[86]

Plasma-sheet thinning is not associated with compression of the plasma near the neutral-sheet region, which implies that there is a transport of plasma out of the mid-tail region. A natural question is, "Where does the

Table 1—Plasma-sheet thinning and recovery velocities.

Speed (km/s)	Reference	Remarks
Plasma-Sheet Thinning		
10 to ≥ 60	Parks et al.[79]	Single event of multiple boundary crossings
23 ± 18	Forbes et al.[80]	Statistical
Plasma Sheet Thickening		
5 to 20	Hones et al.[81,82]	Statistical
90	Buck et al.[83]	Single observation from energetic protons (0.1–1.4 MeV)
63 (Vela) 19 (IMP 4)	Hones et al.[84]	Statistical
40	Pytte et al.[58]	Statistical
11	DeCoster and Frank[85]	Single event
10 to ≥ 60	Parks et al.[79]	Single event of multiple boundary crossings
~ 30 ± 20	Forbes et al.[80]	Statistical

plasma go at the time of thinning?'' The answer to this question is rather unclear. Results from some studies[87,88] indicate that earthward plasma flow is detected more frequently than tailward flow during plasma-sheet thinning. There are also other studies[89,90] showing that plasma-sheet thinning is accompanied more by tailward flows than earthward flows. A surprising finding is that plasma flows near the neutral sheet during substorms are typically very small in spite of generally large (usually about a few hundred km/s or more) plasma flows in the plasma-sheet boundary.[91,92]

The magnetic field component has been examined extensively to infer the magnetic field configuration in the magnetotail during substorm development. Early studies have shown that at the time of the substorm expansion phase onset, the field becomes southward and remains so for about half an hour. This signature is seen over a wide dawn-dusk sector of the magnetotail, even in the tail lobe.[93] When the other components in the magnetic field are examined as well, it is found that southward B_z intervals are accompanied typically with an increase in the B_x component (the sunward-antisunward component), usually reflecting only a slight (5–10°) dip in the magnetic field vector.[94] In addition, the dawn-dusk component is often abnormally large during the southward field intervals.[95] This indicates a complex three-dimensional structure at the time of the southward field. One possibility for this three-dimensional structure is the occurrence of magnetic flux ropes that have been reported from ISEE 1 and 2 measurements.[96] In the neutral-sheet region the magnetic field is predominantly north-

ward[97] although the occurrence frequency of the southward field increases with increasing substorm activity.[98]

Occurrence of southward magnetic field in the magnetotail during plasma-sheet thinning is occasionally accompanied by tailward plasma flow based on some case studies.[99,100] On the other hand, a systematic survey of this association between field and flow over substorm intervals does not show the association to be frequent.[97] If only flows above 300 km/s are considered, then the association appears to be reasonably good.[90] However, as pointed out by Huang,[101] the number of cases constituting good correlation between southward magnetic field and tailward flow is very small, about 2% of the time.

Strong electric fields are often detected in association with thinning and expansion of the plasma sheet. The magnitude is usually 5–10 mV/m lasting typically 5–15 min.[102] The electric field is often observed in the plasma-sheet boundary layer and the associated plasma convection is typically in a direction toward the equatorial plane. On a finer time scale, larger amplitude electric fields up to about 30 mV/m lasting only fractions of a second are observed.[103,104] It is emphasized that these large electric fields are likely to be of limited dawn-dusk extent.

An increase in the density of ionospheric oxygen ions in the magnetotail is often seen about 0.5 hr after substorm expansion phase onset.[105] These particles are usually found to stream tailward and are detected at energies of several keV or even > 100 keV.[106,107] The increase in oxygen ions during active times can be large enough to make oxygen become a dominant ionic species in the magnetotail plasma.[108,109]

Near the end of the substorm expansion phase, the plasma sheet in the mid-tail region is seen to become thick again.[110] The reported thickening speed is given in Table 1. This thickening of the plasma sheet (sometimes called plasma-sheet recovery) is accompanied typically by the occurrence of fast plasma flow (from several hundred to more than a thousand km/s). Note that the recovery of the plasma sheet occurs in the substorm expansion phase and not in the substorm recovery phase. Thus plasma-sheet recovery and substorm recovery do not occur simultaneously. Figure 4 is a plot of the plasma flow during plasma-sheet thickening obtained by IMP 6 measurements.[111] The plasma flow, in general, has a large component along the magnetic field line. There is, in fact, a general trend that the larger the flow, the closer the plasma flow will be along the field line (as shown in Fig. 5). The plasma flow occurs in a thin layer about 1–2 R_e thick at the plasma-sheet boundary.[85,111] More detailed later observations from ISEE 1 and 2 indicate that the layer is a semipermanent feature of the plasma sheet and is named the plasma-sheet boundary layer.[79,88,112,113] The magnetic field at this time of plasma-sheet thickening is generally found to have a northward magnetic field component.

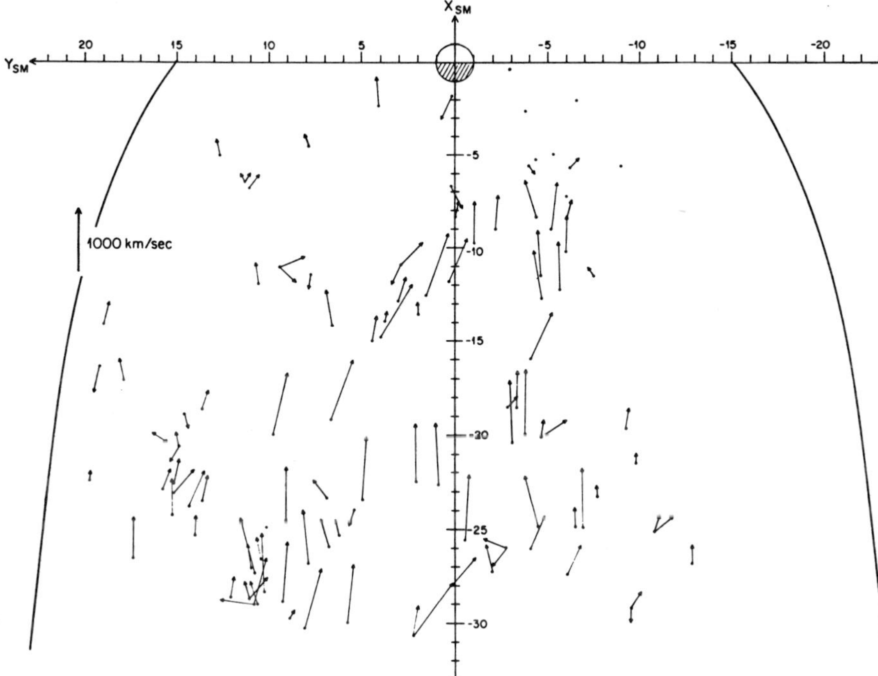

Figure 4—A flow pattern constructed from a statistical study of plasma flow during plasma-sheet recovery from the IMP 6 spacecraft. Earthward flows in the speeds of hundreds of kilometers/second to more than a thousand kilometers/second are observed in an extended region of the magnetotail during plasma-sheet thickening.

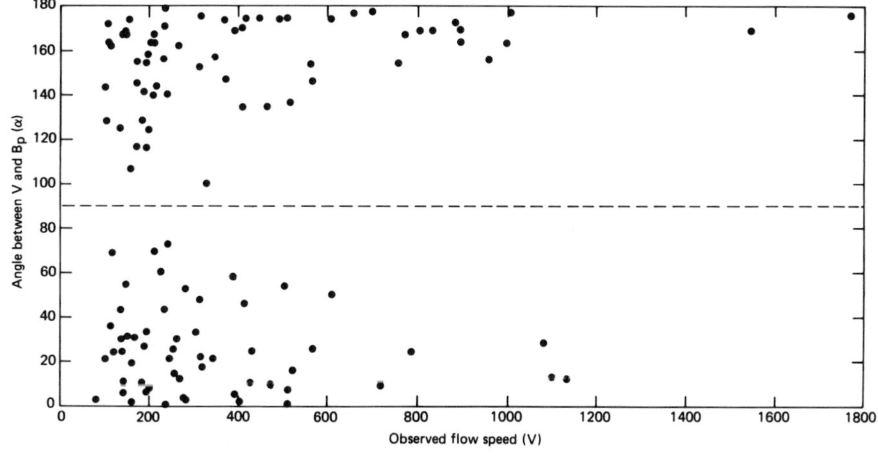

Figure 5—A plot of the angle between the plasma flow and the magnetic field versus the observed flow speed, based on measurements from IMP 6. There is a trend that the higher the speed, the closer the flow with respect to the magnetic field-line direction ($\alpha = 0°$ or $\alpha = 180°$).

Distant Tail

In the distant tail, there appears to be large-amplitude magnetopause motions during substorms.[114] Most of the distant tail activities occur in the late phase of a substorm and are recently revealed by the ISEE-3 measurements.[115-117] After about half an hour from the onset of the substorm expansion phase, observations suggest that the magnetotail cross section appears to be enlarged momentarily. In the tail lobe, the magnetic field is observed to increase in magnitude. The north-south component is first seen northward and then southward. Within the plasma sheet, enhanced tailward flow is seen. The magnetic field usually develops a northward component momentarily and then a fluctuating southward component. This is interpreted as a plasmoid 50–100 R_e long traveling downstream at 300–1000 km/s. Evidence of flux ropes are observed in the distant magnetotail that are believed to be associated with folds on the plasma sheet boundary.[118]

SUBSTORM MODELS

Plasmoid Model

A prevalent model in organizing the set of magnetotail observations during substorms is illustrated in Fig. 6.[119] In this model, southward IMF erodes the dayside magnetic flux by magnetic reconnection, and magnetic flux is built up in the magnetotail. This causes a strengthening of the magnetic field, a stretching of the field into a tail-like configuration, and plasma-sheet thinning in the near-earth region.

At substorm expansion phase onset a neutral line is formed, initially over only a narrow local time sector near the earth at about 10–15 R_e. This causes strong sunward plasma flow in the plasma sheet earthward of the neutral line. Substorm injection and rotation of the magnetic field toward a more dipolar orientation in the near-earth region are related to the onset of magnetic reconnection at the neutral line. Downstream of the neutral line location, strong tailward flow and plasma-sheet thinning are observed. Occurrence of southward B_z is also anticipated tailward of the neutral line. This process presumably results also in the diversion of the cross-tail current into the ionosphere forming the westward auroral electrojet.

The reconnection creates a large-scale magnetic island called a plasmoid that is then released down the tail when reconnection reaches magnetic field lines in the tail lobe. In the distant tail, the plasmoid is associated with a bulge in the tail cross section and is believed to be associated with a compressed magnetic field region known as the traveling compression region.[117] Recovery of the plasma sheet in the mid-tail occurs when the near-earth neutral line moves down the tail.

A large number of observations may be considered as consistent with this model, albeit with considerable variations of this simple model. Case studies conducted for the Coordinated Data Analysis Workshops 6 and 7 serve as good examples.[37,120,121] One freedom of the model is that the near-earth neutral line can be located anywhere in the downstream distance from 10–40 R_e.[122] This uncertainty in the location of the substorm neutral line precludes a definitive test for the predicted topological field change from measurements of most spacecraft surveying the magnetotail, including IMP 6, IMP 7, IMP 8, ISEE 1, and ISEE 2. Other freedoms include the provisions that the near-earth neutral line can be very localized in the dawn-dusk extent, and there may be several neutral lines. This approach adopts the viewpoint that the primary ingredient of the substorm process is a conglomeration of localized and transient activities.

On the other hand, if the primary process of a substrom is a large-scale one, then localized activities are merely secondary phenomena that may or may not occur in association with the substorm process.[123] In this case, the approach of the tearing-mode model to substorm investigation would not lead to a proper identification of the primary process for a substorm. Observations that favor the notion that a substorm is a large-

Plasma sheet configuration changes during a substorm

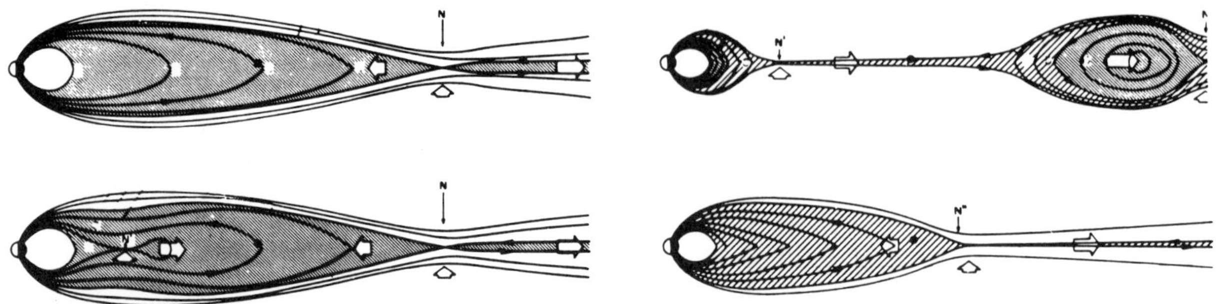

Figure 6—A schematic diagram to illustrate the plasmoid model for a substorm. The open arrows represent plasma flow direction. At the time of substorm expansion phase onset (panel b), a near-earth neutral line is formed and is subsequently ejected down the tail (panel c). The neutral line retreats to the distant tail after staying in the near-earth region for about 0.5 hr.

scale process are the wide local time coverage in the poleward expansion of auroras during substorms,[45] the east-west extent of the substorm westward electrojet,[47] the large spatial extent of the near-earth injection boundary,[63] and the occurrence of plasma sheet thinning phenomenon over a large portion of the magnetotail.[77] As seen in Plate III-1, embedded within the large-scale envelope of substorm activities in the auroral bulge are several spatially localized bright auroral spots that are rather transient events lasting for about a few minutes. It is important in the future to investigate whether these transient localized auroral features are the ionospheric images of the transient localized phenomena in the magnetotail that are regarded in this model as the key element of the substorm process. Are there bright auroral spots during the development of a small substorm? Do bright spots occur prior to the poleward expansion of the discrete auroras? Answers to these questions are important in resolving the issue of whether a substorm is a large-scale process or a conglomeration of small-scale activities.

An important step toward understanding the acceleration mechanism of charged particles in the magnetotail is to identify and investigate the acceleration regions. Such regions can be inferred from a reversal in the particle streaming direction; e.g., observations of tailward flow followed by earthward flow have been interpreted as a neutral line moving past the spacecraft.[124] However, detailed case studies of such a streaming reversal region often show features not readily explicable by the simple plasmoid model.[125] One such example is given in Plate III-2. Strong tailward streaming of energetic ions (0.29–0.5 MeV; second panel) starts at 1143 UT and continues to ~ 1201 UT. This tailward streaming is accompanied by a southward field. The streaming direction gradually changes from tailward to duskward and finally to sunward at 1209 UT. However, Fig. 7 shows that the thermal population detected by the plasma instrument (measuring from 215 eV–45 keV) was flowing tailward from 1143–1209 UT and reversed to flow sunward only at ~ 1210 UT. There is a substantial (~ 8 min) time difference between the times of reversal of tailward to earthward streaming of energetic ions and of reversal from tailward to earthward flow. There are also other features in this event that present difficulties to the plasmoid model (see Ref. 125 for details). In addition, there are other studies that yield results questioning the validity of this model or the importance of reconnection energetically in accelerating particles.[21,37,91,126,128] It is unfortunate that observations that seem to be consistent with the plasmoid model are often emphasized while difficulties with the model are often not, or even ignored.

Boundary-Layer Model

A recently proposed model alternative to the preceeding one for a substorm is the boundary-layer model.[113,129] In this model, illustrated in Fig. 8, the occasional observation of southward B_z is interpreted as a signature of field-aligned currents at the plasma-sheet boundary. This model additionally accounts for

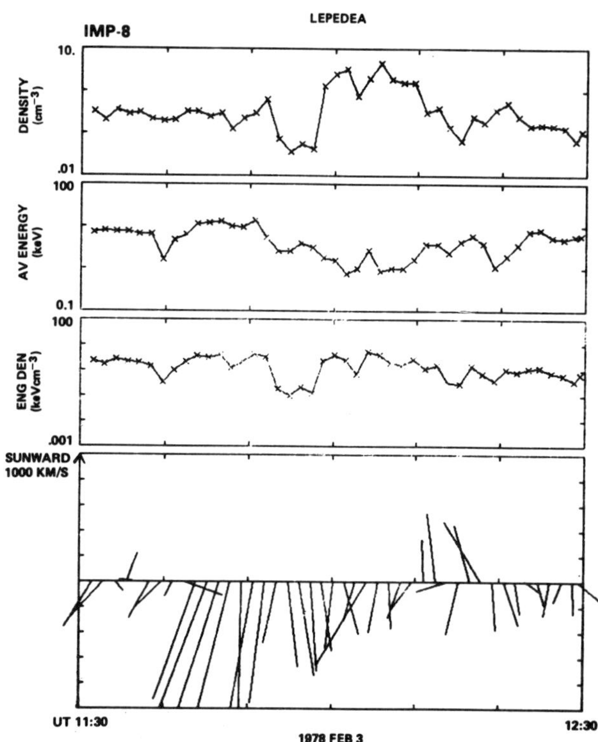

Figure 7—The plasma measurement for the interval showing tailward streaming of energetic ions reversing to earthward streaming (Plate III-1). A close examination shows that there is a large time difference between the streaming reversal of energetic ions and the flow reversal of plasma.

the large B_y component being observed frequently in association with southward B_z. The substorm is due to a Kelvin-Helmholtz instability developed in the velocity shear zone separating the central plasma sheet and the regions of the plasma-sheet boundary layer and the low-latitude boundary-layer. The localized occurrence is explained naturally as the limited local time span influenced by the field-aligned currents flowing out of the ionosphere through the magnetotail. This model incorporates the importance of the plasma-sheet boundary layer and allows a natural explanation for both directly driven and unloading aspects of substorm activity. One uncertainty about this model is how the dynamics of the plasma-sheet boundary layer can be coupled to the outer magnetosphere and form an injection boundary as observed.

Other Models

Another approach to substorm modeling has been carried out by Atkinson[130-133] in a series of papers. By making the assumptions that electrons carry the field-aligned currents and that currents are closed by gradient and curvature drifts of ions in the outer magnetosphere, he has used the observed field-aligned current pattern and convection in the ionosphere to infer regions of enhancement and depletion of plasma in magnetic flux tubes that connect to the auroral regions. The existence of regions of depletion in the outer magnetosphere leads

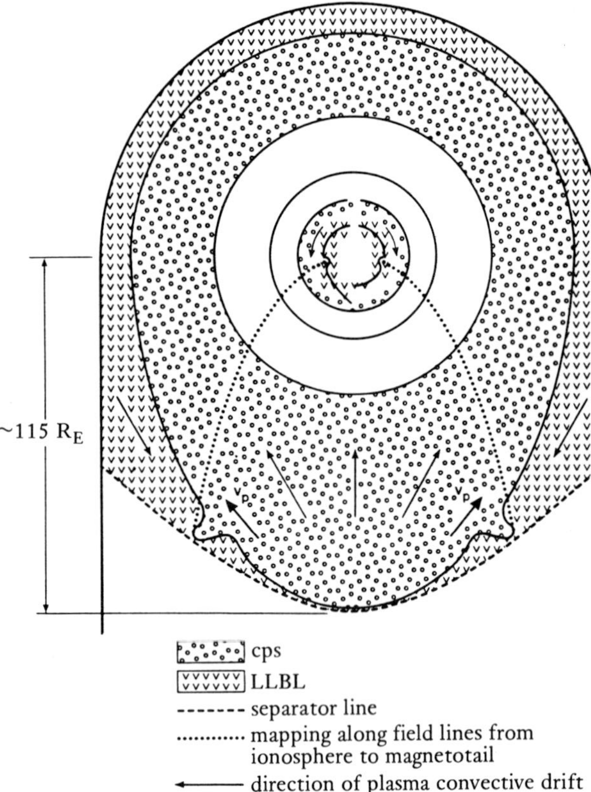

Figure 8—A schematic diagram to illustrate the boundary-layer model for a substorm. Kelvin-Helmholtz instability at the interface between the central plasma sheet and the low-latitude boundary layer is proposed to account for the field-aligned currents generated during a substorm. The field-aligned currents generate strong perturbations in the north-south and dawn-dusk components of the magnetic field (after Rostoker and Eastman[129]).

to thinning of the plasma sheet. This model bears some resemblance to the rarefaction wave model for plasma-sheet thinning proposed by Chao et al.[134] In this model, the formation of the near-earth neutral line is not a necessary part of the process but its presence does not alter the main features of the model.

Another very preliminary model that does not invoke neutral line formation as an essential feature was recently proposed by Smith et al.[135] In this "thermal catastrophe" model, continuous Alfvénic fluctuations heat the plasma-sheet boundary layer. Thermal equilibrium of the plasma-sheet boundary layer is normally set up by heat transfer to the central plasma sheet. The thermal equilibrium is offset at the substorm expansion phase onset and the temperature of the plasma-sheet boundary layer is increased abruptly. There is no indication what plasma flow, magnetic field, and plasma-sheet configurational changes will result from this model.

The substorm activity in the near-earth region has been described by the injection boundary model.[62,66,67] This assumes that at the time of substorm expansion phase onset, particles of all energies appear tailward of a sharp boundary. The particle injections seen after the expansion phase onset are consequences of drift motions of particles behind the injection front. Moore et al.[135] recently proposed that the injection front is associated with a compressional wave launched further downstream. The compression may be produced by a convection surge.[137,138] The injection phenomenon has also been described in terms of enhanced magnetospheric convection.[65] A description of the issues involved is given elsewhere.[139]

LIMITATIONS ON FLUID TREATMENT

Although the fluid description of the magnetotail dynamics is often convenient and tractable, it is not always appropriate. There are observations that clearly indicate that kinetic or single particle consideration is necessary to achieve a better understanding of the processes taking place in the magnetotail. Four examples are given below to illustrate this point.

Velocity Distribution

Although the bulk velocity of a particle distribution identifies the velocity of the center of mass of the population, it does not provide clues to the mechanism by which the bulk velocity of the population is acquired. On the other hand, a detailed examination of the velocity distribution of the population can be extremely informative. The velocity distribution of fast earthward-flowing plasma during plasma-sheet recovery was first presented by DeCoster and Frank[85] who have shown that the velocity distribution has a crescent shape and is not a convecting Maxwellian. Later studies indicate that the shape of the velocity distribution could be reproduced well by particle acceleration in the current sheet.[140] Eastman et al.[112] have discussed further how different acceleration mechanisms can be differentiated by a close examination of the shape of the velocity distribution.

Cleft Ion Fountain

Observations have indicated that the ionosphere is a continuous source of particles to the polar magnetosphere and nightside magnetotail (e.g., Refs. 141-144). Ionospheric ions are often accelerated at low altitudes (e.g., Refs. 136 and 145); some of them acquire more than the escape velocity and subsequently reach the high altitude magnetosphere. The geomagnetic field acts like a giant mass spectrometer[146] and ions with different velocities are separated spatially as they move up the field lines and form a source of particle population in the magnetotail. This effect has been successfully modeled with the upwelling ions treated as single particles.[147-149]

Energetic Ion Beams

The plasma-sheet boundary layer exhibits many dynamic features. At energies higher than the thermal energies, the behavior of these particles often indicates that their motions are consistent with no significant modification from any collective processes. Such an effect is

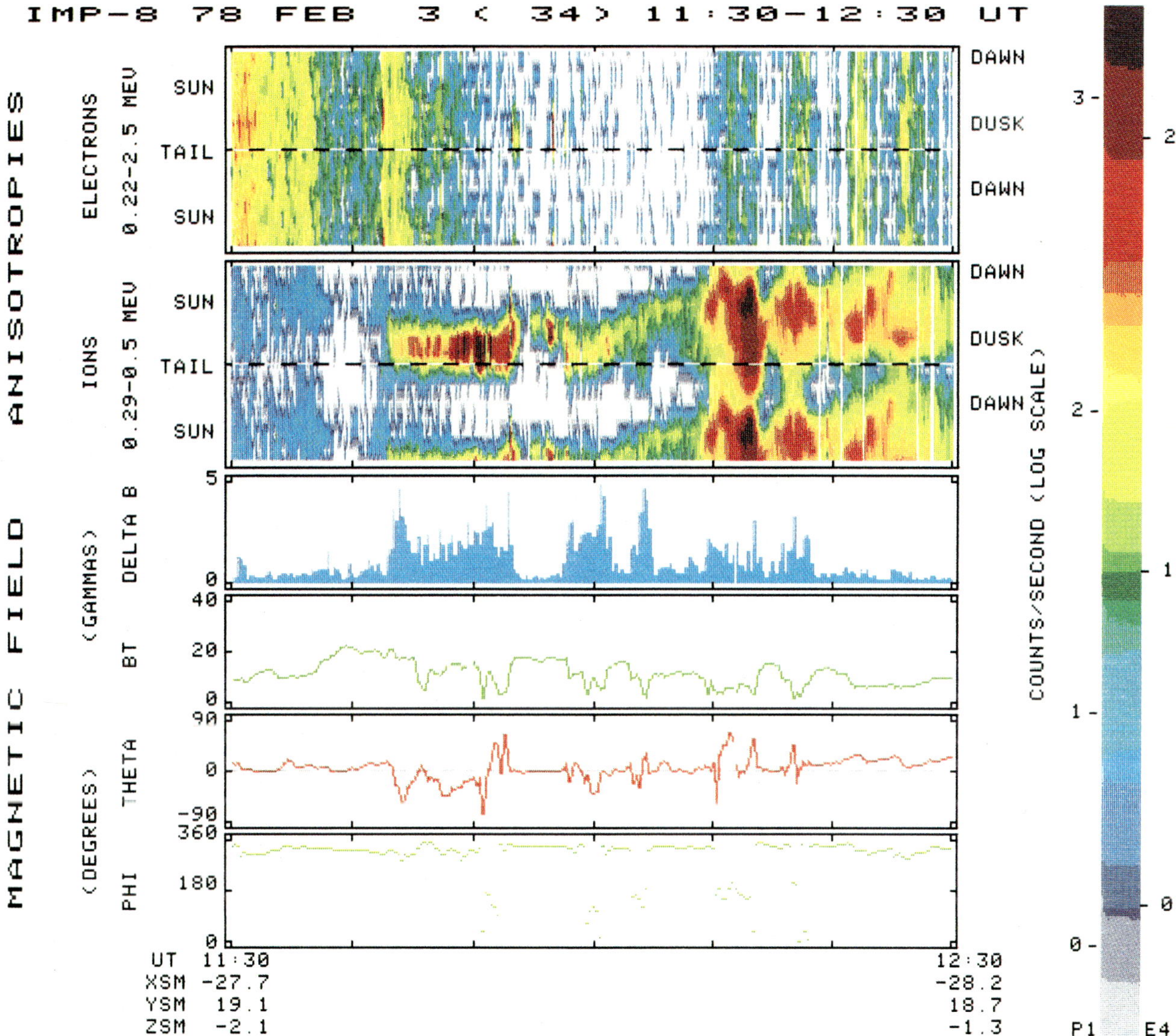

Plate III-2—Anisotropies of energetic electrons (first panel) and energetic ions (second panel) at 10.24 s resolution and magnetic field measurements at 15.36 s averaging intervals from IMP 8 in the magnetotail. For the anisotropy spectrogram, the ordinate is the azimuthal angle near the ecliptic plane. Directional intensities of particles are color coded with blue, green, yellow, orange, and red representing progressively higher intensity, as defined by the color bar at the right. Angular measurements of the particles are repeated to provide visual continuity of intensity variation in the sunward direction. From the third panel down, the magnetic field parameters on the same spacecraft are the standard deviation of magnetic field measurements within the 15.36 s averaging interval (DELTA B), the magnetic field magnitude (BT), the latitude angle (THETA), and the azimuthal angle (PHI) in solar magnetospheric coordinates. Tailward streaming of energetic ions reverses gradually to earthward streaming from ~1201–1209 UT. Intermittent tailward streaming of energetic electrons and southward magnetic field are seen during the interval of tailward streaming of energetic ions.

well documented by Williams,[150] who, by examining the arrival times of ions at different energies and pitch angles, has been able to infer the source location of the observed energetic particles in the near-earth plasma-sheet boundary layer to be at about 80–100 R_e downstream. Later observations of energetic particle phenomena at the plasma-sheet boundary have shown that velocity dispersion effects are very common. Möbius et al.[147] have presented data to show the presence of a thin layer of energetic alpha particles inside a layer of energetic protons. This feature again can be readily understood by invoking a velocity filter effect (imposed by a convection electric field) on single particle motions.

Counterstreaming

The relationship between the motion of the thermal population and the motion of the energetic particles is often quite complex in the magnetotail. In some cases, energetic particles can be treated as a high energy tail of the thermal population[152,153] whereas in other cases, the energetic particles appear to be unrelated to the thermal particles.[125,150,154]

In addition, ionospheric ions are accelerated up along magnetic field lines to the plasma sheet, particularly during substorm intervals.[105] The number density of ionospheric oxygen ions can reach values quite comparable to the hot plasma density in the plasma sheet and its boundary layer. The upward accelerated oxygen is streaming tailward. Since the mass of oxygen is sixteen times that of protons, the bulk velocity of the entire population may be dominated by the tailward flowing oxygen in spite of an earthward moving component of the hot plasma population.[155] Thus, a single fluid treatment of the situation may be easily misinterpreted as the hot plasma streaming tailward.

CLOSING REMARKS

A brief overview on the fluid aspects of the magnetotail dynamics is presented. It is clear that there is a host of dynamic phenomena in the magnetotail, some of which are related to the substorm process and some are not. The intent here is to provide a bird's eye view on the subject and not to engage in a full description of all observed magnetotail phenomena for which other review papers would be more appropriate.[30,101,127,156]

The quest for understanding substorm dynamics in the magnetotail has dominated our research effort, and a proper perspective is therefore important in making progress in this discipline. In assessing the stage of development for the topic of solar-wind/magnetosphere coupling, Siscoe[157] has compared the development of this field with meteorology and noted a number of parallel evolutions. He has emphasized that if we carry the analogy with meteorology development to our effort in achieving a successful account of the substorm process, we are in the phase that requires synthesis of substorm models rather than dogmatically adopting a single prevalent model, as appropriate. The identification of all the essential features for a substorm is crucial for the final stage of mathematical formulation of the substorm process.

Finally, it is important to note that the magnetotail is not a closed system. The boundary conditions imposed by the solar wind and the ionosphere play key roles in many magnetotail phenomena. There is clear observational evidence that the solar wind, particularly its embedded magnetic field, influences the structure and dynamics of the magnetotail. The ionosphere definitely cannot be ignored as a particle source for the magnetotail plasma. It may also play an active role in initiating substorm activities through magnetosphere-ionosphere coupling.

ACKNOWLEDGMENT—This work is supported by the Atmospheric Sciences Section of the National Science Foundation, grant ATM-8611354 to The Johns Hopkins University. This work is carried out to take the place of an undelivered manuscript from the Kick-Off Speaker on this topic in the Chapman Conference on Magnetotail Physics.

REFERENCES

[1] A. T. Y. Lui, "Roadmap to Magnetotail Domains" (this volume, 1987).

[2] V. M. Vasyliunas, "A Survey of Low-Energy Electrons in the Evening Sector of the Magnetosphere with OGO 1 and OGO 3," *J. Geophys. Res.* **73**, 2839 (1968).

[3] F. M. Ipavich and M. Scholer, "Thermal and Suprathermal Protons and Alpha Particles in the Earth's Plasma Sheet," *J. Geophys. Res.* **88**, 150 (1983).

[4] A. Hasegawa, K. Mima, M. Duong-Nam, "Plasma Distribution Function in Superthermal Radiation Field," *Phys. Rev. Lett.* **54**, 2608 (1985).

[5] G. M. Erickson and R. A. Wolf, "Is Steady Convection Possible in the Earth's Magnetotail?" *Geophys. Res. Lett.* **1**, 897 (1980).

[6] K. Schindler and J. Birn, "Self-Consistent Theory of Time-Dependent Convection in the Earth's Magnetotail," *J. Geophys. Res.* **87**, 2263 (1982).

[7] A. T. Y. Lui and A. Hasegawa, "Implications of a Steady-State Magnetospheric Convection," *Planet Space Sci.* **34**, 315 (1986).

[8] E. W. Hones, Jr., J. Birn, S. J. Bame, and C. T. Russell, "New Observations of Plasma Vortices and Insights into Their Interpretation," *Geophys. Res. Lett.* **10**, 674 (1983).

[9] M. A. Saunders, D. J. Southwood, E. W. Hones, Jr., and C. T. Russell, "A Hydromagnetic Vortex Seen by ISEE 1 and 2," *J. Atm. Terr. Phys.* **43**, 927 (1981).

[10] M. A. Saunders, D. J. Southwood, T. A. Fritz, and E. W. Hones, Jr., "Hydromagnetic Vortices – I. The 11 December, 1977 Event," *Planet. Space Sci.* **31**, 1099 (1983).

[11] M. A. Saunders, D. J. Southwood, and E. W. Hones, Jr., "Hydromagnetic Vortices – II. Further Dawnside Events," *Planet. Space Sci.* **31**, 1117 (1983).

[12] T. R. Sanderson, P. W. Daly, K.-P. Wenzel, E. W. Hones, Jr., and S. J. Bame, "Energetic Ion Observations of a Large Scale Vortex in the Distant Geotail," *Geophys. Res. Lett.* **11**, 1094 (1984).

[13] D. G. Sibeck, J. A. Slavin, and E. J. Smith, "ISEE 3 Magnetopause Crossings: Evidence for the Kelvin-Helmholtz Instability," *Proc. Chapman Conf. on Magnetotail Physics, Oct 28-31, 1985*, A. T. Y. Lui, ed., The Johns Hopkins University Applied Physics Laboratory (1985).

[14] N. F. Ness, "The Earth's Magnetic Tail," *J. Geophys. Res.* **70**, 2989 (1965).

[15] D. J. McComas, C. T. Russell, R. C. Elphic, and S. J. Blame, "The Near-Earth Cross-Tail Current Sheet: Detailed ISEE 1 and 2 Case Studies," *J. Geophys. Res.* **91**, 4287 (1986).

[16] J. D. Mihalov, D. S. Colburn, R. G. Currie, and C. P. Sonnett, "Configuration and Reconnection of the Geomagnetic Tail," *J. Geophys. Res.* **73**, 943 (1968).

[17] T. W. Speiser and N. F. Ness, "The Neutral Sheet in the Geomagnetic Tail: Its Motion, Equivalent Currents, and Field Line Connection Through It," *J. Geophys. Res.* **72**, 131 (1967).

[18] K. Schindler and N. F. Ness, "Internal Structure of the Geomagnetic Neutral Sheet," *J. Geophys. Res.* **77**, 91 (1972).

[19] T. W. Speiser, "Magnetospheric Current Sheets," *Radio Sci.* **8**, 973 (1973).

[20] S. B. Bowling, "Transient Occurrence of Magnetic Loops in the Magnetotail," *J. Geophys. Res.* **80**, 4741 (1975).

[21] A. T. Y. Lui, and C.-I. Meng, "Relevance of Southward Magnetic Fields in the Neutral Sheet to Anisotropic Distribution of Energetic Electrons and Substorm Activity," *J. Geophys. Res.* **84**, 5817 (1979).

[22] J. W. Dungey, "Interplanetary Magnetic Field and the Auroral Zones," *Phys. Res. Lett.* **6**, 47 (1961).

[23] H. E. Petschek, "Magnetic Field Annihilation," AAS-NASA Symposium on the Physics of Solar Flares, NASA Spec. Pub. SP-50, p. 425 (1964).

[24] V. M. Vasyliunas, "Theoretical Models of Magnetic Field Line Merging, 1," *Rev. Geophys. Space Phys.,* **13**, 303 (1975).

[25] W. C. Feldman, D. N. Baker, S. J. Bame, J. Birn, E. W. Hones, Jr., S. J. Schwartz, and R. L. Tokar, "Power Dissipation at Slow-Mode Shocks in the Distant Geomagnetic Tail," *Geophys. Res. Lett.* **11**, 1058 (1984).

[26] W. C. Feldman, S. J. Schwartz, S. J. Bame, D. N. Baker, J. Birn, J. T. Gosling, E. W. Hones, Jr., D. J. McComas, J. A. Slavin, E. J. Smith, and R. D. Zwickl, "Evidence for Slow-Mode Shocks in the Deep Geomagnetic Tail," *Geophys. Res. Lett.* **11**, 599 (1984).

[27] E. J. Smith, J. A. Slavin, B. T. Tsurutani, W. C. Feldman, and S. J. Bame, "Slow Mode Shocks in the Earth's Magnetotail: ISEE 3," *Geophys. Res. Lett.* **11**, 1054, (1984).

[28] F. L. Scarf, F. V. Coroniti, C. F. Kennel, E. J. Smith, J. A. Slavin, B. T. Tsurutani, S. J. Bame, and W. C. Feldman, "Plasma Wave Spectra Near Slow Mode Shocks in the Distant Magnetotail," *Geophys. Res. Lett.* **11**, 1050 (1984).

[29] W. C. Feldman, D. N. Baker, S. J. Bame, J. Birn, T. Gosling, E. W. Hones, Jr., S. J. Schwartz, and R. D. Zwickl, "Slow Mode Shocks: A Semi-Permanant Feature of the Distant Geomagnetic Tail," *J. Geophys. Res.* **90**, 233 (1985).

[30] S.-I. Akasofu, *Physics of Magnetospheric Substorms,* D. Reidel Pub. Co., Dordrecht-Holland (1977).

[31] L. J. Zanetti, T. A. Potemra, T. Iijima, W. Baumjohann, and P. F. Bythrow, "Ionospheric and Birkeland Current Distributions of Northward Interplanetary Magnetic Field: Inferred Polar Convection," *J. Geophys. Res.* **89**, 7453 (1984).

[32] T. A. Potemra, L. J. Zanetti, P. F. Bythrow, A. T. Y. Lui, and T. Iijima, "B_y-Dependent Convection Patterns During Northward Interplanetary Magnetic Field," *J. Geophys. Res.* **89**, 9753 (1984).

[33] N. U. Crooker, "Dayside Merging and Cusp Geometry," *J. Geophys. Res.* **84**, 951 (1979).

[34] N. U. Crooker, "The Magnetospheric Asymmetries Associated with the y-Component of the IMF," *Planet. Space Sci.* **29**, 79 (1981).

[35] L. A. Frank, J. D. Craven, C. Y. Huang, K. L. Ackerson, and E. J. Smith, "Observations Pertaining to the Character of Theta Auroras," *Eos* **64**, 300 (1983).

[36] L. A. Frank, J. D. Craven, D. A. Gurnett, S. D. Shawhan, D. R. Weimer, J. L. Burch, J. D. Winningham, C. R. Chappell, J. H. Waite, R. A. Heelis, N. C. Maynard, M. Sugiura, W. K. Peterson, and E. G. Shelley, "The Theta Aurora," *J. Geophys. Res.* **91**, 3177 (1986).

[37] C. Y. Huang, L. A. Frank, W. K. Peterson, D. J. Williams, W. Lennartsson, D. G. Mitchell, R. C. Elphic, and C. T. Russell, "Filamentary Structures in the Magnetotail Lobes," *J. Geophys. Res.* (submitted 1986).

[38] C.-I. Meng, "Polar Cap Arcs and the Plasma Sheet," *Geophys. Res. Lett.* **8**, 273 (1981).

[39] J. S. Murphree, C. D. Anger, and L. L. Cogger, "The Instantaneous Relationship Between Polar Cap and Oval Auroras at Times of Northward Interplanetary Magnetic Field," *Can. J. Phys.* **60**, 349 (1982).

[40] C.-I. Meng and K. Makita, "Dynamic Variations of the Polar Cap," in *Solar Wind/Magnetosphere Coupling,* Y. Kamide and J. A. Slavin, eds., Terra Scientific Pub. Co., p. 605 (1986).

[41] K. W. Behannon, "Mapping of the Earth's Bow Shock and Magnetic Tail by Explorer 33," *J. Geophys. Res.* **73**, 907 (1968).

[42] J. A. Slavin, B. T. Tsurutani, E. J. Smith, D. E. Jones, and D. G. Sibeck, "Average Configuration of the Distant ($>200\ R_e$) Magnetotail: Initial ISEE 3 Magnetic Field Results," *Geophys. Res. Lett.* **10**, 973 (1983).

[43] D. G. Sibeck, G. L. Siscoe, J. A. Slavin, E. J. Smith, B. T. Tsurutani and R. P. Lepping, "The Distant Magnetotail's Response to a Strong Interplanetary Magnetic Field B_y: Twisting, Flattening, and Field Line Bending," *J. Geophys. Res.* **90**, 4011 (1985).

[44] S. W. H. Cowley, "Asymetry Effects Associated with the x-Component of the IMF in a Magnetically Open Magnetosphere," *Planet. Space. Sci.* **29**, 809 (1981).

[45] S.-I. Akasofu, "The Development of the Auroral Substorm," *Planet. Space Sci.* **12**, 273 (1964).

[46] S.-I. Akasofu, *Polar and Magnetospheric Substorms,* D. Reidel Pub. Co., Dordrecht, Holland (1968).

[47] G. Rostoker, S.-I. Akasofu, J. Foster, R. A. Greenwald, Y. Kamide, K. Kawasaki, A. T. Y. Lui, R. L. McPherron, and C. T. Russell, "Magnetospheric Substorms—Definition and Signatures," *J. Geophys. Res.* **85**, 1663 (1980).

[48] Y. I. Feldstein and G. V. Starkov, "Dynamics of Auroral Belt and Polar Geomagnetic Disturbances," *Planet. Space Sci.* **15**, 209 (1967).

[49] A. T. Y. Lui and C. D. Anger, A Uniform Belt of Diffuse Auroral Emission Seen by the ISIS 2 Scanning Photometer," *Planet. Space Sci.* **21**, 799 (1973).

[50] R. L. McPherron, "Substorm Related Changes in the Geomagnetic Tail: The Growth Phase," *Planet. Space Sci.* **20**, 1521 (1972).

[51] R. H. Holzworth and C.-I. Meng, "Mathematical Representation of the Auroral Oval," *Geophys. Res. Lett.* **2**, 377 (1975).

[52] Y. Kamide and J. D. Winningham, "A Statistical Study of Instantaneous Nightside Auroral Oval: The Equatorward Boundary of Electron Precipitation as Observed by ISIS 1 and 2 Satellites," *J. Geophys. Res.* **82**, 5573 (1977).

[53] C. D. Anger, J. S. Murphree, A. Valence-Jones, R. A. King, A. L. Broa, L. L. Cogger, F. C. Crentzberg, R. L. Gattinger, G. Gustafsson, F. R. Harris, J. W. Haslett, E. J. Llewellyn, J. C. McConnell, D. J. McEwen, E. H. Richardson, G. Rostocker, W. R. Sandel, G. G. Shepherd, D. Venkatesan, D. D. Wallis, and G. Witt, "Scientific Results from the Viking Ultraviolet Imager: An Introduction," *Geophys. Res. Lett.* (in press, 1987).

[54] A. T. Y. Lui, D. Venketesan, G. Rostoker, J. S. Murphree, C. D. Anger, L. L. Cogger, and T. A. Potemra, "Dayside Auroral Intensifications During an Auroral Substorm," *Geophys. Res. Lett.* (in press, 1987).

[55] W. G. Tighe and G. Rostoker, "Characteristics of Westward Travelling Surges During Magnetospheric Substorms," *J. Geophys.* **50**, 51 (1981).

[56] J. D. Craven and L. A. Frank, "The Temporal Evolution of a Small Auroral Substorm as Viewed from High Altitudes with Dynamics Explorer 1," *Geophys. Res. Lett.* **12**, 465 (1985).

[57] R. G. Wiens and G. Rostoker, "Characteristics of the Development of the Westward Electrojet During the Expansive Phase of Magnetospheric Substorms," *J. Geophys. Res.* **80**, 2109 (1975).

[58] T. Pytte, R. L. McPherron, M. G. Kivelson, H. I. West, and E. W. Hones, Jr., "Multiple-Satellite Studies of Magnetospheric Substorms: Radial Dynamics of the Plasma Sheet," *J. Geophys. Res.* **81**, 5921 (1976).

[59] G. Rostocker, A. Valence-Jones, R. L. Gattinger, C. D. Anger, and J. S. Murphree, "The Development of the Substorm Expansive Phase: The 'Eye' of the Substorm," *Geophys. Res. Lett.* (in press, 1987).

[60] D. N. Baker, E. W. Hones, Jr., P. R. Higbie, R. D. Belian, and P. Stauning, "Global Properties of the Magnetosphere During a Substorm Growth Phase: A Case Study," *J. Geophys. Res.* **86**, 8941 (1981).

[61] S. E. DeForest and C. E. McIlwain, "Plasma Clouds in the Magnetosphere," *J. Geophys. Res.* **76**, 3587 (1971).

[62] C. E. McIlwain, "Substorm Injection Boundaries," in *Magnetospheric Physics,* B. M. McCormac, ed., D. Reidel, Hingham, Mass., p. 143 (1974).

[63] B. H. Mauk and C. E. McIlwain, "Correlation of Kp with the Substorm-Injected Plasma Boundary," *J. Geophys. Res.* **79**, 3193 (1974).

[64] A. Konradi, C. L. Senar, and T. A. Fritz, "Substorm-Injected Protons and Electrons and the Injection Boundary Model," *J. Geophys. Res.* **80**, 543 (1975).

[65] M. G. Kivelson, S. M. Kaye, and D. J. Southwood, "The Physics of Plasma Injection Events," in *Dynamics of the Magnetosphere*, S.-I. Akasofu, ed., D. Reidel, Hingham, Mass., p. 385 (1979).

[66] B. H. Mauk and C.-I. Meng, "Characterization of Geostationary Particle Signatures Based on the 'Injection Boundary' Model," *J. Geophys. Res.* **88**, 3055 (1983).

[67] B. H. Mauk and C.-I. Meng, "Dynamical Injections as the Source of Near Geostationary Quiet Time Particle Spatial Boundaries," *J. Geophys. Res.* **88**, 10011 (1983).

[68] R. L. Arnoldy, "Fine Structure and Pitch Angle Dependence of Synchronous Orbit Electron Injections," *J. Geophys. Res.* **91**, 13411 (1986).

[69] F. L. Scarf, R. W. Fredericks, C. F. Kennel, and F. V. Coroniti, "Satellite Studies of Magnetospheric Substorms on August 15, 1968: 8, Ogo 5 Plasma Wave Observations," *J. Geophys. Res.* **78**, 3119 (1973).

[70] G. G. Shepherd, R. Bostrom, H. Derblom, C.-G. Fälthammar, R. Gendrin, K. Kaila, A. Korth, A. Pedersen, R. Pellinen, and G. Wrenn, "Plasma and Field Signatures of Poleward Propagating Auroral Precipitation Observed at the Foot of the GEOS-2 Field Line," *J. Geophys. Res.* **85**, 4587 (1980).

[71] T. L. Aggson, J. P. Heppner, and N. C. Maynard, "Observations of Large Magnetospheric Electric Fields During the Onset Phase of a Substorm," *J. Geophys. Res.* **88**, 3981 (1983).

[72] A. T. Y. Lui, C.-I. Meng, and S. Ismail, "Large Amplitude Undulation on the Equatorward Boundary of the Diffuse Aurora in the Afternoon-Evening Sector," *J. Geophys. Res.* **87**, 2385 (1982).

[73] A. Roux, S. Perrault, A. Pedersen, R. Delliness, and D. Rodgers, "Instability of the Plasmasheet in Relation to Substorms," *Phys. Scripta* (submitted, 1986).

[74] K. Maezawa, "Magnetotail Boundary Motion Associated with Geomagnetic Substorms," *J. Geophys. Res.* **80**, 3543 (1975).

[75] D. H. Fairfield, R. P. Lepping, E. W. Hones, Jr., S. J. Bame, and J. R. Asbridge, "Simultaneous Measurements of Magnetotail Dynamics by IMP Spacecraft," *J. Geophys. Res.,* **86**, 1396 (1981).

[76] E. W. Hones, Jr., J. R. Asbridge, S. J. Bame, and I. B. Strong, "Outward Flow of Plasma in the Magnetotail Following Geomagnetic Bays," *J. Geophys. Res.* **72**, 5879 (1967).

[77] A. T. Y. Lui, E. W. Hones, Jr., D. Venkatesan, S.-I. Akasofu, and S. J. Bame, "Complete Dropouts at Vela Satellites During Thinning of the Plasma Sheet," *J. Geophys. Res.* **80**, 4649 (1975).

[78] J. Dandouras, H. Rème, A. Saint-Marc, J. A. Savaud, G. K. Parks, K. A. Anderson, and R. P. Lin, "A Statistical Study of Plasma Sheet Dynamics Using ISEE 1 and 2 Energetic Particle Flux Data," *J. Geophys. Res.* **91**, 6861 (1986).

79 G. K. Parks, D. Leaver, C. S. Lin, K. A. Anderson, R. P. Lin, and H. Rème, "ISEE 1/2 Particle Observations of Outer Plasma Sheet Boundary," *J. Geophys. Res.* **84**, 6471 (1979).

80 T. G. Forbes, E. W. Hones, S. J. Bame, J. R. Asbridge, G. Paschmann, N. Sckopke, and C. T. Russell, "Evidence for the Tailward Retreat of a Magnetic Neutral Line in the Magnetotail During Substorm Recovery," *Geophys. Res. Lett.* **8**, 261 (1981).

81 E. W. Hones, Jr., S.-I. Akasofu, P. Perreault, S. J. Bame, and S. Singer, "Poleward Expansion of the Auroral Oval and Associated Phenomena in the Magnetotail During Auroral Substorms, 1," *J. Geophys. Res.* **75**, 7060 (1970).

82 E. W. Hones, Jr., S.-I. Akasofu, S. J. Bame, and S. Singer, "Poleward Expansion of the Auroral Oval and Associated Phenomena in the Magnetotail During Auroral Substorms, 2," *J. Geophys. Res.* **76**, 4402 (1971).

83 R. M. Buck, H. I. West, Jr., and R. G. D'Arcy, Jr., "Satellite Studies of Magnetospheric Substorms on August 15, 1968, 7. Ogo 5 Energetic Proton Observations – Spatial Boundaries," *J. Geophys. Res.* **78**, 3103 (1973).

84 E. W. Hones, Jr., J. R. Asbridge, and S. J. Bame, "Time Variations of the Magnetotail Plasma Sheet at 18 R_e Determined from Concurrent Observations by a Pair of Vela Satellites," *J. Geophys. Res.* **76**, 4402 (1971).

85 R. J. DeCoster and L. A. Frank, "Observations Pertaining to the Dynamics of the Plasma Sheet," *J. Geophys. Res.* **84**, 5099 (1979).

86 T. Toichi and T. Miyazaki, "Flapping Motions of the Tail Plasma Sheet Induced by the Interplanetary Magnetic Field Variations," *Planet. Space Sci.* **24**, 147 (1976).

87 A. T. Y. Lui, C.-I. Meng, and S.-I. Akasofu, "Search for the Magnetic Neutral Line in the Near-Earth Plasma Sheet, 2. Systematic Study of IMP 6 Magnetic Field Observations," *J. Geophys. Res.* **82**, 1547 (1977).

88 T. E. Eastman, "Boundary Layers of the Earth's Outer Magnetosphere," in *Magnetic Reconnection*, E. W. Hones, Jr., ed. (1984).

89 E. W. Hones, Jr. and K. Schindler, "Magnetotail Plasma Flow During Substorms: A Survey with IMP 6 and IMP 8 Satellites," *J. Geophys. Res.* **84**, 7155 (1979).

90 H. Hayakawa, A. Nishida, E. W. Hones, Jr., and S. J. Bame, "Statistical Characteristics of Plasma Flow in the Magnetotail," *J. Geophys. Res.* **87**, 277 (1982).

91 C. Y. Huang, L. A. Frank, and T. E. Eastman, "Plasma Flows Near the Neutral Sheet of the Magnetotail," *Proc. Chapman Conf. on Magnetotail Physics, Oct. 28-31, 1985*, A.T.Y. Lui, ed., The Johns Hopkins University Applied Physics Laboratory (1985).

92 C. Y. Huang, L. A. Frank, and T. E. Eastman, "Plasma Flows Near the Neutral Sheet of the Magnetotail" (this volume, 1987).

93 A. Nishida and N. Nagayama, "Synoptic Survey for the Neutral Line in the Magnetotail During the Substorm Expansion Phase," *J. Geophys. Res.* **78**, 3782 (1973).

94 A. T. Y. Lui, C.-I. Meng, and S.-I. Akasofu, "Search for the Magnetic Neutral Line in the Near-Earth Plasma Sheet, 1. Critical Reexamination of Earlier Studies on Magnetic Field Observations," *J. Geophys. Res.* **81**, 5934 (1976).

95 S.-I. Akasofu, A. T. Y. Lui, C.-I. Meng, and M. Haurwitz, "Need for a Three-Dimensional Analysis of Magnetic Fields in the Magnetotail During Substorms," *Geophys. Res. Lett.* **5**, 283 (1978).

96 R. C. Elphic, C. A. Cattell, K. Takahashi, S. J. Bame, and C. T. Russell, "ISEE 1 and 2 Observations of Magnetic Flux Ropes in the Magnetotail: FTE's in the Plasma Sheet?" *Geophys. Res. Lett.* **13**, 648 (1986).

97 A. T. Y. Lui, L. A. Frank, K. L. Ackerson, C.-I. Meng, and S.-I. Akasofu, "Plasma Flows and Magnetic Field Vectors in the Plasma Sheet During Substorms," *J. Geophys. Res.* **83**, 3849 (1978).

98 M. N. Caan, D. H. Fairfield, and E. W. Hones, Jr., "Magnetic Fields in Flowing Magnetotail Plasmas and Their Significance for Magnetic Reconnection," *J. Geophys. Res.* **84**, 1971 (1979).

99 L. A. Frank, K. L. Ackerson, and R. P. Lepping, "On the Hot Tenuous Plasmas, Fireballs, and Boundary Layers in the Earth's Magnetotail," *J. Geophys. Res.* **81**, 5859 (1976).

100 A. Nishida, H. Hayakawa, and E. W. Hones, Jr., "Observed Signatures of Reconnection in the Magnetotail," *J. Geophys. Res.* **86**, 1370 (1981).

101 C. Y. Huang, "Quadrennial Review of the Magnetotail," *Rev. Geophys. Space Phys.* (in press, 1986).

102 A. Pedersen, C. A. Cattell, C.-G. Fälthammar, K. Knott, P.-A. Lindqvist, R. H. Manka, and F. S. Mozer, "Electric Fields in the Plasma Sheet and Plasma Sheet Boundary Layer," *J. Geophys. Res.* **90**, 1231 (1985).

103 C. A. Cattell, M. Kim, R. P. Lin, and F. S. Mozer, "Observations of Large Electric Fields Near the Plasma Sheet Boundary by ISEE 1," *Geophys. Res. Lett.* **9**, 539 (1982).

104 C. A. Cattell and F. S. Mozer, "Substorm Electric Fields in the Earth's Magnetotail," in *Magnetic Reconnection*, E. W. Hones, Jr., ed., p. 208 (1984).

105 W. Lennartsson and R. D. Sharp, "Relative Contributions of Terrestrial and Solar Wind Ions in the Plasma Sheet," *Adv. Space Res.* **5**, 411 (1985).

106 F. M. Ipavich, A. B. Galvin, G. Gloeckler, D. Hovestadt, B. Klecker, and M. Scholer, "Energetic (>100 keV) O + Ions in the Plasma Sheet," *Geophys. Res. Lett.* **11**, 504 (1984).

107 F. M. Ipavich, A. B. Galvin, M. Scholer, G. Gloeckler, D. Hovestadt, and B. Klecker, "Suprathermal O + and H + Ion Behavior During the March 22, 1979 (CDAW 6) Substorms," *J. Geophys. Res.* **90**, 1263 (1986).

108 R. D. Sharp, W. Lennartsson, W. K. Peterson, and E. G. Shelley, "The Origins of the Plasma in the Distant Plasma Sheet," *J. Geophys. Res.* **87**, 10420 (1982).

109 R. D. Sharp, R. G. Johnson, W. Lennartsson, W. K. Peterson, and E. G. Shelley, "Hot Plasma Composition Results from the ISEE 1 Spacecraft," in *Energetic Ion Composition in the Earth's Magnetosphere*, p. 231 (1983).

110 S.-I. Akasofu, E. W. Hones, Jr., M. D. Montgomery, S. J. Bame, and S. Singer, "Association of Magnetotail Phenomena with Visible Auroral Features," *J. Geophys. Res.* **76**, 5985 (1971).

111 A. T. Y. Lui, E. W. Hones, Jr., F. Yasuhara, S.-I. Akasofu, and S. J. Bame, "Magnetotail Plasma Flow During Plasma Sheet Expansions: Vela 5 and 6 and IMP 6 Observations," *J. Geophys. Res.* **82**, 1235 (1977).

112 T. E. Eastman, L. A. Frank, and C. Y. Huang, "The Boundary Layers as the Primary Transport Regions of the Earth's Magnetotail," *J. Geophys. Res.* **90**, 9541 (1985).

113 T. E. Eastman, R. J. DeCoster, and L. A. Frank, "Velocity Distributions of Ion Beams in the Plasma Sheet Boundary Layer," in *Proc. Chapman Conf. on Ion Acceleration*, T. Chang, ed. (1986).

114 N. U. Crooker and G. L. Siscoe, "Large Amplitude Substorm Motion of the Magnetotail Boundary," *Geophys. Res. Lett.* **6**, 105 (1979).

115 D. N. Baker, S. J. Bame, R. D. Belian, W. C. Feldman, J. T. Gosling, P. R. Higbie, E. W. Hones, Jr., D. J. McComas, and R. D. Zwickl, "Correlated Dynamical Changes in the Near-Earth and Distant Magnetotail Regions: ISEE 3," *J. Geophys. Res.* **89**, 3855 (1984).

116 E. W. Hones, Jr., D. N. Baker, S. J. Bame, W. C. Feldman, J. T. Gosling, D. J. McComas, R. D. Zwickl, J. A. Slavin, E. J. Smith, and B. T. Tsurutani, "Structure of the Magnetotail at 220 R_e and Its Response to Geomagnetic Activity," *Geophys. Res. Lett.* **11**, 5 (1984).

117 T. A. Slavin, E. J. Smith, B. T. Tsurutani, D. G. Sibeck, H. G. Singer, D. M. Baker, J. T. Gosling, E. W. Hones, Jr., and F. L. Scarf, "Substorm Associated Traveling Compressions Regions in the Distant Tail: ISEE 3 Geotail Observations," *Geophys. Res. Lett.* **11**, 657 (1984).

118 D. G. Sibeck, G. L. Siscoe, J. A. Slavin, E. J. Smith, S. J. Bame, and F. L. Scarf, "Magnetotail Flux Ropes," *Geophys. Res. Lett.* **11**, 1090 (1984).

119 E. W. Hones, Jr., "Plasma Flow in the Magnetotail and Its Implication for Substorm Theories," in *Dynamics of the Magnetosphere*, S.-I. Akasofu, ed., D. Reidel, Hingham, Mass., p. 545 (1979).

120 G. Paschmann, N. Sckopke, and E. W. Hones, Jr., "Magnetotail Plasma Observations During the 1054 UT Substorm on March 22, 1979 (CDAW 6)," *J. Geophys. Res.* **90**, 1217 (1985).

121 T. E. Eastman, G. Rostoker, L. A. Frank, C. Y. Huang, and D. G. Mitchell, "Particle and Field Signatures in the Earth's Magnetotail During Substorm Acitivity: Evidence for and Explanation Involving Boundary Layer Dynamics," *J. Geophys. Res.* (submitted, 1987).

122 J. A. Slavin, E. J. Smith, D. G. Sibeck, D. N. Baker, R. D. Zwickl, and S.-I. Akasofu, "An ISEE 3 Study of Average and Substorm Conditions in the Distant Magnetotail," *J. Geophys. Res.* (in press, 1986).

123 A. T. Y. Lui, "Observations on Plasma Sheet Dynamics During Magnetospheric Substorms," in *Dynamics of the Magnetosphere*, S.-I. Akasofu, ed., D. Reidel Pub. Co., Dordrecht, Holland, p. 563 (1979).

124 A. Nishida and E. W. Hones, Jr., "Association of Plasma Sheet Thinning with Neutral Line Formation in the Magnetotail," *J. Geophys. Res.* **79**, 535 (1974).

125 A. T. Y. Lui, D. J. Williams, T. E. Eastman, L. A. Frank, and S.-I. Akasofu, "Streaming Reversal of Energetic Particles in the Magnetotail During a Substorm," *J. Geophys. Res.* **89**, 1540 (1984).

126 L. R. Lyons, "Dialog on the Phenomenological Model of Substorms in the Magnetotail" (this volume, 1987).

127 L. A. Frank, "Plasmas in the Earth's Magnetotail," in *Space Plasma Simulations*, M. Ashour-Abdalla and D. A. Dutton, eds. (1985).

128 G. Rostoker, "Deninition of a Substorm, Physical Processes in a Substorm, and Sources of Discomfort," in *Magnetic Reconnection in Space and Laboratory Plasmas*, E. W. Hones, Jr., ed., p. 380 (1984).

129 G. Rostoker and T. E. Eastman, "A Boundary Layer Model for Magnetospheric Substorms," *J. Geophys. Res.* (submitted, 1986).

130 G. Atkinson, "Thick Current Sheets in the Renovated Model of the Magnetosphere," *J. Geophys. Res.* **89**, 8949-8955 (1984).

131 G. Atkinson, "Field-Aligned Currents as a Diagnostic Tool. Result: A Renovated Model of the Magnetosphere," *J. Geophys. Res.* **89**, 217-226 (1984).

132 G. Atkinson, "Time-Dependent Flows in the Renovated Model of the Magnetosphere," *J. Geophys. Res.* **90**, 10843-10850 (1985).

133 G. Atkinson, "Magnetospheric Models from Field-Aligned Currents," in *Solar Wind/Magnetosphere Coupling*, Y. Kamide and J. A. Slavin, eds., Terra Scientific Pub. Co., Tokyo, p. 563 (1986).

134 J. K. Chao, J. R. Kan, A. T. Y. Lui, and S.-I. Akasofu, "A Model for Thinning of the Plasma Sheet," *Planet. Space Sci.* **25**, 703 (1977).

135 R. A. Smith, C. K. Goertz, and W. Grossman, "Thermal Catastrophe in the Plasma Sheet Boundary Layer," *Geophys. Res. Lett.* **13**, 1380 (1986).

[136] T. E. Moore, R. L. Arnoldy, J. Feynman, and D. A. Hardy, "Propagating Substorm Injection Fronts," *J. Geophys. Res.* **86**, 6713 (1981).

[137] J. M. Quinn and D. J. Southwood, "Observations of Parallel Ion Energization in the Equatorial Region," *J. Geophys. Res.* **87**, 10536 (1982).

[138] B. H. Mauk, "The Convection Surge Mechanism of Ion Acceleration During Substorms" (this volume, 1987).

[139] M. G. Kivelson, J. Feyman, B. H. Mauk, and R. A. Wolf, "Dialog on Injection Boundary versus Alfvén Layer Model" (this volume, 1987).

[140] L. R. Lyons and T. W. Speiser, "Evidence for Current Sheet Acceleration in the Geomagnetic Tail," *J. Geophys. Res.* **87**, 2276 (1982).

[141] E. G. Shelley, R. D. Sharp, and R. G. Johnson, "Satellite Observations of an Ionospheric Acceleration Mechanism," *Geophys. Res. Lett.* **3**, 654 (1976).

[142] M. Lockwood, J. H. Waite, Jr., T. E. Moore, J. F. E. Johnson, and C. K. Chappell, "A New Source of Suprathermal O^+ Ions Near the Dayside Polar Cap Boundary," *J. Geophys. Res.* **90**, 1611 (1985).

[143] J. H. Waite, Jr., T. Nagai, J. F. E. Johnson, C. R. Chappell, J. L. Burch, T. L. Killeen, P. B. Hays, G. R. Carrigan, W. K. Peterson, and E. G. Shelley, "Escape of Suprathermal O^+ Ions in the Polar Cap," *J. Geophys. Res.* **90**, 1619 (1985).

[144] J. L. Horwitz, M. Lockwood, J. H. Waite, Jr., T. E. Moore, C. R. Chappell, and M. O. Chandler, "Transport of Accelerated Low-Energy Ions in the Polar Magnetosphere" in *Ion Acceleration in the Magnetosphere and Ionosphere*, T. Chang, ed., p. 56 (1986).

[145] B. A. Whalen, W. Bernstein, and P. W. Daly, "Low Altitude Acceleration of Ionospheric Ions," *Geophys. Res. Lett.* **5**, 55 (1978).

[146] J. H. Waite, Jr., T. E. Moore, M. O. Chandler, M. Lockwood, A. Persoon, and M. Sugiura, "Ion Energization in Upwelling Ion Events," in *Ion Acceleration in the Magnetosphere and Ionosphere*, T. Chang, ed., p.61 (1986).

[147] J. L. Horwitz, "Features of Ion Trajectories in the Polar Magnetosphere," *Geophys. Res. Lett.* **11**, 1111 (1984).

[148] J. L. Horwitz and M. Lockwood, "The Cleft Ion Fountain: A Two-Dimensional Kinetic Model," *J. Geophys. Res.* **90**, 9749 (1985).

[149] J. B. Cladis, "Parallel Acceleration and Transport of Ions from Polar Ionosphere to Plasma Sheet," *Geophys. Res. Lett.* **13**, 893 (1986).

[150] D. Williams, "Energetic Ion Beams at the Edge of the Plasma Sheet: ISEE 1 Observations plus a Simple Explanatory Model," *J. Geophys. Res.* **86**, 5507 (1981).

[151] E. Möbius, F. M. Ipavich, M. Scholer, G. Gloeckler, D. Hovestadt, and B. Klecker, "Observations of a Nonthermal Ion Layer at the Plasma Sheet Boundary During Substorm Recovery," *J. Geophys. Res.* **85**, 5143 (1980).

[152] E. C. Roelof, E. P. Keath, C. O. Bostrom, and D. J. Williams, "Fluxes of ≥ 50 keV Protons and ≥ 30 Electrons at $\sim 35\ R_e$, 1, Velocity Anisotropies and Plasma Flow in the Magnetotail," *J. Geophys. Res.* **81**, 2304 (1976).

[153] A. T. Y. Lui and S. M. Krimigis, "Energetic Ion Beam in the Earth's Magnetotail Lobe," *Geophys. Res. Lett.* **10**, 13 (1983).

[154] E. T. Sarris, S. M. Krimigis, C. O. Bostrom, and T. P. Armstrong, "Simultaneous Multi-Spacecraft Observations of Energetic Proton Bursts Inside and Outside the Magnetosphere," *J. Geophys. Res.* **83**, 4289 (1978).

[155] T. E. Eastman, L. A. Frank, W. K. Peterson, and W. Lennartsson, "The Plasma Sheet Boundary Layer," *J. Geophys. Res.* **89**, 1553 (1984).

[156] R. L. McPherron, "Magnetospheric Substorms," *Rev. Geophys. Space Phys.* **17**, 657 (1979).

[157] G. L. Siscoe, "Coupling Between the Solar Wind and the Earth's Magnetosphere: Summary Comments," in *Solar-Wind/Magnetosphere Coupling*, Y. Kamide and J. A. Slavin, eds., Terra Scientific Pub. Co., p. 793 (1986).

SUBSTORM-ASSOCIATED LOWER HYBRID WAVES
IN THE PLASMA SHEET OBSERVED BY ISEE 1

C. A. Cattell* and F. S. Mozer*[†]

Observations of the electric field at frequencies from 2–128 Hz, using the burst mode of the spherical double probe on ISEE 1, have been examined for a time period previously identified as containing the traversal of a near-earth neutral line past the satellite. Intense waves (3 to >30 mV/m) at approximately half the lower hybrid frequency were observed throughout the plasma sheet from the neutral sheet to the boundary, but only during the period of the large DC electric field and $E \times B$ velocity associated with the substorm neutral line. The wave number deduced from linear fits of the data was comparable to the inverse electron gyroradius ($k\rho_e < 1$). These results are consistent with the lower hybrid drift instability. Although peaks between the ion and electron plasma frequency were sometimes observed simultaneously, the integrated power below 100 Hz was usually at least one to two orders of magnitude greater than that above 100 Hz. Although theoretical work has suggested that the instability would be suppressed at the neutral sheet, the largest waves observed occurred right at the neutral sheet when the southward component of the magnetic field was 6γ. The observed waves could provide an anomalous resistivity of $\sim 3 \times 10^{-7}$ to 1×10^{-4} s (compared to the classical value of 1×10^{-18} s).

INTRODUCTION

The occurrence, amplitude, and type of plasma waves in the plasma sheet, particularly at and near the neutral sheet, are important problems in the physics of the magnetotail. Most theories of reconnection require that anomalous dissipation be provided by waves. However, neither observations nor theory have yet provided convincing evidence about which instabilities are important or that the required wave amplitudes occur.[1]

The first observations of waves in the plasma sheet were presented by Scarf et al.[2] Gurnett et al.[3] presented a detailed description of both electrostatic and electromagnetic waves in the plasma sheet, their location, and association with plasma parameters. In light of the study made herein, it is important to note that their typical spectrum from filter data (Fig. 5 in Ref. 3) continued to rise to the lowest frequency measured (40 Hz) and the broadband data (their Fig. 6) showed that the noise generally peaked below 40 Hz. Anderson[4] showed that the waves were generally suppressed right at the neutral sheet. Theoretical explanations of the waves have been provided by two very different mechanisms: (a) the lower hybrid drift instability (applied to the magnetotail by Huba et al.[5-7]) which excites waves near the lower hybrid frequency with wavelengths comparable to the electron gyroradius; and (b) ion beam driven instabilities,[8,9] which drive waves at much higher frequencies from the ion plasma frequency, f_{pi}, to the electron plasma frequency, f_{pe}, with wavelengths comparable to the Debye length.

In this report, we present measurements of the low frequency (2–128 Hz) electric field from ISEE 1 in the plasma sheet at ~ 19 R_e during one substorm period during which a neutral line was inferred to have formed earthward of the satellite (see Nishida et al.,[10] particu-

larly the description of Figs. 1 and 3). This is the first time that waves in this frequency range have been studied in this region of the magnetosphere. The observed waves can be identified as the low-frequency end of broadband electrostatic noise, as can be seen in Fig. 1, which shows composite electric field spectra for several times during this study. The points at ≤ 100 Hz are from the UCB instrument and the points at ≥ 100 Hz are from the University of Iowa sweep frequency receiver (R. Anderson, private communication, 1986). There may be some time-aliasing effects since the swept frequency receiver takes 32 s to produce a spectrum from 10^2 to 4×10^5 Hz, whereas the spectrum below 128 Hz is obtained in 0.5 s, somewhere within that period. It can be seen that there is a definite peak near the lower hybrid frequency and that the power below ~ 100 Hz is several orders of magnitude greater than that above 100 Hz. (Both of these examples were chosen because they also had peaks at high frequencies; however, this was not always the case.) Computer simulations have suggested that the high-frequency waves have only a weak effect on the particles.[11]

In the next section, the instrumentation is described and examples of the electric field data are shown. The third section presents a statistical analysis of the data and linear fits of the observed spectral peaks to determine the frequencies and wavelengths of the waves. The fourth section compares the observations to theories of broadband electrostatic waves, and discusses the importance of the observations to substorm dynamics. The final section presents our conclusions.

DATA

The electric field data presented in this paper were obtained from the University of California, Berkeley double probe experiment on the ISEE 1 satellite.[12] The low-frequency waves were obtained from the small on-board burst memory that sampled 128 points in 0.5 s every 128 s. In addition, the 3-s spin fit DC electric field

*Space Sciences Laboratory, University of California, Berkeley, California 94720.

[†]Also Physics Department.

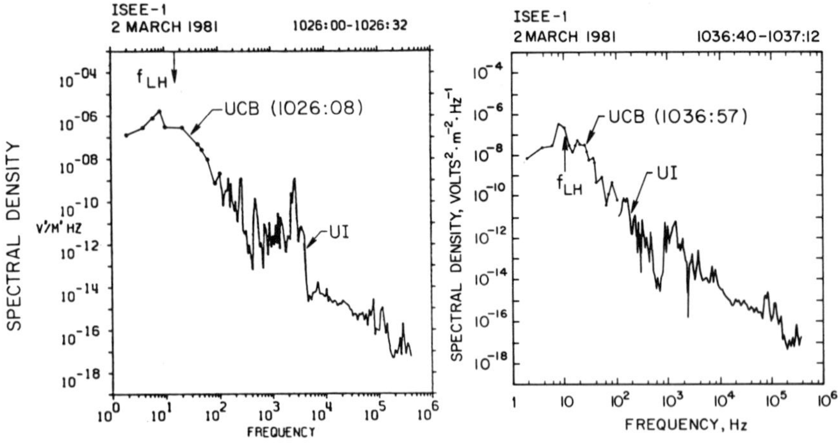

Figure 1—Composite electric field spectra from the UCB burst mode (≤ 100 Hz) and UI SFR data (≥ 100 Hz).

was used to calculate the $E \times B/B^2$ velocity using the assumption that $E \cdot B = 0$. To minimize errors, this calculation was only made when B_y/B_z and $B_x/B_z \leq 5$. Thirty-second averages of the velocities were used in the fits described below. Errors in the electric field measurements and comparison with measured plasma velocities are described by Mozer et al.[13] and Pedersen et al.[14] The measurement of the spacecraft potential, $-V_{2s}$, yields an estimate of the plasma density since $V_{2s} \propto \log nT_e^{1/2}$ in this plasma regime.[13]

The magnetic field data were obtained by the University of California, Los Angeles fluxgate magnetometers on ISEE 1.[15] (ISEE 2 data were also examined to better understand the plasma-sheet structure.) Three-second averages were used to determine the $E \times B$ velocities. For calculating frequencies (cyclotron and lower hybrid) for comparison to the measured frequencies, the high-time-resolution magnetic fields were used. Estimates of the ion composition and temperature were provided from the Lockheed Palo Alto Research Laboratory ion mass spectrometer on ISEE 1.[16] The high-frequency-electric-field data were examined using the sweep frequency receiver data from the University of Iowa plasma wave experiment.[17]

An overview of the DC electric and magnetic field data and selected burst waveforms during the event are shown in Fig. 2. (The DC field data, $E \times B/B^2$ velocity, and ground magnetograms for this event were previously described by Nishida et al.[10]) Panels *a* and *b* plot the GSE x and z components of the magnetic field. Several neutral sheet crossings ($B_x = 0$) can be seen. The GSE y component of the electric field is plotted in panel *c*. The anticorrelation of E_y and B_z produces fast tailward $E \times B$ flow from ~0954 UT (a few minutes after substorm onset) to ~1010 UT, and the correlation of E_y and B_z from ~1010 UT to ~1046 UT produces fast earthward flow. V_{2s}, the negative of the spacecraft potential, is shown in panel *d*. Increases (decreases) in V_{2s} correspond to increases (decreases) in the plasma density. Six selected 0.5-s waveforms from the burst memory are shown in the bottom panels. The first was taken before the substorm onset within the plasma sheet when the DC electric field was small ($\leq \frac{1}{2}$ mV/m) and

the $E \times B$ velocity was <50 km/s and shows that the electric field at higher frequencies was also very small. In the second example, taken after the onset of tailward flow when $|E \times B/B^2| \approx 175$ km/s and $|B| \approx 20\gamma$, the electric field reached amplitudes of ~2 mV/m. The third example was obtained while the satellite was right at the neutral sheet ($B_x < 0.1\gamma$), B_z was 6γ southward, and $|E \times B|/B^2 \approx 1180$ km/s. The waves, which were the most intense of all the burst samples, were strongly saturated (± 7 mV/m) with an estimated amplitude of >30 mV/m. This indicates that, at least during active periods, low-frequency waves are not always suppressed at the neutral sheet, in contrast to previous experimental[4] and theoretical[6] studies. In both the fourth and fifth examples, there were a few saturated points. The fourth example was obtained near the plasma-sheet boundary in a region where the $E \times B$ velocity was not calculated because B_x/B_z or $B_y/B_z > 5$ (this is the only sample in the study for which the velocity was not calculated). The fifth example occurred within the plasma sheet and $|E \times B|/B^2 \approx 580$ km/s. The electric field in the sixth sample, which was near the boundary when $|E \times B|/B^2 \approx 300$ km/s, was lower amplitude. Note that there are large variations in the magnitude of the magnetic field and in the density (as indicated by V_{2s} and the ion densities from the Lockheed instrument (W. Lennartsson, personal communication, 1985)) throughout the event. Generally, regions with high densities had a small magnetic field magnitude; therefore, the values of f_{pe}, f_{pi}, f_{ce}, f_{ci}, and f_{LH} (where f_{pe} and f_{pi} are the electron and ion plasma frequencies, f_{ce} and f_{ci} are the electron and ion cyclotron frequencies, and f_{LH} is the lower hybrid frequency) and the ratios of various frequencies, e.g., f_{pe}/f_{LH}, are quite different for different samples. The samples cover the entire plasma sheet region from the lobe to the neutral sheet.

Since characteristics of the spectra associated with the burst samples will be examined in this study, examples of two spectra and the associated waveforms are presented in Fig. 3. The first, taken at 1004:58 UT, shows the problems involved in determining the wave mode. There are many peaks at frequencies below ~34 Hz which have

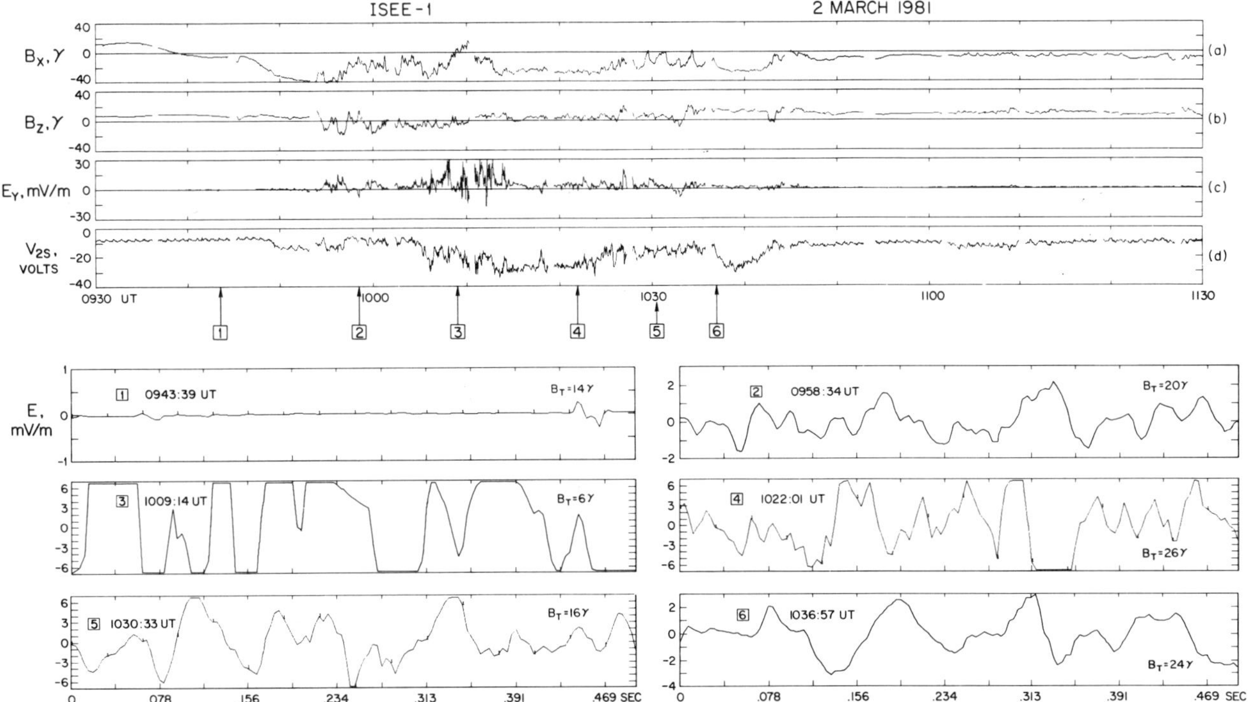

Figure 2—The DC electric and magnetic field data and selected 0.5 s electric field waveforms from the burst memory. (a) the GSE *x*-component of the magnetic field (3-s avg.); (b) the GSE *z* component of the magnetic field (3-s avg.); (c) the GSE *y*-component of the electric field (3-s avg.); (d) the negative of the spacecraft potential, indicating density.

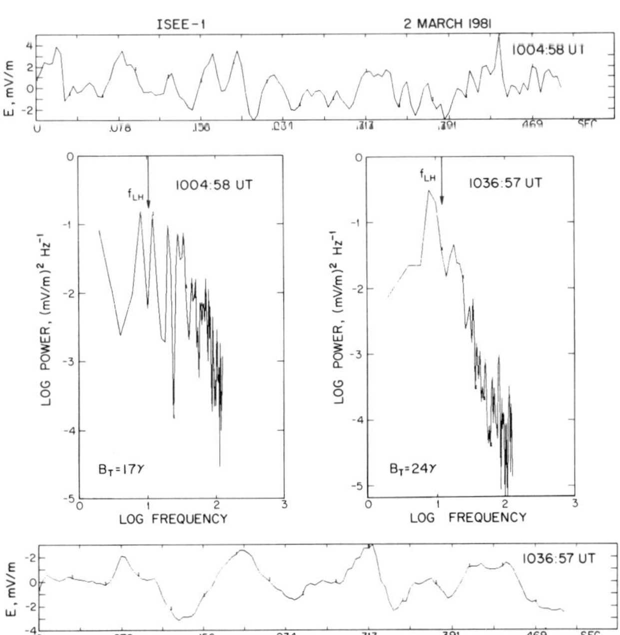

Figure 3—Two 0.5-s electric field waveforms and the associated power spectra. The lower hybrid frequency (assuming 100% H$^+$) is indicated.

comparable power. The second example, taken at 1036:57, is a case where there was a very clear dominant frequency in the spectrum. This occurred in more than half the spectra.

Spectra with peak power $\geq 3 \times 10^{-2}$ (mV/m)^2 Hz^{-1} were observed only in the region where the DC electric field and $E \times B/B^2$ velocity were large ($\sim$0954–1046 UT). Within this time period, there were only four burst samples that were below that power level and all occurred when E_y was small. The first two occurred near 1015–1020 UT in a region that would be classified as lobe (low-density, high magnetic field, primarily in the $-x$ direction) where $E \times B/B^2$ could not be calculated. The second two also occurred in a region of low density, but B_z was large enough that $E \times B/B^2$ could be calculated and it was <50 km/s.

STATISTICAL ANALYSIS AND FITS OF THE DATA

The first step in determining the wave mode of importance is to compare the observed frequencies at peak power for all the burst samples to the lower hybrid frequency, which, in this regime, is $f_{LH} \approx (f_{ce}f_{ci})^{1/2}$. This comparison is plotted in Fig. 4. The frequency that had the maximum power for a given burst spectrum is circled. Dotted lines connect the frequencies that had a power within a factor of 2 of the maximum; broad peaks are connected with solid vertical lines. Samples during which one or more electric-field points were saturated are indicated with an "X." Lines are drawn to indicate where the data points would lie if the measured frequency were equal to the lower hybrid frequency assuming 100% H$^+$, 50% H$^+$ and 50% O$^+$, and 100% O$^+$. Most of the peaks lie below the lower hybrid frequency

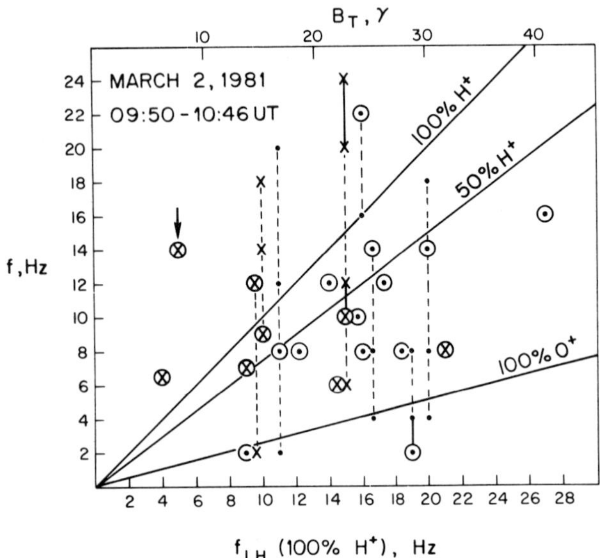

Figure 4—The frequency at peak power for each of the burst waveforms is plotted versus the lower hybrid frequency. The circled points had the maximum power in a given spectrum; "X" indicates that 1 or more points were saturated. The lines indicate where the data points would lie if the frequency were equal to the f_{LH} for 100% H⁺, 50% H⁺ and 50% O⁺, and 100% O⁺.

assuming 100% H⁺ and above the lower hybrid frequency assuming 100% O⁺. The correlation coefficient and fit parameters for a linear fit between the circled points and f_{LH} (100% H⁺) are shown in Table 1. Using preliminary composition data (W. Lennartsson, personal communication, 1985), a fit was made to the actual lower hybrid frequency. As can be seen in Table 2, this improves the fit somewhat, accounting for an additional 5% of the variance. We expect that use of higher time-resolution composition would improve the fit further, since there were some points in composition averaging intervals that had large uncertainties (in particular, the uncertainty in one interval was so large that an average between adjacent intervals was used). (Note that in making these fits, one point (marked with an arrow in Fig. 4) was omitted since it was many standard deviations from the subsequent fits.) The important point is that the frequencies are associated with the lower hybrid frequency and not with either the ion or electron plasma frequencies which are 1–3 orders of magnitude larger.

Implicit in the two fits described above was the assumption that Doppler shifting of the wave frequency was not important. Since this left a large unexplained variance, further analysis was made to include Doppler shifting, i.e., that

$$f_{\text{measured}} = f + k \cdot v/2\pi.$$

Table 1—Correlation coefficients and fit parameters.**

Correlation	Fit	R	Confidence Level
f vs. f_{LH}	$f = 5\ \text{Hz} + 0.3\, f_{LH}$	0.39	90%
f vs. f_{LH}^c *	$f = 5\ \text{Hz} + 0.5\, f_{LH}^c$	0.46	95%
f/B vs. $|v_\perp|$	$f = 0.3\, f_{LH} + 7.7 \times 10^{-3} \left(\dfrac{T_e^{1/2}}{\rho_e} \right) \left(\dfrac{v}{2\pi} \right)$ $f = 0.3\, f_{LH} + (0.4/\rho_e)(v/2\pi)$***	0.57	99%
$\dfrac{f - f_{LH}^c}{B}$ vs. $|v_\perp|$	$f = f_{LH}^c - 0.3\, f_{LH} + 9.4 \times 10^{-3} \left(\dfrac{T_e^{1/2}}{\rho_e} \right) \left(\dfrac{v}{2\pi} \right)$ $f = f_{LH}^c = 0.3\, f_{LH} + (0.5/\rho_e)(v/2\pi)$***	0.68	99.9%
$\dfrac{f - 0.6\, f_{LH}^c}{B}$ vs. $|v_\perp|$	$f = 0.6\, f_{LH} + 9.7 \times 10^{-3} \left(\dfrac{T_e^{1/2}}{\rho_e} \right) \left(\dfrac{v}{2\pi} \right)$ $f = 0.6\, f_{LH} + (0.5/\rho_e)(v/2\pi)$***	0.64	99.8%

*f_{LH}^c is f_{LH} including composition.
**Using $B = 1.5\, f_{LH}$ for 100% H⁺ and $B = 2.4\, T_e^{1/2}/\rho_e$ where T_e is in eV, B is in γ, v is in km/s, and ρ_e is in km.
***for $T_e = 3$ keV.

Table 2—Comparison with Theory.

Theories	f (max. γ)	k (max. γ)	Parameters	Author
Lower hybrid drift	$\sim 0.8\text{--}1\, f_{LH}$	$k\rho_e \sim 0.7\text{--}0.9$	$L_N \sim \rho_i,\ \beta = 1$ $T_e/T_i = 0.5$	Huba et al. (1978)
Ion-beam instability	$\sim 0.1\text{--}0.2\, f_{pe}$ $(\sim 10^2\, f_{LH})^*$	—	$v_b/v_{b,th} = 50$	Grabbe and Eastman (1984)
	$\sim 0.02\text{--}0.9\, f_{pe}$ $(0.2 - 9 \times 10^2\, f_{LH})^*$	—	$v_b/v_{b,th} = 50$	Omidi (1985)
	$\sim 0.6\text{--}4 \times 10^{-2}\, f_{pe}$ $(6\text{--}40\, f_{LH})^*$	$k\lambda_D \sim 0.1\text{--}2$ $(k\rho_e \approx 2\text{--}70)^*$	$v_b/v_{b,th} = 20$	Omidi (1985)
Observations	f (max. power)	k (max. power)		
	$0.3\, f_{LH} \lesssim f \lesssim f_{LH}$ $f \approx 0.6\, f_{LH}$ (inc. comp.)	$k\rho_e \sim 0.3\text{--}0.5$ (for $T_e \sim 1\text{--}3$ keV)		

*Using $f_{ce}/f_{pe} \approx 0.03 - 0.05$, $f_{pe} \approx 10^3\, f_{LH}$, and $\lambda_D \approx 0.03\text{--}0.05\, \rho_e$

(Note that some of the variance may be due to differences in the density gradient upon which the frequency of lower hybrid drift waves depends.) The satellite velocity is negligible compared to the plasma velocity in these events, so the plasma velocity is the appropriate "v." Using the electric field instrument, one can only determine the $E \times B/B^2$ velocity (which is probably the largest perpendicular component of v). Since there is no way to determine the direction and magnitude of k, it was assumed that $k_\parallel \approx 0$ (consistent with theoretical studies of lower hybrid drift waves) and the angle between $v_\perp$ and $k_\perp$ was ignored (it will add to the scatter in the fit). In addition, because of the correlation between f_{measured} and f_{LH}, it was assumed that both the frequency and the wave vector depended on the magnitude of the magnetic field. The following fit was therefore made:

$$f_{\text{measured}}/|B| = a + b|v_\perp|,$$

which, in physical units, can be written as:

$$f_{\text{measured}} = a'f_{LH} + (b'/\rho_e)(|v_\perp|/2\pi).$$

This fit, plotted in Fig. 5, and the fit parameters, listed in Table 1, show that the inclusion of a Doppler shift term improves the fit (explains an additional 12% of the variance). The data are, therefore, consistent with the hypothesis that the observed waves are Doppler shifted lower hybrid drift waves.

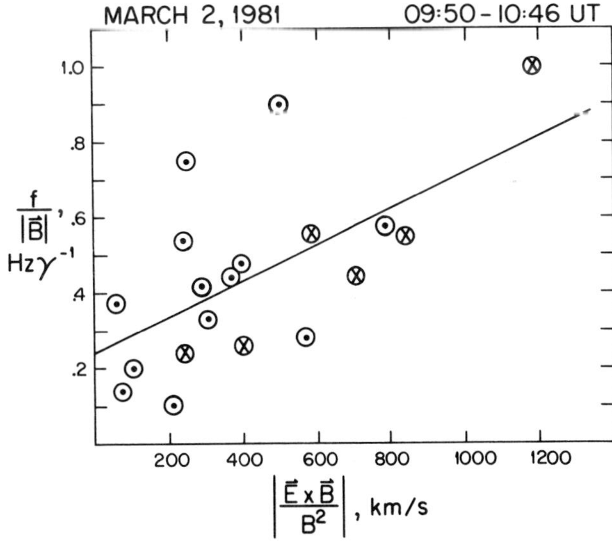

Figure 5—The frequency at peak power for each of the burst waveforms normalized by the magnetic field (i.e., f_{LH}) versus the magnitude of the calculated $E \times B/B^2$ velocity (30-s avg.) The best fit line is shown (parameters are in Table 1).

Even though the parallel plasma velocity is not measured, we can make a plausibility argument to show that the observed waves are not Doppler shifted waves with $k\lambda_D \sim 1$. Previous studies of the plasma velocity and $E \times B$ velocity[18] have shown that $v_\parallel$ is usually comparable to $|v_\perp|$ during active periods of the type described herein. One can estimate, therefore, that $v_\| \approx 100\text{--}1200$ km/s. For $n = 0.1\text{--}1$ cm^{-3} and $T_e = 0.5\text{--}1$ keV, $\lambda_D \approx 0.1\text{--}0.6$ km. The expected Doppler shifts, given by $\Delta f \approx v_\| / 2\pi\lambda_D$, would be $\sim 25\text{--}1100$ Hz. Since the observed peaks are all at frequencies less than 22 Hz and often have widths less than 20 Hz, and since the expected Doppler shifts are so variable, it is extremely unlikely that they could be explained by Doppler shifting of a high frequency wave. On the other hand, the widths of the weaker peaks that are sometimes observed at higher frequencies (see Fig. 1) are consistent with such Doppler shifts.

An additional 14% of the variance is explained by subtracting the lower hybrid frequency (including the measured composition) before fitting, as can be seen in Table 1. For this fit, the slope (i.e., $|k_\perp|$ modified by the cosine of the angle between $k_\perp$ and $v_\perp$) is determined to within 50% at the 95% confidence level. In addition, the hypothesis that the wavelength is comparable to the Debye length can be rejected with a high degree of confidence. When $0.6\, f_{LH}$ is subtracted from the experimental data, the fit is comparably good and the $|B|$ coefficient is <0.01. Given the many possible sources of scatter in the data (problems in choosing the peak frequency, errors in the estimate of ion composition, variable angles between $k_\perp$ and $E \times B/B^2$, differences between $v_\perp$ and $E \times B/B^2$, varying angles between the magnetic field and the electric field booms for each burst, etc.) and the fact that the samples were taken in regions with widely varying densities and magnetic fields (and probably temperatures), the goodness of the fits is quite satisfactory. All the fits were yielded values of k comparable to the electron gyroradius and not to the Debye length.

DISCUSSION

Two theories have been proposed to explain the observed broadband electrostatic noise in the tail: (a) the lower hybrid drift instability (Huba et al.,[5-7] Davidson et al.,[19] Gary,[20] and many others); and (b) ion beam driven instabilities (see Refs. 8, 9, and other papers in this volume). Table 2 presents the theoretically predicted frequencies and wave vectors at maximum growth rate and the parameters deduced from the observations. Note that the theoretical results were obtained for a restricted range of parameters (for example, fixed β) compared to the experimental observations. In addition, since the theories examine only the linear dispersion relations, they cannot provide information on what the final spectral shape should be. However, from examining Table 2 it is clear that the observed frequencies and wavelengths are consistent with the lower hybrid drift instability although ion composition effects must be included. Finite β corrections to the theory, however, increase the real frequency and wavenumber[19] and therefore worsen the agreement with data.

The fact that the growth rate is also reduced by finite β has presented a problem for theoretical explanations of reconnection using lower hybrid drift waves to provide dissipation (see, for example, Refs. 6 and 7). In contrast, the observations described herein showed that intense waves were observed even right at the neutral sheet. This may be reconciled with theory in the following ways: (a) the observed z-component of the magnetic field was large ($\sim 5\text{--}6\ \gamma$) at the neutral sheet so that β was probably always less than 10–20 and the plasma sheet was very turbulent, so there may have been small scale size gradients (as suggested by Huba et al.[6]) large enough to overcome the finite β stabilization and drive the waves; or (b) there could be nonlinear penetration of the waves to the neutral sheet. This has been observed in simulations (Tanaka and Sato,[21] Winske,[22] Brackbill et al.[23]) although they observed electromagnetic waves with only small electric components whereas we observed large electric fields.

The dissipation which could be provided by the observed waves has been calculated from the expression for the resistivity due to lower hybrid drift waves (equation on page 1200 of Ref. 20) using the measured wave electric fields instead of the theoretically derived saturation values:

$$\eta \approx \frac{(8\pi)^{1/2}\,(1 - 0.3\,(T_i/T_e)^{1/4})}{1 + T_i/T_e}\ \left(\frac{1}{f_{LH}}\right)\left(\frac{|E|^2}{nT_e}\right).$$

For $T_i/T_e = 2.5$, $T_i = 5$ keV, and $n = 0.5$/cm^3, η is $\sim 3 \times 10^{-7}$ to 1×10^{-4} s for the observed range of f_{LH} and electric field wave amplitudes. The classical resistivity is only $\sim 4 \times 10^{-18}$ s. The anomalous resistivity could support a duskward electric field of 0.05–18 mV/m for a cross-tail current of $\sim 2 \times 10^{-8}$ A/m^2.

Table 2 also shows that the frequencies expected from ion-beam instabilities are 1 to 3 orders of magnitude larger than the frequencies at peak power observed during this active interval and that the predicted wavenumbers are also very different from those inferred from the data. Ion-beam instabilities probably explain the much weaker waves observed at high frequencies. This is consistent with a recent study of plasma waves in the tail[24] that suggested that BEN may be a superposition of several different wave modes and showed that there are often enhancements at the upper hybrid frequency and/or the electron cyclotron frequency.

CONCLUSIONS

Low frequency (≤ 128 Hz) electric field data have been presented for one substorm event in the magnetotail at ~ 19 R$_e$ near midnight when the DC field data were interpreted as being due to the traversal of a near-earth neutral line past the satellite. Most of the power occurred at frequencies $\leq 20\text{--}50$ Hz. The observed frequencies ($\leq f_{LH}$) and deduced wavenumber ($k\rho_e < 1$) were consistent with the lower hybrid drift instability, although finite β corrections to the theory worsen the agreement. The ion composition was found to have an important effect on the observed frequencies. In approximately half the observed spectra, the peak was very narrow suggesting that often only a narrow range of frequencies and wavelengths was excited. The waves were found throughout the plasma sheet from the plasma-sheet boundary to the neutral sheet, and, in contrast to linear theory, which predicts that the waves would be suppressed at high β, the most intense waves were observed right at the neutral sheet.

The observed strong waves ony occurred simultaneously with the large DC electric fields and $E \times B/B^2$ velocities associated with the near-earth neutral line. This observation is consistent with the suggestion that lower hybrid drift waves (although originally suppressed in the central plasma sheet) become unstable when the plasma sheet thins prior to substorm onset. They may, therefore, provide the dissipation required for reconnection to occur at a new, near-earth neutral line. The difference between the conclusions herein and those of Anderson[4] may be due to this association between intense lower hybrid drift waves and the large flows associated with a neutral line, since the neutral sheet crossings described therein were not associated with large plasma velocities. In addition, sensitivities of the two instruments differ. Further studies are being made to determine the occurrence frequency and intensity of lower hybrid drift waves at the neutral sheet during other time periods, as well as the association of this wave mode with large $E \times B$ velocities, the plasma sheet boundary layer, and the plasma sheet.

ACKNOWLEDGMENT—The authors thank C. T. Russell for use of the ISEE 1 magnetic field data, W. Lennartsson for the use of the ISEE 1 ion composition data, D. Gurnett and R. Anderson for the ISEE 1 SFR wave data, and M. Colby for production of the manuscript. The research has been carried out under NASA grant NAG5-375.

REFERENCES

[1] B. U. O. Sonnerup et al., "Reconnection of Magnetic Fields," in *Solar Terrestrial Physics: Present and Future,* NASA Ref. Pub. 1120, 1(1) (1984).

[2] F. L. Scarf, L. A. Frank, K. L. Anderson, and R. P. Lepping, "Plasma Wave Turbulence at the Distant Crossings of the Plasma Sheet Boundaries and Neutral Sheet," *Geophys. Res. Lett.* **1**, 109 (1974).

[3] D. A. Gurnett, L. A. Frank, and R. P. Lepping, "Plasma Waves in the Distant Magnetotail," *J. Geophys. Res.* **81**, 6059 (1976).

[4] R. R. Anderson, "Plasma Waves at and near the Neutral Sheet," *Proc. Conf. Ach. of IMS,* ESA SP-217, 199 (1984).

[5] J. D. Huba, N. T. Gladd, and K. Papadopoulos, "The Lower-Hybrid-Drift Instability as a Source of Anomalous Resistivity for Magnetic Field Line Reconnection," *Geophys. Res. Lett.* **4**, 125 (1977).

[6] J. D. Huba, N. T. Gladd, and K. Papadopoulos, "Lower-Hybrid-Drift Wave Turbulence in the Distant Magnetotail," *J. Geophys. Res.* **83**, 5217 (1978).

[7] J. D. Huba, N. T. Gladd, and J. F. Drake, "On the Role of the Lower Hybrid Drift Instability in Substorm Dynamics," *J. Geophys. Res.* **86**, 5881 (1981).

[8] C. L. Grabbe and T. E. Eastman, "Generation of Broadband Electrostatic Noise by Ion Beam Instabilities in the Magnetotail," *J. Geophys. Res.* **89**, 3865 (1984).

[9] N. Omidi, "Broadband Electrostatic Noise Produced by Ion Beams in the Earth's Magnetotail," *J. Geophys. Res.* **90**, 12330 (1985).

[10] A. Nishida, Y. K. Tulunay, F. S. Mozer, C. A. Cattell, E. W. Hones, Jr., and J. Birn, "Electric Field Evidence for Tailward Flow at Substorm Onset," *J. Geophys. Res.* **88**, 9109 (1983).

[11] M. Ashour-Abdalla and H. Okuda, "Theory and Simulations of Broadband Electrostatic Noise in the Geomagnetic Tail," *J. Geophys. Res.* **91**, 6883 (1986).

[12] F. S. Mozer, R. B. Torbert, U.V. Fahleson, C.-G. Fälthammar, A. Gonfalone, and A. Pedersen, "Measurements of Quasistatic and Low Frequency Electric Fields with Spherical Double Probes on the ISEE 1 Spacecraft," *IEEE Trans. Geosci. Electron.* **GE-16**, 258 (1978).

[13] F. S. Mozer, E. W. Hones, Jr., and J. Birn, "Comparison of Spherical Double Probe Measurements with Plasma Bulk Flows in Plasmas Having Densities Less than 1 cm^{-3}," *Geophys. Res. Lett.* **10**, 737 (1983).

[14] A. Pedersen, C. A. Cattell, C.-G. Fäthammar, V. Formisano, P.-A. Lindquist, F. Mozer, and R. Torbert, "Quasistatic Electric Field Measurements with Spherical Double Probes on the GEOS and ISEE Satellites," *Space Sci. Rev.* **37**, 269 (1984).

[15] C. T. Russell, "The ISEE-1 and-2 Fluxgate Magnetometers," *IEEE Trans. Geosci. Electron.* **GE-16**, 239 (1978).

[16] E. G. Shelley, F. D. Sharp, R. G. Johnson, J. Geiss, P. Eberhardt, H. Balsiger, G. Haerendel, and H. Rosenbauer, "Plasma Composition Experiment on ISEE-A," *IEEE Trans. Geosci. Electron.* **GE-16**, 266 (1978).

[17] D. A. Gurnett, F. L. Scarf, R. W. Fredricks, and E. J. Smith, "ISEE 1 and ISEE 2 Plasma Wave Investigation," *IEEE Trans. Geosci. Electron.* **GE-16**, 225 (1978).

[18] C. A. Cattell, F. S. Mozer, E. W. Hones, Jr., R. R. Anderson, and R. D. Sharp, "ISEE Observations of the Plasma Sheet Boundary, Plasma Sheet, and Neutral Sheet: 1. Electric Fields, Magnetic Field, Plasma, and Ion Composition," *J. Geophys. Res.* **91**, 5663 (1986).

[19] R. C. Davidson, N. T. Gladd, C. W. Wu, and J. D. Huba, "Effects of Finite Plasma Beta on the Lower Hybrid Drift Instability," *Phys. Fluids* **20**, 301 (1977).

[20] S. Gary, "Wave-Particle Transport from Electrostatic Instabilities" *Phys. Fluids* **23**, 1193 (1980).

[21] M. Tanaka and T. Sato, "Simulations on Lower Hybrid Drift Instability and Anomalous Resistivity in the Magnetic Neutral Sheet," *J. Geophys. Res.* **86**, 5541 (1981).

[22] D. Winske, "Current-Driven Microinstabilities in a Neutral Sheet," *Phys. Fluids,* **24**, 1069 (1981).

[23] J. U. Brackbill, D. W. Forslund, K. B. Quest, and D. Winske, "Nonlinear Evolution of the Lower Hybrid Drift Instability," *Phys. Fluids* **27**, 2682 (1984).

[24] C. A. Cattell, F. S. Mozer, E. W. Hones, Jr., R. R. Anderson, and R. D. Sharp, "ISEE Observations of the Plasma Sheet Boundary, Plasma Sheet, and Neutral Sheet: 2. Waves," *J. Geophys. Res.* **91**, 5681 (1986).

PLASMA FLOWS NEAR THE NEUTRAL SHEET OF THE MAGNETOTAIL

C. Y. Huang, L. A. Frank*, and T. E. Eastman[†]

Using the University of Iowa Lepedea on board ISEE 1, we investigate the bulk flow of plasma in the neutral-sheet region (defined as the area where $B_x \simeq 0$) of the magnetotail. For the majority of crossings, there is no appreciable change in the macroscopic plasma parameters, i.e., the density, temperature, and velocity of the plasma remain constant through the neutral sheet. This is true even during active periods, when $AE \gtrsim 100$ nT. However, for a small number of crossings, all during disturbed times, large plasma bulk velocities ($|V| \gtrsim 300$ km s^{-1}) are observed. The velocity distributions during these events are qualitatively similar to those of the plasma-sheet boundary layer that is usually observed at higher latitudes. We suggest that the acceleration mechanism that creates the plasma-sheet boundary layer extends to relatively small radial distances during these active periods.

INTRODUCTION

In most studies of substorm effects in the earth's magnetotail, a great deal of attention is focused on the near-earth neutral-sheet region. Ideally the point at which $|B| \simeq 0$, the neutral-sheet region is the most likely place for magnetic reconnection, via tearing-mode instability, to occur.[1,2] However, statistical studies of the central plasma sheet during quiet and active times show that the near-earth neutral-sheet region is not characterized by the large bulk flows that would result from neutral line formation.[3,4] In sharp contrast the plasma-sheet boundary layer usually consists of flowing plasma.[5,6] The exact role of the boundary layer in substorm theory is not well understood, but it seems clear that energy and momentum flow primarily within this region and not in the central plasma sheet itself.[7,8]

In this study we examine in detail an event in which large bulk flow is detected near the center of the magnetotail. We should emphasize at the outset that events of this kind are extremely rare. The chosen example is one of the more convincing events in which we detect large bulk velocities when $B_x \simeq 0$. In the vast majority of cases, streaming plasmas are confined to the edges of the plasma sheet.

INSTRUMENTATION

The observations presented here are primarily from the University of Iowa Lepedea on board ISEE 1. Samples are taken in 32 passbands in the energy-per-charge range from 1 V to 45 kV in high-bit-rate (HBR) mode, or 215 V to 45 kV in low-bit-rate (LBR) mode. For each energy passband 12 (16) azimuthal sectors are covered, while 7 polar angles are sampled simultaneously in each azimuthal sector in HBR (LBR) mode. A complete sample is obtained in 128 or 512 seconds.

Energy-time (E-t) spectrograms are constructed from these data. An example is shown in Plate III-3A for the period 0600 UT to 1200 UT on May 25, 1978 (day 145). This spectrogram displays detector responses with the color-coded scale at right. The upper four panels show the detector responses to ion intensities in four solid-angle segments near the solar ecliptic plane. From top to bottom, these are for ion velocities directed antisunward, dawnward, sunward, and duskward, respectively. The fifth panel shows the electron intensities averaged over all azimuthal sectors, in a plane near the solar ecliptic.

Energy-phase angle (E-ϕ) spectrograms provide details of the velocity distribution in terms of energy versus azimuthal angle for each of the seven polar angles during each instrument cycle. Plate III-3A,C shows examples of such spectrograms. Responses are displayed for each of the seven ion and electron detectors and of the Geiger-Mueller (GM) tube. The detectors are numbered 1P through 7P for ions, where the look angles vary from 13° from the spacecraft spin axis (1P) through 90° (4P) to 167° (7P). A similar polar range is displayed for the electron channels numbered 1E through 7E. The GM tube responds primarily to high-energy electrons ($E \gtrsim$ 45 keV). The intensity maximum in the solar direction (panel center) is due to solar x-rays. Each detector sweeps through all azimuthal angles at a given energy before stepping up to the next energy level. From left to right across the azimuthal range displayed in each panel, the velocity directions correspond to velocities directed sunward, duskward, antisunward, dawnward, and again sunward. For more details the reader is referred to Frank et al.[9]

OBSERVATIONS

An example of a typical quiet-time neutral sheet is shown in Plate III-3. On this day (May 25, 1978, i.e., day 145) the reversal in B_x occurs at 1046 UT, but it is undistinguished by any change in the macroscopic plasma parameters, such as density, bulk velocity, or temperature. Plates III-3B,C show the detailed three-dimensional velocity distributions observed during

*Department of Physics and Astronomy, The University of Iowa, Iowa City, Iowa 52242.
[†]NASA Headquarters, Space Plasma Physics Branch, Washington, D.C. 20546.

this crossing, but no discernible signature can be detected when $B_x \simeq 0$. The moments are shown in Fig. 1, with the neutral-sheet crossing indicated by the dashed line. Throughout the two-hour period the central plasma sheet can be characterized as isotropic with high thermal energy. This is almost always true of the central plasma sheet even during active times. In this case the *AE* index varies between 70 and 100 nT. During high geomagnetic activity the plasma sheet thins[7,10] and thermal energies increase, but quasi-isotropic velocity distributions are persistently observed.[3]

In Plate III-4A,B shows examples of velocity distributions observed in the plasma sheet boundary layer during a period of low activity (*AE* ~ 60–70 nT) on March 25, 1980. The ion velocity distributions are highly anisotropic with high bulk speeds. This is evident in the plasma parameters plotted in Fig. 2. The magnetic field (not shown) has a magnitude of approximately 30 nT and is generally directed earthward during this interval. The measured speed during the two instrument cycles shown in Plate III-4A,B is 300–500 km s^{-1}. The fluctuations in the measured velocities shown in Fig. 2 are due to multiple transitions between the boundary layer and the lobe or central plasma sheet. However, the velocities are generally high, and reach a maximum of 900 km s^{-1} in the absence of any significant geomagnetic activity. Such dynamic behavior is characteristic of the plasma-sheet boundary layer[5,7,11] and indicates that an acceleration process operates continuously.

Estimates of the location of the acceleration region place it approximately 100 R_e from the earth.[12] As the flow direction observed at the ISEE spacecraft is usually earthward, the ion beams are presumably generated tailward of the orbit apogee at 23 R_e. A survey of the ISEE 3 magnetic field observations in the deep tail confirms that the boundary layer flow pattern undergoes a change at 100 R_e.[13] The essential difference between

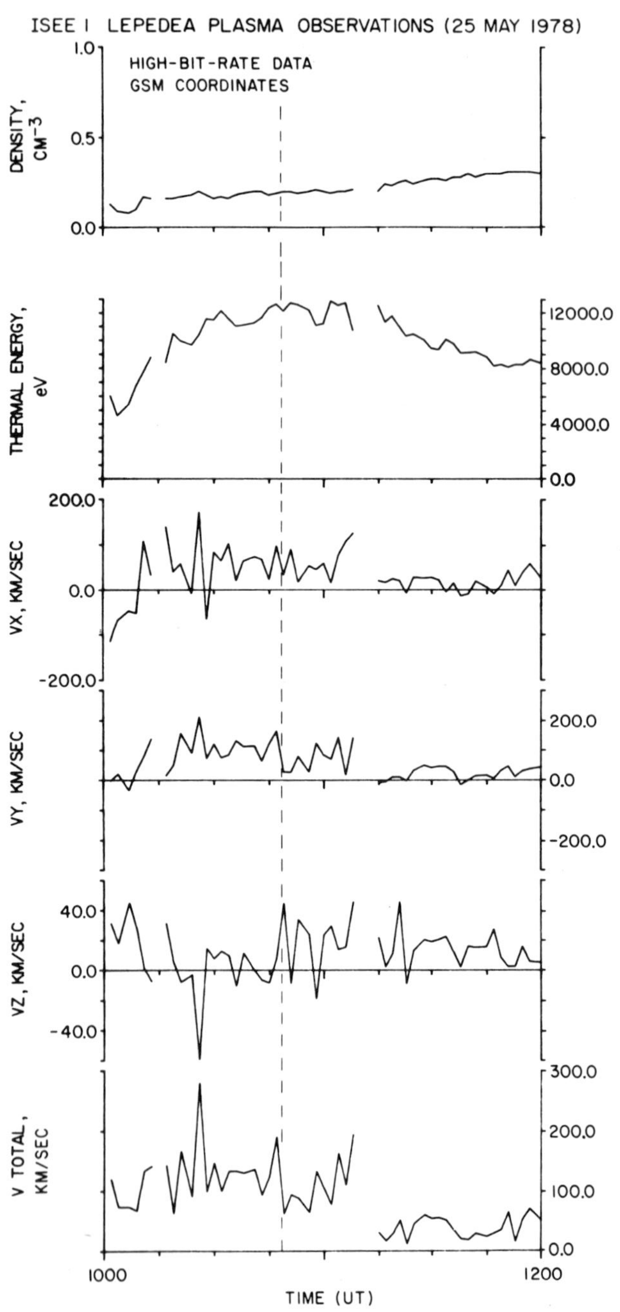

Figure 1 — The plasma parameters on May 25, 1978 are shown here during a typical encounter with the central plasma sheet. The neutral-sheet crossing at 1046 UT is indicated by a dashed line.

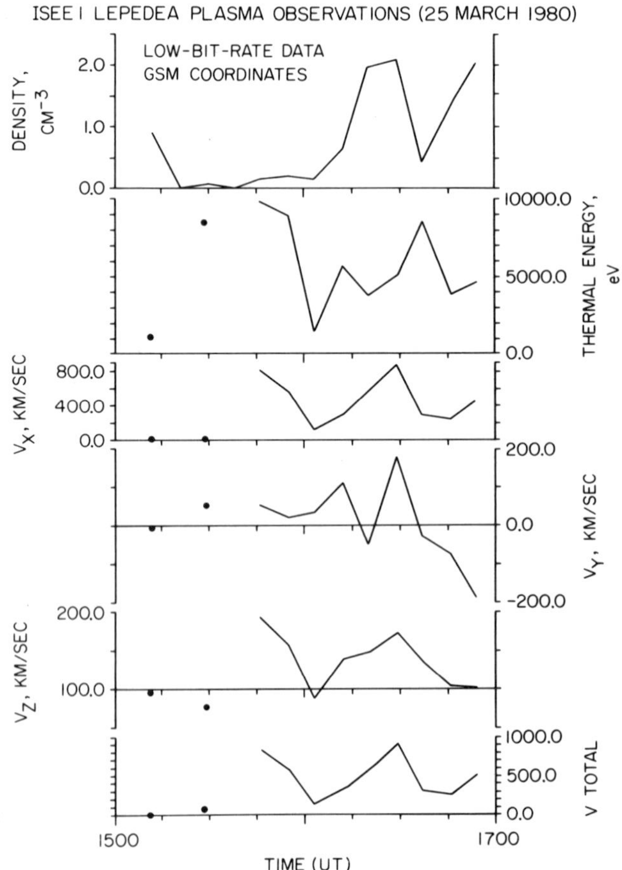

Figure 2 — The bulk properties of the plasma are plotted here for a typical boundary-layer crossing on March 25, 1980. Note the large earthward-directed velocities throughout the event.

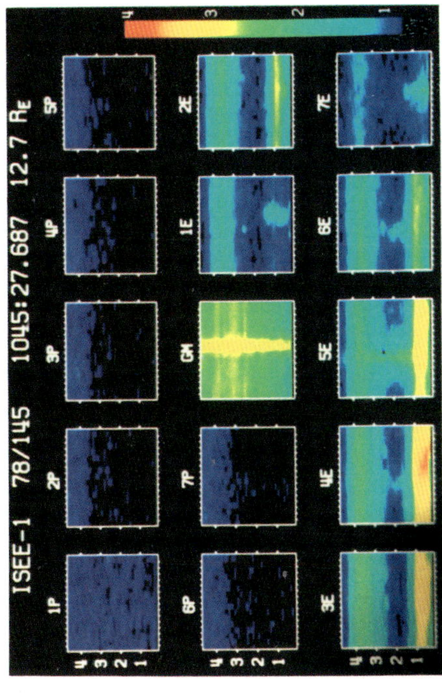

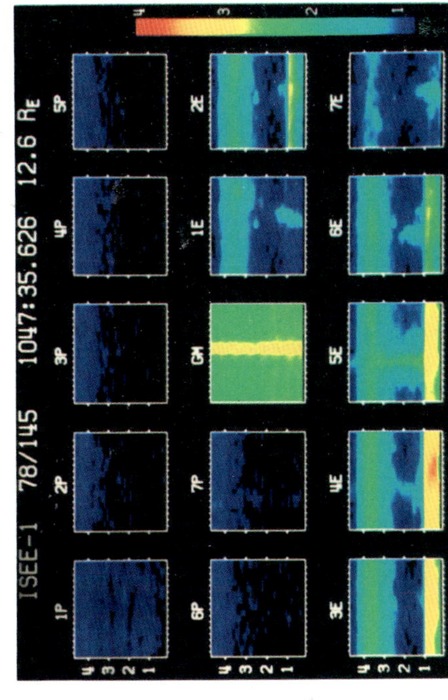

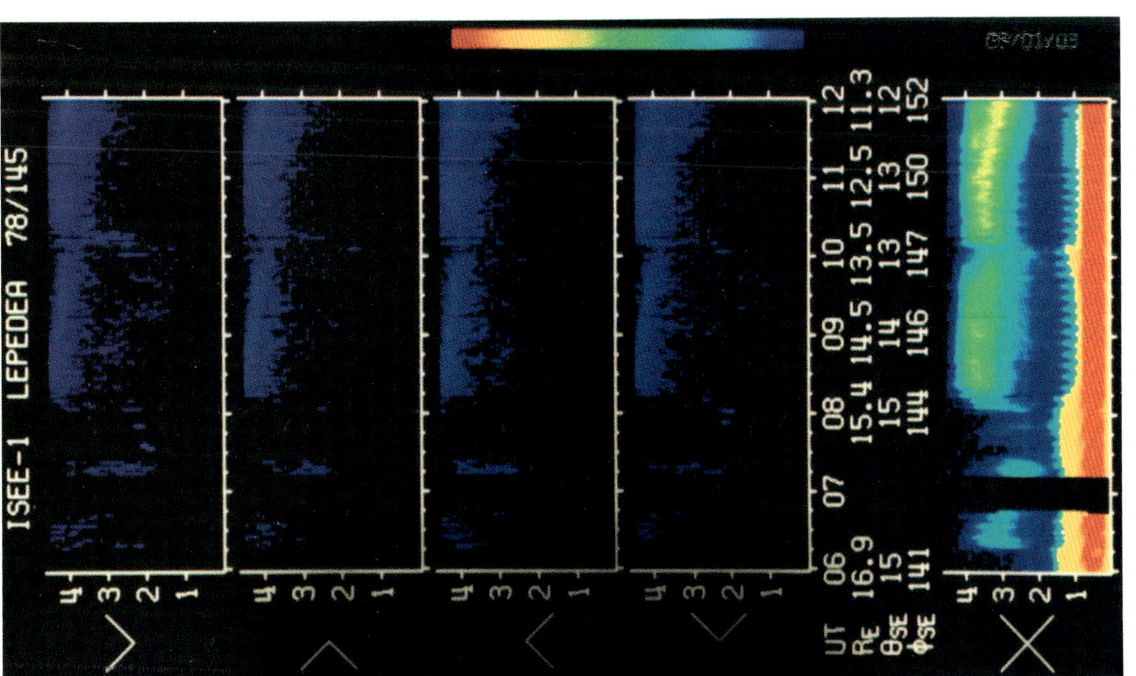

Plate III-3.—(A) This energy-time (*E-t*) spectrogram for the period 0300–1200 UT on May 25, 1978 shows typical observations within the central plasma sheet. The neutral-sheet crossing at 1046 UT is indicated by the dashed line. (B) During this instrument cycle, the neutral sheet is encountered. The velocity distributions shown in this energy phase angle (*E-φ*) spectrogram are isotropic. (C) This E-φ spectrogram shows the plasma velocity distribution on the consecutive cycle. There is no change in any of the macroscopic plasma parameters during the neutral-sheet crossing.

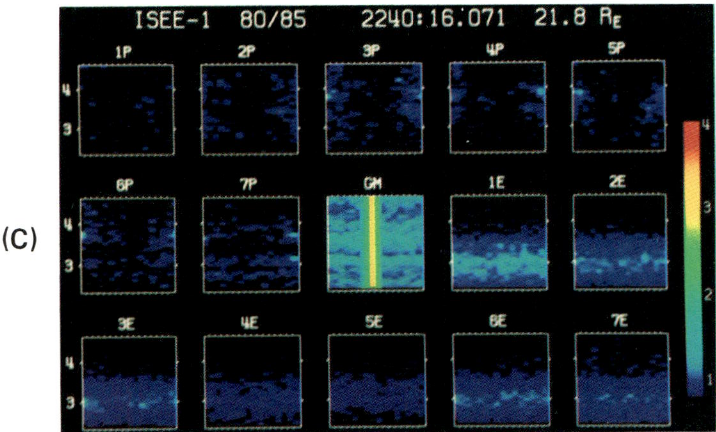

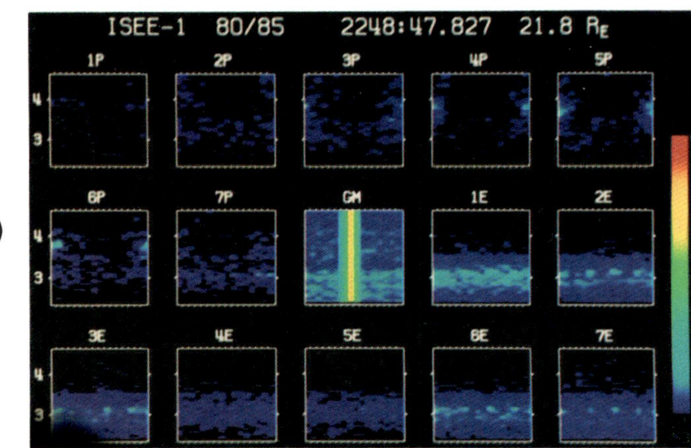

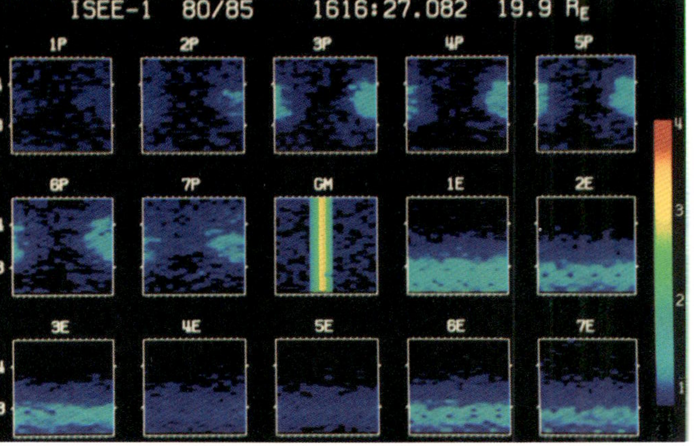

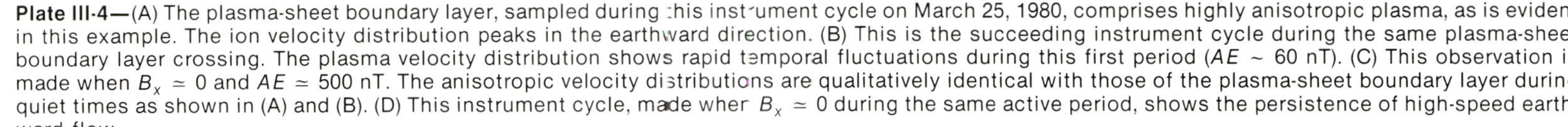

Plate III-4—(A) The plasma-sheet boundary layer, sampled during this instrument cycle on March 25, 1980, comprises highly anisotropic plasma, as is evident in this example. The ion velocity distribution peaks in the earthward direction. (B) This is the succeeding instrument cycle during the same plasma-sheet boundary layer crossing. The plasma velocity distribution shows rapid temporal fluctuations during this first period ($AE \sim 60$ nT). (C) This observation is made when $B_x \simeq 0$ and $AE \simeq 500$ nT. The anisotropic velocity distributions are qualitatively identical with those of the plasma-sheet boundary layer during quiet times as shown in (A) and (B). (D) This instrument cycle, made when $B_x \simeq 0$ during the same active period, shows the persistence of high-speed earthward flow.

the central plasma sheet and the plasma-sheet boundary layer is evident in these two examples, which are typical of many hundreds of events observed with the ISEE spacecraft. Under all levels of geomagnetic activity the plasma-sheet boundary layer consists of streaming plasma while the central plasma sheet remains quasi-isotropic.

We now consider an event in which streaming-velocity distributions occur near the neutral sheet. This occurs later on March 25, 1980, and is shown in Plate III-4C,D. The plasma and magnetic field values are plotted in Figs. 3 and 4. The spacecraft is in the tail lobe prior to 2203

UT, and isotropic velocity distributions typical of the central plasma sheet are observed shortly after 2300 UT. During the entire hour there is a high level of activity detected on the ground. The *AE* index is at least 500 nT throughout.

The measured velocities show flow directed persistently earthward at high speed. The magnitude varies from more than 1000 km s^{-1} at the start to 600 km s^{-1} at the end of the event at 2240 UT. Both in magnitude and direction, these velocities are typical of the plasma-sheet boundary layer, a fact that can be confirmed by comparing Figs. 2 and 3. Further, the velocity distributions are indistinguishable from those common to the boundary layer, as can be seen in Plate III-4. It seems highly unlikely that the processes that operate in the case of the 2200–2300 UT interval are qualitatively different from those that produce the plasmas observed from 1500–1700 UT. The essential difference between the two sets of observations lies in the magnetic field. Between 2200 and 2300 UT, B_x reverses sign several times (see Fig. 4), while during the 1500–1700 UT interval, B_x is large and positive. It appears that during the second period the spacecraft is located in the midplane of the plasma sheet, while observing plasma with boundary layer characteristics.

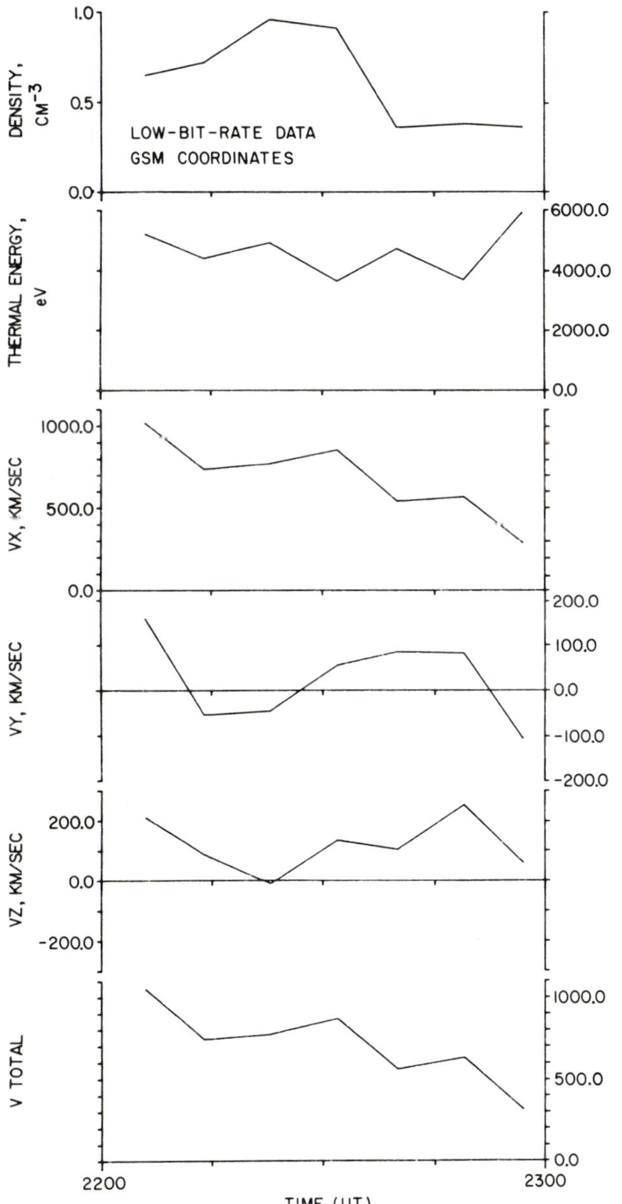

Figure 3 — During this period the spacecraft is in the vicinity of the neutral sheet. However the bulk flow is typical of the boundary layer.

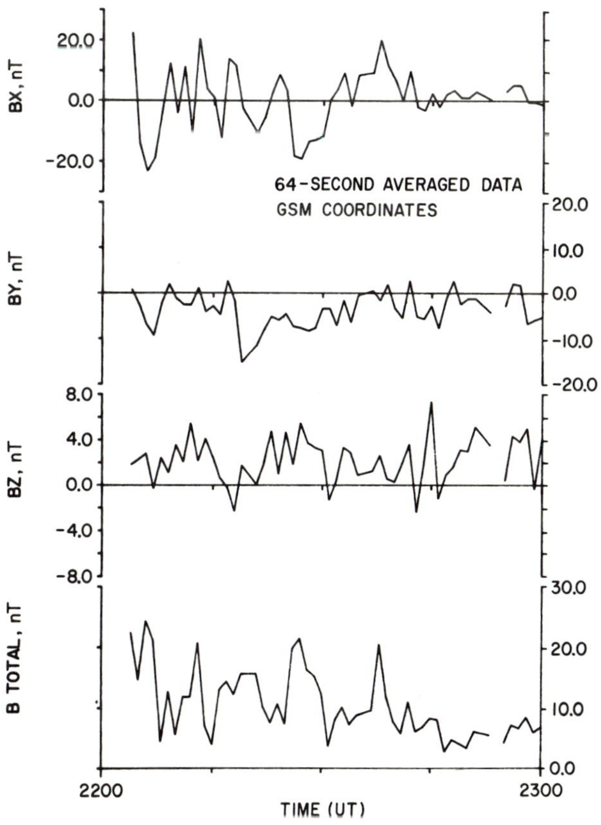

Figure 4 — The magnetic field components during the 2200–2300 UT interval on March 25, 1980 are shown. The B_x component reverses sign several times during this period.

DISCUSSION

A schematic diagram of the geomagnetic tail during quiet times is given in Fig. 5a. The main regions of interest are (a) the acceleration region in the deep tail, indicated by cross-hatching, which generates the streaming plasmas in the boundary layer; (b) the plasma-sheet boundary layer, indicated by the dotted region in which accelerated plasmas are detected; and (c) the central plasma sheet, interior to the boundary layer, consisting of plasma with high thermal energy but low bulk speeds. During active periods the tail thins, causing the boundary to move toward the neutral-sheet region. The essential question in this case is whether the thinning results in formation of a plasmoid, with a magnetic neutral line close to the earth, or whether thinning is accompanied by the earthward extension of the deep-tail acceleration region. This latter possibility is sketched in Fig. 5b, while near-earth neutral-line formation is shown in Fig. 5c. The essential difference between these possibilities is that in case 5b no magnetic flux is disconnected from the tail, while in Fig. 5c a plasmoid is ejected from the magnetosphere.

While a boundary-layer dynamics model is being developed,[14] it is still in a preliminary stage and does not provide for quantitative predictions that can be compared with observations. The neutral-line model, however, can be compared with our observations. The bulk speed of the outflowing plasma is predicted to be the local Alfvén speed, V_A, with a maximum velocity of 2 V_A.[15] Strictly speaking, the predicted outflow velocity is directed perpendicular to the magnetic field. During the two-hour interval, the magnetic field undergoes a number of fluctuations (see Fig. 4), but as B_x reverses sign many times, and the bulk speed is generally earthward, it is safe to assume that V is essentially perpendicular to B. In our case the local Alfvén speed varies from 360 km s^{-1} to 170 km s^{-1}, while the measured bulk speed is approximately two to three times this value. Further, the predicted speed at which the plasmoid retreats is $\sim V_A$. This is inconsistent with the duration of these flows for 40 minutes, since the neutral line would have retreated tailward by ~ 100 R$_e$ in this interval. It should be noted that during the entire time, the observed density and thermal energy indicate that the spacecraft never enters the lobe but remains in the plasma-sheet boundary layer.

We can make some rough estimate of the amount of magnetic flux that must reconnect in order to supply the measured energetic plasma flow in a steady-state reconnection model. The average density, thermal energy, and bulk speed are approximately $N \simeq 0.6$ cm^{-3}, $kT \simeq$ 4.5 keV, $V \simeq 800$ km s^{-1}. Energy transport per unit area for the 40-minute period is $\simeq 8.3 \times 10^2$ erg cm^{-2}. This must be supplied by the lobe magnetic field. Assuming an average magnetic field strength of 40 nT, the thickness, Δz, of the reconnected flux region is given by

$$\left(\frac{B^2 \, \Delta z}{8\pi} \right) A_{xy} = (8.3 \times 10^2) \, A_{yz}$$

where A_{xy} is the cross-sectional area in the x-y plane over which reconnection occurs, and A_{yz} is the cross-sectional area in the y-z plane threaded by streaming

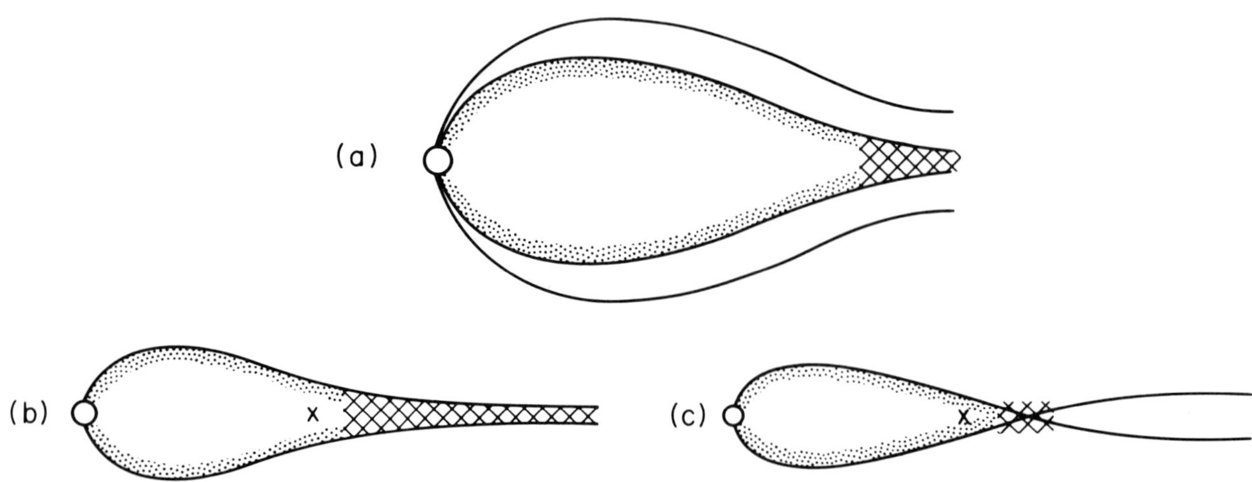

Figure 5 — This sketch shows two possible interpretations of the observed data. The cross-hatched area shows acceleration regions, while the dotted area indicates the location of the plasma-sheet boundary layer. (a) Figure corresponds to the observations on May 25, 1978 of a central plasma sheet bordered by the boundary layers. (b) The active magnetotail is sketched under conditions of extreme thinning. The acceleration process extends to lower equatorial radial distances close to the spacecraft, which is marked by an X. (c) An alternative explanation in which an X-line forms tailward of the spacecraft. A new acceleration region is created near the spacecraft, while the distant acceleration region is presumably still located in the deep tail.

plasma. The spatial extent in the x direction of the neutral line is small,[16] $\Delta x \leq 1$ R_e. The thickness of the boundary layer is estimated to be 1–2 R_e.[5] The value of Δz is then ≥ 200 R_e.

In order to maintain the observed plasma flow away from the reconnection region, the entire tail would be eroded of magnetic flux. Even in nonsteady-state simulations[17] there is no set of conditions that can yield energetic flows over such an extended period.

A more fundamental problem in near-earth reconnection is the exclusion of the plasma-sheet boundary layer in virtually all existing models. It is unclear how this region will affect neutral-line formation at small radial distances. If the acceleration process that forms the boundary layer is provided by deep tail reconnection it would be reasonable to expect a distinctive change in the velocity distributions as a near-earth neutral line becomes dominant. Yet no such qualitative change is observed, as we have pointed out above.

An interpretation based on a boundary layer dynamics model such as that sketched in Fig. 5b is still speculative. The cause of the extreme thinning of the plasma sheet is not supplied by the model. The acceleration process in the boundary layer dynamics model is provided by nonadiabatic motions of ions and electrons in the cross-tail current sheet.[18] Observational evidence confirming that the relative motion of ions and electrons produces the neutral sheet current is shown by Frank et al.[19] with analysis of ISEE plasma measurements near the neutral sheet. Preliminary calculations of the ion velocity distributions near the current sheet show that observations such as those presented in this study can be approximated by this acceleration mechanism (Speiser, private communication). Further investigation of this possibility is being pursued.

SUMMARY

In a detailed examination of the velocity distributions observed near the neutral sheet during a period of extended activity, we find that the observations are qualitatively identical with those of the plasma sheet boundary layer. The processes that produce this streaming plasma during substorms must be similar to those that create the boundary layer during quiet times. Comparison between the measured plasma and magnetic field values and those predicted by the near-earth neutral-line model reveals a number of discrepancies. A simple observation of strong flow in the neutral sheet during substorms

does not in itself constitute verification of near-earth reconnection. One possible explanation for such events is that the acceleration region, normally located at approximately 100 R_e from the earth, can extend to the near-earth region.

ACKNOWLEDGMENT—We wish to thank Dr. C. T. Russell for use of the ISEE 1 magnetic field data. This research was supported in part by the National Aeronautics and Space Administration under contract NAS5-26257 and grants NGL-16-001-002 and NAG5-295 and by the Office of Naval Research under grant N00014-76-C-0016.

REFERENCES

[1] K. Schindler, "A Theory of the Substorm Mechanism," *J. Geophys. Res.* **79**, 2803 (1974).

[2] A. A. Galeev, F. V. Coroniti, and M. Ashour-Abdalla, "Explosive Tearing Mode Reconnection in the Magnetospheric Tail," *J. Geophys. Res.* **5**, 707 (1978).

[3] C. Y. Huang and L. A. Frank, "A Statistical Study of the Central Plasma Sheet," *Geophys. Res. Lett.* (submitted, 1986).

[4] A. T. Y. Lui, "Observations on Plasma Sheet Dynamics During Magnetospheric Substorms," in *Dynamics of the Magnetosphere,* S. Akasofu, ed., p. 563 (1979).

[5] R. J. DeCoster and L. A. Frank, "Observations Pertaining to the Dynamics of the Plasma Sheet," *J. Geophys. Res.* **84**, 5099 (1979).

[6] G. K. Parks, C. S. Lin, K. A. Anderson, R. P. Lin, and H. Reme, "ISEE 1 and 2 Particle Observations of the Outer Plasma Sheet Boundary," *J. Geophys. Res.* **84**, 6471 (1979).

[7] T. E. Eastman, L. A. Frank, and C. Y. Huang, "The Boundary Layers as the Primary Transport Regions of the Earth's Magnetotail," *J. Geophys. Res.* **90**, 9541 (1985).

[8] L. R. Lyons and D. S. Evans, "An Association Between Discrete Aurora and Energetic Particle Boundaries," *J. Geophys Res.* **89**, 2395 (1984).

[9] L. A. Frank, D. M. Yeager, H. D. Owens, K. L. Ackerson, and M. R. English, "Quadrispherical Lepedeas for ISEE's -1 and -2 Plasma Measurements," *IEEE Trans. Geosci. Electron.* **16**, 221 (1978).

[10] E. W. Hones, Jr., J. R. Asbridge, and S. J. Bame, "Time Variations of the Magnetotail Plasma Sheet at 18 R_e Determined from Concurrent Observations by a Pair of Vela Satellites," *J. Geophys. Res.* **76**, 4402 (1971).

[11] T. E. Eastman, L. A. Frank, W. K. Peterson, and W. Lennartsson, "The Plasma Sheet Boundary Layer," *J. Geophys. Res.* **89**, 1553 (1984).

[12] D. J. Williams, "Energetic Ion Beams at the Edge of the Plasma Sheet: ISEE 1 Observations Plus a Simple Explanatory Model," *J. Geophys. Res.* **86**, 5507 (1981).

[13] B. T. Tsurutani, D. E. Jones, J. A. Slavin, D. G. Sibeck, and E. J. Smith, "Plasma Sheet Magnetic Fields in the Distant Tail," *Geophys. Res. Lett.* **11**, 1062 (1984).

[14] G. Rostoker, T. E. Eastman, C. Y. Huang, L. A. Frank, A. T. Y. Lui, D. G. Mitchell, L. R. Lyons, and G. K. Parks, *Eos* **65**, 1046 (1984).

[15] S. W. H. Cowley, "Plasma Populations in a Simple Open Model Magnetosphere," *Space Sci. Rev.* **26**, 217 (1980).

[16] E. W. Hones, Jr., "Substorm Processes in the Magnetotail," *J. Geophys. Res.* **82**, 5633 (1977).

[17] J. Birn, "Computer Studies of the Dynamic Evolution of the Geomagnetic Tail," *J. Geophys. Res.* **85**, 1214 (1980).

[18] L. R. Lyons and T. W. Speiser, 'Evidence for Current Sheet Acceleration in the Geomagnetic Tail," *J. Geophys. Res.* **87**, 2276 (1982).

[19] L. A. Frank, C. Y. Huang, and T. E. Eastman, "Currents in the Earth's Magnetotail," in *Magnetospheric Currents,* T. A. Potemra, ed., Geophysical Monograph 28, American Geophysical Union (1984).

PLASMA AND MAGNETIC FIELD VARIATIONS IN THE DISTANT MAGNETOTAIL ASSOCIATED WITH NEAR-EARTH SUBSTORM EFFECTS

D. N. Baker, S. J. Bame, D. J. McComas, and R. D. Zwickl*

J. A. Slavin and E. J. Smith[†]

Examination of many individual event periods in the ISEE 3 deep-tail data set has suggested that magnetospheric substorms produce a characteristic pattern of effects in the distant magnetotail. During the growth, or tail energy storage phase of substorms, the magnetotail appears to grow diametrically in size, often by many earth radii (R_e). Subsequently, after the substorm expansive phase onset at earth, the distant tail undergoes a sequence of plasma, field, and energetic particle variations as large-scale plasmoids move rapidly down the tail following their disconnection from the near-earth plasma sheet. ISEE 3 data are appropriate for the study of these effects since the spacecraft remained fixed within the nominal tail location for long periods. Using newly available auroral electrojet indices (*AE* and *AL*) and geostationary orbit particle data to time substorm onsets at earth, we have performed superposed epoch analyses of ISEE 3 and near-earth data prior to, and following, substorm expansive phase onsets. These analyses quantify and extend substantially our understanding of the deep-tail pattern of response to global substorm-induced dynamical effects.

INTRODUCTION

The ISEE 3 extended mission allowed a detailed exploration of the earth's magnetotail in a geocentric distance range (80–240 R_e) not previously studied. Examination of the ISEE 3 data has shown that the magnetotail beyond lunar orbit (~ 60 R_e) retains its qualitative near-earth structure,[1,2] but a significant and systematic quantitative evolution of tail properties also occurs with increasing radial distance.[3,4] These latter, systematic studies show that the magnetotail is well ordered out to at least ~ 240 R_e. In the near-earth magnetotail one identifies a plasma sheet separating two tail lobes that are nearly devoid of measurable plasma. Very close to the near-earth magnetopause one finds boundary-layer plasmas with densities, temperatures, and bulk velocities approaching those of the adjacent magnetosheath. As demonstrated, for example, in the analyses of Zwickl et al.,[3] the magnetotail plasma evolves such that strong, tailward bulk flow becomes the dominant characteristic of all of the distant tail plasmas. The average plasma sheet density decreases while the average lobe density increases such that the density distributions of these two regions largely overlap in the distant tail. The distant plasma sheet remains characteristically hotter than the distant lobes so that the current reversal region in the tail midplane remains a relatively high-β region.[4] Nonetheless, the boundary layers of the distant tail appear to have largely subsumed the empty lobes by ~ 240 R_e.

Among the more notable results from the ISEE 3 geotail mission were the recurring sequences of plasma and field variations found during enhanced geomagnetic activity at earth. During the substorm energy storage (growth) phase, the distant tail was inferred to grow substantially in diameter.[5] Following substorm expansive phase onsets at earth, the ISEE 3 data were often found to be consistent with the passage of hot plasma blobs containing closed magnetic field loops. These self-contained plasma structures, called plasmoids,[6,7] were found to be traveling tailward at high speeds (500–1000 km/s). They also were inferred to produce strong compressive effects in the tail lobes, dubbed traveling compression regions (TCRs), as they traveled through the distant tail.[8]

On the basis of many individual case studies as referenced above, a composite picture of the distant geotail response to substorms has emerged. To date, however, there has not been a systematic assemblage of data from near-earth and distant-tail sources to characterize fully the global magnetospheric behavior during substorms. In this paper, we use the method of superposed epoch analysis in order to determine the average plasma and field properties of the distant tail during substorms, and we relate these averaged data to similarly superposed near-earth measures of substorm effects.

THE NEAR-EARTH SUBSTORM SEQUENCE

A phenomenological model of the substorm sequence is shown in the noon-midnight cross-sectional diagram of Fig. 1 (see also, e.g., Russell and McPherron[9]). In this model, enhanced geomagnetic activity is initiated by the southward turning of the interplanetary magnetic field (IMF) (see Dungey[10]). Interplanetary field lines merge on the dayside with terrestrial field lines to form "open" magnetic flux tubes. These open field lines are then dragged by the shocked solar wind flow into the nightside magnetosphere where they form the tail lobes. Since the open flux tubes contain shocked solar wind plasma and thus represent a significant amount of magnetic and plasma energy, the dayside merging and subsequent addition of flux tubes to the nightside represents a large increase in the stored energy content of the magnetotail.

*University of California, Los Alamos National Laboratory, Los Alamos, New Mexico 87545.

[†]California Institute of Technology, Jet Propulsion Laboratory, Pasadena, California.

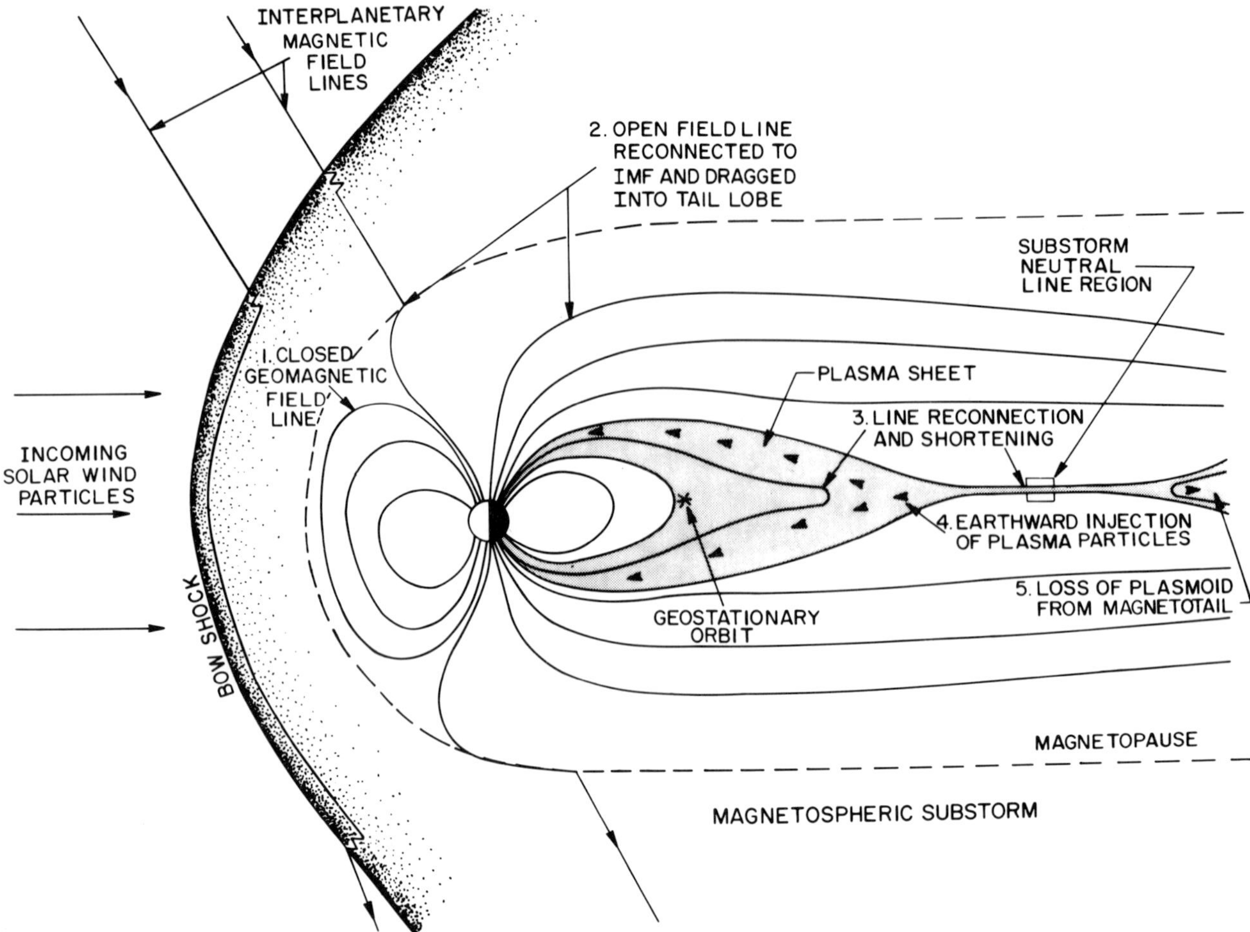

Figure 1—Features of magnetic substorms inferred from many years of near-earth observations. Interplanetary field lines interconnect with terrestrial field lines near the front of the magnetosphere and are dragged back to form the tail lobes, carrying plasma and energy with them. The net effect is to increase the magnetic field energy density in the lobes. Far down the tail (off the figure to the right), lobe field lines reconnect to form the plasma sheet. When the midsection of the plasma sheet becomes thin enough (due to increased magnetic pressure from the tail lobes), field lines reconnect there also, forming the magnetic neutral region shown. The release of energy that results from the reconnection heats and accelerates magnetotail plasma, driving part of the plasma back along field lines into the earth's ionosphere near the poles. Plasma is also ejected to the right (as a plasmoid), flowing at great velocity down the tail.

In the above picture, the occurrence of dayside merging must eventually be balanced by reconnection of magnetic flux on the nightside. During relatively quiet (near-equilibrium) conditions it is assumed that the requisite nightside reconnection would occur far down the tail (far to the right of Fig. 1) and would form the quiet-time plasma sheet. During periods of high geomagnetic activity, with the rapid addition of magnetic energy to the tail lobes, the midsection of the plasma sheet thins substantially. This produces a very taillike field even as close to earth as 6.6 R_e (the geostationary orbit), as shown in Fig. 1.[11]

When the plasma sheet thins sufficiently, the model suggests that a near-earth reconnection region (neutral line) forms and essentially pinches off an azimuthally limited portion of the plasma sheet. This closed magnetic bubble is the plasmoid. IMP 8 measurements, studied by means of superposed epoch methods,[12] provided rather convincing evidence that this sequence of events actually occurs.

The substorm neutral-line region is thought to occur typically at 10–20 R_e in the nightside plasma sheet. There the reconnection process is hypothesized to heat plasma and to jet it both earthward and tailward at high speeds. On the earthward side of the neutral line, field lines rapidly snap back toward the earth due to magnetic tension and in the process carry hot plasma and energetic particles deep into the outer trapping region (near 6.6 R_e) giving rise to strong "injection" events near local midnight. Also, the magnetotail dissipation at the neutral line is thought to produce strong auroral effects at the foot of field lines where they map into the ionosphere.

The near-earth features of the above model have been rather well-established by many and varied observations (see Ref. 13 and references therein). The formation of plasmoids, however, was a more speculative aspect of the model, which required a systematic analysis of data from the distant tail. The ISEE 3 data studied here are well-suited for this purpose.

A SUPERPOSED EPOCH DATA ANALYSIS

A convenient period of time was identified from January 26 to February 8, 1983 when ISEE 3 was relatively fixed in location in the nominally aberrated tail at ~220 R_e geocentric distance.[5,6] During this 14-day interval there were some 20 plasmoid events detected by ISEE 3 of which ~15 were "clean" cases without substantial data gaps and were such that we had a geostationary orbit spacecraft in the midnight sector during the substorm sequence. We use these events in this study to determine average plasmoid and near-earth substorm properties.

The data used in this analysis are the ISEE 3 plasma electron moments (n, V, T_e) provided in 1-min bins, plus ~1-min averages of the ISEE 3 magnetic field components (B_x, B_y, B_z, and $|B|$). We have also averaged data from the Los Alamos geostationary orbit spacecraft 1977-007, 1981-025, and 1982-019 for 1-min intervals. For the present study we have used energetic electron data from whichever of these spacecraft was in the midnight sector during each of the plasmoid events. Finally, we used the 1-min resolution AE and AL indices (World Data Centers A and C) which have been specially prepared for the ISEE 3 deep tail mission interval.

The types of data that have gone into our study are shown in Fig. 2. They are common timescale plots of ISEE 3 bulk speed (V), magnetic field strength (B), and the north-south field components (B_z). In the lower panel of Fig. 2 we show the AL index as a measure of substorm activity at earth. The AL data show that the substorm expansive phase commenced at ~1315 UT. Approximately 25 minutes later ISEE 3 saw a large increase in V as it rapidly entered the plasma sheet (note the reduced, fluctuating $|B|$). B_z first went strongly northward (positive) beginning at 1340 UT and then at ~1400 UT went strongly southward. The period after 1340 UT is therefore identified as a plasmoid in the ISEE 3 data.

Our method of analysis has been to identify plasmoid onset at ISEE 3 by the rapid increase in the plasma bulk flow. This is taken as the zero epoch time for each of our plasmoid events. Such a zero epoch is shown by the vertical dashed line in Fig. 2. For all of our plasmoid events, we have then taken the zero epoch time and used all available data for each event interval, where an event interval is defined as ±2 h from the zero epoch time.

The results of our superposed epoch analysis are shown in Fig. 3. Generally, from $t = -120$ min to $t \simeq -60$ min the plasma flow speed (top panel) is tailward (direction not shown) at ~350 km/s in the event-averaged superposition units. The average flow speed then gradually diminishes until $t = 0$ at which time the flow speed increases substantially. This is the onset of the plasmoid, and the large increase in V at $t = 0$ is expected since this is the signature which was used to define the zero epoch.

All other averaged values (panels 2–5) in Fig. 3 were determined only by reference to the zero epoch time defined by V in the top panel. The average values of $|B|$ are low and fluctuating from $t = -120$ min to

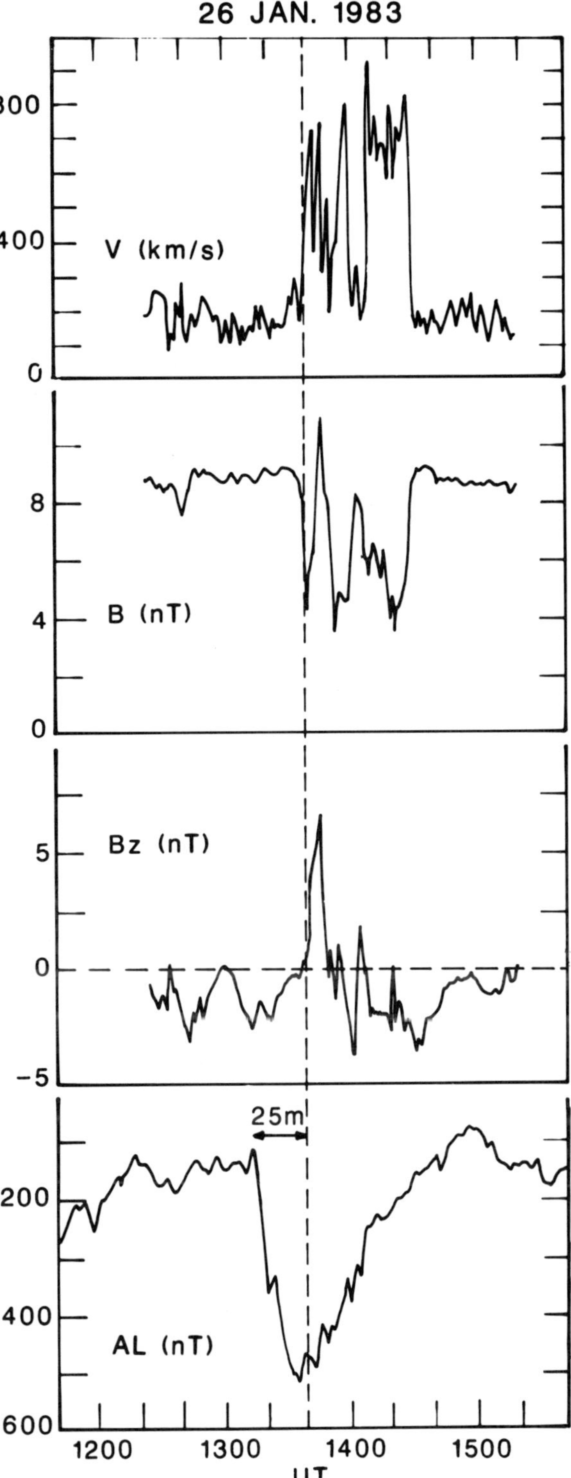

Figure 2—One-minute averages of ISEE 3 data (top three panels as labeled) and the AL index (bottom panel) as used in the superposed epoch analysis discussed in the text. The data here cover the period 1140–1540 UT on January 26, 1983. The vertical dashed line marks a plasmoid arrival at ISEE 3 at ~1340 UT.

$t = -60$ min. For the interval $-60 < t < 0$ min, $|B|$ values are large and steady as in the distant lobes.[2] At

$t = 0$, $|B|$ drops abruptly, consistent with entry into the plasma sheet. The B_z variation in Fig. 3 is notable by the positive rotation of the field orientation near $t = 0$ and by the long period of strongly southward field

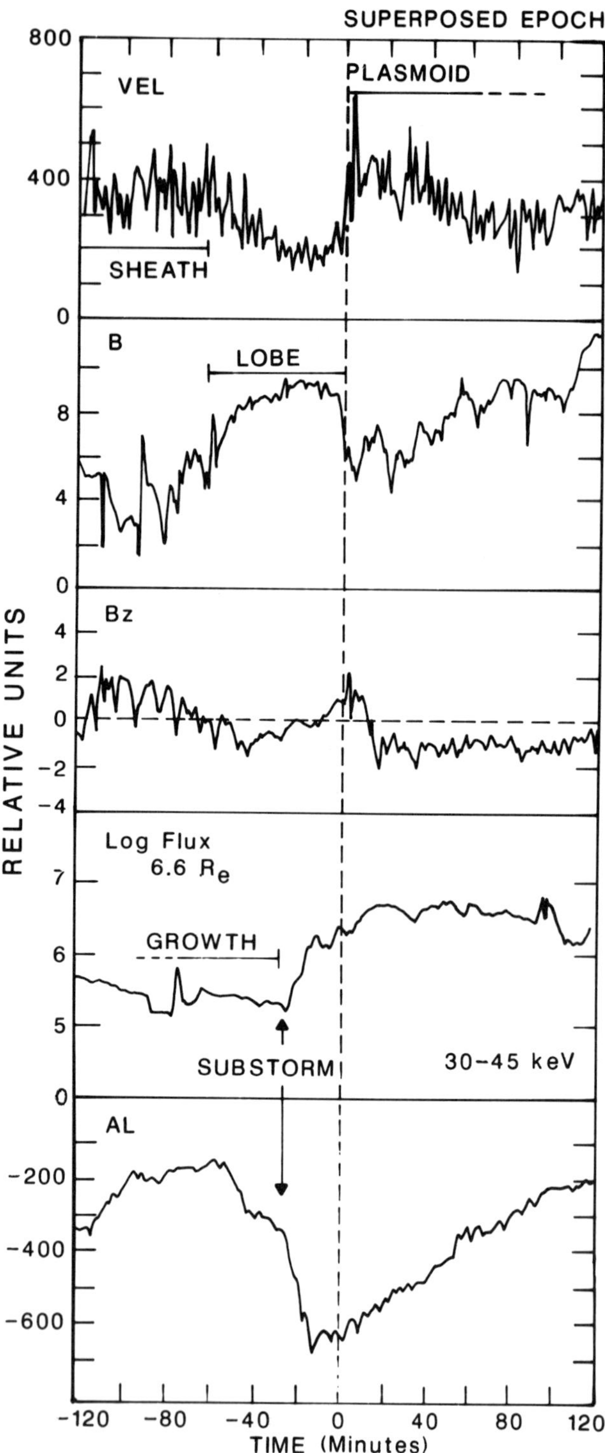

Figure 3—A superposed epoch analysis (as described in the text) for ISEE 3 plasma and magnetic field data (top three panels) and for electron data at 6.6 R_e and the AL index obtained from ground magnetic records. Zero epoch time corresponds to plasmoid arrival at ISEE 3, which was at ~ 220 R_e in the geotail for all of the superposed events.

beyond $t = +20$ min. This is an expected plasmoid feature.[6]

Although not shown here because of space limitations, the superposed values of electron number density (n) and electron temperature (T_e) also show systematic behavior. From $t = -120$ min to $t = -60$ min, the values of n are relatively high and the values of T_e are low as is characteristic of the distant magnetosheath.[1,5] From $t = -60$ min to $t = 0$ min, the density decreases (on average) and the temperature gradually increases. After $t = 0$ (i.e., within the identified plasmoid), the density increases slightly while the temperature jumps up markedly. These latter plasma characteristics are consistent with the statistically determined properties in the distant plasma sheet.[3]

Of substantial interest in Fig. 3 are the lower two panels, which are completely independent of measurements taken at ISEE 3. The fourth panel from the top shows the average 30–45 keV electron flux (log (e⁻/cm²-s-sr-keV)) as a function of epoch time. The flux diminishes gradually but systematically from $t \simeq -80$ min to $t \simeq -25$ min. At $t \sim -25$ min there is a large, rapid increase in the electron flux by about an order of magnitude, or more. This is the average signature of a strong flux injection, i.e., the substorm expansive phase onset. The flux decrease for $t \lesssim -25$ min is consistent with a growth phase feature.[5,13]

Finally, in the bottom panel of Fig. 3. we show the superposed values of the AL index. After being reasonably steady from $t \simeq -100$ min to $t \simeq -60$ min, the AL index drops fairly rapidly in value until $t \simeq -25$ min. At $t \simeq -25$ min, AL begins to drop much more rapidly until reaching a broad minimum at $t \sim 0$. Thereafter, AL gradually and systematically recovers. The superposed AL data would be consistent with a growth phase interval ($-60 \lesssim t \lesssim -25$ min) and an expansive phase onset ($t \sim -25$ min). This latter time comports well with the average "injection" onset time seen in the energetic electron data at 6.6 R_e (panel 4).

DISCUSSION AND CONCLUSIONS

Based on examination of individual plasmoid cases and the superposed results of Fig. 3, we conclude that the distant tail responds in a very systematic way in association with substorms observed at earth. In the cases analyzed here, we see evidence that on average some 40 to 60 minutes prior to (near-earth) substorm onset there are features in the energetic electrons (6.6 R_e) and in the AL index which indicate the storage of energy in earth's magnetotail.[14] This is the storage, or growth, phase of the substorm. For the cases examined here, at the beginning of the growth phase, we find that on average ISEE 3 was located in a sheath-like environment. After roughly 20 minutes of growth phase development near earth (see Fig. 3), we see that (on average) ISEE 3 enters the tail lobe. Finally, about 25 minutes after the near-earth expansive phase onset, ISEE 3 enters the plasmoid structure that was released from the near-earth plasma sheet at substorm expansive phase onset. We have labeled Fig. 3 to indicate these interpretations.

These results have led us to picture the growth phase as follows (top portion, Fig. 4). A southward turning of the interplanetary field lines enhances the rate at which magnetic reconnection occurs on the dayside of the magnetosphere. As these newly connected lines are dragged into the tail, magnetic and plasma energy are added at an enhanced rate, and the tail grows in diameter. During our late January and early February observations, ISEE 3 was positioned in relatively stationary fashion near the distant magnetotail surface so that growth of the tail would eventually cause it to envelop the spacecraft.

Of course, with a single spacecraft it is virtually impossible to specify one's exact location relative to the center of the tail or the magnetopause boundary. As-

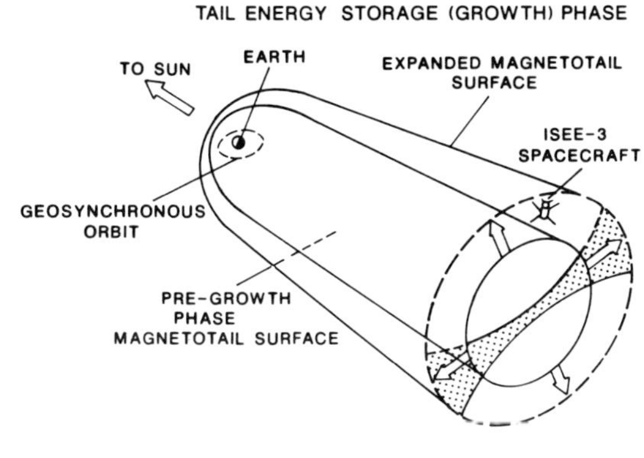

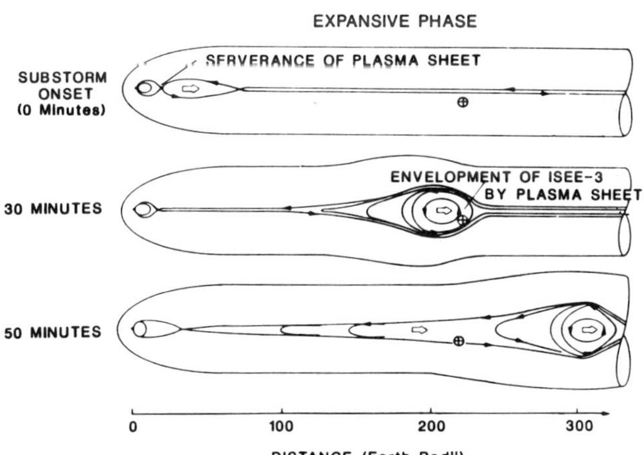

Figure 4—Growth-phase model (top). Before the growth phase of a substorm begins, the magnetotail is relatively small in cross section and, quite often, ISEE 3 resides outside. As plasma and energy are added to the tail, the cross section apparently grows until ISEE 3 is enveloped. The data show this as an inward crossing from outside the magnetosphere to one of the tail lobes. Plasmoid model (bottom) (after Hones et al.[6]). After plasma and energy have built up in the magnetotail during the growth phase, a portion of the plasma sheet is severed by reconnection close to earth, causing substorm onset. The plasmoid thus formed moves rapidly down the tail, eventually enveloping ISEE 3 about 25 minutes later at ~200 R_e from earth. The envelopment is seen by instruments in the spacecraft as a crossing into the plasma sheet.

suming a nominally aberrated tail, during the period used here for our superposed epoch analysis, ISEE 3 was ~10–20 R_e from the expected center of the tail.[5] Magnetic field data[2] suggest a tail radius at $X \sim -200\ R_e$ of $r \sim 30\ R_e$. The fact that ISEE 3 commonly was in the magnetosheath on nearly every day of our analysis interval clearly reinforces the interpretation that the distant tail must flap around substantially due to solar-wind changes. If, in addition to this flapping effect, we assume that there is a 10–20% increase in tail magnetic flux during a substorm growth phase,[5] we then calculate that the distant tail radius can easily increase by 2–4 R_e. This diametrical growth seems to be the pattern revealed in Fig. 3.

The results of our present analysis also confirm another important feature, namely the envelopment of ISEE 3 by plasmoids. This occurs about 20 to 30 minutes after the start of a magnetic disturbance at the earth. Further, numerous examples of data recorded by ISEE 3 show the spacecraft passing from one lobe into the plasma sheet and then back into the same lobe again as if a bulge in the plasma sheet had passed by.

We believe that the composite behavior described above is the signature of plasmoid release in the manner depicted in the lower panel of Fig. 4.[6] A central portion of the plasma sheet is severed close to earth, flows tailward as a plasmoid, and reaches ISEE 3 about 20 to 30 minutes later. The delay in arrival is consistent with the observed bulk flow speeds for the plasma of about 500 to 1000 km/s. The plasmoid's dimensions increase substantially as it is ejected down the tail because it is moving through regions of decreasing magnetic pressure. The large size of the plasmoid also results in a bulge in the magnetopause that travels along with it, and this feature also produces TCRs.

ISEE 3 provided two other pieces of information that support the plasmoid model as the correct interpretation of the substorm-related encounters with the plasma sheet. First, during these encounters the magnetic field in the newly expanded plasma sheet usually points steeply northward for a few minutes, then points southward (often steeply) for a longer period of time. Examination of Fig. 4 shows that this is the expected magnetic signature of a passing plasmoid—at the front of the plasmoid the field lines point northward, just past its center they reverse and point southward, and toward the back they gradually resume a more radial direction. This feature is confirmed by our superposed epoch results (Fig. 3). ISEE 3 energetic particle data also support the plasmoids interpretation (e.g., Ref. 7).

Alternative models to the near-earth neutral-line/plasmoid picture are currently being proposed (e.g., Ref. 15). These models suggest that the plasma-sheet boundary layer is of dominant dynamical importance in the near-earth region and that substorm processes generally are controlled by the boundary layer. At present, it is not clear to us how the boundary-layer dynamics model would account for the kind of systematic plasma and magnetic field features that emerge from our superposed epoch analysis. It would appear to be an interesting and important test of any alternatives to the plasmoid model to determine if they can account in a natural way for the observed deep-tail sequence.

Based on our current understanding, we conclude from our superposed epoch analysis that:

1. The substorm energy storage (growth) phase commences 30–60 min before the expansive onset and produces clear effects at ~ 220 R_e;
2. The *AL* and 6.6 R_e injection criteria give similar expansive phase onset times;
3. The plasmoid reaches ISEE 3 at ~ 220 R_e at a time ~ 25 min. after substorm onset near earth; and
4. The velocity, magnetic field magnitude, and B_z superpositions at ISEE 3 give a good average characterization of plasmoids at ~ 220 R_e.

Thus, examination of the average responses in the ISEE 3 deep-tail data set has shown that magnetospheric substorms regularly produce a quite consistent pattern of effects in the distant magnetotail. During the growth, or tail energy storage, phase of substorms, the distant magnetotail appears to grow in diametrical size, often by many earth radii (R_e). Subsequently, after the substorm expansive phase onset at earth, the distant tail undergoes a clear sequence of plasma, field, and energetic particle variations as large-scale plasmoids move rapidly down the tail following their disconnection from the near-earth plasma sheet. These analyses confirm and extend our understanding of the deep-tail pattern of response to substorm-induced dynamical effects.

ACKNOWLEDGMENT — This research was supported at Los Alamos and JPL by grants from NASA. Work at Los Alamos was done under the auspices of the U.S. Department of Energy.

REFERENCES

[1] S. J. Bame, R. C. Anderson, J. R. Asbridge, D. N. Baker, W. C. Feldman, J. T. Gosling, E. W. Hones, Jr., D. J. McComas, and R. D. Zwickl, "Plasma Regimes in the Deep Geomagnetic Tail: ISEE 3," *Geophys. Res. Lett.* **10**, 912 (1983).

[2] J. A. Slavin, B. T. Tsurutani, E. J. Smith, D. E. Jones, and D. G. Sibeck, "Average Configuration of the Distant (≤ 220 R_e) Magnetotail: Initial ISEE 3 Magnetic Field Results," *Geophys. Res. Lett.* **10**, 973 (1983).

[3] R. D. Zwickl, D. N. Baker, S. J. Bame, W. C. Feldman, J. T. Gosling, E. W. Hones, Jr., D. J. McComas, and E. J. Smith, "Evolution of the Earth's Distant Magnetotail: ISEE 3 Electron Plasma Results," *J. Geophys. Res.* **89**, 11007 (1984).

[4] J. A. Slavin, E. J. Smith, D. G. Sibeck, D. N. Baker, R. D. Zwickl, and S. −I. Akasofu, "An ISEE 3 Study of Average and Substorm Conditions in the Distant Magnetotail," *J. Geophys. Res.* (in press, 1985).

[5] D. N. Baker, S. J. Bame, R. D. Belian, W. C. Feldman, J. T. Gosling, P. R. Higbie, E. W. Hones, Jr., D. J. McComas, and R. D. Zwickl, "Correlated Dynamical Changes in the Near-Earth and Distant Magnetotail Regions: ISEE 3," *J. Geophys. Res.* **89**, 3855 (1984).

[6] E. W. Hones, Jr., D. N. Baker, S. J. Bame, W. C. Feldman, J. T. Gosling, D. J. McComas, R. D. Zwickl, J. A. Slavin, E. J. Smith, and B. T. Tsurutani, "Structure of the Magnetotail at 220 R_e and Its Response to Geomagnetic Activity," *Geophys. Res. Lett.* **11**, 5 (1984).

[7] M. Scholer, G. Gloeckler, B. Klecker, F. M. Ipavich, and D. Hovestadt, "Fast Moving Plasma Structures in the Distant Magnetotail," *J. Geophys. Res.* **89**, 6717 (1984).

[8] J. A. Slavin, E. J. Smith, B. T. Tsurutani, D. G. Sibeck, G. L. Siscoe, H. J. Singer, D. N. Baker, J. T. Gosling, E. W. Hones, Jr., and F. L. Scarf, "Substorm Associated Traveling Compression Regions in the Distant Tail: ISEE 3 Geotail Observations," *Geophys. Res. Lett.* **11**, 657 (1984).

[9] C. T. Russell and R. L. McPherron, "The Magnetotail and Substorms," *Space Sci. Rev.* **15**, 205 (1973).

[10] J. R. Dungey, "Interplanetary Magnetic Field and the Auroral Zones," *Phys. Rev. Lett.* **6**, 47 (1961).

[11] R. L. McPherron, "Growth Phase of Magnetospheric Substorms," *J. Geophys. Res.* **28**, 5591 (1970).

[12] J. W. Bieber, E. C. Stone, E. W. Hones, Jr., D. N. Baker, and S. J. Bame, "Plasma Behavior During Energetic Electron Streaming Events: Further Evidence for Substorm-Associated Magnetic Reconnection," *Geophys. Res. Lett.* **9**, 664 (1982).

[13] D. N. Baker, S.-I. Akasofu, W. Baumjohann, J. W. Bieber, D. H. Fairfield, E. W. Hones, Jr., B. H. Mauk, R. L. McPherron, and T. E. Moore, "Substorms in the Magnetosphere," Chapter 8 in *Solar Terrestrial Physics—Present and Future,* NASA Pub. 1120, Washington, D.C. (1984).

[14] D. N. Baker, T. A. Fritz, R. L. McPherron, D. H. Fairfield, Y. Kamide, and W. Baumjohann, "Magnetotail Energy Storage and Release During the CDAW-6 Substorm Analysis Intervals," *J. Geophys Res.* **90**, 1205 (1985).

[15] G. Rostoker and T. Eastman, "A Boundary Layer Model for Magnetospheric Substorms," *J. Geophys. Res.* (submitted, 1986).

A CASE STUDY OF THE LARGE-SCALE RESPONSE OF THE MAGNETOSPHERE TO A SOUTHWARD TURNING OF THE IMF

J. A. Sauvard, J. P. Treilhou, A. Saint-Marc, J. Dandouras, H. Rème[*]

A. Korth, G. Kremser[†]

G. K. Parks[‡]

A. N. Zaitzev, V. Petrov[§]

L. Lazutine[‖]

R. Pellinen[#]

The magnetospheric response of the directional changes of the interplanetary magnetic field (IMF) that occurred on March 4, 1979 was monitored by particle and field instruments at the nightside geostationary orbit, in the geomagnetic tail, and on a high-altitude balloon and on the ground in the auroral zone. This coordinated data set showed that, coincident with the southward turning of the IMF at $\simeq 2117$ UT, (a) the electrojet intensity increased and the aurora moved equatorward, (b) the H component of the geostationary magnetic field decreased while both V and D components increased, a consequence of enhanced dawn-dusk current, (c) trapped particle fluxes at the geostationary orbit decreased adiabatically, and (d) weak fluxes of apparently trapped energetic electrons of average energy about 100 keV precipitated into the auroral ionosphere. Approximately 40 minutes later, the IMF turned abruptly northward for a few minutes. Upon returning to the southward direction, also abruptly, the same magnetospheric response was again observed by the various instruments. A series of weak, short-lived substorms is superimposed on these large-scale changes during the period of southward directed IMF. A major substorm then occurred when ϵ was at maximum. Finally, magnetic and auroral activity subsided as soon as the IMF B_z component reached positive values. This study suggests that for the March 4 event, the bulk of the energy dissipated in the magnetosphere and auroral ionosphere was directly extracted from the solar wind.

INTRODUCTION

The exact response of the magnetosphere to changes of the interplanetary magnetic field (IMF) still remains a puzzle. This point was emphasized in a recent article by Rostoker et al.[1] who compared the responses of the electrojet to the changes of the IMF direction and found that, while southward turnings of the IMF lead to electrojet intensifications and low level substorm activity, northward turnings of the IMF appear capable of triggering large substorm disturbances.

To better identify the dynamical changes of the entire magnetosphere associated with the southward turning of the IMF, our study will include a broader database. A period of enhanced magnetic activity that occurred on March 4, 1979 was extensively covered by spacecraft in the solar wind (ISEE 3, IMP 8), in the nightside magnetosphere, at synchronous altitude (GEOS 2) and in the plasma sheet (ISEE 1 and 2), by balloon borne experiments (SAMBO 79 campaign), by three meridional chains of ground based magnetometers, and by all-sky cameras. A detailed study of this coordinated data set reveals that the effects of the southward turning of the IMF can be detected with time delays of less than about 5 minutes in the auroral ionosphere and inner magnetosphere. The observations are consistent with an interpretation that IMF changes can directly drive a series of magnetospheric processes including the intensification of auroral electrojets, auroral equatorward motions, changes of equatorial particle distributions induced by magnetic reconfiguration of the inner magnetosphere, and relativistic electron precipitation. Superimposed on these changes, a series of low-level substorms is detected soon after the turning of the southward IMF.

DATA ANALYSIS

In the sections that follow we present the relationship between the IMF and the ground magnetic activity; then we focus on the associated particles and fields dynamical changes observed at the geostationary orbit; following, we relate the B field reconfiguration of the inner magnetosphere to the relativistic electron precipitation observed in the nightside auroral zone; and finally we describe the plasma-sheet behavior pertaining to this period.

---

[*]CESR, CNRS-Toulouse University, Toulouse, France.
[†]Max-Planck Institut für Aeronomie, Katlenburg-Lindau, Federal Republic of Germany.
[‡]Geophysics Program AK-50, University of Washington, Seattle, Washington.
[§]IZMIRAN, USSR Academy of Sciences, Moscow Region, USSR.
[‖]PGI, Kola Branch of the USSR Academy of Sciences, Apatity, USSR.
[#]Finnish Meteorological Institute, Helsinki, Finland.

Electrojets

The variations of the *AU* and *AL* indices[2] from 1800 UT on March 4, 1979, to 0600 UT on March 5 are shown in Fig. 1.

The north-south component of the IMF displayed in Fig. 2 has been obtained from the IMP 8 and ISEE 3 spacecraft. The ISEE 3 B_z profile has been shifted by 65 minutes in order to fit the available IMP 8 data.

At earth orbit, the IMF turns southward at $\simeq 2117$ UT. A brief northward turning takes place for some minutes around 2200 UT while the final northward turning occurs during the 2300–2320 UT interval. The eastward electrojet, as monitored by the horizontal component of the geomagnetic field measured at Fort Churchill, exhibits two strong increases in the time interval 2100–2230 UT (14–16 MLT) that are clearly related to the IMF changes. A time delay of less than about 5 minutes is observed between the IMF southward turning and the strengthening of the eastward electrojet. A similar correlated feature can be seen between the IMF B_z component variations and the westward electrojet, as monitored by the *H* component of the geomagnetic field at Kanin-Nos in the time interval 2100–2235 UT ($\simeq 0.5$–2.0 MLT). The intense geomagnetic bay beginning at 2237 is related to an auroral breakup identified from all-sky camera data, which will be discussed later.

The detailed behavior of the westward electrojet in the near midnight sector has been deduced from three meridional magnetometer chains located in the Soviet Union, at corrected geomagnetic longitudes of 145°, 120° and 115° (Fig. 3). Here the westward electrojet is assumed to be a single horizontal current sheet flowing at 115 km height. Induction effects are taken into account and field-aligned currents are neglected.[3,4] When the electrojets become too wide or shifted from the view of the stations, the uncertainty in the calculations becomes large and this is indicated by a dashed line on the intensity curves.

The substorm onset at $\simeq 2237$ UT is associated with a sudden northward shift of the electrojet boundaries (Fig. 3). Other fast motions of the electrojet boundaries, indicating substorm intensifications, are then seen at $\simeq 2247$, $\simeq 2254$, and $\simeq 2310$ UT ($\phi_m = 115°$, 120°). The preceding period corresponds to a general southward motion of the electrojet and an increase in the horizontal current intensity, which is first detected at the easternmost stations at the time of the IMF B_z reversal. It must be stressed that the southward motion of the electrojet is in good agreement with a transfer of magnetic flux to the tail lobe.[5-8] The electrojet behavior

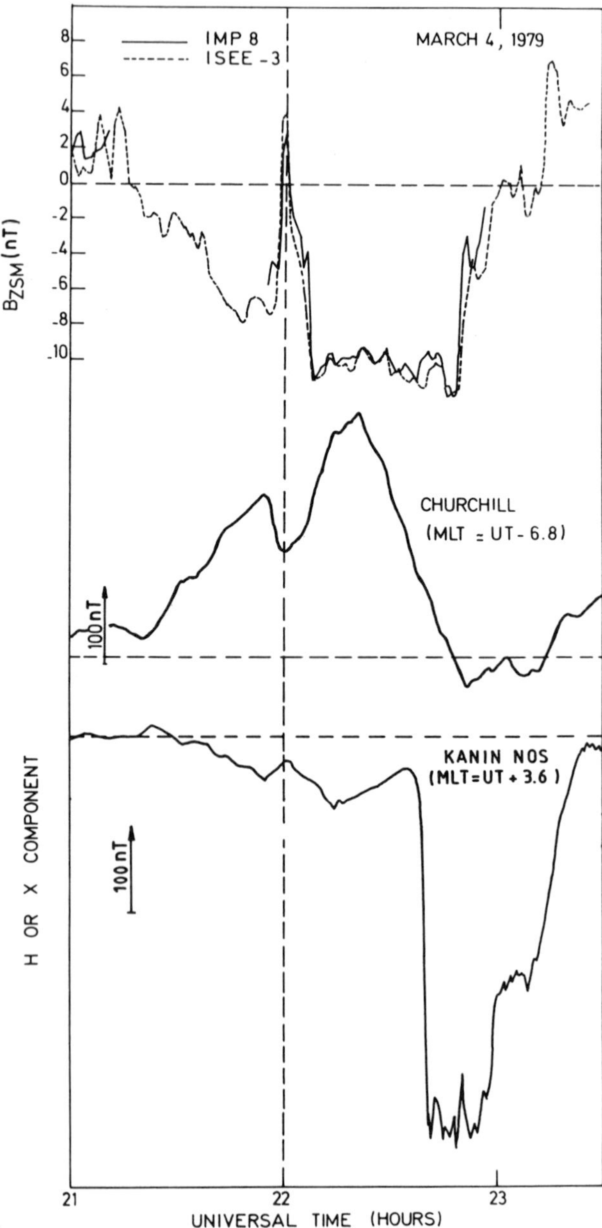

Figure 2—From top to bottom: North-south component of the IMF from the IMP 8 and ISEE 3 spacecraft, in solar magnetospheric coordinates; X component of the ground magnetic field at Fort Churchill; H component of the ground magnetic field at Kanin-Nos.

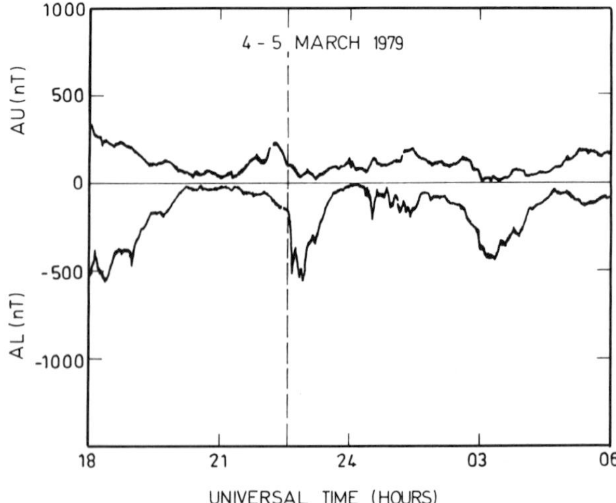

Figure 1—*AU* and *AL* indices from 1800 UT on March 4, 1979 to 0600 UT on March 5, 1979.

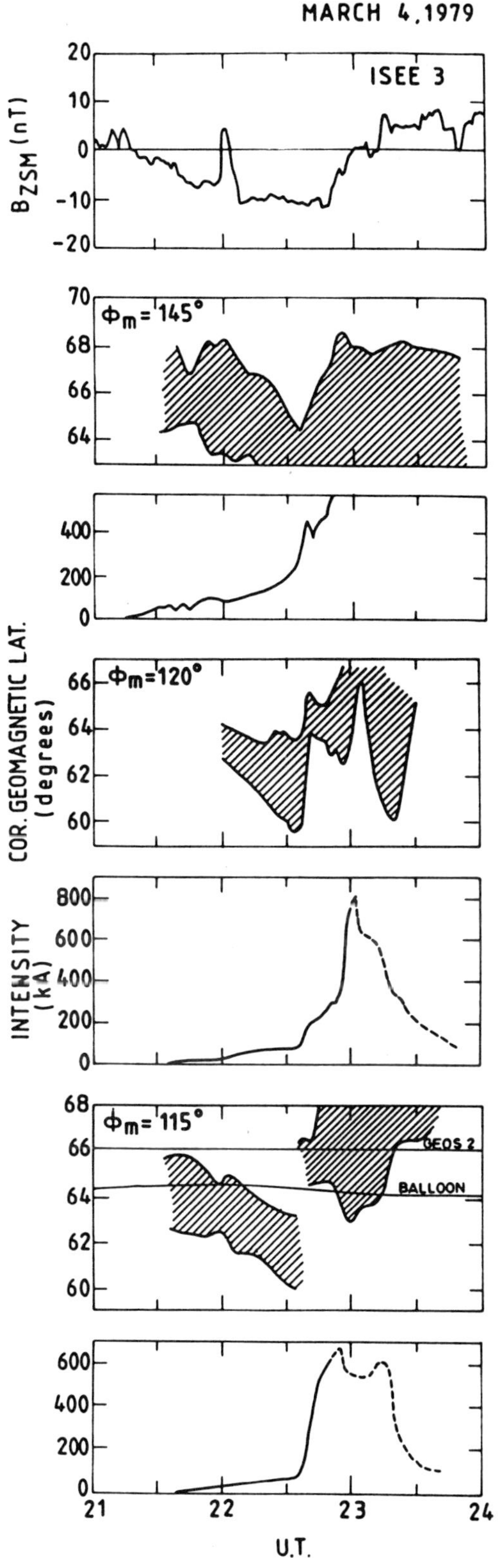

Figure 3—From top to bottom: B_z component of the IMF and computed longitudinal extent and intensity of the westward electrojet for three magnetometer chains located in the post-midnight sector, at corrected geomagnetic longitudes of 145°, 120°, and 115°. The GEOS 2 position, projected along the magnetic field lines, has been computed using the Olson and Pfitzer (1974) magnetospheric model.

here shows that the ionospheric current responds as a whole to the change of the IMF B_z component. Note that several disturbances are superposed on the overall motion of the westward electrojet. The first one, registered at $\simeq 2145$ UT, consists of a fast poleward motion of the northern electrojet boundary followed by an equatorward shift of its southern boundary (Fig. 2, $\phi_m = 145°$). Such features have been described as characteristic of a substorm expansion phase by Kamide and Akasofu.[9] Other weak perturbations, localized in longitude, can be seen at about 2200 UT ($\phi_m = 145°$, $\phi_m = 115°$), about 2213 UT ($\phi_m = 145°$), and about 2220 UT ($\phi_m = 120°$). This type of localized auroral activation has been shown to occur frequently in the DE 1 and Viking auroral oval imaging data (J. Craven and C. Anger, personal communication, 1986).

Outer Radiation Zone

As already shown by Baker et al.,[10] the magnetic field topology at the nightside geostationary orbit is very sensitive to the orientation of the IMF. For the event presented here the magnetic field at the GEOS 2 orbit (MLT $\simeq$ UT + 2.4 H) becomes increasingly tail-like after the IMF southward turning until the strong particle injection is observed at 2237 UT (Fig. 4). This injection, as usual, is associated with a rapid return of the local magnetic field to a dipole-like configuration and coincided with a substorm onset (Fig. 4). It must be stressed that this onset occurs while the IMF is still southward directed. The increasing tail-like B field configuration during the preinjection period has already been interpreted as the result of the increase of the cross-tail current.[11-13] The energetic particles drifting near the equator ($\alpha \simeq 90°$) adiabatically respond to the slow development of this tail-like magnetic field topology by moving earthward, keeping their total energy and first adiabatic invariant constant. The inward motion of the outer radiation zone boundary causes the observed depletion of the 90° pitch angle electron flux.[14] The response of low pitch angle particles to the change of B field due to IMF reversal will be examined in the next section.

We presented data above that showed that this tail-like deformation of the B field at 6.6 R_e occurred in coincidence with the appearance of weak substorms. ULF observations of waves in the Pi 2 range onboard GEOS 2 are shown in Fig. 5. Bursts of activity observed in the V component at $\simeq 2143$, 2215, and 2237 UT are in good agreement with onset times deduced from the westward electrojet changes (Fig. 3). This figure also shows that the wave activity begins as soon as the IMF turns southward. (The activity in the D component at 2154 UT may be associated with the northward turning of the IMF, but the origin remains unknown.) Pi 2 activity is associated with the auroral expansion phase.[15] Hence we conclude that the weak events that occur after $\simeq 2117$ are indeed substorms.

Relativistic Electron Precipitation

Figure 6a shows that as soon as the B field at 6.6 R_e starts to develop a tail-like configuration, the integral

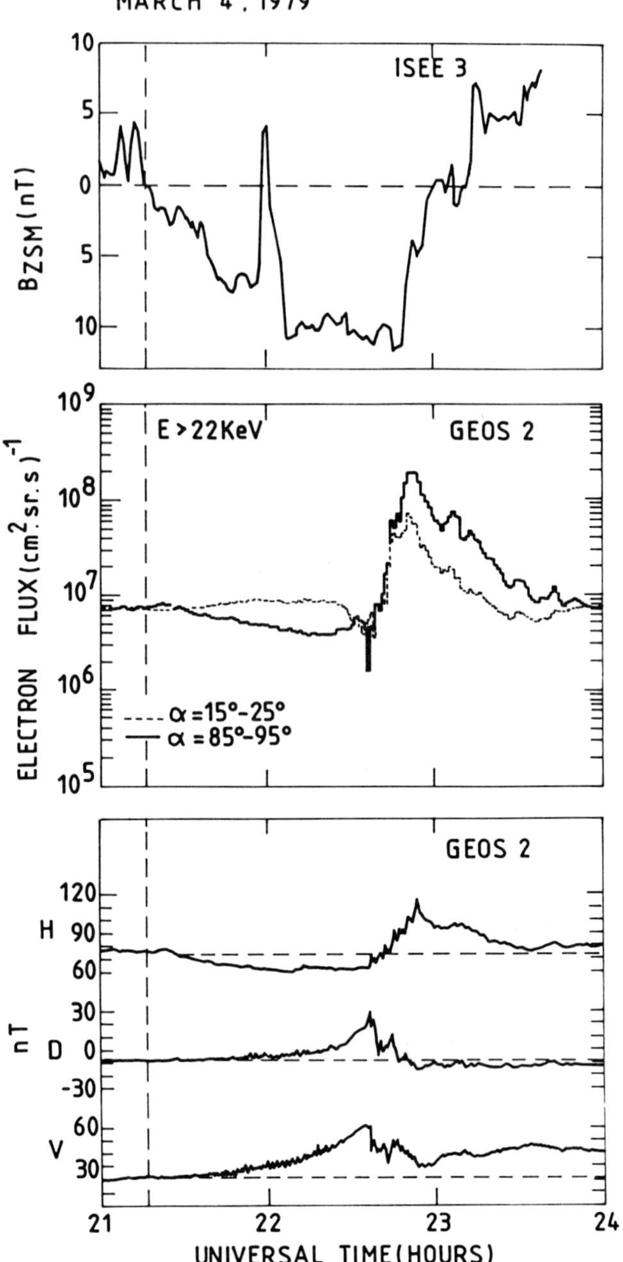

Figure 4—Top panel: IMF B_z component. Middle panel: Integral flux of electrons with energies >22 keV, for two pitch angles outside the loss cone, measured onboard GEOS 2 in the night sector. Bottom panel: The three components of the magnetic field measured onboard GEOS 2. H is northward directed, V is directed toward the earth, and D is eastward directed. The dashed vertical line refers to the southward turning of the IMF.

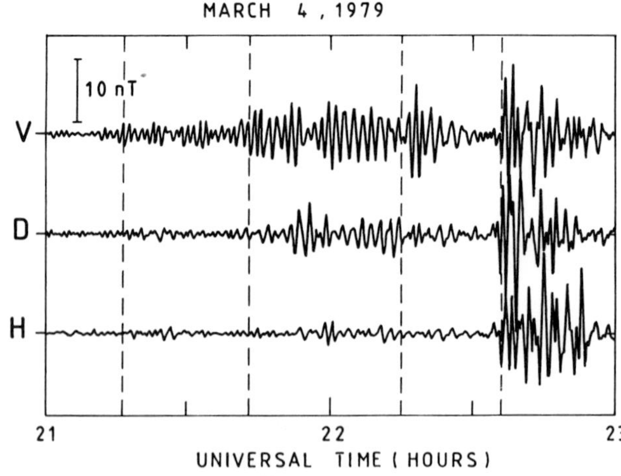

Figure 5—GEOS 2 magnetometer filtered Pi 2 data taken on March 4, 1979 between 2100 and 2300 UT. The vertical solid line indicates the time when the IMF turns southward while the dashed lines refer to bursts of Pi 2 activity related to the electrojet perturbations displayed in Fig. 3.

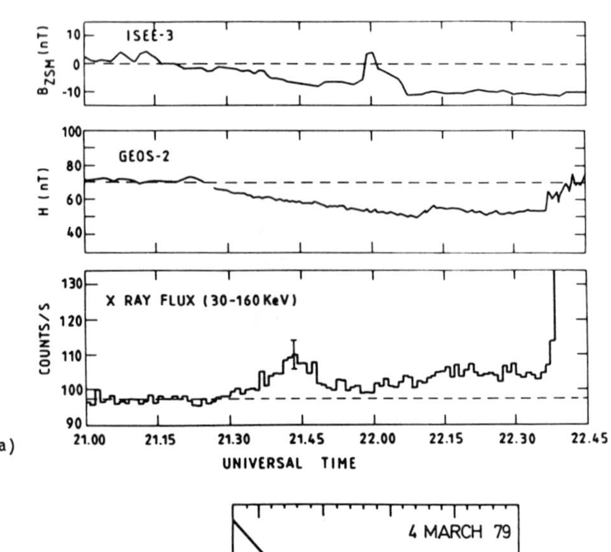

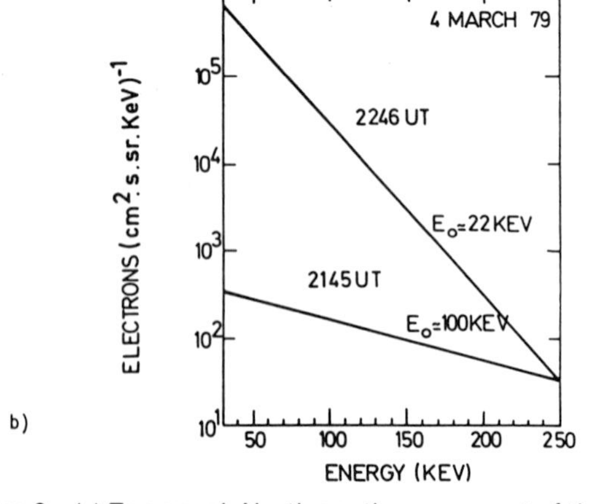

Figure 6—(a) Top panel: North-south component of the IMF measured onboard the ISEE 3 spacecraft. Middle panel: Northward component of the B field measured onboard GEOS 2. Bottom panel: Integral flux of 30–160 keV X-rays measured onboard a balloon near the footprint of the GEOS 2 field line. (b) Computed differential energy spectrum of the electrons responsible for the X-ray emission observed during the pre-breakup phase (2145 UT) and after substorm onset (2246 UT).

flux of X-rays between 30 and 160 keV measured onboard a SAMBO 79 balloon located 1.8° southward of the GEOS 2 magnetic line footprint started to increase above the cosmic background level. As no injection of energetic electrons was detected onboard GEOS 2 during this presubstorm period (Fig. 4), the X-ray increase is interpreted as bremsstrahlung X-rays from precipitation of the ambient population of trapped energetic electrons. This conclusion is supported by the fact that the parent electron energy spectrum responsible for the X-rays[16] has an *e*-folding energy E_0 of about 100 keV during this preinjection period (Fig. 6b). This value is about five times the *e*-folding energy usually found during precipitation associated with the expansion phase of substorms. As no ELF waves were observed during this period onboard GEOS 2 (N. Cornilleau-Wehrlin, personal communication, 1985), the precipitation here must be due to a process other than wave-particle interaction. An alternative explanation of the precipitation can be proposed on the basis of dynamical drift-shell splitting due to the conservation of the two first adiabatic invariants.[17]

Contrast the hard precipitation observed before 2237 UT with the intense X-ray fluxes observed after 2237 UT during the main substorm expansion phase (Fig. 6a,b). Here, the characteristic energy of the precipitating electrons is about 20 keV and the precipitation is associated with intense ELF waves on GEOS 2. Cornilleau-Wehrlin et al.[18] have shown that this precipitation can be caused by the interaction between waves and newly accelerated particles.

Plasma Sheet

At 2200 UT, GEOS 2 was located near local midnight and the ISEE spacecraft were located in the morning sector of the plasma sheet, at a radial distance of 22.35 R_e. The ISEE 1/ISEE 2 separation was about 150 km at this time and both were located south of the neutral sheet. Between 2100 and 2200 UT, the magnetic local time of the ISEE satellites is almost constant and equal to about 2.3 H while the geomagnetic local time of the easternmost magnetometer chain varies from 2.14 to 3.14 H.

Figure 7 shows, from top to bottom, the IMF B_z component, the H component of the magnetic field at 6.6 R_e, and the 1.5 keV and 6 keV electron fluxes measured onboard ISEE between 2000 and 2400 UT. The 1.5 keV electron flux onboard ISEE accompanying the substorm onset at about 2237 UT decreased to below the threshold level of our detector. From 2100 to 2145 UT, the electron flux measured onboard ISEE 2 (data gap for ISEE 1) was relatively stable. The first indication of flux decrease was observed on ISEE 1 at 2145, in coincidence with the weak substorm onset detected on the ground by the easternmost nightside magnetometer chain ($\phi_m = 145°$, Fig. 3) and at 6.6 R_e (Pi 2 pulsations). A second (transient) electron flux decrease is detected at 2200 UT and a third around 2213 UT, in conjunction with a small change of the H component at the GEOS 2 orbit (see also Figs. 4 and 6). This last feature indicates a partial return of the inner magnetosphere B field topology toward a dipolar configuration and is typical of a substorm onset at this orbit.[14,19,20] We shall now examine more precisely the different signatures of the 2213 UT event in the magnetotail and in the ionosphere.

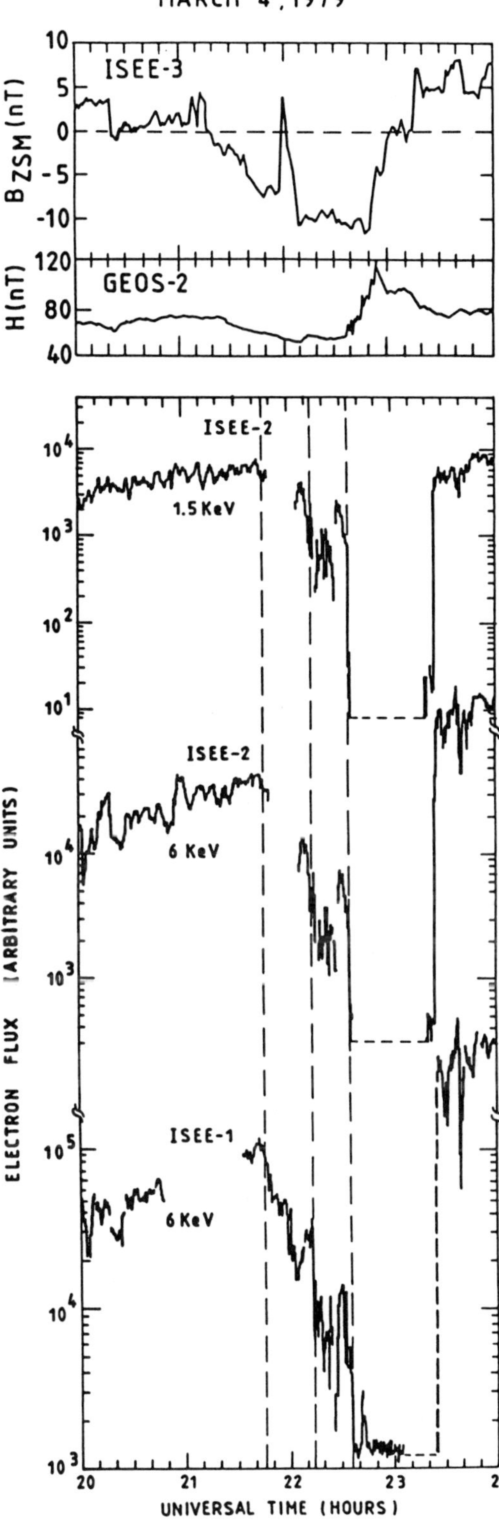

Figure 7—Comparison between various parameters from 2000 UT to 2400 UT in the interplanetary medium, at geostationary orbit near local midnight, and in the distant plasma sheet. From top to bottom: North-south component of the IMF (IMP 8 and ISEE 3 spacecraft); H component of the magnetic field measured at the geostationary orbit; 1.5 and 6 keV electron flux measured onboard ISEE spacecraft in the distant plasma sheet, at $\approx 22\ R_e$. The light vertical dashed lines refer to the 2145, 2213, and 2237 UT events (see text).

Figure 8 illustrates the relationship between the plasma sheet particle variations at 2213 UT and the ionospheric electric field measured by a balloon-borne instrument located 1.8° south of the computed GEOS 2 magnetic line footprint. The 1.5 keV and 6 keV electron flux decreases recorded onboard ISEE 1 coincide with a reversal toward the north and a large intensification of the ionospheric electric field that reached a maximum value of $\simeq 60$ mV/m at 2220 UT. Furthermore, all-sky camera data from Ivalo (not shown) showed a diffuse precipitation starting at about 2215 UT, following the appearance and the development of auroral arcs just after 2210 UT. The appearance and the southward motion of auroral structures lasted until $\simeq 2233$ UT. These dynamical events coincide with the northward excursion of the northern boundary of the westward electrojet, observed at $\simeq 2220$ UT at $\phi_m = 120°$ (Fig. 3). This substorm event occurred in association with a particle injection at the geostationary orbit affecting the lowest (16–22 keV) energy particles detected onboard GEOS 2 by the S-321 experiment (Fig. 9). Note that this local injection began at about 2220 UT, at the time the largest northward electric field was observed onboard the SAMBO balloon.

To conclude this part, all plasma-sheet thinnings observed onboard the ISEE spacecraft near apogee ($\simeq 22$ R_e) are impulsive and are associated with substorm onsets. We found no evidence of the slow plasma sheet thinning usually described as a "growth phase" feature.[5]

Note finally that plasma-sheet fluxes recover around $\simeq 2325$ UT (Fig. 7). This recovery coincides with the poleward shift of the equatorward boundary of the auroral electrojet observed in longitudinal interval 115°–120° (Fig. 3).

Relation to the ϵ Parameter

The relationship of the various events discussed above to the parameter[21] is shown in Fig. 10 (events labeled A through G). The increase (A) of the ϵ parameter associated with the southward turning of the IMF at 2117 UT is quickly followed by an increase of the AE index. The sharp decrease of ϵ around 2300 UT, associated with the northward turning of the IMF, is followed by a return of the AE index to quiet time values.

It can be seen that the main substorm onset (E), at $\simeq 2237$ UT, occurs when ϵ has already reached maximum values. The first weak event (B) occurs at 2145 UT while ϵ is still increasing. The second weak event (D) occurs at $\simeq 2213$ UT just after the ϵ parameter reached a maximum value ($> 1.2 \times 10^{12}$ W). Note also the occurrence at 2247 UT (F) and 2254 UT (G) of two AE intensifications following the main substorm onset, each being associated with small increases of the ϵ parameter.

Another salient feature (C) is the correlation between the ϵ and AE decreases around 2200 UT. It thus appears that, with the exception of the main substorm onset, the ionospheric currents are mainly controlled by the interplanetary medium.

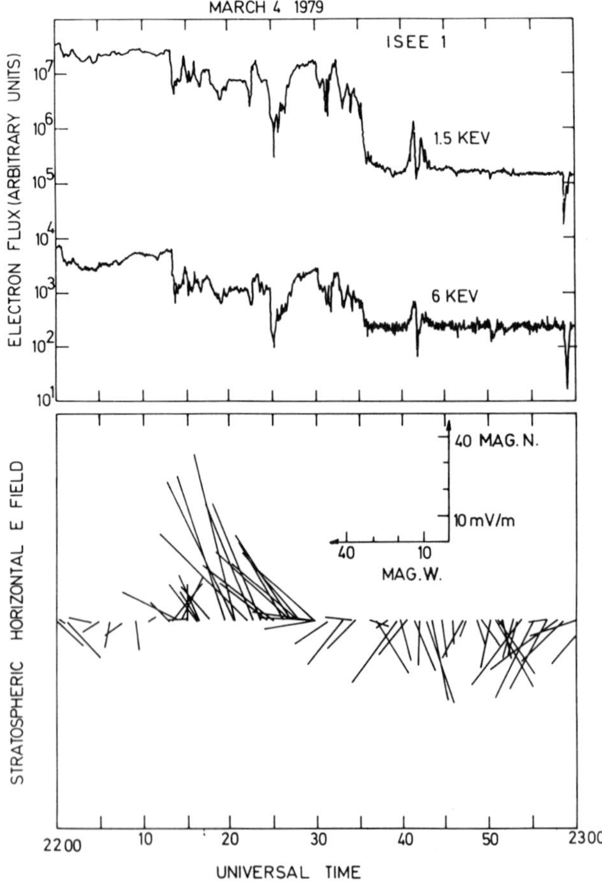

Figure 8—Comparison between the 1.5 keV and 6 keV electron fluxes measured onboard ISEE 1 in the 2200–2300 UT interval and the ionospheric vector electric field measured onboard the SAMBO balloon.

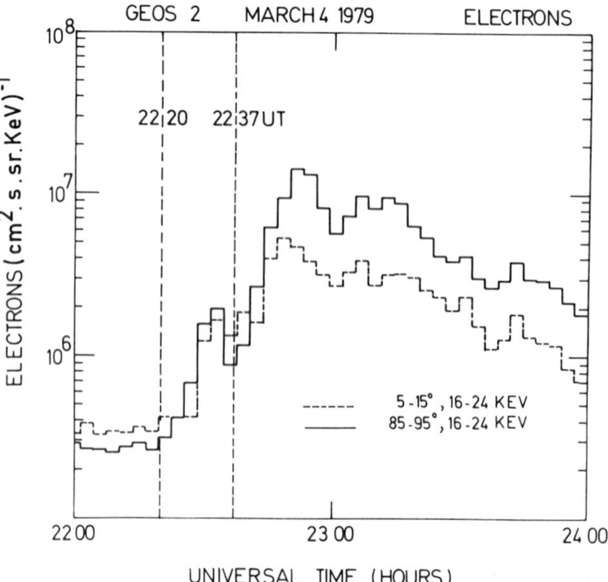

Figure 9—Low-energy electron fluxes (16–23 keV) recorded between 2200 and 2400 UT onboard GEOS 2 for two different pitch angles.

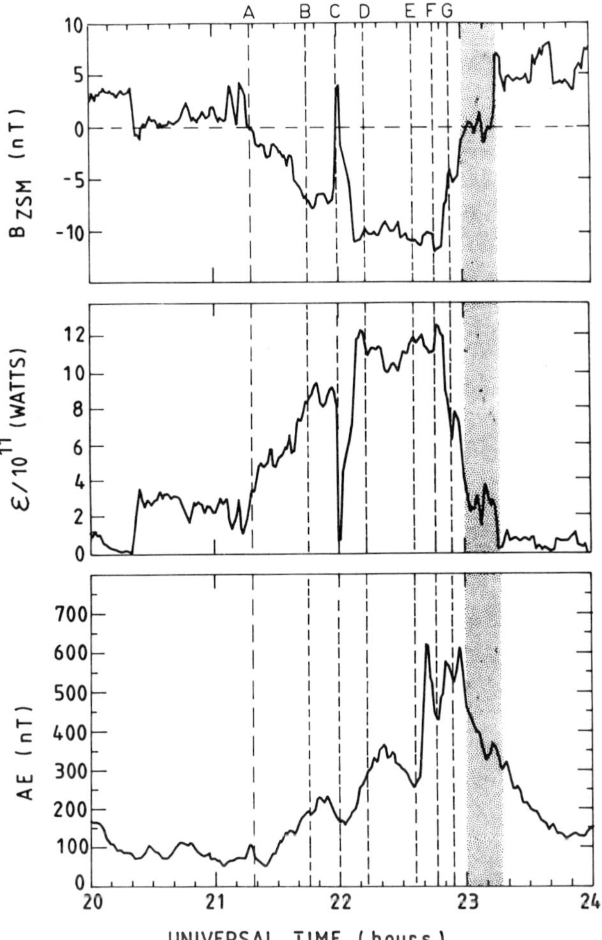

Figure 10—From top to bottom: North-south component of the interplanetary magnetic field; solar-wind/magnetosphere energy coupling function ϵ; AE index. The dashed vertical lines refer to the events discussed in the text. The shaded area indicates the interval when the IMF turns northward.

DISCUSSION

The March 4, 1979 event indicates that a turning of the interplanetary magnetic field to the southward direction drives various magnetospheric processes affecting both the day and night sectors of the magnetosphere. The main results of this case study can be summarized as follows:

1. The change of the interplanetary magnetic field direction is associated with a time delay of less than about 5 minutes with an increase of the electrojets and precedes by ~10 minutes the beginning of a tail-like deformation of the B-field geometry at 6.6 R_e near midnight.
2. After the IMF reversal and before the main substorm onset, the general motion of the westward electrojet is southward directed.
3. At 6.6 R_e, the high energy particles adiabatically respond to the change of the local B-field configuration. The adiabatic response of small pitch angle energetic electrons induces their precipitation

in the auroral ionosphere, causing a weak but very energetic X-ray emission.
4. Short-lived substorms occur soon after the IMF reversal. They are characterized by all or part of the following phenomena: partial plasma-sheet flux decreases at the ISEE orbit ($\simeq 22$ R_e), weak dipolar reconfiguration of the magnetic field at 6.6 R_e, change of the ionospheric electric field, limited impulsive motions of the boundaries of the westward electrojet, injection of low-energy particles at 6.6 R_e, and Pi 2 bursts.
5. The main breakup, occurring about 1.3 hours after the first B_z reversal, is associated with energetic particle injection toward the inner magnetosphere and complete particle flux dropout observed by the ISEE satellites in the plasma sheet.
6. The AE index, which is a rough measure of the total current of the electrojets, decreases to quiet time values when the IMF B_z component reaches positive values (Fig. 10).

Our correlated observations clearly show that in the distant tail, while the IMF is southward, impulsive plasma sheet thinnings occur in time coincidence with the onset of weak substorms corresponding to impulsive energy dissipation periods.

The good general agreement found between the ionospheric currents and the interplanetary parameter (B_z, ϵ) variations suggests that the energy for driving the auroral electrojet and tail current systems comes first directly from the solar wind, before the main substorm onset. This result is in agreement with previous work by Pellinen,[22] Rostoker et al.,[1] and Ahn et al.[23] However it must be stressed that the main substorm onset corresponds to a period where the solar-wind/magnetosphere energy coupling function has already reached maximum values and does not show a consequent new increase. If the parameter is a real measure of the energy input from the solar wind to the magnetosphere, we can conclude that the weak substorms preceding the main one did not dissipate all the available energy and consequently that a part of the energy dissipated during the main substorm was stored in the magnetotail during the preceding southward IMF episode.

SUMMARY

The extensive data set for the March 4, 1979 event was used to better understand the cause of substorms. In particular, our original intention was to use it to determine whether the solar wind energy is directly accessible to drive substorms or whether some of the energy is stored in the magnetotail. Our study of a very large data set indicates that there is evidence that substorms can directly derive their energy from the solar wind (as exemplified by the near-immediate response of the various data to IMF changes). This result is in agreement with the "driven model" of substorms (see e.g., Akasofu[6,21]). However the main substorm occurred independently of the solar-wind IMF changes. The cause of this substorm appears to reside in the magnetosphere and it is not inconsistent with the idea that stored energy is being dissipated during this event (see e.g., McPherron[12,24]).

ACKNOWLEDGMENT—The authors would like to thank S. Mariani and the GEOS magnetometer group for providing the geostationary magnetic field measurements used in this study. The ISEE 3 IMF and solar-wind velocity data come from the Data Pool; we thank the principal investigators, respectively, E. J. Smith and S. J. Bame.

REFERENCES

[1] G. Rostoker, M. Mareschal, and J. C. Samson, "Response of Dayside Net Downward Field Aligned Current to Changes in the Interplanetary Magnetic Field and to Substorm Perturbations," *J. Geophys. Res.* **87**, 3489 (1982).

[2] T. Kammei and H. Maeda, "World Data Center C2 for Geomagnetism Data Books, Data Analysis Center for Geomagnetism and Space Magnetism," Faculty of Science, Kyoto Univ. (1982).

[3] Y. Kamide and A. Brekke, "Auroral Electrojet Current Density Deduced from the Chatanika Radar and from the Alaska Meridian Chain of Magnetic Observatories," *J. Geophys. Res.* **80**, 587 (1975).

[4] V. G. Petrov, "The Modeling of Induction in the Conducting Earth for the Polar Electrojets' Dynamics" (in Russian), *Geomagn. Aeron.* **22**, 159 (1982).

[5] R. L. McPherron, C. T. Russel, and M. P. Aubry, "Satellite Studies of Magnetospheric Substorms on August 15, 1968," *J. Geophys. Res.* **78**, 3131 (1973).

[6] S.-I. Akasofu, "The Roles of the North-South Components of the Interplanetary Magnetic Field on Large-Scale Auroral Dynamics Observed by the DMSP Satellite," *Planet. Space Sci.* **23**, 1349 (1975).

[7] R. E. Holzer and J. A. Slavin, "Magnetic Flux Transfer Associated with Expansions and Contractions of the Dayside Magnetosphere," *J. Geophys. Res.* **83**, 3831 (1978).

[8] R. E. Holzer and J. A. Slavin, "A Correlative Study of Magnetic Flux Transfer in the Magnetosphere," *J. Geophys. Res.* **84**, 2573 (1979).

[9] Y. Kamide and S.-I. Akasofu, "Latitudinal Cross-Section of the Auroral Electrojet and Its Relation to the Interplanetary Magnetic Field Polarity," *J. Geophys. Res.* **79**, 3755 (1974).

[10] D. N. Baker, E. W. Hones, Jr., R. D. Belian, P. R. Higbie, R. P. Lepping, and S. Stauning, "Multiple-Spacecraft and Correlated Riometer Study of Magnetospheric Substorm Phenomena," *J. Geophys. Res.* **87**, 6121 (1982).

[11] R. L. McPherron, "Substorm Related Changes in the Geomagnetic Tail: The Growth Phase," *Planet. Space Sci.* **20**, 1521 (1972).

[12] R. L. McPherron, "Magnetic Variations During Substorms," in *Dynamics of the Magnetosphere*, S.-I. Akasofu, ed., D. Reidel Pub. Co., Dordrecht, Holland, p. 631 (1980).

[13] K. N. Erickson and J. R. Winckler, "Auroral Electrojets and Evening Sector Electron Dropouts at Synchronous Orbit," *J. Geophys. Res.* **78**, 8373 (1973).

[14] J. A. Sauvaud and J. R. Winckler, "Dynamics of Plasma, Energetic Particles, and Fields Near Synchronous Orbit in the Nighttime Sector During Magnetospheric Substorms," *J. Geophys. Res.* **85**, 2043 (1980).

[15] G. Rostoker and J. C. Samson, "Polarization Characteristics of Pi 2 Pulsations and Implications for their Source Mechanisms: Location of Source Regions with Respect to the Auroral Electrojets," *Planet. Space Sci.* **29**, 225 (1981).

[16] J. P. Treilhou, "A Semi-Empirical Method to Obtain the Precipitated Electron Energetic Spectrum from the X-ray Spectrum," IV Scientific Assembly, IAGA, Edinburgh (Aug 1981).

[17] J. A. Sauvaud, J. P. Treilhou, A. Saint-Marc, H. Rème, J. Dandouras, A. Korth, G. Kremser, G. K. Parks, A. N. Zaitzev, V. Petrov, and L. Lazutine, "A Multipoint Study of a Substorm on March 4, 1979," *Proc. of the Int. Conf. on Results of the ARCAD-3 Project and of the Recent Programs in Magnetospheric and Ionospheric Physics, Toulouse, May 1984*, Cepadues, ed., p. 775 (1985).

[18] N. Cornilleau-Wehrlin, J. Solomon, A. Korth, and G. Kremser, "Experimental Study of the Relationship Between Energetic Electrons and ELF Waves Observed On Board GEOS: A Support to Quasi-Linear Theory," *J. Geophys Res.* **90**, 4141 (1985).

[19] J. W. Lezniak and J. R. Winckler, "Experimental Study of Magnetospheric Motions and the Acceleration of Energetic Electrons During Substorms," *J. Geophys. Res.* **75**, 7075 (1970).

[20] A. T. Y. Lui, "Estimates of Current Changes in the Geomagnetotail Associated with a Substorm," *Geophys. Res. Lett.* **5**, 853 (1978).

[21] S.-I. Akasofu, "The Solar-Wind/Magnetosphere Energy Coupling and Magnetospheric Disturbances," *Planet. Space Sci.* **28**, 495 (1980).

[22] R. J. Pellinen, W. Baumjohann, W. J. Heikkila, V. A. Sergeev, A. G. Yahnin, G. Marklund, and A. D. Melnikov, "Event Study of Pre-Substorm Phases and their Relation to the Energy Coupling Between Solar Wind and Magnetosphere," *Planet. Space Sci.* **30**, 371 (1982).

[23] B. H. Ahn, Y. Kamide, and S.-I. Akasofu, "Global Ionospheric Current Distributions During Substorms," *J. Geophys. Res.* **89**, 1613 (1984).

[24] R. L. McPherron, "Magnetospheric Substorm," *Rev. Geophys. Space Phys.* **17**, 657 (1979).

DISCUSSION

D. Baker: Your data support the interpretation that a low level of magnetospheric activity commences very rapidly in response to solar wind input changes. On a substantially longer time scale one sees a much larger dissipation onset which persists long after the input (ϵ) has diminished to near zero. Consequently, both a "driven" and an "unloading" component is clear. Thus, I believe you should emphasize the duality of the magnetospheric response to solar wind changes. As a comment, the growth-phase precipitation events detected in your X-ray data are probably not new since similar features have been well documented in ground-based riometer data by Ranta and Ranta among others.

G. Parks: You raise two questions. One is about the fact that toward the end of the event the activity still persisted while ϵ had diminished. I did mention that we had noticed this feature and interpreted that the magnetosphere was at this time possibly dissipating stored energy. However, the main feature is the close correlation observed between the temporal behavior of ϵ and AE. This feature is better interpreted in terms of a directly driven model. The second point is about observing electron precipitation before the major onset. Although this has been observed before, energy spectra were not available. Consequently, the riometer data could not establish with certainty as we did that trapped particles were being modulated by IMF.

V. Vasyliunas: You treat the ϵ function as though it were established as the solar wind energy input into the magnetosphere. In fact, it is a function constructed to match the energy output as closely as possible and is not otherwise proved to be the input. Therefore, only if one assumes a priori that input = output can one take as the input; if that assumption is being questioned or tested, there is no longer a logical basis for assigning a specific meaning to ϵ.

G. Parks: We're not making any assumptions about "input = output". We computed the parameter and compared to AE. It is from such good correlation that we infer that the solar wind appears to drive the substorm phenomenon (deduced by means of AE directly).

AN INTERPRETATION OF MAGNETOSONIC WAVES IN THE BOUNDARY REGIONS OF THE PLASMA SHEET AS SEEN BY THE ISEE 3 SPACECRAFT

P. R. Smith,* K. I. Hopcraft,[†] and N. Murphy[‡]

We discuss recent calculations that derive the normal mode spectrum of an idealized magnetic current sheet. The Harris neutral sheet equilibrium is perturbed with an ideal magnetohydrodynamic displacement u of the form $u = u(z) \exp [i(kx - \omega t)]$. We calculate the longitudinal polarization of the fundamental modes as a function of the position in the sheet. Using data from the energetic ion instrument aboard ISEE 3, we estimate the thickness of the plasma sheet in the deep geomagnetic tail. This parameter enables a quantitative comparison between the boundary oscillations reported by Tsurutani and Smith and the normal mode oscillations derived by Hopcraft and Smith to be performed. The normal mode solutions are found to be consistent with observation. Finally we point out further aspects of the MHD wave spectrum that may lead to an observable variation of the mode character across the boundary of the plasma sheet.

INTRODUCTION

The appearance of large-amplitude magnetosonic waves in the boundary layers of the plasma sheet has been one of the interesting and intriguing observations of the ISEE 3 geotail mission. Hopcraft and Smith[1] have recently suggested that ideal MHD normal modes of a magnetic neutral sheet account for many of the characteristics of these waves, which were first described by Tsurutani and Smith.[2] The waves derived in Ref. 1 propagate along the ambient magnetic field and are magnetosonic. A discrete spectrum of modes exists whose eigenfunctions exhibit a strong variation in polarization as a function of the position in the sheet. The dispersion properties of these waves have been described in Ref. 1, although a real comparison with observations requires knowledge of the plasma-sheet width. This parameter has only recently been quantifiable in the distant geomagnetic tail.

The waves reported in Ref. 2 may be summarized as follows. The oscillations occur in the plasma-sheet boundary layers and are approximately field aligned with a propagation speed in excess of the local Alfvén speed. The characteristic frequency is near the proton cyclotron frequency and the waves are generally right-hand circularly or elliptically polarized. There is usually a significant compressional component whose amplitude varies between 30 and 90% of the total wave amplitude.

In this paper we demonstrate the consistency between the waves observed and the MHD modes derived in Ref. 1. In order to do this we use data to calculate the plasma-sheet width. This latter parameter is estimated using a remote sensing technique similar to that used by Daly.[3] We can thereby narrow the theoretical spectrum of waves that are appropriate for real plasma-sheet thicknesses and compare with measured values. Finally we discuss other properties of the sheet modes that may inspire further experimental study.

*Department of Mechanical Engineering, University College London, London WC1E 7JE.

[†]Wheatstone Laboratory, Department of Physics, King's College (KQC) London, London WC2R 2LS.

[‡]Blackett Laboratory, Department of Physics, Imperial College of Science and Technology, London SW7 2BZ.

MODEL SOLUTIONS

The one-dimensional equilibrium configuration we use to model the current sheet is that derived by Harris[4] for which the magnetic field (B^0) and plasma pressure (p^0) distributions are

$$B^0 = B_0 \tanh(z/h)x$$

$$p^0 = p_0 \operatorname{sech}^2(z/h)$$

The magnetic field is in the x-direction only and reverses at $z = 0$, the center plane of the current sheet. The sheet width is characterized by h. Taking a displacement of the form $u = u(z) \exp [i(kx - \omega t)]$, the linearized momentum balance equations for ideal MHD may be reduced to a second-order differential equation (see Ref. 1) in the z-displacement (u_z) as follows:

$$A(z,\omega,k)u_z'' + B(z,\omega,k)u_z' + C(z,\omega,k)u_z = 0 \qquad (1)$$

where primes denote differentiation with respect to z and

$$A(z,\omega,k) = \sinh^2(z) - (G\omega^2)/f$$

$$B(z,\omega,k) = 2 \tanh(z)$$
$$\{1 - (G\omega^2 R/f)[1 + (G-1)k^2/2f]\}$$

$$C(z,\omega,k) = \omega^2/R - k^2 \sinh^2(z)$$

$$R = [\operatorname{sech}(z)]^{2(G-1)/G}$$

$$f = GRk^2 - \omega^2$$

and $G = 5/3$ is the ratio of specific heats. Note that z and k are normalized to the characteristic width h, and ω is normalized to the average Alfvén frequency ω_A given by:

$$(\omega_A)^2 = (B_0)^2/(\mu_0 r_0 h^2)$$

where r_0 is the plasma density at the center of the sheet, and B_0 is the magnitude of the magnetic field outside the sheet.

Treating Eq. 1 as an eigenvalue equation for u_z, solutions may be obtained by a finite-difference "shooting" technique for eigenfunctions satisfying the correct boundary conditions at $z = 0$ and infinity. For a full derivation of the boundary conditions and the method for solving Eq. 1 see Ref. 1. Both symmetric and antisymmetric solutions exist for u_z and in each case an infinite sequence of eigensolutions is obtained. Here we investigate the properties of the more important fundamental harmonics. Before discussing these eigenmodes, it is convenient to parameterize the regimes into which the ISEE 3 measurements fall with respect to our theoretical model. This is achieved by estimating the plasma-sheet width, which is performed in the next section.

DATA ANALYSIS

To estimate a typical plasma-sheet thickness, data from the ISEE 3 Energetic Proton Anisotropy Spectrometer (EPAS) are used. This experiment measures the three-dimensional ion distribution function (for all ion species) in eight logarithmically spaced, contiguous energy channels from 35 to 1600 keV. This is achieved by means of three independent particle telescopes inclined at 30, 60, and 135° to the spacecraft spin axis, which is nearly perpendicular to the ecliptic. Output from these telescopes is binned into eight azimuthal sectors. A more detailed description of the experiment can be obtained from Ref. 5.

The EPAS experiment has six telescope/sector combinations that view in the GSE y-z plane. These sectors will detect ions with pitch angles of 90° when the magnetic field is parallel or antiparallel to the GSE x-axis, as is the case in the lobes of the earth's magnetotail. It is therefore possible to use these ions to sense remotely the velocity and orientation of the plasma sheet during the passage of ISEE 3 from one tail lobe to the other.[3,6,7] Several such crossings were documented in Ref. 8 and five of these events have been used to obtain a thickness by using the time of flight of ISEE 3 through the plasma sheet together with the mean velocity as deduced above. The average plasma-sheet thickness obtained in this way is 1.5 ± 1.1 earth radii, i.e., 9600 ± 7040 km. The dimension that characterizes the plasma-sheet width in our present study is, in fact, the half width, and so we take $h = 0.75$ earth radii, or 4800 km.

WAVE PROPERTIES

Figure 1 shows the solutions for the perturbed magnetic field components in x and z together with the perturbed pressure for the fundamental symmetric (Fig. 1a.) and antisymmetric (Fig. 1b) modes, where the symmetry refers to the z-displacements. As the wave analysis is linear, the amplitudes of the perturbed quantities are not determined. For the sake of convenience the solutions have been normalized to B_0 and p_0 where appropriate. Although it is not shown here, note that higher harmonic solutions will exhibit more structure inside the sheet ($-1 < z < 1$) but the boundary and external regions will

be much the same. In this sense all the waves are "guided" by the neutral sheet. In Ref. 1 the authors show that the perturbed pressure is either purely in phase or antiphase with the compressional component (x) of the magnetic field. Thus from Fig. 1, note that the symmetric mode has the phase character of a fast magnetosonic wave while the antisymmetric mode can be fast or slow, depending on the position in the sheet.

In Fig. 2 the polarization ellipticity $e = B_x/B_z$ has been plotted as a function of z. There is a large variation in e, as the boundary region of the neutral sheet is traversed, in response to changes in background field gradient at the edge of the system. As our analysis is for field aligned waves only, the y-component of the magnetic field is not determined. Thus we are not able to discuss the transverse polarization of the waves and not able therefore to make a direct comparison with the polarization reported in Ref. 2, which is approximately B_y/B_z (i.e., for field aligned waves). However it is clear from Figs. 1 and 2 that the amplitude of the compressional component (B_x) varies strongly with z, which is consistent with there being observed (in Ref. 2) a large scatter in the ratio of the compressional component to

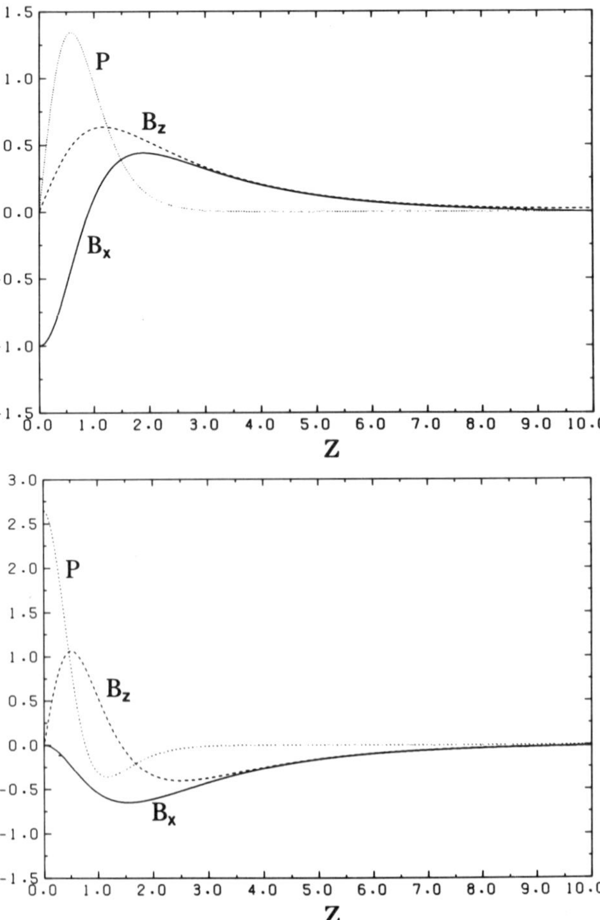

Figure 1—Perturbed magnetic field components in x and z together with the perturbed pressure for the fundamental symmetric (a) and antisymmetric (b) displacements in z. The amplitudes are normalized values.

152

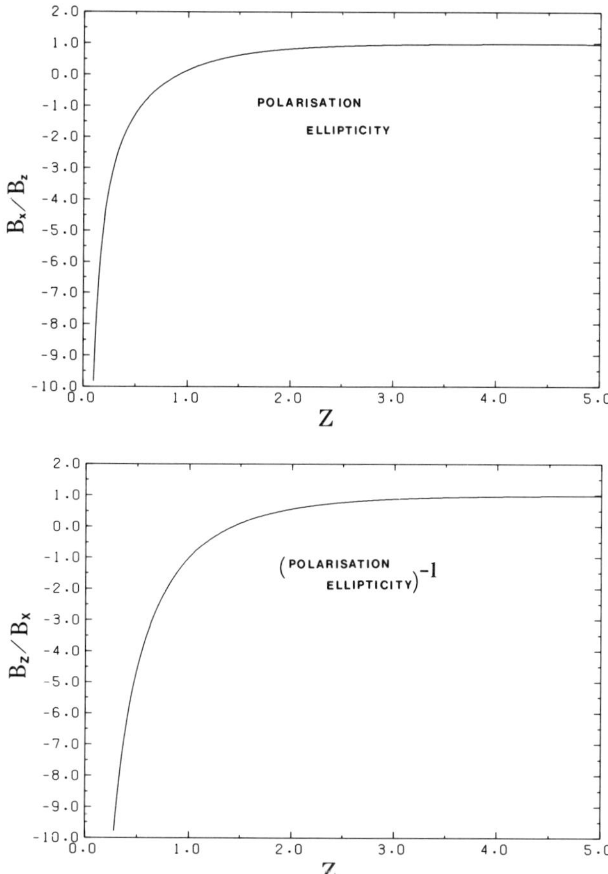

Figure 2—Polarization ellipticity $e = B_x/B_z$ as a function of z for the fundamental symmetric (a) and antisymmetric (b) solutions.

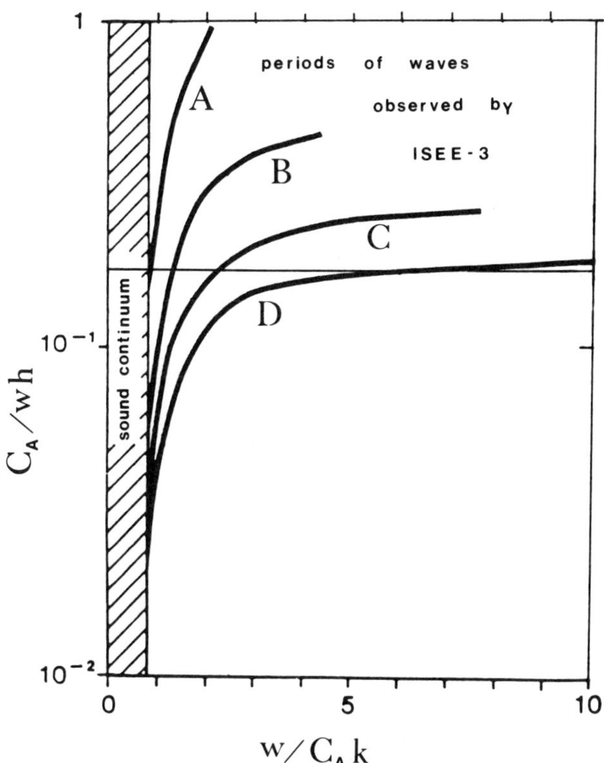

Figure 3—The normalized period of the zeroth and first harmonic of the symmetric and antisymmetric modes plotted as a function of the normalized phase speed.

the wave peak-to-peak transverse amplitude. A direct comparison is only possible if the waves are allowed to propagate across the field, in which case all magnetic field components can be determined. This calculation has recently been performed,[9] and if it were possible to analyze the data so that all the field components are measured independently from the propagation direction, then a more thorough comparison could be made.

We now address the dispersive properties of the waves in Fig. 3. There the normalized wave period is plotted against the normalized phase speed for the symmetric and antisymmetric zeroth and first harmonics. We have chosen this interval of wave period in the light of our earlier discussion of the plasma-sheet width and the range of wave frequencies recorded in Ref. 2. Thus the periods fall into the range plotted if we make the reasonable assumption that the characteristic sheet width h lies between 1000 and 50,000 km. All modes have a cutoff when the phase speed is equal to the normalized sound speed ($\omega/k = (G/2)^{1/2}c_A$). For phase speeds less than the sound speed Eq. 1 has a singularity when $A = 0$, thereby indicating a continuum of solutions,[10] as opposed to the discrete spectrum at higher speeds. Further examination of Fig. 3 correctly suggests that the solutions extend unbounded to infinite phase speed. This is

an artifact of the infinite equilibrium, which allows the local Alfvén speed to increase without limit as the distance from the neutral sheet increases. In reality this would not occur because of the presence of the lobe magnetoplasma.

Let us now examine some specific examples of the wave data from days 24 and 27 (1983) of the deep tail mission. Data are taken from values published in Tsurutani et al.,[11] where these two days are chosen for special study. Averaging all the right-hand polarized wave events (on days 24 and 27) we obtain a typical wave period = 11 seconds, and a local Alfvén speed = 480 km.s^{-1}. These values together with the estimated plasma-sheet width are used to define the horizontal line in Fig. 3. Thus it is clearly evident that for a fast-mode magnetosonic wave (phase speed > Alfvén speed) the fundamental ideal MHD modes are entirely in agreement with directly measured quantities.

DISCUSSION

We have shown the consistency between the properties of magnetosonic waves as observed in the boundary of the plasma sheet and ideal MHD modes of the Harris magnetic neutral sheet. Since the waves are often observed in association with substorms, it is reasonable to conjecture that the ideal modes are driven by forces resulting from the onset of magnetic field line reconnection. In Ref. 2 the authors stress the importance of high-energy particle beams at the outer edge of the current sheet and the possibility of a linear instability giving rise to the waves observed. This is clearly an interesting approach that may be explored in parallel to our own. At the present stage, both explanations are consistent with the measurements and therefore further investigation is required. One clear way of verifying the applicability of our work is to obtain measurements of the plasma-sheet width during magnetosonic wave events, which would necessitate finding a plasma-sheet crossing associated with a wave event in one or both of the boundary layers. Any systematic correlation between properties of the magnetosonic waves and the width of the plasma sheet would support the explanation due to ideal MHD modes of the current sheet.

A more extensive analysis of the wave data has very recently been made,[11] in which a careful study of the periods, polarizations, and propagation directions of the waves is performed over 44 days of the ISEE 3 deep tail mission. Although more examples are presented than in Ref. 2, the analysis of the wave data is identical and therefore no new information is available to compare with the mechanism suggested here. As in Ref. 2, Tsurutani et al.[11] calculate the rest frame frequencies and phase velocities of the waves by assuming, a priori, a particular dispersion relationship that is a function of directly measurable quantities. This dispersion relationship is very similar to that of the resonant ion beam instability,[12] as indeed the authors point out. We are therefore unhappy with one of their conclusions that states that two "unambiguous" examples of the resonant ion-beam instability are presented, as their conclusion is based, at least in part, on the assumption that the instability is in operation. More convincing is the authors' analysis and discussion of the particle data, which clearly indicate a strong correlation between the presence of ion beams and the magnetosonic boundary-layer waves. Both of these phenomena are therefore substorm related.[11] At this stage it is not possible to prove unambiguously a direct relationship between the particle beams and the magnetosonic waves, although it is probably not coincidental that they occur together, since plasma and field phenomena associated with substorms are complex and multifaceted.

It has been the intention of this paper to present an alternative interpretation of the plasma-sheet waves from that proposed in Ref. 2. We do not discount the interpretation due to a resonant ion-beam instability, as there is clearly important evidence for this mechanism. However the evidence is not "unambiguous," as we have shown that the wave properties are also consistent with inhomogeneous ideal MHD modes of a magnetic neutral sheet.

REFERENCES

[1] K. I. Hopcraft and P. R. Smith, "Guided Modes in the Vicinity of the Earth's Geomagnetic Tail," *Phys. Lett.* **113A**, 75 (1985).

[2] B. T. Tsurutani and E. J. Smith, "Magnetosonic Waves Adjacent to the Plasma Sheet in the Distant Magnetotail," *Geophys. Res. Lett.* **11**, 331 (1984).

[3] P. W. Daly, "Remote Sensing of Energetic Particle Boundaries," *Geophys. Res. Lett.* **9**, 1329 (1982).

[4] E. Harris, "On a Plasma Sheath Separating Regions of Oppositely Directed Magnetic Field," *Il Nuovo Cimento* **23**, 115 (1962).

[5] A. Balogh, G. vanDijen, J. vanGenechten, J. Henrion, R. Hynds, G. Korfmann, T. Iversen, J. vanRooijen, T. Sanderson, G. Stevens, and K.-P. Wenzel, "The Low Energy Proton Experiment on ISEE-C," *IEEE Trans. Geosci. Electron.* **GE-16**, No. 3, 176 (1978).

[6] N. Murphy, J. A. Slavin, D. N. Baker, and W. J. Hughes, "Enhancements of Energetic Ions Associated with Travelling Compression Regions in the Deep Geomagnetic Tail," *J. Geophys. Res.* (in press, 1986).

[7] I. G. Richardson and S. W. H. Cowley, "Plasmoid-Associated Energetic Ion Bursts in the Deep Geomagnetic Tail: Properties of the Boundary Layer," *J. Geophys. Res.* **90**, 12133 (1985).

[8] J. T. Gosling, D. N. Baker, S. J. Bame, W. C. Feldman, R. D. Zwickl, and E. J. Smith, "North-South and Dawn-Dusk Plasma Asymmetries in the Distant Tail Lobes: ISEE-3," *J. Geophys. Res.*, (in press, 1986).

[9] K. I. Hopcraft and P. R. Smith, "Magnetohydromagnetic Waves in a Neutral Sheet," *Planet. Space Sci.* (in preparation, 1986).

[10] H. Grad, "Magnetofluid Dynamic Spectrum and Low Shear Stability," *Proc. Nat. Acad. Sci.* **70**, 3277 (1973).

[11] B. T. Tsurutani, I. G. Richardson, R. M. Thorne, W. Butler, E. J. Smith, S. W. H. Cowley, S. P. Gary, S.-I. Akasofu, and R. D. Zwickl, "Observations of the Right-Hand Resonant Ion Beam Instability in the Distant Plasma Sheet Boundary Layer," *J. Geophys. Res.* **90**, 12159 (1985).

[12] A. Barnes, "Theory of Generation of Bow-Shock-Associated Hydromagnetic Waves in the Upstream Medium," *Cosmic Electrodynam.* **1**, 90 (1970).

AN EVALUATION OF THE TOTAL MAGNETOSPHERIC ENERGY OUTPUT PARAMETER, U_T

R. D. Zwickl,* L. F. Bargatze,*‡ D. N. Baker,* C. R. Clauer,† and R. L. McPherron‡

Over the last few years the relationship between U_T, the magnetospheric energy consumption or output rate, and ϵ, a commonly used solar-wind/magnetosphere energy input function, has been explored in some detail. Very high correlations between U_T and ϵ are found during periods of strong activity, and by using linear prediction filtering techniques a "δ-function" impulse response was found for filter elements representing essentially zero delay. In light of these remarkable results, the derivation of U_T for these intervals is reexamined. We find that U_T is dominated in each event interval by the term containing τ_R, the ring current decay time, and that when τ_R is defined as a function of ϵ the δ-function impulse response is present. If a constant τ_R is assumed, the δ-function part of the filter disappears completely. Thus, this δ-function, which has been taken as being indicative of the directly driven component, is an artifact of the earlier analysis, and it is due to the dependence of U_T on ϵ. Our results imply that until U_T can be derived independently from ϵ, these two quantities cannot be compared in a meaningful way, and that results obtained in previous studies are not valid.

INTRODUCTION

Solar-wind/magnetosphere coupling studies are concerned with the issues of how solar wind mass, momentum, and energy are transferred to the magnetosphere, the physical processes that affect the transfer, and the time scales on which the processes operate. Over the years direct in situ measurements of the relevant solar-wind parameters have been cross correlated with various combinations of geomagnetic activity indices in an attempt to understand the coupling process (e.g., Refs. 1–4). Lately, the more powerful method of linear prediction filtering has been used in studies involving solar-wind/magnetosphere coupling.[5-10]

A recent application of linear prediction filtering to solar-wind/magnetosphere coupling presented a comparison of ϵ, an estimator of solar-wind energy input, and U_T, an estimator of total magnetospheric energy output.[11] A principal result of Akasofu et al.[11] was that the filter or impulse response function, relating ϵ and U_T, often resembles a narrow rectangular spike near zero lag time. They concluded from this result that the "magnetosphere is primarily a directly driven system during disturbed periods." Given the scale size of the magnetosphere, the finite signal propagation time that must exist within such a system, and the uncertainty in correcting for propagation of the solar wind from ISEE 3 to earth, the zero lag effect found by Akasofu et al. was quite surprising. This result prompted us to re-examine in detail the relationship between ϵ and U_T to help us understand the origin of this seemingly unphysical result.

ANALYSIS TECHNIQUE

In order to understand fully the Akasofu et al. result, it is necessary to start at the beginning and recalculate all necessary quantities, being careful not to combine the various products until their relative importance is understood. The solar-wind energy input parameter, ϵ, is given by

$$\epsilon = VB^2 l_0^2 \sin^4(\theta/2) \qquad (1)$$

where $l_0 = 7\,\mathrm{R_e}$, θ is the angle between the z-axis and the projection of the interplanetary magnetic field, B, into the y-z plane of the GSM coordinate system, and V is the solar-wind speed.[12] The total magnetospheric energy output parameter, U_T, is given by the sum of the ring current injection rate, U_R, the ionospheric Joule heating rate, U_J, and the energy dissipation rate of auroral particle precipitation, U_A. According to Perreault and Akasofu:

$$U_T = 4 \times 10^{20} \left(\frac{\partial \overline{D_{st}}}{\partial t} + \frac{\overline{D_{st}}}{\tau_R} \right) + 3 \times 10^{15}\, AE \qquad (2)$$

where U_R is the sum of the first two terms, and the last term represents $U_J + U_A$. τ_R is the lifetime of the ring-current particles, AE is the auroral electrojet index, and $\overline{D_{st}}$ is the pressure corrected D_{st} index defined by

$$\overline{D_{st}} = |D_{st}| - D_{st}(P) \qquad (3)$$

where

$$D_{st}(P) = aP^{1/2} - b\,,$$

and $a = 0.2\ \mathrm{nT/(eV\ cm^{-3})}$, $b = 20\ \mathrm{nT}$, and P is the solar-wind dynamic pressure.[13] U_T has units of ergs.

*University of California, Los Alamos National Laboratory, Los Alamos, New Mexico 87545.
†Space, Telecommunications and Radioscience Laboratory, Stanford University, Stanford, California 94305.
‡Institute of Geophysics and Planetary Physics, University of California, Los Angeles, California 90024.

The most uncertain parameter in the above formulation is τ_R, and here we adopt the same conventions as outlined in Ref. 14. Three different functional forms of the ring-current decay time are considered here: (a) a constant, (b) a multi-variable function of ϵ, or (c) a continuous function of ϵ.

All data presented here were obtained from the same combined data set used in the Baker et al.[4] and Akasofu et al.[11] studies. Values of U_T that use the continuous function of ϵ were calculated at the Geophysical Institute, University of Alaska (S.-I. Akasofu and co-workers).

RESULTS

Three storm intervals were selected for detailed examination from among those published in Ref. 11. While results from only one of those intervals will be discussed here, all three contained similar features. The basic 5-minute averaged solar-wind plasma and geomagnetic-index data for the September 29, 1978 storm interval are shown in Fig. 1. All solar-wind data (measured at ISEE 3) have been time shifted to correspond to the arrival of plasma at the subsolar point. Beginning at ~0700 UT, the nonpressure-corrected D_{st} index decreased rapidly, indicating very rapid growth in the ring current and the onset of an intense geomagnetic storm. During the initial phase of the storm the pressure correction term, $D_{st}(P)$, was comparable to the magnitude of the D_{st} index.

The first step in the analysis is to calculate each of the terms of U_T as outlined above. The various parts of each term will initially be kept separate to allow a comparison of its relative importance. The first attempt at calculating the individual components revealed a minor problem—the partial derivative terms contained a large noise contribution that obscured the results. To help minimize this problem, the original data were passed through a 15-minute running average filter, and then the components of U_T were recalculated. A similar procedure has been used in the past (e.g., Ref. 14). The D_{st}-related terms using a constant ring-current decay time of one hour are shown in the top four panels of Fig. 2. Two things are immediately apparent. The D_{st}/τ_R component (panel 4) is the dominant term, while the partial derivative pressure correction term (panel 1) is composed largely of noise. For comparison the bottom two panels show the data for the ring-current decay terms using the variable step decay time shown in Table 1. In this case the magnitude of the components is larger and distinct steps are seen whenever a new τ_R step is encountered. This latter effect is due directly to a change in the magnitude of ϵ.

The pressure-corrected terms are subtracted from their parent $|D_{st}|$ terms as prescribed by Eq. 3 above, and the results are shown in the top two panels of Fig. 3. The total U_R term is shown in the third panel. A quick glance at Fig. 3 is all that is needed to see that the term containing the ring-current decay parameter dominates U_R, while the superimposed fine structure is due mainly to the derivative term. This same effect is seen in all cases

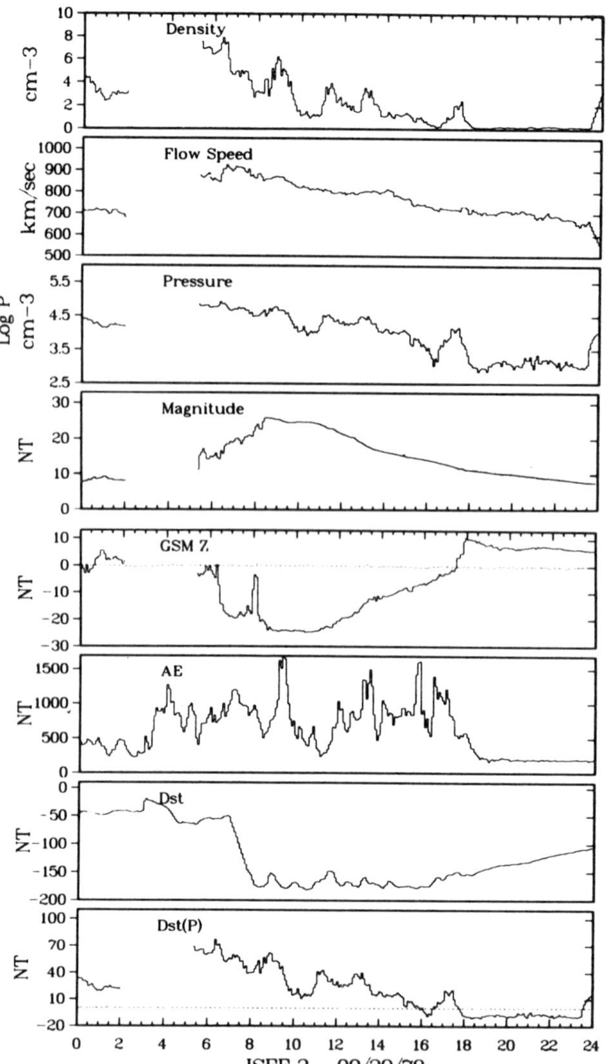

Figure 1—The interplanetary and magnetospheric parameters for the intense geomagnetic storm on September 29, 1978. The top five panels illustrate the one-minute-averaged, transit-time corrected, solar-wind plasma and magnetic-field data obtained by ISEE 3. The bottom three panels present the one-minute-averaged AE and D_{st} indices and the pressure correction term, $D_{st}(P)$, which is used to correct the D_{st} index.

studied to date. The fourth panel from the top displays the contribution of the AE index ($U_J + U_A$) to the total U_T, shown in the bottom panel. The AE contribution to U_T, as shown in Fig. 3, is unimportant compared to U_R even though the AE index was extremely large. A similar result is obtained in all events studied to date. It is possible that the ring-current injection term is overstated or, alternatively, that the AE contribution is understated in the Perreault and Akasofu formulation of U_T. For example, Davis and Parthasarathy[15] found $\tau_R = 10$ hours, empirically, while simulation studies suggest that the U_R and ($U_J + U_A$) terms should be closer to unity (e.g., Harel et al.[16]).

The data presented in Figs. 2 and 3 show that U_T is

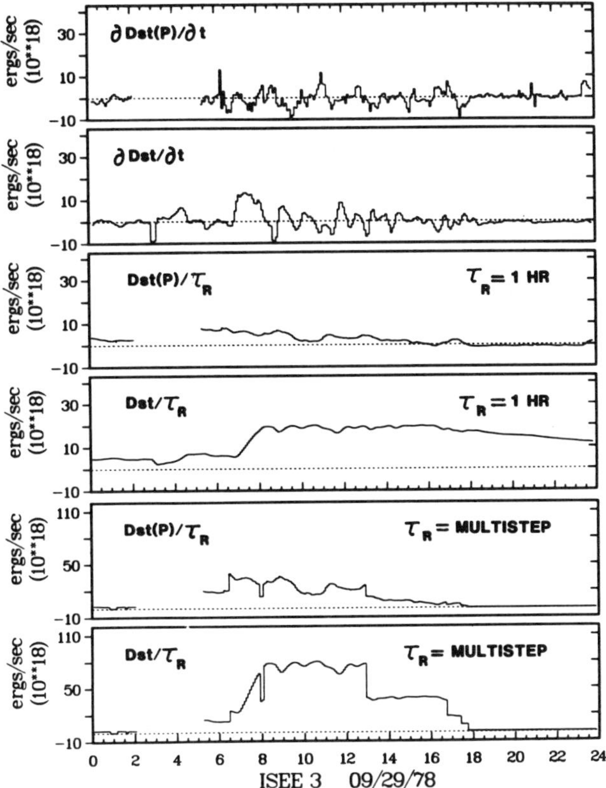

Figure 2—The top two panels show components of the first term of U_T that involve the derivative of the pressure-corrected D_{st} Index. The next two panels show the components of D_{st}/τ_R term for $\tau_R = 1$ hour. The bottom two panels show the ring-current decay term using a variable multistep τ_R. All terms have been multiplied by the appropriate constants in Eq. 2. The D_{st}/τ_R term is by far the largest, while the components of the derivative term are small and contain a large degree of noise. The data have been passed through a five-point running-average filter to help reduce the high-frequency noise.

Table 1—Variable step ring current decay time. (after Akasofu[14] and personal communication, 1982)

ϵ (erg/s)	τ_R $(hours)$
$\epsilon \leq 10^{18}$	20
$10^{18} \leq \epsilon \leq 5 \times 10^{18}$	2
$5 \times 10^{18} < \epsilon \leq 10^{19}$	1
$10^{19} < \epsilon \leq 5 \times 10^{19}$	½
$\epsilon > 5 \times 10^{19}$	¼

dominated by the single term containing the ring-current decay parameter τ_R, at least when it is set equal to one hour. In order to study the impact of different functional forms of τ_R on U_T, we have compared U_T calculated by Akasofu using a nearly continuous function of τ_R on ϵ, U_T calculated using $\tau_R = 1$ hour, and U_T calculated from a multistep function (Table 1 and bottom of Fig. 2). These results are shown in Fig. 4 along with the ϵ time profile for comparison. Since everything was held

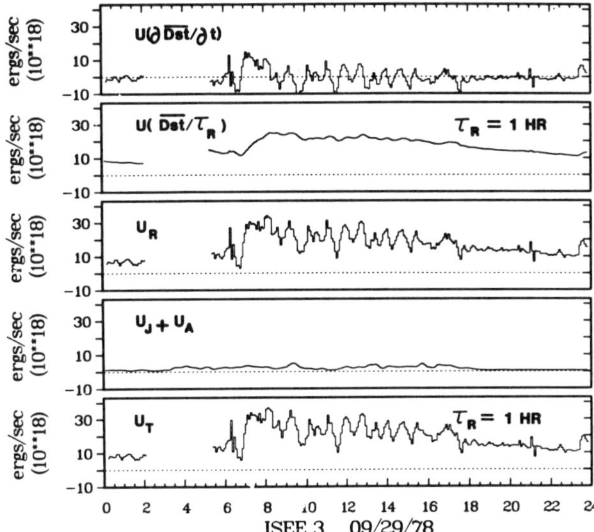

Figure 3—The top two panels represent the individual D_{st} contributions to U_R for the September 29 event, based on a constant (c) ring current decay time of one hour. The third panel shows the complete U_T term, formed by combining the data in the top two panels. The $U_J + U_A$ contribution to U_T is shown in the next panel. The total calculated U_T is presented in the last panel. This step-by-step analysis shows clearly that U_T is dominated by the D_{st}/τ_R contribution, the fine structure is due to the D_{st} derivative term, and $(U_J + U_A)$ is small and does not contribute in any meaningful way.

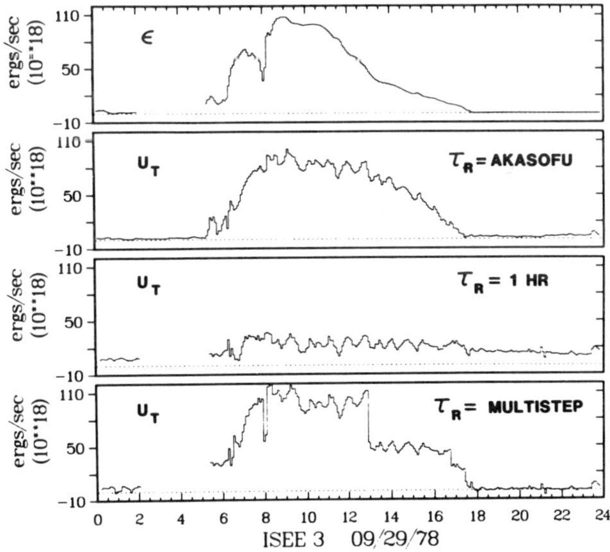

Figure 4—Comparison of all three methods of calculating U_T, based on the same input data. All three calculations use the same ϵ time history shown in the top panel. The second panel contains U_T as calculated by Akasofu. The bottom two panels show our calculated U_T values based on published work by Akasofu in 1981,[14] and based on a program supplied by him in 1982. While our multistep calculation of U_T is somewhat similar to that of Akasofu, the obvious step-like decreases are not present in his curve. To remove these decreases it is necessary to make τ_R a smoother function of ϵ. These data indicate clearly that the final form of U_T is strongly dependent upon the functional form chosen for the ring-current decay time.

constant except for τ_R, all differences seen in Fig. 4 are directly due to the different functions of τ_R on ϵ. It is very clear that the final form of U_T is strongly dependent upon τ_R, which in turn depends directly upon ϵ in the second and fourth panels from the top in Fig. 4. Only the Akasofu et al. version of U_T shows a very obvious correlation with ϵ. The U_T derived from the multistep function contains some unphysical step-like changes, and U_T derived with a constant τ_R shows no evident correlation. In the above cases, it is clear that when τ_R depends upon ϵ, U_T also become a function of ϵ. Thus, it is not surprising that U_T and ϵ are highly correlated in the Akasofu et al. study.[11]

A principal result of the Akasofu et al. study[11] was that the impulse response function between ϵ and U_T often resembled a narrow rectangular impulse near zero lag time. This result led them to conclude that the magnetosphere is primarily a directly driven system during disturbed periods. Given the results shown above concerning the dependency of U_T on ϵ, we felt that the major result of Akasofu et al. might be due to an artifact in the analysis procedure. To check this hypothesis, a linear prediction filtering program[9] was applied to the data taken during all four storm intervals for each of the τ_R functional forms. The results for the September 29 storm for τ_R = constant and for a continuously variable τ_R[11] are shown in Fig. 5. The input data and the predicted U_T based on the derived filter are shown on the left and the derived impulse response function is shown on the right. The upper response function

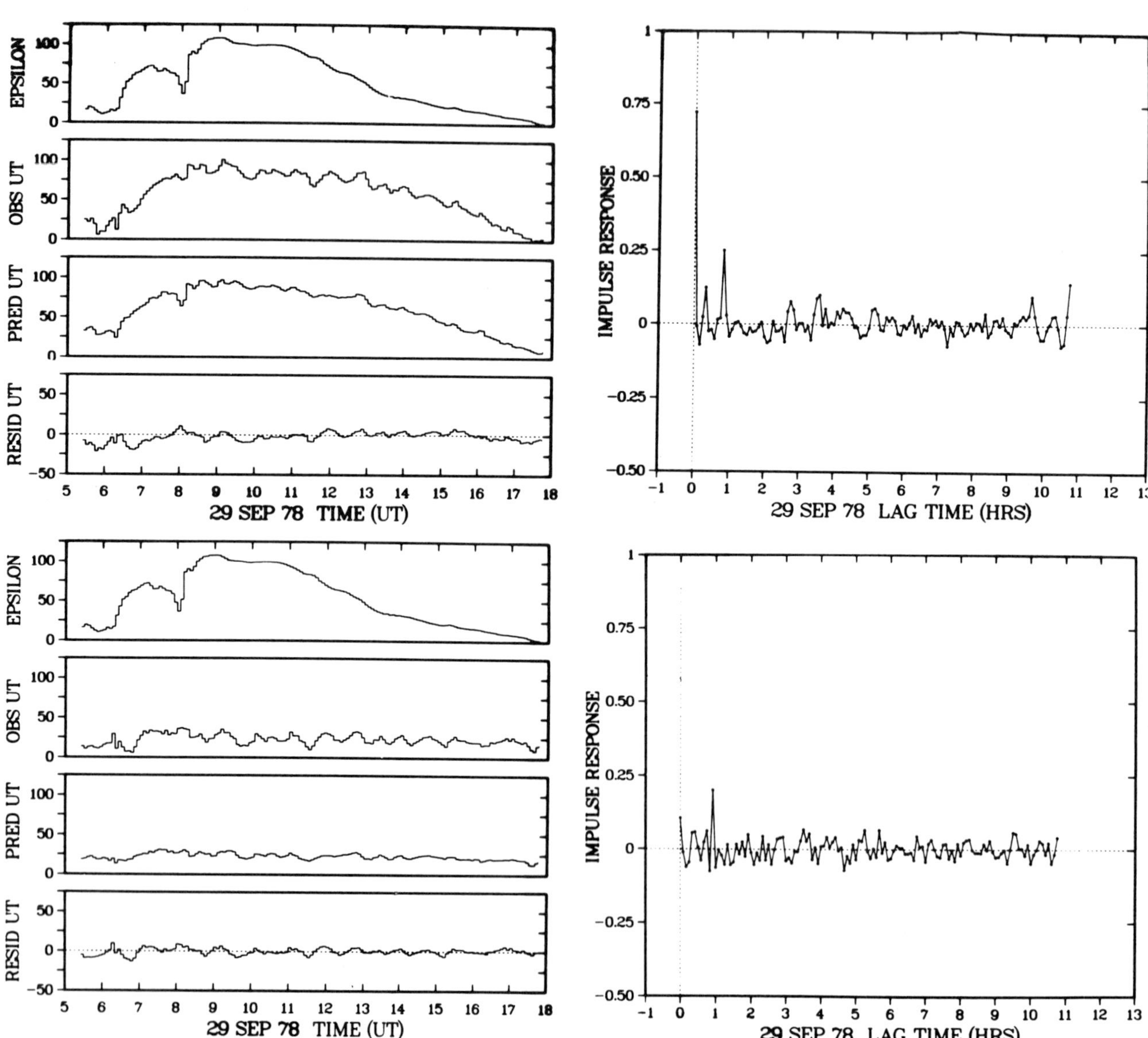

Figure 5—Comparison of the impulse response functions and their products for U_T calculated by Akasofu in the upper panels and U_T calculated using a constant value of the ring-current decay time in the lower panels. The plots on the left-hand side display the actual derived linear filters. The upper filter is similar to that published in Ref. 11. The major difference in the two filters shown in the right-hand panels is the presence of a very large first point in the upper filter and the absence of such a point in the lower filter.

contains a spike at near zero lag, exactly as Akasofu et al.[11] found, while the response function shown in the bottom panel contains no such peak. Note, however, that the peak seen just inside one hour is still present.

All four cases studied to date show a result similar to that seen in Fig. 5. Thus, we confirm that the peak at zero lag is indeed present, as Akasofu et al. claim, but only under certain conditions: only when τ_R is defined as a function of ϵ will the spike at zero lag appear. Hence, the spike can only be due to an artifact in the analysis and, consequently, no physical significance can be attached to it.

CONCLUSIONS

In our reevaluation of U_T, starting at the most basic level, we have confirmed many of the earlier statements concerning how to calculate the most accurate U_T. It is necessary to filter out high-frequency noise from the derivative terms by passing a running-average filter over the data before any calculation is performed. In all cases studied and for all three forms of τ_R used, U_T was dominated by the D_{st}/τ_R term while the AE term did not contribute significantly. Because of the importance of this single term, any change in the definition of τ_R causes a comparable change in U_T, as seen in Fig. 4. In those cases where τ_R is defined as a function of ϵ, U_T must therefore become a direct and sensitive function of ϵ. Thus, statements made in many previous publications concerning the strong correlation of ϵ with U_T (e.g., Refs. 4, 11 and 14) are accurate but meaningless. Physical insight about the relationship between two quantities can be gained only by comparisons between quantities that are independently determined.

Akasofu et al.[11] conclude that the narrow impulse response function found near zero lag shows that ϵ and U_T are highly correlated with little or no time delay, and that these results are consistent with the model of a directly driven system. We have shown here that this δ-function response is an artifact of the analysis procedure, and hence all of these conclusions are without physical basis. Because of the facts described above, it is not possible to make any definitive statement about any magnetospheric model from their analysis.

ACKNOWLEDGMENT—This work was done under the auspices of the U.S. Department of Energy with support from NASA and the University of California Institute of Geophysics and Planetary Physics. Support for this work at Stanford was provided by the National Aeronautics and Space Administration under grant NAGW-235 and also with partial support from the Air Force Geophysics Laboratory under contract F19628-85-K-001. Support for this work at UCLA was provided by grants from the National Science Foundation, ATM 83-18200, and the Office of Naval Research, ONR N0014-C-0158.

REFERENCES

[1] R. L. Arnoldy, "Signature in the Interplanetary Medium for Substorms," *J. Geophys. Res.* **76**, 5189 (1971).

[2] C.-I. Meng, B. Tsurutani, K. Kawasaki, and S.-I. Akasofu, "Cross-Correlation Analysis of the AE Index and the Interplanetary Magnetic Field B_z Component," *J. Geophys. Res.* **78**, 617 (1973).

[3] R. E. Holzer and J. A. Slavin, "An Evaluation of Three Predictors of Geomagnetic Activity," *J. Geophys. Res.* **87**, 2558 (1982).

[4] D. N. Baker, R. D. Zwickl, S. J. Bame, E. W. Hones, Jr., B. T. Tsurutani, E. J. Smith, and S.-I. Akasofu, "An ISEE 3 High Time Resolution Study of Interplanetary Parameter Correlations with Magnetospheric Activity," *J. Geophys. Res.* **88**, 6230 (1983).

[5] T. Iyemori, H. Maeda, and T. Kamei, "Impulse Response of Geomagnetic Indices to Interplanetary Magnetic Field," *J. Geomagn. Geoelectr.* **31**, 1 (1979).

[6] C. R. Clauer, R. L. McPherron, C. Searls, and M. G. Kivelson, "Solar Wind Control of Auroral Zone Geomagnetic Activity," *Geophys. Res. Lett.* **8**, 915 (1981).

[7] C. R. Clauer, R. L. McPherron, and C. Searls, "Solar Wind Control of the Low Latitude Asymmetric Magnetic Disturbance Field," *J. Geophys. Res.* **88**, 2123 (1983).

[8] R. L. McPherron, R. A. Fay, C. R. Garrity, L. F. Bargatze, D. N. Baker, C. R. Clauer, and C. Searls, "Coupling of the Solar Wind to Measures of Magnetic Activity," *Proc. Conf. Achievements of the IMS,* Graz, Austria, ESA SP-217, 161 (Sep 1984).

[9] L. F. Bargatze, D. N. Baker, R. L. McPherron, E. W. Hones, Jr., "Magnetospheric Impulse Response for Many Levels of Geomagnetic Activity," *J. Geophys. Res.* **90**, 6387 (1985).

[10] C. R. Clauer, "The Technique of Linear Prediction Filters Applied to Studies of Solar Wind-Magnetosphere Coupling," in *Solar Wind-Magnetosphere Coupling,* Y. Kamide and J. A. Slavin, eds., Reidel, Dordrecht (in press 1985).

[11] S.-I. Akasofu, C. Olmsted, E. J. Smith, B. Tsurutani, R. Okida, and D. N. Baker, "Solar Wind Variations and Geomagnetic Storms: A Study of Individual Events Based on High Time Resolution ISEE-3 Data," *J. Geophys. Res.* **90**, 325 (1985).

[12] P. Perreault and S.-I. Akasofu, "A Study of Geomagnetic Storms," *Geophys. J. R. Astron. Soc.* **54**, 547 (1978).

[13] R. K. Burton, R. L. McPherron, and C. T. Russell, "An Empirical Relationship Between Interplanetary Conditions and D_{st}," *J. Geophys. Res.* **80**, 4204 (1975).

[14] S.-I. Akasofu, "Energy Coupling Between the Solar Wind and the Magnetosphere," *Space Sci. Rev.* **28**, 121 (1981).

[15] T. N. Davis and R. Parthasarathy, "The Relationship Between Polar Magnetic Activity DP and Growth of the Geomagnetic Ring Current," *J. Geophys. Res.* **72**, 5825 (1967).

[16] M. Harel, R. A. Wolf, R. W. Spiro, P. H. Reiff, C.-K. Chen, W. J. Burke, F. J. Rich, and M. Smiddy, "Quantitative Simulation of a Magnetospheric Substorm 2. Comparison with Observations," *J. Geophys. Res.* **86**, 2242 (1981).

COMMENTS ON "AN EVALUATION OF THE TOTAL MAGNETOSPHERIC ENERGY OUTPUT PARAMETER, U_T" BY R. D. ZWICKL, L. F. BARGATZE, D. N. BAKER, C. R. CLAUER, AND R. L. McPHERRON

S.-I. Akasofu and C. Olmsted*

In Zwickl et al.[1] the authors make specific reference to Akasofu et al.[2] and assert that the existence of a δ-shaped response function relating ϵ, the solar wind parameter, and the total energy dissipation rate, U_T, "can only be due to an artifact in the analysis." This conclusion is unwarranted and overstated. The results of Akasofu et al.[2] are not erroneous. The procedures and assumptions are clearly documented.

Zwickl et al.[1] point out that τ_R, the ring current decay time, is not available as data and must be estimated; this fact was pointed out earlier by Akasofu (see Ref. 3, Section 9, p. 187). Actually, it has long been known that τ_R varies considerably during a geomagnetic storm, from less than a few hours to several tens of hours.[4] It has been pointed out that the assumption of a single value for τ_R will result in the violation of the total energy conservation; that is, the input energy and the output energy integrated throughout a storm period will differ significantly if it is simply assumed $\tau_R = $ constant.[3,5] In particular, compare Fig. 4b and Fig. 32 and read Section 9 in Ref. 3. It has also been pointed out that τ_R should depend on ϵ, since the plasma sheet distance should depend on the cross-tail potential drop ϕ_T and ϕ_T should, in turn, depend on ϵ.[2,3] Indeed, Reiff et al.[6] demonstrated that ϕ_T and ϵ are related by $\phi_T \propto \sqrt{\epsilon}$. The estimate of τ_R is critical to the analysis, especially as it relates to ϵ. Very simply, if $\tau_R \sim 1/\epsilon$ when ϵ is large, $U_T \sim 1/\tau_R \sim \epsilon$. Thus, far from being an "artifact of the analysis," such as a spurious root or Gibbs-type overshoot, this item is a crucial element of the analysis. In summary, all the past analyses of the energy coupling with the assumption of $\tau_R = $ constant violate the total energy conservation principle and thus are invalid.

A correct approach (variable τ_R) should not be blamed to support a wrong approach (constant τ_R) that leads to the violation of a basic physical principle.

In their introduction, Zwickl et al.[1] describe the response function analysis method as "powerful." We must remember, however, that it also has limitations. Thus, in processing our U_T with their ϵ, Zwickl et al.[1] arrive at a similar response function as shown in their Fig. 5 (upper right). There they note a "peak seen just inside one hour." In our response function, computed from statistically equivalent data, we see, however, no such peak (see Fig. 3b, bottom). Fluctuations are quite uniform and low after the initial spike. This is the kind of feature, unexplained by any specific aspect or assumption of the method, that can properly be called an artifact.

*Geophysical Institute, University of Alaska, Fairbanks, Alaska 99775.

REFERENCES

[1] R. D. Zwickl, L. F. Bargatze, D. N. Baker, C. R. Clauer, and R. L. McPherron, "An Evaluation of the Total Magnetospheric Energy Output Parameter, U_T," *Chapman Conf. Proc.,* Laurel, Md. (1986).

[2] S.-I. Akasofu, C. Olmsted, E. J. Smith, B. Tsurutani, R. Okida, and D. N. Baker, "Solar Wind Variations and Geomagnetic Storms: A Study of Individual Storms Based on High Time Resolution ISEE 3 Data," *J. Geophys. Res.* **90,** 325 (1985).

[3] S.-I. Akasofu, "Energy Coupling Between the Solar Wind and the Magnetosphere," *Space Sci. Rev.* **28,** 121 (1981).

[4] S.-I. Akasofu, S. Chapman, and D. Venkatesan, "The Main Phase of Great Magnetic Storms," *J. Geophys. Res.* **68,** 3345 (1968).

[5] P. Perreault and S.-I. Akasofu, "A Study of Geomagnetic Storms," *Geophys. J. R. Astron. Soc.* **54,** 547 (1978).

[6] P. H. Reiff, R. W. Spiro, and T. W. Hill, "Dependence of Polar Cap Potential Drop on Interplanetary Parameters," *J. Geophys. Res.* **86,** 1639 (1981).

SUPERPOSED EPOCH ANALYSIS OF MAGNETOSPHERIC SUBSTORMS USING SOLAR WIND, AURORAL ZONE, AND GEOSTATIONARY ORBIT DATASETS

L. F. Bargatze,[*††] D. N. Baker,[*] and R. L. McPherron[††]

A primary goal of studies of solar-wind/magnetosphere interaction is to understand how the solar wind controls the temporal sequence of events during substorms at many widely separated regions within the magnetosphere. This paper examines the average relationship between definitive solar wind, auroral zone, and geostationary orbit parameters during isolated substorms. High time resolution (1 minute) measurements of two solar-wind quantities, B_z and VB_s, two auroral electrojet indices; AE and AL, and three parameters that define the energetic (30–300 keV) electron distribution at geostationary orbit from 13 events are analyzed, using the superposed epoch technique. The zero epoch times used to organize the analysis were defined as the beginning of electron injection events at synchronous orbit. The average variations of the auroral zone and geostationary orbit parameters and their relation to the solar wind are discussed in context of the three-phase (growth, expansion, and recovery) model of substorms. Notably, we find that particle injection at synchronous orbit lags about 6 minutes behind expansion phase onset in the auroral zone. A possible explanation for this time lag is briefly discussed.

INTRODUCTION

A magnetospheric substorm has been defined as the collection of physical processes that occurs when energy input from the solar wind into the magnetosphere is dissipated in the ring current and the ionosphere.[1] A model has been developed that describes the substorm as a sequence of three phases: growth, expansion, and recovery.[2] In the growth phase, the solar-wind/magnetosphere energy coupling rate is greatly enhanced due to dayside magnetic reconnection. Magnetic flux is added to the tail as the reconnected magnetic field lines are drawn antisunward by the solar wind. As a consequence, energy is stored in the magnetotail in the form of increased magnetic field energy density. Later at expansion phase onset (EPO), magnetic reconnection starts in the near-earth magnetotail. This releases the stored energy that is dissipated at the earth by Joule heating in the ionosphere, by energetic particle precipitation, and by injection of energetic particles into the ring current. The recovery phase is defined as the period of time during which the magnetosphere relaxes to its presubstorm configuration.

The simple model just described applies well to what is called an "isolated" substorm. However, more complicated situations occur in which multiple onsets follow one another without intervening intervals of magnetospheric calm. In this study, strict criteria will be enforced in order to select only isolated substorm events for analysis. That is, only events with one clear EPO will be studied. We will adhere to a strict definition of EPO defined by ground magnetic signatures: expansion phase onset corresponds to the beginning of markedly enhanced electrojet activity as evidenced by large, rapid increases in the magnitudes of the auroral electrojet indices.

As discussed above, the north-south component (B_z) of the interplanetary magnetic field plays a crucial role in determining the response of the magnetosphere to the solar wind. A southward directed IMF leads to enhanced geomagnetic activity; a northward directed IMF leads to little geomagnetic activity. Such evidence led Burton et al.[3] to propose the "half-wave" rectifier model for solar-wind/magnetosphere energy transfer. They suggested that VB_s, which is equal to the solar-wind bulk speed multiplied by the magnitude of the southward IMF component (B_s) measured in geocentric solar magnetospheric (GSM) coordinates, is a good parameter to use for correlating solar wind and geomagnetic activity variations.

Other workers have suggested that ϵ, which is closely related to the Poynting flux of the solar wind incident upon the magnetosphere, is also an accurate predictor of geomagnetic activity.[4,5] Based on high correlations between ϵ and geomagnetic index time series, Akasofu[6] has suggested that magnetospheric substorms occur primarily in direct response to solar-wind variations. There is still support for the driven concept of magnetospheric substorms within the community of scientists studying solar-terrestrial interaction. However, since recent work[7] strongly suggests that substorm activity results from the superposition of activity driven directly by the solar wind and activity driven by release of stored energy in the earth's magnetotail, we will discuss the results of this study in the context of the growth-expansion-recovery phase substorm model outlined above.

In order to quantify the effects of magnetospheric substorms as measured by ground-based magnetic observatories, geomagnetic activity indices have been defined. These include the auroral electrojet indices (AE, AL, and AU) that crudely approximate the density of currents flowing in the westward and eastward electrojets[8] and the D_{st} index that is indirectly related to the total kinetic energy of the ring current. Recently, advanced numerical techniques and realistic ionospheric conductivity

---

[*]University of California, Los Alamos National Laboratory, Los Alamos, New Mexico 87545.

[†]Institute of Geophysics and Planetary Physics, University of California, Los Angeles, California 90024.

[‡]Department of Earth and Space Sciences, University of California, Los Angeles, California 90024.

models have been combined with ground-based magnetometer data to calculate such quantities as the ionospheric electric field and current strengths, the field-aligned current strength, and the Joule heat production rate.[9] At present, such quantities are not readily available since a large amount of computer time is required to obtain them. These quantities have been calculated only for a few, select substorm events. For these reasons, statistical analyses must still rely on indices such as AE and D_{st} to measure geomagnetic substorm variations.

At geostationary orbit (GSO), particle and magnetic field signatures of substorms have been frequently observed.[10] These signatures include the stretching and subsequent dipolarization of tail magnetic field lines[11,12] and the decrease or dropout of energetic particle fluxes followed by a large particle flux increase. They can be understood as consequences of magnetic reconnection on the dayside magnetopause and, subsequently, in the near-earth magnetotail. The second-order anisotropy of the energetic electron distribution (C_2) is also seen to undergo substorm associated variations as the trapped particle distributions respond to the distortion of magnetic field lines.[13] Prior to substorm EPO, C_2 usually increases and becomes positive representing a "cigarlike," field-aligned distribution. Following EPO, C_2 tends to decrease rapidly and become negative representing a "pancakelike" trapped distribution.

Our present study can be seen as an extension of the work of Foster et al.[14] who performed a superposed epoch analysis using 54 time intervals when both AE and B_z were available. The list of 54 substorm events was compiled by scanning seven months of AE index data from late 1967. To be selected, an event had to satisfy two criteria: the AE index was required to reach a value of at least 200 nT and the substorm event had to be separated by at least four hours from any previous or subsequent intervals of substorm activity. The last criterion ensured that the substorm events could be characterized as isolated. Foster et al.[14] chose to organize their superposed epoch analysis by defining the zero epoch time (T_0) as the time of EPO. They determined T_0 individually for each substorm event using both auroral zone and mid-latitude magnetometer records. They found that the solar wind B_z component turned southward about 70 minutes prior to T_0 and remained southward for a total of 3 hours. Shortly following the southward turning, and preceding T_0, the superposed AE index values increased gradually showing evidence of a growth phase. From T_0 until T_0 plus 40 minutes, the average AE index values increased greatly. This interval corresponds to the substorm expansion phase. In the two subsequent hours, the average AE index values recovered to presubstorm values.

The purpose of this paper is to determine the average correlated behavior of several solar wind, auroral zone, and GSO parameters during isolated magnetospheric substorms. Seven parameters are examined in all, using 13 isolated substorm events and the superposed epoch analysis technique. The parameter list includes two solar-wind quantities, VB_s and B_z; two auroral electrojet indices, AE and AL; and three GSO parameters that describe the distribution function of the energetic electron population: the flux of 30–300 keV electrons, the second-order anisotropy magnitude C_2, and θ_B, which is an indirect measure of the local magnetic field inclination. C_2 and θ_B are determined from electron distribution using a spherical harmonic analysis.[15]

DATA SETS, DATA-SET PREPARATION, AND EVENT SELECTION

The data elements utilized in this study include IMP 8 solar-wind plasma and magnetic-field measurements, AE and AL indices provided by the World Data Center, and GSO parameters from two satellites designated 1976-059 and 1977-007. All data sets were either originally sampled at, or were subsequently averaged to, 1-minute time resolution.

Before analysis, the solar-wind time series was shifted in time to account for the expected solar-wind travel time from IMP 8 to the dayside magnetopause. The technique utilized accounts for solar-wind convection both in the X- and Y-GSM directions;[16] further details of the time shifting model are discussed in Ref. 7.

In the first step of the event selection procedure, GSO parameters were examined whenever IMP 8 was expected to be monitoring solar-wind conditions. The objective was to identify and to perform a cursory evaluation of energetic electron flux variations sensed by the Los Alamos energetic particle instruments onboard 1976-059 and 1977-007. A preliminary event list was compiled by scanning data from 1979 and selecting events subject to two criteria. An event was selected if and only if the monitoring satellite was within four hours of local midnight and only if there was no evidence of substorm activity at the satellite within two hours prior or subsequent to the event under consideration. The first criterion was enforced in order to ensure reliable measurements of the energetic electron flux, C_2, and θ_B. The motivation for the second criterion lies in our desire to study only isolated substorm events. At this stage, 94 possible study events were identified.

Next, the AE indices for the 94 events were examined to address the isolated substorm restriction. Once again, events were discarded if there was evidence of complex substorm activity within two hours prior to, or subsequent to, the substorm under consideration. Both multiple onset and continuous activity substorms were culled at this step. Finally, World Data Center plots of IMP 8 VB_s time series were checked to ensure that solar-wind conditions were indeed monitored over at least 75% of the four hour time interval centered roughly on the time of electron injection at GSO. Due to the severe restrictions imposed, only 13 events remained on the final event list (see Table 1).

ZERO EPOCH TIME DETERMINATION AND EVENT NORMALIZATION

In contrast to Foster et al.[14] who used ground station magnetograms to determine T_0, we have chosen to

Table 1—Event list summary.

Date	Geo. Sat.	Onset Time (UT)	Local Time
Jan 15	1977-007	0838	2343
Jan 29	1977-007	1216	0325
Feb 10	1977-007	1059	0205
Mar 22	1977-007	1103	0205
Apr 17	1976-059	0552	0120
May 21	1977-007	0810	2314
Jun 17	1976-059	0246	2159
Aug 3	1976-059	0738	0305
Aug 6	1976-059	0736	0303
Oct 6	1976-059	0206	2129
Oct 6	1976-059	0604	0127
Nov 24	1977-007	2113	0145
Dec 8	1976-059	0336	0112

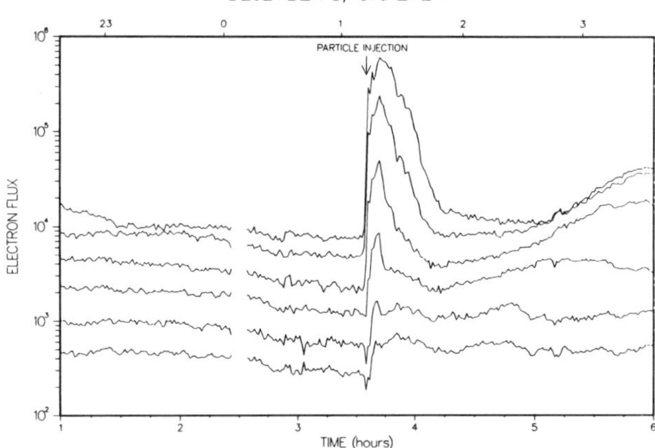

Figure 1—The differential electron flux intensity for six channels spanning the 30–300 keV energy range is shown for the December 8, 1979 event. The individual channels span the following energy ranges: 30–45 keV (top trace until about 0500 UT), 45–65 keV, 65–95 keV, 95–140 keV, 140–200 keV, and 200–300 keV (bottom trace). The electron flux is plotted using units of counts/(cm^2-s-sr-keV). The start time of the particle injection (T_0) is at 0335 UT. The local time position of spacecraft 1976-059 is listed every hour at the top of the panel.

define T_0 as the time of particle injection at GSO. In particular, we define T_0 as the first, large, identifiable energetic electron flux increase at GSO. Figure 1 demonstrates our method. It displays the differential energetic electron flux variations observed by detectors onboard 1976-059 for the 0100–0600 UT time interval on December 8, 1979. In all, six flux channels are plotted, which span the 30–300 keV energy range. Prior to 0335 UT one can see a gentle flux decrease in all six channels that is a repeatable signature of the substorm growth phase. At 0335 UT, all channels show nearly instantaneous, large flux increases that signify that a large particle injection has occurred. Thus, the zero epoch time is 0335 UT. For this event, the integrated 30–300 keV electron flux (not shown in Fig. 1) increased by about a factor of 45 in seven minutes. The values of flux-increase rates for all of the 13 events span the range from a minimum of a fivefold increase in 20 minutes up to a maximum of a fiftyfold increase in 5 minutes. The events with the smaller flux-increase rates are those with satellite positions between 0200 and 0400 local time. Table 1 lists all of the dates and onset times of the 13 events chosen for analysis. The table also lists the designation of the observing geostationary satellite and the satellite's local time position at T_0.

In order to properly weight the contributions of each event to the final superposed epoch time series, a normalization procedure was carried out on five parameters. These parameters are the integrated 30–300 keV electron flux, B_z, VB_s, the AE index, and the AL index. All of these quantities were scaled event by event such that the largest, absolute data value within each individual time series was normalized to 1.0. Since the superposed epoch analysis will be performed only over the time interval corresponding to T_0 plus or minus 2 hours, only these time intervals were considered during normalization.

The θ_B and C_2 time series were not normalized prior to analysis. It was felt that absolute average variations of these two quantities would be physically more interesting to determine. For instance, no normalization

allows one to determine how much the magnetic field inclination at GSO changes at the time of dipolarization.

RESULTS OF THE SUPERPOSED EPOCH ANALYSIS

Figure 2 displays the average or superposed epoch traces of the seven solar-wind, auroral-zone, and GSO parameters for the 2 hours preceding and following T_0. T_0 is designated by the vertical lines drawn at $T_0 = 0.0$ minutes. The five normalized parameters are plotted as a fraction of their maximum values which, of course, is equal to 1.0 after scaling.

Let us first focus on the 30–300 keV electron flux time series which is the parameter we chose to determine T_0. In the two hours preceding T_0, the electron flux decreases slowly by approximately 25%. The electron flux increases sharply at about T_0 minus 1 minute and it reaches a peak near T_0 plus 15 minutes. The total flux increase in this time interval is nearly tenfold. After T_0 plus 15 minutes, the electron flux decreases slowly yet steadily; however, it does not decrease to the normalized background level (0.1) by the end of the analysis interval.

Two points require mention before proceeding to discussion of the other parameters. First, as discussed above, the slow decrease prior to T_0 is one of the signatures of the substorm growth phase; we note, however, that it is difficult to identify when the growth phase commences from examining the average electron flux time series alone. Second, we expect, a priori, that the flux increase signifying the beginning the average electron injection event would begin exactly at T_0. Instead, the electron injection starts one minute before the desired

165

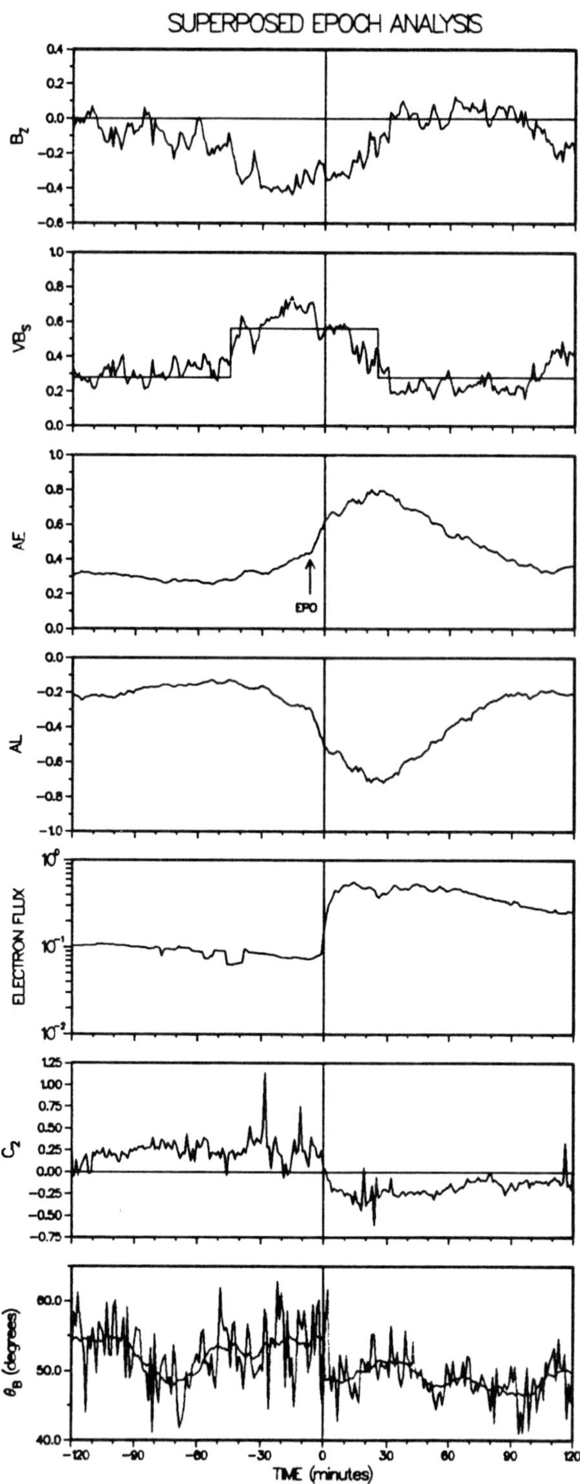

Figure 2—Superposed epoch times series for solar wind, auroral zone, and GSO parameters. Each trace represents the average behavior of that quantity during an isolated substorm.

time. This discrepancy reveals a small, systematic error in our timing of T_0.

Next we focus on the solar wind variables VB_s and B_z. The interval of time between T_0 minus 45 minutes and T_0 plus 25 minutes has been highlighted by the addition of a squarewave to the VB_s panel. These two times correspond roughly to when the IMF turns strongly southward and then strongly northward. During this interval the average B_z and VB_s time series reach values exceeding -0.4 and 0.7, respectively. Note that while the baseline value for B_z outside of this interval is near zero (indicating little solar-wind energy input), the VB_s baseline is about 0.25 (indicating at least some energy input). This apparent discrepancy is easily explainable. It is due to "one-sided" averaging of the VB_s time series whose individual data values are always greater than or equal to zero.

Now examine the superposed epoch time series for the AE index and the AL index. They are nearly identical in form except that AL reaches larger values relative to its baseline. We infer from this that the AL index is less noisy than the AE index, which includes AU index variations ($AE = AU - AL$). Upon comparison of AL and AE with the solar-wind time series, one detects that the gradual growth-phase-associated increases of the index magnitudes begin roughly when the solar-wind energy input enhances at T_0 minus 45 minutes. At T_0 minus seven minutes, the slopes of the AE and AL index curves change markedly: AE increases rapidly and AL decreases rapidly. These signatures indicate that the substorm expansive phase has begun in the auroral zone. At about T_0 plus 25 minutes, the auroral electrojet indices begin to recover toward baseline values. This time is nearly coincident with the northward turning of the IMF as evidenced by the correlated B_z variations. It is interesting that AL appears to recover more quickly reaching presubstorm values some 20 minutes before AE.

Finally, examine the remaining GSO parameters. First, note the high degree of noise exhibited by the average θ_B time series. The high noise level is probably due to the inherent uncertainty in the calculation of the anisotropy direction from the 30–300 keV electron distribution.[17] This fact must be considered when using the θ_B values to infer the local magnetic field orientation. A low-pass averaged version of the θ_B curve has been added to the plot panel in order to better reveal the general behavior of this parameter. The curve was obtained by calculating 19-minute running averages of the θ_B values; the averages for the first and second halves of the curve were calculated independently and connected across T_0. The slope of the smoothed curve suggests that θ_B does increase (an increase of θ_B corresponds to a stretching of the magnetic field lines into a more tail-like orientation) between T_0 minus 45 minutes and T_0 (the inferred growth phase interval at GSO). However, the field inclination at T_0 minus 120 minutes is just as large as at T_0. The field orientation dipolarizes at T_0; on the average, the field inclination decreases by about 5 degrees.

The C_2 superposed epoch time series is similar to the θ_B time series in that it does not clearly reflect changes in response to the southward turning of the IMF. Still, it does reflect the expected variation at the time of particle injection; it changes from positive (field-aligned distribution) to negative (pancake distribution) values. We are presently testing whether normalization of the individual C_2 time series prior to superposed epoch analysis would produce the previously reported C_2 response discussed in the introduction.

The time lag of approximately 6 minutes between EPO in the auroral zone and particle injection at GSO is a surprising result of this analysis. The significance of this feature and possible implications of its existence are discussed below in the final section of this paper.

DISCUSSION

It is interesting to compare and contrast the results of the superposed epoch analysis presented here with the results of Foster et al.[14] First, the average AE index response to the solar wind found here confirms the Foster et al. observations. That is, there is evidence for growth, expansion, and recovery phase features seen in the average AE index trace that appear to be in direct response to the average B_z time series variations. One difference between the two studies is that B_z remains strongly negative for almost 3 hours in the Foster et al. study while it remains negative for only about 1.5 hours in the present study. Also, the return of the AE index to presubstorm values takes nearly an hour longer in their study. These facts may be related to the more strict selection criteria used here to identify isolated substorm events.

Now let us focus on the apparent time delay of particle injection at GSO in relation to EPO in the auroral zone. This time delay (6 minutes) is larger than the expected error in determining the time of particle injection at GSO. We estimate that this error is about ± 2 minutes, including systematic error. Preliminary attempts to determine the time delay individually, event-by-event, utilizing ground-based magnetogram records and additional GSO data sets seem to suggest that the time delay is always positive. This basic result compares favorably with the results of Kamide and McIlwain[18] who found that GSO particle injections occur about 10 minutes later than the time of EPO. They examined 10 events occurring within a 36 hour interval beginning on December 2, 1970 using ATS-5 energetic proton data and H-component ground magnetogram records from auroral zone and mid-latitude observatories. In 9 of 10 cases, they found a positive time delay (auroral zone ahead of GSO). Examining Fig. 1a of their paper, it is possible to assert that all 10 events possessed a positive time delay. In contrast to the present study, their events were spread randomly throughout all local time; our events occurred at GSO within about 3 hours of local midnight.

Kamide and McIlwain[18] offer two possible explanations to account for the systematic lag of particle injection at GSO in relation to EPO in the auroral zone. One explanation is related to the method of defining EPO. In their study, they defined EPO by determining the time at which the first recognizable signature of onset was recorded at a ground magnetic observatory. The authors suggest that choosing the time when the amplitude of the magnetic disturbance reaches half its maximum value would be more consistent with satellite-determined times of particle injection. As another possible explanation, they suggest that the "simplistic 'extrapolation back to infinite energy' method which is used along with satellite data to determine the time of particle injection from the magnetotail" is systematically in error. They contend that this method neglects delays related to radial particle motions caused by east-west electric fields.

Another possible explanation for the systematic time delay may be postulated based on physical arguments. Consider for a moment that a near-earth neutral line forms in the magnetotail at expansion phase onset.[2] The information that the line has formed is transmitted both along magnetic field lines connecting the neutral line to the the auroral ionosphere and across field lines radially inward toward GSO. The information is carried to the ionosphere both by energetic electrons (transit time on the order of a few seconds) and by the propagation earthward of a Pi2 pulsation (transit time of about 1 minute if the wave speed is approximately equal to the Alfvén speed). The information is carried radially to GSO by an inward moving compression wave that is launched as newly reconnected field lines begin to dipolarize. It has been suggested that this inward-moving compressional wave could be responsible for the injection of energetic particles at GSO.[19] Assuming that the compression wave moves at the speed of a fast mode MHD wave and that it moves primarily through the plasma sheet, which is characterized by a low magnetic field strength and a high plasma density (relative to the lobes), it would take the compression wave several minutes to travel from the neutral line to the vicinity of GSO. Thus, it is plausible, based on these simple arguments, that the observed time delay is a physically relevant time lag related to the different speeds at which the information of near-earth neutral line formation travels to reach these two regions.

SUMMARY

The average response of the GSO energetic (30–300 keV) electron distribution and of the auroral electrojet indices AE and AL in relationship to solar-wind B_z and VB_s variations has been studied by using the superposed epoch analysis technique. The results of this analysis agree favorably with an earlier analysis performed by Foster et al.,[14] which focused on the relationship between B_z and the AE index. The average variations of the auroral electrojet indices and the GSO parameters are easily understood when they are considered in the context of the three-phase (growth, expansion, and recovery) isolated substorm model.

An interesting result of the present study is the apparent time lag of particle injection at GSO in relation to expansion-phase onset in the auroral zone. The time lag, which is equal to about 6 minutes, appears to be systematic in that expansion phase onset occurred first in the auroral zone for all of the 13 events studied here. At the moment, we consider this result to be preliminary; a new and larger study is required to test its validity.

ACKNOWLEDGMENT—The authors thank R. C. Anderson for significant computer programming support. Work at Los Alamos was done under the auspices of the United States Department of Energy with support from the University of California Institute of Geophysics and Planetary Physics. Work at the University of California, Los Angeles was supported by the National Science Foundation (grant ATM 83-18200 A01), the Office of Naval Research (grant ONR N00014-85-K-0556), and the National Aeronautics and Space Administration (Grant NGL 05-007-004).

REFERENCES

[1] D. N. Baker, S.-I. Akasofu, W. Baumjohann, J. W. Bieber, D. H. Fairfield, E. W. Hones, Jr., B. Mauk, R. L. McPherron, and T. E. Moore, "Substorms in the Magnetosphere," in *Solar Terrestrial Physics: Past and Future, NASA Ref. Publ. 1120*, Chapter 8 (1984).

[2] C. T. Russell and R. L. McPherron, "The Magnetotail and Substorms," *Space Sci. Rev.* **15**, 205 (1973).

[3] R. K. Burton, R. L. McPherron, and C. T. Russell, "An Empirical Relationship Between Interplanetary Condition and D_{st}," *J. Geophys Res.* **80** 4204 (1975).

[4] P. Perreault and S.-I. Akasofu, "A Study of Geomagnetic Storms," *Geophys. J. R. Astron. Soc.* **54**, 547 (1978).

[5] S.-I. Akasofu, "Interplanetary Energy Flux Associated with Magnetospheric Substorms," *Planet. Space Sci.* **27**, 425 (1979).

[6] S.-I. Akasofu, "Energy Coupling Between the Solar Wind and the Magnetosphere," *Space Sci. Rev.* **28**, 121 (1981).

[7] L. F. Bargatze, D. N. Baker, R. L. McPherron, and E. W. Hones, Jr., "Magnetospheric Impulse Response for Many Levels of Geomagnetic Activity," *J. Geophys. Res.* **90**, 6387 (1985).

[8] Y. Kamide and S.-I. Akasofu, "Notes on Auroral Electrojet Indices," *Rev. Geophys. Space Phys.* **21**, 1647 (1983).

[9] Y. Kamide and W. Baumjohann, "Estimation of Electric Fields and Currents from International Magnetospheric Study Magnetometer Data for the CDAW-6 Intervals: Implications for Substorm Dynamics," *J. Geophys. Res.* **90**, 1305 (1985).

[10] D. N. Baker, "Particle and Field Signatures of Substorms in the Near Magnetotail," in *Magnetic Reconnection in Space and Laboratory Plasmas*, E. W. Hones, Jr., ed., American Geophysical Union, Washington, D.C., p. 193 (1984).

[11] R. L. McPherron, "Growth Phase of Magnetospheric Substorms," *J. Geophys. Res.* **28**, 5592 (1970).

[12] R. L. McPherron, "Substorm Related Changes in the Geomagnetotail: The Growth Phase," *Planet. Space Sci.* **20**, 1521 (1972).

[13] D. N. Baker, P. R. Higbie, E. W. Hones, Jr., and R. D. Belian, "High-Resolution Energetic Particle Measurements at 6.6 R_e 3. Low-Energy Electron Anistropies and Short-Term Substorm Predictions," *J. Geophys. Res.* **83**, 4863 (1978).

[14] J. C. Foster, D. H. Fairfield, K. W. Ogilvie, and T. J. Rosenberg, "Relationship Between Interplanetary Parameters and Occurrence of Magnetospheric Substorms," *J. Geophys. Res.* **76**, 6971 (1971).

[15] D. N. Baker, R. D. Belian, P. R. Higbie, and E. W. Hones, Jr., "High-Energy Magnetospheric Protons and Their Dependence on Geomagnetic and Interplanetary Conditions," *J. Geophys. Res.* **84**, 7138 (1979).

[16] R. D. Zwickl, J. R. Asbridge, S. J. Bame, W. C. Feldman, and J. T. Gosling, "Convective Nature of Transient Solar Wind Phenomena Observed by ISEE 3 and ISEE 1 (abstract)," *Eos Trans. AGU* **61**, 354 (1980).

[17] P. R. Higbie, R. D. Belian, and D. N. Baker, "High-Resolution Energetic Particle Measurements at 6.6 R_e 1. Electron Micropulsations," *J. Geophys. Res.* **83**, 4851 (1978).

[18] Y. Kamide and C. E. McIlwain, "The Onset Time of Magnetospheric Substorms Determined from Ground and Synchronous Satellite Records," *J. Geophys. Res.* **79**, 4787 (1974).

[19] T. E. Moore, R. L. Arnoldy, J. Feynman, D. A. Hardy, "Propagating Substorm Injection Fronts," *J. Geophys Res.* **86**, 6713 (1981).

THE KELVIN-HELMHOLTZ INSTABILITY AND ITS ROLE IN THE GENERATION OF THE ELECTRIC CURRENTS ASSOCIATED WITH Ps 6 AND WESTWARD TRAVELING SURGES

G. Rostoker*

Surges and omega bands are wave-like auroral forms associated with substorm expansive phase activity that form in the poleward portion of the auroral oval and that propagate in the sunward direction in the evening and morning sectors, respectively. Both forms are associated with three-dimensional current systems that produce large (up to several hundred nT) magnetic perturbations with a positive D-component spike at their leading edges being a feature common to both disturbances. In this paper we present a scheme for mapping the auroral oval regimes into the distant magnetotail. In particular we suggest that the electric field transition from dawn to dusk across the polar cap to dusk to dawn at lower latitudes marks the presence of velocity shear zones that map to the interface between the low-latitude boundary layer and central plasma sheet as viewed in a plane parallel to the neutral sheet. In this framework, westward-traveling surges and omega bands are manifestations of the Kelvin-Helmholtz instability in this velocity shear zone at times when bursts of magnetic-field-line reconnection lead to significantly enhanced momentum density of the central plasma sheet.

INTRODUCTION

A magnetospheric substorm is a transient enhancement of energy transfer from the solar wind to the magnetosphere during which time there is enhanced energy dissipation in the auroral oval and storage of energy in the electromagnetic field of the magnetotail and ring current as well as in the kinetic energy of the magnetospheric plasma. Deposition of energy in the ionosphere that was previously stored in the tail often takes place episodically with each outburst involving the development of field-aligned potential drops on auroral oval field lines threading regions of convective velocity shear. These potential drops, found at altitudes of 1–2 R_e, accelerate electrons to energies of several keV and change their pitch angles to the extent that the energized electrons precipitate into the ionosphere leading to the formation of discrete auroral features.

Each outburst described above represents an expansive phase of the overall magnetospheric substorm[1] and their auroral signature is that of the auroral substorm defined by Akasofu.[2] In the auroral substorm, the region of brightened arcs expands poleward and the western edge of the disturbed region develops a characteristic S-shaped contour that expands westward at velocities ranging from zero to a few km per second. Virtually all substorm expansive phases feature the characteristic surge form, and sequential expansive phase intensifications can lead to multiple surge forms that can coexist with one another.[3]

In contrast to the surge, which seems to characterize each expansive phase, only a few magnetospheric substorms appear to feature clearly identifiable omega bands. When they are observed, they tend to occur close to the dawn meridian and most often are observed in bunches. Both the group velocity and phase velocity of the magnetic waveforms associated with omega bands are eastward with velocities ranging from 0.5–2 km/s.[4]

While the westward velocity of surges is highly variable (viz., a surge may start to propagate westward at a few km/s but may then come to a complete stop) omega bands seem to propagate a relatively constant velocity over their lifetimes. Nonetheless, both surges and omega bands have a common characteristic when it comes to velocity of propagation, namely, both forms propagate sunward with somewhat similar average velocities.

The magnetic variations associated with omega bands have been labeled Ps 6 (Ref. 5) and these magnetic variations look very similar to those associated with westward traveling surges both in magnitude (tens to hundreds of nT) and the duration of the characteristic positive D-component pulse ($\sim$ 15 minutes). Possible ionospheric and magnetospheric current systems have been proposed for Ps 6 by Kawasaki and Rostoker[6] and Andre and Baumjohann[7] and for westward traveling surges by Rostoker and Hughes[8] and Tighe and Rostoker.[9] All models involve an intense equatorward ionospheric current within the waveform that is responsible for the positive D-component perturbation.

In this paper we shall propose a mapping between the auroral ionosphere and the magnetotail in which the sunward velocity of surges and omega bands translate into the sunward motion of surface waves at the interface between the low-latitude boundary layer (LLBL) and the central plasma sheet (CPS). We shall then identify the physical processes that lead to the development of the magnetosphere-ionosphere current systems whose magnetic signatures are those of surges and Ps 6 variations.

MAPPING FROM THE IONOSPHERE TO THE MAGNETOSPHERE

Our scheme for mapping begins by noting that the energetic electron characteristics in the topside ionosphere dictate that the particles precipitating into the auroral oval be subdivided into two distinct regions—the central plasma sheet (CPS) and boundary plasma sheet (BPS) as defined by Winningham et al.[10] The CPS fea-

*Institute of Earth and Planetary Physics and Department of Physics University of Alberta, Edmonton, Alberta, Canada T6G 2J1.

tures power-law electron spectra (suggesting no field-aligned electric fields affect those particles) while the BPS features localized bursty energetic electron fluxes characterized by drifting Maxwellians (suggesting the presence of a parallel electric field on these field lines). The polarity reversal of the north-south component of the ionospheric electric field (identified with the Harang discontinuity in the premidnight quadrant) is found in the heart of the BPS.[11,12] The locations of the CPS and BPS relative to the directly driven auroral electrojets are shown in Fig. 1. Here it can also be seen that the equatorward edge of the CPS and auroral electrojets maps to the inner edge of the magnetospheric CPS.

We next note that there are two clearly identified electric field transitions found in the magnetotail. One is found in the heart of the plasma-sheet boundary layer (PSBL)[14] and the other is found in the low latitude magnetotail at the interface between the CPS and the

low-latitude boundary layer (LLBL).[15] Our mapping demands that the E-field transition in the ionospheric BPS and those in the PSBL and at the LLBL/CPS interface are one and the same. That is, if one follows a magnetic field line out of the region of the E-field transition in the ionospheric BPS, one will find oneself threading the PSBL in the "high-latitude" magnetotail with the (closed) field line crossing the neutral sheet at the LLBL/CPS interface as shown in Fig. 1. The evidence that the ionospheric BPS maps to the LLBL close to the earth is considerable (cf. Heelis et al.[16] and Bythrow et al.[17]) with strong evidence being available suggesting that BPS field lines are threaded by Region 1 field-aligned currents (e.g., McDiarmid et al.[18]). Our mapping scheme proposes that this relationship between BPS and LLBL exists to the antiearthward edge of the closed-field-line region. These relationships pertain, of course, only to the closed-field-line regime.

Dayside merging processes lead to an open magnetosphere demanding that our model involve a separator line defining the outer boundary of the closed-field-line regime. This separator line, shown in Fig. 1 by the projection on a plane parallel to the neutral sheet, coincides with the magnetopause for some distance downtail. When the closed LLBL field lines can be extended no further downtail, the separator line no longer follows the magnetopause and converges toward the center of the tail. In the plane of the neutral sheet, this separator line would be seen as the neutral line. The average position of the neutral line in the center of the tail is ~ 115 R_e behind the earth based on the rate at which the tail-lobe magnetic field falls off with distance down the magnetotail.[19] Reconnection of the magnetotail field lines may be expected to occur across the neutral line during episodes of magnetospheric activity when magnetic flux added to the tail through frontside merging is returned to the nightside inner magnetosphere.

KELVIN-HELMHOLTZ INSTABILITY AND SUBSTORM EXPANSIVE PHASE

From our mapping shown in Fig. 1, it is evident that there is a velocity shear zone at the interface between the LLBL and CPS. We propose that a burst of reconnection across the neutral line will result in earthward "jetting" of plasma. This jetting will tend to be field aligned in the PSBL and convective adjacent to the neutral sheet. The mathematical description of the Kelvin-Helmholtz instability (KHI) has been given in general by Southwood[20] and applied to our specific geometry by Rostoker et al.[21] Assuming that the antiearthward momentum flux of the boundary layer plasma adjacent to the CPS is less than that of the CPS plasma, KHI theory predicts an earthward phase velocity V_P of waves generated in the shear zone of

$$V_P = \frac{1}{\rho_{PS} + \rho_{BL}} (\rho_{PS} V_{PS} - \rho_{BL} V_{BL}) \quad (1)$$

where the subscripts *PS* and *BL* indicate the CPS and LLBL, respectively, and $k \parallel (\rho_{PS} V_{PS} + \rho_{BL} V_{BL})$.

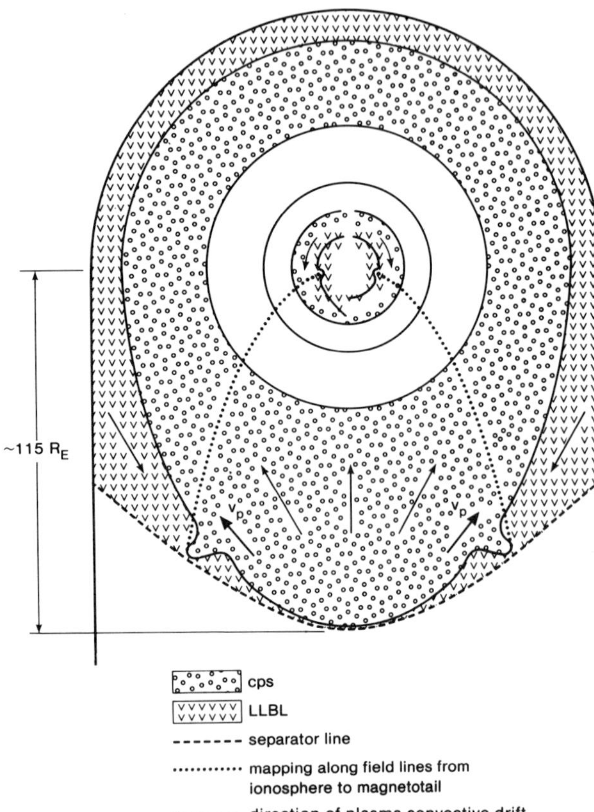

Figure 1—Mapping of the particle distributions in the topside ionosphere (CPS and BPS) into the magnetospheric equatorial plane/plane of the neutral sheet emphasizing the corresponding tail particle regimes (CPS and LLBL, respectively). Transient bursts of field-line reconnection across the separator line near -115 R_e (on the average) leads to fast earthward convective flow in the distant CPS. This results in enhanced velocity shear at the interface between the CPS and the boundary layer, leading to the growth of a Kelvin-Helmholtz instability. The resultant waveforms propagate sunward along the distant shear zones and are manifested in the ionosphere by westward-traveling surges in the evening sector and auroral omega bands in the dawn sector (after Rostoker and Eastman[13]).

Mapped into the ionosphere, sunward velocities reflect westward motion in the dusk sector and eastward motion in the dawn sector (which is precisely what is observed for surges and omega bands, respectively). The plasma-sheet convective velocity V_{PS} should be of the order of the Alfvén velocity computed from the magnetic field adjacent to the reconnection region and the number density of plasma in the reconnection region.[22] Choosing these parameters as $B = 10$ nT and $n \sim 1$ cm^{-3}, respectively, the velocity at which the plasma jets earthward from the reconnection region is ~ 225 km·s^{-1}. As the plasma convects earthward, it will gradually slow down and the density will decrease due to the gradual increase of the cross section of the plasma sheet (in the plane normal to the axis of the tail) and the need for mass flux to be conserved (viz., $\Phi_M = \rho VA =$ constant). The decrease in kinetic drift energy is consistent with that energy being available to drive portions of the Birkeland current systems.[23]

The development of the waveform seen at the LLBL/CPS interface (Fig. 1) features a growth rate given by γ where

$$\gamma^2 = \frac{\rho_{PS}\rho_{BL}}{(\rho_{PS}+\rho_{BL})^2} \, [k \cdot (V_{BL} - V_{PS})]^2$$

$$- \frac{(B_{PS} \cdot k)^2 + (B_{BL} \cdot k)^2}{\mu_0 \, (\rho_{PS} + \rho_{BL})} \qquad (2)$$

and k is the wave number of the growing wave.[21] Within the CPS adjacent to the neutral sheet we consider $B_{PS} \perp k$, $V_{PS} \| k$, and $V_{BL} \| -k$. A major unknown is the orientation of k to B_{BL}. That is, we do not know if the closed field line LLBL has become so spatially confined as to be inconsequential close to the portion of the separator line across which reconnection is occurring. If the LLBL is still to be reckoned with, then $B_{BL} \perp k$ and the growth rate can be calculated from the first term of Eq. 2. For $\rho_{PS} \doteq \rho_{BL}$ (see Slavin et al.[24]) and $|V_{PS}| > |V_{BL}|$ we find the characteristic growth time $\gamma^{-1} \simeq$ 2–3 minutes where $|V_{PS}| \simeq |V_A|$ in accordance with Petschek.[22] In this calculation $k = V_P/\omega$ and ω corresponds to the 1 mHz frequency which dominates Ps 6 disturbances[4] and surges. If, on the other hand, the boundary layer adjacent to the CPS neutral sheet region involves open field lines (viz. $k \| B_{BL}$), then Eq. 2 reduces to

$$\gamma^2 = \frac{k^2}{2} \, (2V_{PS}^2 - V_A^2)$$

and one would expect reduced wave growth due to the stabilizing effect of the magnetic shear across the velocity shear zone. For our KHI concept to be valid, there cannot be too much magnetic shear across the LLBL/CPS interface, namely, the closed field line LLBL must be consequential at least along part of that interface. (This criterion may, in fact, determine at what local time (as seen in the ionosphere) waveforms generated

through the action of the KHI can develop.) We shall proceed on that basis to show how the substorm current wedge is a logical consequence of the development of the KHI described above.

Our approach is to consider the MHD analog of the hydrodynamic problem involving the treatment of a velocity shear zone surface of discontinuity described by a vortex sheet.[25] By considering a perturbation of such a vortex sheet (including second-order terms) Rosenhead followed the evolution of the resulting waveform (Fig. 2) and noted that the elementary vortices redistributed themselves in a manner involving concentrations and depletions relative to the pre-existing uniform distribution of the vortices. In the MHD approach, each vortex element can be considered as an element of space charge; that is, from $\rho = \epsilon_0 \nabla \cdot E$ and $E = -v \times B$, one can show that a portion of the total space charge is given

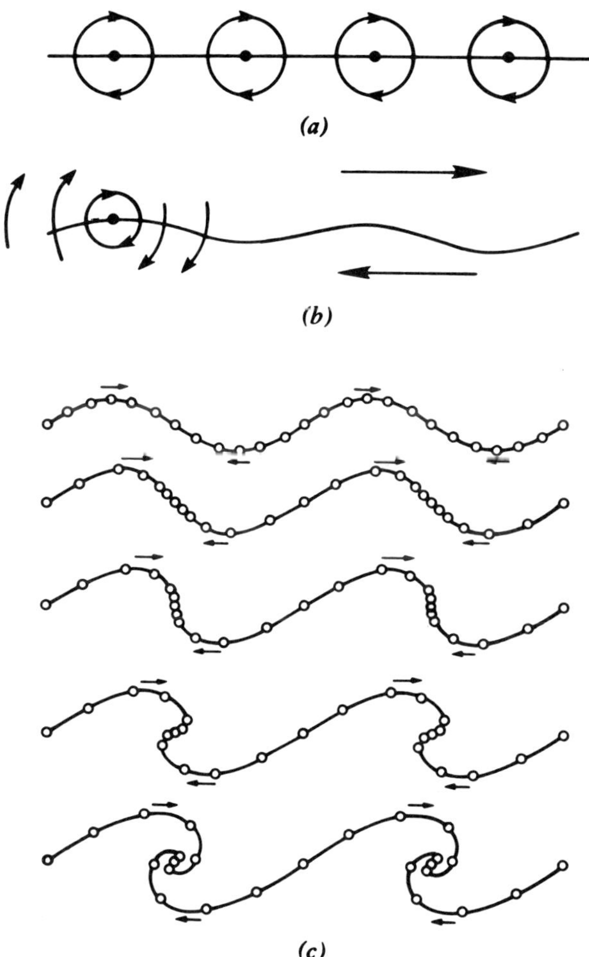

Figure 2—Representation of the velocity shear zone at the CPS/boundary layer interface by a vortex sheet (after the hydrodynamic approach of Rosenhead[25]). The figure shows the unperturbed vortex (a), the initial perturbation of the sheet (b), and the evolution of the instability (c). Note the alternating concentrations and depletions of the vortex elements. Since each vortex element can be considered as an element of space charge (see text), the Kelvin-Helmholtz instability leads to changes in space charge density as a consequence of which field-aligned currents are caused to flow.

by $\rho = -\epsilon_0 \boldsymbol{B} \cdot \boldsymbol{\Omega}$ where $\boldsymbol{\Omega}$ ($= \nabla \times \boldsymbol{v}$) is the vorticity. For the velocity shear configuration shown in Fig. 1, it can be seen that the evening sector shear zone should be populated by negative space charge and the morning sector shear zone by positive space charge (consistent with the expected dawn-to-dusk electric field across the CPS). The redistribution of elementary vortices (i.e., space charge) leads to the growth of spatially alternating concentrations of space charge with concentrations of negative space charge on the evening side and positive space charge on the morning side. One should expect field-aligned current to flow in response to the changing space charge distributions as shown in Fig. 3, with field-aligned current flowing out of the ionosphere into the core of the waveform in the evening sector and into the ionosphere from the core of the waveform in the morning sector. (Theoretical evaluations of field-aligned currents that develop through the action of the KHI can be found in Refs. 26 and 27.) The complete three-dimensional current system associated with the KHI would be solenoidal in character with ionospheric Pedersen current converging toward (away from) the region of upward (downward) field-aligned current in the evening (morn-

ing) sector waveform. Closure is effected through downward (upward) field-aligned currents in the regions surrounding the core field-aligned currents in the evening (morning) sector and transverse currents antiparallel to the perturbation electric field in the velocity shear zone adjacent to the neutral sheet. Of course, a ground observer would not detect the magnetic effects of the solenoidal current system described above, but would rather see the signature of the Hall current vortices in the ionosphere (shown in Fig. 3) driven by the perturbation electric field transverse to the background magnetic field (as suggested in Ref. 28). We note, in passing, that based on the Hall current patterns shown in Fig. 3, one should detect positive D-component magnetic perturbations (due to concentrated southward current flow) at the leading edge of westward-traveling surges and omega bands as is, indeed, the case. Finally we note the schematic elongation of the surge Hall current pattern relative to that of the omega band, which is intended to reflect the rather different ionospheric conductivity distribution in the two cases. That is, the ionosphere is highly conducting behind the surge in the evening sector so one would expect the poleward current to be more distributed in longitude behind the surge than behind the omega band (which is embedded in a lower background conducting ionosphere).

SUMMARY

We have presented, in this paper, the contention that the surge and omega band waveforms associated with substorm expansive phase activity originate through the action of a Kelvin-Helmholtz instability at the outer fringe of the CPS in the deep magnetotail. We suggest that the waveforms arise through enhanced earthward CPS convective flow caused by bursts of reconnection at the neutral line (nominally located near $\sim 115\ R_e$). Based on the MHD analog to the hydrodynamic treatment of the KHI by Rosenhead,[25] we expect the growth of the KHI to lead to field-aligned current flow. As suggested by Rostoker and Samson,[29] the current flows so as to stabilize the instability by removing free energy from the site of the instability (or, as viewed in an MHD context, by creating a transverse magnetic field leading to the development of a stabilizing magnetic shear in the velocity shear zone). The development of the concentrations of negative (positive) space charge in the dusk (dawn) sector shear zone lead to the development of solenoidal current systems (Fig. 3), with the ground magnetic effects being due to associated Hall current flow in the ionosphere. The sunward propagation of both surge and omega bands seen at ionospheric levels simply reflects the fact that the momentum density of the sunward convecting CPS plasma exceeds that of antisunward flowing boundary layer plasma under normal circumstances.

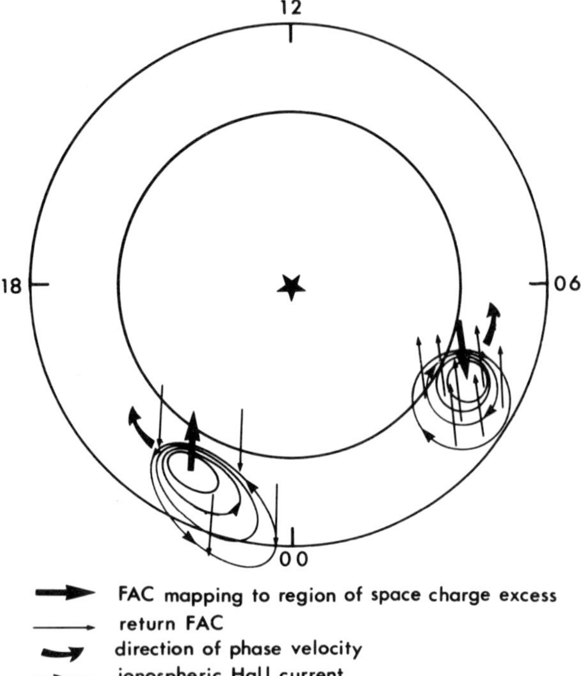

Figure 3—Polar view of surge and omega-band currents (the star marking the pole). Field-aligned currents flow out of the ionosphere at the core of a westward-traveling surge in the dusk sector and into the ionosphere at the core of an omega band in the dawn sector. The three-dimensional current system that develops involving Pedersen closure currents in the ionosphere is solenoidal in character so that its magnetic perturbations are not visible below the ionosphere. The ground observer detects the magnetic effects of the Hall currents that circulate around the cores of the waveforms. The key magnetic signatures of both waveforms are positive D-component perturbations associated with concentrated equatorward ionospheric current flow at the leading edges of the sunward propagating auroral forms.

ACKNOWLEDGMENT—This research has profited by many useful discussions with J. C. Samson. The vortex sheet approach was suggested to me by A. D. M. Walker. The research was supported by the Natural Sciences and Engineering Research Council of Canada.

REFERENCES

[1] G. Rostoker, S.-I. Akasofu, J. Foster, R. A. Greenwald, Y. Kamide, K. Kawasaki, A. T. Y. Lui, R. L. McPherron, and C. T. Russell, "Magnetospheric Substorms—Definition and Signatures," *J. Geophys. Res.* **85**, 1663 (1980).

[2] S.-I. Akasofu, "The Development of the Auroral Substorm," *Planet. Space Sci.* **12**, 273 (1964).

[3] R. G. Wiens and G. Rostoker, "Characteristics of the Development of the Westward Electrojet during the Expansive Phases of Magnetospheric Substorms," *J. Geophys. Res.* **80**, 2109 (1975).

[4] G. Rajaram, G. Rostoker, and J. C. Samson, "Wave Characteristics of Ps 6 Magnetic Variations and Their Implications for Convective Flow in the Magnetotail," *Planet. Space Sci.* (in press, 1986).

[5] T. Saito and A. Morioka, "Dawn-Dusk Asymmetry of Magnetospheric Substorm," in *Proc. 3rd Magnetospheric Symp.,* ISAS Pub., Tokyo, pp. 66-71 (1971).

[6] K. Kawasaki and G. Rostoker, "Perturbation Magnetic Fields and Current Systems Associated with Eastward Drifting Auroral Structure," *J. Geophys. Res.* **84**, 1464 (1979).

[7] D. Andre and W. Baumjohann, "Joint 2-D Observations of Ground Magnetic and Ionospheric Electric Fields Associated with Auroral Currents 5. Current Systems Associated with Eastward Drifting Omega Bands," *J. Geophys.* **50**, 194 (1982).

[8] G. Rostoker and T. J. Hughes, "A Comprehensive Model Current System for High Latitude Magnetic Activity 2. The Substorm Component," *Geophys. J.* **58**, 571 (1979).

[9] W. G. Tighe and G. Rostoker, "Characteristics of Westward Travelling Surges during Magnetospheric Substorms," *J. Geophys.* **50**, 51 (1981).

[10] J. D. Winningham, F. Yasuhara, S.-I. Akasofu, and W. J. Heikkila, "The Latitudinal Morphology of 10-eV to 10-keV Electron Fluxes during Magnetically Quiet and Disturbed Times at the 2100-0300 MLT Sector," *J. Geophys. Res.* **80**, 3148 (1975).

[11] J. D. Winningham, K. Kawasaki, and G. Rostoker, "Energetic Particle Precipitation into the High-Latitude Ionosphere and the Auroral Electrojets 1. Definition of the Electrojet Boundaries Using Energetic Electron Spectra and Ground Based Magnetometer Data," *J. Geophys. Res.* **84**, 1993 (1979).

[12] R. A. Heelis, "Ionospheric Convection at High Latitudes," in *Proc. Magnetospheric Boundary Layers Conference,* Alpbach, Austria, 11-15 Jun 1979, B. Battrick, ed., European Space Agency, Paris, France, p. 175 (1979).

[13] G. Rostoker and T. E. Eastman, "A Boundary Layer Model for Magnetospheric Substorms," *J. Geophys. Res.* (submitted for publication, 1986).

[14] S. Orsini, M. Candidi, V. Formisano, H. Balsiger, A. Ghielmetti, and K. W. Ogilvie, "The Structure of the Plasma Sheet–Lobe Boundary in the Magnetotail," *J. Geophys. Res.* **89**, 1573 (1984).

[15] T. E. Eastman, B. Popielawska, and L. A. Frank, "Three-Dimensional Plasma Observations near the Outer Magnetospheric Boundary," *J. Geophys. Res.* **90**, 9519 (1985).

[16] R. A. Heelis, J. D. Winningham, W. B. Hanson, and J. L. Burch, "The Relationships Between High-Latitude Convection Reversals and the Energetic Particle Morphology Observed by Atmospheric Explorer," *J. Geophys. Res.* **85**, 3315 (1980).

[17] P. F. Bythrow, R. A. Heelis, W. B. Hanson, R. A. Power, and R. A. Hoffman, "Observational Evidence for a Boundary Layer Source of Dayside Region 1 Field-Aligned Currents," *J. Geophys. Res.* **86**, 5577 (1981).

[18] I. B. McDiarmid, J. R. Burrows, and M. D. Wilson, "Comparison of Magnetic Field Perturbations at High Latitudes with Charged Particle and IMF Measurements," *J. Geophys. Res.* **83**, 681 (1978).

[19] J. A. Slavin, B. T. Tsurutani, E. J. Smith, D. E. Jones, D. G. Jones, and D. G. Sibeck, "Average Configuration of the Distant ($<220\ R_e$) Magnetotail; Initial ISEE-3 Magnetic Field Results," *Geophys. Res. Lett.* **10**, 973 (1983).

[20] D. J. Southwood, "Magnetopause Kelvin-Helmholtz Instability," in *Magnetospheric Boundary Layers,* B. Battrick and J. Mort, eds., ESA Sci. Tech. Publ. Branch, Noordwijk, The Netherlands, p. 357 (1979).

[21] G. Rostoker, I. Spadinger, and J. C. Samson, "Local Time Variation in the Response of Pc 5 Pulsations in the Morning Sector to Substorm Expansive Phase Onsets near Midnight," *J. Geophys. Res.* **89**, 6749 (1984).

[22] H. E. Petschek, "Magnetic Field Annihilation," in *Proc. AAS-NASA Symp. Solar Flares,* NASA SP-50, W. N. Hess, ed., Washington, D. C., pp. 425-439 (1964).

[23] G. Rostoker and R. Bostrom, "A Mechanism for Driving the Gross Birkeland Current Configuration in the Auroral Oval," *J. Geophys. Res.* **81**, 235 (1976).

[24] J. A. Slavin, E. J. Smith, D. G. Sibeck, D. N. Baker, R. D. Zwickl, and S.-I. Akasofu, "An ISEE-3 Study of Average and Substorm Conditions in the Distant Magnetotail," *J. Geophys. Res.* **90**, 10,875 (1985).

[25] L. Rosenhead, "The Formation of Vortices from a Surface of Discontinuity," *Proc. R. Soc. A* **134**, 170 (1931).

[26] A. Miura and P. L. Pritchett, "Nonlocal Stability Analysis of the MHD Kelvin-Helmholtz Instability in a Compressible Plasma," *J. Geophys. Res.* **87**, 7431 (1982).

[27] W. B. Thompson, "Parallel Electric Fields and Shear Instabilities," *J. Geophys. Res.* **88**, 4805 (1983).

[28] G. Gustafsson, W. Baumjohann, and I. B. Iversen, "Multi-Method Observations and Modeling of the Three-Dimensional Currents Associated with a Very Strong Ps 6 Event," *J. Geophys.* **49**, 138 (1981).

[29] G. Rostoker and J. C. Samson, "Can Substorm Expansive Phase Effects and Low Frequency Pc Magnetic Pulsations be Attributed to the Same Source Mechanism?" *Geophys. Res. Lett.* **11**, 251 (1984).

MHD ASPECTS OF MAGNETOTAIL DYNAMICS

F. V. Coroniti*

The equations of ideal MHD represent the conservation of mass, momentum, and energy; thus, MHD must, in some broad sense, provide a correct description of large-scale structure of the earth's magnetosphere. However, steady-state, ideal MHD (the simplest description) allows only a static interaction between the solar wind and the magnetosphere. Hence an MHD theory of magnetospheric dynamics must involve unsteady flow turbulence, such as might be generated by Kelvin-Helmholtz turbulence at the magnetopause, and dissipative transport, such as reconnection, which violates ideal MHD. This brief review summarizes recent theoretical efforts to explore and extend the early MHD magnetospheric models of Axford and Petschek. Analytic calculations show that steady convection in the magnetotail is nearly impossible. Global magnetospheric simulations exhibit substorm reconnection phenomena for southward IMF and predict a novel convection pattern for northward IMF. Reconnection simulations show plasmoid formation and field-aligned current flows in the tail. The Kelvin-Helmholtz instability can provide a strong anomalous viscosity which can drive the viscous convection system.

INTRODUCTION

The early viscous[1] and reconnection[2,3] models of the earth's magnetosphere used the principles of magneto-hydrodynamics (MHD) to describe the large-scale stresses and convective flows within the magnetosphere and the dissipative interactions of the geomagnetic fields and plasmas with the solar wind. Although these models proved to be amazingly perceptive and accurately predictive, they were only conceptual solutions to the MHD equations, a qualitative image of what an MHD solution for the magnetosphere might be. During the last decade, analytical calculations and numerical simulations have established the essential rectitude of these conceptual MHD solutions, and have extended the early models to more complex interactions between the magnetosphere and solar wind, and to time-dependent internal magnetospheric dynamics. The purpose of this short and highly selective review is to summarize these recent accomplishments, to remark upon their successes, and to indicate future potentialities for MHD investigations.

Before starting, however, a few general comments about MHD are perhaps in order. The magnetospheric plasma is collisionless, and typically contains distributions that are highly structured in velocity space. Since MHD sums over these various distributions, an immediate question is whether or not the MHD equations can in principle adequately describe the magnetosphere. The answer is not altogether clear. Since MHD essentially represents the conservation of the total mass, momentum, and energy in the plasma, and since the macroscopic stresses are directly related to the global internal and external boundary conditions, MHD should provide a reasonable, zeroth order description of magnetospheric structure and dynamics; although the local plasma distributions are complex, they must still integrate to yield the correct macroscopic stresses.

The more serious difficulty with MHD is in its description of dissipation. In classical transport theory, plasma dissipation takes the analytical form of resistance, viscosity, and thermal conduction, each with a locally determined diffusion coefficient. Both the early conceptual models and the recent MHD simulations implicitly or explicitly assume that dissipation processes are locally diffussive. Whether the collisionless plasma dissipation in the magnetosphere can be adequately represented by local transport relations of classical form is definitely uncertain and even dubious. Advocates for MHD will argue that the total dissipation is determined by the large-scale boundary conditions and stresses, and that locally the collisionless plasma will somehow adjust to provide the required dissipation; hence, by describing the plasma dissipation with classical transport coefficients, some local structural details may not be precisely modeled but the global configuration is. Although plausible, this argument remains to be validated.

The next section reviews several analytic calculations for the MHD structure of a quiet, slowly convecting geomagnetic tail. These models are based on ideal, adiabatic MHD, and demonstrate that, under these assumptions, steady convection is impossible. Global convection requires dissipation, which is most readily studied with MHD simulations. The third section discusses some of the recent MHD simulations that support both the viscous and reconnection models of the magnetosphere. The global simulations do not resolve the detailed flow and stress configurations of the reconnection and viscous processes. However, these are accessible to small-scale MHD simulations that have been examined for both free and driven reconnection, with particular attention to the generation of field-aligned currents, and Kelvin-Helmholtz turbulence which provides an anomalous viscosity discussed in the fourth section. The final section concludes with some comments on the future of the MHD studies of the magnetosphere.

QUASISTATIC MODELS OF THE MAGNETOTAIL

The magnetosphere is often observed to be quiescent. Low rates of dayside reconnection and viscous dissipa-

---

*Department of Astronomy and Department of Physics, University of California, Los Angeles, California 90024.

tion slowly transport flux into the tail. The plasma sheet is thick and weakly stressed, and plasma flow velocities are low and often disordered. The structure of a quiet tail should be adequately described by ideal MHD since local dissipation would only slightly affect the global structure. In a series of papers, Birn and Schindler have investigated two- and three-dimensional MHD models of a long tail for both static and slowly convecting configurations. Recently Erickson[4] has extended these analytical models to include a two-dimensional dipole field. These calculations have provided invaluable insight on the structure and temporal evolution of the quiet magnetosphere.

Since the flow velocities are subsonic and sub-Alfvénic, the plasma and magnetic fields must be in quasistatic stress balance so that $0 = -\nabla p + J \times B/c$. A given tail model can then be constructed by specifying the spatial dependence of the pressure in the plane of symmetry; in principle, the midplane pressure in the plasma sheet must balance the magnetosheath dynamic and static pressure that is exerted at the tail magnetopause. Rather than solve the full MHD energy equation, it is usually assumed that the plasma obeys the polytropic equation of state P/ρ^γ = constant. For $\gamma = 5/3$, the entropy is conserved along flow streamlines, which should be valid for very low dissipation rates; however, $\gamma < 5/3$ can usefully model entropy and/or particle losses during convection.[5] Perpendicular flow velocities are governed by the frozen-in condition $E + v \times B/c = 0$, and parallel flows are determined by the continuity equation. Finally, specifying the magnetic vector potential at some boundary, such as the tail magnetopause, completes the model and permits a calculation of the tail configuration.

Schindler and Birn[5] studied the structure of a two-dimensional tail with steady convection and showed that unless γ was less than unity, the magnetic field lines were concave near the magnetopause. Subsequent three-dimensional tail models confirmed this topology.[6] Since the tail field is observed to be convex, they concluded that steady convection was inconsistent with an adiabatic ($\gamma < 5/3$) flow, and could only occur if the plasma sheet suffered severe particle losses ($\gamma < 1$). The difficulty with loss-free convection is that flux tubes at large radial distances have huge volumes and particle content. As a distant flux tube convects to a new location near the earth, adiabatic compression causes the pressure in the convecting tube to increase above the pressure in the flux tube that originally occupied that location. Hence, unless particle losses are very large ($\gamma < 1$), earthward convecting flux tubes rapidly destroy local stress balance, and the flow must accelerate to Alfvénic speeds. Since the quiet tail should be nearly adiabatic, Schindler and Birn[5] conclude that even at quiet times, convection must be unsteady, a conclusion first reached by Erickson and Wolf.[7]

Time-dependent tail models have been investigated by Schindler and Birn,[5] Birn and Schindler,[6] and by Erickson,[4] who included a two-dimensional dipole field and retained full magnetic tension stresses. Time-

dependence is introduced by varying (increasing) the value of the vector potential at the magnetopause or of the pressure in the tail's midplane; thus, a time-sequence of quasistatic equilibria are calculated (see Fig. 1). Although flux is added to the tail, the electric field in the plasma sheet remains much smaller than that in the lobes; plasma-sheet flux tubes can adiabatically convect earthward only at the rate allowed by the small increase in magnetic stresses associated with the compression of the plasma sheet. Since flux tubes, which convect out of the earthward end of the plasma sheet, are replaced by adiabatically compressed flux tubes with a much higher pressure, a deep minimum develops in the magnetic field in the near-earth region of the tail. For some parameters, distant flux tubes actually convect tailward, stretching the tail field lines and further reducing the normal magnetic component in the plasma sheet.

The above calculations treat the evolution of the plasma sheet as a succession of quasistatic equilibria. In the context of the theoretical model for an isolated substorm posed by Coroniti and Kennel,[8,9] the onset of dayside reconnection following a southward IMF shift launches tailward propagating fast compression and rarefaction waves that dynamically reconfigure the plasma sheet in exactly the same manner as found in the quasistatic calculations.[10] As in the Coroniti and Kennel model,[8] Schindler and Birn[5] argue that the structural changes drive the plasma sheet toward the instability threshold for the onset of tail reconnection. However, the quasistatic calculations demonstrate that the compression of the plasma sheet only decreases its thickness by about a factor of two from its quiet value. Therefore, when tail reconnection starts, the $J \times B$ forces in the

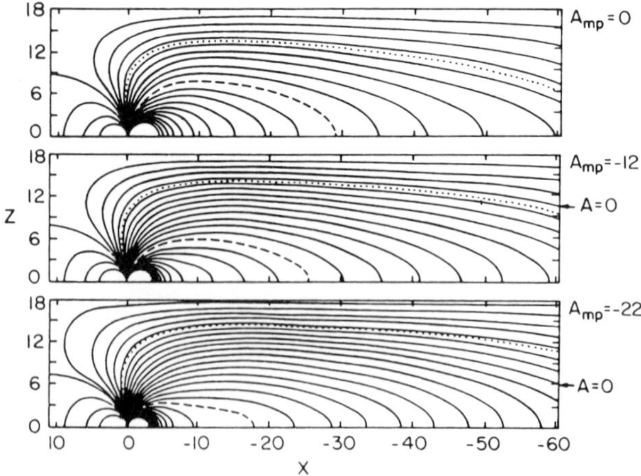

Figure 1—Quasistatic evolution of the magnetotail (after Erickson[4]). Flux is added to the tail by decreasing the vector potential (A_{mp}) at the magnetopause. Lobe field lines convect toward the plasma sheet as the tail field strength increases. In the plasma sheet, field lines cannot convect rapidly earthward since the resulting adiabatic compression would result in violating the local stress balance. As field lines leave the near earth plasma sheet, a deep minimum develops in the strength of the normal magnetic field component. Note that the plasma-sheet thickness decreases by only about a factor of two.

176

plasma sheet are too small to accelerate the plasma to the Alfvén speed. Thus, reconnection starts slowly, and accelerates only as the plasma sheet thins and the normal magnetic field component increases. The reconnection collapse of the initially thick plasma sheet can proceed explosively in time so that rapid reconnection dissipation still occurs on the 5–10 min time scale of substorm breakup.[10] The explosive reconnection model, however, lies outside the domain of MHD since the dissipation required for reconnection is fundamentally kinetic in nature.

GLOBAL MHD SIMULATIONS OF THE MAGNETOSPHERE

Substorms involve changes in the entire configuration of the magnetosphere that are not yet amenable to complete analytic modeling. The magnetosphere's global structure and dynamics have been investigated in a series of increasingly sophisticated numerical MHD simulations. Although the global simulations are constrained by limited spatial resolution and either numerical or ad hoc dissipative transport, they provide essential confirmation of the conceptual and analytic models of the magnetosphere.

The earliest MHD simulations studied the interaction of the solar wind with a two-dimensional magnetic dipole.[11,12] In two dimensions, a southward IMF must reconnect with the earth's field, and the high numerical dissipation rates in the early codes resulted in the formation of a very short magnetic tail. Nevertheless, the expected substorm reconnection behavior of the tail was clearly discernable.

MHD simulations began to yield new results with the advent of three-dimensional numerical codes. Wu et al.[13] studied the tear-drop magnetosphere produced by the flow of an unmagnetized solar wind over a three-dimensional dipole. They found that the internal magnetospheric field in the simulation agreed, to a remarkable degree, with the observed quiet-time field. Thus, the quiet magnetospheric plasma and $J \times B$ stresses are indeed determined by a steady MHD interaction with minimal dissipative coupling to the solar wind.

Three-dimensional substorm dynamics were revealingly investigated by Brecht et al.[14] using a code in which the numerical dissipation depended only on the time derivatives in the MHD equations; hence, steady reconnection could not occur. Starting from a quiet, nonreconnecting magnetospheric configuration, a southward IMF shift resulted in immediate nose reconnection followed about an hour later by the formation of a tail neutral line in the near-earth plasma sheet. The unsteady reconnection created successive plasmoids (or magnetic islands) that were accelerated tailward by the Maxwell stress of the open field lines; as the X-type neutral line retreated tailward, a new neutral line formed near the earth. Hence, the phenomenological substorm model of Hones[15,16] and Russell and McPherron[17] is fully consistent with the MHD behavior of an unsteady reconnecting magnetosphere.

The Brecht et al.[14] simulation showed that during the strong reconnection events, the three-dimensional structure of the tail was highly variable in space and time. Although strong tailward flows with both positive and negative B_z fields dominate the central region of the plasma sheet, in the region on the dawn and dusk sides of the reconnection flow, the plasma velocity could be directed toward the noon-midnight meridian. Thus a spacecraft that was located to either side or well away from the central reconnection flow channel would observe a magnetic field and flow configuration that exhibited little or no evidence of reconnection. Brecht et al.[14] made the reasonable suggestion that the spatially variable flow structure could explain the often conflicting observations of plasma-sheet dynamics that have been reported during substorms.

Global MHD simulations have achieved a new plateau with the recent work of Ogino and his collaborators.[18-20] Previous simulations did not permit the flow of field-aligned currents that are essential for coupling plasma stresses between the magnetosphere and ionosphere. Ogino's simulations have an artificial ionosphere located at ~ 3 R_e, that can absorb and emit field-aligned current; although clearly not completely self-consistent, the numerical magnetospheric models can now be tested against the thoroughly observed field-aligned current system. Figure 2 shows several of Ogino's diagnostics[20] for the interaction of the magnetosphere with a strong southward IMF. The Region 1 field-aligned currents are primarily generated by the plasma flow parallel vorticity $\Omega_{\parallel} = \boldsymbol{B} \cdot (\nabla \times V)$ which

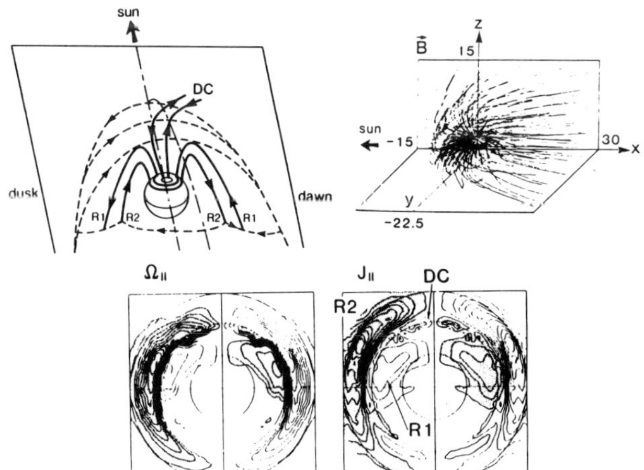

Figure 2—The distribution of Region 1 (R1) and Region 2 (R2) currents, the magnetic field topology, the parallel vorticity ($\Omega_{\parallel}$), and field-aligned current ($J_{\parallel}$) projected into the ionosphere from a global MHD simulation for a southward IMF, steady-state magnetosphere (after Ogino[20]). The vorticity and field-aligned current are parallel (antiparallel) to the magnetic field in the shaded (open) regions. The dayside polar cusp (DC) current system results from the twisting of open field lines. The high-latitude $\Omega_{\parallel}$ is associated with the Region 1 currents and the antisunward polar cap convection pattern. The low-latitude $\Omega_{\parallel}$ is generated by a viscous interaction along the flank magnetopause and the return flow of reconnected tail field lines.

results from the solar-wind torsion imposed on newly reconnected tail field lines; thus, the Region 1 currents are directly coupled to the cross-tail current system. Some parallel vorticity is also produced along the flanks by the viscous interaction with the magnetosheath flows. The Region 2 currents are not generated by vorticity but by plasma pressure gradients that develop nearer the earth; some Region 1 current also couples directly into the Region 2 system.

Conceptual and analytical models of the magnetosphere have typically concentrated on the ideal case of strong southward IMF and maximal spatial symmetry. Global simulations have now begun to consider more realistic configurations with large B_y IMF fields.[19] The most striking "unconventional" simulation was done by Ogino and Walker,[18] who considered a reconnecting magnetosphere with strong northward IMF B_z. This simulation showed that northward interplanetary field lines reconnect with terrestrial field lines that are located tailward of the polar cusp (Fig. 3). After reconnection, the solar wind "peels away" a tail field line; the stresses on the lines of force, which are earthward of the reconnection site, accelerate the plasma toward the dayside magnetopause. These field lines then join the viscous convection system and are dragged into the tail. The reconnection removal of tail field lines creates a pressure minimum in the lobes, causing plasma to flow away from the tail midplane. Thus, some plasma-sheet field lines in the viscous convection system flow poleward and tailward, resulting in the north-south bifurcation of the central plasma sheet. When projected to the ionosphere, those field lines flow sunward across the polar cap until they reach the original reconnection point just tailward of the cusp. The implied precipitation of plasma sheet particles along the sunward moving convection path actually predicted the recently discovered quiet-time polar cap theta aurorae.[21]

The level of detail in Ogino's simulations is truly astounding. The convincing agreement between the numerical models and the observed structure of the magnetosphere demonstrates that MHD does describe the most important large-scale features of the magnetosphere. Clearly global MHD simulations are evolving into a powerful tool for investigating magnetospheric dynamics with the exciting possibility of explaining and/or revealing complex configurations that are extremely difficult to discern in spacecraft observations.

LOCAL MHD SIMULATIONS

The global magnetospheric simulations contain both viscosity and reconnection, but do not possess the spatial resolution required to examine the detailed flow patterns associated with those dissipation processes. Furthermore, the dissipation rates are usually either numerical or scaled to a spatially constant diffusion coefficient; thus the global simulations have only a limited ability to adjust the dissipation rates to the local flow conditions. In order to provide greater insight into their local behavior, several groups have carried out MHD simulations of reconnection and viscous shear flows in

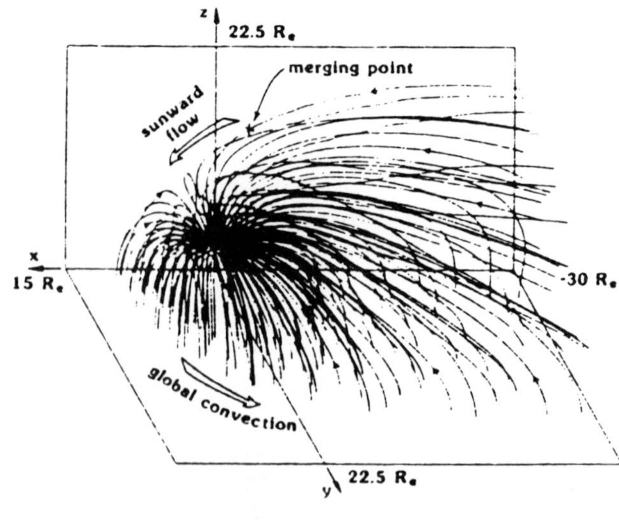

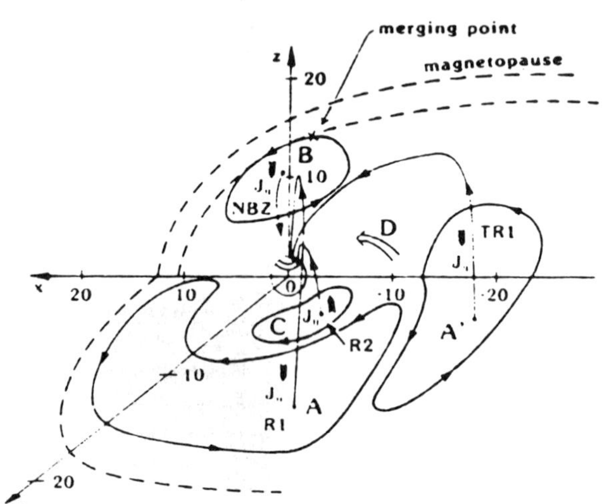

Figure 3—The magnetic field topology and a schematic of the internal convection system from a global MHD simulation for a northward IMF (after Ogino and Walker[18]). Field lines on the tailward side of the polar cusp reconnect with the northward IMF. The newly formed closed field line moves toward the nose and enters the viscous convection system (*A*). The plasma sheet has three convection systems: (a) *A* returns flux in the viscous system to the nose; (b) *A'* is a region of closed, circulating flux; and (c) *D* is the return of flux to the polar cusp reconnection region by sunward flow across the polar cap.

which the dissipative transport is modeled with greater realism and the boundary conditions are more precisely controlled. This section will briefly summarize and comment upon some of these local simulation effects.

Tail Reconnection

The four major attempts to simulate tail reconnection have been led by Birn,[22,23] Sato,[24-26] Ugai,[27,28] and recently Lee,[29] each using somewhat different MHD codes and studying somewhat different problems. The results of Birn and Sato are thoroughly reviewed in the proceedings of the Chapman Conference on Reconnection[30,31] and thus will only be briefly described here.

Birn's simulations model the onset of reconnection in a plasma sheet with initially closed field lines surrounded by open flaring tail lobes. His code solves the two-dimensional or three-dimensional MHD equations, assuming a constant resistivity and an isothermal plasma. Since no external (or driving) electric field is imposed, reconnection evolves as a single long-wavelength, resistive tearing-like mode. An X-type neutral line forms near the earth, creating a magnetic island or plasmoid on the tailward side. The pressure stress due to flaring produces a tailward acceleration of the plasmoid. In the three-dimensional simulation with a plasma-sheet thickness that increases toward dawn and dusk, the tail reconnection flow is concentrated near the noon-midnight meridian, and significant east-west flows develop into the reconnection region. The dawn-dusk gradients of the magnetic field and pressure generate field-aligned currents directly away from (toward) the earth on the dawn (dusk) side of the tail, which is the sense of the Region 2 currents.

Sato and coworkers have emphasized reconnection flows that are driven by an external electric field; the flow geometry is symmetric about the X-type neutral line and thus resembles the classic Petschek[32] configuration. In the three-dimensional calculations,[26,33] the injection velocity at the z-edge of the simulation box is assumed to peak above the X-line and decrease to zero at the x and y edges as a cosine function. In an attempt to model the expected behavior of an anomalous resistance, the resistivity is assumed to scale as a quadratic function of the difference between the local current drift velocity and the critical drift speed, typically assumed to be three times the Alfvén speed. Since Sato's code solves the full MHD energy equation, the nonlinear resistivity results in strong local Joule heating in the diffusion region surrounding the X-line and at the standing slow switch-off shocks that bound the outflow regions.

The simulations start with a thick current sheet that subsequently thins and commences reconnection when the anomalous resistance threshold is exceeded. The nonlinear resistivity sustains a large inductive electric field that leads to a rapid or explosive conversion of magnetic energy into heat and kinetic energy of outflowing plasma. Since the code neglects heat conduction, the ohmic temperature rise combined with the boundary condition of almost constant total pressure in the x-direction forces the density to become highly rarefied in the current sheet. The spatial profile of the driving velocity at the z-boundary causes the magnetic pressure to peak just above the neutral line and decrease in both $|x|$ and $|y|$ away from the origin. Since the pressure in the outflow region must balance the external magnetic pressure, a reverse current (opposite to the current in the slow shocks) is forced to flow in the center of the neutral region;[34] the x-component of the field reverses sign, and the Maxwell stress from the reversed current decelerates the flow and balances the pressure gradient.

On either side of the x-z plane (noon-midnight meridian) field-aligned currents flow near the region of the neutral sheet; these currents have the sense of the Re-

gion 1 system toward (away from) the earth on the dawn (dusk) side. The field-aligned currents close across the B_z field in the neutral sheet, where the $J \times B_z$ stress confines the high pressure near the origin and drives east-west plasma flows into the central reconnection region. Thus, the Region 1 sense of the field-aligned currents is solely a consequence of the injection velocity profile at the z-edge of the simulation box, which forces the magnetic pressure to maximize just above the origin. Presumably the opposite, Region 2, sense of the field-aligned currents in Birn's simulations is also a consequence of the imposed boundary conditions that allow the field strength to decrease just above the X-line as the plasmoid accelerates tailward.

Ugai's[27,28] simulations have also investigated symmetric reconnection flows, but differ from Sato's work in two important aspects: (a) the anomalous resistivity is spatially confined, with a Gaussian profile to the diffusion region; and (b) the reconnection flow is not externally driven, but is allowed to develop freely once the current sheet thins sufficiently to trigger strong resistivity. In addition, a return current at large $|z|$ models the tail magnetopause. Once initiated, reconnection proceeds in a fast or explosive fashion, plasma rapidly accelerates along the outflow region, and the "open" magnetic flux is quickly reduced. Hence, the explosive behavior of reconnection is not the result of an externally imposed electric field or flow, but is a consequence of the Maxwell stresses that develop to accelerate the plasma out of the X-line region. In a recent simulation,[35] the resistivity model was changed so that ohmic dissipation could occur anywhere in the flow. After the fast reconnection collapse of the current sheet, the neutral line splits into a pair of X-lines on either side of a central O-line. Apparently the high value of the resistivity and the small current-sheet thickness permit the resistive tearing mode to grow at a rate that is competitive with the flow-time scale. The bifurcation of the initial X-line suggests an alternative possibility for plasmoid formation in the tail.

Lee et al.'s simulations[29] are discussed elsewhere in these proceedings, and therefore only a few comments will be made here. Lee's code solves the incompressible, resistive, two-dimensional MHD equations. The code is very resistive in that the analytically prescribed magnetic Reynolds length (for the runs shown in the paper) is 10 times smaller than the cell size in the z-direction; thus, the resistivity is actually numerical, which explains why the reconnection time scale for plasmoid formation is essentially independent of the prescribed value of the resistivity. The resistive diffusion time in the z-direction, based on the numerical resistivity, is comparable with the convective time scale, defined as the time required for the flux in the tail to double.

For a steady imposed electric field at the tail magnetopause, Lee et al.[29] found that reconnection proceeds by the successive formation and tailward ejection of plasmoids. The time interval between successive plasmoid formation, the numerical resistive diffusion time scale, and the convection time scale are all comparable. Since the exit flow velocity on the earthward side of the simulation box is held fixed, the earthward flow from the reconnection neutral line is choked. Thus, a steady-state reconnection flow is prevented from developing, which therefore forces successive plasmoids to form in order to remove the injected plasma. Presumably, if the earthward outflow velocity were permitted to adjust to the reconnection rate, which would occur if the fast mode were allowed in the code, a steady reconnection flow would be possible.

Kelvin-Helmholtz Simulations

The global MHD simulations exhibit a component of the internal convection system that is driven by viscous stresses at the near-earth magnetopause. In the magnetosphere, the anomalous viscosity is undoubtedly the consequence of turbulent Maxwell and Reynolds stresses that are generated by the Kelvin-Helmholtz instability.

Miura[36] has simulated the MHD Kelvin-Helmholtz instability for symmetric shear flow configurations in which the flow is either perpendicular or parallel to the magnetic field. For both configurations, the shear boundary layer develops the familiar vortex flows that eventually cascade to the longest wavelength modes that can be contained in the simulation system. The parallel flow configuration also results in a very distorted and twisted magnetic topology that is relaxed by reconnection. The anomalous viscous stress is estimated to be about one percent of the flow dynamic pressure. For magnetospheric parameters, the corresponding electric potential drop in the viscous convection system is of order 10–20 kV, in reasonable accord with observations.

Recently, the Kelvin-Helmholtz instability has been examined for flow configurations that more nearly approximate the dayside magnetopause. Miura[37] studied a nonsymmetric shear layer in which the flow velocity asymptotes to zero within the magnetosphere, and the magnetic field changes direction and increases in strength from the magnetosheath to the magnetosphere. The turbulent viscous stresses transport momentum into the magnetosphere so that plasma flow develops inside the magnetopause current layer; analogous plasma flows are often observed within the low-latitude boundary layer. Wu[38] has studied the symmetric Kelvin-Helmholtz configuration but has employed such a long simulation system that the instability convectively propagates rather than being periodic. Wu[38] finds that the saturation kinetic energy of the convective instability can be 10–100 times larger than for the periodic case. Since the Kelvin-Helmholtz instability should convectively amplify at the magnetopause, the strength of the anomalous viscosity may exceed the modest value obtained by Miura.[36]

CONCLUSIONS

From the above selective summary of recent research on the application of MHD to the magnetosphere, several broad conclusions stand out:

1. Slow, adiabatic steady convection is not possible in the geomagnetic tail. With the addition of magnetic flux, unsteady convection drives the plasma sheet toward the threshold for the formation of a near-earth neutral line.
2. Global simulations yield magnetospheric topologies, current systems, and convection patterns that are familiar and similar to both observations and theoretical preconceptions; thus MHD does adequately describe the magnetopshere's large-scale properties.
3. For a tail geometry, reconnection does produce plasmoids that are accelerated tailward by the self-consistent stresses.
4. Three-dimensional reconnection flows generate field-aligned currents, but their sense depends on the imposed boundary conditions.
5. Kelvin-Helmholtz turbulence does provide an effective anomalous viscosity; more simulations are needed in order to firmly establish the strength of the viscous convection system.

The present MHD simulations have already demonstrated an impressive ability to extend theoretical understanding of the magnetosphere. Ogino's simulations of convection during northward B_z and large B_y in the IMF have greatly clarified the magnetosphere's field and flow topologies under conditions that are difficult to conceptualize. Clearly, the next step in this direction is to investigate how the magnetosphere evolves from one convection state to another as the IMF changes direction.

In the future, MHD simulations should become an even more powerful technique for investigating magnetospheric dynamics. A frequent criticism of the current simulations is that they are not really quantitative. The simulations do produce the correct order of magnitude for the magnetosphere's magnetic fields, plasma flows, and field-aligned currents; however, these quantities are basically either input parameters or the result of dimensional analysis. Realistic simulations will require accurate estimates of the magnetosphere's local and global dissipation rates, which are, presumably, primarily associated with plasma kinetic processes. The self-consistent incorporation of collisionless plasma dissipation into the MHD codes may not be straightforward since the dissipation processes are likely not to be describable by local transport coefficients. However, even observationally the magnetosphere's dissipation processes and rates are highly uncertain. Perhaps progress will come from making detailed comparisons between observations and the predictions of simulations that contain adjustable transport rates. Such comparisons might prove to be the best method for actually determining the magnetosphere's dissipation rates.

In a somewhat related area, local spacecraft observations are often so complex and variable that their relation to global phenomena is difficult, if not impossible, to determine. Simulations may become the essential technique for uniquely interpreting the local measurements. If the ISTP mission is approved, global simulations may provide the only reliable method of correlating observations obtained by widely separated spacecraft.

Finally, there is an interesting theoretical question—how much of the magnetosphere's structure and dynamics is adequately described by MHD? MHD should break down in the thin boundary layers, in situations where single particle dynamics are important, or when nonlocal transport, such as global heat conduction by energetic particles, dominates the dissipation. However, since the magnetosphere is basically controlled by global stresses, MHD may prove to be a more rugged description than plasma kinetic theorists expected.

ACKNOWLEDGMENT—This work was partially supported under National Science Foundation grant ATM-85-03424 and NASA grant NGL-05-007-190.

REFERENCES

[1] W. I. Axford and C. O. Hines, "A Unifying Theory of High Latitude Geophysical Phenomena and Geomagnetic Storms," *Can. J. Phys.* **39**, 1433 (1961).

[2] J. W. Dungey, "Interplanetary Magnetic Field and the Auroral Zones," *Phys. Rev. Lett.* **6**, 47 (1961).

[3] W. I. Axford, H. E. Petschek, and G. L. Siscoe, "Tail of the Magnetosphere," *J. Geophys., Res.* **70**, 1231 (1965).

[4] G. M. Erickson, "On the Range of X-Line Formation in the Near-Earth Plasma Sheet: Results of Adiabatic Convection of Plasma Sheet Plasma," in *Magnetic Reconnection in Space and Laboratory Plasmas*, E. W. Hones, Jr., ed. American Geophysical Union, Washington, D. C., p. 296 (1984).

[5] K. Schindler and J. Birn, "Self-Consistent Theory of Time-Dependent Convection in the Earth's Magnetotail," *J. Geophys. Res.* **87**, 2263 (1982).

[6] J. Birn and K. Schindler, "Self Consistent Theory of Three-Dimensional Convection in the Geomagnetic Tail," *J. Geophys. Res.* **88**, 6969 (1983).

[7] G. M. Erickson and R. A. Wolf, "Is Steady Convection Possible in the Earth's Magnetotail," *Geophys. Res. Lett.* **7**, 887 (1980).

[8] F. V. Coroniti and C. F. Kennel, "Can the Ionosphere Regulate Magnetospheric Convection?" *J. Geophys. Res.* **78**, 2837 (1973).

[9] F. V. Coroniti and C. F. Kennel, "Magnetospheric Reconnection, Substorm, and Energetic Particle Acceleration," in *Particle Acceleration Mechanisms in Astrophysics*, J. Arons, C. Max, and C. McKee, eds., American Institute of Physics, New York, p. 169 (1979).

[10] F. V. Coroniti, "Explosive Tail Reconnection: The Growth and Expansion Phases of Magnetospheric Substorms," *J. Geophys. Res.* **90**, 7427 (1985).

[11] J. M. LeBoeuf, T. Tajima, C. F. Kennel, and J. M. Dawson, "Global Simulations of the Time-Dependent Magnetosphere," *Geophys. Res. Lett.* **5**, 609 (1978).

[12] J. G. Lyon, S. H. Brecht, J. D. Huba, J. A. Fedder, and P. J. Palmadesso, "Computer Simulation of a Geomagnetic Substorm," *Phys. Rev. Lett.* **46**, 1038 (1981).

[13] C. C. Wu, R. J. Walker, and J. M. Dawson, "A Three-Dimensional MHD Model of the Earth's Magnetosphere," *Geophys. Res. Lett.* **8**, 523 (1981).

[14] S. H. Brecht, J. G. Lyon, J. A. Fedder, and K. Hain, "A Time-Dependent Three-Dimensional Simulation of the Earth's Magnetosphere: Reconnection Events," *J. Geophys. Res.* **87**, 6098 (1982).

[15] E. W. Hones, Jr., "Plasma Flow in the Plasma Sheet and Its Relation to Substorms," *Radio Sci.* **8**, 879 (1973).

[16] E. W. Hones, Jr., "Transient Phenomena in the Magnetotail and Their Relation to Substorms," *Space Sci. Rev.* **23**, 393 (1979).

[17] C. T. Russell and R. L. McPherron, "The Magnetotail and Substorms," *Space Sci. Rev.* **15**, 205 (1973).

[18] T. Ogino and R. J. Walker, "A Magnetohydrodynamic Simulation of the Bifurcation of Tail Lobes During Intervals with a Northward Interplanetary Magnetic Field," *Geophys. Res. Lett.* **11**, 1018 (1984).

[19] T. Ogino, R. J. Walker, M. Ashour-Abdalla, and J. M. Dawson, "An MHD Simulation of B_y-Dependent Magnetospheric Convection and Field-Aligned Currents during Northward IMF," *J. Geophys. Res.* **90**, 10835 (1985).

[20] T. Ogino, "A Three-Dimensional MHD Simulation of the Interaction of the Solar Wind with the Earth's Magnetosphere," *J. Geophys. Res.* (submitted, 1985).

[21] L. A. Frank, J. D. Cravan, J. L. Burch, and J. M. Winningham, "Polar Views of the Earth's Aurora with Dynamics Explorer," *Geophys. Res. Lett.* **9**, 1001 (1982).

[22] J. Birn, "Computer Studies of the Dynamic Evolution of the Geomagnetic Tail," *J. Geophys. Res.* **85**, 1214 (1980).

[23] J. Birn and E. W. Hones, Jr., "Three Dimensional Computer Modeling of Dynamic Reconnection in the Geomagnetic Tail," *J. Geophys. Res.* **88**, 6969 (1981).

[24] T. Sato, "Strong Plasma Acceleration by Slow Shocks Resulting from Magnetic Reconnection," *J. Geophys. Res.* **84**, 7177 (1979).

[25] T. Sato and T. Hayashi, "Externally Driven Magnetic Reconnection and a Powerful Magnetic Energy Converter," *Phys. Fluids* **22**, 1189 (1979).

[26] T. Sato, T. Hayashi, R. H. Walker, and M. Ashour-Abdalla, "Neutral Sheet Current Interpretation and Field-Aligned Current Generation by Three-Dimensional Driven Reconnection," *Geophys. Res. Lett.* **10**, 221 (1983).

[27] M. Ugai, "Spontaneously Developing Magnetic Reconnection in a Current-Sheet System under Different Sets of Boundary Conditions," *Phys. Fluids* **25**, 1027 (1982).

[28] M. Ugai, "Self-Consistent Development of Fast Magnetic Reconnection with Anomalous Plasma Resistivity," *Plasma Phys.* **26**, 1549 (1984).

[29] L. C. Lee, Z. F. Fu, and S.-I. Akasofu, "A Simulation Study of Forced Reconnection Processes and Magnetospheric Storms and Substorms," *J. Geophys. Res.* **90**, 10,896 (1985).

[30] J. Birn, "Three-Dimensional Computer Modeling of Dynamic Reconnection in the Magnetotail; Plasmoid Signatures in the Near and Distant Tail," in *Magnetic Reconnection in Space and Laboratory Plasmas*, E. W. Hones, Jr., ed. American Geophysical Union, Washington, D. C., p. 264 (1984).

[31] R. J. Walker and T. Sato, "Externally Driven Magnetic Reconnection," in *Magnetic Reconnection in Space and Laboratory Plasmas*, E. W. Hones, Jr., ed., American Geophysical Union, Washington, D. C., p. 272 (1984).

[32] H. E. Petschek, "Magnetic Field Annihilation," NASA Spec. Pub., SP-50, 425 (1964).

[33] T. Sato, R. J. Walker, and M. Ashour-Abdalla, "Driven Magnetic Reconnection in Three-Dimensions; Energy Conversion and Field-Aligned Current Generation," *J. Geophys. Res.* **89**, 9761 (1984).

[34] H. E. Petschek and R. M. Thorne, "The Existence of Intermediate Waves and Neutral Sheets," *Astrophys. J.* **147**, 1157 (1967).

[35] M. Ugai, "Temporal Evolution and Propagation of a Plasmoid Associated with Asymmetric Fast Magnetic Reconnection," *J. Geophys. Res.* **90**, 9576 (1985).

[36] A. Miura, "Anomalous Transport by Magnetohydrodynamic Kelvin-Helmholtz Instabilities in the Solar Wind-Magnetosphere Interaction," *J. Geophys. Res.* **89**, 801 (1984).

[37] A. Miura, "Kelvin-Helmholtz Instability at the Magnetospheric Boundary," *Geophys. Res. Lett.* **12**, 635 (1985).

[38] C. C. Wu, "Kelvin-Helmholtz Instability of the Magnetopause Boundary," *J. Geophys. Res.* (submitted, 1985).

A MAGNETOHYDRODYNAMIC SIMULATION OF RECONNECTION IN THE MAGNETOTAIL DURING INTERVALS WITH SOUTHWARD INTERPLANETARY MAGNETIC FIELD

R. J. Walker,* T. Ogino,*[†] and M. Ashour-Abdalla*[‡]

We have simulated the interaction between the solar wind and the earth's magnetosphere by using a time-dependent three-dimensional magnetohydrodynamic model. The calculation was performed for several orientations of the interplanetary magnetic field (IMF) between dawnward pointing and southward. When the IMF has a dawnward component, the plasma sheet rotates northward on the dawnside of the tail and toward the south on the duskside. As the southward component becomes larger, the plasma sheet becomes thinner and develops a wavy cross section because of patchy or localized tail reconnection. The field-aligned currents associated with this localized reconnection have a filamentary layered structure. When projected onto the polar cap the filamentary currents are located in the same region as the tail region 1 currents. At lower latitudes strong region 2 sense currents that originate in the plasma sheet are found. Field-aligned currents are found on field lines that map to the polar cap even for southward IMF. These currents have many of the properties of the observed polar-cusp currents. The polar-cusp field-aligned currents evolve from the polar-cap (NB_z) field-aligned currents as the IMF is rotated from northward to southward.

INTRODUCTION

It has been well established observationally that the interplanetary magnetic field (IMF) controls magnetospheric dynamics. Most studies have concentrated on the effects of a southward IMF because the occurrence of magnetospheric substorms strongly correlates with intervals of southward IMF (see McPherron[1] and Akasofu[2] for reviews).

The IMF orientation governs the structure and dynamics of the convection of plasma and magnetic field within the magnetosphere. For a southward IMF, two-celled convection patterns are observed in the polar cap in which the convective flow is tailward over the pole while a sunward return flow occurs around the dawn and dusk sides at lower latitudes.[3-5] The geometry of these large-scale convection cells is strongly influenced by the B_y (east-west) component of the IMF. For example, on the northern polar cap, the dusk convection cell becomes crescent-shaped and the dawn cell expands to higher latitudes for $B_{yIMF} < 0$ and $B_{zIMF} < 0$.

The magnitude and distribution of field-aligned currents also depend on the IMF.[6,7] The intensity of both the region 1 and region 2 currents increases as the southward IMF component becomes larger.[6] The polar-cusp field-aligned currents are observed at higher latitudes than the region 1 currents and flow into the ionosphere in the postnoon sector and away from the ionosphere prenoon. These currents strengthen, too, for a stronger southward IMF.

In this study, we have used a three-dimensional magnetohydrodynamic (MHD) simulation code to model the earth's magnetospheric configuration when the IMF has components both in the southward and east-west directions. Our results demonstrate that the IMF orientation

has a strong effect on the plasma-sheet configuration, magnetospheric convection, field-aligned current generation, and the magnetic reconnection process as was suggested by the observational studies.

SIMULATION MODEL

Our simulation model has been described in great detail elsewhere,[8,9] so we will describe only the main features here. We have solved the MHD and Maxwell's equations as an initial-value problem by using the two-step Lax-Wendroff scheme. The normalized resistive MHD equations that we solved are written as follows:

$$\frac{\partial \rho}{\partial t} = -\nabla \cdot (v\rho) + D\nabla^2 \rho \tag{1a}$$

$$\frac{\partial v}{\partial t} = -(v \cdot \nabla)v - \frac{1}{\rho}\nabla p$$
$$+ \frac{1}{\rho}J \times B + g + \frac{1}{\rho}\phi \tag{1b}$$

$$\frac{\partial p}{\partial t} = -(v \cdot \nabla)p - \gamma p \nabla \cdot v + D_p \nabla^2 p \tag{1c}$$

$$\frac{\partial B}{\partial t} = \nabla \times (v \times B) + \eta \nabla^2 B \tag{1d}$$

$$J = \nabla \times (B - B_d) \tag{1e}$$

where ρ is the plasma density; v, the flow velocity; p, the plasma pressure; B, the magnetic field; J, the current density; g, the gravity force; $\phi \equiv \mu \nabla^2 v$, the viscosity; $\gamma = 5/3$, the ratio of the specific heats; $\eta = \eta_0(T/T_0)^{-3/2}$, the resistivity; and T/T_0 is the temperature normalized by its value in the ionosphere. The units for distance, velocity, and time are the earth's radius, $R_e = 6.37 \times 10^6$ m, the Alfvén speed at one earth radius on the equator, $V_A = 6.80 \times 10^6$ m/s, and the Alfvén transit time, $t_A = R_e/V_A = 0.937$ s.

*Institute of Geophysics and Planetary Physics, University of California, Los Angeles, California 90024.
[†]Also at Research Institute of Atmospherics, Nagoya University, Toyokawa, Japan.
[‡]Also at Department of Physics, University of California, Los Angeles, California 90024.

The numerical parameters are $\eta_0 = 0.01$, $\mu/\rho_{sw} = D = D_p = 0.005$ where ρ_{sw} is the density of the solar wind. The magnetic Reynold's number, which is the magnetic diffusion time divided by the Alfvén transit time, is $S = \tau_\eta/\tau_A = 100 - 1000$. The diffusion and viscosity terms in Eqs. 1a, 1b, and 1c were added to suppress start-up fluctuations associated with the numerical scheme.

The solar magnetospheric coordinate system used in the calculation is the same as that shown in Plate III-5 of Ogino et al.[8] A uniform solar wind with $n_{sw} = 5/\text{cm}^3$, $V_{sw} = 300$ km/s, and $T_{sw} = 2 \times 10^5$ K flows into a simulation box of dimensions $-x_0 \leq x \leq x_0$, $-y_0 \leq y \leq y_0$, and $0 \leq z \leq z_0$ at $x = x_0$ where $x_0 = y_0 = z_0 = 24.5$ R$_e$. The uniform IMF is initially given by B_{IMF} ($= 0$, $B_{\mathrm{IMF}} \cos \theta$, $B_{\mathrm{IMF}} \sin \theta$) in the whole situation region where the angle θ is determined by rotating in a counterclockwise direction from the y-axis. We used free boundary conditions through which plasmas and waves can freely enter or exit the system at $x = -x_0$, $y = \pm y_0$, and $z = z_0$. A form of mirror boundary conditions was used at $z = 0$ such that quantities in the northern dusk (dawn) quadrant map into the southern dawn (dusk) quadrant. The ionospheric boundary condition imposed near the earth was determined by requiring static equilibrium.[10] We kept all of the parameters (ρ, v, p, B) fixed for $\xi \equiv (x^2 + y^2 + z^2)^{1/2} < 3.5$ R$_e$ and all perturbations were damped out by using a smoothing function near the ionosphere ($\xi \leq 5.5$ R$_e$). Therefore, the field-aligned currents also disappear near the earth and cannot close in the ionosphere. The MHD equations were solved on a 48 $\times$ 48 $\times$ 24 point grid not including the boundary grid points. The mesh size was $\Delta x = \Delta y = \Delta z = 1$ R$_e$ and the time step was $\Delta t = 3.8$ s.

SIMULATION RESULTS

The code was run for 512 time steps or 32 minutes. The model is intrinsically time dependent; however, in this study we have analyzed the results at only one instant in time. All of the results presented below represent a "snapshot" of the magnetosphere taken after 32 minutes. In that time interval, the solar wind with $V_{sw} = 300$ km/s travels a distance of 90 R$_e$ (approximately twice the length of the simulation box).

All of the parameters of the model were evaluated at each grid point at each time. Plate III-5 shows one way of presenting three-dimensional information in a two-dimensional format. In the two left-hand columns the parallel vorticity ($\Omega = \nabla \times v$, $\Omega_\parallel = \Omega \cdot B/B$) and the field-aligned current ($J_\parallel = J \cdot B/B$) have been projected along magnetic field lines into the northern ionosphere. The projected values were determined by calculating $\int (f/B)\, dl / \int (1/B)\, dl$ for each parameter, f, along the magnetic field lines. Only that part of open field lines within the magnetosphere was included in the integral. The $\Omega_\parallel$ and $J_\parallel$ values are color coded with red and yellow parallel to B and blue antiparallel. Yellow and light blue represent larger magnitudes than red and dark blue, respectively. From top to bottom, the IMF orientation

changes from $\theta = 180°$ (dawnward) to $\theta = 270°$ (southward).

There are four major convection systems evident: large-scale viscous convection cells (A), tail lobe cells (A$'$), high-latitude merging cells (B), and low-latitude cells (C). The largest system is the global system (A). The flow corresponding to this system is clockwise at dusk (i.e., antisunward over the polar cap and sunward at lower latitudes) while it is counterclockwise in the cell at dawn. This is mainly a viscous system since it is found even when $B_{\mathrm{IMF}} = 0$ (Ogino[10]). The nightside extension of the A system (A$'$) differs from the A system in that it is primarily driven by reconnection. The C system is found in the inner magnetosphere.

The final convection system, the polar cap or high latitude system, is most readily seen when $B_{y\mathrm{IMF}} \neq 0$. For an IMF with a dawnward pointing component, the polar cap flow is dominated by antiparallel vorticity and appears as an extension of the blue dawn side A cell into the polar cap. The corresponding parallel vorticity is the small red cell on the dawn side.

There also are four types of field-aligned currents: region 1 currents (R1), region 2 currents (R2), tail region 1 currents (TR1), and high-latitude currents composed of the dayside magnetopause currents and polar cap currents (PC). It is easiest to locate the current systems for the case where $\theta = 270°$ because of the symmetries present when $B_{y\mathrm{IMF}} = 0$. The dayside region 1 currents are closely correlated with the A convection cell. These currents are driven by the flow.[8,11] The region 2 currents, on the other hand, while associated with the C convection cells, are primarily driven by pressure gradients in the inner magnetosphere.

For southward IMF the currents in the polar cap region have a region 2 sense (i.e., earthward on the duskside and tailward on the dawnside). Just equatorward of the polar currents there are two additional currents (CC) which also have the region 2 sense. For $B_{y\mathrm{IMF}} < 0$ the earthward current, which is on the duskside for $\theta = 270°$, moves into the center of the polar cap. The corresponding upward current moves to lower latitude and toward dusk. This current eventually becomes indistinguishable from the duskside region 1 current. For dawnward IMF this leads to ring-like distributions of current with an upward (blue) ring at auroral latitudes and a downward (red) ring at lower latitudes.

The tail region 1 currents are also most clearly seen for southward IMF. This current undergoes a major change when $B_{y\mathrm{IMF}} < 0$, becoming very structured and almost filamentary. An additional current system appears in the tail at lower altitudes for $B_{y\mathrm{IMF}} < 0$. This current has a region 2 sense and is labeled TR2. The earthward flowing part of the current is clearly seen as a yellow and red region near midnight for $\theta = 210°$ and 240°. In subsequent runs this current was found for purely southward IMF as well but at slightly later times.

The plots in the right-hand column in Plate III-5 delimit in orange and yellow the region of the polar cap on open field lines. The yellow region indicates those field lines that pass through the top of the simulation

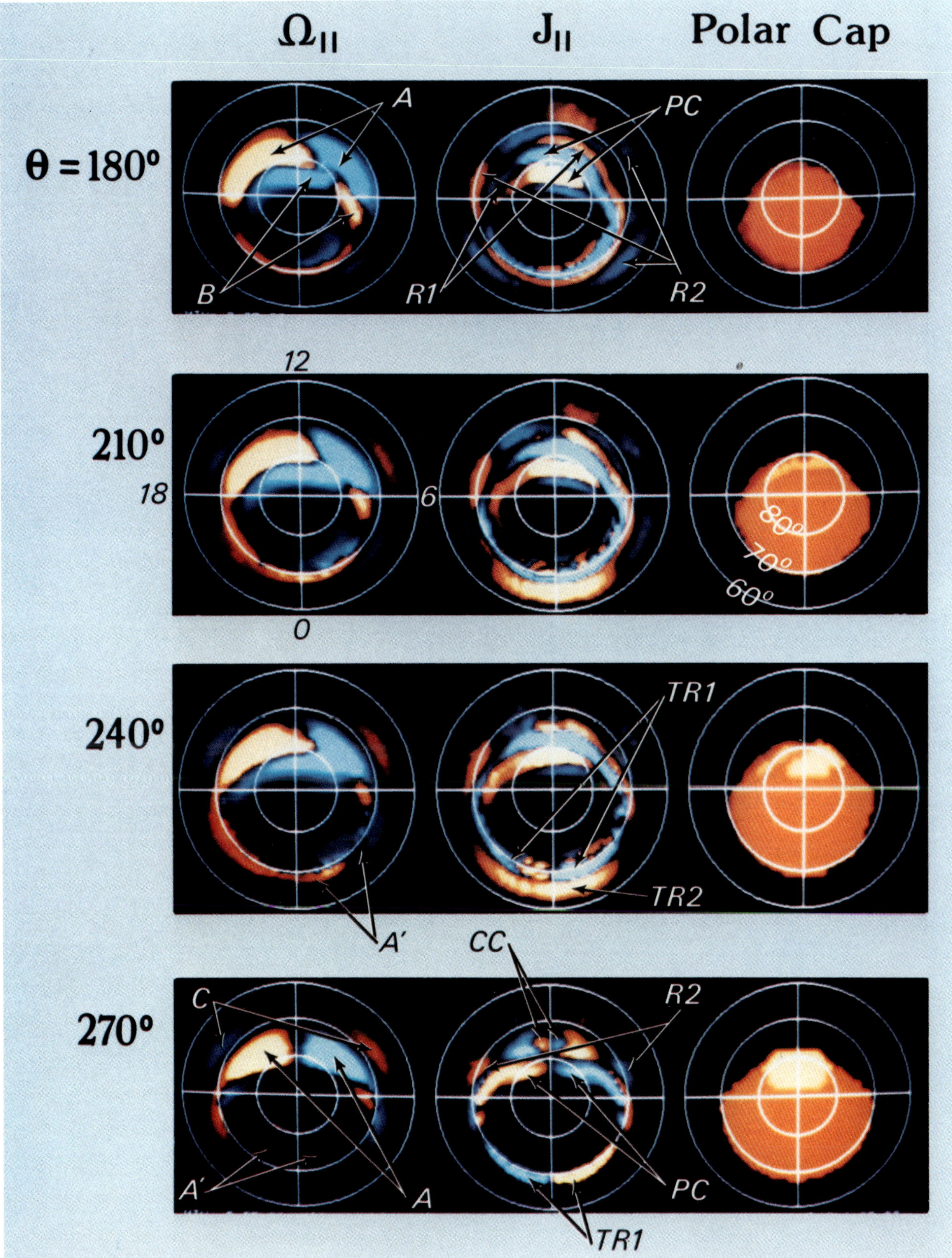

Plate III-5—Northern polar projections of the field-aligned vorticity, $\Omega_{\parallel}$, the field-aligned currents, $J_{\parallel}$, and a plot of the polar cap. Red and yellow indicate $\Omega_{\parallel}$ and $J_{\parallel}$, which are parallel to **B**, while blue indicates antiparallel quantities. In the polar-cap plots yellow indicates the region from which open field lines exit the top of the simulation box, θ gives the direction of the IMF with $\theta = 180°$ representing dawnward IMF and $\theta = 270°$ representing southward IMF.

box. Note how the size of the polar cap (open-field-line region) increases between $\theta = 180°$ and $\theta = 270°$. The polar cap currents are on open field lines while the cusp currents (CC) are on closed field lines. Thus there are region 2 type currents on both the open and closed field lines in the polar cusp. The tail region 1 currents occur at the boundary between open and closed field lines, while the tail region 2 currents occur on closed plasma-sheet field lines.

Contours of plasma pressure (P) and field-aligned velocity ($v_\parallel$) have been plotted in the y-z plane at $x = -18.5\ R_e$ for IMF angles between $\theta = 150°$ and $\theta = 270°$ in Fig. 1. For $B_{y\text{IMF}} < 0$ the plasma sheet has been rotated out of the equatorial plane. The rotation is northward on the dawnside of the noon-midnight meridian and southward on the duskside. In addition, as the IMF goes from northward ($\theta = 150°$) to southward ($\theta > 180°$) the plasma sheet becomes thinner. When $B_{z\text{IMF}} < 0$ and $B_{y\text{IMF}} \neq 0$ ($\theta = 240°$) pressure islands form in the plasma sheet. The parallel velocity

reverses sign on either side of the pressure islands (arrows in the right panel). This sharp gradient in the $v_\parallel$ profile is a characteristic sign of localized reconnection.

Cross-sectional patterns showing the field-aligned currents in the same plane at $x = -18.5\ R_e$ have been plotted in Fig. 2. Currents parallel to B have been contoured in the right panel while antiparallel currents are represented in the left panel. For southward IMF ($\theta = 270°$) the currents in the center of the tail have the region 1 sense. These are the tail region 1 currents in Plate III-5. For $\theta = 210°$ and $\theta = 240°$, the currents have a great deal more structure and those near the outer edge of the plasma sheet are filamentary. These are the filamentary currents that we observed in Plate III-5 for $B_y \neq 0$. The negative current near the center of the plasma sheet appears as a region of enhanced outward current (blue) just north of the TR2 currents in Plate III-5.

The magnetic field configuration has been plotted in three dimensions in Fig. 3. Only the northern hemisphere part of the field lines has been included. The draping of IMF field lines over the magnetosphere can be seen. Magnetospheric field lines are frequently bent or twisted. In particular, note the wave-like structure of tail field lines. This results from localized or patchy reconnection.

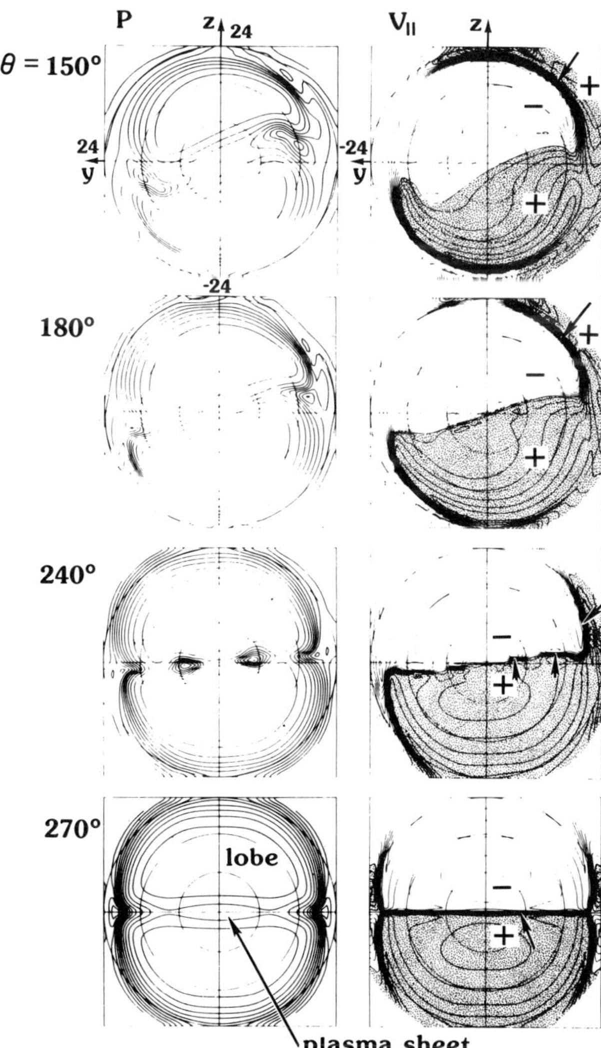

Figure 1—Cross-sectional patterns in the y-z plane at $x = -18.5\ R_e$ of the pressure (p) and the parallel velocity ($v_\parallel$).

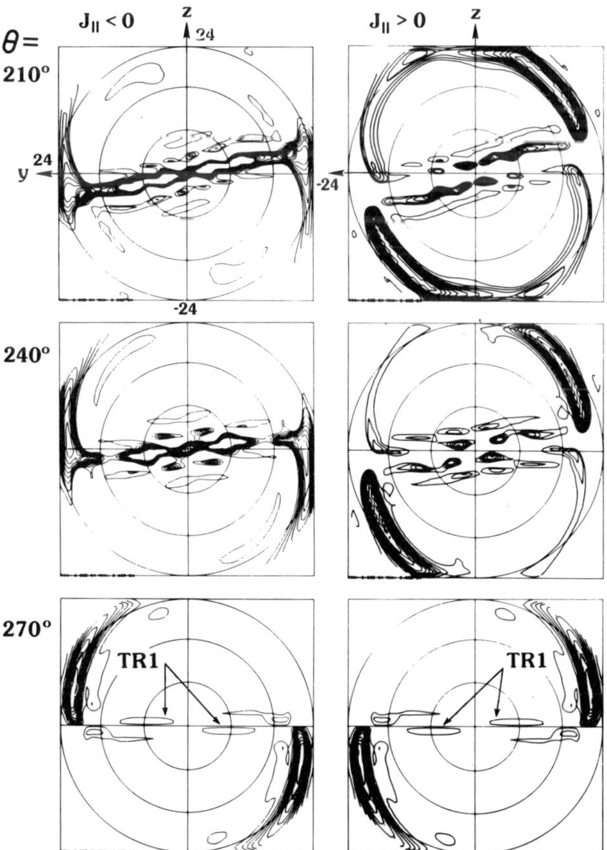

Figure 2—Cross-sectional patterns of the field-aligned currents at $x = -18.5\ R_e$ in the tail region for $\theta = 210°$, $240°$, and $270°$. The left column shows $J_\parallel$ opposite to B ($J_\parallel < 0$) while the right panel contains $J_\parallel$ along B ($J_\parallel > 0$).

187

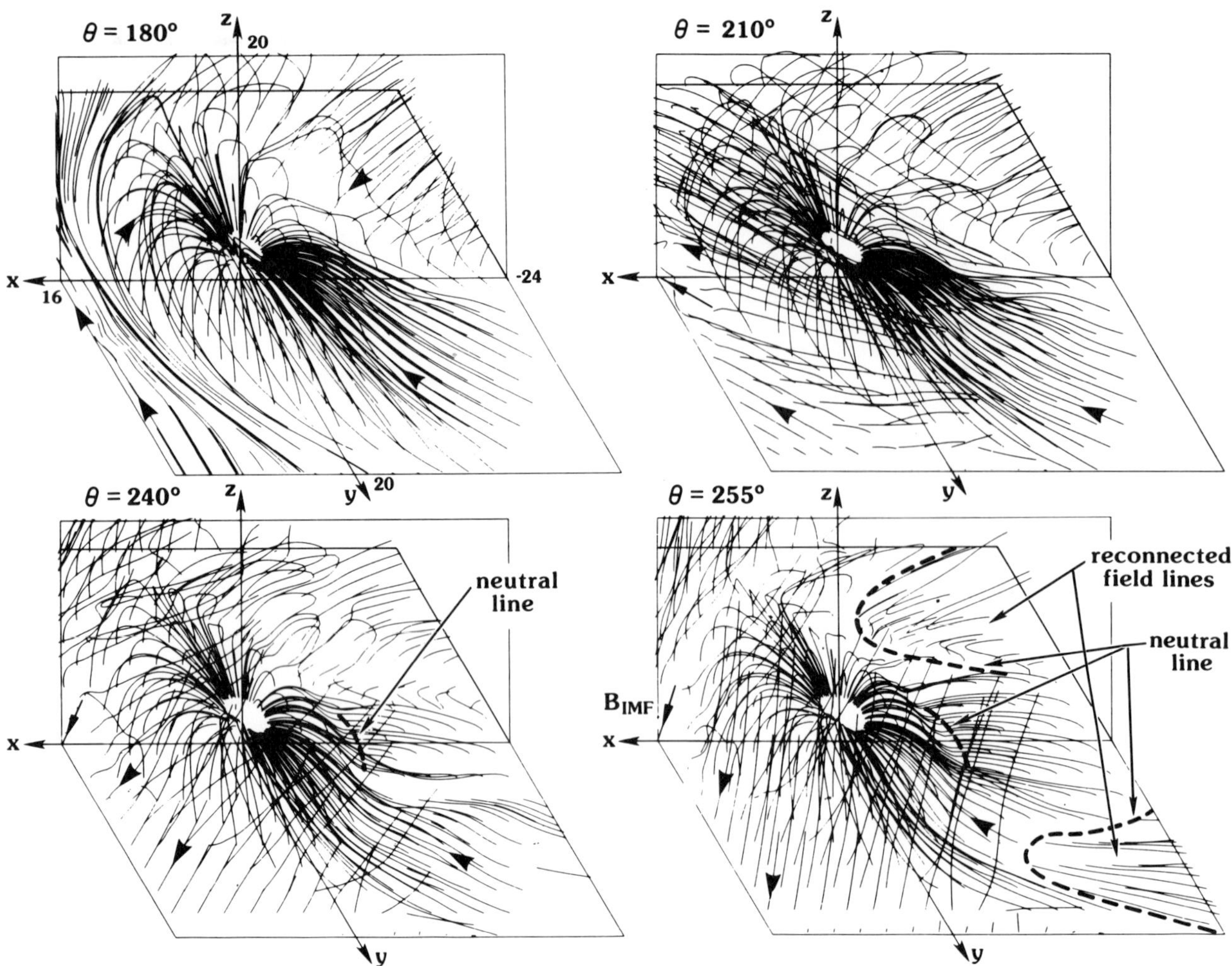

Figure 3—Three-dimensional plots of the magnetic field configuration for 180° ≤ θ ≤ 255°. The dashed lines show the magnetic neutral lines.

We have indicated neutral lines (dashed) and regions of reconnected field lines for $\theta = 240°$ and 255°. The positions of the neutral lines change in time and are associated with the plasma pressure islands in Fig. 1.

DISCUSSION

The IMF B_y component introduces a dawn-dusk asymmetry into the magnetospheric configuration. This asymmetry can be seen in the dayside polar cusp, the plasma sheet, and the tail lobes. In the northern hemisphere the dayside cusp shifts toward dawn for $B_y <$ 0.[9] In the tail, the plasma sheet rotates toward the northern lobe on the dawnside and toward the southern lobe on the duskside (Fig. 1). This twisting is consistent with the earlier simulation results of Brecht et al.[12] and the theoretical model of Cowley.[13] In the tail earthward of the neutral lines, the magnetic field lines are twisted toward dusk for a dawnward-pointing IMF. Both the motion of the polar cusp and the twisting of the tail field lines are caused by the imposition of a dawnward B_y on the magnetosphere. That such a B_y component is superimposed on the magnetosphere has been demonstrated observationally by Fairfield[14] and Cowley and Hughes.[15]

There are four major convection systems and four major field-aligned current systems in the model results. The large-scale convection cells (A) and the low-latitude convection cells (C) are primarily viscous effects. They are found even when there is no IMF. When the IMF is northward the potential difference across the A cells is about 70 kV with the electric field pointing from dawn to dusk. That this is somewhat higher than the 12 kV to 60 kV determined from polar-orbiting satellites[16-18] indicates that the model is too viscous. In contrast, the tail lobe convection cells (A') and the high-latitude merging cells (B) depend very strongly on the IMF orientation. For the case when the IMF is entirely southward, the flow driven by reconnection adds to that of the viscous cells (Fig. 1 lower left panel). Then the total potential difference across the A system is about 180 kV. The difference between the northward IMF case and the southward IMF case (110 kV) is caused by reconnection. This potential difference is consistent with the 40 kV to 170 kV inferred from the observations.

In order to better see how the IMF affects the flow pattern and current pattern, the regions in which reconnection may be occurring have been sketched in Fig. 4. In Fig. 4 the shaded areas indicate regions in which the magnetic field is small ($|B| < 0.3\ B_{IMF}$). For each IMF orientation, the top panel shows the projection in the noon-midnight meridian and the bottom panel shows the projections onto the equatorial plane. For $\theta = 180°$ reconnection occurs on the nightside magnetopause at high latitudes. For an increasing southward IMF component the reconnection moves to the dayside and to lower latitudes. On the dayside, the location of reconnection is consistent with that expected if merging occurs primarily in regions in which the magnetospheric field at the magnetopause is antiparallel to the IMF.[9] The reconnection region also moves within the magnetosphere when the IMF has a southward as well as a dawnward component. For $\theta = 240°$, there are three regions of reconnection in the plasma sheet.

The high-latitude polar-cap cells are located in the region of open field lines and are associated with high-latitude nightside or dayside reconnection. For $B_{yIMF} < 0$ and $B_{zIMF} < 0$ most of the field lines which reconnect on the dawn magnetopause are pulled tailward

around the dusk magnetopause by the solar wind. This convection produces a strong field-aligned vorticity with $\Omega_\parallel < 0$ throughout the whole dayside polar cap region in Plate III-5. A small fraction of the open field lines convects tailward around the dawn magnetopause producing the small dawnside convection cell.

The tail-lobe convection cells (A′) are driven by tail reconnection. These convection cells consist of tailward flow and flow that converges toward the sun-earth meridian along the plasma-sheet boundary. This is followed by flows expanding earthward away from the reconnection regions in the plasma sheet. The low-latitude boundaries of the tail-lobe convection cells occur at the boundary between open and closed field lines (see Plate III-5) which is the outer boundary of the plasma sheet.

Two field-aligned current generation mechanisms, which may be applicable to the magnetosphere, involve the parallel vorticity and pressure gradients. The field-aligned vorticity can lead to field-aligned currents in two ways. First currents such as the region 1 currents can be generated by large-scale convection on closed field lines. Second currents can be generated by the twisting of reconnected field lines. The tail region 1 (TR1) currents that occur at the boundary between open and closed field lines are generated by the reconnection. As noted previously, the region 2 currents are primarily generated by pressure gradients associated with the inner edge of the plasma sheet. The tail region 2 currents (TR2) occur on closed plasma-sheet field lines and are probably caused by pressure gradients in the plasma sheet earthward of the magnetic neutral line.

The dayside cusp currents (CC) are located on closed field lines in the polar cusp. They can be most easily distinguished when the IMF orientation is nearly south ward. The polar-cap currents occur on open field lines and are located just tailward of the cusp currents for southward IMF. These currents always have a region 2 sense. The location of the polar-cap field-aligned currents changes smoothly as the IMF direction and hence the location of reconnection on the magnetopause changes.[9] In particular, the polar cap currents evolve from the northward B_z currents[7,11] to the cusp currents when the IMF is rotated from northward to southward.[9] In all cases, the currents are caused by the twisting caused by reconnection.

There are several limitations to the model. First, the boundary condition near the earth is fixed. This boundary condition ensures a steady-state ionosphere in which all perturbations are damped to zero. Therefore, the field-aligned currents do not form a closed circuit in the ionosphere. Remember that the currents in Plate III-5 are averages along the field line and thus represent magnetospheric values projected onto the earth, not values in the ionosphere. Despite this limitation, it is important to emphasize that the field-aligned currents in the magnetosphere are calculated self-consistently and have a distribution similar to that of the observed currents.

The second limitation is caused by the length of the tail. The tail was cut off at $x = -24.5\ R_e$, and a free

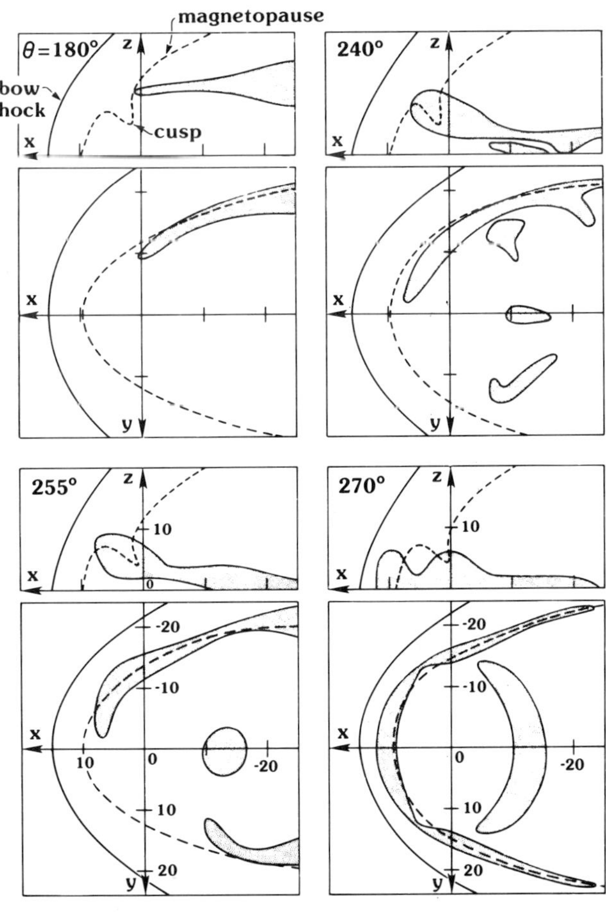

Figure 4—Areas of magnetic reconnection for $180° \leq \theta \leq 270°$. The shaded areas indicate regions projected onto the noon-midnight (*x-z*) and equatorial (*x-y*) planes for which $|B| < 0.3\ B_{IMF}$.

boundary condition, in which the spatial derivative for all the quantities was set to zero, was used. We chose this boundary condition because it minimizes the effect of the boundary on phenomena inside the boundary. Recently, we have had the opportunity to run the case with a southward IMF using a larger grid in which the boundary was at $x = -130\ R_e$. All of the features discussed in this report were confirmed in the calculations with the longer tail.[19] Therefore, we are confident that the length of the tail did not seriously influence any of the results shown here.

Finally, the interaction mechanism in the calculation is the numerical resistivity, η. We chose this to vary in a classical way on temperature (see second section). We made this selection because it represents a simple model for the resistivity. We feel that we should wait before trying any more sophisticated models until kinetic studies of the plasma instabilities thought to be important in the magnetosphere can be used to drive the appropriate scaling laws.

ACKNOWLEDGMENT—This work was supported by the NASA Solar Terrestrial Theory Program, grants NAGW-78 and NSF ATM 82-18746.

REFERENCES

[1] R. L. McPherron, "Magnetospheric Substorms," *Rev. Geophys. Space Phys.* **17**, 657 (1979).

[2] S.-I. Akasofu, "What is a Magnetospheric Substorm?" in *Dynamics of the Magnetosphere,* S.-I. Akasofu, ed., D. Reidel Pub. Co., Dordrecht, Holland, p. 447 (1980).

[3] K. Maezawa, "Magnetic Convection Induced by the Positive and Negative Z Components of the Interplanetary Magnetic Field: Quantitative Analysis Using Polar Cap Magnetic Records," *J. Geophys. Res.* **81**, 2289 (1976).

[4] J. L. Horwitz and S.-I. Akasofu, "On the Relationship of the Polar Cap Current System to the North-South Component of the Interplanetary Magnetic Field," *J. Geophys. Res.* **84**, 2567 (1979).

[5] R. A. Heelis, "The Effects of Interplanetary Magnetic Field Orientation on Dayside High Latitude Ionospheric Convection," *J. Geophys. Res.* **89**, 2873 (1984).

[6] T. Iijima and T. A. Potemra, "Field-Aligned Current in the Dayside Cusp Observed by Triad," *J. Geophys. Res.* **81**, 5971 (1976).

[7] T. Iijima, T. A. Potemra, L. J. Zanetti, and P. F. Bythrow, "Large-Scale Birkeland Currents in the Dayside Polar Region During Strongly Northward IMF – A New Birkeland Current System," *J. Geophys. Res.* **89**, 7441 (1984).

[8] T. Ogino, R. J. Walker, M. Ashour-Abdalla, and J. M. Dawson, "An MHD Simulation of B_y-Dependent Magnetospheric Convection and Field-Aligned Currents During Northward IMF," *J. Geophys. Res.* **90**, 10835 (1985).

[9] T. Ogino, R. J. Walker, M. Ashour-Abdalla, and J. M. Dawson, "An MHD Simulation of the Effects of the Interplanetary Magnetic Field B_y Component on the Interaction of the Solar Wind with the Earth's Magnetosphere During Southward IMF" (preprint PPG-855), *J. Geophys. Res.* (submitted, 1986).

[10] T. Ogino, "A Three-Dimensional MHD Simulation of the Interaction of the Solar Wind with the Earth's Magnetosphere: The Generation of Field-Aligned Currents" (preprint PPG-796), *J. Geophys. Res.* (in press, 1986).

[11] T. Ogino and R. J. Walker, "A Magnetohydrodynamic Simulation of the Bifurcation of Tail Lobes During Intervals with a Northward Interplanetary Magnetic Field," *Geophys. Res. Lett.* **11**, 1018 (1984).

[12] S. H. Brecht, J. G. Lyon, J. A. Fedder, and K. Hain, "A Simulation Study of East-West IMF Effects on the Magnetosphere," *Geophys. Res. Lett.* **8**, 397 (1981).

[13] S. W. H. Cowley, "Magnetospheric Asymmetries Associated with the Y-Component of the IMF," *Planet. Space Sci.* **29**, 79 (1981).

[14] D. H. Fairfield, "On the Average Configuration of the Geomagnetic Tail," *J. Geophys. Res.* **84**, 1950 (1979).

[15] S. W. H. Cowley and W. J. Hughes, "Observation of an IMF Sector Effect in the Y Magnetic Field Component at Geostationary Orbit," *Planet. Space Sci.* **31**, 73 (1983).

[16] P. H. Reiff, R. W. Spiro, and T. W. Hill, "Dependence of Polar Cap Potential Drop on Interplanetary Parameters," *J. Geophys. Res.* **86**, A9, 7639 (1981).

[17] J. R. Wygant, R. B. Torbert, and F. S. Mozer, "Comparison of S3-3 Polar Cap Potential Drops with the Interplanetary Magnetic Field and Models of Magnetopause Reconnection," *J. Geophys. Res.* **88**, A7 (1983).

[18] M. A. Doyle and W. J. Burke, "S3-2 Measurements of the Polar Cap Potential," *J. Geophys. Res.* **88**, 9125 (1983).

[19] T. Ogino, R. L. Richard, R. J. Walker, and M. Ashour-Abdalla, "A Global MHD Simulation of the Earth's Magnetotail Dynamics" (abstract), *Eos, Trans. AGU* **66**, 1044 (1985).

MULTIPLE *X*-LINE RECONNECTION IN THE EARTH'S MAGNETOSPHERE

L. C. Lee, Z. F. Fu, and S.-I. Akasofu*

A comprehensive review of the multiple *X*-line reconnection process is made. It is found that magnetic reconnection occurs either with single or multiple *X*-lines, depending on several conditions. The multiple *X*-line reconnection process tends to occur when the system length is long, the reconnection rate is high, or the resistivity is small. The multiple *X*-line reconnection is a time-dependent process and may lead to the repeated formation and convection of plasmoids (or magnetic flux tubes). The multiple *X*-line reconnection may lead to the occurrence of flux transfer events observed at the earth's dayside magnetopause and the formation of substorms and plasmoids observed in the magnetotail. It appears that multiple *X*-line reconnection occurs more frequently than the classical single *X*-line reconnection. A criterion for the transition from a single *X*-line to a multiple *X*-line reconnection is obtained.

INTRODUCTION

The open magnetosphere model was first proposed by Dungey.[1] In his open model, the interplanetary magnetic field (IMF) and the geomagnetic field reconnect along a single neutral line (*X*-line) near the subsolar region of the dayside magnetopause as shown in Fig. 1a. The reconnected magnetic field lines are then convected toward the magnetotail lobes. The lobe field lines reconnect along the *X*-line in the equatorial plane of the magnetotail. The antisolar part of the reconnected field lines is then convected further in the antisolar direction and eventually becomes the IMF field lines. The earthward part of the reconnected field lines is convected earthward and eventually returns to the dayside magnetosphere. In principle, a quasisteady state of the magnetic field configuration could be reached.

This open model was further developed by Levy et al.,[2] who predicted the existence of rotational discontinuities and high speed plasma flows at the dayside magnetopause. However, the predicted rotational discontinuities and high-speed plasma flow are not observed with the expected frequency. Further, the "flux transfer events" (FTEs) first reported by Russell and Elphic[3] indicated that the dayside magnetic reconnection geometry is different from that in Dungey's quasisteady-state model. The identification of FTEs is mainly based on the signature in B_N (the component of the magnetic field in the direction normal to the local magnetopause).

Recently, Lee and Fu[4] proposed a multiple *X*-line reconnection model for the dayside magnetopause as illustrated in Fig. 1b, which can explain reasonably well the occurrence and the observed properties of FTEs.

The multiple *X*-line reconnection can be considered as a magnetic reconnection process between two approaching magnetized plasmas in which magnetic field lines reconnect at multiple sites.

The multiple *X*-line reconnection is intrinsically a time-dependent process. Magnetic flux tubes are expected to be formed intermittently and repeatedly. The multiple

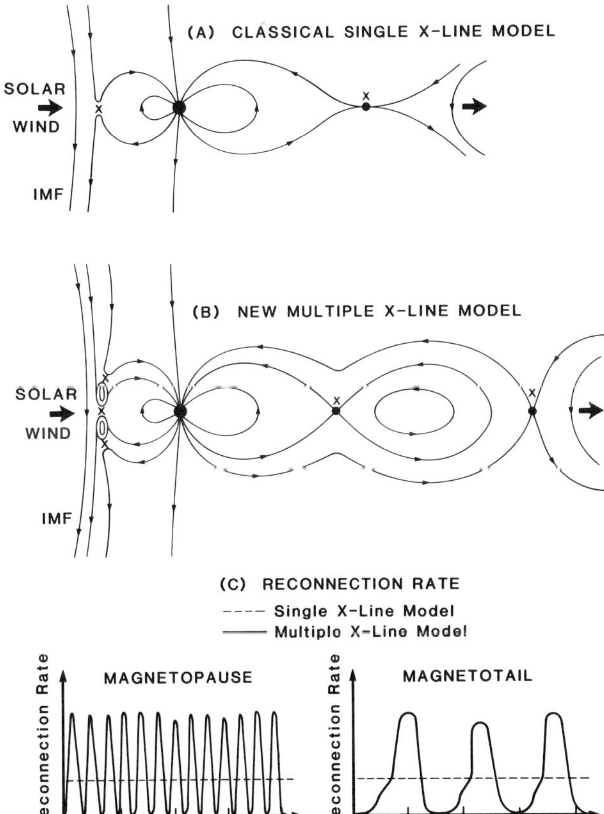

Figure 1—(a) Dungey's single *X*-line open magnetosphere model, (b) the new multiple *X*-line model, and (c) the reconnection rates at the dayside magnetopause and in the magnetotail for the two models when the IMF turns southward for a long time.

X-line reconnection process is different from the steady-state single *X*-line reconnection proposed by a number of workers, including Petschek,[5] Sonnerup,[6] and Yeh and Axford.[7] A criterion for the transition from a single *X*-line reconnection to a multiple *X*-line reconnection was recently obtained by Lee and Fu.[8]

*Geophysical Institute and Physics Department, University of Alaska, Fairbanks, Alaska 99775-0800.

The multiple *X*-line reconnection process may also occur in the magnetotail as shown in Fig. 1b. A great variety of substorm-associated magnetotail phenomena observed by satellite has been interpreted in terms of (in sequence) the initial plasma-sheet thinning, appearance of a new *X*-line, formation of plasmoids, further stretching and thinning of the plasma sheet, and tailward convection of the new *X*-line and plasmoids (e.g., Hones,[9] Birn,[10] Lyon et al.,[11] and Brecht et al.[12]).

Recently, Lee et al.[13] simulated the multiple *X*-line reconnection processes of a long magnetotail (~ 200 R_e) under a constant and time-varying driving force delivered from the solar wind. It is found that under a constant driving force the magnetic fields in the magnetotail tend to reconnect impulsively and the formation of *X*-line and plasmoids occurs intermittently and repeatedly every 2-4 hours.

Biskamp[14] and Leboeuf et al.[15] also discussed some aspects of the multiple *X*-line reconnection process. However, they did not discuss the repeated formation and convection of plasmoids. The readers are referred to Lee and Fu[8] for more details on the multiple reconnection process.

The purpose of the present paper is to give a brief review of the multiple *X*-line reconnection process that takes place at the dayside magnetopause and in the nightside magnetotail. In the next section a criterion for the transition from a single *X*-line to a multiple *X*-line reconnection will be examined. In the third section, we present the multiple *X*-line reconnection process at the dayside magnetopause, which leads to the occurrence of flux transfer events. In the fourth section, the multiple *X*-line reconnection process in the magnetotail, which leads to the occurrence of substorms, will be discussed.

A CRITERION FOR THE MULTIPLE X-LINE RECONNECTION

In an attempt to understand the occurrence of multiple *X*-line reconnections, Lee and Fu[8] have carried an extensive simulation study of magnetic reconnections based on a two-dimensional incompressible MHD code. Both single and multiple *X*-line reconnections are observed in the simulation.

The simulation model is briefly described as follows. The simulation length in the *x* direction is $L_x \equiv a$ and the simulation length in the *z* direction is L_z. The magnetic field $\boldsymbol{B}$ at time $t = 0$ is given by

$$\boldsymbol{B} = -B_0 \tanh \left(\frac{x}{d}\right) \boldsymbol{z} \qquad (1)$$

where *d* is the characteristic thickness of the current layer, and *z* is the unit vector in the *z* direction.

Along the boundary $x = a$, the magnetic field B_z is maintained at the initial value as given in Eq. 1 and the normal velocity is set to be $V_1 = R_0 V_A$, where V_A is the Alfvén speed based on B_0, and the parameter R_0 can be considered as the imposed reconnection rate. A "free" boundary is imposed at $z = \pm L_z$.

An example of the single *X*-line reconnection is shown in Fig. 2, in which $L_z = 2a$ and $R_0 = 0.1$. The final quasisteady configurations of the field lines and streamlines and the contours of current density (*J*) are shown in the top three panels. The bottom two panels show the profiles of the current density *J* and velocity v_z / V_A

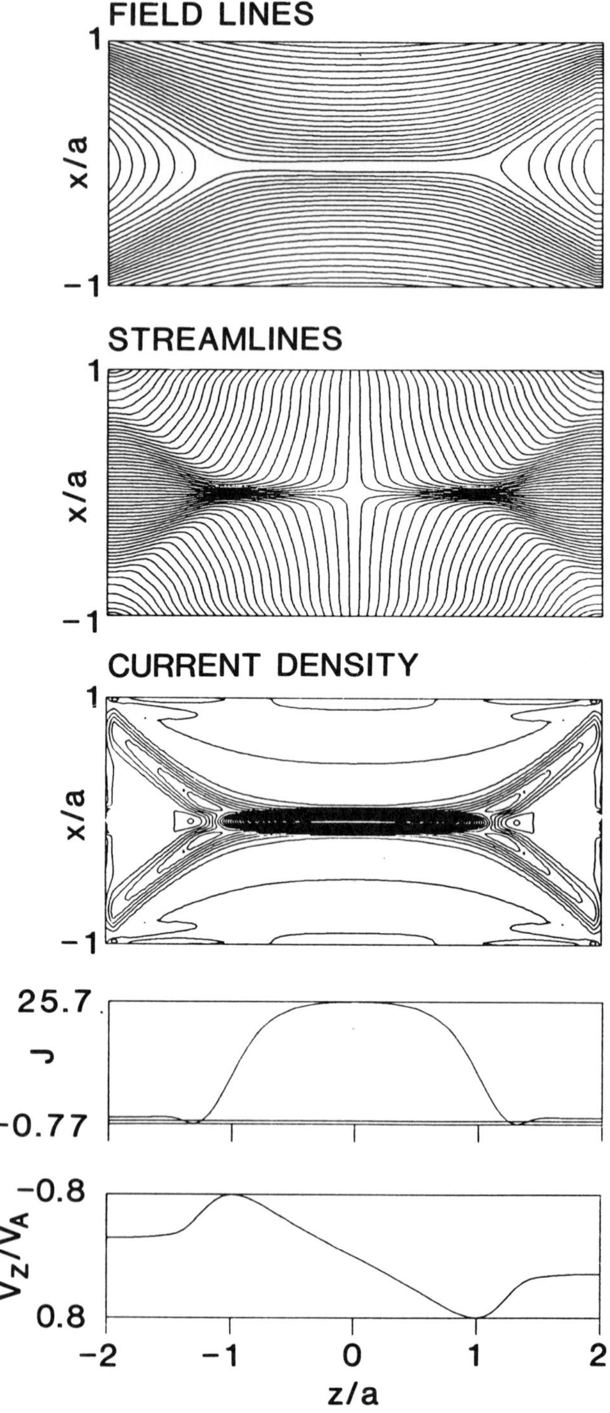

Figure 2—The magnetic field lines, streamlines, contours of the current density *J*, the current profile, and the v_z profile for case A. The current density *J* is expressed in units of $B_0/\mu_0 a$.

along the z axis ($x = 0$). As shown in Fig. 2, magnetic field lines are convected inward from $x = \pm a$ and reconnected in the central diffusion region. The reconnected field lines are then convected outward through the boundary at $z = \pm L_z$. It can be seen from the streamlines that four large-amplitude standing Alfvén waves are attached to the central diffusion region. The current density in the diffusion region is much higher than those along the four wave fronts.

From the v_z/V_A profile, it is found that the flow speed is highest at the two ends of the diffusion region and is $U_2 \sim 0.8\ V_A$. The outflow speed at $z = \pm L_z$ is $V_2 \sim 0.35\ V_A$. The length of the diffusion region ($2l$) can be determined from the current density J profile. The half-length of the diffusion region is found to be $l \approx a$. The half-width δ of the diffusion region is defined to be the half-width of the current profile along the x axis and is found to be $\delta = 0.065a$.

The criterion for the multiple X-line reconnection to occur depends on the development of tearing instability in the diffusion region, which in turn depends on the resistivity η, the half-length l, and half-width δ of diffusion region (e.g., Furth et al.[16]). As the ratio l/δ

becomes larger, multiple X-lines will appear in the diffusion region, leading to the formation of magnetic islands as shown in Fig. 3, which is to be further discussed in the next section.

Based on the simulation of 50 cases, the ratio l/δ is found to be

$$l/\delta \simeq C_0^2\ (L_z/a)^{2\alpha}\ R_0^{2\beta-1}\ (\eta/\mu_0 a V_A)^{2\gamma}$$

$$\simeq 14\ (L_z/a)^{1.2}\ R_0^{1.0}\ (\eta/\mu_0 a V_A)^{-0.25} \qquad (2)$$

where $C_0 \simeq 3.7$, $\alpha = 0.6$, $\beta \simeq 1.0$, $2\gamma = -0.25$. The 50 cases used in the study are plotted in Fig. 4, in which each circle in the l-δ plane represent the case with a single X-line reconnection, while the points marked with an asterisk are cases that eventually evolve into an unsteady multiple X-line reconnection. The length and width of the diffusion region in the multiple X-line case are those measured before the development of magnetic islands. The straight line in Fig. 4 is $l = 10\delta$. It can be seen that multiple X-line reconnection occurs when

$$l/\delta \gtrsim 10\ . \qquad (3)$$

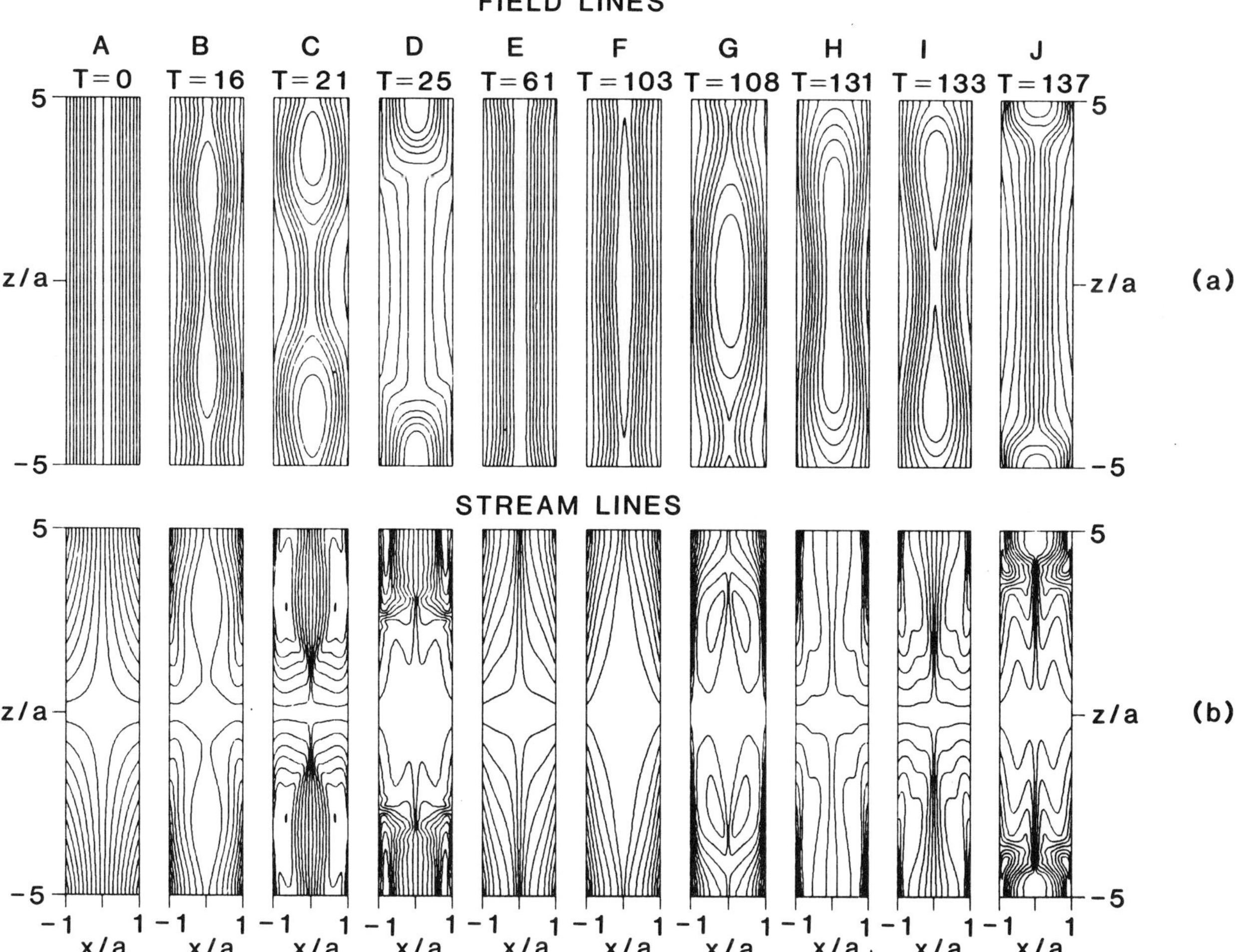

Figure 3—An example of the multiple X-line reconnection at the dayside magnetopause: (a) the magnetic field lines, and (b) the streamlines at different simulation times.

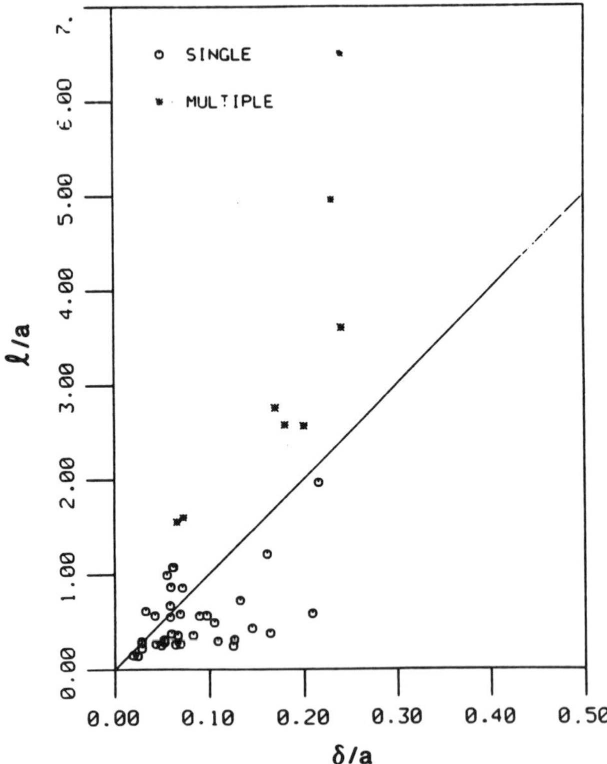

Figure 4—The half-length and half-width of the 50 cases simulated in the present study are plotted in the l-δ plane. The straight line is $l = 10\delta$. The points marked with a circle are for the single X-line reconnection and the points marked with an asterisk are for the multiple X-line reconnection.

The numerical criterion in Eq. 3 can be explained as follows. Consider the development of tearing instability in a current layer with a length of $2l$ and a thickness of 2δ. The following condition must be met for the formation of magnetic islands: the e-folding time t_g for the tearing instability must be shorter than the convection time $t_c \simeq 2l/V_A$. In estimating the convection time t_c, we assume the average convection speed to be $V_A/2$. The growth time is found to be $t_g \simeq 20\,\delta/V_A$.[8] The requirement that $t_g < t_c$ leads to the condition in Eq. 3 for the occurrence of multiple X-line reconnections.

Therefore, the multiple X-line reconnection process tends to occur when the system is long, the reconnection rate R_0 is high, or the resistivity is small.

It may be noted that without the driving force imposed at $x = \pm a$ the reconnection rate R_0 generally increases with η (cf., Ref. 5) so that l/δ in Eq. 2 may be proportional to η^s where s is a positive number. However, in the presence of a driving force the situation is different and the reconnection rate actually depends only on the imposed electric field at the boundary (cf., Ref. 7). This can be understood as follows. If the initial reconnection rate in the current sheet is small due to a small resistivity η, the current sheet may be pinched or the magnetic field outside the current sheet may pile up, which leads

to an enhanced reconnection rate, matching the imposed driving rate.

At the dayside magnetopause and the magnetotail, the system length is long and the anomalous resistivity is expected to be smaller than the simulation values. Therefore, the multiple X-line reconnection may dominate the reconnection processes in the magnetosphere, leading to the occurrence of the flux transfer events and substorms. For example, if we take $L_z = 5\ \mathrm{R_e}$, $a = 0.5\ \mathrm{R_e}$, $R_0 = 0.1$, $V_A = 100\text{–}200$ km, and $\eta < 0.1\ \mu_0 a V_A$, we then obtain, from Eq. 2, $l/\delta > 40$, which leads to the multiple X-line reconnection. On the other hand, if we choose $L_z = 200\ \mathrm{R_e}$, $a = 20\ \mathrm{R_e}$, $R_0 = 0.05$ (corresponding to a cross-tail potential $\phi_{ct} = 100$ kV), and $\eta < 0.1\ \mu_0 a V_A$ for the magnetotail, we have $l/\delta > 20$ and hence the multiple X-line reconnection in the magnetotail.

FLUX TRANSFER EVENTS AT THE DAYSIDE MAGNETOPAUSE

An example of the multiple X-line reconnection at the dayside magnetopause is shown in Fig. 3a and 3b, in which the magnetic field lines and streamlines are plotted. The parameters used are $R_0 = 0.1$, $L_z/a = 5$, and $a \simeq 1500$ km.

It can be seen from Fig. 3a that under a driving force applied at the boundary $x = \pm a$, magnetic islands (magnetic flux tubes) are repeatedly formed and convected out of the boundary at $z = \pm L_z$. Magnetic-flux pile-up near the current sheet is often observed. The first pair of islands is formed at $t = 16t_A$ and grow to a size of $S_z \simeq 5a$ in the z direction and $S_x \simeq 1.5a$ in the x direction. A second pair of islands is formed by splitting the large central island at $z = 0$ and the splitting is nearly completed at $t = 133t_A$. Note that the magnetic field configuration and the streamline pattern at $t = 61t_A$ are similar to those at $t = 0$. A third pair of islands are formed at $t \simeq 230t_A$. Figure 3b shows that vortices in plasma flow are formed during the multiple X-line reconnection process.

It is seen from Fig. 3 that magnetic flux tubes are repeatedly formed and convected out of the simulation region at a time interval $\tau \simeq 100t_A$. At the dayside magnetopause near the subsolar region, the Alfvén speed is observed to be $V_A \simeq 200$ km/s and $t_A \simeq a/V_A \simeq 7.5$ s. Therefore, we have $\tau \simeq 750$ s. The cross-section of the flux tubes is given by $S_z \simeq 5a \simeq 1.25\ \mathrm{R_e}$ and $S_x \simeq 2a \simeq 0.5\ \mathrm{R_e}$. The time interval τ and the size of the flux tubes obtained in Fig. 3 are consistent with satellite observations.[17]

Figure 5 shows B_x and the magnetic field magnitude B at $x = 0.8a$ and $z = 3a$ as a function of time t. It is seen that the B_x profile shows the typical B_N signature observed with spacecraft. The field magnitude B increases inside the flux tubes following by a dip that may correspond to a rarefaction region behind the flux tubes. Thus, the simulation results shown in Figs. 3 and 5 can reproduce most of the observed features of FTEs.

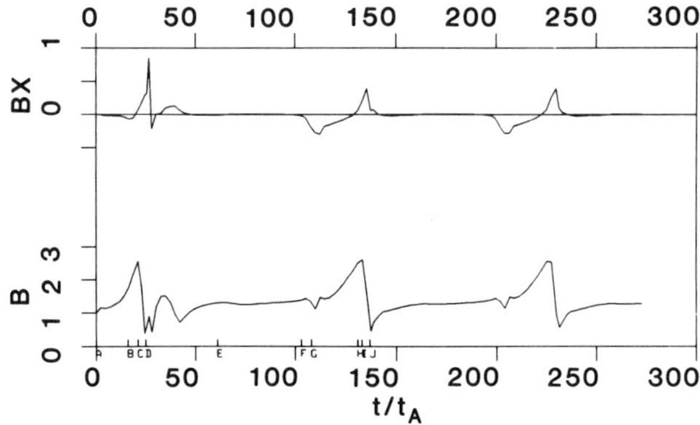

Figure 5—B_x and B as a function of time t for the case shown in Fig. 3.

A three-dimensional MHD simulation of the reconnection at the dayside magnetopause by Sato (presented at this conference) also shows that the magnetic field lines can reconnect at multiple sites, leading to the formation of helical flux tubes as proposed by Lee and Fu.[4]

MULTIPLE *X*-LINE RECONNECTION IN THE MAGNETOTAIL

In this section, we present the simulation results of the magnetotail dynamics obtained by Lee at al.[13] Figure 6 shows a case of the forced reconnection in the magnetotail, in which the convection electric field imposed at the magnetopause boundary at $z = \pm 20$ R$_e$ is $E \simeq$ 0.48 mV/m. The tail length in the simulation region is $L_x - 200$ R$_e$. The Alfvén time t_A shown in the figure is $\simeq 10$ s. Note that the coordinate system in this section is different from that in the previous section.

The magnetic field lines at different simulation times are shown in Fig. 6. It can be seen in the figure that plasma-sheet thinning, appearance of the new *X*-line, formation of magnetic islands (plasmoids), and covection of plasmoids occur in sequence repeatedly, which can be identified as substorms.

The left panels in Fig. 6 show the sequence of the first event. At $t = 529$ t_A, the plasma sheet near the earthward end starts the initial thinning process. At $t = t_1$ = 610, a new *X*-line appears at $x = -35$ R$_e$, and a large magnetic island starts to form. At $t = 772$, the magnetic reconnection rate is very high, and the plasma sheet near the new *X*-line is further stretched and becomes very thin. As can be seen in Fig. 6, the plasma sheet after the onset becomes much thinner than before the onset. At $t = 1113$, a small second magnetic island is formed in the region earthward of the large plasmoid. At $t = 1381$, the large plasmoid is convected out of the far-earth simulation boundary, while the second magnetic island grows to a large size. The newly formed *X*-line remains at the location of its formation for about 100 t_A and then moves tailward. At $t = 1583$, the newly formed *X*-line has moved to $x = -130$ R$_e$, and the first substorm event is completed. The tailward convec-

tion speed of the *X*-line is $v_{x1} \approx 50$ km/s. Both the magnetic reconnection at the new *X*-line and the tailward convection of the *X*-line may lead to the formation of the poleward expanding bulge after the onset of substorms.

The right panels in Fig. 6 show that three more events occur at $t_2 = 1704$, $t_3 = 3232$, and $t_4 = 4763$. It is noted that the new *X*-lines for these three events are located at $x = -27$, -28, and -30 R$_e$, respectively. The time intervals between successive events are $\tau = 610$, 1094, 1528, and 1531 t_A, which correspond to 3–5 hr. The size of plasmoids is $\simeq 100$ R$_e$ in the x direction and ~ 40 R$_e$ in the z direction, which is consistent with the ISEE 3 observations.[18] It is also found that plasma flows in the tail lobe are dominantly field-aligned and super-Alfvén high-speed flows occur at $x \simeq -100$ R$_e$, which are consistent with recent ISEE observations.[19]

Plasmoids associated with substorms have been identified on the basis of magnetic field signatures observed by ISEE 3 satellite at $x \simeq -200$ R$_e$.[18-20] About 20–50 minutes after the onset of substorms, large northward magnetic fields followed by large southward fields are often observed. Figure 7 shows the z component of the magnetic field, B_z, as a function of time at several points in the equatorial plane from the simulation data. The north-south magnetic field signatures associated with the onset of the first three substorms can be clearly seen in the figure. It is found that at $x = -200$ R$_e$ the north-south magnetic field signatures will appear approximately at a time 250 t_A ($\simeq 40$ min) after the onset of substorms, in agreement with ISEE observations. It is interesting to note that the north-south magnetic field signature associated with plasmoids in the distant tail is similar to the B_N signature of FTEs as shown in Fig. 5.

The repeated occurrence of plasmoids under a constant driving force is confirmed by the observations of two large storms, namely, (a) the storm of December 31, 1969 to January 1, 1970 (Akasofu[21]) and (b) the storm of October 30, 1978 (Lee et al.[13]). These two storms were characterized by an almost constant and high ϵ value (or the magnitude of the southward component of the IMF) for almost 24 hours. It was found that the AE index had impulsive changes superposed on a highly enhanced level. Auroral substorms were identified during the impulsive changes of the AE index. The recurrence time for the auroral substorms was observed to be 3–5 hours. We suggest that some of such impulsive changes are associated with the repeated occurrence of plasmoids, as our simulation study suggests.

SUMMARY

From the above study, it appears that when the system length is long compared to its width, magnetic reconnections tend to take place at multiple sites, which leads to the time-dependent multiple *X*-line reconnection. Magnetic islands (plasmoids or flux tubes) are formed intermittently and repeatedly under driving force.

At the dayside magnetopause, the multiple *X*-line reconnection process will lead to the occurrence of flux transfer events, which are observed with spacecraft ev-

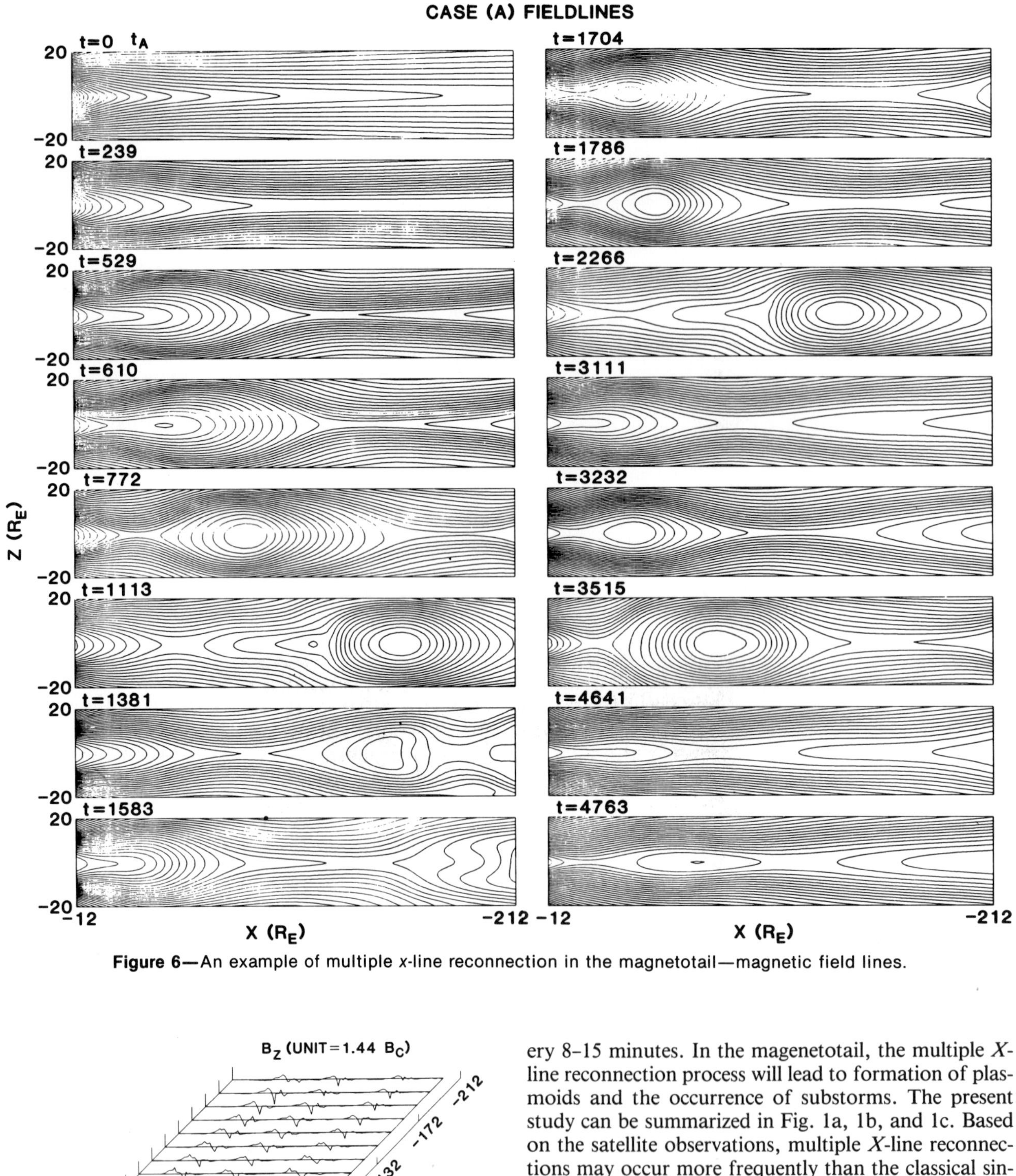

CASE (A) FIELDLINES

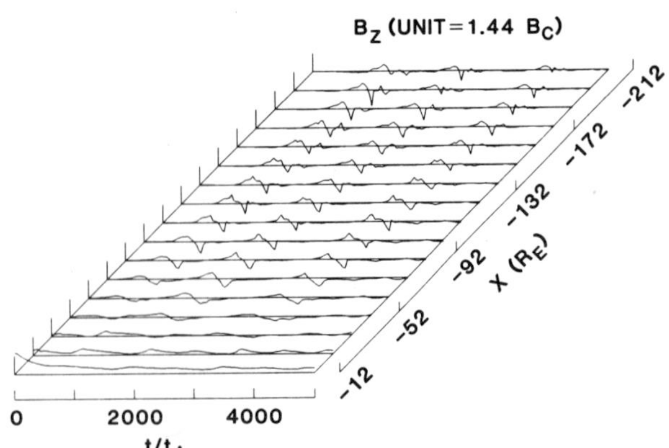

Figure 6—An example of multiple *x*-line reconnection in the magnetotail—magnetic field lines.

Figure 7—The *z* component of magnetic field B_z as a function of time *t*, observed at different positions in the equatorial plane for the case shown in Fig. 6.

ery 8–15 minutes. In the magenetotail, the multiple *X*-line reconnection process will lead to formation of plasmoids and the occurrence of substorms. The present study can be summarized in Fig. 1a, 1b, and 1c. Based on the satellite observations, multiple *X*-line reconnections may occur more frequently than the classical single *x*-line reconnections.

ACKNOWLEDGMENT—This work is supported by Toray Science Foundation of Japan, NSF grant ATM 85-51115 and DOE grant DE-AT06-76ER0005 at the University of Alaska.

REFERENCES

[1] J. W. Dungey, "Interplanetary Field and the Auroral Zones," *Phys. Rev. Lett.* **6**, 47 (1961).

[2] R. H. Levy, H. E. Petschek, and G. L. Siscoe, "Aerodynamic Aspects of the Magnetospheric Flow," *AIAA J.* **2**, 206 (1964).

[3] C. T. Russell and R. C. Elphic, "ISEE Observations of Flux Transfer Events at the Dayside Magnetopause," *Geophys. Res. Lett.* **6**, 33 (1979).

[4] L. C. Lee and Z. F. Fu, "A Theory of Magnetic Flux Transfer at the Earth's Magnetopause," *Geophys. Res. Lett.* **12**, 105 (1985).

[5] H. E. Petschek, "Magnetic Field Annihilation," *NASA Spec. Pub.* SP-50, 425 (1964).

[6] B. U. O. Sonnerup, "Magnetic Field Reconnection in a Highly Conducting Incompressible Fluid," *J. Plasma Phys.* **4**, 161 (1970).

[7] T. Yeh and W. I. Axford, "On the Reconnection of Magnetic Field Lines in Conducting Fluids," *J. Plasma Phys.* **4**, 207 (1970).

[8] L. C. Lee and Z. F. Fu, "Multiple X-line Reconnection: 1. A Criterion for the Transition from a Single X-line to a Multiple X-line Reconnection," *J. Geophys. Res.* (in press, 1986).

[9] E. W. Hones, "Plasma Flow in the Magnetotail and Its Implications for Substorm Theories," in *Dynamics of the Magnetosphere*, S.-I. Akasofu, ed., D. Reidel, Hingham, Mass., p. 545 (1979).

[10] J. Birn, "Computer Studies of the Dynamic Evolution of the Geomagnetic Tail," *J. Geophys. Res.* **85**, 1214 (1980).

[11] J. Lyon, S. H. Brecht, J. D. Huba, J. A. Fedder, and P. J. Palmadesso, "Computer Simulation of a Geomagnetic Substorm," *Phys. Rev. Lett.* **46**, 1038 (1981).

[12] S. H. Brecht, J. G. Lyon, J. A. Fedder, and K. Hain, "A Time Dependent Three-Dimensional Simulation of the Earth's Magnetosphere: Reconnection Events," *J. Geophys. Res.* **87**, 6098 (1982).

[13] L. C. Lee, Z. F. Fu, and S.-I. Akasofu, "A Simulation of Forced Reconnection Processes and Magnetospheric Substorms and Storms," *J. Geophys. Res.* **90**, 10896 (1985).

[14] D. Biskamp, "Effect of Secondary Tearing Instability on the Coalescence of Magnetic Islands," *Phys. Lett.* **87A**, 357 (1982).

[15] J. N. Leboeuf, T. Tajima, and J. M. Dawson, "Dynamic Magnetic X Points," *Phys. Fluids* **25**, 784 (1982).

[16] H. P. Furth, J. Killeen, and M. N. Rosenbluth, "Finite Resistivity Instabilities of a Sheet Pinch," *Phys. Fluids* **6**, 459 (1963).

[17] M. A. Saunders, C. T. Russell, and N. Sckopke, "Flux Transfer Events: Scale Size and Interior Structure," *Geophys, Res. Lett.* **11**, 131 (1984).

[18] E. W. Hones, D. N. Baker, S. J. Bame, W. C. Feldman, J. T. Gosling, D. J. McComas, R. D. Zwickl, J. A. Slavin, E. J. Smith, and B. T. Tsurutani, "Structure of the Magnetotail at 200 R_e and Its Response to Geomagnetic Activity," *Geophys. Res. Lett.* **11**, 5 (1984).

[19] J. A. Slavin, E. J. Smith, D. G. Sibeck, D. N. Baker, R. D. Zwickl, and S.-I. Akasofu, "An ISEE 3 Study of Average and Substorm Conditions in the Distant Magnetotail," *J. Geophys. Res.* **90**, 10875 (1985).

[20] B. T. Tsurutani, D. E. Jones, J. A. Slavin, D. G. Sibeck, and E. J. Smith, "Plasma-Sheet Magnetic Fields in the Distant Tail," *Geophys. Res. Lett.* **11**, 1062 (1984).

[21] S.-I. Akasofu, "A Magnetospheric Storm with a Nearly Constant Input Rate for about 24 Hours," *Planet. Space Sci.* **33**, 81 (1985).

[22] Z. F. Fu and L. C. Lee, "Simulation of Multiple X-line Reconnection at the Dayside Magnetopause," *Geophys. Res. Lett.* **12**, 291 (1985).

AURORAL ARCS AND TEARING-MODE INSTABILITIES IN THE MAGNETOTAIL

D. W. Swift*

It is argued that discrete auroral arcs are caused by electric potentials as a result of tearing-mode instability in the magnetotail current sheet that occurs in the presence of a finite normal component of the magnetic field. It is suggested that the presence of a conducting ionosphere on magnetic field lines passing through the current sheet may have a destabilizing effect.

INTRODUCTION

Auroral substorms are manifested by a brightening of auroral arcs and expansion of the width of the auroral oval.[1] Substorms are also associated with significant morphological changes in the magnetotail[2,3] which have been attributed to tearing-mode instabilities and reconnection processes.[4]

One of the most obvious features of the auroral display is the banded structure of auroral arcs. It has been known for a long time that auroral arcs are caused by precipitating electrons. More recently there have been observations of strong electric fields at altitudes above 4000 km in association with the aurora.[5,6] The fact that such strong fields are not seen at ionospheric heights led Swift et al.[7] and Mozer et al.[5] to conclude that electrons responsible for discrete arcs must have been accelerated by field-aligned potential differences existing between altitudes above 4000 km and the ionosphere. These field-aligned potentials can be related to potentials applied across magnetic field lines above the ionosphere. Since the ionosphere will draw field-aligned currents, current-limiting mechanisms described by Kan and Lee,[8] Lyons,[9] Chiu et al.,[10] and Swift[11] can lead to the maintenance of field-aligned potential differences.

The purpose of this paper is to propose and provide supporting evidence for a mechanism for the generation of potentials across magnetic field lines by tearing-mode instabilities in a way that is consistent with auroral arc structure. Further, it will be suggested that the ionosphere at the base of the magnetic field lines connecting into the magnetotail current sheet may have a destabilizing effect on tearing-mode instabilities.

GENERATION OF ELECTROSTATIC POTENTIALS BY TEARING MODES

A current sheet, such as exists in the earth's magnetotail, may be envisioned as being composed of a large number of small current filaments. The current filaments mutually attract each other. The tearing mode is therefore manifested by the bunching of the plasma-sheet current filaments into a series of O-type neutral lines. These O-type neutral lines, which carry a more concentrated current, also have a strong mutual attraction, so they tend to coalesce with a substantial release of magnetic energy. This coalescence process represents a transfer of energy from shorter to longer wavelength modes. During this nonlinear phase the growth rate of the long wavelength modes becomes faster than exponential; this is sometimes referred to as the explosive phase.

Swift[12] conducted a series of numerical simulations using a magnetoinductive particle code that included the electrostatic interaction between ions and electrons. The results showed the bunching of the current sheet into a series of O-type neutral lines, as predicted by linear theory. The O-type lines coalesced into a single large O-type line. The diagnostics indicated that most of the electrons were effectively entrained inside the O-type lines while many of the ions, because of the higher pressure and larger gyroradius, were not. Hence the O-lines became negatively charged, and potential differences on the order of the plasma-sheet-ion kinetic energy were generated. Similar results were also obtained by Leboeuf et al.[13] in a simulation of driven reconnection. The results of these simulations mean that tearing-mode instabilities are capable of generating potential differences across magnetic field lines on the order of several kV.

These simulations had no equilibrium magnetic field normal to the current sheet. As a result, the potentials generated could never be communicated to the ionosphere to drive field-aligned currents. However, Swift[14] carried out a simulation in a single-component plasma with no electrostatic coupling to electrons, in which the normal z-component was 1/5 the magnitude of the asymptotic x-component. The regions outside the current sheet were assumed to be populated by a tenuous, cold component that was convected into the current sheet by an assumed dawn-to-dusk convection electric field. Electron dynamics were not included in the model. The x-component of the Maxwell stress was approximately balanced by the momentum transferred to particles being convected into the neutral sheet. Figures 1 and 2, reproduced from Ref. 14, show results of the simulation. The important features are the magnetic fluctuations at $t = 25$ and the transfer of energy to longer wavelengths, typical of tearing-mode behavior, at $t = 55$. There are even small regions of field reversal in which B_z becomes negative. The point illustrated here is that tearing-mode instabilities can occur in a finite B-normal, if the electrostatic interaction among particles is neglected.

---

*Geophysical Institute, University of Alaska-Fairbanks, Fairbanks, Alaska 99775-0800.

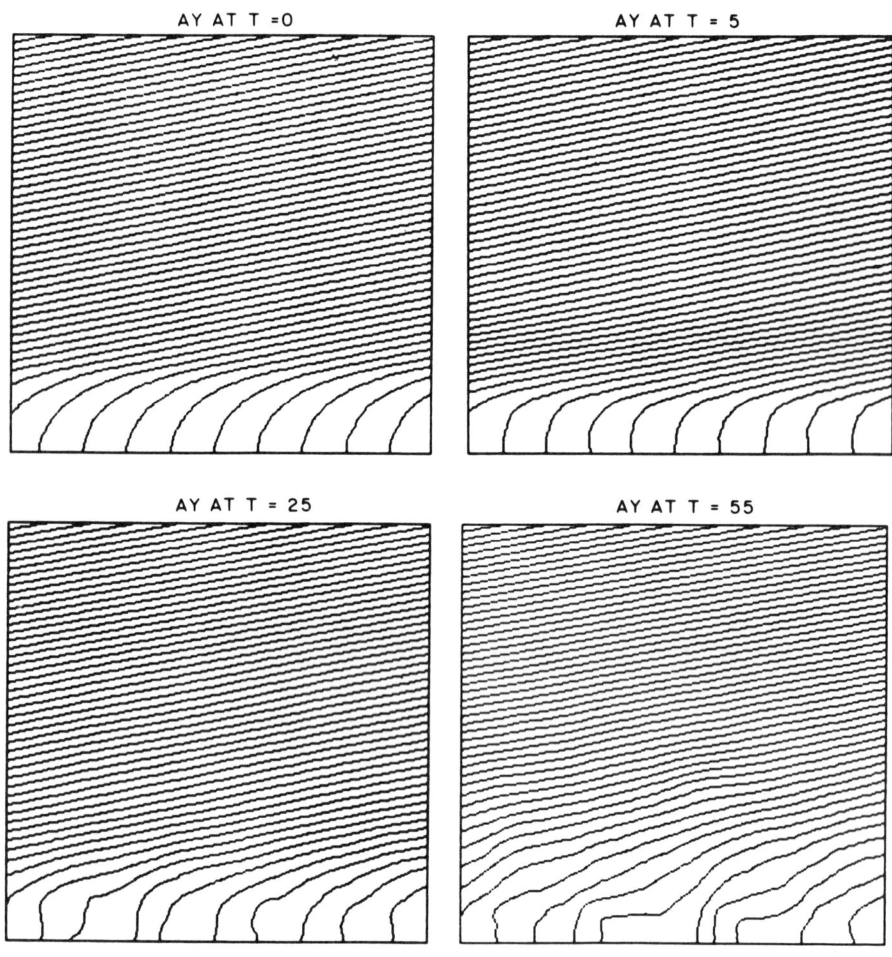

Figure 1—Contours of constant magnetic vector potential during a simulation of a single-component plasma with finite B-normal. This is a single-component model. Earthward is to the right (from Ref. 14). Time units are inverse gyrofrequency of ions in the asymptotic field.

However, Lembege and Pellat[15] point out that when B-normal becomes strong enough to magnetize the electrons, the tearing mode is inhibited. These results were confirmed when the simulation of Swift[14] was repeated with the inclusion of electron dynamics and electrostatic coupling between ions and electrons. The simulation (Swift, unpublished) indicated no tearing-mode growth.

The reason that electrons inhibit tearing-mode growth is that as magnetic field lines bunch and become locally compressed, the highly magnetized electrons must similarly be compressed. Since the ions are electrostatically bound to the electrons, they must also be compressed. However, the energy cost of compressing the ions is higher than the magnetic energy released by the instability, so the instability is inhibited.

EFFECT OF THE IONOSPHERE

If the ions were not electrostatically coupled to the electrons, then provided the electron β is small enough, the ions could drive the tearing mode. The presence of a B-normal allows a connection of the current sheet to the ionosphere, as depicted in Fig. 3. The importance of a connection between the ionosphere and the current sheet was first pointed out by Goldstein and Schin-dler,[16] who were interested in the role of tearing-mode instabilities in driving Birkeland currents. The ionosphere is also a source of cold electrons that are capable of depolarizing the electrostatic fields generated by the tearing mode, thus allowing decoupling of the ions from electrons and allowing the mode to grow. Since the electron β in the current sheet is not small, the work compressing the trapped electrons could inhibit the growth of the tearing mode. However, a small compression of the field lines could induce an anisotropy of the electrons, that would drive them whistler-mode unstable. This would allow the escape of the hot electrons and significantly mitigate the stabilizing effect of finite electron pressure. It is therefore of interest to examine the destabilizing effect of the ionosphere when the effect of the finite electron pressure is neglected.

This was done by Swift,[17] who did an analytical study of the effect of ionospheric conductivity on tearing-mode growth rates. The study was based on a model in which the pressure of the magnetospheric x-component of the magnetic field on either side of the neutral sheet was balanced by the ion pressure. A normal component of the field was assumed, which left the hot ions in the current sheet unmagnetized and the electrons strongly magnetized. With the neglect of electron

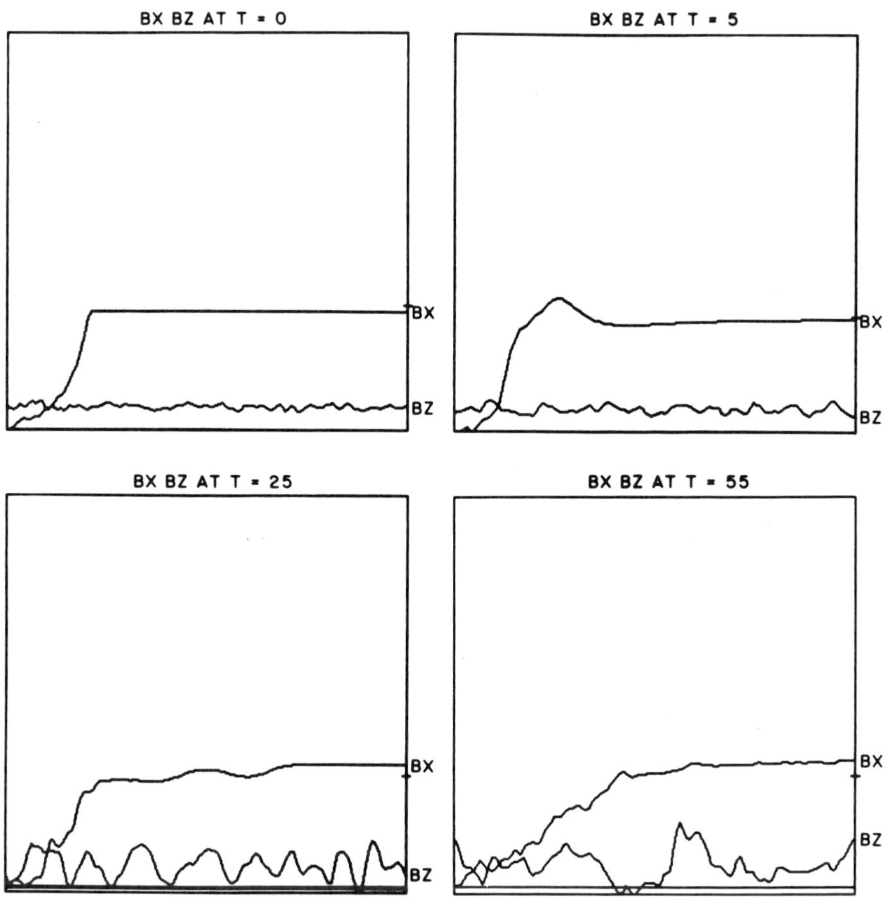

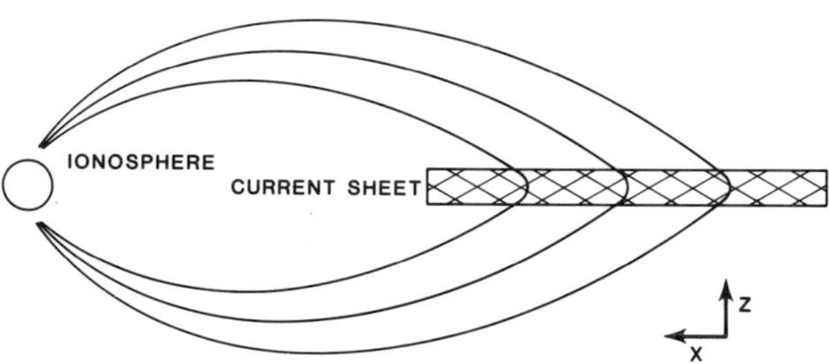

Figure 2—Plots of the *x*- and *z*-components of the magnetic field at times corresponding to the contour plates shown in Fig. 1. The *x*-component is shown as a function of *z*. The *z*-component is shown as a function of *x* along the neutral plane.

pressure, the electrons simply responded to the $E \times B$ drift. The lobe region of the magnetotail was assumed to be populated by a more tenuous cold plasma in which both the ions and electrons were assumed to be strongly magnetized.

The problem was solved by calculating the perturbed ion distribution function in which the ions are affected by both the induction and electrostatic electric fields, as well as the magnetic field. The perturbed ion distribution function was used to calculate the perturbed ion density and current density. The current density was used to calculate a perturbed vector potential. The electron density was calculated from the continuity equation in which the motion perpendicular to the magnetic field was the $E \times B$ drift in the induction electric field. The elec-

Figure 3—A sketch of the current-sheet ionosphere system and connecting field lines.

tron density was also determined by the field-aligned drift due to currents connecting to the ionosphere. These field-aligned currents were determined by the ionospheric conductivity and the potential across the field lines, which were assumed to be equipotentials. The potential field was determined by the quasineutrality condition, namely that the electron and ion densities be the same. This gave rise to a dispersion relation which determined the growth rate.

Figure 4a and b shows plots of the computed tearing-mode growth rates as a function of the ratio of the wavelength to the half thickness of the current sheet. For the H^+ model a normal field of 2γ was assumed, while a 5γ normal field was assumed in the O^+ model. The asymptotic tangential field was assumed to be 25γ. The assumed current-sheet half widths were 1720 and 2580 km, respectively, for the H^+ and O^+ models. The particle density in the current sheet was assumed to be 0.78 cm^{-3}, and the density in the lobe was assumed to be 0.078 cm^{-3} for both models. The difference in shape of the curves between Fig. 4a and b is related to the fact that in the H^+ model, the real and imaginary parts of the growth rate are about equal, whereas the growth rate in the O^+ model is mostly real. This, in turn, is related to the change in phase of the potential as a function of z in going from the middle to the edge of the current sheet. This phase change is large in the H^+ model because of the smaller value of the normal field assumed.

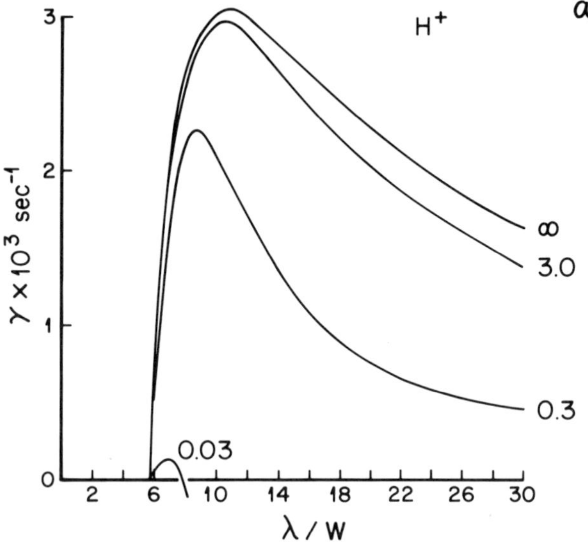

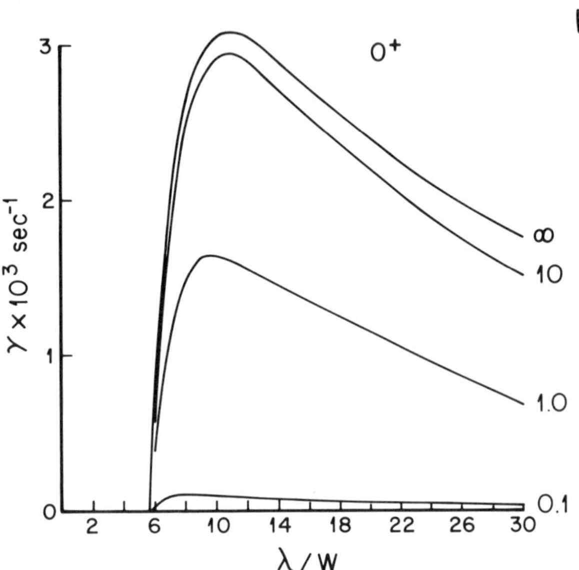

Figure 4—(a) Real part of the growth rate as a function of wavelength for various assumed ionospheric conductivities assuming an H + plasma. Wavelengths are expressed in half thicknesses of the current sheet and height-integrated ionospheric conductivities in mhos. (b) Same as (a) except for an O + plasma.

Typical auroral ionospheric height-integrated conductivities are in the 1–10 mho range.[18] From Fig. 4a and b it can be seen that the depolarizing effects of the ionosphere are sufficient to allow growth of the tearing mode as if the current sheet behaved as a one-component plasma. However, the maximum *e*-folding times are comparable to the Alfvén transit time, so actual growth rates would likely be dependent on the distance between the ionosphere and the magnetospheric current sheet.

DISCUSSION

The basic thesis put forth here is that tearing-mode instabilities can generate potentials across magnetic field lines comparable to the plasma-sheet ion temperature, but a magnetic field component normal to the current sheet is needed to communicate the potential difference to the ionosphere in order that the generated potentials may be effective in accelerating auroral electrons. A component of the magnetic field normal to the current sheet strong enough to magnetize electrons will inhibit the growth of the tearing mode. However, a conducting ionosphere on field lines passing through a current sheet may enable the tearing mode to grow. The major question yet to be answered is whether pitch angle diffusion processes would be sufficient to overcome the stabilizing effect of the finite electron β in the current sheet.

The model presented here further suggests that reconnection would be favored on field lines that connect to the ionosphere, and that reconnection would be manifested by the appearance of auroral arcs or field lines linking the ionosphere to the reconnection region. This is consistent with the observations that reconnection is initiated in the near-earth plasma sheet[3] where there is a relatively short field-line path to the ionosphere.

ACKNOWLEDGMENT — This work was supported by grant ATM 83-17459 through the Atmospheric Sciences Section of the National Science Foundation.

REFERENCES

[1] J. D. Craven, L. A. Frank, and T. E. Eastman, "Latitudinal Motions of Aurora in the Midnight Section During Substorms" (this conference, 1985).

[2] S.-I. Akasofu, *Physics of Magnetospheric Substorms*, D. Reidel, Dordrecht-Holland (1977).

[3] E. W. Hones, Jr., "Substorm Processes in the Magnetotail: Comments in 'On Hot Tenuous Plasma Fireballs, and Boundary Layers in the Earth's Magnetotail' by L. A. Frank, K. L. Ackerson, and R. P. Lepping," *J. Geophys. Res.* **82**, 5633 (1977).

[4] B. Coppi, G. Loval, and R. Pellat, "Dynamics of the Geomagnetic Tail," *Phys. Rev. Lett.* **1207** (1966).

[5] F. S. Mozer, C. A. Cattell, M. K. Hudson, R. L. Lysak, M. Termerin, and R. B. Torbert, "Satellite Measurements and Theories of Low Altitude Auroral Particle Acceleration," *Space Sci. Rev.* **27**, 155 (1980).

[6] E. M. Wescott, H. C. Stenbaek-Nielsen, T. J. Hallinan, T. N. Davis, and H. M. Peek, "The Skylab Barium Plasma Injection Experiments, 2 Evidence for a Double Layer," *J. Geophys. Res.* **81**, 4495 (1976).

[7] D. W. Swift, H. C. Stenbaek-Nielsen, and T. J. Hallinan, "An Equipotential Model for Auroral Arcs," *J. Geophys. Res.* **81**, 3931 (1976).

[8] J. R. Kan and L. C. Lee, "Theory of Imperfect Magnetosphere-Ionosphere Coupling," *Geophys. Res. Lett.* **7**, 633 (1980).

[9] L. R. Lyons, "Discrete Auroras as a Direct Result of an Inferred High-Altitude Generating Potential Distribution," *J. Geophys. Res.* **86**, 1 (1981).

[10] Y. T. Chiu, A. L. Neuman, and J. M. Cornwall, "On the Structure and Mapping of Auroral Electrostatic Potentials," *J. Geophys. Res.* **86**, 10079 (1981).

[11] D. W. Swift, "Mechanisms for Auroral Precipitation: A Review," *Rev. Geophys. Space Phys.* **19**, 185 (1981).

[12] D. W. Swift, "Numerical Simulations of Tearing-Mode Instabilities," *J. Geophys. Res.* **91**, 219 (1986).

[13] J. N. Leboeuf, T. Tajima, and J. M. Dawson, "Dynamic Magnetic X-points," *Phys. Fluids*, **25**, 784 (1982).

[14] D. W. Swift, "A Two-Dimensional Simulation of the Interaction of the Plasma Sheet with Lobes of the Earth's Magnetic Field," *J. Geophys. Res.* **88**, 125 (1983).

[15] B. Lembege and R. Pellat, "Stability of a Thick Two-Dimensional Quasi-Neutral Sheet," *Phys. Fluids* **22**, 1995 (1982).

[16] H. Goldstein and K. Schindler, "On the Role of the Ionosphere in Substorms: Generation of Field-Aligned Currents," *J. Geophys. Res.* **83**, 2574 (1978).

[17] D. W. Swift, "The Effect of the Ionosphere on the Growth of Tearing Mode Instabilities," *J. Geophys. Res.* **91**, 4256 (1986).

[18] P. H. Reiff, "Models of Auroral-Zone Conductions," in *Magnetospheric Currents*, T. A. Potemra, ed., American Geophysical Union, Washington, D.C. (1984).

REMARKS ON THE MHD PROBLEM OF GENERIC MAGNETOSPHERES AND MAGNETOTAILS

D. Montgomery*

It is argued that a high-priority item for magnetosphere and magnetotail physicists is the understanding of the plasma physics problem of MHD flow around a magnetized sphere or cylinder in different Reynolds number regimes, and that much can be learned from experimental and computational hyrodynamic precedents.

The flow of an electrically conducting fluid around a magnetized sphere is a natural MHD generalization of the hydrodynamic problem of the flow of a neutral fluid around a sphere or cylinder. The earth's magnetosphere and magnetotail are only special cases of a wide class of possible magnetospheres that might exist for other values of the magnetofluid flow parameters. These include, as another special case, the case of zero magnetic field strength, which is exactly the hydrodynamic problem.

The (simpler) hydrodynamic problem remains unsolved, except for the relatively uninteresting case of low Reynolds numbers. However, a great deal of information exists from extensive experimental investigation, and what we do know is classified most usefully according to which phenomena characterize which Reynolds number regimes (e.g., Panton[1]). Much of our understanding of flow around more complicated obstacles derives from information gleaned from the sphere and the cylinder (e.g., Batchelor[2]). A parallel set of plasma physics investigations of flows around magnetic obstacles seems likely to yield knowledge of more general scientific interest than that obtained by continuing to focus on the minutiae of single-point measurements for the case of the earth.

Consider a rigid sphere or infinite cylinder centered at the origin (for the cylinder, the axis may be assumed to coincide with the y-axis), with radius $r = a$. A uniform-density incompressible magnetofluid moves with average speed U_0 in the x-direction, so that far from the obstacle, the fluid velocity is $v \to U_0 \hat{e}_x$. The magnetic field B at the surface of the sphere (cylinder) is that of a three-dimensional (two-dimensional) dipole, with maximum field strength B_0.

Measure all velocities in units of U_0, all magnetic fields in units of B_0, all lengths in units of $L \equiv 2a$, and all times in units of L/U_0. In these units, the relevant dimensionless magnetofluid equations are

$$\frac{\partial v}{\partial t} + v \cdot \nabla v = -\nabla p^* + A\, B \cdot \nabla B + \frac{1}{R}\, \nabla^2 v \qquad (1)$$

$$\nabla \cdot v = 0 \qquad (2)$$

$$\frac{\partial B}{\partial t} + v \cdot \nabla B = B \cdot \nabla v + \frac{1}{R_M}\, \nabla^2 B \qquad (3)$$

$$\nabla \cdot B = 0 \qquad (4)$$

In Eq. 1, p^* is the ratio of the pressure (mechanical plus magnetic) to the density, in dimensionless units, and is determined by taking the divergence of Eq. 1, using $\nabla \cdot \partial v/\partial t = 0$, and solving the resulting Poisson equation for p^*. There are three dimensionless numbers in Eqs. 1–4: the Alfvén number $A \equiv (B_0^2/4 \pi \rho_0)U_0^{-2}$, where ρ_0 is the mass density; the Reynolds number $R \equiv U_0 L/\nu$, where ν is the kinematic viscosity; and $R_M \equiv U_0 L/\eta$, where η is the magnetic diffusivity, $c^2/4\pi\sigma$ in cgs units (c = the speed of light, σ = the electrical conductivity). The boundary conditions are that $B \to 0$ and $v \to \hat{e}_x$ far from the sphere (cylinder), while $v \to 0$ at the surface of the sphere (cylinder) and $B \to$ its dipole form there.

The hydrodynamics problem is recovered by taking only Eqs. 1 and 2 and setting $A = 0$; there is, then, only one variable physical number left in the problem, the Reynolds number R. In the MHD problem there are three, strictly speaking: A, R, and R_M. If the temperature and density of the impinging magnetofluid are regarded as given, ν and η are thereby determined, so that picking two of three numbers A, R, R_M determines the third, and there are only two independent ones. If $A \gg 1$, the fluid will, in effect, be colliding with a magnetic barrier much larger in dimension than the sphere, and if $A \ll 1$, the magnetic barrier will not greatly alter the flow; A of order unity, or somewhat greater, may be the regime of greatest interest. For the earth's magnetosphere, R and R_M are $\gg 1$.

It is not difficult to make a list of important physical effects in the earth's magnetosphere and magnetotail that are omitted from the above description: the bow shock (because of the incompressibility); interaction with the atmosphere and ionosphere; effects of irregularities in the incoming solar-wind stream; and all Vlasov and single-particle effects. It is believed, however, that the problem posed is still very rich and is the simplest one that contains the dynamics of a true magnetosphere and magnetotail in a global sense. It is not unimportant that it contains the simpler, but nevertheless analytically intractable, hydrodynamics problem of flow around a

*Physics and Astronomy Department, Dartmouth College, Hanover, New Hampshire 03755.

sphere. Any analytical or numerical method that purports to be able to deal with the MHD problem faces the unforgiving test of the well-measured hydrodynamic situation as a special case (see Panton's discussion[1] and some pertinent remarks by Rostoker[3]).

The hydrodynamic problem is analytically tractable only for Stokes flow, $R < 1$. It can be dealt with numerically up to $R \sim$ a few hundred.[4] It is unstable above $R \sim$ a few hundred, and continues to develop and enter interesting new turbulent regimes as R is raised to tens of thousands or more. Significant features are asymmetrical vorticity shedding, separation and re-attachment of the flow, turbulent boundary layers, and a general lack of equivalence between the instantaneous streamlines of velocity and vorticity and their time averages.[1,2,5]

It is these last exotic features that one may expect also to characterize the MHD problem. Highly symmetric, laminar models of the magnetosphere and magnetotail (even ones that are perhaps subject to an occasional decorous interruption by an "instability" that soon cleans itself up) seem unlikely in view of the high Reynolds numbers involved. An extensive program of laboratory measurement and numerical computation seems desirable to supplement the wealth of existing spacecraft data. A brave beginning on the computational side of the problem has recently been reported[6,7] and will become even more interesting as viscosity is added and fewer restrictive assumptions of symmetry are imposed.

ACKNOWLEDGMENT—The work was supported in part by NASA grant NAG-W-7I0 and the U.S. Dept. of Energy grant DE-FG02-85ER53I94.

REFERENCES

[1] R. L. Panton, *Incompressible Flow*, Wiley-Interscience, Inc. Chaps. 15 and 23 (1984).

[2] G. K. Batchelor, *Introduction to Fluid Dynamics*, Cambridge University Press. Chap. 5 (1967).

[3] G. Rostoker, "Definition of a Substorm, Physical Processes in a Substorm, and Sources of Discomfort," in *Geophysical Monograph 30*, American Geophysical Union, Washington, D.C., pp. 380-384 (1984).

[4] B. Fornberg, "Steady Viscous Flow Past a Circular Cylinder up to Reynolds Number 600," *J. Comput. Phys.* **61**, 297-320 (1985).

[5] M. Van Dyke, *An Album of Fluid Motion*, Parabolic Press (1982) (contains valuable flow visualizations).

[6] T. Ogino, R. L. Richard, R. J. Walker, and M. Ashour-Abdalla, "Structure and Dynamics of the Plasmoids in the Distant Magnetotail," *Bull. Am. Phys. Soc.*, Ser. II, **30**, 1467 (paper 4RI3) (1985).

[7] R. L. Walker, T. Ogino, and M. Ashour-Abdalla, "An MHD Simulation Showing the Location of Magnetospheric Reconnection," *Bull. Am. Phys. Soc.*, Ser. II, **30**, 1469 (paper 4R22) (1985).

DISCUSSION

W. Gekelman: Comment: I completely agree with David Montgomery's call for some laboratory experimental work. The paper I am giving later in the week is the only laboratory effort presented at this meeting. Although it is probably not possible to scale completely the entire magnetosphere in our lab, very meaningful experiments can be conducted. For example in our experiment we can measure the magnetic field at several thousand spatial locations and several thousand time intervals during a reconnecting event in the simulation of a neutral sheet. Jetting has been observed, flow velocity maps have been generated, and distribution functions measured. Experiments can be reconfigured and done (rather cheaply) in a matter of weeks (as opposed to satellite experiments, which take years). Aside from observing phenomena familiar to most people in this room, some unexpected things that should be of interest to this community have been seen as well. In my opinion a stronger interaction with this community and laboratory efforts is highly desirable and would be fruitful.

S. Peter Gary: Your pessimistically large Reynolds numbers are based on collision-dominated transport coefficients. In a collisionless plasma, collective effects will probably dominate transport, providing a possibly increased resistivity and therefore smaller Reynolds numbers. If this is true, a turbulent magnetotail may not be as inevitable as your talk has suggested.

D. Montgomery: This is quite possible. It would be a matter of putting down some specific numbers for the anomalous resistivity and viscosity, and recalculating the Reynolds numbers with the appropriate velocities and length scales. But it would have to pull the Reynolds numbers down quite a bit, since values of a few thousand are probably already adequate for turbulence to set in, assuming the MHD model has any validity at all.

T. Sato: I'd like to point out one important thing that is missing in the kinetic treatment of the collisionless tearing, as applied to the magnetospheric substorms, but may connect the microscopic treatment to the macroscopic treatment. The point is that the tail configuration is not steady but changing with time. The growth time of the ion collisionless tearing instability is of the order of minutes, which is not ignorable time in the configurational change. The point I wish to emphasize is the fact that the kinetic analysis starts with an arbitrary Harris type equilibrium, which includes a free parameter of the sheet width. The question is then how such an unstable equilibrium is established. For the analysis to be meaningful, the equilibrium must be established with a time scale much shorter than the growth time. At this point the MHD treatment becomes meaningful. The driven reconnection process requires a continuous thinning of the plasma sheet, that eventually leads to a condition where reconnection process does not require any particular triggering mechanism of reconnection.

IV. KINETIC ASPECTS OF MAGNETOTAIL DYNAMICS

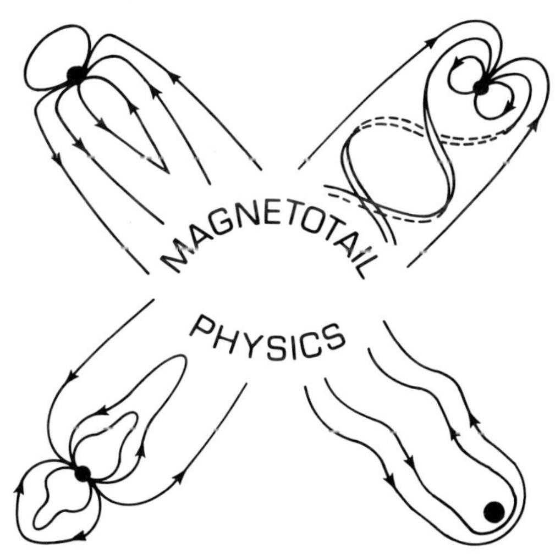

KINETIC ASPECTS OF MAGNETOTAIL DYNAMICS – OBSERVATIONS

D. G. Mitchell*

We discuss observed kinetic (as opposed to fluid) aspects of particle behavior in the geomagnetic tail. Such behavior has been observed from thermal plasma energies to energies greater than or approximately equal to a few MeV. The importance in recognizing and analyzing the kinetic behavior of tail particles is that it provides a powerful diagnostic tool for inferring the state of the local and remote environment, the mechanisms responsible for observed tail dynamics, and a way of determining when fluid analyses (relying upon moments such as bulk velocity and temperature) may be misleading. Thus kinetic analyses provide a valuable complement to fluid-dynamics-based studies of the collective behavior of geomagnetic tail particles. Observed kinetic effects can be analyzed to examine, for example: (a) spatial gradients; (b) convection electric fields; (c) distances to source regions; (d) source region characteristics; (e) magnetic field topology; (f) heating/acceleration mechanism (e.g., wave-particle interactions, beam-plasma instabilities); (g) boundary characteristics; and (h) remote mechanisms affecting local particle distributions.

By presenting examples of observed kinetic effects throughout the energy range observed in the tail, we will demonstrate the importance of recognizing and analyzing these effects in understanding the dynamics of the geomagnetic tail.

INTRODUCTION

We will discuss magnetotail dynamics in the context of the plasma and energetic particle observations with emphasis on two aspects: (a) the importance of particle kinetics to the physical processes that govern magnetotail dynamics, and (b) the importance of a kinetic approach in the understanding of observations made in the tail. These aspects are obviously related. Certainly if kinetic effects are important in a physical process, then it will be useful to employ a kinetic approach in interpreting the observations. On the other hand, even a process that can be modeled adequately using a fluid approach may be understood far more completely if it is analyzed using a kinetic treatment. By the "kinetic approach" we mean an approach in which the plasma and its transport properties are describe on the basis of the configuration and velocity space distribution of the particles, the correlation between particles, and the microfields produced by the particles (see Krall and Trivelpiece, Ch. 7.)[1] Here, and throughout this paper, we will take the "fluid approach" to be synonymous with the "moment approach," by which in turn we mean the use of the first several (≤ 4) moments of the particle distribution to describe the plasma characteristics.

It is always possible to calculate the moments of a physical distribution such as the density, velocity, pressure, and energy flux. However, the crucial questions are (a) whether the transport of mass, energy, and momentum can be functionally related to those moments, as is assumed in the fluid approach, and (b) if that assumption is correct, whether the fluid approach alone is sufficient to determine the proper choices of constraints defining the system. We will show that in many cases, the moments alone are grossly inadequate to analyze magnetotail observations. Of course, for a system where a simple drifting Maxwellian distribution prevails,

these moments are often fully descriptive and a kinetic approach is unnecessary. But under what conditions does a Maxwellian arise? It generally arises in a system in which the particle mean-free-path is short compared with the size of the system. In the noncollisional magnetized plasmas in the magnetotail, magnetic confinement constrains the motion of particles perpendicular to the magnetic field. This provides an effective collision process in the sense that the plasma acts (more or less) in a collective manner in the direction perpendicular to B. This process does not directly transfer momentum between particles, so it is not entirely analogous to collisions in a gas. However, it is the "localization" of the action of individual particles that often allows for the formulation of a moments approach. Such a localization makes it more likely, but still not certain, that the transport processes can be related to the moments and their derivatives. Parallel to B, the constraints on the particle motion can violate the short mean-free-path or localization conditions considerably. Although quasineutrality must be maintained parallel to B (except under unusual circumstances), and wave-particle interactions may provide effective collisions and momentum transfer, there are many regions in the magnetotail in which free streaming of particles parallel to B is common, and a fluid approach will give an incomplete, potentially misleading picture of the physical situation.

If the particles that make up a distribution measured at a point in space and at an instant in time are streaming parallel to B, then the moments of the distribution generally will not be useful in telling where the particles came from or where they are going. The moments will still be useful in instantaneously determining quantities such as the local current or the plasma β. But without a kinetic treatment the history and the predictive information in the distribution will be lost, as will details of the shape of the distribution, which can significantly modify both local and remote physical processes that are contributing to the overall dynamics of the system.

*The Johns Hopkins University Applied Physics Laboratory, Laurel, Maryland 20707.

In the remainder of this paper, we will illustrate a variety of conditions in the magnetotail for which a kinetic approach to the observations is recommended. We will discuss these conditions first through a simple treatment of charged particle dynamics in relation to the measurements made by spacecraft, and second through a number of examples of plasma and energetic particle observations from the literature.

TUTORIAL FRAMEWORK

For complete knowledge of the history, present state, and future of a system, one would need to follow in detail the trajectory of every particle in the system, i.e., to know the mass, charge, and energy of each particle, as well as know all the forces operating throughout the system. This is not a practical approach. However, statistical mechanics provides a framework in which we can describe the ensemble of particles by a distribution function, which includes a statistical description of all of the quantities listed above, except for the externally applied forces operating throughout the system. In principle, the kinetic approach utilizes the full distribution function and includes all external forces (magnetic field, B, electric field, E, and gravitation when necesary) as well as statistical descriptions of the effects of internal forces that may lead to modification of the distribution function through interactions between particles, either direct (e.g., Coulomb collisions) or indirect (e.g., polarization potentials, wave-particle interactions). In many cases, particularly in the high energy tails of distributions, a single particle (or test particle) kinetic approach may be used.

In Fig. 1, we illustrate with simple idealized distribution functions a variety of common situations encountered in space plasmas. We will show that even in these very simple cases, kinetic considerations are required to fully understand the physics of the situation. We will also point out the signatures of these situations in the measurements commonly obtained on spacecraft. Figure 1a shows a one-dimensional cut through a Maxwellian, or thermal distribution function, under the influence of perpendicular E and B fields. The well known $E \times B$ drift velocity results in the displacement of the peak of the distribution from $V_\perp = 0$. Using the fluid approach, one can easily calculate the moments of the distribution to find the density (n), drift velocity (V_0), and pressure (P), from the zeroth, first, and second moments of $f(v)$. For the case of a drifting Maxwellian, these completely describe the plasma (or at least one species in a plasma) and no more detailed approach is necessary. The primary signature of this situation in a spacecraft measurement would be an anisotropy in the flux at a given energy in the direction of $E \times B$, which becomes smaller toward high energy. This signature will be discussed more fully in the next section.

Figure 1b shows a Maxwellian modified by a smooth spatial gradient in the plasma pressure perpendicular to B ($\nabla P_\perp$) with $E = 0$. In this case, the fluid approach can be misleading. At high velocities, the distribution function is not symmetrical about zero velocity, while

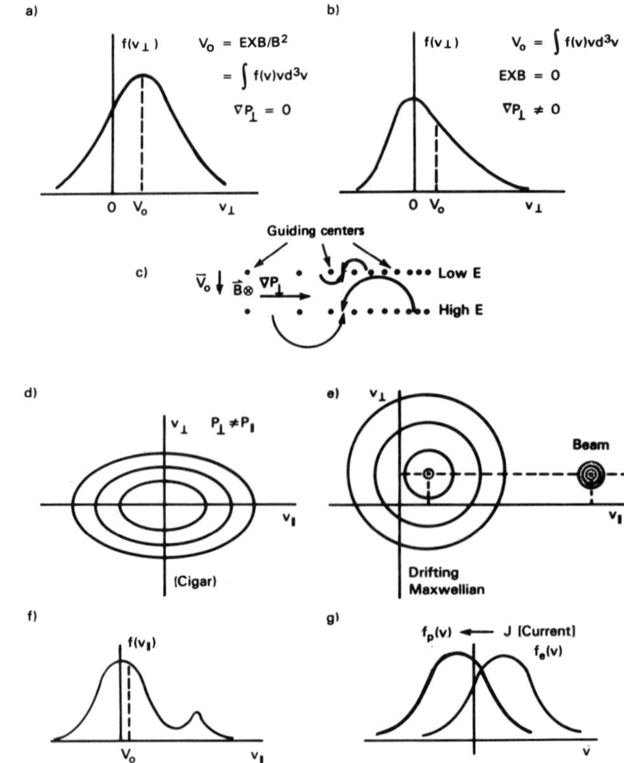

Figure 1—A series of schematics to illustrate a variety of conditions common in the magnetotail plasma. (a) A drifting Maxwellian in perpendicular magnetic and electric fields. (b) A skewed distribution, representative of a pressure gradient. (c) Schematic to demonstrate the source of a pressure gradient anisotropy. The density of guiding centers is represented by the spacing of points (increasing density, or pressure, to the right). Notice that larger gyroradius particles detected at one location have guiding centers displaced parallel to the gradient a greater distance than smaller gyroradius particles. Therefore, the flux anisotropy at the detector is higher for higher energy particles, as in (b). (d) Cigar distribution typical of closed, elongated field lines. Other typical distributions on closed field lines are isotropic (circles) and pancake ($P_\perp > P_\parallel$). (e) Two-component distribution — a low-energy, warm plasma and a high-energy cold beam. (f) Cut along the $V_\parallel$ axis through a two-component plasma comprised of a Maxwellian at rest and a high parallel velocity beam. (g) Schematic to indicate that a current is the result of the velocity space displacement of the ion distribution relative to the electron distribution in a plasma.

at low velocities it is nearly symmetrical. This skewness arises because the particle gyroradii increase with velocity, and so the gyro-centers of particles measured perpendicular to B and perpendicular to $\nabla P_\perp$ can be separated significantly in a direction parallel to $\nabla P_\perp$ (see illustration, Fig. 1c). This leads to an increasing anisotropy in the spacecraft measured flux or distribution function, $f(v)$, toward higher energies in the direction $B \times \nabla P_\perp$. In the illustration, $\nabla P_\perp$ is proportional to $\nabla n_\perp$, a gradient in the plasma density perpendicular to B. Clearly, this distribution also has a nonzero first moment, with the dimensions of velocity. While this first moment does relate directly to the current at the measurement point, it does not have anything to do with an electric field ($E = 0$), and in the fluid approach no distinction between the cases in Fig. 1a and 1b may be apparent in the first moment. A finite third moment would quantify the skewness, but would not reveal the functional dependence of the skewness on v. In the extreme case of a sharp boundary (e.g., the magnetopause), the skewness would appear as a discontinuous break in $f(v)$ at some v. In magnetotail plasmas and fields, this ambiguity inherent in the fluid approach can become problematic at energies $\gtrsim 10$ keV $(Z^2 m_p/M)$ where m_p is a proton mass, and M and Z are the mass and charge number of the species being considered. This expression simply scales other species by gyroradius to a 10 keV proton. Most pressure gradients will have scales ranging from much greater than to the order of a thermal ion gyroradius, so gradient effects in the distribution function only become important above the typical thermal energy for tail plasmas of several keV. Notice that by this formula (and in practice), one rarely need be concerned about this source of ambiguity for magnetotail electrons; the fluid approach is usually unambiguous for electrons in the direction perpendicular to B.

In Fig. 1d we show in the $(v_\parallel, v_\perp)$ two-dimensional space a contour plot of a cigar distribution, one in which the pressure parallel to B ($P_\parallel$) is greater than that perpendicular to B ($P_\perp$). The reverse ($P_\perp > P_\parallel$) is called a pancake or trapping distribution. Both cigar and pancake distributions (as well as isotropic, $P_\parallel \cong P_\perp$) are characteristic of charged particles in closed field topologies in the tail, during quiet periods in the case of ions, and under most conditions in the case of electrons $\gtrsim 100$ eV in energy. This type of distribution can often be well described by moment calculations; however, details such as loss cones, butterfly distributions (which have maximum pressure at an intermediate pitch angle), and evolving distributions (e.g., cigar at low energies to pancake at high energies) require a kinetic treatment for interpretation.

Figure 1e shows (also using a contour plot of constant-phase space density) a two-part distribution function, comprised of a slowly drifting, warm plasma and a fast, cool beam traveling parallel to B relative to the warm plasma. Such a situation can arise due to spatial inhomogeneity in the magnetotail system (particularly in the plasma-sheet boundary layer), coupled with velocity dispersion. We will expand on this situation later. If we take a one-dimensional cut through the center of the distribution illustrated in Fig. 1e, we will see something like Fig. 1f. Again, a moment approach will yield a velocity V_0 that is really a poor description of the physical situation. It is a weighted average of two essentially independent populations, neither of which alone is characterized by that V_0. The kinetic approach, i.e., examination of the distribution function, yields much more physically useful information.

Finally, in Fig. 1g we illustrate an ion and an electron distribution, each with different first moments. Such a situation results in a net current flow, and can be well determined by applying the fluid approach to both the ions and the electrons. There are, however, situations in which currents can be more readily measured with a kinetic approach, examples of which will be shown in the next section.

Having presented a number of idealized cases in Fig. 1 (by no means complete—the possibilities are many), we will in the next section relate them to examples of published observations over energy ranges from tens of eV to hundreds of keV, that have been interpreted using a kinetic approach. Before leaving this section though, we would like to illustrate a few simple concepts to explain one frequently ignored reason that a kinetic approach is often necessary to interpret observations.

The primary difficulty with the fluid approach to particle observations in the magnetotail is that central to the formalism is the assumption that each particle in the ensemble of particles that make up the fluid or plasma is in close communication with the entire ensemble. If this communication is not present or is weak, then the moments of the distribution will not accurately reflect either the past or the future of the distribution nor even the past or future of the moments themselves. The result is the introduction of uncertainty in the physical interpretation of the causes and the consequences of the observations. For a more theoretical and in depth development of this statement, see Chew, Goldberger, and Low.[2]

Much of this difficulty stems from the large physical size of the magnetotail system, coupled with the large mean-free-path of the plasma particles and the broad range of velocities of the particles that make up a distribution. A further complication is the finite temporal resolution of spacecraft instrumentation.

We will discuss these points in the context of the schematic presented in Fig. 2. In this figure, we show a cartoon of the magnetotail region, locating a spacecraft at about $X_{SM} = -22$ R$_e$, $Z_{SM} = +1$ R$_e$ in the midnight meridian. We then ask ourselves two (related) questions: (a) Given an instantaneous measurement of the proton distribution at the spacecraft, where were those protons located 60 seconds earlier? (b) If an instrument takes 60 seconds to make a complete characterization of the particle distribution, over what regions of the magnetotail system do the measured protons travel? The choice of a 1 minute time scale is taken to be representative (within an order of magnitude) of the typical resolution of present day detectors. It is equally representative of the minimum period over which energetically important phenomena in the tail evolve. Figure 2 then outlines the spatial regions in the tail that can be considered as possible sources for the protons measured at the indicated point during a 1 minute period, as a function of energy. We have included finite gyroradius effects, parallel streaming, global dawn-to-dusk electric field values of $0 - 0.2$ mV/m, and nonadiabatic motion in the current sheet. For a 10 nT magnetic field, these electric field values correspond to drift speeds of $0 - 20$ km/s, or $0 - 1200$ km displacement in 60 seconds.

The first point to notice is that not only do the protons cover a lot of space in a minute, but protons at different energies, all of which contribute to a distribu-

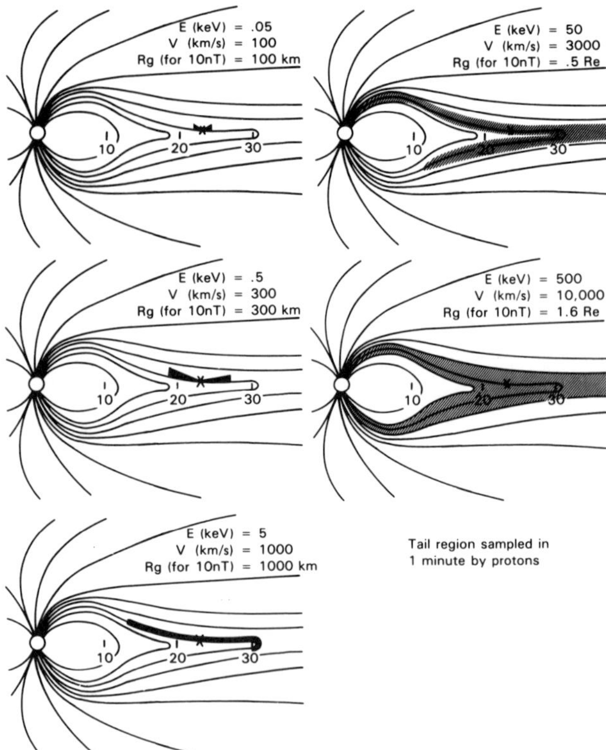

Figure 2—Schematic of the magnetotail (midnight meridional cut) to illustrate the broad range of spatial scales which are represented by a point measurement of a distribution of protons with energies ranging from 50 eV (a) to 500 keV (e). Drifts toward the central plasma sheet due to dawn-dusk electric fields of 0 – 0.2 mV/m are included. Each case assumes 1 minute of propagation time prior to detection. A distribution measured over a broad range of energies will include protons with radically different recent histories.

tion function that in the fluid approach is treated as a locally coherent ensemble, come from widely different spatial regions in the tail, differing both in location and characteristics.

Starting with the lowest energy (50 eV, the lower range of most plasma instruments), we see that during 1 minute these protons travel about 1 R_e parallel to B and between 0.0 and 0.5 R_e perpendicular to B, depending upon the value of E. Their velocity is ~ 100 km/s and their gyroradius is $1000/B(nT)$ km, which is usually negligible when compared with a typical drift distance of 1200 km. The 1 R_e traveled parallel to B may or may not be considered "local," depending on conditions. At 500 eV protons traveling at ~ 300 km/s cover 3 R_e in 1 minute, with a $3000/B(nT)$ km gyroradius. Again, with a 0.2 mV/m cross-tail electric field the $E \times B$ drift displacement in 1 minute (1200 km in a 10 nT B field) is much greater than the gyroradius, which for most cases is still negligible. The 3 R_e parallel travel may be significant depending on the scale of and distance to trajectory—and/or energy-modifying phenomena. Note that even at these lowest energies, two particles at 50 eV and 500 eV measured at a given location and time were physically in very different locations 1 minute earlier.

This difference becomes more and more profound as we consider higher energies, and it is increasingly likely that protons measured as part of a single distribution in fact have histories (and futures) having little or nothing to do with other protons at different velocities and pitch angles in the same distribution. Still, in some cases, the measurement of 500 eV protons may be considered a local measurement.

Going to 5.0 keV (still in the heart of the sensitivity of most plasma instruments, and a typical "thermal plasma" energy), protons have a velocity of 1000 km/s, a gyroradius of $10,000/B(nT)$ km, and in 1 minute travel 10 R_e parallel to B. Now the gyroradius is becoming significant (comparable to the $E \times B$ drift distance in 1 minute, and comparable to the scale of rather steep density gradients in the tail), and the distance traveled is becoming a sizable fraction of the scale of the system. In 1 minute, 5 keV protons at different pitch angles have traveled from source regions that are topologically quite different from each other, and the physical processes operating in those regions are likely to be different. For most purposes, this can no longer be considered a local measurement, although the effects of locally operating phenomena (e.g., $E \times B$ drifts, gradients perpendicular to B) may be well observed at these and higher energies.

At 50 keV (the upper range of a plasma instrument, the lower range of a solid-state detection system) protons travel 30 R_e parallel to B in 1 minute, exhibit nonadiabatic motion in the tail current sheet, and sample regions from the ionosphere, to the conjugate hemisphere, to the deep tail. Their gyroradius, $5/B(nT)$ R_e, is comparable to the scale of many gradients in pressure perpendicular to B and greater than the typical $E \times B$ drift distance over 1 minute.

Finally at 500 keV, in the middle range of solid-state detector-based magnetospheric instruments, the protons move at 10,000 km/s, have gyroradii of $16/B(nT)$ R_e, and travel 100 R_e in 1 minute, covering a large fraction of the plasma-sheet region.

From this discussion we can see that a distribution function of magnetotail protons covering a broad range in velocity will not in general be comprised of protons with common sources or destinations. This discussion has assumed single-particle motion in time-stationary electromagnetic fields. To the extent that there is significant communication via wave-particle interactions, the fluid approach may become a better approximation. In the next section, however, we will show numerous examples for which a kinetic approach is essential to interpret the data correctly.

Table 1 gives the relationships between various energy protons (and elecrons) discussed above in the context of Fig. 2. Notice that virtually all the electrons measured by a plasma instrument ($E_e \geq 25$ eV) in 1 minute cover a region that is large compared with the dimension of the tail. For electron energies $E_e \geq 10$ keV the electrons travel ≥ 100 R_e in 10 seconds, which makes them especially useful as tracers of field topology—if the field is not closed (and in the absence

Table 1—Magnetotail dynamics scales for protons (electrons)
(*Rg* calculated for 10 nT field).

E (keV)	0.05 (0.00003)	0.47 (0.00025)	5.2 (0.0028)	47 (0.25)	520 (0.28)
Rg (km)	104 (2.4)	316 (7.4)	1040 (24)	3160 (74)	10400 (240)
$X(R_e)/64s$	1	3	10	30	100
v(km/s)	100	300	1000	3000	10000

(Electrons 0.3 – 10 keV $\geq$ 10000 – 60000 km/s)

of strong scattering), these nearly relativistic electrons will show a unidirectional (streaming) anisotropy parallel to $\boldsymbol{B}$.

To summarize this section, we would like to make two points. First, with the exception of a very few, very simple cases, many details of the distribution function and therefore of the local physics require a kinetic approach to observations of magnetotail plasmas. Second, ambiguities regarding the history leading to a given observation can greatly compromise a fluid interpretation of plasma measurements (particularly at higher energies). The various particles that make up a velocity distribution at a given point in the magnetotail system may have participated in an array of unrelated processes (energization, scattering, drifts, etc.) in a diversity of magnetotail regions prior to their observation; therefore, any analysis (such as the fluid approach) that treats the observations as a homogeneous distribution may seriously distort the interpretation of the observations. The kinetic approach contains the means to differentiate among the various contributions to a given observation.

This is not to say the fluid approach is without merit. Certainly the moments of a distribution have physical meaning independent of the details of that distribution, and they are useful in many inquiries that do not require detailed knowledge of the processes leading up to a given observation.

EXAMPLES

We now turn our attention to a series of examples chosen from the literature to focus both on the importance of a kinetic approach to the study of a broad range of magnetotail dynamical situations and on the role of kinetics in magnetotail dynamics. A large number of these observations are from the plasma-sheet boundary layer (PSBL). This is not coincidental; the characteristics of the PSBL are such that a kinetic approach is generally the only appropriate choice in analysis of observations in this region. These characteristics include: limited extent perpendicular to $\boldsymbol{B}$, leading to structure perpendicular to $\boldsymbol{B}$ of the order of an ion gyroradius; plasmas of very different properties bounding it on either side (lobe on one side, plasma sheet on the other), leading to multiple possibilities for cross-field gradients and plasma

sources and sinks; $\boldsymbol{E} \times \boldsymbol{B}$ drifts due to the cross-tail electric field, leading to velocity-filter effects; and a topology that involves magnetic connection with the ionosphere at one end and with a variety of possible regions including the current sheet, the reconnection neutral line region, and the solar wind at the other end, all leading to a variety of sources of plasma in the PSBL.

Although our selected examples are dominated by the PSBL, there are examples from other tail regions (lobe, plasma sheet, low-latitude boundary layer, plasma mantle) that illustrate the ubiquitous nature of particle kinetics throughout the magnetotail system at all energies.

Our first example (Fig. 3) is a figure from Forbes et al.,[3] which shows four plasma instrument distribution functions from successive time periods, representing the evolution in the structures of the PSBL from near the lobe boundary (Fig. 3a) to the plasma sheet (Fig. 3d). The first panel (a) shows an ion beam traveling earthward, parallel to $\boldsymbol{B}$. With time (b and c), the velocity of the peak density of this beam decreases, and a second, higher velocity beam appears traveling tailward. Finally (d), the distribution becomes nearly isotropic, typical of the central plasma sheet. In Fig. 3b and 3c, a moment calculation would produce a first moment that would be a weighted average over the two peaks, and would seriously misrepresent the phenomenology in the distributions. It was only through a kinetic treatment that the authors correctly concluded they were observing a

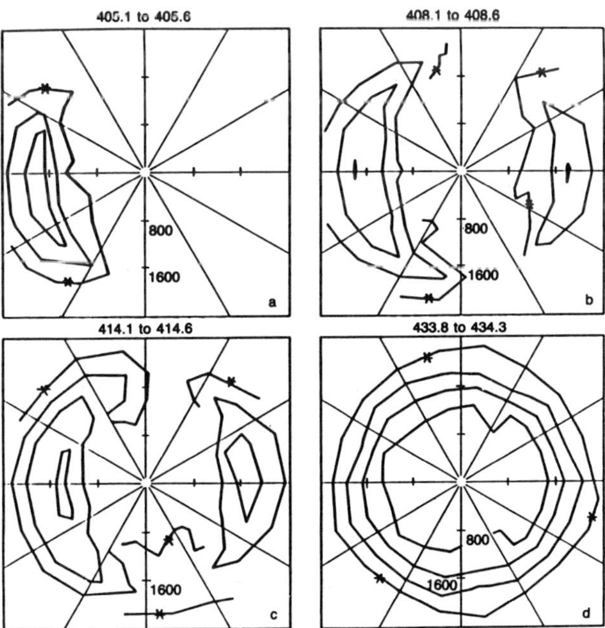

Figure 3—Contours of the $\log_{10}$ of the proton distribution function (f) obtained from a two-dimensional analyzer on ISEE 2 from 04:05 to 04:34 on March 1, 1978. The coordinates are solar ecliptic V_x and V_y. The sun and earth are to the left ($+V_x$ axis) and the dusk direction is at the bottom ($+V_y$ axis). The contours with asterisks correspond to an f value of $10^{-27.75}$ s^3 cm^{-6}, and the log f values of adjacent contours increase by steps of 0.5. The average magnetic field for this time period is aligned within 1° of the $+V_x$ axis. From Ref. 3.

beam and its mirrored return, both at free-streaming velocities and indicating a velocity filter effect due to an $E \times B$ drift toward the central plasma sheet.

In Fig. 4 (from Cladis and Francis[4]) the authors schematically show the kinetic trajectories of various velocity ionospheric ions as they populate the storm-time plasma sheet and ring current. This kinetic property of ionospheric ion propagation, which includes their parallel velocities coupled with the $E \times B$ filter effect, has been shown explicitly in ISEE 1 and 2 observations by Orsini et al.[5]

In another investigation of ionospheric ion streaming in the tail, Orsini et al.[6] used a kinetic approach to derive the north-south electric field structure in the PSBL region. Figure 5 (from Ref. 6) shows the reversal of the $E \times B$ deflection of ionospheric O^+ ions relative to the B field direction (center dashed line) as the spacecraft crosses the PSBL. From this the authors derive the reversal of E_z and the presence of a charge layer in the PSBL.

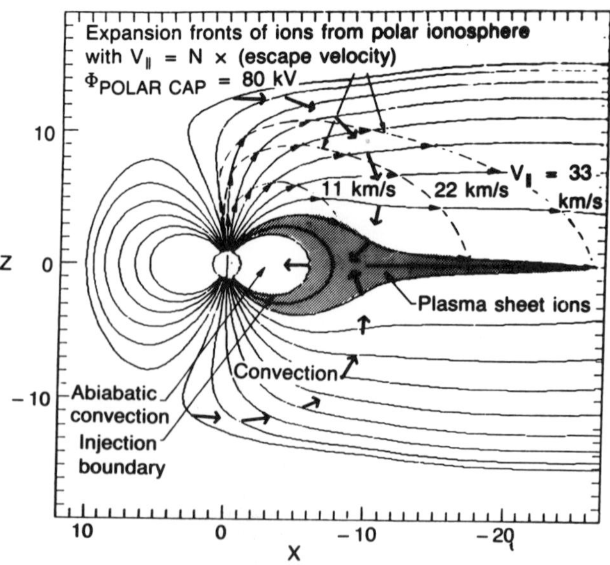

Figure 4—Illustration of ions escaping from polar ionosphere and transported to inner magnetosphere. From Ref. 4.

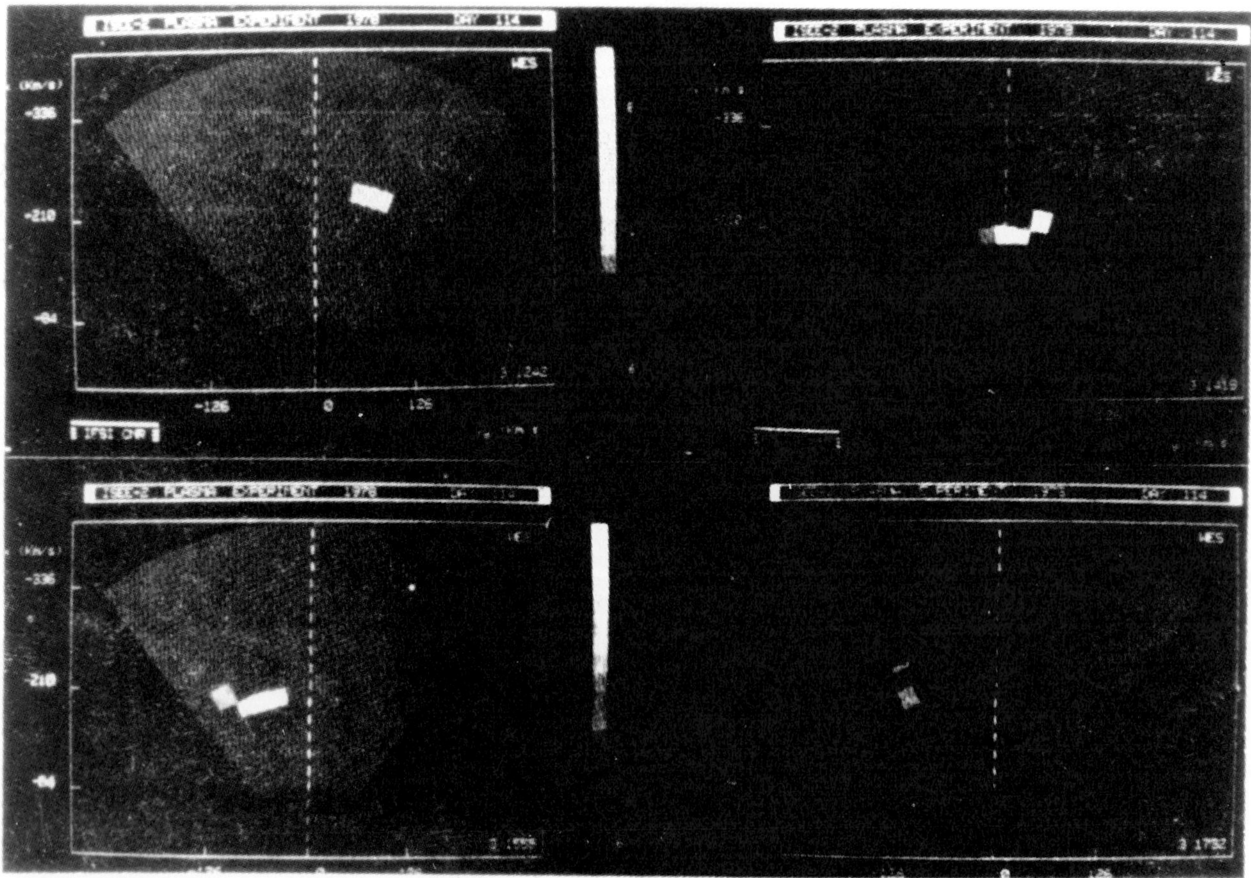

Figure 5—A time sequence of four time-contiguous angular distributions of a typical O^+ ion beam detected by the ISEE 2 plasma experiment on day 114, 0312–0317 UT, during a plasma-sheet/lobe boundary crossing in a coordinate system V_x, V_y SE. In this type of display, all particles are attributed the same mass, and equivalent H^+ velocities are computed. Every angular distribution is taken over 96 s. The gray tones are proportional to the log of the number fluxes according to the scales on the right of the single panels. The light areas are the $-34°$ to $+34°$ solar ecliptic longitude tailward flow sectors in which the high angular resolution data are taken. The broken lines are the ISEE 2 X, Y_{SE} magnetic field projections. From Ref. 6.

Figure 6 shows plasma observations by Eastman and Frank[7] in the PSBL that show energetic plasma-sheet ions flowing earthward in conjunction with lower energy ionospheric ions flowing tailward. They argue that a moment calculation over these distributions, without breaking them down into components based on composition and source regions, would yield misleading results with regard to the causes of observed flows, as well as the true values of energy and mass transport. Whichever component dominates in density will likely dominate higher moments and mask the counterstreaming character of the distribution, which can be crucial to local physics (e.g., wave-particle interactions—see Dusenbury and Lyons[8]) as well as to the diagnostics of the possibly distant sources of these distributions.

Figure 7 shows data from Frank et al.[9] that the authors used in a kinetic treatment to quantify both field-aligned and cross-field currents in the tail. By a kinetic approach, they were able to use the electron distribution on the left to determine that the primary current carriers were cold ionospheric electrons. On the right they show, for another event, that the anisotropy perpendicular to $\boldsymbol{B}$ of the high-energy electrons is opposite that

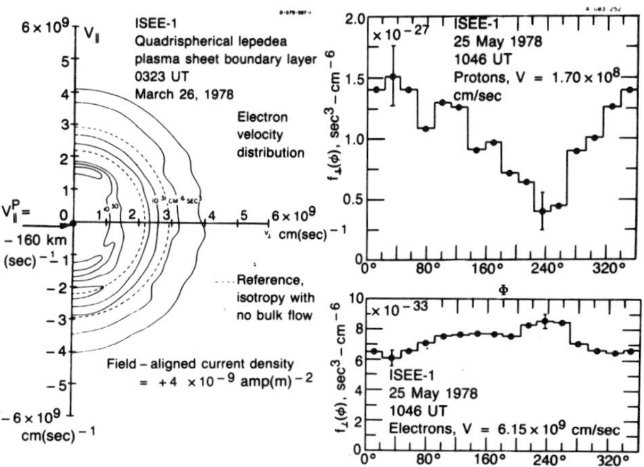

Figure 7—Isodensity contours (left) for electron velocity distributions within the first field-aligned current sheet encountered at 0323 UT by ISEE 1 as the plasma sheet expands over the spacecraft position. The current density is directed into the ionosphere. The proton velocity distribution (right, top) perpendicular to $\boldsymbol{B}$, $f_\perp(\phi)$, during the crossing of the neutral sheet at 1046 UT. The angle ϕ is the azimuthal angle about $\boldsymbol{B}$ and the proton speed is 1.70×10^8 cm/s. (right, bottom) The electron velocity distribution (right, bottom) $f_\perp(\phi)$ measured at 1046 UT and to be compared with that for the proton distribution shown in Plate 1. The electron speed for this cut of the velocity distribution is 6.15×10^9 cm/s. From Ref. 9.

Plasma sheet boundary near onset

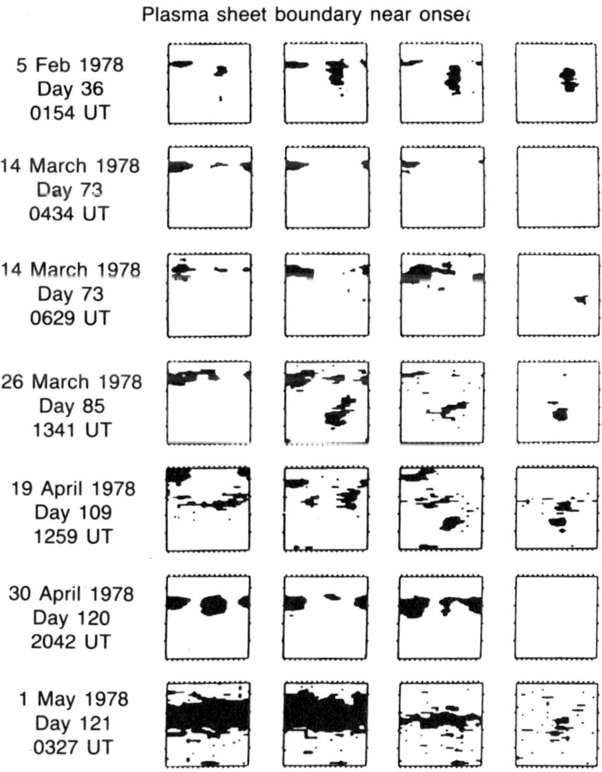

5 Feb 1978
Day 36
0154 UT

14 March 1978
Day 73
0434 UT

14 March 1978
Day 73
0629 UT

26 March 1978
Day 85
1341 UT

19 April 1978
Day 109
1259 UT

30 April 1978
Day 120
2042 UT

1 May 1978
Day 121
0327 UT

Figure 6—Energy-phase (E-ϕ) plots of ion responses from the LEPEDEA equatorial plane detector for seven examples of plasma sheet "dropout." All cases include the last three instrument cycles in high-data-rate (128-s instrument cycles) sampled within the plasma-sheet boundary layer. The first sample of lobe plasma is shown in the right-hand column and the time of this transition is listed for each event. The center of each plot corresponds to antisunward flow whereas sunward flow corresponds to $\phi = 0°$. Ion flow with duskward or dawnward components occurs with phase angle ranges of $0 \leq \phi \leq 360°$, respectively. From Ref. 7.

of the ions, and when interpreted in a kinetic treatment they calculate the magnitude of the cross-field current. These results required the examination of the full distribution function; neither a fluid nor a single particle approach would have yielded the information.

Before moving to higher energy, we will show two examples of the moment (fluid) treatment that are applied under circumstances appropriate to the fluid technique and that yield results that would be more painstaking to obtain with a kinetic approach. The first example is shown in Fig. 8, from Ref. 10. In this figure the authors record what they interpret to be the passage of a plasmoid in the deep tail. Important to their interpretation is the field topology, which they infer in part on the basis of electron flows. They indicate these flows in two ways: at low energy they use the third-order moment of the plasma electron distribution, or heat flux (fifth panel from top), and for energetic electrons they use the flux anisotropy, which is a kinetic detail of the electron distribution function. These are seen to be in general agreement, with the periods of high heat flux and large energetic electron anisotropy consistent with an open topology and the consequent unidirectional electron flux out of the tail into the solar wind, while periods of low heat flux and roughly isotropic energetic electrons are consistent with a closed field topology.

Another example of a fluid technique (or a moment calculation) is shown in Fig. 9 from Birn et al.[11] Using plasma ion moments in the plasma sheet, where large streaming anisotropies and strong gradients are rare, they measure the rotation of the first moment velocity vector at ISEE 1 and ISEE 2 and derive the properties of "vortices" in the plasma sheet. Because the plasma sheet

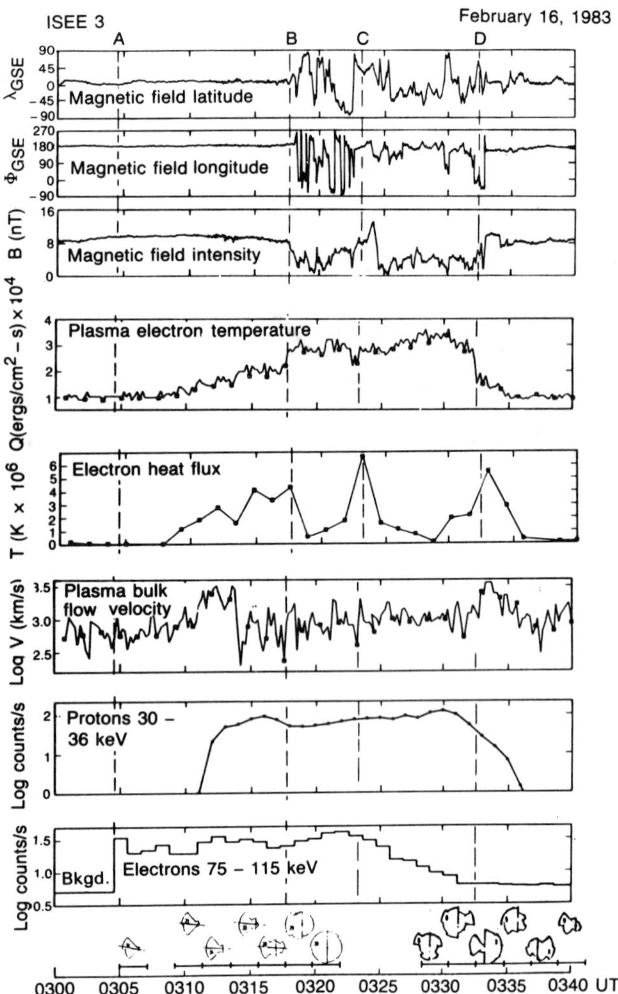

Figure 8—The top three panels show 3 second averages of the magnetic field latitude, longitude, and intensity. The next three panels show the temperature, heat flux, and bulk velocity of plasma electrons (E_e = 10–1000 eV). The bottom two panels show the intensities of energetic protons and electrons. At the bottom are azimuthal angular distributions of the energetic electrons with lines indicating the 128-second time intervals over which each was measured. Leftward pointing sectors (as in the first distribution, for example) designate tailward-flowing electrons. From Ref. 10.

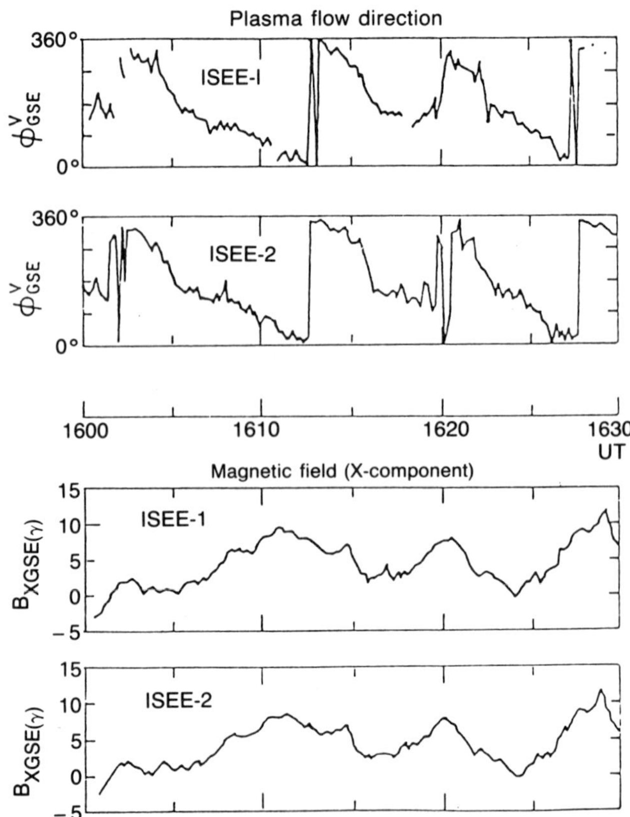

Figure 9—Variations in the longitude of plasma flow and in the x-component of the magnetic field measured simultaneously at ISEE 1 and 2 during a plasma vortex encounter on March 2, 1978 (event 6). From Ref. 11.

plasma is presumably roughly homogenous and acting in response to a global phenomenon, the bulk plasma properties adequately reflect the physics of this phenomenon.

We will now focus attention on the higher energy particles. The interpretation of energetic particle observations is routinely carried out using a kinetic (usually single particle) approach, necessitated by the high velocity and large gyroradii relative to the scale of the magnetotail system.

Again, the PSBL is a region of intense interest and fruitful application of the kinetic approach. The nature of energetic particle detection systems and the rapidly falling density toward high energies make the most convenient representation of the data differential flux

(counts/cm^2-s-sr-keV) versus energy or energy per charge (as a function of angle or look direction), rather than distribution function, in order to reduce the large dynamic range required for display of the data.

Angular distributions taken at a particular energy are equivalent to cuts through a distribution function at constant velocity, so that by including the flux at all energies, the information is the same as that contained in the distribution function. In practice, one often needs only a few energies to apply a kinetic approach to the energetic particle data.

We have already briefly discussed the use of energetic electron angular distributions in determining field topology (Fig. 8). Now, in Plate IV-1, we present a time series of energetic-ion three-dimensional angular distributions in the PSBL that show bidirectional streaming, with beams at low energies and conics at higher energies. Using these data, Williams[12] derived a location of $X_{SE} \sim -100\ R_e$ for a source region in the tail and an $E \times B$ drift toward the central plasma sheet causing energy (or parallel velocity) dispersion. In this example the author uses the kinetic aspects of the observations to infer the characteristics not only of the propagation, but of the source region itself. By using a kinetic ap-

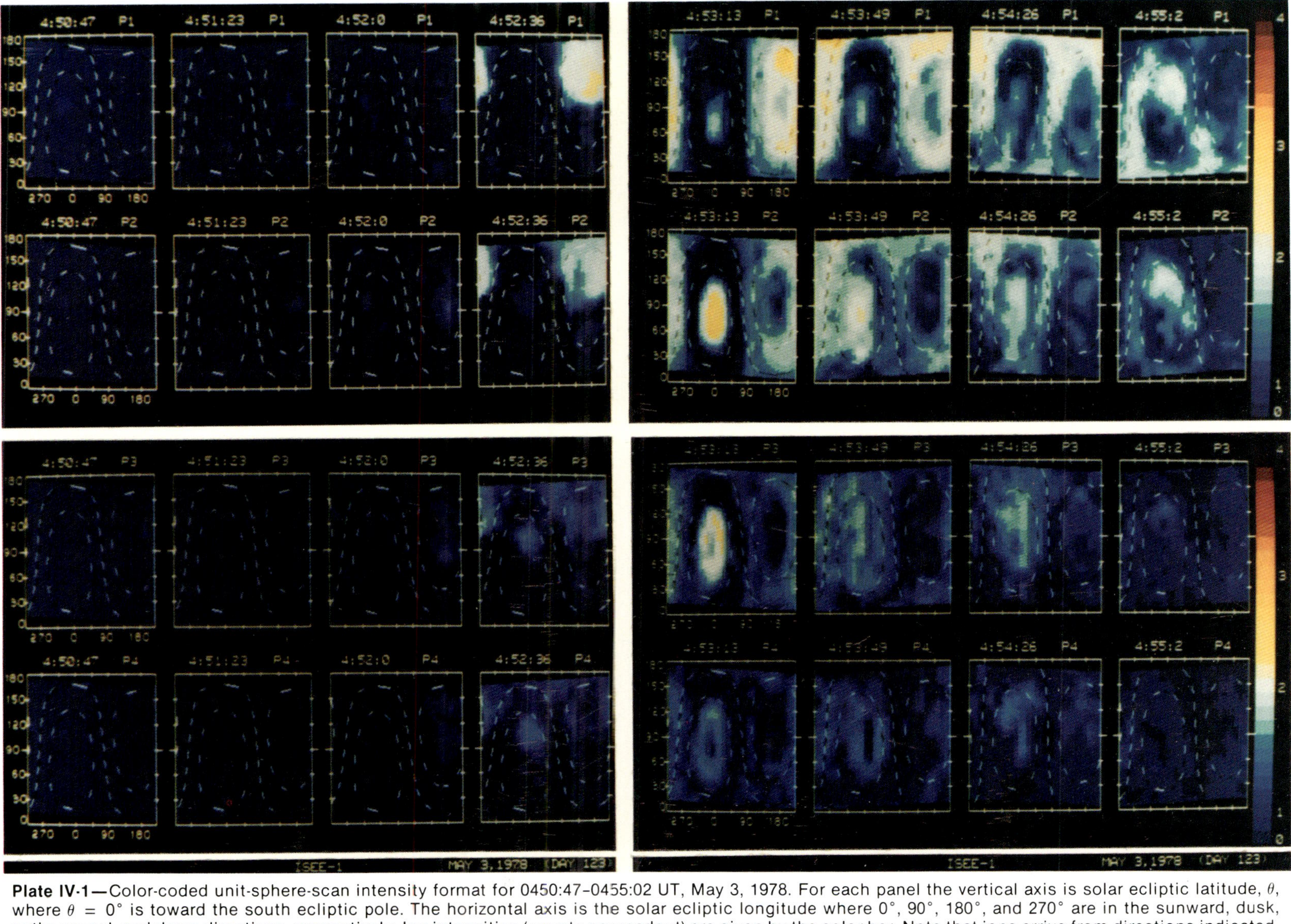

Plate IV-1—Color-coded unit-sphere-scan intensity format for 0450:47–0455:02 UT, May 3, 1978. For each panel the vertical axis is solar ecliptic latitude, θ, where $\theta = 0°$ is toward the south ecliptic pole. The horizontal axis is the solar ecliptic longitude where 0°, 90°, 180°, and 270° are in the sunward, dusk, antisunward, and dawn directions, respectively. Ion intensities (counts per readout) are given by the color bar. Note that ions arrive from directions indicated. The center time of each unit sphere scan is given above the individual color panel. A time sequence of P1 (24–34.2 keV), P2 (34.2–44.4 keV), P3 (44.4–54.8 keV), and P4 (54.8–65.3 keV) intensities is shown. All intensities can be converted to differential flux values through multiplication by 69.3. Thus comparison of the intensities in these channels shows directly the spectral characteristics in this ion energy range. From Ref. 12.

proach, one can reconstruct and probe the dynamics of distant regions in the system. This technique works in the following way. If a source of particles turns on at some distance from the observer, then at a given energy (or velocity) low-pitch-angle particles arrive earlier than high- pitch-angle particles, due to a larger $V_\parallel$. By timing the arrival of different pitch angles at a given energy (or different energies at the same pitch angle), the distance to the source region can be inferred. As shown by Williams and Speiser,[13] this technique can be applied to either a time-varying source or a steady-state source in the presence of significant convection electric fields. In the latter case, the ratio of the parallel particle velocity (a function of energy and pitch angle) to the convection velocity separates the streaming particles spatially in a predictable fashion. For details on the use of these techniques, the reader is referred to the above references.

Notice that early in the event, the fluxes at higher energy exceed those at lower energy. This is caused not by an acceleration process but simply by the velocity dispersion due to the propagation of these fluxes from the distant source, coupled with the $E \times B$ velocity filter effect. This process alone can result in a positive slope of the distribution function. In Fig. 10 we show an example from Scholer et al.[14] of two spectra, one from the plasma sheet, which falls monotonically, and the other from the PSBL, which has a local maximum in flux at ~ 80 keV. This is also a local maximum in the distribution function, and such a condition may be unstable to wave growth. An example of wave generation in the PSBL, attributed by Tsurutani et al.[15] to the generation of the right-hand resonant ion-beam instability by PSBL ion streaming, is shown in Fig. 11 (from the same work). This is an example of the importance of kinetic effects to the local physics of the PSBL. In this manner it is possible that ions energized in one region in the magnetotail might develop, due to propagation alone, into an unstable distribution at some distant point, as discussed in Ref. 12. The instability then drives waves, which may interact with any local, cool plasmas, heating them and so transporting energy over large distances kinetically. The importance of wave-particle interactions (a kinetic effect) to magnetotail dynamics cannot be overemphasized. Scarf et al.[16] and Gurnett et al.[17] showed the association of the boundary plasma sheet with broadband electrostatic noise (BEN) and whistler noise, while the neutral-sheet region exhibited occasional bursts of electron cyclotron harmonic emissions. BEN has been associated more directly with the PSBL and the streaming ion beams therein by Grabbe and Eastman,[18] who also give a theoretical treatment, and Parks et al.[19] Further theoretical work has been contributed by Dusenburg and Lyons,[8] Omidi,[20] and Akimoto and Omidi.[21]

As discussed in the previous section, plasmas are constrained perpendicular to B so that they often exhibit collective behavior in that plane. Electric field drifts introduce anisotropy in energetic ion observations that can most easily be understood as follows. We are usually measuring an energy spectrum that decreases in flux monotonically with increasing energy, and we use constant-energy-window detectors. Any motion of the de-

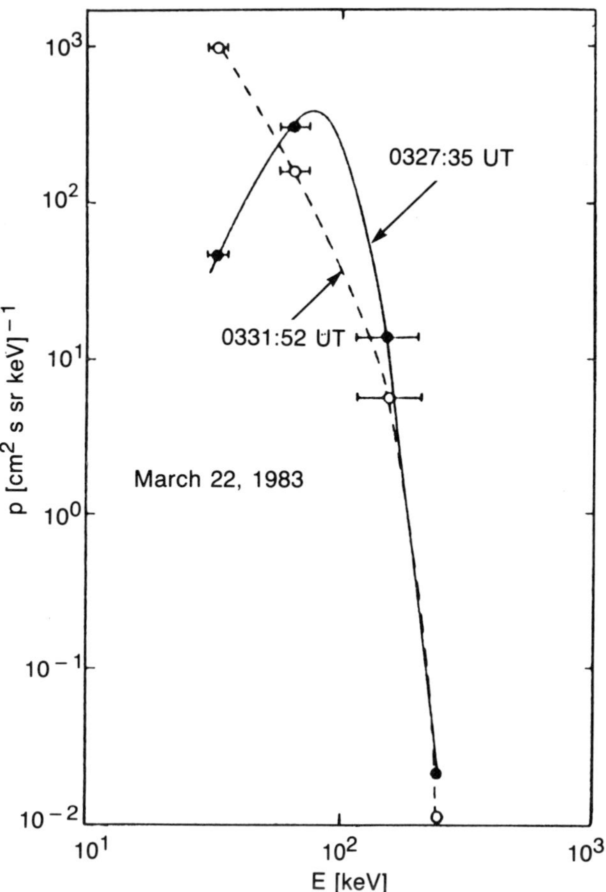

Figure 10—Omnidirectional differential proton spectra during two time periods. The time given is the start time of the 128-s averaging interval. Lines are drawn to guide the eye. Spectrum indicated by solid line is obtained when the LANL electron plasma instrument indicates a lobe-like environment. From Ref. 14.

tector relative to an isotropic distribution will add a little energy to ions entering the detector when it is looking in the direction of its relative motion (i.e., into the flow or upstream). Since there is a higher flux of ions at lower energy, and since the relative motion increases the energy of all the ions viewed in this direction, the apparent flux at a given energy increases. This increase is a function of both the relative velocity of the detector and the plasma and of the slope of the spectrum (a larger effect for a steeper slope). The reverse effect is seen in the downstream direction. This results in an anisotropy in the flux, which decreases toward higher energy due to the decreasing fraction of the particle energy, which depends on the relative motion of the detector. This is known as the Compton-Getting effect from its original application to cosmic rays by Compton and Getting.[22]

Figure 12 is an example from Mitchell et al.[23] of a series of energy spectra taken at various azimuths perpendicular to B. The $E \times B$ convective flow produces the azimuthal flux variation, which is seen to decrease as a function of energy. A frame transformation reveals a velocity of ~ 600 km/s of the energetic ion rest frame relative to the spacecraft, which is located in the sheath at this time.

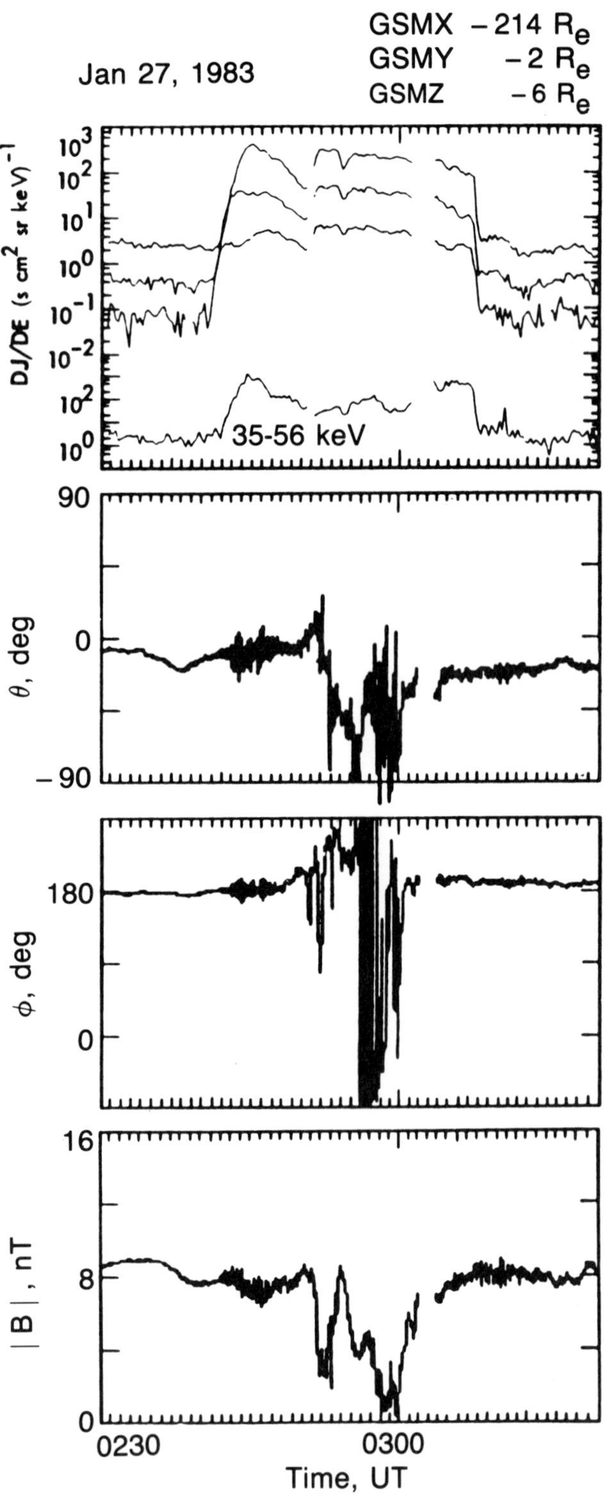

Figure 11—Energetic (>35 keV) ion flux, magnetic fields, and magnetosonic waves (high-frequency oscillations) for a plasma-sheet boundary layer and plasma-sheet encounter on January 27, 1983. From Ref. 15.

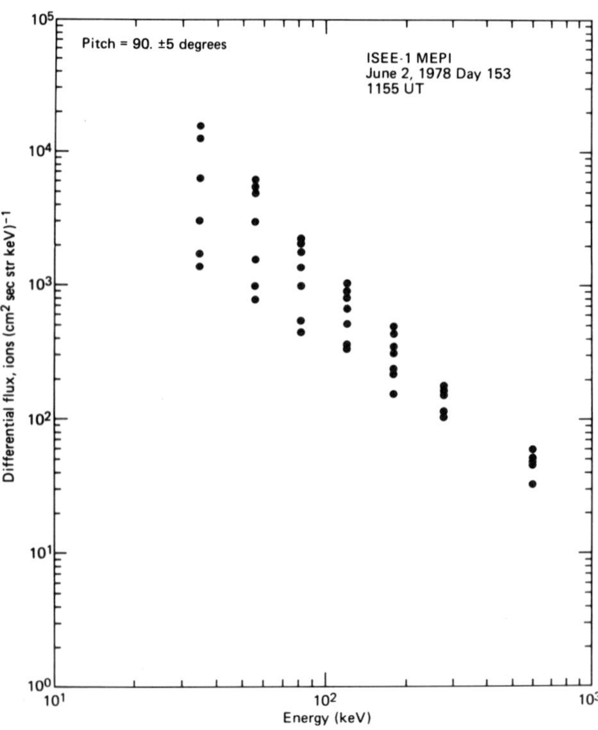

Figure 12—Energy spectrum of all unit sphere samples centered within ±5° of 90° pitch angle, for a period when the anisotropy perpendicular to *B* is dominated by *E* × *B* convection. From Ref. 23.

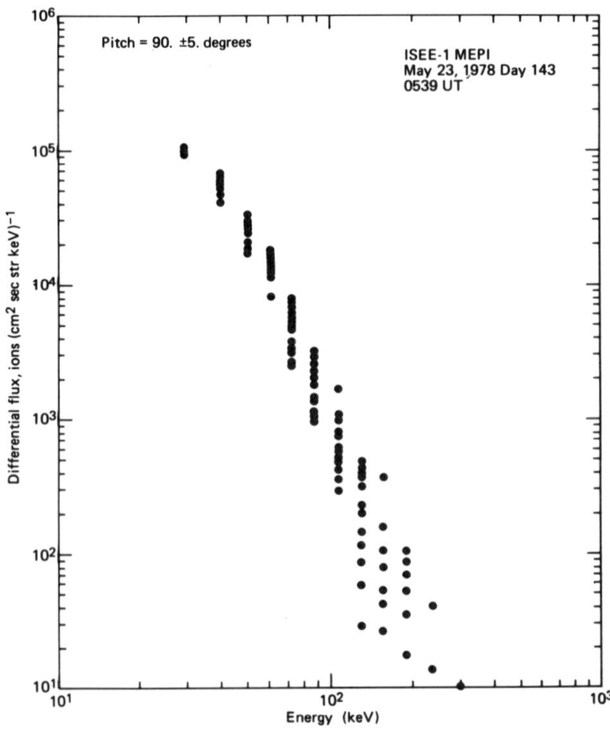

Figure 13—Energy spectrum of all unit sphere samples centered within ±5° of 90° pitch angle, for a period when the anisotropy perpendicular to *B* is dominated by a pressure gradient. From Ref. 23.

An anisotropy perpendicular to *B* may also arise (as stated earlier) from pressure gradients. An example of spectra at various azimuths perpendicular to *B* for a gradient is shown in Fig. 13, from Mitchell et al.[23] Notice that, contrary to the convective case, the anisotropy increases with energy, due to the larger gyroradii of higher energy ions.

At any particular energy, the signatures in the data in Figs. 12 and 13 are qualitatively the same. It is only through a kinetic approach that these different phenomena are identified and quantified.

Figure 14 is taken from Daly et al.[24] Here they compare energetic ion bulk velocities with plasma electron measurements. Their technique for determining the energetic ion bulk velocity is qualitatively similar to that discussed with Fig. 12, except that they do not restrict the process to the plane perpendicular to *B*. Since, as discussed earlier, energetic particles may often stream parallel to *B* in essentially single particle motion, the usefulness of the concept of a distribution rest frame parallel to *B* is highly dependent on such conditions as closed fields and/or strong scattering. In the portion of Fig. 14 where the velocities agree, independent considerations suggest the spacecraft is in the plasma sheet where we expect the ions to behave in a more-or-less collective manner. In the region where they disagree, the authors conclude that the spacecraft is in the PSBL (which is similar to the lobe), where we expect free streaming of the ions and where the "bulk velocity" determined from the energetic ions exhibited large uncertainties. A similar comparison between energetic ion and plasma ion velocities by Sarris et al.[25] produced comparable results.

In an earlier application of the same technique as that discussed above regarding Daly et al.,[24] Daly et al.[26] show a preponderance of tailward flow for X_{SE} positions ≤ -100 R_e. They also show a persistent anistropy perpendicular to *B*, which they attribute to a gradient toward the center of the plasma sheet in the energetic ions, with a scale of 1 to 2 R_e. In both findings, a kinetic treatment was essential to the correct interpretation of the observations.

Also in the deep tail, Richardson and Cowley[27] use the streaming characteristics of the energetic ions to identify the regions the spacecraft passes through (see Fig. 15). In this figure, as in Figs. 14 and 11 and typical of many other of the data discussed here, the regions of strong (tailward) ion streaming are identified with lobe, or PSBL (depending upon one's definition), while the regions of more moderate tailward flow are essentially identical in the regions labeled "plasmoid" and "plasma sheet," which were differentiated from one another

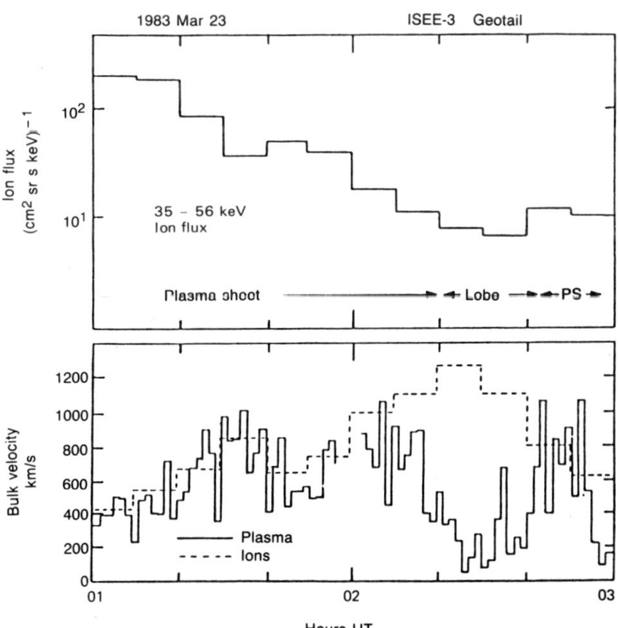

Figure 14—The omnidirectional ion flux in the lowest energy channel and the calculated ion bulk velocity (dashed line) plotted against time on March 23, 1983, from 0100 – 0300 UT, in 10-min intervals. The ISEE 3 spacecraft was 120 R_e antisunward of the earth at this time. The solid line in the lower panel shows the plasma bulk velocity, provided by the Los Alamos National Laboratory (R.D. Zwickl, private communication, 1984). The lobe region from 0220 – 0245 UT is an energetic ion boundary layer, containing high-speed ions and low-speed plasma. From Ref. 24.

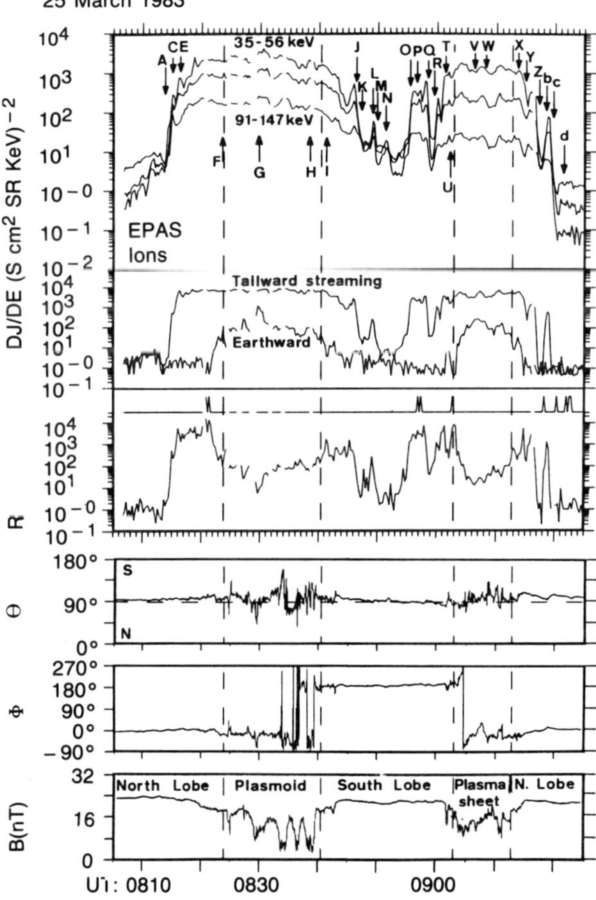

Figure 15—ISEE 3 energetic ion and magnetic field data for the plasmoid of 0824–0840, 25 March 1983, and a subsequent plasma sheet encounter at 0904–0915. In both cases, a tailward-streaming ion boundary layer is observed in the lobe adjacent to the plasma sheet on entrance and exit. In addition, brief burst-like ion enhancements occur in the lobe which are more evident in the lower energy channels and show strong tailward streaming. From Ref. 27.

on the basis of magnetic field properties. While a moment calculation might distinguish between these regions in the magnitude of the flow velocity, it would not indicate the qualitative difference between the regions, namely, that the lobe regions show no earthward flowing ions, while the other regions do.

Leaving the deep tail region, we go to the low-latitude boundary layer. Plate IV-2 shows four nearly consecutive times that highlight the energetic particle signatures as the spacecraft (ISEE 1) travels through the closed field region of the LLBL (trapped electrons, anisotropic ions that indicate tailward $E \times B$ convection), to open field boundary layer in the second period (with reverse-draped magnetic field as in Hones et al.[28] and unidirectional streaming electrons and ions, with some convection still present in the ions). The third panels are in the magnetopause layer, where the ion and electron streaming continues and rotates with the field toward the sheath orientation. The last panels are in the sheath, where there are no energetic electrons and only a few convecting energetic ions. All these signatures are kinetic details of the distribution that probably have little effect on the local physics, but that are essential to the interpretation of the phenomenology. The analysis of this period in terms of the energetic particle anisotropies is discussed in Mitchell et al.[29]

In a superposed epoch analysis of substorms, Bieber et al.[30] used tailward streaming energetic (>200 keV) electrons as a fiducial time (see Fig. 16). This kinetic signature was interpreted as the point of passage of the spacecraft from closed to open magnetic fields in the separatrix layer associated with a neutral line. Ordered by this signature, the density, velocity, B_z, and AE index were shown to exhibit characteristics commonly observed in association with substorm onset. In addition, the plasma electron temperature often increased and developed a bump-on-tail spectrum, a kinetic feature from which the authors infer acceleration due to a parallel electric field. This is a good example of kinetics providing diagnostics of local phenomena, distant source characteristics, and field-line topology.

As a final example (Fig. 17 from Baker[31]), we show an example of the use of energetic electron anisotropies used to infer the morphology of closed magnetic field lines. The author uses the development of a cigar distribution in the drifting electrons to infer the development of tail-like field morphology at geosynchronous orbit, and the storage of energy in the tail. With the dipolarization of the magnetic field at substorm onset and the injection of newly energized plasma into the geosynchronous region, the electron distributions return suddenly to a pancake or trapping distribution.

In this section we have attempted to present a broad range of observations from different regions in the magnetotail and different energy ranges and species. The common thread in this selection is that a kinetic treatment was necessary to the interpretation of the data in each case. Furthermore, in many cases the particle kinetics behavior played a significant role in modifying the local environment.

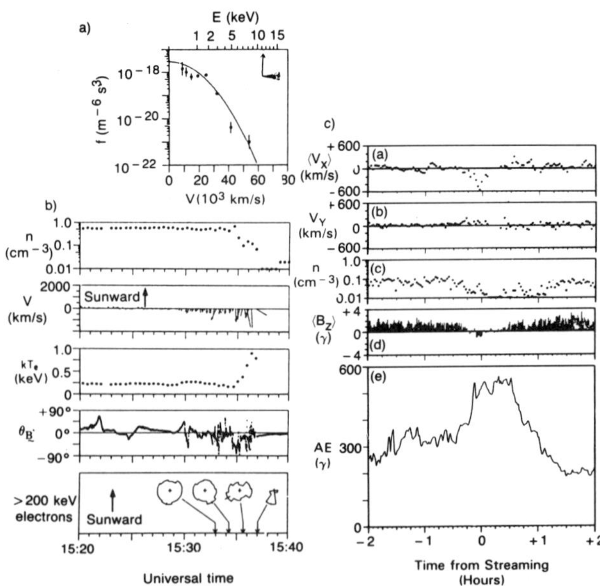

Figure 16—(a) Example of a bump-in-tail electron velocity distribution observed at the time of peak electron heating for the event shown in Fig. 2. The curve is a fitted Maxwellian. Error bars on the points are due to counting statistics. At upper right, the arrow shows the azimuth of the magnetic field, and the shaded fan shows the azimuthal range of electron velocity vectors included in the distribution, where earthward is up and dawnward is to the right. The distribution required ~0.25 s to accumulate. (b) Example of an energetic electron streaming event observed on IMP 8, located in the central plasma sheet 32 R_e tailward of earth, on 14 November 1973. From top to bottom are shown plasma density n, bulk flow velocity V as calculated both from ion data (arrow-tipped vectors) and electron data (untipped vectors), electron temperature kT_e, elevation angle of magnetic field θ_B (GSM coordinates), and polar plots of angular distribution >200 keV electrons (normalized to a constant size of peak sector). (c) Superposed epoch analysis of: X_{GSE}-component of plasma flow velocity; Y_{GSE}-component of plasma flow velocity; plasma density (plasma parameters: 2-min averages for protons of energy 84 eV, to 2-min averages for protons of energy 84 eV to 15 keV); Z_{GSM}-component of magnetic field (1-min averages); AE index (1-min values). Plotted values are the median over 16 events. Zero epoch time is determined by the first observation of intense streaming of >200 keV electrons. From Ref. 30.

DISCUSSION AND SUMMARY

We have illustrated the importance of kinetics in observations of magnetotail dynamics. In simulations of the magnetotail system and its dynamics, a fluid approach is probably necessary in order to make the problem tractable. We have had two notes of caution on this point at this conference. F. Coriniti characterized the MHD approach as one that maximizes one's ignorance; nevertheless, given the enormous complication of a kinetic approach on the scale of the whole tail, MHD remains a compromise worth making and he suggested directions to be taken for improving this approach. D. Montgomery cautioned that with the large Reynolds number of the system, one must be wary of the results obtained from simulations based upon techniques that do not include the full range of scale sizes. From our

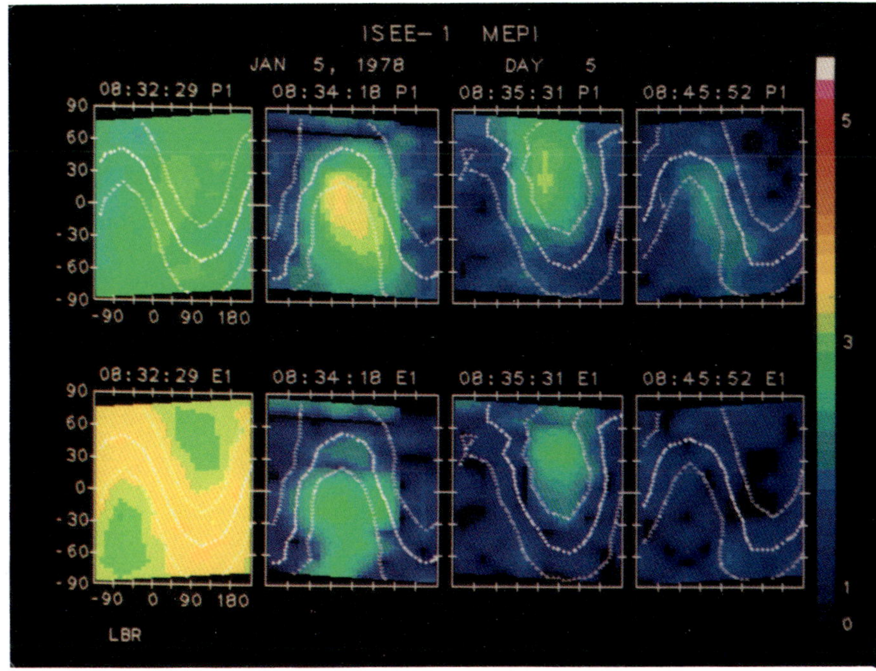

Plate IV·2—Detailed ISEE 1 MEPI ion and electron three-dimensional distributions. Top panels are 24–44.5 keV ions and bottom panels are 22.5–39 keV electrons. This display is a flat projection of the unit sphere, with polar angle running from $-90°$ to $+90°$ and azimuth from $-180°$ to $+180°$ (offset by $45°$). The convention for the angles is the GSE look direction of the detector; e.g., $\theta = 0, \phi = 0$ represents particles coming from the sunward direction, traveling tailward. Flux in log (counts/cm^2-s-sr-keV) is keyed to the color bar to the right. Measured contours of look directions corresponding to particle pitch-angles of 60° (bottom curve in first panel), 90° (center curve), and 120° (top curve) are overlaid in white. The time given is the center-point time of the 36 s scan.

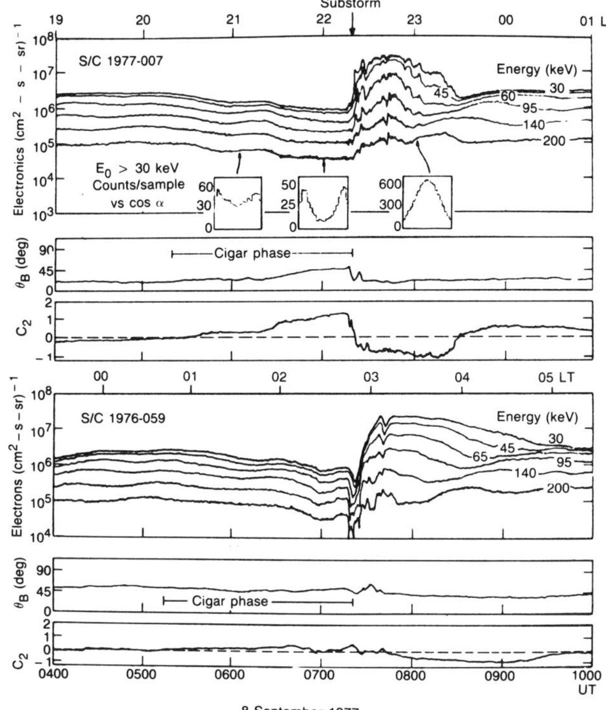

Figure 17—A detailed plot of the geostationary orbit spin-averaged energetic electron fluxes, local magnetic field line tilt angle (θ_B), and second-order anisotropy amplitude (C_2) on September 8, 1977. All electron channels (energies as labeled) have a common upper cutoff energy of 300 keV. The upper panels show data for spacecraft 1977-007, while the lower panels show data for spacecraft 1976-059. Universal time is shown along the bottom of the figure, while geographical local time is shown for each satellite. A substorm injection event is seen at ~0720 UT, preceded by a substorm growth (cigar) phase of ~2 hour duration. From Ref. 31.

discussion of kinetics in an observational context, we have seen that at some times, in some places, the microscale phenomena can indeed be important to the system as a whole.

The regions in which kinetics are particularly important are:

1. The plasma-sheet boundary layer,
2. The current-sheet region of the plasma sheet,
3. The lobe and mantle,
4. The low latitude boundary layer, and
5. Any neutral lines that may be present.

These regions share at least one of two aspects that make a kinetic approach important in interpreting observations: spatial scales of the order of an ion gyroradius and/or high-speed parallel streaming flows. Neither condition is handled properly in a fluid or moment approach.

We have pointed out some observations in the magnetotail that suggest the importance of collective, kinetic effects to the transport of energy and the modification of particle phase space distributions. Parks et al.[19] shows correlations between BEN and PSBL signatures,

and Grabbe and Eastman[18] also correlate these and provide a theoretical framework to explain their findings. It is likely that particles carry energy from the tail toward the earth in the form of beams, and that these beams generate waves through beam-plasma instabilities and are ultimately thermalized by interaction with the wave field as they mirror and convect into the central plasma sheet, thereby adding to and heating the central plasma sheet. These ions may originate in a thin current sheet in the tail, accelerated across the tail current sheet by the cross-tail electric field in the manner described by Speiser[32] and verified by Lyons and Speiser.[33] They may also originate in a neutral-line merging region in the tail current sheet. The signature of such a region in terms of observations is not established, and the neutral line region itself remains elusive. However, kinetic effects such as ion and electron inertia, nonadiabatic motion, and anomalous resistivity due to interaction of the plasma with electrostatic or electromagnetic waves/noise would certainly play important roles in such a region. Large electric fields due to $\partial B/\partial t$ would be expected to accelerate particles to high energies such as those often observed in association with substorm-related events.

An essential point that we made in the first section is that the distribution in phase space results from the history of each volume element in phase space. Only a kinetic treatment allows for the reconstruction of the history of the distribution. Also, in the likely situation where the transport processes cannot be functionally related to the moments, a kinetic approach is required to reconstruct the history of even the moments themselves and not just that of the detailed distribution. In addition, a kinetic treatment is necessary to investigate many details of the distribution that contribute to the local microscale physics.

For a static system or a system dominated by collisional processes, the fluid approach will be adequate. For interpretation of observations in the inhomogeneous dynamic magnetotail system, a kinetic approach is usually essential.

ACKNOWLEDGMENT—I am grateful to D. J. Williams and B. H. Mauk for many fruitful discussions in the course of this work. I also thank G. W. Snyder for typing the manuscript. This work was supported in part by the National Science Foundation Atmospheric Sciences Section, grant ATM 8219436 to The Johns Hopkins University, and the National Aeronautics and Space Administration contract to The Johns Hopkins University Applied Physics Laboratory and the Department of the Navy under Task I2UOS10 under contract N00024-85-C-5301.

REFERENCES

[1] N. A. Krall and A. W. Trivelpiece, *Principles of Plasma Physics,* McGraw-Hill, New York, p. 674 (1973).
[2] G. F. Chew, M. L. Goldberger, and F. E. Low, "The Boltzmann Equation and the One-Fluid Hydromagnetic Equations in the Absence of Particle Collisions," *Proc. R. Soc.* (*London*) **A236**, 112 (1956).
[3] T. G. Forbes, E. W. Hones, S. J. Bame, J. R. Asbridge, G. Paschmann, N. Sckopke, and C. T. Russell, "Evidence for the Tailward Retreat of a Magnetic Neutral Line in the Magnetotail During Substorm Recovery," *Geophys. Res. Lett.* **8**, 261 (1981).
[4] J. B. Cladis and W. E. Francis, "The Polar Ionosphere as a Source of the Storm Time Ring Current," *J. Geophys. Res.* **90**, 3465 (1985).
[5] S. Orsini, M. Candidi, K. Altwegg, and H. Balsiger, "Ionospheric H$^+$ and O$^+$ at ISEE 1 and ISEE 2 Flowing Tailward with Same Parallel Velocity" (this publication, 1986).

223

[6]S. Orsini, M. Candidi, V. Formisano, H. Balsiger, A. Ghielmetti, and K. W. Ogilvie, "The Structure of the Plasma Sheet-Lobe Boundary in the Earth's Magnetosphere," *J. Geophys. Res.* **89**, 1573 (1984).

[7]T. E. Eastman and L. A. Frank, "Boundary Layers of the Earth's Outer Magnetosphere," in *Magnetic Reconnection in Space and Laboratory Plasmas*, E. W. Hones, Jr., ed., Geophysical Monograph 30, American Geophysical Union, p. 249 (1984).

[8]P. B. Dusenbury and L. R. Lyons, "The Generation of Electrostatic Noise in the Plasma Sheet Boundary Layer," *J. Geophys. Res.* **90**, 10935 (1985).

[9]L. A. Frank, C. Y. Huang, and T. E. Eastman, "Currents in the Earth's Magnetotail," in *Magnetospheric Currents*, T. A. Potemra, ed., Geophysical Monograph 28, American Geophysical Union, p. 147 (1984).

[10]E. W. Hones, Jr., J. Birn, D. N. Baker, S. J. Bame, W. C. Feldman, D. J. McComas, R. D. Zwickl, J. A. Slavin, E. J. Smith, and B. T. Tsurutani, "Detailed Examination of a Plasmoid in the Distant Magnetotail with ISEE 3," *Geophys. Res. Lett.* **11**, 1046 (1984).

[11]J. Birn, E. W. Hones, Jr., and S. J. Bame, "Analysis of 16 Plasma Vortex Events in the Geomagnetic Tail," *J. Geophys. Res.* **90**, 7449 (1985).

[12]D. J. Williams, "Energetic Ion Beams at the Edge of the Plasma Sheet: ISEE 1 Observations Plus a Simple Explanatory Model," *J. Geophys. Res.* **86**, 5507 (1981).

[13]D. J. Williams and T. W. Speiser, "Sources for Energetic Ions at the Plasma Sheet Boundary: Time Varying or Steady State?" *J. Geophys. Res.* **89**, 8877 (1984).

[14]M. Scholer, D. N. Baker, S. J. Bame, W. Baumjohann, G. Gloeckler, F. M. Ipavich, E. J. Smith, and B. T. Tsurutani, "Correlated Observations of Substorm Effects in the Near-Earth Region and in the Deep Magnetotail," *J. Geophys. Res.* **90**, 4021 (1985).

[15]B. T. Tsurutani, I. G. Richardson, R. M. Thorne, W. Butler, E. J. Smith, S. W. H. Cowley, S. P. Gary, S.-I. Akasofu, and R. D. Zwickl, "Observations of the Right-Hand Resonant Ion Beam Instability in the Distant Plasma Sheet Boundary Layer," *J. Geophys. Res.* **90**, 12159 (1985).

[16]F. Scarf, L. A. Frank, K. L. Ackerson, and R. P. Lepping, "Plasma Wave Turbulence at Distant Crossings of the Plasma Sheet Boundaries and Neutral Sheet," *Geophys. Res. Lett.* **1**, 189 (1974).

[17]D. A. Gurnett, L. A. Frank, and R. P. Lepping, "Plasma Waves in the Distant Magnetotail," *J. Geophys. Res.* **8**, 6059 (1976).

[18]C. L. Grabbe and T. E. Eastman, "Generation of Broadband Electrostatic Noise by Ion Beam Instabilities in the Magnetotail," *J. Geophys. Res.* **89**, 3865 (1984).

[19]G. K. Parks, M. McCarthy, R. J. Fitzenreiter, J. Etcheto, K. A. Anderson, R. R. Anderson, T. E. Eastman, L. A. Frank, D. A. Gurnett, C. Huang, R. P. Lin, A. T. Y. Lui, K. W. Ogilvie, A. Pedersen, H. Reme, and D. J. Williams, "Particle and Field Characteristics of the High-Latitude Plasma Sheet Boundary Layer," *J. Geophys. Res.* **89**, 8885 (1984).

[20]N. Omidi, "Broadband Electrostatic Noise Produced by Ion Beams in the Earth's Magnetotail," *J. Geophys. Res.* **90**, 12330 (1985).

[21]K. Akimoto and N. Omidi, "The Generation of Broadband Electrostatic Noise by an Ion Beam in the Magnetotail," *Geophys. Res. Lett.* **13**, 97 (1986).

[22]A. H. Compton and I. A. Getting, "An Apparent Effect of Galactic Rotation on the Intensity of Cosmic Rays," *Phys. Rev.* **47**, 818 (1935).

[23]D. G. Mitchell, R. Lundin, and D. J. Williams, "Analyses of Convective Flow and Spatial Gradients in Energetic Ion Observations," *J. Geophys. Res.* **91**, 8827 (1986).

[24]P. W. Daly, T. R. Sanderson, and K.-P. Wenzel, "A Method to Measure the Bulk Velocity of an Energetic Ion Distribution in the Presence of Ion Composition Mixing," *J. Geophys. Res.* **90**, 1499 (1985).

[25]E. T. Sarris, D. J. Williams, and S. M. Krimigis, "Observations of Counterstreaming Between Plasma and Energetic Particles in the Magnetotail," *J. Geophys. Res.* **83**, 8655 (1978).

[26]P. W. Daly, T. R. Sanderson, and K.-P. Wenzel, "Survey of Energetic (E > 35 keV) Ion Anisotropies in the Deep Geomagnetic Tail," *J. Geophys. Res.* **89**, 10,733 (1984).

[27]I. G. Richardson and S. W. H. Cowley, "Plasmoid-Associated Energetic Ion Bursts in the Deep Geomagnetic Tail: Properties of the Boundary Layer," *J. Geophys. Res.* **90**, 12133 (1985).

[28]E. W. Hones, Jr., B. U. O. Sonnerup, S. J. Bame, G. Paschmann, and C. T. Russell, "Reverse Draping of Magnetic Field Lines in the Boundary Layer," *Geophys. Res. Lett.* **9**, 523 (1982).

[29]D. G. Mitchell, F. J. Kutchko, D. J. Williams, T. E. Eastman, L. A. Frank, and C. T. Russell, "An Extended Study of the Low Latitude Boundary Layer on the Dawn and Dusk Flanks of the Magnetosphere," *J. Geophys. Res.* (in press, 1987).

[30]J. W. Bieber, "Streaming Energetic Electrons in Reconnection Events," in *Magnetic Reconnection in Space and Laboratory Plasmas,"* E. W. Hones Jr., ed. Geophysical Monograph 30, American Geophysical Union, p. 185 (1984).

[31]D. N. Baker, "Particle and Field Signatures of Substorms in the Near Magnetotail," in *Magnetic Reconnection in Space and Laboratory Plasmas,* E. W. Hones, Jr., ed., Geophysical Monograph 30, American Geophysical Union, p. 193 (1984).

[32]T. W. Speiser, "Particle Trajectories in Model Current Sheets, 2, Applications to Auroras Using a Geomagnetic Tail Model," *J. Geophys. Res.* **72**, 3919 (1967).

[33]L. R. Lyons and T. W. Speiser, "Evidence for Current Sheet Acceleration in the Geomagnetic Tail," *J. Geophys. Res.* **87**, 2276 (1982).

DISCUSSION

J. Birn: I agree with you that MHD theory has its limitations. This does not mean, however, that the calculation of moments from a local distribution function becomes meaningless. In fact, usually only the moments calculated from a full distribution function make physical sense (even when the distribution function is very complicated). They fulfill for instance the continuity equation, while a subpopulation within a certain energy channel does not obey such a law.

D. J. Mitchell: The moments are not meaningless, provided one applies them only to the equations that they do fulfill. My argument is that in the cases I've presented, and in many more, they provide little physical insight, and if they are applied in a broader sense than that for which they are strictly defined, they become misleading. In the same spirit if one tries to use one energy channel to answer questions about the continuity equation, one will not be successful. Other aspects of a problem may be better addressed by one channel than by moments of the distribution (e.g., unidirectional streaming 100 keV electrons).

V. M. Vasyliunas: The fluid equations, if correctly applied, are always valid and provide applicable constraints. The only limitation is that they may not always be fully useful and may not answer the questions one is asking.

D. J. Mitchell: I agree. My point is that there are many questions, some critical to the understanding of observations in a dynamic magnetotail, which do benefit from a kinetic interpretation. See also my response to J. Birn.

A STATISTICAL STUDY OF ION PITCH-ANGLE DISTRIBUTIONS

D. G. Sibeck, R. W. McEntire, A. T. Y. Lui, and S. M. Krimigis*

Preliminary results of a statistical study of energetic (34–50 keV) ion pitch-angle distributions (PADs) within 9 R_e of earth provide evidence for an orderly pattern consistent with both drift-shell splitting and magnetopause shadowing. Normal ion PADs dominate the dayside and inner ($R < 6$ R_e) magnetosphere. Butterfly PADs typically occur in a narrow (1–2 R_e wide) belt stretching from dusk to dawn through midnight, where they approach within 6 R_e of earth. While those ion butterfly PADs that typically occur on closed drift paths are mainly caused by drift-shell splitting, there is also evidence for magnetopause shadowing in observations of more frequent butterfly PAD occurrence in the outer magnetosphere near dawn than dusk. Isotropic and gradient boundary PADs terminate the tailward extent of the butterfly ion PAD belt.

INTRODUCTION

No statistical survey of the various energetic ion PADs in the near-earth ($R \leq 9$ R_e) magnetosphere exists. In this paper, we report the preliminary results of such a study for 34–50 keV ions and compare them to previous electron and ion PAD observations.

West et al.[1] describe the three most common energetic ($E > 79$ keV) electron PADs: normal (or pancake), butterfly (or cigar), and isotropic. The particle flux peaks perpendicular to the magnetic field in a normal PAD, but at pitch angles α and $180° - \alpha$ (with $0 < \alpha < 90°$) in a butterfly PAD. There are no clear flux minima and maxima in isotropic PADs. To these categories we add the gradient PADs reported at the plasma-sheet/lobe interface by Buck et al.[2] and at the magnetopause by Williams.[3] Gradient PADs differ from the others in that the observed particle flux is not symmetric about the magnetic field, but depends instead on the phase of the arriving particles. These definitions will be used throughout our study.

Normal PADs are formed by radial and pitch-angle diffusion and the loss of low pitch-angle particles to the atmosphere. West et al. show that normal electron PADs dominate the dayside and inner ($R \leq 6$ R_e) magnetosphere.

There are three proposed generation mechanisms for butterfly PADs. West et al. show that the combined effects of drift-shell splitting and a negative radial flux gradient can transform normal dayside PADs into butterfly nightside PADs. Particles with 90° pitch angles drift inward toward the earth as they move from the dayside magnetosphere to the nightside, while particles with lower pitch angles follow more circular drift paths. Since the particles with 90° pitch angles come from a lower flux region, the resulting nightside PADs may have a relative flux minimum at 90° pitch angles—the definition of a butterfly PAD.

More extreme butterfly PADs may occur on the drift paths of 90° pitch-angle particles that intersect the mag-netopause. Since particles are lost to the magnetosheath at the magnetopause, they will be absent from these drift paths. Particles with lower pitch angles may still be present, since their drift paths differ from those of the 90° particles. Thus, butterfly PADs may be formed on the drift paths for 90° particles that originate at, or lie in the shadow of, the magnetopause. One expects westward drifting protons to be shadowed in the dawnside magnetosphere and eastward drifting electrons to be shadowed in the duskside magnetosphere.

Finally, Luhmann[4] noted that particles with 90° pitch angles drift faster than those with lower pitch angles. Thus, a group of injected particles disperses so that a stationary satellite might observe butterfly PADs whose flux maxima gradually shift to lower pitch angles.

The regions in which butterfly electron PADs typically occur agree very well with the models based on drift-shell splitting. West et al. report that they are found along the dusk magnetopause, throughout the pre-midnight quadrant, on occasion in the post-midnight quadrant, and in a narrow belt some 5–8 R_e from earth in the nightside magnetosphere. More recently, Garcia and Spjeldvik[5] have determined that butterfly ion PADs with $E < 50$ keV occur in a belt near 8 R_e from 1700 to 0500 LT. Thus, they found butterfly ion PADs at this energy to be more likely at dusk than dawn, in disagreement with the shadowing model.

West et al. noted that isotropic PADs can be created from normal and butterfly PADs that encounter a region of magnetic turbulence or one in which the magnetic field changes sharply within one gyroradius. These conditions may be encountered in the magnetotail plasma sheet by westward drifting protons and eastward drifting electrons. One therefore expects isotropic PADs tailward of the butterfly belt. This prediction has been verified for electrons by West et al., but Garcia and Spjeldvik report quasinormal ion PADs tailward of the butterfly belt.

Gradient PADs are observed when the spacecraft nears the boundary between an energetic particle-rich region and a particle-void region. Two such boundaries are the magnetopause and plasma-sheet/lobe interface,

*The Johns Hopkins University Applied Physics Laboratory, Laurel, Maryland 20707.

where particles arrive from the magnetosphere (plasma sheet) but not the magnetosheath (lobe).

DATA

We present observations of energetic ($34 \leq E \leq 50$ keV) ion PADs obtained by the Medium Energy Particle Analyzer (MEPA) on the AMPTE Charge Composition Explorer (CCE) satellite. The CCE's orbit lies nearly in the equatorial plane and has an apogee of 8.8 R_e. The satellite's geomagnetic latitude does not exceed 16.2°. The MEPA[6] measures ion flux in 32 contiguous 11.25° sectors perpendicular to the spacecraft spin axis, which lies near the earth-sun line. The local pitch angle of each sector is determined by simultaneous magnetic-field measurements by the magnetometer[7] on the CCE.

We inspected 6.4 min accumulations of MEPA ion observations to determine the PAD observed at each point along each CCE orbit. Only the four PADs discussed above (normal, butterfly, isotropic, and gradient) persistently recur, and Fig. 1 shows typical examples of each. This study groups isotropic and gradient PADs together since we expect (and find) that the regions in which isotropic plasma-sheet and gradient plasma-sheet/lobe interface PADs occur are nearly coincident, and it is easy to distinguish this magnetotail region from that in which the dayside magnetopause gradient PADs occur. The occurrence ratio of gradient to isotropic magnetotail PADs is about 8:1.

Three problems arise in interpreting the data. First, if several different PADs occur during a 6.4 min accumulation, the accumulation may not reflect the most typical PAD within the interval. A high flux butterfly PAD might dominate 15 other lower flux 24 s normal PADs in the 6.4 min period. Since PAD types endure for 30 min or longer and over 1 R_e or more, this is a problem only at sharp spatial boundaries or sudden temporal changes and does not modify the results of this statistical study.

Second, when the CCE is not at the geomagnetic equator, the large-pitch-angle ions of equatorial PADs cannot reach the spacecraft to be recorded by the MEPA. Assuming a dipolar magnetic field, the MEPA observes a range of equatorial pitch angles that varies from 0° to 70° or 90°, depending on the CCE's magnetic latitude. In contrast, when the magnetic field has a component parallel to the spacecraft's spin axis, the MEPA cannot observe ions with low pitch angles. This situation occurs when the spacecraft encounters tail-like magnetic fields. Thus the latter two factors conspire to eliminate low- and high-pitch-angle coverage of equatorial PADs when the spacecraft is off the geomagnetic equator and observing tail-like magnetic fields. Within the limited pitch-angle coverage (at worst a few tens of degrees), gradient PADs may easily be distinguished from other PADs. If coverage is limited to the lowest pitch angles, the particle flux depression near 90° equatorial pitch angles that distinguishes butterfly from normal PADs cannot be observed. Like previous authors, we assign such PADs to the normal category, thereby underestimating butterfly PAD occurrence.

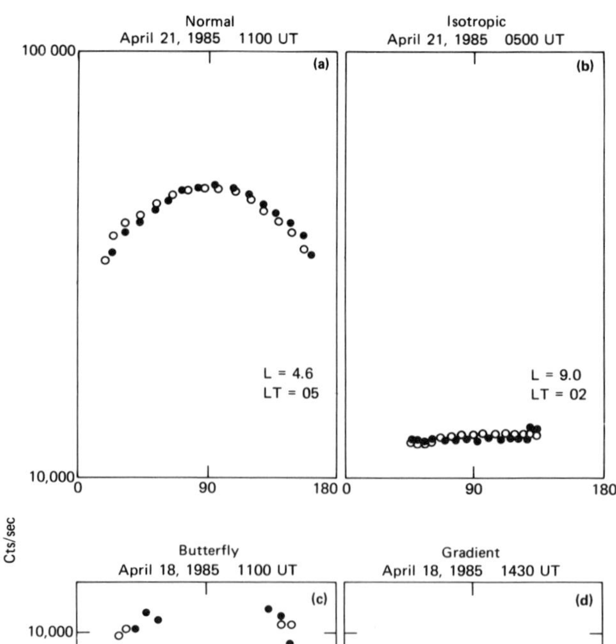

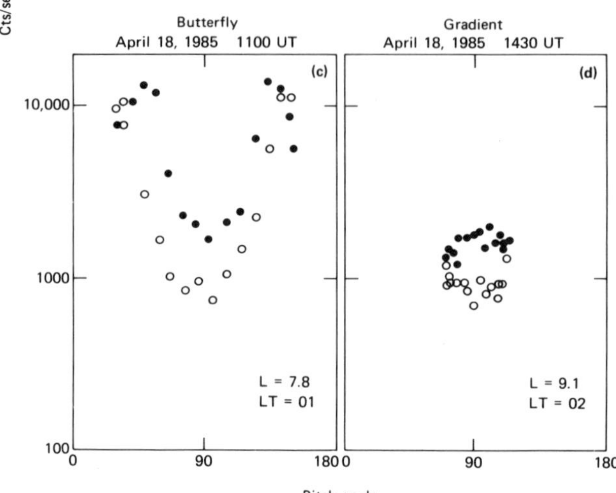

Figure 1—Typical normal (a), isotropic (b), butterfly (c), and gradient (d) PADs. Solid and open circles at each pitch angle mark MEPA sectors receiving particles separated by phases of 180°.

Isotropic PADs are easily distinguished by their refusal to fall to lower flux levels at the lowest pitch angles like normal PADs or at the highest pitch angles like butterfly PADs. Those few cases in which almost no pitch-angle coverage is available are assigned to the isotropic category, although they could well be normal or butterfly PADs. Reassigning them to either category would not affect the results of this study, since the isotropic PAD occurrence rate is about an order of magnitude less than that for other PADs, even in the magnetotail region where they are most frequent.

Figure 2 shows the extent of the processed MEPA coverage. We divided the equatorial plane into 36 longitudinal sectors of 10° each, and established seven radial ranges from 2–9 R_e. The figure shows the number of passes through each of the resulting 252 bins. There is sufficient coverage for a statistical study everywhere except near the dusk meridian, and this will be remedied as further data are processed.

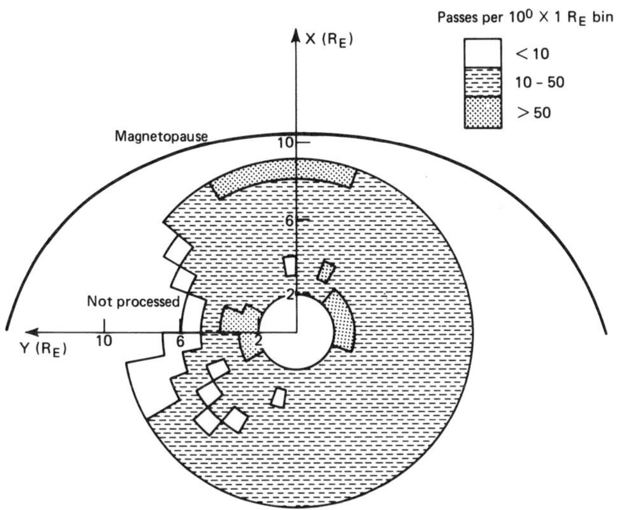

MEPA data coverage, August 25, 1984 to August 7, 1985.

Figure 2—The processed coverage of MEPA observations during the first year of CCE operations. The sun lies at the top of the figure, at local noon. The *y*-axis corresponds to local dusk.

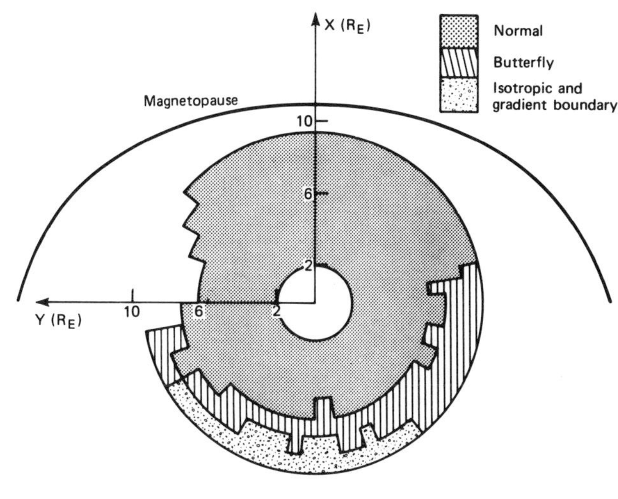

Most likely PAD.

Figure 3—The PAD that occurs most frequently in each 10° by 1 R_e bin.

MOST LIKELY PAD

We tabulated the PADs on each orbit and binned them in the grid. Figure 3 shows the ion PAD with the highest occurrence rate in each bin. When two PADs have equal occurrence rates, the bin has been divided in half longitudinally. The figure shows a very regular pattern in which normal PADs dominate the entire dayside magnetosphere as well as the nightside magnetosphere within 6 R_e of earth. The nightside region lies tailward, dominated by butterfly PADs that stretch in a band from where processed coverage begins in the premidnight sector to shortly after dawn. The butterfly band lies beyond 7 R_e at dawn where its thickness is at least 2 R_e, but moves inward to 6 R_e from earth at midnight where its thickness is about 1 R_e. Tailward of the butterfly PADs lies a region of plasma-sheet and gradient PADs, lying beyond 7 R_e at midnight and 8 R_e toward dusk and dawn. The figure suggests that butterfly PADs are somewhat more common on the dawnside but that isotropic and gradient PADs occur more frequently on the duskside.

Figure 3 is basically consistent with the electron PADs reported and the ion PADs predicted by West et al. However, in contrast to the results of Garcia and Spjeldvik, we find that isotropic and gradient PADs terminate the tailward extent of a nightside band of butterfly PADs extending from dusk through dawn. They found that the band of butterfly PADs extends from dusk through midnight to a predawn local time and that quasinormal PADs terminate the band's tailward extent. We feel the differing results can be explained by our larger data set and observations closer to the equatorial plane. Garcia and Spjeldvik used observations made up to 40° from the geomagnetic equator (for a ratio of observed local to model equatorial magnetic field strengths of B/B_0 = 1.5), while our observations were made no more than

16.2° corresponding to B/B_0 = 1.1 in a dipole field. Since equatorial butterfly PADs map to normal distributions off the equator, they may have overestimated the extent and occurrence of normal PADs at the expense of butterfly PADs.

DAWN-DUSK ASYMMETRY IN BUTTERFLY ION PAD OCCURRENCE

West et al. predicted that butterfly ion PADs would be more prevalent on the dawnside of the magnetosphere than the duskside since the particles drift westward and are shadowed by the dawnside magnetopause. We might also understand a relative absence of duskside butterfly ion PADs by invoking a mechanism for producing isotropic PADs from butterfly PADs in the magnetotail. A final alternative is that injected isotropic PADs supplant duskside butterfly PADs.

Figure 4 depicts a detailed study of butterfly PAD occurrence. The figure shows that butterfly PADs occur in the morning quadrant as an extension of the nightside belt, but at a rate that diminishes toward noon. These dayside butterfly PADs probably result from magnetopause shadowing, as they are observed within the CCE orbit when K_p is high, i.e., when the magnetopause is compressed so that the CCE crosses the open drift paths of the outer magnetosphere. The frequency of butterfly PAD observations increases from noon toward dawn because the open drift paths lie nearer earth at dawn and dusk than at noon. Butterfly PADs were also observed on occasion in the dayside magnetosphere within 3 R_e in contrast to the nightside, where they were never observed nearer earth than 4 R_e. These near-earth dayside butterfly PADs, first reported by Lyons and Williams,[8] will be the subject of a separate study.

Here we concentrate on showing that butterfly ion PADs are more prevalent in the post- rather than premidnight quadrants. To do this, we compare the but-

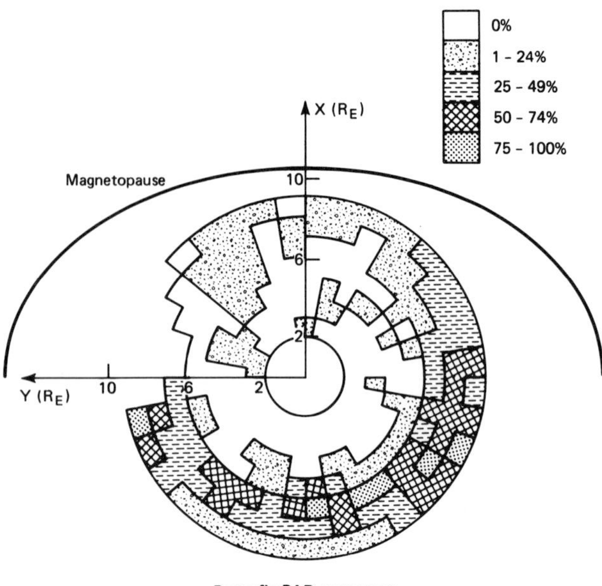

Figure 4—The occurrence of butterfly PADs. They are more common in the post- rather than premidnight quadrant.

terfly occurrence rate of each premidnight quadrant bin with its symmetric partner in the postmidnight quadrant. For example, butterfly PADs occur more frequently (75–100%) in the bin between $R = 6$ and 7 just after midnight local time than in the bin at the same radial distance just prior to midnight (50–74%). On this basis, we find that of 61 comparable bins, the postmidnight occurrence rate is greater in 21 bins, the premidnight rate is greater in 3 bins, and the two are roughly equal in 37 bins. Thus, the postmidnight butterfly ion PAD occurrence rate is greater than the premidnight rate, confirming shadowing model predictions.

As further evidence for the shadowing mechanism, we have examined the relationship between butterfly PADs and magnetopause crossings. Magnetospheric drift paths of outer-magnetospheric ions with 90° pitch angles terminate at the postnoon magnetopause and begin at the prenoon magnetopause. Thus ions with all pitch angles may be supplied to the region inside the postnoon magnetopause, while 90° pitch-angle particles cannot arrive from the prenoon magnetopause. We therefore expect to find butterfly PADs just inside prenoon magnetopause crossings, but not inside postnoon magnetopause crossings. In fact, butterfly PADs were observed in the outer magnetosphere on 11 of 17 prenoon orbits that crossed the magnetopause, but on only 2 of 14 postnoon orbits that crossed the magnetopause, confirming the prediction.

BUTTERFLY PADs ON CLOSED DRIFT PATHS

Although some butterfly PADs in the nightside magnetosphere result from magnetopause shadowing, we can show that others occur on average inside closed drift paths and therefore must be caused by the drift shell

splitting and radial flux gradient mechanism. Figure 5 compares regions in which butterfly PADs occur more than 50% of the time with contours of constant magnetic field strength reported by Fairfield.[9] The comparison depends on the reasonable assumption that the solar-wind pressure was not significantly different on average in 1984-85 when the MEPA observations were made than in 1963-67 when Fairfield's data were obtained. The region of butterfly observations straddles the closed 70 nT and open 60 nT contours at dawn and dusk, but lies well within the closed nightside 70 nT contour. Since substorm particle injection and drift cannot be a steady cause of butterfly PADs and shadowing only occurs on open contours, most of the steady butterfly PADs within the 70 nT contour must be caused by the drift-shell splitting and radial flux gradient mechanism. Note that although shadowing by a compressed magnetopause could produce occasional nightside butterfly PADs within 6 R_e of earth, it cannot account for their presence under typical conditions.

CONCLUSION

We have presented preliminary results of the first statistical energetic (34–50 keV) ion PAD study within 9 R_e of earth. The results agree with a "mirror image" projection of the electron PADs presented by West et al., as expected, since the particles drift in opposite directions. Normal ion PADs dominate the dayside and in-

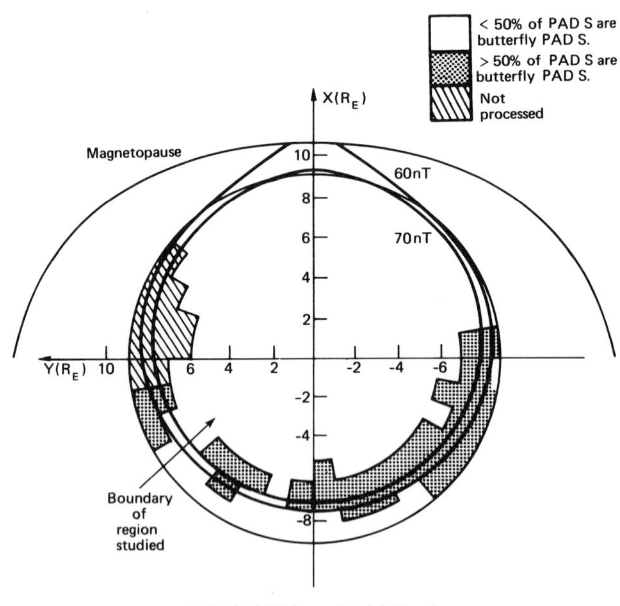

Figure 5—A comparison of butterfly PAD occurrence and particle drift paths. Nightside butterfly PADs are commonly observed within the closed 70 nT drift path. These PADs are caused by drift-shell splitting and a negative radial flux gradient. Butterfly PADs are occasionally (see Fig. 4), but not typically (Fig. 5), observed in the morning quadrant. This is consistent with the occasional CCE traversal of compressed open drift paths from which 90° particles are absent. The drift-shell splitting mechanism cannot produce butterfly PADs in this dayside region of negative radial flux gradient.

ner magnetosphere. Butterfly ion PADs typically occur in a narrow belt that passes from dusk to dawn through midnight, where they approach within 6 R_e of earth. In contrast to Garcia and Spjeldvik, we find that butterfly PADs occur more frequently on the dawnside of the magnetosphere than the duskside. The dawnside butterfly PADs are probably caused by magnetopause shadowing since they occur on open dayside drift paths. However, butterfly PADs near midnight typically occur on closed drift paths, implying that they are caused by the drift-shell splitting and flux gradient mechanism. Finally, isotropic and gradient boundary ion PADs lie beyond the butterfly ion PAD belt.

ACKNOWLEDGMENT—This work has been supported by NASA under Task I of contract N00024-85-C-5307. We thank D. H. Fairfield for helpful comments.

REFERENCES

[1] H. I. West, Jr., R. M. Buck, and J. R. Walton, "Electron Pitch Angle Distributions Throughout the Magnetosphere as Observed on Ogo 5," *J. Geophys. Res.* **78**, 1064-1081 (1973).

[2] R. M. Buck, H. I. West, Jr., and R. G. D'Arcy, Jr., "Satellite Studies of Magnetospheric Substorms on August 15, 1968. 7. Ogo 5 Energetic Proton Observations–Spatial Boundaries," *J. Geophys. Res.* **78**, 3103-3119 (1973).

[3] D. J. Williams, "Magnetopause Characteristics Inferred from Three-Dimensional Energetic Particle Distributions," *J. Geophys. Res.* **84**, 101-104 (1979).

[4] J. G. Luhmann, "The Effect of Transient Localized Sources on Equatorial Pitch Angle Distributions in the Magnetosphere," *J. Geophys. Res.* **82**, 1943-1946 (1977).

[5] H. A. Garcia and W. N. Spjeldvik, "Anisotropy Characteristics of Geomagnetically Trapped Ions," *J. Geophys. Res.* **90**, 347-358 (1985).

[6] R. W. McEntire, E. P. Keath, D. E. Fort, A. T. Y. Lui, and S. M. Krimigis, "The Medium Energy Particle Analyzer (MEPA) on the AMPTE/CCE Spacecraft," *IEEE Trans. Geosci. Remote Sensing* **GE-23**, 230-233 (1985).

[7] T. A. Potemra, L. J. Zanetti, and M. H. Acuna, "The AMPTE/CCE Magnetic Field Experiment," *IEEE Trans. Geosci. Remote Sensing* **GE-23**, 246-249 (1985).

[8] L. R. Lyons and D. J. Williams, "The Storm and Post-Storm Evolution of Energetic (35-560 keV) Radiation Belt Electron Distributions," *J. Geophys. Res.* **80**, 3985-3994 (1975).

[9] D. H. Fairfield, "Average Magnetic Field Configuration of the Outer Magnetosphere," *J. Geophys. Res.* **73**, 7329-7338 (1968).

ACCELERATION OF IONS OF IONOSPHERIC ORIGIN
IN THE PLASMA SHEET DURING SUBSTORM ACTIVITY

E. Möbius, M. Scholer, B. Klecker, D. Hovestadt*

G. Gloeckler, F. M. Ipavich[†]

The ionic mass and charge composition of the suprathermal (10–230 keV/e) population of the plasma sheet has been measured using the Suprathermal Energy Ionic Charge Analyzer (SULEICA) on AMPTE/IRM before and after a substorm onset on April 8, 1985. Simultaneously with the substorm onset and strong earthward plasma flows, a flux increase of the suprathermal ions and a significant hardening of the spectra are observed that are most pronounced for O^+. The enhancement ratios can be best organized in terms of energy per charge. The observations are interpreted as an indication of particle acceleration by electric fields and additional injection of ionospheric plasma into the plasma sheet after substorm onset.

INTRODUCTION

It has been recognized in recent years that the ionosphere is an important source for the plasma-sheet population; e.g., Balsiger et al.,[1] Lennartsson et al.,[2] and Peterson et al.[3] have shown that O^+ ions can represent more than half of the number density in the plasma sheet at energies below 17 keV/e. The contribution of ionospheric ions to the plasma sheet depends on geomagnetic activity; e.g., Sharp et al.[4] found that the O^+/H^+ ratio varied from 2% during magnetically quiet times ($AE < 100\ \gamma$) to $\approx 40\%$ after major substorms ($AE > 500\ \gamma$).

First observations of O^+ in the plasma sheet at higher energies, i.e., above 100 keV, were reported by Ipavich et al.[5] They found O^+/H^+ abundance ratios at 130 keV as high as 0.35. Ipavich et al.[6] investigated in detail the intensity-time behavior of the 130 keV O^+ ion flux during the CDAW 6 substorm on March 22, 1979. They found (a) tailward-directed O^+ ions appear as early as ≈ 10 min. before substorm onset, (b) a second increase of the O^+/H^+ ratio in the tailward direction ≈ 12 min after onset, and (c) a decrease of the O^+/H^+ ratio in the late phase of the substorm when the H^+ flow was earthward. Ipavich et al.[6] suggested that the O^+/H^+ ratio increase after substorm onset was due to ionospheric ions flowing into and being accelerated at the neutral line after all plasma-sheet field lines were being reconnected.

Ipavich et al.[5,6] presented differential fluxes of O^+ ions from one energy channel only. With the improved instrumentation on the AMPTE/IRM and CCE spacecraft it has now become possible to analyze the various ionic species in the suprathermal energy range in greater detail. We will report in this paper for the first time spectral measurements of suprathermal O^+, He^+, and CNO ($Q > 2$) ions in the plasma sheet in the course of a geomagnetic substorm that occurred on April 8, 1985. These measurements are supplemented by spectral measurements of H^+ and He^{++}. The data have

been obtained with the Suprathermal Energy Ionic Charge Analyzer (SULEICA) on AMPTE/IRM. The instrument has been described in detail by Möbius et al.[7] Studying the spectral variations of the various ionic species in the course of a substorm will not only reveal the relative contribution of the ionosphere and the solar wind as a source for the plasma sheet, but should result in new insight into the acceleration process itself.

OBSERVATIONS

The time period of interest was geomagnetically active as indicated in the ground magnetometer reading from the EISCAT magnetometer stations (not shown here). Major variations of the magnetic field occurred at about 1850 UT ($\Delta B \approx 100$ nT), 2200 UT ($\Delta B \approx 300$ nT), and 2305 UT ($\Delta B \approx 800$ nT). According to the AMPTE/IRM plasma measurements the satellite entered the plasma sheet at ≈ 1850 UT and stayed there for the whole time period. Between 2100 and 2300 UT several neutral sheet crossings were observed, indicating that the satellite was near the center of the plasma sheet at that time. In this paper we will concentrate on the strongest event at 2305 UT.

Figure 1 shows the 1-hour period from 2300 to 2400 UT when the satellite was at a distance between 14.5 and 15.2 R_e and at ≈ 2300 LT on the outbound part of the orbit. From top to bottom, the panels display the plasma flow velocity and its direction in azimuth as obtained by the three-dimensional-plasma instrument[8] and the omnidirectional differential fluxes of 160 keV protons, 160 keV O^+ ions, and 1 MeV protons as obtained by the SULEICA instrument. The 160 keV H^+ and O^+ ions are sampled every 19th spin according to the voltage stepping scheme of the electrostatic analyzer.[7] Strong plasma flows into the earthward direction with velocities up to 1000 km/s are observed for short time periods of ≈ 1 min each starting at 2307, 2318, and 2321 UT. The first flow event coincides according to Fig. 1 with a strong flux increase of the 160 keV protons and O^+ ions. Such a flux increase is actually seen over a wide energy range. The flux of the 1 MeV protons increases slowly, starting at ≈ 2312 UT and reaching its maximum around 2320 UT, seeming to coincide with

*Max-Planck Institut für extraterrestrische Physik, 8046 Garching, Federal Republic of Germany.

[†]Department of Physics and Astronomy, University of Maryland, College Park, Maryland 20742.

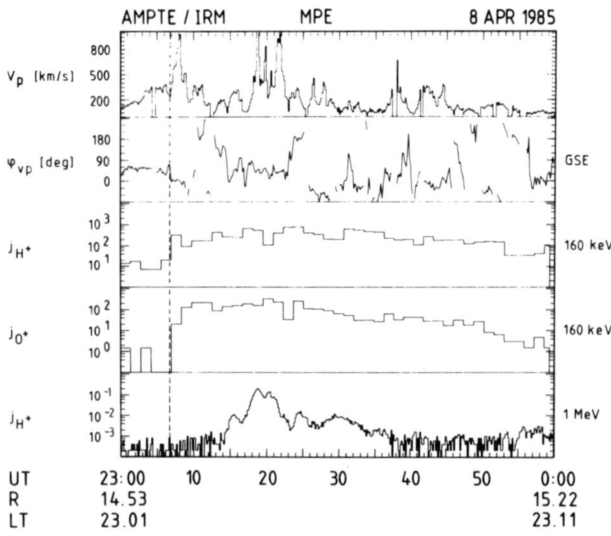

Figure 1—From top to bottom: plasma flow velocity v_p, azimuthal flow direction φ_{v_p} in GSE coordinates, differential flux of 160 keV protons, 160 keV O$^+$ ions, and 1 MeV protons (in particles/s-cm^2-sr-keV/e) versus time.

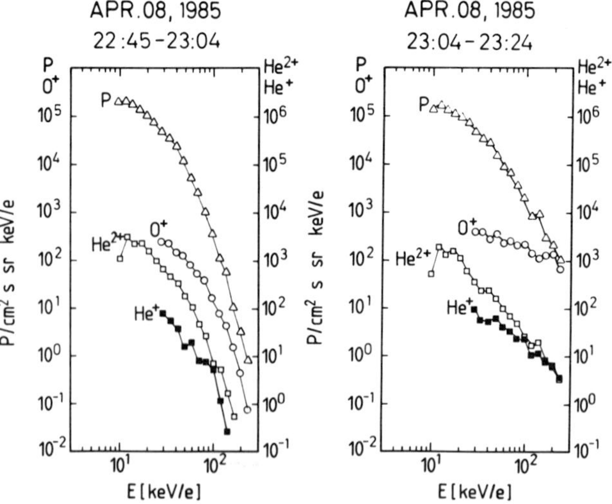

Figure 2—Energy spectra of H$^+$, O$^+$, He^{2+}, and He$^+$ ions. The representation of He$^+$ and He^{2+} is shifted by one order of magnitude. Left: before substorm onset. Right: after substorm onset.

the second series of plasma flow events. We note in this respect that the first velocity increase was predominantly parallel to the magnetic field, while the second flow increase was almost entirely perpendicular to the magnetic field resulting in strong convection electric fields. The energetic ion events observed in situ in the plasma sheet are well correlated with the large substorm ($\Delta B \approx 800$ nT), which occurred at ≈ 2305 UT.

Figure 2 shows energy spectra before and after the flux increase. In order to increase the statistical significance, 20 min averages were taken. The fluxes stay relatively constant during the averaging intervals. The differential flux of H$^+$, He$^+$, He^{2+}, and O$^+$ is shown versus energy per charge. For a better separation of the individual species the fluxes of He$^+$ and He^{2+} have been multiplied by 10^{-1}. The data for He$^+$ and O$^+$ below 28 keV/e have been omitted in this representation, since they would require a substantial background subtraction in order to remove the time-of-flight pile up at low energies, where no solid-state detector signal can be used for the identification. During both time periods the He^{2+} spectrum is similar to the proton spectrum, while the spectra of He$^+$ and O$^+$ ions are significantly harder than those of H$^+$ and He^{2+}. Between the two time periods there is little or no change at the low-energy end of the spectra. However, after the substorm onset at 2307 UT there is a considerable flux increase at higher energies, which reaches three orders of magnitude at ≈ 200 keV/e, i.e., all the spectra are significantly harder than before the substorm.

In order to present a more quantitative description of the spectral changes, the ratios $K_i(E/Q)$ of the ion fluxes after and before the substorm onset have been calculated for each individual species, now including also ions of the CNO group with an ionic charge $Q > 2$. The results are shown in Fig. 3 as a function of (a) rigidity R, (b) energy per mass unit E/A, and (c) energy per charge E/Q. The increase of the ratios with energy is due to the hardening of the energy spectra. There is a large separation between the individual curves in a representation ratio versus rigidity and ratio versus energy per mass. However, the ratios for the various species exhibit the same relative behavior as a function of energy per charge, i.e., for each ion i, $K_i(E/Q)$ can be written as $K_i(E/Q) = D_i \cdot K_{H^+}(E/Q)$ where D_i is the enhancement factor of each ion species. The factors D_i are compiled in Table 1. They represent the mean enhancement (respective depletion) of the ion flux ratios (after and before substorm onset) as compared to the proton flux ratio. He^{2+} and CNO ions are depleted compared to H$^+$, while He$^+$ and O$^+$ ions are enhanced. As can be seen from Fig. 3, the depletion of He^{2+} and CNO indeed corresponds to a flux decrease during the substorm at low energies (< 50 keV/e). Above 50 keV/e, He^{2+} and CNO ions are also enhanced after the onset of the substorm. Note that for all species but O$^+$ and He$^+$ the fluxes are depleted at low energies after the substorm onset.

DISCUSSION

We have shown in this paper that after the onset of a major substorm the fluxes of various ionic species in the plasma sheet increase sharply in the suprathermal energy range (up to ≈ 230 keV/e). After the onset, large earthward directed plasma flows have been observed concurrent with energetic protons up to ≈ 1 MeV. The ratios of the ion flux after and before substorm onset increase with energy. Their spectral behavior can be best organized for all species in terms of energy per charge. The contents of ionospheric ions, like O$^+$ and He$^+$ during the substorm, is significantly enhanced compared to solar-wind ions like He^{2+} and CNO with $Q > 2$. From the plasma observations and the presence of the

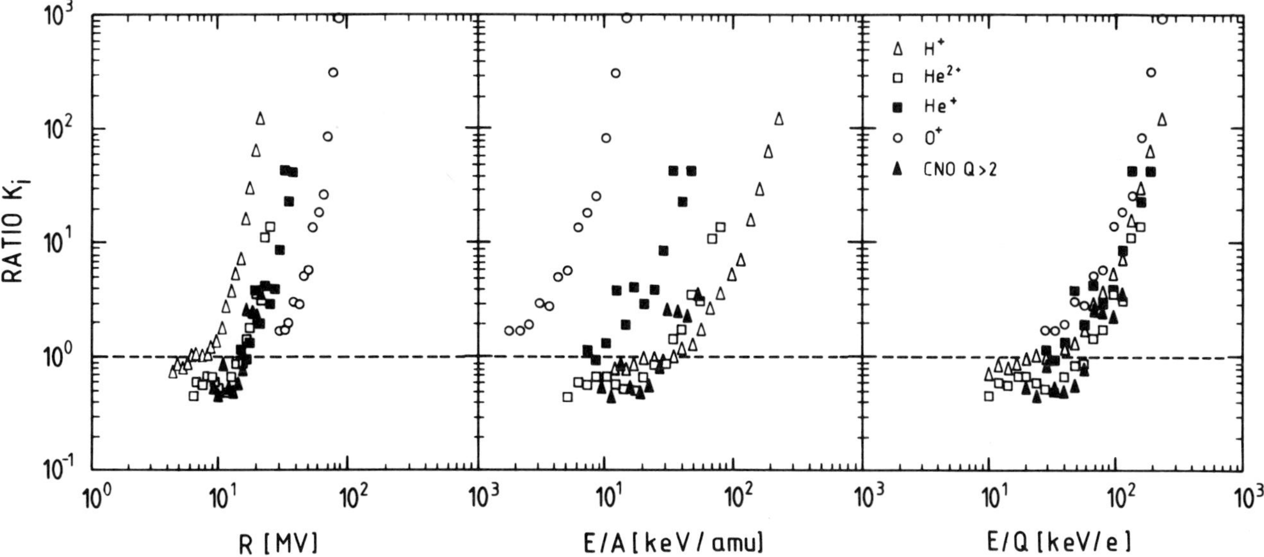

Figure 3—Ratios K_i of the ion fluxes after and before substorm onset for H$^+$, He^{2+}, He$^+$, O$^+$, and CNO with $Q > 2$ as a function of rigidity (left), energy/mass (middle), energy/charge (right).

Table 1—Mean enhancement factors D_i compared to protons.

He^{2+}	CNO with $Q > 2$	He$^+$	O$^+$
0.61	0.65	1.37	2.83

suprathermal ions it can be concluded that the spacecraft was during the whole time period well within the plasma sheet.

We will interpret our observations in terms of the classical picture of a substorm (e.g., Hones[9]) in which a new magnetic neutral line is formed in the thinning plasma sheet and most of the plasma sheet is ejected tailward as a plasmoid. The observation of strong earthward flows at the satellite during substorm onset is consistent with the spacecraft being earthward of the neutral line within the remainder of the old plasma sheet. In this model the observed flux increase of suprathermal ions is due to acceleration at the neutral line and subsequent injection into the plasma sheet.

The ordering of the flux variations in energy per charge for species with different ionic charge (varying from $Q = 1$ for H$^+$, He$^+$, O$^+$ to $Q \approx 6$ for CNO) strongly suggests an electric-field acceleration mechanism for these ions. In a representation ratio versus rigidity, on the other hand, different species exhibit a different behavior. This rules out a strong rigidity dependence of the acceleration process. There is also no ordering of the ratio change as a function of velocity. The acceleration by electric fields is in accordance with the concept that strong electric fields are induced in the neutral-line region during substorm activity (e.g., Axford[10]). However, the variation of the energy spectra from the prestorm situation to the time period after substorm onset is not a mere shift in energy per charge that would be observed in the case of a linear acceleration in an electric field. An electric field acceleration process, where the particles reside for a time of the order of $1/\omega_{ci}$ in the acceleration region, can also be ruled out, since different species with the same rigidity behave differently. The enhancement of the ion fluxes increases exponentially with energy per charge. This is similar to heating of a thermal distribution, so that acceleration in electric fields together with strong scattering by waves may explain the observations.

Although the acceleration process works in the same way for all species, there is a significant overall enhancement of ionospheric ions like O$^+$ and He$^+$ over solar wind ions like He^{2+} and CNO with $Q > 2$, as can be seen from Table 1. This result cannot be explained in terms of preferential heating or acceleration for different species, since there is no unique ordering of the enhancement factors with any of the parameters like rigidity or energy per mass. In addition, it can be shown that the total density of suprathermal ions (in the energy range 28–226 keV/e) has increased for He$^+$ and O$^+$ after substorm onset, while the corresponding value of He^{2+} and CNO has decreased. The results are compiled in Table 2. The suprathermal proton density shows only a small variation.

These results are a clear indication for a preferential injection of ionospheric material into the acceleration region during substorm activity. In order to reach the magnetic neutral line, the ionospheric ions first have to travel either along the lobe field lines close to the plasma sheet or in the plasma-sheet boundary layer. Such a model is sketched in Fig. 4—ionospheric plasma is heated at the foot points of the last open field lines, i.e., in the auroral zones. This causes the outflow of thermal O$^+$, He$^+$, and H$^+$ ions along the magnetic field lines. Such ions have been observed in the boundary layer

Table 2—Densities of suprathermal ions.

28–226 keV/e	H^+	He^{2+}	CNO with $Q > 2$	He^+	O^+
2245–2305 UT n_1 (cm^{-3})	4.16×10^{-2}	8.49×10^{-4}	3.15×10^{-5}	1.57×10^{-4}	1.3×10^{-3}
2305–2324 UT n_2 (cm^{-3})	5.21×10^{-2}	6.44×10^{-4}	1.96×10^{-5}	3.33×10^{-4}	5.34×10^{-3}
n_2/n_1	1.25	0.76	0.62	2.12	4.11

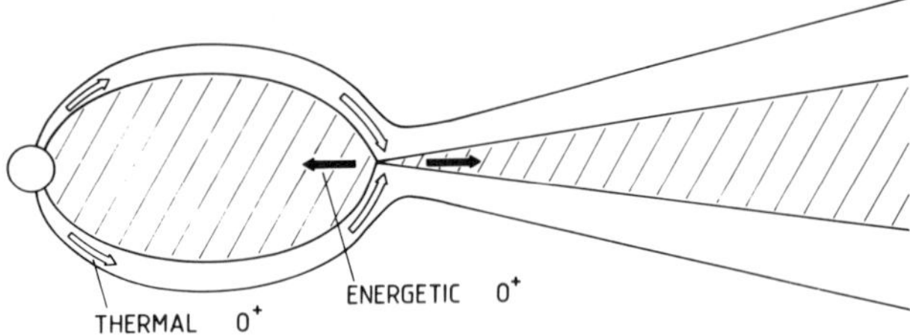

Figure 4—Schematic view of the magnetotail and the flow of ionospheric ions.

and the lobe by Sharp et al.[11] and Candidi et al.[12] The thermal ions are accelerated in the neutral line region and finally are injected into the plasma sheet.

Contrary to the ions of ionospheric origin, those of pure solar-wind origin as He^{2+} and CNO with $Q > 2$ are depleted compared to protons. This behavior seems to indicate that solar-wind ions play only a minor role in supplying the plasma sheet during substorm activity compared to ionospheric ions. However, we observe in addition, a depletion in absolute flux at low energies. Expansion of the plasma sheet earthward of the new neutral line in the course of the dipolarization of the magnetic field may cause such a flux depletion. On the other hand, injection from the ionospheric source seems to exceed the effect of the expansion. The H^+ ions partly originate in the solar wind and partly in the ionosphere. Therefore, the variation of the proton density is between that of ions originating in the ionosphere and in the solar wind.

ACKNOWLEDGMENT—The authors are grateful to the many individuals at the Max-Planck Institut and the University of Maryland who contributed to the success of the SULEICA instrument and the AMPTE/IRM spacecraft. We thank G. Paschmann for the plasma data, H. Lühr for the ground magnetometer data, and R. P. Lin for the high-energy-proton data.

REFERENCES

[1] H. Balsiger, P. Eberhardt, J. Geiss, and D. T. Young, "Magnetic Storm Injection of 0.9-16 keV/e Solar and Terrestrial Ions into the High-Altitude Magnetosphere," *J. Geophys. Res.* **85**, 1645 (1980).

[2] W. Lennartsson, R. D. Sharp, E. G. Shelley, R. G. Johnson, and H. Balsiger, "Ion Composition and Energy Distribution During 10 Magnetic Storms," *J. Geophys. Res.* **86**, 4628 (1981).

[3] W. K. Peterson, R. D. Sharp, E. G. Shelley, and R. G. Johnson, "Energetic Ion Composition of the Plasma Sheet," *J. Geophys. Res.* **86**, 761 (1981).

[4] R. D. Sharp, W. Lennartsson, W. K. Peterson, and E. G. Shelley, "The Origins of the Plasma in the Distant Plasma Sheet," *J. Geophys. Res.* **87**, 10,420 (1982).

[5] F. M. Ipavich, A. B. Galvin, G. Gloeckler, D. Hovestadt, B. Klecker, and M. Scholer, "Energetic (>100 keV) O^+ Ions in the Plasma Sheet," *Geophys. Res. Lett.* **11**, 504 (1984).

[6] F. M. Ipavich, A. B. Galvin, M. Scholer, G. Gloeckler, D. Hovestadt, and B. Klecker, "Suprathermal O^+ and H^+ Ion Behavior During the March 22, 1979 (CDAW 6) Substorms," *J. Geophys. Res.* **90**, 1263 (1985).

[7] E. Möbius et al., "The Time-of-Flight Spectrometer SULEICA for Ions of the Energy Range 5-270 keV/charge on AMPTE/IRM," *IEEE Trans. Geosci. Remote Sensing* **GE-23**, 274 (1985).

[8] G. Paschmann, H. Loidl, P. Obermayer, M. Ertl, R. Laborenz, N. Sckopke, W. Baumjohann, C. W. Carlson, and D. W. Curtis, "The Plasma Instrument for AMPTE/IRM," *IEEE Trans. Geosci. Remote Sensing* **GE-23**, 262 (1985).

[9] E. W. Hones, Jr., "Substorm Processes in the Magnetotail: Comments On 'Hot Tenuous Plasmas, Fireballs, and Boundary Layers in the Earth's Magnetotail' by L. A. Frank, K. L. Ackerson, and R. P. Lepping," *J. Geophys. Res.* **82**, 5633 (1977).

[10] I. Axford, "Magnetic Field Reconnection," in *Magnetic Reconnection in Space and Laboratory*, Geophysical. Monograph 30, American Geophysical Union, Washington, D.C., p. 1 (1984).

[11] R. D. Sharp, D. L. Carr, W. K. Peterson, and E. G. Shelley, "Ion Streams in the Magnetotail," *J. Geophys. Res.* **86**, 4639 (1981).

[12] M. Candidi, S. Orsini, and V. Formisano, "The Properties of Ionospheric O^+ Ions as Observed in the Magnetotail Boundary Layer and Northern Plasma Lobe," *J. Geophys. Res.* **87**, 9097 (1982).

DISTINGUISHING AMONG ELECTRON INJECTION TYPES

J. Feynman*

Three types of electron injections taking place in the near-earth region of the magnetotail have been distinguished previously using SCATHA particle and field data. Defining characteristics are given here for each type of event, and the positions of the magnetosphere where they are expected to occur are discussed. These three event types can be difficult to distinguish in data sets that are more limited than the SCATHA set that carried instruments detecting magnetic fields and charged particles over an energy range from eVs to MeVs. It is suggested that determining the magnetospheric regions at which each of these event types occurs will considerably clarify the phenomenological description of substorms available for theoretical analysis.

INTRODUCTION

The purpose of this note is to differentiate among three types of electron injections seen in the near-earth magnetotail. These types of injections are clearly distinguishable in data from the near-geosynchronous satellite, SCATHA, but are difficult or even impossible to differentiate in data available from some other spacecraft. As a result, it is often impossible to tell from the published data which type of event is being described, and the three essentially different types of events are often assumed to be different examples of the same type of event. The events represent different aspects of substorm and/or increased convection phenomena, and the failure to distinguish among event types may have, and probably has, resulted in a confused description of near-earth magnetotail phenomena. Because these phenomenological descriptions guide the development of theory, it may well be that studies of when and where these different event types occur will lead to considerable clarification of the substorm problem as presented to the theorist for analysis.

Before proceeding to the description of these events it should be stressed that these three types are not at all intended to exhaust the different types of phenomena seen in this region of the magnetosphere.

The data shown here are from the AFGL rapid-scan particle detector (SC5) aboard SCATHA. The satellite was launched in January 1979 into an orbit covering the earth-satellite distance range from 5.3–7.9 R_e. The satellite drifts 6° per day in longitude toward later local time. The electron detectors were sensitive to electrons in the range from 0.05 keV to 1 MeV. The electron data used are 1 min averages of the count rate of detectors with look directions nominally perpendicular to the magnetic field (i.e., pitch angles near 90° unless the magnetic field was highly distorted). The ion data in Fig. 2 are in the same format and from the same instrument. SCATHA also carried a NASA/Goddard magnetometer built and operated under the direction of B. G. Ledly. Data from that instrument are the basis of some of the qualitative statements concerning magnetic field changes accompanying the events.

---

*Jet Propulsion Laboratory, Pasadena, California 91109.

DEFINITIONS

Figure 1 shows schematic representations of the earthward regions of the magnetotail in which 40 keV and 20 keV 90° pitch-angle electron fluxes are observed. (The label "90° pitch angle" will be omitted but should be understood in the remainder of this comment.) These schematics represent the midnight situation for observation in the near geosynchronous region. The equatorward boundary of the plasma sheet for 40 keV particles is shown at $\geq 6\ R_e$. This value was chosen for illustrative purposes only. The solid line between the shaded and stippled regions separates the region of the magnetosphere in which 40 keV electrons are always present (shaded) from the higher latitude region in which fluxes of 40 keV electrons are intermittent (stippled). The position of this boundary was determined from an extensive set of SCATHA observations at near geosynchronous.[1] Some 352 days of almost continuous data were used for this determination and the position shown for the boundary is that for $K_p \leq 4+$. The boundary between the region of intermittent fluxes and the region in which 40 keV fluxes are almost never seen is shown as a dashed line. The actual position of this boundary could not be determined from SCATHA observations and is crudely estimated here. The situation for 20 keV electrons is shown on the right-hand side of the figure, which is meant to be a crude representation based on the notions that the 20 keV electrons can penetrate closer to earth than the 40 keV electrons in the equatorial region and

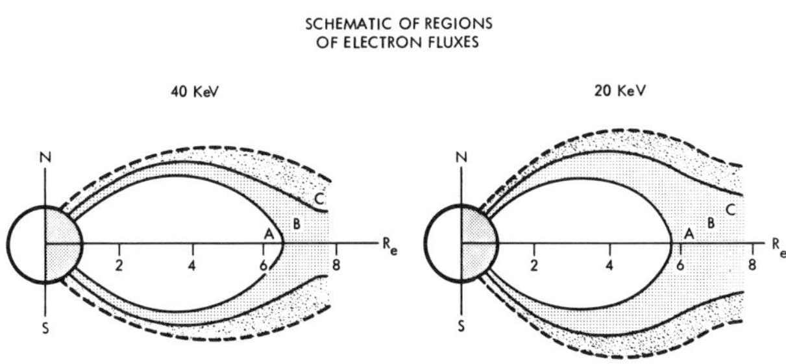

Figure 1—Schematic representation of earthward regions of the magnetotail in which 40 keV and 20 keV electron fluxes are observed. See text for description.

also will be trapped to higher latitudes since lower energy electrons can follow field lines with smaller radii of curvature in the tailward part of the plasma sheet than can higher energy particles.

Three regions at which satellites can observe events have been labeled A, B, and C. In the figure, which represents the situation at the equinoxes, these three positions seem to be at clearly distinguishable latitudes. However, as discussed later, the B and C latitudes are often very difficult (but not impossible) to tell apart in practice.

Plasma Sheet Entry (Region A)

The position of the equatorward extent of 40 keV electron penetration drifting toward the earth from the more distant tail will depend on K_p or, more correctly, the cross-tail electric field. Lower energy electrons will come closer to earth. If the cross-tail electric field remains constant for a period of time, the positions of the equatorward boundaries will remain unchanged. A geosynchronous satellite will move across these quiescent boundaries and, as it moves to later local time and deeper into the plasma sheet, it will cross the boundaries for one energy after another. As it crosses the boundaries, the fluxes of the pertinent electron energies will increase. The observational result will be a highly dispersed "electron injection," using the term "electron injection" in the sense defined by Kivelson et al.,[2] i.e., an increase in electron fluxes.

Figure 2 shows SCATHA electron and high-energy ion data for February 12, 1979. The data for this day were discussed in Ref. 3. A rise in low-energy electron fluxes was seen by ATS 6 at about 0145 local time (1545 UT), some 1¾ h before it was seen at SCATHA at about 0335 UT (1615 local time) at 5.9 R_e. This was interpreted as being due to the spacecraft crossing quiescent boundaries and, of course, no substorm was reported accompanying these particle increases.

There is a somewhat different but related process in which the boundaries corresponding to the most equatorward extents of plasma-sheet electrons of different energies move across the spacecraft. In this type of event an enhancement of the convection electric field deflects the sunward flow of the plasma-sheet electrons causing the boundary for a particular energy to come closer to the earth in the equatorial regions so that higher energy electrons may now reach the satellite. These events show a large energy dispersion, with higher energies appearing later and little or no magnetic perturbation at the satellite[2,4,5] and have been referred to as "electric substorms."

Dynamic Injections (Region B)

A second type of event, "dynamic," injection occurs when the spacecraft is in Region B, within the region of high 40 keV particle fluxes. Two examples are seen at about 0500 and 0730 UT in Fig. 2. This type of event and these specific examples were previously discussed in Ref. 3. They are accompanied by a change in the magnetic field to a more dipolar configuration and show little

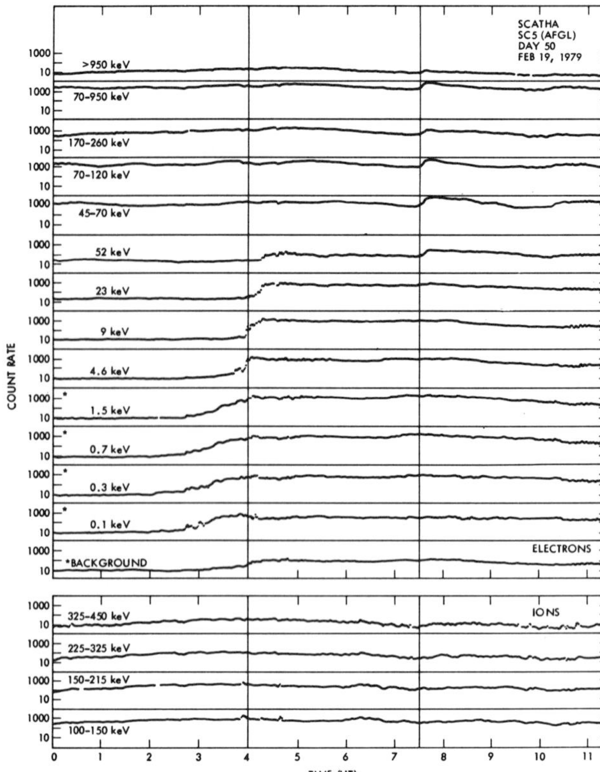

Figure 2—Particle count rates observed on SCATHA using AFGL rapid-scan particle detector. The count rate labeled "background" should be subtracted from the count rates marked with a star (electrons at 0.1–1.5 keV) to find the measured count rate at those energies. An event of type A occurs at about 0345 UT. See text for satellite positions. B events are seen at about 0500 UT and 0730 UT. At 0500 UT SCATHA was at 1915 local time and 5.6 R_e, ATS 6 was at 2030 LT and 6.6 R_e. At 0730 UT SCATHA was at 2230 LT and 5.7 R_e and ATS 6 was at 2115 LT and 6.6 R_e.

or no dispersion. An important feature to note is that the event occurs in an energy range that is limited from both above and below. For example the 0500 event involves mainly electrons between 1.5 and 52 keV. It is also not seen in the high energy ions shown in the bottom panels of Fig. 2. In Ref. 3 these events were interpreted as being due to the passage of a propagating injection front across the satellite. The magnetospheric positions at which events of this type occur have not yet been adequately mapped.

Return from Dropout (Region C)

The third type of event, seen in Region C, is shown in Fig. 3 at 0305 and 0315 UT and between 0800 and 0900 UT. It is a return from particle dropout. The examples shown occur at distances greater than geosynchronous; however, events of this type are also commonly seen at geosynchronous (cf. Bogott and Mozer[6]). Return events are defined[1] so that particle fluxes in some selected energy band must be below some chosen threshold before the event takes place and the particle flux at the energy must increase at event time.

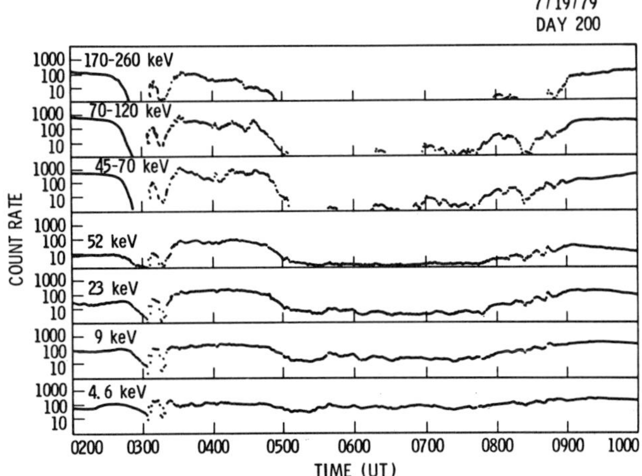

Figure 3—Electron count rates using the same format as in Fig. 2. C events occur at about 0305 UT, 0315 UT, and from ~0700 UT to ~0900 UT. At 0300 UT SCATHA was at 2300 LT, 7.8 R_e, and 12° solar magnetic latitude. At 0800 UT SCATHA was at 0220 LT, 7.1 R_e, and 18° solar magnetic latitude.

The events are also defined by the requirement that they involve the highest energy electrons and ions to which the detector is sensitive. Both electrons and ions take part in the event and the event is more marked at the highest observed energies than at lower energies. There may or may not be dispersion, but if they are dispersed the lower energy fluxes increase first. They may or may not be accompanied by a clear change in the magnetic field. In our study of SCATHA "return" events, the 40 keV electron count rate was used as the energy in the definition of the pre-event threshold and the pre-event count level was required to be less than 10. Since electrostatic analyzers often have an upper energy cutoff of 20–40 keV for electrons, the selection of these criteria for energy and flux levels results in the dropout and return events corresponding to the passage of the "high latitude boundary of trapping" across the satellite.[6] A survey of 352 days of SCATHA data collected between March 15, 1979 and March 14, 1980 revealed 103 days on which returns were observed.[1]

DISCUSSION

The three types of events described are difficult to distinguish when using a limited data set. For example, the main observational differences between electric substorms (Region A) and dynamic injections (Region B) are in the magnetic field observations in space and/or the observation of a nearly simultaneous substorm signature on surface magnetometers, and/or the amount of dispersion in the electrons. It may well be that no space or surface magnetic field observations are available, and in practice the dispersion criterion is difficult to apply because there is only a quantitative difference in dispersion between type A and B events.

A common difficulty occurs in attempts to distinguish dynamic injections from return events. In Fig. 3, each of the returns follows a rather rapid dropout, but this is not always the case. The dropout can be obscured because it is gradual or, more often, incomplete; i.e., the pre-event-measured fluxes do not decline to values as low as those used in the definition. It may also be that ion observations are not available and/or only a limited range of electron energies has been observed. For example, if the only data available were for electrons at less than 40 keV (as is common for electrostatic analyzers) it would be very difficult to distinguish between the dynamic injection seen at 0730 in Fig. 2 and the return seen at about 0800 in Fig. 3. There are many reports in the literature where the data presented do not permit such a distinction to be made, but such data are often interpreted as if they occurred in Regions A or B.

As mentioned above, it may seem that the differences of latitudes at which B and C events occur should be sufficient to distinguish them, but this is not so.[1] Figure 1 shows the geometrically simple case of the equinoxes. At this time no particle returns (or dropouts) are seen near the equator when $K_p \leq 4+$. The equator is at the same place whether it is defined in terms of solar magnetic (SM) coordinates or geocentric solar magnetospheric (GSM) coordinates.[7] However, at other seasons of the year, returns (and dropouts) were seen in the SCATHA data on the equator for $K_p \leq 4+$ for both coordinate systems. This is discussed in detail in Ref. 1, and the discussion will not be repeated here because of lack of space. The result, however, is important—that particle returns due to crossing the high-latitude boundary of trapping occur in unexpected regions of the magnetosphere if the magnetosphere is described in the usual coordinate systems. It was found, however, that a new coordinate system was easily constructed that was halfway between SM and GSM. Using this coordinate system, returns (and dropouts) occurred only at high latitudes for $K_p \leq 4+$. This coordinate system takes into account the fact that the near-geosynchronous region is partially influenced by the hinging of the tail.

CONCLUSION

The point to be stressed in this comment is that the similarity of essentially different types of electron injections, when observed with a restricted data set, can result, and most probably has resulted, in a confusing description of substorm events seen at geosynchronous altitudes. A redefinition of event types and a new determination of where types A and especially B events occur (in an appropriate coordinate system) may well result in clarification of the description of substorms available to theorists, providing the theorist with a clearer and simpler picture with which to compare theoretical results.

ACKNOWLEDGMENT—This work was supported in part by Air Force contract F19628-82-K-0011.

REFERENCES

[1] J. Feynman, D. A. Hardy, and E. G. Mullen, "The 40 keV Electron Durable Trapping Region," *J. Geophys. Res.* **89,** 1517 (1984).

[2] M. G. Kivelson, S. M. Kaye, and D. J. Southwood, "The Physics of Plasma Injection Events," in *Dynamics of the Magnetosphere,* S.-I. Akasofu, ed., D. Reidel, Hingham, Mass, p. 385 (1980).

[3] T. E. Moore, R. L. Arnoldy, J. Feynman, and D. A. Hardy, "Propagating Substorm Injection Fronts," *J. Geophys. Res.* **86,** 6713 (1981).

[4] S. M. Kaye and M. G. Kivelson, "Time Dependent Convection Electric Fields and Plasma Injection," *J. Geophys. Res.* **84,** 4183 (1979).

[5] B. Hultqvist, B. Aparicio, H. Borg, R. Arnoldy, and T. E. Moore, "Decrease of keV Electron and Ion Fluxes During the Early Phase of Magnetospheric Disturbances," *Planet. Space Sci.* **28,** 107 (1981).

[6] F. H. Bogott and F. S. Mozer, "Nightside Energetic Particle Decreases at Synchronous Orbit," *J. Geophys. Res.* **78,** 8119 (1973).

[7] C. T. Russell, "Geophysical Coordinate Transformations," *Cosmic Electrodynam.* **2**(2), 184 (1971).

COMPOSITION AND VELOCITY OF IONS STREAMING IN THE PLASMA MANTLE AND IN THE LOBE

S. Orsini,* M. Candidi,* and H. Balsiger[†]

The ISEE 1 Ion Composition Experiment detects ionospheric and solar-wind ions streaming tailward with the same mean parallel velocity in the open magnetic field regions of the geomagnetotail, at distances ranging from 10 R_e to 20 R_e. This is interpreted to be an effect of the so-called "magnetospheric mass spectrometer" which organizes these streams via the inward $E \times B$ drift, allowing only those ions having a particular flow velocity to get to a certain point down the tail. This effect mixes solar-wind and ionospheric particles, thus filling the plasma mantle and the lobe region with a multicomponent plasma which is cold and collimated and flows with several different energies, one for each ion species involved in this process.

INTRODUCTION

The distinction between the mantle and lobe regions in the magnetotail is generally made, based on the plasma density.[1] The plasma that flows tailward has much higher density in the mantle, close to the magnetopause, than in the lobes, inside the tail. The transition from the mantle to the lobe is gradual; sometimes the lobe plasma is so dilute that plasma experiments cannot easily detect it, and this region has therefore often been defined as "empty." Nevertheless, during disturbed periods cold streams of ionospheric plasma have been detected in the lobe regions, flowing tailward with energies from a few eV to about 1 keV.[2,3] Another criterion to distinguish between mantle and lobe regimes is the presence of solar-wind ions; it has been used to define the mantle as having its source in the solar wind, while the lobe plasma should be of pure ionospheric origin.[4,5]

The studies carried out on the effect of the "magnetospheric mass spectrometer" (in the colorful expression of Lockwood et al.[6]), usually referred to as the $E \times B$ drift filter effect, have helped better define some of these features. They have revealed that ionospheric ions (mainly O^+ and H^+) and solar-wind ions (mainly H^+ and He^{++}) flow with roughly the same velocities in the mantle.[7-9] This is true at large distances from the earth only, while at distances of the order of 1 R_e (where the $E \times B$ drift filter has not had time enough to affect strongly the properties of the streams) the ions appear with equal energies.[10-12] At distances of the order of 10–20 R_e, it has been inferred that the slight differences observed between the flow velocities of the ionospheric ions and of the solar-wind ions may be accounted for by slight differences in travel path inside the mass spectrometer itself.[8]

The ISEE 2 EGD experiment has previously been used to study both the O^+ and H^+ streams in the plasma mantle,[7] but the low-energy cutoff of this experiment (50 eV/q) is too high to measure the bulk of the H^+ ions in the lobes. The ion streams that have been observed by this experiment in the lobes have been identi-fied to be mostly O^+ by comparison with simultaneous data from the ISEE 1 Ion Composition Experiment (ICE). The typical mean velocities of these O^+ ions are less than 100 km/s. Similar velocities of the H^+ ions correspond to energies below the energy range of the EGD experiment but may be sufficient to bring the H^+ ions within the range of the ICE (10 eV/q).

The validity of the magnetospheric mass spectrometer model in all the open-field line regions of the geomagnetotail, including the lobes, will be tested here in a statistical study of these streams as observed by using the ISEE 1 ICE in 1978-79. In order to better qualify the data collected by ICE with its low time resolution and wide energy channels at low energy, the periods have been selected on the basis of the detailed observation of these streams by the ISEE 2 EGD experiment. The data from the ISEE 2 EGD experiment have been used as a monitor only. We will focus on the properties of H^+ and O^+ ions, streaming tailward in the mantle and lobe regions of the magnetotail during geomagnetically disturbed periods; it will be shown that these ions are observed to flow with the same parallel velocity in the lobes as they were already shown to be doing in the plasma mantle.[7] The mixing of plasma from different sources will be shown to be a possible reason for confusion about region nomenclature.

INSTRUMENTATION

The Ion Composition Experiment on ISEE 1 uses the same basic ion optics as the one flown on GEOS 1 and 2, Dynamic Explorer (DE-1), and on the AMPTE CCE spacecraft. A description of the experiment along with the modes of operation is given in Ref. 13, and a more thorough analysis of the ion optics can be found in Ref. 14. We describe here only the major characteristics and the mode of operation that have been used in our study. The one of two identical ion mass spectrometers that have been used for our analysis has its field of view centered 5° below the satellite spin plane; the spin plane is approximatively parallel to the solar ecliptic plane, and the spin period is 3 s. The field of view is 10° in the spin plane and varies with energy in the direction perpendicular to the spin plane from 30° at 100 eV/e to 10° above

*Instituto di Fisica dello Spazio Interplanetario, CNR, C. P. 27, 00044 Frascati (Roma), Italy.
[†]Physikalisches Institut, University of Bern, 3012 Bern, Switzerland.

10 keV/e. The mass spectrometer consists of an energy analyzer (cylindrical electrostatic analyzer) followed by a mass analyzer (crossed electrostatic and magnetic fields) and an electron multiplier as detector. The data that are used in this investigation are collected in a mode where five major species (H^+, He^{++}, He^+, O^+, and O^{++}) and a background channel are sampled in sequence; at each mass channel the energy distribution is measured by stepping the energy analyzer between 0.01 and 18 keV/e. A full cycle in this mode lasts about 10 min. The ISEE 2 EGD experiment is described in Ref. 15.

DATA ANALYSIS

During 1978 and 1979 ISEE 1 and 2 spent three to four months with their apogees in the earth's magnetotail; from these periods 133 hours of data have been singled out, during which tailward flowing O^+ streams were observed by both experiments at both satellites; these data are uniformly distributed over 40 different days. The ISEE 2 EGD experiment can show the temporal trend of the stream parameters with a time resolution of tens of seconds and allows the determination of periods over which the ion stream parameters are constant. During these periods the ISEE 1 ICE data show the presence of O^+ at some energy, whenever the O^+ mass step is sampled, and H^+ at another energy, whenever the H^+ mass step is sampled. The gaps can be as large as 10 min. The simultaneous observation by ISEE 2 EGD of persistent ion streams with the same energy as the O^+ in the lobe or persistent ion streams at both the O^+ and H^+ energies in the mantle (where the generally higher velocities draw H^+ streams into the ISEE 2 EGD experiment energy range) allows the inference that the streams are continuously detected at ICE as well. The separation between the two spacecraft does not impair this inference since the observed ion mean velocities (above 20 km/s) would introduce a delay of about 300 s over distances of the order of 1 R_e, which is short with respect to the minimum integration time on the ICE data (about 40 min).

Due to the variability of the stream properties (as monitored by the ISEE 2 EGD experiment), appropriate averages over the ICE data have been made over different time spans to produce a database of 72 points for the statistical analysis. Whenever a persistent stream was observed in the EGD data with constant velocity (oscillations within 10%) over a certain time interval, data from ICE were used to define the O^+ and H^+ stream parameters for that interval. A moment analysis was performed over the distribution function in velocity space. Densities and velocities were thus derived based on the velocity and angular distribution of the ions as detected by ICE; errors correspond to 1 σ. Cold H^+ was detected by ICE in all 72 cases, together with the O^+, but appropriate care had to be used to select periods during which no contamination from plasma sheet H^+ occurred; this hot H^+ would have introduced a bias in the H^+ stream parameter computation. In order to avoid this contamination the proton-second velocity moment in the energy per charge range 200 eV/q to 16 keV/q was used, and cases when it was in excess of 1.1 keV were excluded. This moment is indicative of the plasma sheet H^+ temperature and is not the temperature of the ionospheric streams. This selection reduced the number of cases for H^+ to roughly one-half. Figure 1 shows the spatial distribution of the 72 events in a coordinate system that has proved useful for this kind of study;[7,16] i.e., GSM X on the horizontal axis, and "s", the square root sum of the squared Y and Z GSM coordinates; i.e., the distance from the GSM X axis (or the distance from "the central axis of the magnetotail") on the vertical axis.

An example of a single event is given in Fig. 2. The energy-per-charge spectra of the different ion species detected by ICE are shown for two time periods on day 129, 1978, 1002–1100 UT (~ 16 R_e distance, ~ 2015 GSM LT, $\sim 26°$ GSM latitude) in the upper panel and 2100–2200 UT (~ 22 R_e distance, ~ 2150 GSM LT, $\sim 38.5°$ GSM latitude) in the lower panel. In the lower panel, four ion species are present, H^+, He^{++}, He^+, O^+, all roughly flowing with the same mean velocity. The presence of solar wind He^{++} in this spectrum suggests the detection of the plasma mantle. In the upper panel, only ionospheric ions are observed (H^+, He^+, O^+), again flowing with the same mean velocity, so that we assume that the satellite is in the lobe region. This individual case is given here to show how, during disturbed periods, all the main ion species detected in

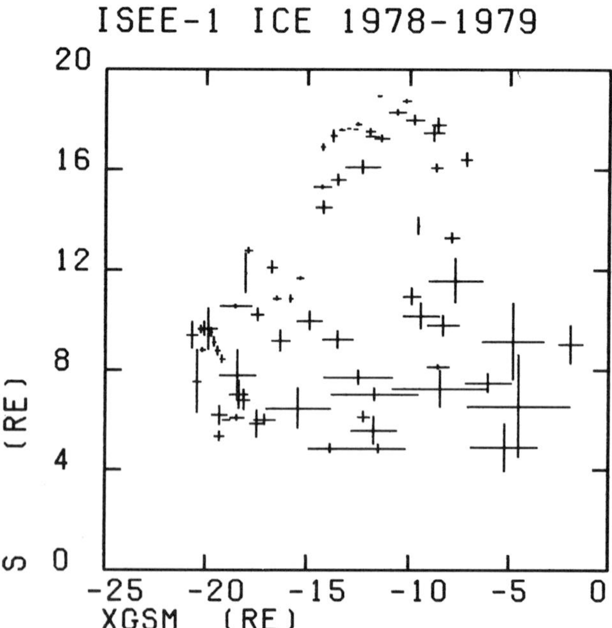

Figure 1—The spatial distribution of the 72 events detected in 1978 and 1979 is shown. The coordinate system is s, the square root sum of the squared Y and Z GSM coordinates versus GSM X. The crosses represent the extension of the region for which each single data point has been computed. The larger extent of these regions at low X and s values is determined by the larger velocity of the satellites when they are closer to the earth.

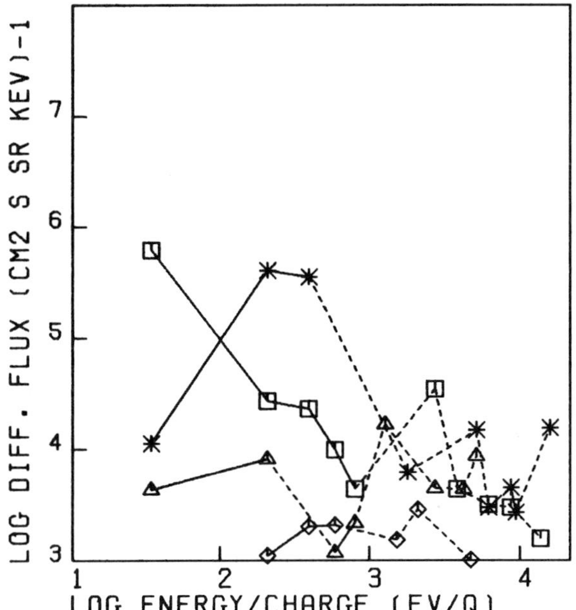

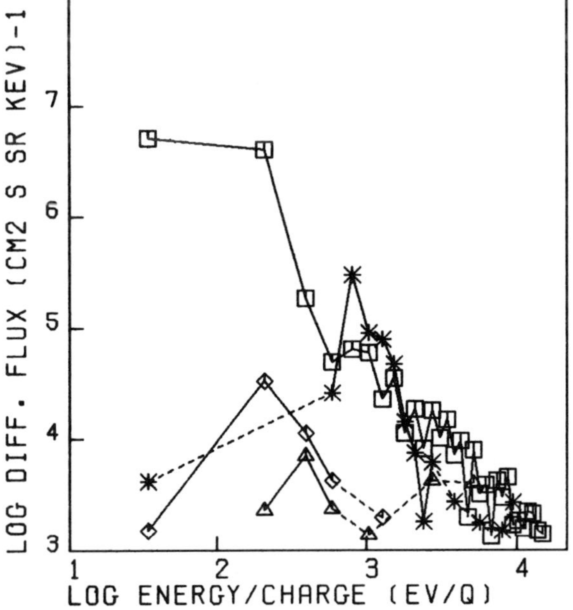

Figure 2—Energy per charge spectra for the various ion species detected by ICE at ISEE 1 on day 129, 1978. The data refer to the tailward flow direction (90° sector) and show two periods: between 1002 and 1100 UT, upper panel, and between 2100 and 2300 UT, lower panel. The range of distances from the earth (*R*), GSM local time (LT), and latitude angle (LAT) are also listed.

the magnetotail flow together with similar mean velocities, at a given location, in both the two regions, mantle and lobe.

Figures 3 and 4 show the dependence of the O^+ and H^+ stream velocity on the two coordinates *X* GSM and *s*, as defined above. In Fig. 3, panels A and B refer to the dependence of the velocity of the two ion species on *X* GSM for *s* smaller than 13 R_e (panel A) and *s* larger than 13 R_e (panel B); it is shown that the velocities are lower for low values of *s* than for high values of *s*. At the same time the velocity has a similar dependence on *X* GSM, and decreases when *X* GSM varies from $-22\ R_e$ to $-2\ R_e$ (panel A). Panels C and D show the same effects when the data are selected based on *X* GSM, from 0 R_e to $-12\ R_e$ in panel C, and from $-12\ R_e$ to $-25\ R_e$ in panel D; again the range of lower values of *X* GSM corresponds to lower velocities, at corresponding values of *s*; similarly, for a given range of *X* GSM, increasing *s* corresponds to increasing stream velocity. In Fig. 4 the number of points for H^+ is lower than those available for O^+ in Fig. 3, and the trends are less clear; still the points at small *X* values (panel B) indicate larger H^+ velocity at closer distances than the points at larger *X* (panel A), and the same effect with *s* can be inferred if the plot in panel C is compared with the plot in panel D.

The apparent similarity of the curves referring to O^+ and H^+ velocities is shown quantitatively in the scat-

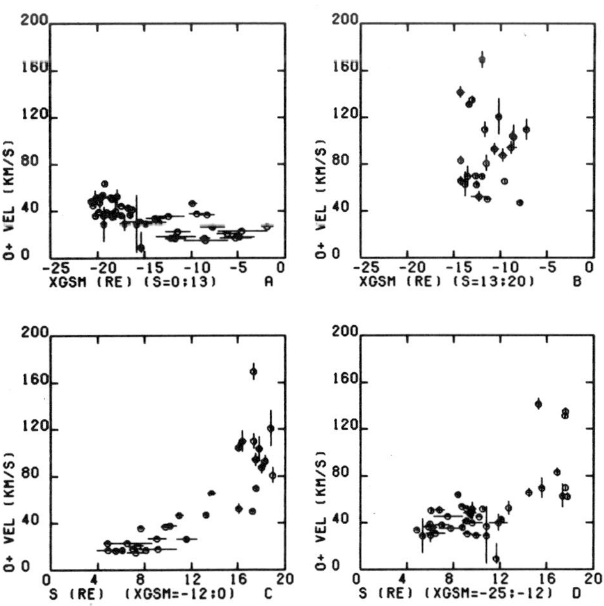

Figure 3—The O^+ ion stream flow velocity is plotted as a function of GSM coordinates. Panels A and B show the O^+ velocity versus *X* GSM for two selected ranges of *s* (see Fig. 1 for its definition); panel A for *s* lower than 13 R_e and panel B for *s* between 13 and 20 R_e. Panels C and D show the O^+ velocity versus *s* for two selected ranges of *X* GSM; panel C for *X* GSM between the earth and $-12\ R_e$; panel D for *X* GSM between $-12\ R_e$ and $-25\ R_e$.

ISEE-1 ICE 1978-1979

EXB FILTER EFFECT (H+ VEL)

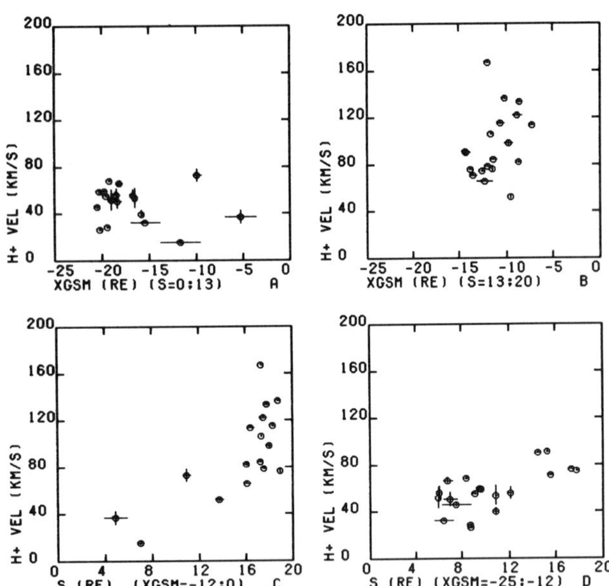

Figure 4—The H$^+$ stream velocity is plotted versus the same coordinates as in Fig. 3.

ter plot of Fig. 5 (panel A); the correlation coefficient of the O$^+$ ion stream velocity with the velocity of the concurrent H$^+$ ion stream is 0.88. The slope of the linear regression is 0.90 ± 0.11, which is consistent with equal flow velocities and not with equal energies (which would imply a slope of 0.25). Panel B shows the same comparison on the velocity component parallel to the ambient magnetic field. This is very similar to the other since the drift velocity of these ion streams is low with respect to their flow velocity.

CONCLUSIONS

The magnetospheric mass spectrometer effect was shown, in the ISEE 2 EGD experiment data, to organize the flow of O$^+$ in the magnetotail.[7] In this analysis of the ISEE 1 ICE data it appears that it affects H$^+$ as well as O$^+$, both in the magnetotail plasma lobes and in the plasma mantle. The ion flow velocity is higher at larger X GSM down the tail, for low values of s (panel A, Fig. 3), and it is higher as well at larger s distances from the tail axis in a certain range of X GSM values (panel C, Fig. 3). This distribution is generated by the fact that higher velocity ions have shorter travel times, and the inward $E \times B$ drift brings the various ions a distance that is proportional to this travel time; at a given distance downstream the faster ions flow closer to the magnetopause, and, on the axis, the faster ions convect to larger distances. As a further consequence of this effect, the ISEE 1 ICE data show that in the cases when the two species are detected together, they are flowing with equal velocity (Fig. 5). Sharp et al.[2] leaned toward the interpretation of nearly equal energies possibly because of the limited time resolution of the ICE data. The

ISEE-1 ICE 1978-1979

O+ VS H+ FLOW VELOCITY

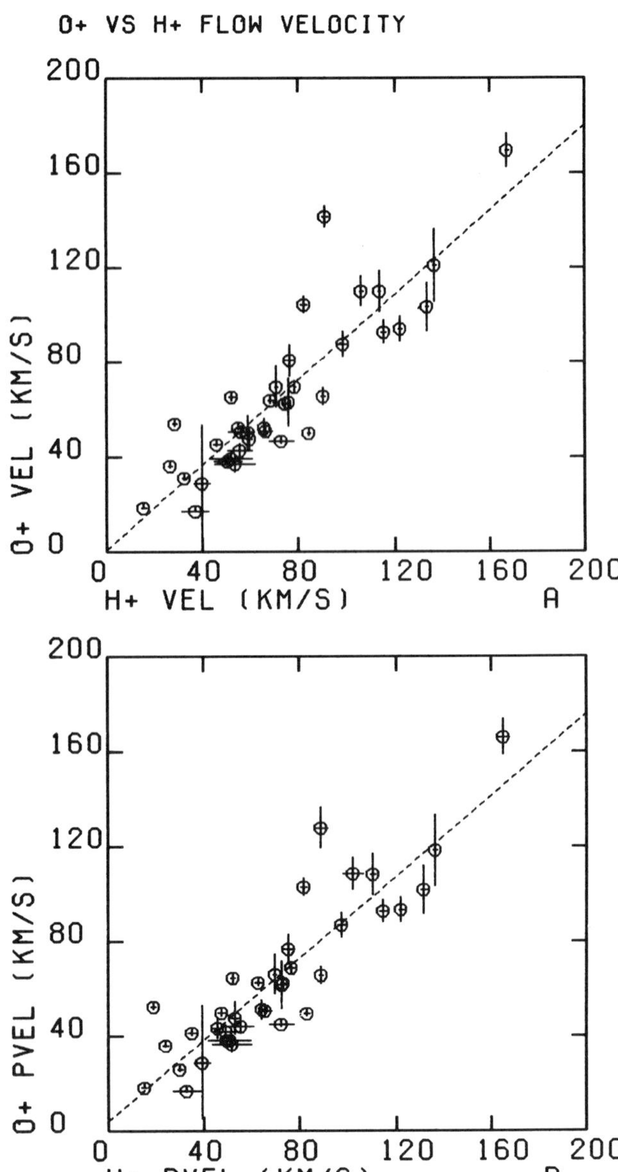

Figure 5—Scatter plot of the flow velocity of the O$^+$ ions versus the flow velocity of the H$^+$ as detected by ICE at ISEE 1 during the same time interval. The best fit line shown has a slope of 0.90 ± 0.11, with a correlation coefficient of 0.88.

ICE data by themselves do not allow the selection of events in which H$^+$ and O$^+$ are present at the same time, but this can be done in the EGD data, used in this study as a monitor to select the periods, while the analysis is carried out on the ICE data that are unequivocal with respect to ion identification. However, the most significant difference between this study and the study of Sharp et al.[2] may have been the energy ranges. Sharp et al. limited their study to energies above 100 eV/q, which would have excluded all H$^+$ ions with velocities below 138 km/s.

The detection of nearly equal H^+ and O^+ ion flow velocity and direction implies that both species emerge from similar locations, possibly the polar cusp region, and are equally dispersed by the $E \times B$ drift filter effect into the plasma lobes and mantle regions. As far as the ionospheric ions are concerned, there is no distinction between lobes and mantle, and the transition from one to the other is to be looked for in other data. The filter mixes species according to the ion velocity; ionospheric ions from the high-energy end of their distribution at the source are intermixed with solar-wind ions from the low-energy end of their distribution outside the magnetosphere, at all positions. A consequence of this effect is that even the criterion that the presence of solar-wind He^{++} is the distinctive factor between the two regions may be loose. As already suggested by Sckopke and Paschmann[1] the determination of a boundary between mantle and lobe seems to be more related to instrument capabilities than to physical constraints. Our study, which takes into account only disturbed periods, when the ionosphere plays a significant role in the geomagnetotail dynamics, suggests that the lobe can be considered as a low-density extension of the plasma mantle towards the plasma-sheet plane. At distances of 10–20 R_e, this region is mainly refilled by ionspheric ions since the magnetospheric mass spectrometer allows these ions to get there; it is reasonable to expect that the lobe region, because of the same effect, would be reached by the more energetic solar-wind ions at higher distances from the earth.

ACKNOWLEDGMENT—Part of this work was supported by the Swiss National Foundation, grant 2.007.83.

REFERENCES

[1] N. Sckopke, and G. Paschmann, "The Plasma Mantle: A Survey of Magnetotail Boundary Layer Observations," *J. Atmos. Terr. Phys.* **40**, 261 (1978).

[2] R. D. Sharp, D. L. Carr. W K. Peterson, and E. G. Shelley, "Ion Streams in the Geomagnetic Tail," *J. Geophys. Res.* **86**, 4639 (1981).

[3] S. Orsini, M. Candidi, H. Balsiger and A. Ghielmetti, "Ionospheric Ions in the Near Earth Magnetotail Plasma Lobes," *Geophys. Res. Lett.* **9**, 163 (1982).

[4] H. Rosenbauer, H. Grunewaldt, M. D. Montgomery, G. Paschmann, and N. Sckopke, "HEOS 2 Plasma Observations in the Distant Polar Magnetosphere," *J. Geophys. Res.* **80**, 2723 (1975).

[5] D. A. Hardy, H. K. Hills, and J. W. Freeman, Jr., "A New Plasma Regime in the Distant Geomagnetic Tail," *Geophys. Res. Lett.* **2**, 169 (1975).

[6] M. Lockwood, T. E. Moore, J. H. Waite, Jr., C. R. Chappell, J. L. Horwitz, and R. A. Heelis, "The Geomagnetic Mass Spectrometer-Mass and Energy Dispersion of Ionospheric Ion Flows into the Magnetosphere," *Nature* **316**, 612 (1985).

[7] M. Candidi, S. Orsini, and V. Formisano, "The Properties of Ionospheric O^+ Ions as Observed in the Magnetotail Boundary Layer and Northern Plasma Lobe," *J. Geophys. Res.* **87**, 9097 (1982).

[8] M. Candidi, S. Orsini, and A. Ghielmetti, "Observation of Multiple Ion Streams in the Magnetotail Low Latitude Boundary Layer. Evidence for a Double H^+ Population," *J. Geophys. Res.* **89**, 2180 (1984).

[9] R. Lundin, B. Hultqvist, E. Dubinin, A. Zackarov, and N. Pissarenko, "Observations of Outflowing Ion Beams on Auroral Field Lines at Altitudes of Many Earth Radii," *Planet. Space Sci.* **30**, 715 (1982).

[10] E. G. Shelley, "Heavy Ions in the Magnetosphere," *Space Sci. Rev.* **23**, 465 (1979).

[11] D. J. Gorney, A. Clarke, D. Croley, J. F. Fennell, J. Luhmann, and P. F. Mizera, "The Distribution of Ion Beams and Conics Below 8000 km," *J. Geophys. Res.* **86**, 83 (1981).

[12] A. W. Yau, B. A. Whalen, W. K. Peterson and E. G. Shelley, "Distribution of Upflowing Ionospheric Ions in the High-Altitude Polar Cap and Auroral Ionosphere," *J. Geophys. Res.* **89**, 5507 (1984).

[13] E. G. Shelley, R. D. Sharp, R. G. Johnson, J. Geiss, P. Eberhardt, H. Balsiger, G. Haerendel, and H. Rosenbauer, "Plasma Composition Experiment on ISEE A," *IEEE Trans. Geosci. Electron.* **GE-16**, 266 (1978).

[14] H. Balsiger, P. Eberhardt, J. Geiss, A. Ghielmetti, H. P. Walker, D. T. Young, H. Loidl, and H. Rosenbauer, "A Satellite Borne Ion Mass Spectrometer for the Energy Range 0 to 16 keV," *Space Sci. Instrum.* **2**, 499 (1976).

[15] C. Bonifazi, P. Cerulli-Irelli, A. Egidi, V. Formisano, and G. Moreno, "The FGD Experiment Positive Ion Experiment on the ISEE-B Satellite," *IEEE Trans. Geosci. Electron.* **GE-16**, 243 (1978).

[16] M. Candidi and S. Orsini, "Estimates of the North-South Electric Field Component in the Low Latitude Boundary Layer," *Geophys. Res. Lett.* **8**, 637 (1981).

ENERGETIC ION AND ELECTRON BEAMS AT THE PLASMA-SHEET BOUNDARY IN THE DISTANT TAIL

M. Scholer, B. Klecker, D. Hovestadt,* G. Gloeckler, F. M. Ipavich, A. B. Galvin,[†]
D. N. Baker,[‡] and B. T. Tsurutani[§]

We have analyzed several energetic particle bursts observed by ISEE 3 in the distant tail. The energetic particle data are supplemented by the electron plasma and magnetic field measurements. These bursts are characterized by large velocity-dispersion effects, with energetic electrons observed first, followed by ions with continuously lower velocities. Both ions and electrons stream in the tailward direction. Protons and alpha particles of the same energy per nucleon are observed at the same time. The dispersion effects are observed in reverse order when the spacecraft leaves from the boundary layer into the lobe. No obvious signature indicating the passage of a plasmoid has been observed in the particle or magnetic field data during or immediately following these bursts. It is concluded that these beams at the plasma-sheet boundary originate at a steady source, presumably close to a neutral line, and not at a time varying source. The velocity filter effect in a dawn-dusk electric field arranges particles with progressively lower velocity further inside the wedge-shaped region of reconnected field lines. The plasma electron distribution exhibits a tailward directed heat flux within this layer.

INTRODUCTION

The plasma-sheet boundary layer in the near-earth region has been extensively studied in recent years by the ISEE 1 and 2 spacecraft. Parks et al.,[1] Möbius et al.,[2] Andrews et al.,[3] and Williams[4] found energetic ion beams at the outer edge of the plasma-sheet boundary. When entering the plasma sheet from the lobe, usually earthward energetic ion beams are observed first. Further inside the boundary layer, earthward as well as tailward streaming is observed. Williams[4] has shown that the various spectral shapes and the counter streaming would follow from considerations of single particle motion from a distant tail source and subsequent adiabatic reflection in the near-earth region. It is, however, not clear from the near-earth observations whether these particle beams are due to a source that is temporarily varying or spatially varying (see Ref. 5).

Energetic particle layers have also been observed in the deep tail by ISEE 3. Scholer et al.[6] and Richardson and Cowley[7] have shown that tailward directed ion beams in the deep tail often precede the passage of plasmoids and are thus thought to be accelerated at a newly created near-earth neutral line. Richardson and Cowley[7] have actually shown that in the majority of cases the spacecraft enters the layer because the tailward propagating plasmoid drives the boundary layer over the spacecraft. This layer is on open field lines, since the energetic electrons stream intensively tailward.[6] There exist, however, also cases where the spacecraft enters the boundary layer due to thickening of the layer by reconnecting field lines to higher latitudes, as originally proposed by Scholer et al.[6] Scholer et al.[8] reported one particular event where following a substorm onset, ISEE

1 observed earthward directed ion beams at the boundary of an expanding plasma sheet and ISEE 3 entered simultaneously an energetic particle layer in the deep tail with tailward directed ion and electron beams. This layer was entered long before the tailward moving plasmoid could be sensed as a traveling compression region.[9] Simultaneous occurrence of earthward directed ion beams in the near-earth region and tailward directed ion beams in the distant tail have also been observed during plasma-sheet recovery.[10]

In this paper we want to investigate further the occurrence of energetic particle layers in the deep tail. We found many examples of an energetic particle layer adjacent to the plasma sheet, i.e., when the spacecraft entered or left the plasma sheet. But quite often the spacecraft entered such a layer, stayed in it for some tens of minutes, and left the layer again without ever encountering the plasma sheet. The magnetic field and electron plasma data showed a lobe-like environment during such time periods. In all these cases no direct signature of a tailward-propagating plasmoid has been found in the particle or magnetic field data.

OBSERVATIONS

In order to demonstrate the various signatures of the energetic-particle boundary layer we will present in the following, three different examples of ISEE 3 encounters with such a layer. Figure 1 shows from top to bottom, magnetic field data (magnitude in nT, latitude and longitude in GSE coordinates) and energetic particle count rates (30–36 keV protons, 215–308 keV ions, 75–115 keV electrons), as obtained by ISEE 3 at $\sim 220\ R_e$. The Max-Planck-Institut/University of Maryland particle sensor measures in eight sectors in the ecliptic plane. Figure 1 shows the count rates in the sunward looking sector containing the magnetic field; this is the sector with the maximum count rate, since both protons and electrons stream highly collimated along the magnetic field. ISEE 3 enters the plasma sheet from the south lobe

*Max-Planck-Institut für Physik und Astrophysik Institut für estraterrestrische Physik, 8046 Garching, Federal Republic of Germany.
[†]University of Maryland, College Park, Maryland 20742.
[‡]Los Alamos National Laboratory, Los Alamos, New Mexico 87545.
[§]Jet Propulsion Laboratory, California Institute of Technology, Pasadena, California 91109.

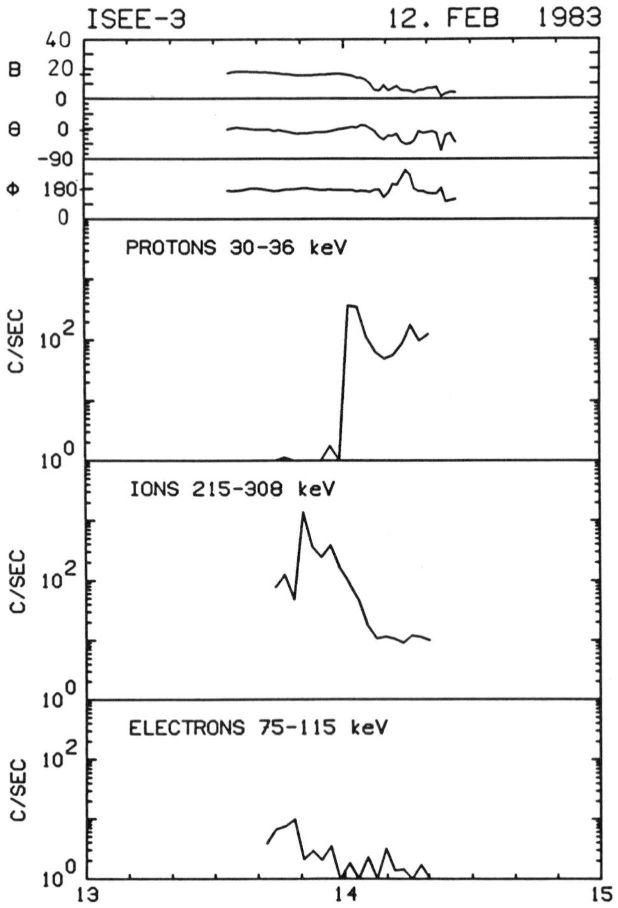

Figure 1—From top to bottom: Magnetic field magnitude (nT), latitude, and longitude, count rate of 30–36 keV protons, 215–308 keV ions, and 75–115 keV electrons for a 2-hour period on February 12, 1983, as measured by ISEE 3 at 220 R_e.

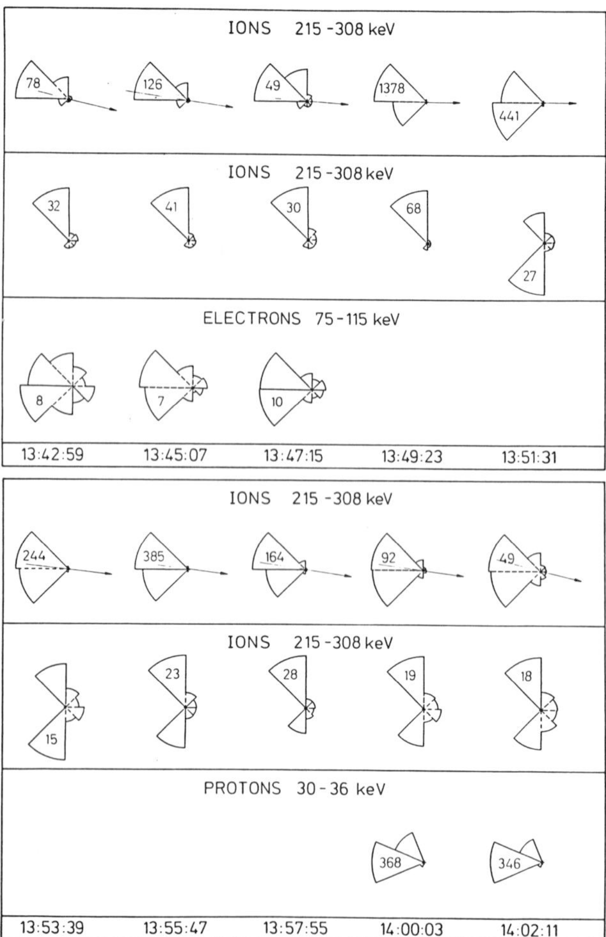

Figure 2—128-s averaged energetic particle angular distributions on February 12, 1983. Shown is intensity versus look direction (the sun is to the left of the figure). In the middle panel the two sunward looking sectors are eliminated in order to emphasize dawn-dusk anisotropies. Time given is the beginning of a 128-s averaging interval.

at ~ 1405 UT (see the depression of the magnetic field magnitude). Before entering the plasma sheet, the spacecraft encounters bursts of energetic particles. Higher velocity particles (electrons) are observed before lower velocity particles (protons). Throughout the occurrence of the electron and high energy ion bursts the magnetic field strength has the lobe value. We will now demonstrate that these bursts are not of a temporal nature but due to a crossing of layers of particles with progressively lower velocity.

Figure 2 shows angular distributions of energetic particles in the ecliptic plane. The intensity is plotted linearly versus instrument look direction; the sun is toward the left and dawn is at the top. The arrow indicates the projection of the magnetic field direction (averaged over the 124 s sampling time of the energetic particles) onto the ecliptic plane. The upper panel shows distributions of 215–308 keV ions. In the middle panel we show this distribution again but have excluded the two sunward sectors in order to demonstrate any dawn-dusk anisotropy resulting from a north-south density gradient. The lower panel shows distributions of either 75–115 keV electrons (upper half of figure) or 30–36 keV protons

(lower half of figure). No electrons are observed above the threshold after 1349:23 UT and no 30–36 keV protons are observed before 1400 UT. From the second panel in the upper half of Fig. 2, it can be seen that although the ions stream intensively along the field there is a dawn-dusk anisotropy, i.e., the spacecraft enters the layer from the south. During the sample between 1349:23 and 1351:31 UT the anisotropy has reversed its direction and the count rate has decreased by a factor of 3. This indicates that the burst is spatial and not temporal in nature. The spacecraft is now above the maximum of the higher energy ion layer but has still not entered the 30–36 keV proton layer. We should note that we observe 215–308 keV ions with 45–90° pitch angle long before we see the 30–36 keV proton layer: ~67° pitch angle ~250 keV protons have the same parallel velocity as field-aligned 36 keV protons. If the layering is due to a velocity filter effect these ions should be observed at the same time. The fact that the large-pitch-angle high-energy particles are observed as early as the field-aligned high-energy particles can be explained by assuming that the large-pitch-angle ions are due to pitch-angle scattering of the field- aligned beam all the way between source and observation point. However, it cannot be excluded that the large-pitch-angle particles constitute a low-intensity background population. The intensity of the $\sim 67°$ pitch-angle 215–308 keV ions at ~ 1400 UT (i.e., when 30–36 keV field-aligned protons are observed) is considerably less (~ 19 c/s) than the intensity in the field-aligned 215–308 keV ion beam (~ 1378 c/s). This indicates that ions are either not injected isotropically at the source position but are injected highly beamed along the magnetic field or that they are strongly adiabatically focussed. The dawn-dusk anisotropy changes within 4 min. From this way we can compute an approximate downward velocity of the layer relative to the spacecraft of ~ 20 km/s, which leads to a total thickness of the energetic particle layer of ~ 4 R_e.

Figure 3 shows magnetic field and particle data between 1800 and 2000 UT on March 20, 1983, when ISEE

3 moved out of the plasma sheet into the north lobe at ~ 130 R_e. In addition to the energetic proton and electron rates, we show the count rates of two alpha particle channels (30–36 keV/e and 58–75 keV/e). When leaving the plasma sheet the dispersion effects are seen in reverse order, i.e., the layers of particles with progressively higher velocity are higher above the plasma sheet. This substantiates further our conclusion that the particle beams at the plasma-sheet boundary are a spatial and not a temporal phenomenon. Furthermore, we are able to prove that the dispersion is indeed a velocity-dependent effect: 30–36 keV protons and 58–75 keV/e alpha particles (both of 0° pitch angle) have the same velocity so that the layers of both these species should overlap. As can be seen from Fig. 3, this is indeed the case. The intensity of the lowest velocity particles, 30–36 keV/e alpha particles, still increases toward the plasma sheet. Unfortunately, there is a data gap and it is not clear whether these lowest velocity particles have their maximum intensity also at the plasma-sheet boundary.

Our final example is an entry of ISEE 3 into an energetic particle boundary layer at 220 R_e on February 12, 1983. Figure 4 shows magnetic field and particle data for a 3-hour period. The spacecraft enters the energetic electron layer at ~ 1104 UT and observes subsequent layers of energetic ions with progressively lower velocity. The spacecraft never enters the plasma sheet proper but stays in the boundary layer for ~ 1 hour. When leaving the boundary layer the velocity dispersion effects are seen again in reverse order. The dashed vertical lines in Fig. 4 mark the maxima of the 30–36 keV proton intensity. The intensity of the ions and electrons in these layers has changed between entry and exit, indicating that the intensity of the boundary layer can vary on time scales at least as short as 40 min. In the lower part of Fig. 4 we show velocity distribution functions of plasma electrons. Below each distribution is the time (in minutes and seconds after 1100 UT) after which each 3 s data acquisition started. The times run from right to left. The logarithm of f is shown for 15 energy levels (10–1000 eV) and for 16 azimuthal angles. The sunward-looking samples are identified by the small peak in the lower energy levels caused by sunlight entering the instrument and are indicated by an arrow. In the samples at 1104:27 UT and later, a tailward streaming population can be seen clearly in the distribution function at the higher energy end, giving rise to a tailward directed heat flux (the distribution at 1105:51 UT is obtained in a photoelectron mode of the instrument).

DISCUSSION

We have presented a limited data set that is representative for many more energetic particle burst events observed by ISEE 3 in the deep tail. The results can be summarized as follows:

1. Energetic electron and ion bursts are observed in the magnetotail lobe adjacent to the plasma sheet.
2. Before entering the plasma sheet, energetic electrons are observed first, followed by ions with suc-

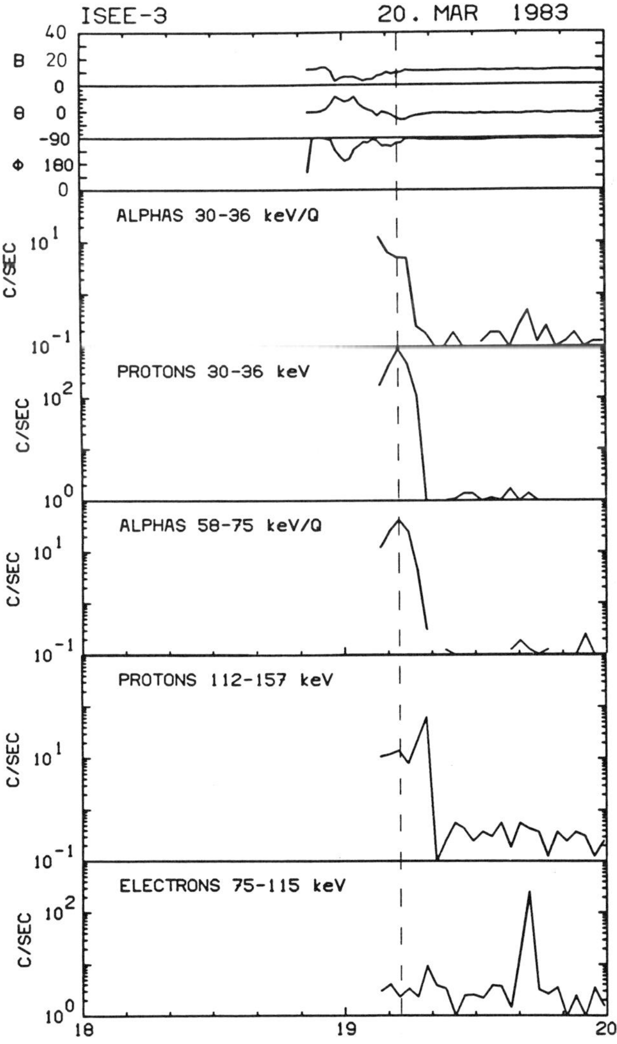

Figure 3—Magnetic field and particle data for a 2-hour period on March 20, 1983, as measured by ISEE 3 at ~ 140 R_e.

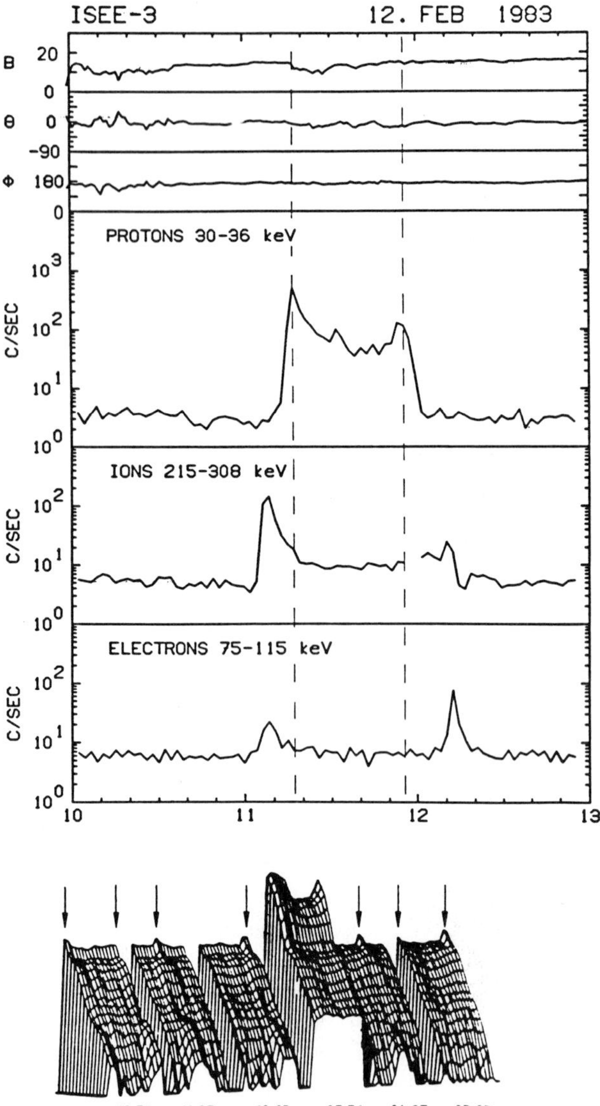

Figure 4—Upper part: Magnetic field and particle data for a 3-hour period on February 12, 1983. Lower part: Velocity distribution functions of plasma electrons. The log of f is shown for 15 energy levels (10–1000 eV) and for 16 azimuthal angles. Below each distribution is the time (in minutes and seconds after 1100 UT) after which each 3-s data acquisition started.

cessively lower velocity. When leaving the plasma sheet the dispersion is observed in the reverse order.

3. Occasionally, energetic particle bursts are observed in the magnetotail lobe although the spacecraft does not subsequently enter the plasma sheet. In these cases bursts exhibiting normal velocity dispersion are observed first, followed by bursts with inverse velocity dispersion.

4. Intensity maxima of proton and alpha particle bursts of ions with the same velocity (same energy per nucleon) occur at the same time.

5. During the occurrence of the energetic particle bursts, the plasma electron distribution shows a population of electrons coming from the sunward

direction at the highest energies (~ 1000 eV). These enhanced fluxes of tailward propagating electrons constitute a tailward directed heat flux (see Ref. 1).

The observations support the neutral line model as given by Hill[12] and Cowley and Southwood [13] (see also Ref. 14). These authors pointed out that ions accelerated near a neutral line will form wedge-shaped regions where the plasma flows away from the neutral line. The ejection velocity is expected to be twice the Alfvén velocity of the incoming lobe plasma. This region does not fill the whole width of the region of reconnected field lines. Higher velocity particles produced or injected near the neutral line will be found at the edge of the plasma beam within the region of reconnected field lines. As this region is entered, energetic electrons are observed first close to the separatrix, followed by ions with progressively lower velocities. This model is illustrated in Fig. 5.

From our ISEE 3 observations we can indeed infer that there are occasionally layers of energetic particles adjacent to the plasma sheet. Comparison between protons and alpha particles shows that these layers are separated due to a velocity filter effect. Remote sensing from above and below as the spacecraft traverses the layer not only indicates that the bursts are steady state and not time varying, but allows a spacecraft velocity to be derived with respect to the layer. We find then a thickness of the whole boundary layer of ~ 5 R_e. However, there are temporal changes of the intensity in these layers on time scales at least as short as ~ 40 min. The energetic ions in these layers are highly field aligned, i.e., intensities at large pitch angles are considerably reduced compared to the field-aligned intensities. This suggests that the ions are either injected at the source and highly beamed along the magnetic field or that they are strongly adiabatically focused relatively near the source.

The question arises as to whether the energetic-particle boundary layer in the deep tail is a permanent feature, only substorm related, or even present only during a particular phase of a substorm. Lui et al.[15] investigated the occurrence of ion beams at the plasma-sheet boundary in the near-earth region in relation to various phases of the substorm and concluded that the activation of the source is not related to any particular phase of the substorm, although the ion beams were more intense and reached higher energies during the substorm recovery phase. We found by inspecting the hourly AE index that the occurrence of the energetic particle beams in the dis-

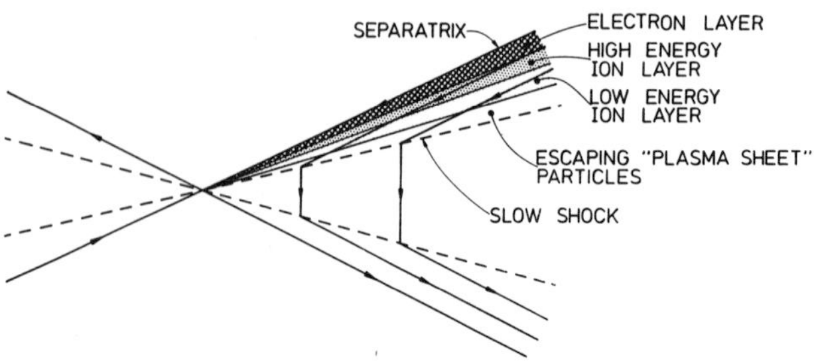

Figure 5—Schematic of the magnetic field topology and of the boundary layer, tailward of a neutral line.

tant tail is generally a high *AE* phenomenon: the *AE* index on February 12, 1983, 1300–1400 UT (Fig. 1), was 734 and 1021; the *AE* index on March 20, 1983, 1800–2000 UT (Fig. 3), was 564 and 885; the *AE* index on February 12, 1983, 1100-1300 UT (Fig. 4), was 548, 1171, and 832.

The features of the boundary layer found here are similar to the ones found by Richardson and Cowley,[7] e.g., the time dispersion effects being due to the spatial motion of the various layers over the spacecraft. Richardson and Cowley[7] concluded that in the majority of cases that they investigated the layer was pushed over the spacecraft by the tailward retreating plasmoid. However, in the cases investigated here no obvious signature could be found in the magnetic field or particle data that marks the passage of a plasmoid before, during or after the observation of the particle layer. Such signatures would be first north, followed by south excursions of the magnetic field or isotropic electron distributions. When entering the plasma sheet (case in Fig. 1) the magnetic field actually turns southward but before leaving the plasma sheet through the boundary layer (Fig. 3) the field was northward. We note that in the cases investigated here energetic electrons are observed above the detector threshold only in the boundary layer and not in the plasma sheet. It seems that entry and exit of the spacecraft into the layer is due to a flapping of the tail, which may be initialized by the substorm process itself.

ACKNOWLEDGMENT – The Max-Planck-Institut portion of this work was supported by the BMFT, Federal Republic of Germany, under contract RV14-B8/74 and the University of Maryland portion of this work was supported by NASA under contract NAS5-26739, grants NAGW101 and NGR21-002-316. The Los Alamos portion of this work was carried out under the auspices of the U.S. Department of Energy with NASA support under S-54496A. The Jet Propulsion Laboratory portion of this work was carried out under contract with NASA.

REFERENCES

[1] G. K. Parks, D. Leaver, C. S. Lin, K. A. Anderson, R. P. Lin, and H. Reme, "ISEE 1/2 Particle Observations of Outer Plasma Sheet Boundary," *J. Geophys. Res.* **84**, 6471 (1979).

[2] E. Möbius, F. M. Ipavich, M. Scholer, G. Gloeckler, D. Hovestadt, and B. Klecker, "Observations of a Nonthermal Ion Layer at the Plasma Sheet Boundary during Substorm Recovery," *J. Geophys. Res.* **85**, 5143 (1980).

[3] M. K. Andrews, P. W. Daly, and E. Keppler, "Ion Jetting at the Plasma Sheet Boundary: Simultaneous Observations of Incident and Reflected Particles," *Geophys. Res. Lett.* **8**, 987 (1981).

[4] D. J. Williams, "Energetic Ion Beams at the Edge of the Plasma Sheet: ISEE-1 Observations Plus a Simple Explanatory Model," *J. Geophys. Res.* **86**, 5507 (1981).

[5] D. J. Williams and T. W. Speiser, "Sources for Energetic Ions at the Plasma Sheet Boundary: Time Varying or Steady State?" *J. Geophys. Res.* **89**, 8877 (1984).

[6] M. Scholer, G. Gloeckler, B. Klecker, F. M. Ipavich, D. Hovestadt, and E. J. Smith, "Fast Moving Plasma Structure in the Distant Magnetotail," *J. Geophys. Res.* **89**, 6717 (1984).

[7] I. G. Richardson and S. W. H. Cowley, "Plasmoid-Associated Energetic Ion Bursts in the Deep Geomagnetic Tail: Properties of the Boundary Layer," *J. Geophys. Res.* **90**, 12133 (1985).

[8] M. Scholer, D. N. Baker, G. Gloeckler, B. Klecker, F. M. Ipavich, T. Terasawa, B. T. Tsurutani, and A. B. Galvin, "Energetic Particle Beams in the Plasma Sheet Boundary Layer Following Substorm Expansion: Simultaneous Near-Earth and Distant Tail Observations," *J. Geophys. Res.* (submitted, 1985).

[9] J. A. Slavin, E. J. Smith, B. T. Tsurutani, D. G. Sibeck, H. J. Singer, D. N. Baker, J. T. Gosling, E. W. Hones, Jr., and F. L. Scarf, "Substorm Related Travelling Compression Regions in the Distant Tail: ISEE 3 Geotail Observations," *Geophys. Res. Lett.* **11**, 657 (1984).

[10] M. Scholer, G. Gloeckler, D. Hovestadt, F. M. Ipavich, B. Klecker, D. N. Baker, W. Baumjohann, B. T. Tsurutani, and R. D. Zwickl, "Simultaneous Observation of the Plasma Sheet in the Near Earth and Distant Magnetotail: ISEE-1 and ISEE-3," *Geophys. Res. Lett.* **11**, 1034 (1984).

[11] W. C. Feldman, D. N. Baker, S. J. Bame, J. Birn, J. T. Gosling, E. W. Hones, Jr., and S. J. Schwartz, "Slow-Mode Shocks: A Semipermanent Feature of the Distant Geomagnetic Tail," *J. Geophys. Res.* **90**, 233 (1985).

[12] T. W. Hill, "Magnetic Merging in a Collisionless Plasma," *J. Geophys. Res.* **80**, 4689 (1975).

[13] S. W. H. Cowley and D. J. Southwood, "Some Properties of a Steady State Geomagnetic Tail," *Geophys. Res. Lett.* **7**, 833 (1980).

[14] S. W. H. Cowley, "The Distant Geomagnetic Tail in Theory and Observations," in *Magnetic Reconnection in Space and Laboratory Plasmas*, E. W. Hones, Jr., ed., Geophysical Monograph No. 30, American Geophysical Union, Washington, D.C., p. 228 (1984).

[15] A. T. Y. Lui, D. J. Williams, T. E. Eastman, and L. A. Frank, "Observations of Ion Streaming during Substorm," *J. Geophys. Res.* **87**, 7753 (1983).

PLASMOID-ASSOCIATED ENERGETIC ION BURSTS IN THE DISTANT GEOMAGNETIC TAIL

I. G. Richardson and S. W. H. Cowley*

The results of a detailed study of plasmoid-associated energetic ion enhancements observed by the EPAS instrument on ISEE 3 in the distant geomagnetic tail are summarized and are illustrated by a representative example. We discuss the overall structure of the events in terms of the neutral line model of substorms and show that the observations are quite consistent with expectations. We also discuss the relationship between the thermal plasma bulk speed and speeds deduced from energetic ion anisotropies, plasmoid lengths deduced therefrom, the dependence of energetic ion fluxes on the thermal plasma temperature, and the motion and orientation of the energetic particle boundary layer surrounding the plasma sheet as deduced from the ion data.

INTRODUCTION

The ISEE 3 geotail mission in 1982-83 provided a wealth of new information about the structure and dynamics of the geomagnetic tail, out to distances from earth of ~ 230 R$_e$.[1] In this paper we summarize results obtained from an extensive study of energetic ions (35–1600 keV) observed by the Imperial College/ESTEC/SRL Utrecht EPAS experiment, only some of which have, as yet, been published.[2,3] Our studies mainly concern energetic ions in the plasma sheet and in the closed-loop "plasmoid" structures that have been observed to travel down the tail following substorm onset.[4-7] It has been suggested that these plasmoids arise from rapid reconnection at a neutral line formed close to the earth at substorm onset, which results in the disconnection of the presubstorm plasma sheet from the earth and its subsequent tailward propagation (Ref. 8 and references therein). This picture has been the subject of considerable controversy during recent years in its application to near-earth data, but in the course of our studies it has become clear that it does provide a consistent framework for understanding the deep-tail data. In this paper we will therefore discuss the data in terms of the neutral line substorm model and would point out that to the best of our knowledge no other detailed interpretation of these data has, as yet, been forwarded. In general terms we would also stress the overwhelming evidence that supports the view that the earth's magnetosphere is a magnetically open structure (Ref. 9 and references therein) such that the reconnection picture of the tail should form a natural starting point. In the next section we thus review these pictures with particular reference to energetic particle observations, preparatory to discussing the data in the third section.

THEORETICAL CONSIDERATIONS

Figure 1 indicates the relationship between field, plasma, and energetic particle regions expected for the simplest case of steady-state reconnection.[10] Solid lines indicate magnetic field lines, while the wedge-shaped hatched regions in Fig. 1a indicate accelerated plasma

*Blackett Laboratory, Imperial College, London SW7 2BZ, United Kingdom.

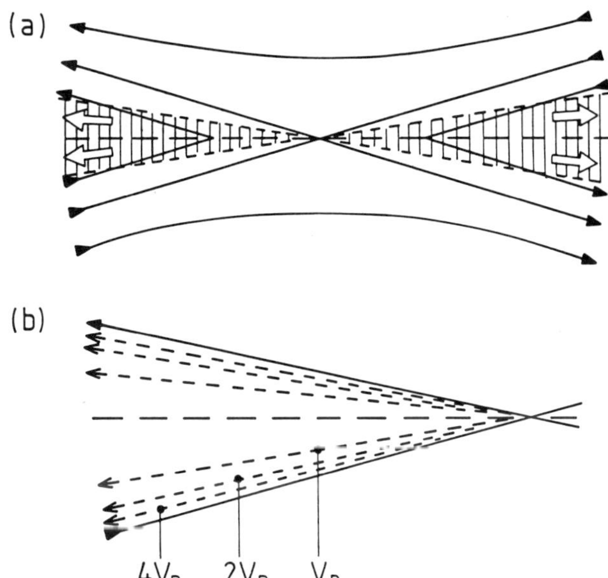

Figure 1—Relationship between magnetic field (solid line), plasma sheet (hatched area), and particle boundaries (dashed lines) for steady-state reconnection. In the lower diagram the dashed line marked V_B represents the plasma-sheet boundary, as also shown in the upper diagram, while the other dashed lines show the outer boundaries of particles having twice and four times the plasma-sheet speed, respectively. (After Cowley[10]).

flows corresponding, in the tail, to the plasma sheet. This population originates from tail-lobe plasma in which $E \times B$ drifts into the current sheet at the center of the system and is accelerated to form nearly field-aligned jets that flow away from the neutral line on either side at approximately the tail-lobe Alfvén speed (typically ~ 500–1000 km/s). Due to the finite speed of the plasma along the reconnected field lines, these jets occupy only the central portion (typically one-half[10,11]) of the reconnected field wedge. Any energetic particles in the "tail" of the jet population, however, will, because of their higher speed, stream out ahead of the main population, forming an energetic particle "boundary layer" between the plasma-sheet edge and the field separator mapping to the neutral line. The dashed lines in Fig. 1b show the outer edge of the plasma-sheet jet of speed V_B, as in Fig. 1a, together with the outer boundaries of particles having speeds twice and four times V_B. The boundary of particles of infinite speed is the separator itself. This picture should be valid to zeroth order in the distant tail, but to understand the properties of the near-earth plasma sheet the effects of particle mirroring must be included.[10]

During substorms the effects of time-dependence must also be considered, according to the picture developed from near-earth studies by Hones.[8] At substorm onset a new neutral line is suggested to form within the near-earth (~ 20 R$_e$) plasma sheet, resulting after a few minutes in the latter's disconnection from earth and propagation down-tail as a closed-loop "plasmoid." Substorm recovery is associated with tailward retreat of the neutral line to quiet-time distances ~ 100 R$_e$ and diminishing reconnection rates. This sequence is shown in Fig. 2.[3] As in Fig. 1 the hatched areas indicate hot plasma-sheet populations, while energetic particles extend in energy-dependent layers out to the field separatrices. Reconnection at the new neutral line results in expanding layers of disconnected flux surrounding the retreating plasmoid and presubstorm down-tail plasma sheet. The hot plasma formed tailward of the new neutral line only extends a finite distance down-tail along these field lines due to its finite down-tail speed. At larger distances only energetic particles of increasingly large speed will be present on these field lines. Since the plasmoid itself travels down-tail at a speed similar to, but

generally a little slower than, the newly formed hot plasma, it is only slowly engulfed by this plasma. Consequently the new hot plasma is mainly confined to a wedge-shaped region on the earthward side of the plasmoid, while on the tailward side only energetic particle layers will be present (Fig. 2c, d). At ISEE 3 distances ~ 200 R$_e$ down-tail, therefore, only expanding energetic particle layers surrounding the old plasma sheet should be present following substorm onset and before plasmoid arrival. An ~ 60 R$_e$ long plasmoid disconnected near the earth and traveling at ~ 500 km/s should then arrive ~ 25 min after substorm onset and depart after a further ~ 15 min, after which the plasma sheet will remain thick (Fig. 2d) until reconnection rates fall and the new neutral line moves tailward to reestablish quiet-time conditions (Fig. 2a). It should be noted that the plasma-sheet thickness is expected to have opposite variations during a substorm in the near-earth and distant tail regimes. In the near-earth tail, it is thick during quiet times and thin during disturbed times, and vice versa in the distant tail.

Energetic particle observations at ~ 200 R$_e$ during substorms will thus depend on spacecraft location relative to the tail center plane. Close to the center plane energetic particles may be observed well before plasmoid arrival, while further away they will first be observed as the disconnected flux layer is swept over the spacecraft by plasmoid motion, the spacecraft then immediately entering the plasmoid. In either case the spacecraft should remain for some time in the "post-plasmoid plasma sheet" after plasmoid departure, until substorm recovery occurs. Energetic particle onset characteristics will be different in these two cases, however, the layer expansion speed being faster and with less energy dispersion in the second case than in the first. In the first case the expansion speed is determined by the reconnection rate at the new neutral line (a few tens of km/s for electric fields of a few $\times 10^{-4}$ V/m), while the dispersion at a given location should just correspond to the particle time-of-flight from the near-earth neutral line (a few minutes over the EPAS energy range). In the second case the expansion speed due to plasmoid motion (~ 100 km/s for an ~ 20 R$_e$ wide structure) will be added to the above, and the observed dispersion reduced by this sweeping of the layer. The relative probability of observing these two types of onset will depend on the thickness of the layer of disconnected flux relative to the plasmoid thickness, and this depends in turn on the reconnection rate at the new neutral line.

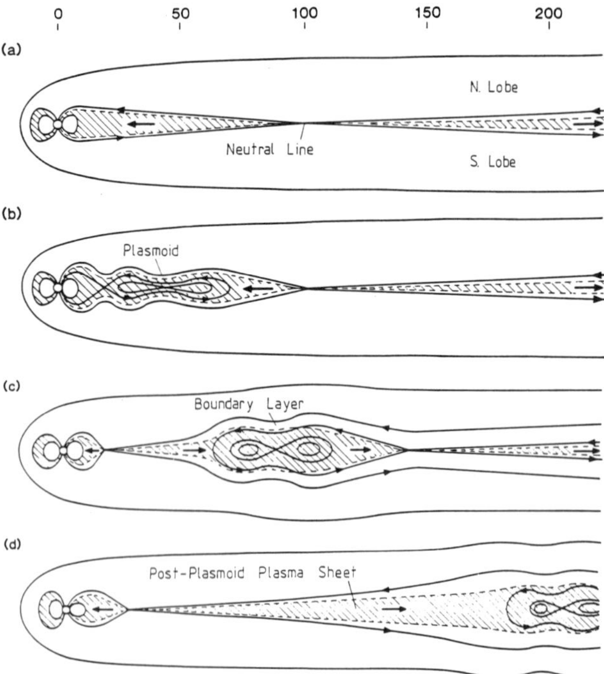

Figure 2—Sketch of field and plasma behavior in the neutral line model of substorms. The hatched areas indicate regions of hot-plasma-sheet plasma, while energetic particles extend further toward the separatrices of the field. (After Richardson and Cowley[3]).

OBSERVATIONS

Energetic ions observed by EPAS during 37 plasma sheet encounters identified as "plasmoids" by Hones et al.[4] and Scholer et al.[6,7] have been studied. Evidence for the presence of closed loops in the magnetic field is provided by the occurrence of characteristic north-south field tilting, combined with energetic electron observations that show a transition from tailward streaming distributions in the boundary layer to isotropic in the plasmoid.[5,6] EPAS[12] measures ions in eight logarithmic energy channels from 35–1600 keV with a time resolution of 16 s. It consists of three identical telescopes viewing at 30°, 60°, and 135° to the spin axis, which is directed northward perpendicular to the ecliptic. As the spacecraft spins, the output from each telescope is binned into eight 45° sectors, giving 24 directional intensity measurements for each energy range.

In Fig. 3 we present a representative plasmoid encounter that occurred on 26 January 1983 when ISEE 3 was at GSM coordinates $(-213, -0.6, -6.6)$ R_e. The top three panels show, respectively, the direction-averaged ion flux in the three lowest EPAS energy channels, the tailward and earthward streaming 35–56 keV fluxes observed by telescope 2 (which views closest to 30° above the ecliptic), and their ratio R. The lower three panels show the direction and strength of the magnetic field from the JPL magnetometer.[13]

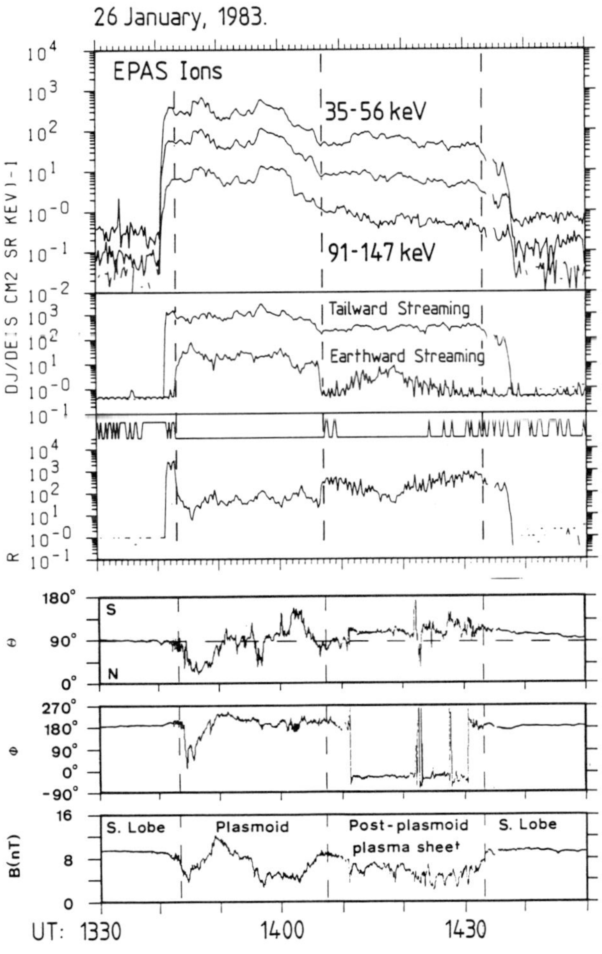

Figure 3—ISEE 3 energetic ion and magnetic field data for the interval 1330–1450 UT on January 26, 1983. The ion data in the top three panels show, respectively, the direction-averaged fluxes at energies 35–56, 56–91, and 91–147 keV, the tailward and earthward-streaming fluxes at 35–56 keV observed by EPAS telescope 2, and their ratio R. When no earthward-directed ions were detected, a count of 1 ion was substituted in the R calculation, as indicated by the flag above, thus giving a lower limit for R. The three lower panels show the direction and strength of the magnetic field (GSM coordinates).

Substorm onset occurred at 1310 UT[4] when the spacecraft was in the southern tail lobe. The energetic ion burst, however, does not begin until 1341 UT, ~ 30 min later. As usual,[3] the burst has four distinct phases. For the first two minutes, tailward-moving ions are observed streaming along tail-lobe field lines (though a slight diamagnetic field depression is also observed), and earthward-moving ions are essentially absent. At the beginning of the burst a cross-field ion flow is superposed on the tailward ion motion, indicative of the motion of a spatial layer across the spacecraft (the flow being due to a gradient anisotropy). The orientation and speed of this layer can be deduced from the gyrophase-dependence of the "switch-on" times of large pitch angle ions as described by Richardson and Cowley,[3] an analysis that in this case yields a layer speed of 106 ± 50 km/s in a direction $24 \pm 2°$ south of duskward. The implied layer thickness is ~ 2 R_e. These characteristics, together with the subsequent direct entry into the plasmoid (see below), indicate that the ion onset in this case was due to layer sweeping by plasmoid motion. This corresponds to the great majority of the events studied, indicating that 20–30 min after substorm onset the layer of disconnected flux remains relatively narrow by comparison with the plasmoid thickness. Average layer characteristics are speeds ~ 95 km/s, durations ~ 6 min, and thicknesses ~ 4.5 R_e, though wide variations do occur. The quartile ranges of these quantities are 45 to 115 km/s, 4–8 min, and 1.8–5.5 R_e, respectively. The outward boundary normals are usually directed approximately north or south (depending on the tail lobe in which the spacecraft is located), as expected on the basis of Fig. 2, though as in the above example, large deviations can also be observed (see Fig. 14 in Ref. 3). These may be due either to the three-dimensional shape of the plasmoid or to large-scale twisting of the tail structure. Another unusual feature of the example discussed here is the lack of any ion energy dispersion on entry to the layer that is discernable within the 16 s resolution of the EPAS data (upper panel, Fig. 3). Usually some dispersion is observed, and, following from considerations similar to those discussed in the previous section, this can be used to deduce the reconnection rate at the substorm neutral line at the time that the boundary is crossed, provided the neutral line location is known.[3] In the present case only an upper limit of ~ 0.15 mV/m can be deduced for the value of the electric field along the neutral line, compared with an average value of ~ 0.45 mV/m, these numbers being derived on the assumption that the neutral line lies in the near-earth tail at the time that the boundary is crossed.

In general, entry into the plasmoid is identified principally from the simultaneous appearance of hot, tailward flowing plasma, isotropic energetic electrons (when available), and generally depressed, variable magnetic fields that show systematic tilts, first to the north then to the south. In this case the plasma (not shown) and magnetic-field data indicate that entry occurred at 1343 UT,[4] the north-south deflections of the field suggesting passage of a two-loop structure over the spacecraft during the next 25 min. Approximately half of our events show evidence of multiple (usually two) rather than single field loops. Earthward-moving ions are observed in this region for the first time in the event, but tailward streaming remains dominant. These ions represent the high energy tail of the thermal plasmoid population and convect with the plasma (see below), such that the bulk velocity can be deduced from their angular anisotropy. In this case the tailward speed is 700 ± 100 km/s, and combining this with the time taken for plasmoid passage yields a total length of ~ 150 R_e. This is long compared with the average 76 R_e obtained by Scholer et al.,[7] but from our general study we find that multiple-loop plasmoids are on average nearly twice as long as single-loop plasmoids (110 ± 46 R_e compared to 59 ± 24 R_e), indicating that individual loops are usually approximately the same length (~ 60 R_e) regardless of number. Now for the single-loop plasmoids, the observed delay in arrival following substorm onset is generally in good accord with expectations based on release of the earthward end of the plasmoid from the near-earth regime at substorm onset, as expected from the "neutral line" substorm picture. For example, the tailward end of a plasmoid ~ 60 R_e long, released with its earthward end located at ~ 20 R_e from earth at substorm onset, would arrive at ISEE 3 ~ 220 R_e from earth after ~ 23 min, if traveling at the observed average plasmoid speed of ~ 660 km/s (based on the data in Fig. 4). This delay agrees well with the observed delays of 20–30 min.[3] However, the large lengths of the multiple-loop structures are harder to accommodate. In the present case, for example, the total plasmoid length and speed combine to give an expected delay of $\lesssim 7$ min, which is hard to reconcile with the observed 31-min delay. However, much better agreement is obtained if only the first loop is taken to represent the pinched off presubstorm plasma sheet, in this case giving an expected delay of ~ 20 min, which can be reconciled with observation given some acceleration of the plasmoid as it moves downtail. Subsequent loops may be formed, e.g., by early tailward retreat of the substorm neutral line followed by further pinching off, possibly associated with multiple-onset substorms. Other possible interpretations also exist however, such that this topic requires further investigation.

Exit from the plasmoid in Fig. 3 occurs at 1408 UT after which ISEE 3 remained within the post-plasmoid plasma sheet until 1433 UT where the field remained depressed in strength and tilted nearly continuously southward. Fluxes are reduced in this region compared with the plasmoid and more anisotropic (higher R value), indicating faster tailward plasma flow (~ 800 km/s). Typically we find flow speeds in the post-plasmoid plasma sheet to exceed those within the plasmoid by $\sim 25\%$. We note that observation of an extended plasma-sheet interval following, but not prior to, entry into the plasmoid is in accord with the discussion in the previous section.

After exit from the plasma sheet, a boundary layer of unidirectionally tailward streaming energetic ions is observed for a further 5 min, also in conformity with expectation. The edge of the layer retreated in a direction $19 \pm 8°$ duskward of north at a speed 25 ± 1 km/s, indicating a layer thickness of 1.2 R_e. The latter two numbers are rather smaller than usual, the quartile ranges being 40–130 km/s for the boundary speed, and 0.7–3.8 R_e for the boundary thickness. "Reverse" velocity dispersion is evident at the edge of the layer, where lower energy ion fluxes fall first. The neutral line electric field deduced from the dispersion is ~ 0.1 mV/m in this case.

We now turn from considering the specific example in Fig. 3 and its implications, to some more general results of our study, obtained by comparing EPAS data and thermal electron data obtained from the LANL plasma experiment on ISEE 3.[14] In Fig. 4 we compare the average flow speed $\bar{V}_p$ deduced from the energetic ion anisotropy using the method of Daly et al.,[15] with that obtained from the thermal electrons (V_e) for each region separately (boundary layer, plasmoid, post-plasmoid plasma sheet) for each of the events in our study. In the boundary layers it can be seen that the inferred ion speed generally considerably exceeds the plasma speed, such that the boundary layer ions clearly form a nonthermal beam flowing through the tail-lobe plasma, as expected. In the plasmoid and post-plasmoid plasma sheet, however, there exist a few examples that appear to show a real difference between bulk speeds deduced from thermal electron and energetic ion data, but a large majority of the events show good agreement, within ~ 200 km/s. This result indicates that the ions represent the high energy tail of the plasma-sheet ion population that convects with the plasma. The implication is that the ions are sufficiently strongly scattered by field irregularities within the plasma sheet that their distribution is maintained close to isotropic in the plasma rest frame. It is for this reason that some earthward-moving ions may be observed within (e.g., the post-plasmoid plasma sheet (see Fig. 3)), though they may not initially be expected in this region in the simplest picture.

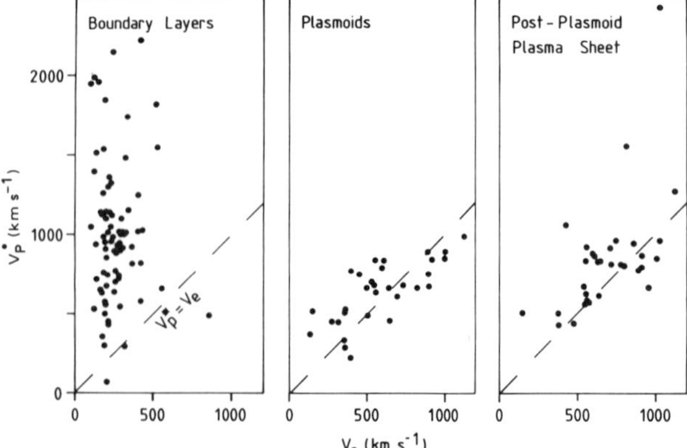

Figure 4—Comparison of the average down-tail flow speeds deduced from energetic ions ($\bar{V}_p$) and thermal electrons (V_e) in each region of the 37 plasmoid-associated ion burst events studied.

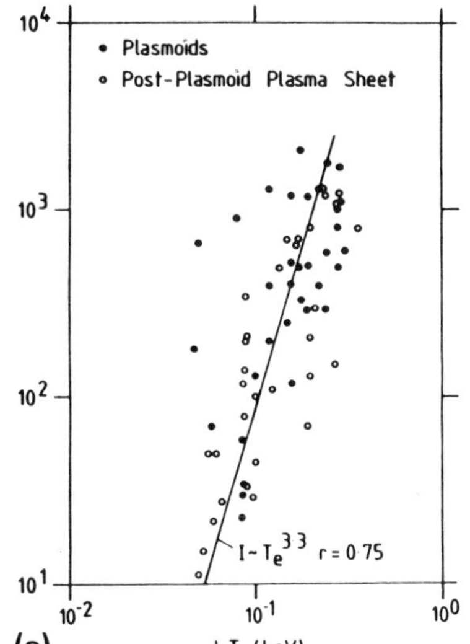

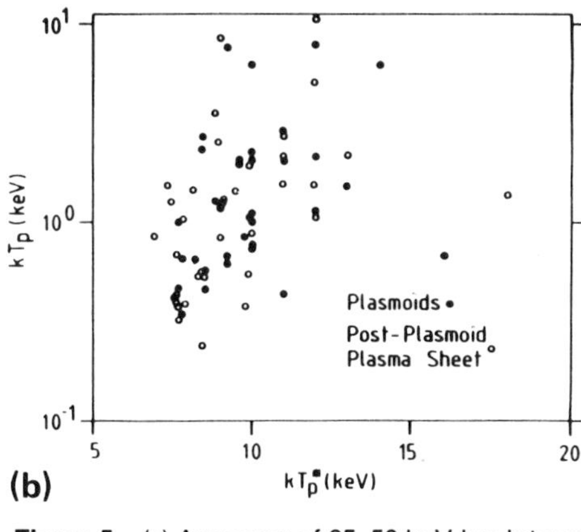

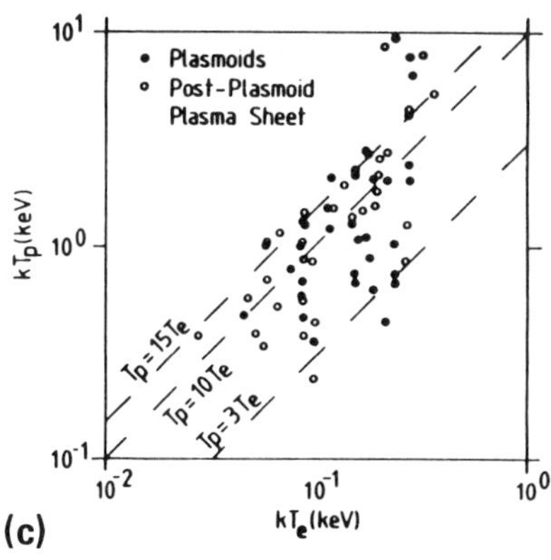

Figure 5—(a) Averages of 35–56 keV ion intensity versus electron temperature in the plasmoids and the post-plasmoid plasma sheet for the events in our study. Note that the ion intensity varies by more than two orders of magnitude from event to event, in concert with electron temperature variations. (b) Effective energetic ion temperature $\bar{T}_p$ versus thermal ion temperature T_p deduced from pressure balance. (c) Thermal ion temperature T_p versus thermal electron temperature T_e.

Another indication of the close relation between the energetic ions and the plasma-sheet populations comes from the positive correlation between energetic ion intensity and electron plasma temperature shown in Fig. 5a. This result suggests a link between thermal ion and electron temperatures and hence to energetic ions, but this cannot be directly tested because of the lack of thermal ion data from ISEE 3 during the geotail mission. However, we have attempted to infer thermal ion properties by assuming balance between the field pressure of the tail lobes and the sum of the pressures in the plasma sheet due to the residual magnetic field, the thermal electrons, and the thermal and energetic ions, the latter two taken to form separate components. The temperature and density of the energetic ion population are obtained by modeling the distribution as an isotropic convecting Maxwellian whose parameters are determined from the EPAS data. The only unknown is then the thermal ion pressure, which having been obtained from pressure balance, then yields the thermal ion temperature, since the number density is known from the difference between thermal electron and energetic ion densities. Results are shown in Fig. 5b and c where we show plots of T_p versus $\bar{T}_p$ (log-linear format) and T_p versus T_e (log-log format), respectively. Positive correlations are seen in both cases, thus substantiating the links suggested above. The correlation coefficients corresponding to the distributions shown in Fig. 5b and c are 0.36 and 0.63, respectively, the correlations in both cases being statistically significant at better than the 99% confidence level. The figures also show that typical energies are $kT_e \simeq 100$ eV, $kT_p \simeq 1$ keV, and $k\bar{T}_p \simeq 10$ keV, and that $T_p \simeq 3 - 15\ T_e$ similar to the situation in the near-earth plasma sheet.

In conclusion we wish to emphasize two points. First, our examination of the proposed plasmoid structures in the distant tail shows good accord with expectations based upon the neutral line model of substorms. Second, more generally, we stress the wealth of informa-tion that can be extracted from careful analysis of energetic ion data in geomagnetic tail studies, particularly when combined with simultaneous magnetic field, plasma, and energetic-electron data.

ACKNOWLEDGMENT—The ISEE 3 EPAS is a joint project of the Blackett Laboratory, Imperial College, London; the Space Research Laboratory, Utrecht; and the Space Science Dept., ESTEC, Noordwijk. We thank E. W. Hones, Jr., S. J. Bame, and R. D. Zwickl (Los Alamos National Laboratory) for providing the electron plasma data and J. A. Slavin and B. T. Tsurutani (Jet Propulsion Laboratory) for providing the magnetic field data. This work was performed while IGR was supported by SERC grant SG/D 09072. The Royal Society is thanked for providing a travel grant to allow attendance at this meeting.

REFERENCES

[1] B. T. Tsurutani and T. T. von Rosenvinge, "ISEE-3 Distant Geotail Results," *Geophys. Res. Lett.* **11**, 1027 (1984).

[2] S. W. H. Cowley, R. J. Hynds, I. G. Richardson, P. W. Daly, T. R. Sanderson, K.-P. Wenzel, J. A. Slavin, and B. T. Tsurutani, "Energetic Ion Regimes in the Deep Geomagnetic Tail: ISEE-3," *Geophys. Res. Lett.* **11**, 275 (1984).

[3] I. G. Richardson and S. W. H. Cowley, "Plasmoid-Associated Energetic Ion Bursts in the Deep Geomagnetic Tail: Properties of the Boundary Layer," *J. Geophys. Res.* **90**, 12133 (1985).

[4] E. W. Hones, Jr., D. N. Baker, S. J. Bame, W. C. Feldman, J. T. Gosling, D. J. McComas, R. D. Zwickl, J. A. Slavin, E. J. Smith, and B. T. Tsurutani, "Structure of the Magnetotail at 220 R_e and Its Response to Geomagnetic Activity," *Geophys. Res. Lett.* **11**, 5 (1984).

[5] E. W. Hones, Jr., J. Birn, D. N. Baker, S. J. Bame, W. C. Feldman, D. J. McComas, and R. D. Zwickl, "Detailed Examination of a Plasmoid in the Distant Magnetotail with ISEE-3," *Geophys. Res. Lett.* **11** (1984).

[6] M. Scholer, G. Gloeckler, B. Klecker, F. M. Ipavich, D. Hovestadt, and E. J. Smith, "Fast Moving Plasma Structures in the Distant Magnetotail," *J. Geophys. Res.* **89**, 6717 (1984).

[7] M. Scholer, G. Gloeckler, D. Hovestadt, B. Klecker, and F. M. Ipavich, "Characteristics of Plasmoid-Like Structures in the Distant Magnetotail," *J. Geophys. Res.* **89**, 8872 (1984).

[8] E. W. Hones, Jr., "Transient Phenomena in the Magnetotail and Their Relation to Substorms," *Space Sci. Rev.* **23**, 393 (1979).

[9] S. W. H. Cowley, "Evidence for the Occurrence and Importance of Reconnection Between the Earth's Magnetic Field and the Interplanetary Magnetic Field," in *Magnetic Reconnection in Space and Laboratory Plasmas*, E. W. Hones, Jr., ed., Geophysical Monograph No. 30, American Geophysical Union, Washington, D.C., p. 375 (1984).

[10] S. W. H. Cowley, "The Distant Geomagnetic Tail in Theory and Observation," in *Magnetic Reconnection in Space and Laboratory Plasmas*, E. W. Hones, Jr., ed., Geophysical Monograph No. 30, American Geophysical Union, Washington, D.C., p. 228 (1984).

[11] T. W. Hill, "Magnetic Merging in a Collisionless Plasma," *J. Geophys. Res.* **80**, 4689 (1975).

[12] A. Balogh, G. van Dijen, J. van Genechten, J. Henrion, R. Hynds, G. Korfmann, T. Iversen, J. van Rooijen, T. Sanderson, G. Stevens, and K.-P. Wenzel, "The Low Energy Proton Experiment on ISEE-C," *IEEE Trans. Geosci. Electron.* **GE-16**, 176 (1978).

[13] A. M. A. Frandsen, B. V. Connor, J. van Amersfoort, and E. J. Smith, "The ISEE-C Vector Helium Magnetometer," *IEEE Trans. Geosci. Electron.* **GE-16**, 195 (1978).

[14] S. J. Bame, J. R. Asbridge, H. E. Felthauser, J. P. Glore, H. C. Hawk, and J. Charez, "ISEE-C Solar Wind Plasma Experiment," *IEEE Trans. Geosci. Electron.* **GE-16**, 160 (1978).

[15] P. W. Daly, T. R. Sanderson, and K.-P. Wenzel, "Survey of Energetic (E > 35 keV) Ion Anisotropies in the Deep Geomagnetic Tail," *J. Geophys. Res.* **89**, 10733 (1984).

GENERAL DISCUSSION ON MAGNETOTAIL KINETICS

M. Scholer*

I would like to concentrate in this introduction to the general discussion on two examples where kinetic aspects of tail dynamics are important: the plasma-sheet boundary layer and high-energy-particle phenomena in the tail.

ISEE 3 observations have demonstrated that beyond $\sim 140\ R_e$ the spacecraft is on the tailward side of the source producing the particle beams in the plasma-sheet boundary layers (e.g., Refs. 1-3). Energetic electrons in the distant plasma-sheet boundary layer stream highly collimated tailward so that either field lines on the tailward side of the source are open, or energetic electrons are nongyrotropic and are lost. Because of the small gyroradius the latter is highly unlikely. If earthward of the source, field lines are closed and tailward of the source they are open, then the source is necessarily identical with a neutral line. Richardson and Cowley[2] and Scholer et al.[4] found that the potential along such a possible neutral line is at most of the order of ~ 100 keV. We observe, on the other hand, in the plasma-sheet boundary layer, energetic proton beams with energies in excess of ~ 300 keV. How are these ions accelerated? Figure 1 shows a typical spectrum of the suprathermal protons in the near-earth plasma sheet (Ipavich and Scholer[5]). Also indicated by open symbols are differential fluxes at two energies of the proton beams in the plasma-sheet boundary layer in the distant tail. The fluxes indicated by a square were obtained in the layer around a plasmoid that was studied extensively by Scholer et al.[6] and Hones et al.[7] The fluxes indicated by a box were obtained at $\sim 130\ R_e$ in the late phase of a substorm. Note that these fluxes are the highest fluxes observed in the layer, i.e., taking into account spatial dispersion effects. At ~ 250 keV the directional fluxes in the deep tail are about two orders of magnitude smaller than typical omnidirectional fluxes in the near-earth plasma sheet. A potential limited to ~ 100 kV and the occurrence of ~ 300 keV protons in the boundary layer in the deep tail is thus not necessarily a contradiction—we have to take into account the three dimensionality of the plasma sheet, i.e., the neutral line has only a limited extent in the dawn-dusk direction and only a limited portion is pinched off in the reconnection process. Figure 2 shows, schematically, cuts through the magnetotail at three different distances. Near the neutral line, high-energy ions from the plasma sheet may have access via nonadiabatic motion to the neutral line region and are ejected along open field lines in the downtail direction. At larger

*Max-Planck-Institut für Physik und Astrophysik, Institut für extraterrestrische Physik, 8046 Garching, Federal Republic of Germany.

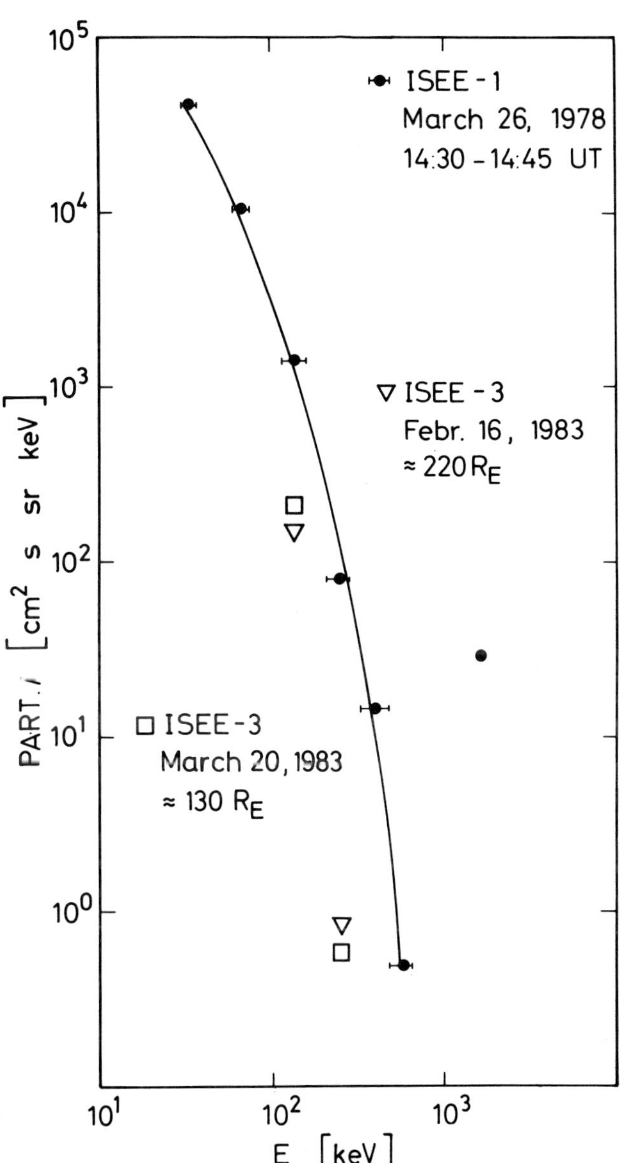

Figure 1—Solid dots: typical omnidirectional suprathermal proton spectrum in the near-earth plasma sheet as obtained by ISEE 1. Open triangles and squares: unidirectional fluxes of the proton beams in the plasma-sheet boundary layer in the distant tail as obtained by ISEE 3.

distances ($\sim 30\ R_e$) a complicated plasma sheet boundary layer may result—earthward beams from the distant neutral line with reflected particles more inside the boundary layer and tailward beams will be found at different longitudes.

"

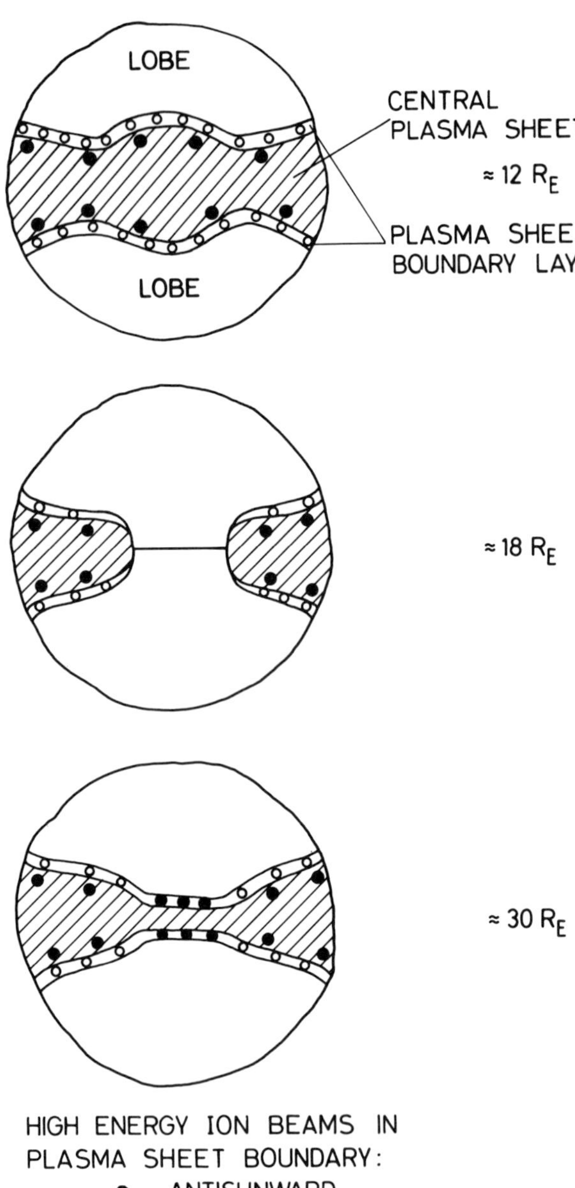

Figure 2—Schematic cuts through the tail at three different distances after a substorm onset.

When and where are the high-energy particles of the plasma sheet accelerated? The highest fluxes are usually observed when the plasma sheet recovers, i.e., ~ 40–60 min after a substorm onset. Baker et al.[8] have explained this in terms of a model where acceleration occurs in a short time period after substorm onset. The particles are injected toward the earth onto closed drift shells. During plasma-sheet recovery they leak out into the expanding plasma sheet. The question is whether this model is able to explain the high fluxes observed after recovery. It may be that the main acceleration occurs during the fast tailward retreat of the substorm-associated neutral line, i.e., between steps 9 and 10 of the Hones[9] model. This phase of the substorm is, by the way, the least understood. What causes the new neutral line to sit close to the earth for about an hour and then to move rapidly tailward? MHD simulators have yet to give a satisfying answer.

Let us come back to the kinetic aspects of the geomagnetic tail. The fast plasma experiment, the three-dimensional plasma experiment, and the energetic particle experiments on ISEE 1 and 2 have provided detailed distribution functions of ions and electrons in the plasma sheet and the plasma-sheet boundary layer. The complexity of these distribution functions exceeds by far present theoretical understanding (see, e.g., the distribution functions reported by Paschmann et al.[10] during the CDAW-6 event). New insight could possibly be gained by performing single particle trajectory calculations in the time-dependent MHD substorm models. Such calculations have been performed by, e.g., Sato et al.[11] and Matthaeus et al.[12]; however, they have not constructed detailed distribution functions. Methods such as the one employed by Lyons and Speiser[13] to predict the distribution function by current-sheet acceleration may be promising.

REFERENCES

[1] S. W. H. Cowley, R. D. Hynds, I. G. Richardson, P. W. Daly, T. R. Sanderson, K.-P. Wenzel, J. A. Slavin, and B. T. Tsurutani, "Energetic Ion Regimes in the Deep Geomagnetic Tail: ISEE-3," *Geophys. Res. Lett.* **11**, 275 (1984).

[2] I. G. Richardson and S. W. H. Cowley, "Plasmoid-Associated Energetic Ion Bursts in the Deep Geomagnetic Tail: Properties of the Boundary Layer," *J. Geophys. Res.* (submitted, 1986).

[3] M. Scholer et al., "Energetic Ion and Electron Beams at the Plasma Sheet Boundary in the Distant Tail" (this volume, 1986).

[4] M. Scholer et al., "Energetic Particle Beams in the Plasma Sheet Boundary Layer Following Substorm Expansion: Simultaneous Near-Earth and Distant Tail Observations," *J. Geophys. Res.* (in press, 1986).

[5] F. M. Ipavich and M. Scholer, "Thermal and Suprathermal Protons and Alpha Particles in the Earth's Plasma Sheet," *J. Geophys. Res.* **88**, 150 (1983).

[6] M. Scholer, G. Gloeckler, B. Klecker, F. M. Ipavich, D. Hovestadt, and E. J. Smith, "Fast Moving Plasma Structures in the Distant Magnetotail," *J. Geophys. Res.* **89**, 6717 (1984).

[7] E. W. Hones, Jr., et al., "Detailed Examination of a Plasmoid in the Distant Magnetotail with ISEE-3," *Geophys. Res. Lett.* **11**, 1046 (1984).

[8] D. N. Baker, R. D. Belian, P. R. Higbie, and E. W. Hones, Jr., "High-Energy Magnetospheric Protons and Their Dependence on Geomagnetic and Interplanetary Conditions," *J. Geophys. Res.* **84**, 7138 (1979).

[9] E. W. Hones, Jr., "Substorm Processes in the Magnetotail: Comments on 'On Hot Tenuous Plasmas, Fireballs, and Boundary Layers in the Earth's Magnetotail' by L. A. Frank, K. L. Ackerson, and R. P. Lepping," *J. Geophys. Res.* **82**, 5633 (1977).

[10] G. Paschmann, N. Sckopke, and E. W. Hones, Jr., "Magnetotail Plasma Observations During the 1054 UT Substorm on March 22, 1979 (CDAW 6)," *J. Geophys. Res.* **90**, 1217 (1984).

[11] T. Sato, H. Matsumoto, and D. Nagai, "Particle Acceleration in the Time Developing Magnetic Reconnection Process," *J. Geophys. Res.* **87**, 6089 (1982).

[12] W. H. Matthaeus, J. J. Ambrosiano, and M. L. Goldstein, "Particle Acceleration by Turbulent Magnetohydrodynamic Reconnection," *Phys. Rev. Lett.* **53**, 1449 (1984).

[13] L. R. Lyons and T. W. Speiser, "Evidence for Current Sheet Acceleration in the Geomagnetic Tail," *J. Geophys. Res.* **87**, 2276 (1982).

DISCUSSION

E. T. Sarris: It appears from some of your viewgraphs that there are boundary layer crossings without any noticeable particle intensity enhancements at these energies. Therefore it seems that the very presence of energetic particles is not a steady phenomenon. Whatever the mechanism producing or ejecting these particles, it must be acting sporadically.

M. Scholer: Your are specifically referring to the time period 1005–1035 UT on February 12, 1983. We have checked the ISEE 3 electron plasma distribution functions and conclude that the spacecraft was not in a plasma sheet boundary layer during this time period but was in the magnetosheath. However, in a recent paper, Scholer et al. (J. Geophys. Res., 1986, in press) present an event during which ISEE 3 was at 130 R_e earthward of an energetic particle source when entering the plasma sheet but did not observe energetic particles when again leaving the plasma sheet. This may indicate that the mechanism producing or ejecting these particles acts indeed not all the time but sporadically.

S. Peter Gary: Comment: Although some very interesting physics has been presented in this session, I have been disappointed with the "kinetic" label. Most of the papers in the session have concerned energetic particles and their responses to global fields. As theoretician, "kinetic" means to me the self-consistent interaction of particles and fields. If we are to address particle acceleration, which Dr. Scholer has listed as a fundamental issue, it is necessary to do fully kinetic physics, which involves thermal as well as energetic plasma, and fluctuating as well as relatively steady-state fields. The paper cited by Dr. Goldstein appears to be a start in this direction. I think that true "kinetic" experimental studies should, like kinetic theories, involve full data sets of particles and fields.

W. I. Axford: The properties of energetic ion beams at the plasma sheet boundary, described by Scholer in particular, indicate that the acceleration mechanism must be simple, "clean" and spatially/temporally limited. Since at times the energies involved were 1 MeV, the acceleration must be extraordinarily efficient. The very large emf's that are sometimes required in this case ($> 10^6$ volts) can only be short-lived since they are "inductive" and the amount of magnetic flux available is limited (because of the small number of observations of dispersed beams at these higher energies). Rapid reconnection requires an X-point rather than a neutral sheet and the short transit time of plasma through this point makes conditions difficult for the growth of turbulence – instead one would expect that the electric resistivity required for reconnection is simply associated with electron inertia and is unrelated to turbulence. Note that the voltage drop available is limited to $V_A LB \simeq 3 \times 10^6$ V if $V_A \sim 3000$ km/s, $L \sim 15\ R_e$, and $B \sim 10\ \gamma$.

M. Scholer: Comment: The number of acceleration processes are probably legion. In laboratory experiments at UCLA, for example, a current filamentation instability within a neutral sheet gives rise to a double layer and particle acceleration. In other experiments if the B_z field is forced to rapidly fluctuate locally, a current interruption can occur that also gives rise to particle acceleration along the sheet direction. A great deal more data are needed before it is unraveled.

MEASUREMENT OF MAGNETIC HELICITY DURING THE DISRUPTION OF A NEUTRAL CURRENT SHEET

W. Gekelman, J. M. Urrutia, and R. L. Stenzel*

A basic physics experiment on magnetic field line reconnection in a laboratory plasma is described. A neutral sheet is formed by applying time-varying opposing magnetic fields to a highly conducting plasma. Complete diagnostics for the three-dimensional magnetic field topology $B(r,t)$, particle distribution $f(v,r,t)$, fluid properties (n,T,v,j,p), and waves $(\delta n, \delta B, \delta E)$ have been developed. Quasistationary reconnection is observed and understood, to first order, by fluid models but, upon closer inspection, modified by kinetic effects (anisotropic in $f(v)$, microinstabilities, and space charge fields). A controlled disruption of the neutral sheet current results in rapid conversion of stored magnetic energy into particle kinetic energy. The disruption is accompanied by a sharp jump in the helicity, consistent with measured linking of magnetic flux tubes.

INTRODUCTION

Magnetic field line reconnection is recognized as an important process in fusion devices,[1] magnetospheric physics,[2] and the description of solar flares.[3] Reconnection involves a breakdown of the frozen-in condition between field and fluid that takes place at magnetic null regions (X points, neutral sheets) where the finite conductivity has to be considered in order to avoid a current singularity. Diffusion permits a flux transfer across the separatrix. According to Faraday's law the flux change implies an electric field along the separator (neutral line) that can be taken as a measure for the reconnection rate. Reconnection may be driven by an externally imposed plasma flow into a magnetic null region (e.g., magnetotail) or by temporally increasing external magnetic fields (e.g., pinch plasmas). Irrespective of its motional or inductive origin, the reconnection electric field drives a plasma current along the neutral line or sheet. The current density is consistent with the magnetic topology ($j = \nabla \times H$) and, according to fluid theory, with Ohm's law ($E + v \times B = \eta j$). The magnetic field associated with the plasma current provides a potential energy source that can be released during a disruption of the current sheet. Such disruptions may arise from tearing modes, resistivity changes due to current driven instabilities, or space charge instabilities (double layers) at large current densities ($v_d \rightarrow v_e$).

Whatever the causes of the disruption, the result will consist of a change in the magnetic field topology (if the field associated with the initial current is comparable to any ambient field) and an energy release. Fundamental questions deal with the conversion of magnetic field energy into particle energy[4] in space physics. A complex chain of events may ensue; for example, the disruption of the crosstail current may give rise to auroras. In fusion plasmas, the importance of reconnection by tearing modes lies in the associated transport processes.

The topological properties of the magnetic field lines are also of fundamental importance. To first order, a magnetized charged particle will spiral about a field line.

If the MHD approximation is valid, the magnetic field is frozen into the fluid plasma. Finally, the magnetic field configuration greatly aids in the visualization of most situations. The magnetic helicity measures the linkage of flux tubes[5,6] and may be associated with changes in the topology.

Detailed measurements of field topologies and current systems are not possible for solar physicists since observations are remote (resolution ~ 500 km, theoretical sheet width 10 m). Magnetotail measurements, although performed in situ, yield data only at a few points and suffer from space-time ambiguities. In hot, dense fusion plasmas, magnetic field measurements are done remotely or inferred from other data (X-rays, cyclotron emissions). Thus, there is a need to establish a case where reconnection can be studied in detail. This is possible in a well diagnosed laboratory experiment.

In the UCLA experiments,[7] the simplest possible configuration has been chosen, i.e., a linear device with an X-type neutral point on axis. A nearly collisionless discharge plasma is generated. By temporally increasing the strength of the antiparallel magnetic fields, a plasma current is induced, which assumes self-consistently the shape of a current sheet. Once established, the current sheet is disrupted with the use of a magnetic switch and is carefully diagnosed using modern digital data processing techniques.

The following section describes the experimental setup. Then the properties of the current sheet prior to disruption are reviewed. The disruption is then discussed in detail with emphasis on the plasma currents and helicity measurements. The summary will include a discussion of this experiment with regard to magnetotail physics.

EXPERIMENTAL ARRANGEMENT

Figure 1 shows schematically the main components of the experiment. A 2 m long discharge plasma is generated between a large cathode (1 m diameter) and anode. The plasma of parameters listed in Fig. 1b is uniform, quiescent, essentially collisionless, and highly reproducible in pulses of duration $t_p \simeq 5$ ms repeated every $t_r = 1.5$ s. The plasma may be immersed in a uniform

*Department of Physics, University of California, Los Angeles, California 90024.

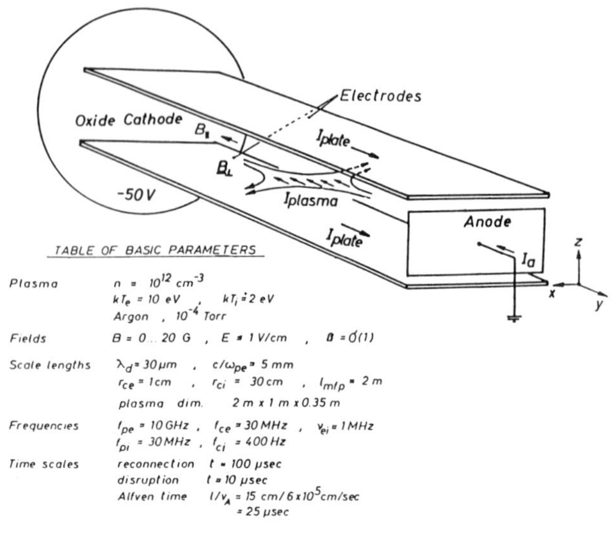

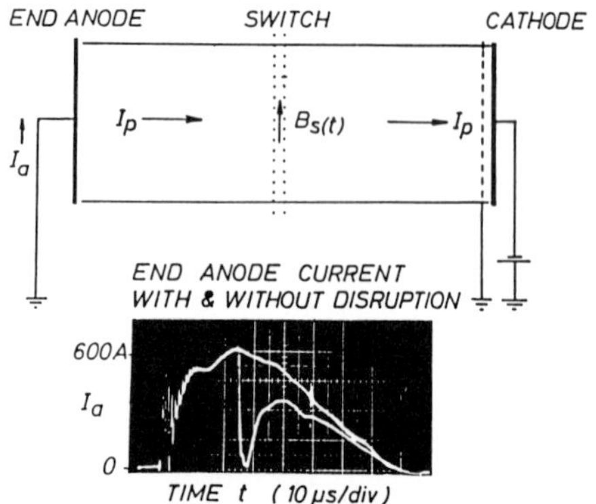

Figure 1—(a) Schematic representation of the experimental setup. (b) Table of basic plasma parameters. (c) Schematic (side view) of chamber and magnetic switch for studying current disruption. (d) Characteristic current waveform.

axial magnetic field of variable strength $B_{y0} = 0 \dots$ 100 G. For the reconnection experiments, a time varying ($t_{rise} = 100$ μs) magnetic field is applied transverse to the plasma column ($|B_\perp| = 0 \dots 20$ G). This field is produced by pulsing currents (I_{plate}) through two parallel-plate electrodes (1.8 m long × 72 cm wide) adjacent to, but insulated from, the plasma column. All return currents flow coaxially on the cylindrical metallic chamber (inner diameter 1.5 m) wall (which is not shown for purposes of clarity).

In vacuum, the transverse field topology $B_{\perp(x,z)}$ contains an X-type neutral point on the axis of the device. However, when the space between the plates is filled with plasma, a secondary current is induced antiparallel to the plate current. This induced plasma current (I_{plasma}) flows preferentially near the X-point but is so large ($\simeq 1000$ A) that the vacuum field topology is greatly modified. A neutral sheet and O-type neutral points may arise.

The machine was designed for accessibility. The vacuum vessel has 50 ports through which a variety of probes may be inserted via pumpdown stations without disturb-

ing the activated (and easily poisoned by oxygen) cathode. The plasma is cool enough to allow probes to be placed in it. The typical density ($10^{12}/\text{cm}^3$) is enough to provide for high β effects but not so large as to make structures of the order of the collisionless skin depth unreasonable or the coulomb collision frequency too high.

Magnetic field measurements are obtained with three orthogonal loops mounted on a coaxial telescoping axial shaft that allows them to be placed anywhere within the plasma volume. In this way, vector magnetic fields $B(r,t)$ are obtained in situ at up to 4000 spatial locations ($\Delta r \cong 2$ cm) and 1000 temporal points ($\Delta t \simeq 100$ ns).

From repeated measurements, statistical averages are formed (mean, standard deviation, correlations). Distribution functions are measured with a modified retarding potential analyzer that filters particles through a passive microchannel plate and thereby obtains high directional sensitivity ($\Delta\Omega/4\pi \simeq 10^{-3}$). The small detector ($\sim 3$ mm radius) can be moved in real space, rotated at each position through the two orthogonal spherical angles θ, ϕ so as to obtain from the differential particle flux the three-dimensional distribution function $f(v,\theta,\phi)$ or $f(v_x,v_y,v_z)$. Electron and ion phase space measurements $f(v,r,t)$ are time resolved to within a few microseconds. Ensemble averages yield fluctuations in velocity space. Measurements of such multidimensional functions produce large data flows ($n > 10^9$ numbers), which can only be handled by digital techniques.

The analog traces are therefore digitized with 10 MHz, 32 kbyte, 8-bit A/D converters housed in a CAMAC system that is serviced by an LSI 11/23 computer. The LSI serves as a slave to a VAX 11/750 computer and is linked to it by a specialized high-speed (230 kbyte/s transfer rate) network. The data are placed in large arrays that are accessed in real time (the experimental repetition rate is ~ 1 Hz) by an array processor that digitally filters and performs correlations and signal averaging. The VAX is in turn linked to a Cray computer for further data analysis and graphics.

SUMMARY OF RESULTS OF RECONNECTION AT A NEUTRAL POINT

The magnetic field topology during the time of external plate current rise is shown in Fig. 2a.[8] Whereas in a vacuum the field vanishes at one single point, the X-point, in the plasma the null region is extended along a separatrix with two contact points, described by two opposing horizontal Y's ($\succ\!\!-\!\!-\!\!\prec$). This topology is that of the classical neutral sheet[9] that arises when the induced plasma currents slow down the penetration and reconnection of magnetic fields at the field reversal region. The result is a pileup of flux although, in contrast to a superconductor, the shielding is incomplete and reconnection does take place.

In Fig. 2b a typical current density profile is shown, obtained by calculating $j = \nabla \times B/\mu_0$. The dashed region of largest current density indicates the shape of the current sheet, which is nearly uniform in the y direction. The thickness $\Delta z \simeq 5$ cm is in the range between

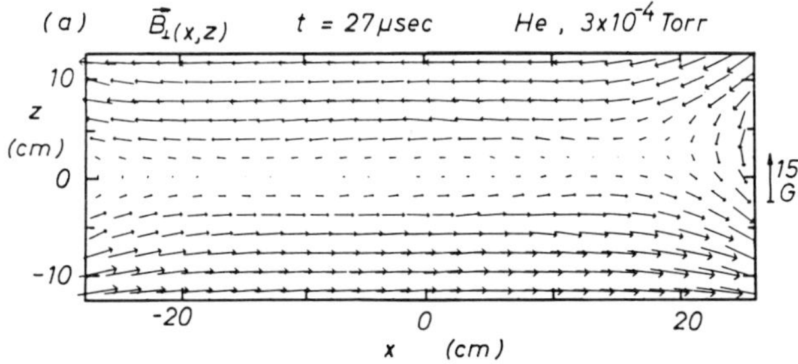

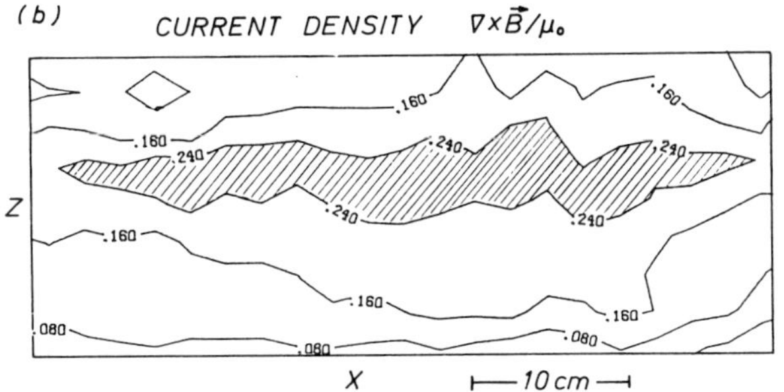

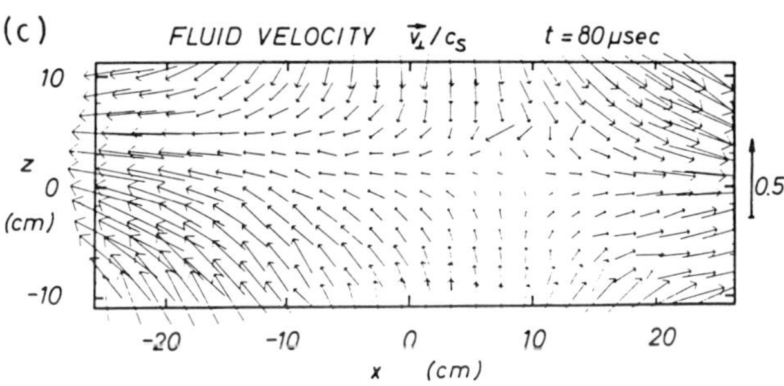

Figure 2—(a) Magnetic field topology in a plane showing the neutral sheet at $z \simeq 0$. (b) Current density in the undisturbed neutral sheet in A/cm^2. (c) Transverse ion flow $v_\perp (x,z)$ normalized to the local sound velocity $c_s = (kT_e/m_i)^{1/2}$. The characteristic fluid flow during reconnection is established, i.e., vertical inflow toward the neutral sheet and horizontal outflow at high velocities $(v/c_s)_{max} \simeq 0.6$. Note that the driving $j \times B$ force acts on the electrons while the ions are accelerated via space charge electric fields.

the electron Larmor radius ($r_{ce} \simeq 1$ cm) and the ion Larmor radius ($r_{ci} \simeq 60$ cm). The current is dominantly carried by electrons that are freely accelerated in the null region and drift in its vicinity with $v_d = E_\perp \times B_\perp/B_\perp^2 > v_{ion}$. The current sheet widens when a magnetic field component B_y is present along the separator. In this three-dimensional case, which will be depicted later, the electrons remain magnetized and carry axial current while drifting along $B_\perp$.

In two-dimensional reconnection models,[8] the rate of flux transfer across the separatrix is measured by the inductive electric field E_y along the separator ($-\partial\phi/\partial t = V, \phi = \int A_y d_y, V = \int E_y dy, \therefore E_y = -\partial A_y/\partial t$). However, the assumption of axial uniformity ($\partial/\partial y = 0$) and absence of space charge electric fields is an unrealistic simplification. From measurements of the plasma potential, we observe that a space charge electric field

$E_s \equiv -\nabla\phi_p$ builds up in the direction opposite to the applied inductive electric field. The net electric field is reduced to a value consistent with the electron supply and plasma resistivity and is of order $1/10$ the inductive electric field.

Space charge electric fields also develop self-consistently in the direction perpendicular to the neutral sheet. For example, in Fig. 2c, the observed ion flow in the transverse x–z plane is shown. The ion flow exhibits the, characteristic fluid flow during reconnection, which is driven by the $j_y \times B_\perp$ force. This body force acts on the current-carrying electrons that set up a space charge electric field and accelerate the unmagnetized ions until a common fluid flow is established. The evolution of the classical fluid flow at a neutral sheet has been studied in detail by comparing the measured acceleration $\rho\partial v/\partial t$ with the total force density $j \times B - \nabla p$.[10] The presence of fluctuating electric and magnetic fields causes an effective drag on the ions and has to be added to the average force densities. The fluid flow does not generate a density maximum on axis. Density and temperature maximize at the edges of the current sheet near the two contact points of the Y-shaped separatrix.[11]

Further investigations of the fluid properties have been concerned with the electrical resistivity η, which is important for energy dissipation, reconnection rates, and current disruptions.[12] From an analysis of the generalized Ohm's law, the effective resistivity is obtained and found to be at least one order of magnitude larger than the classical resistivity, to exhibit large spatial variations and to vary with current density as expected from current driven microinstabilities. This spatial dependence is shown in Fig. 3. Here $\eta = J \cdot E$, where $E = -\nabla\phi - \partial A/\partial t$, was measured with a specialized probe, and $J = 1/\mu_0 \nabla \times B$.

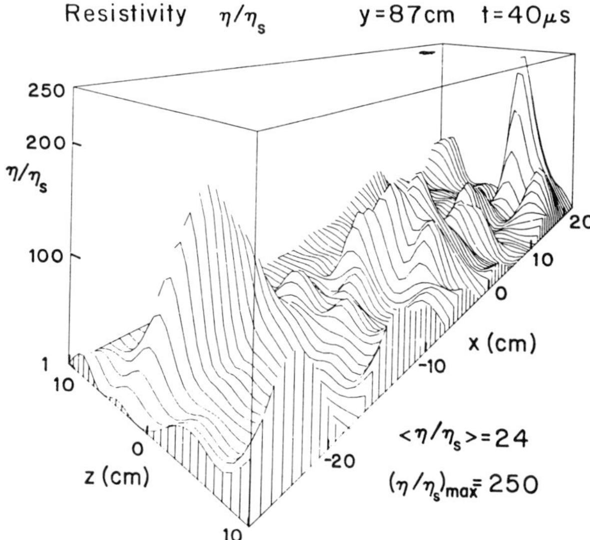

Figure 3—Three-dimensional view of the measured resistivity normalized to the local Spitzer value. Data are taken in the center of the machine before the current disruption. Spatial average $\langle\eta\rangle = 0.22$ Ωcm.

The magnetic field topology shown in Fig. 2, as well as the other parameters discussed so far, are ensemble averages. In the vicinity of the neutral sheet, considerable fluctuating components of the fields ($\delta B/B_\perp \sim 1$, $\delta\Phi/\phi \sim 0.05$) reflecting wave activity have been observed. By correlating the signals from a pair of electrostatic probes[13] tuned to monitor $\delta\Phi$ at a fixed frequency within the broadband turbulence ($f \lesssim f_{pi}$), the cross-spectral function (CSF), $C_{12}(\Delta y)$, may be obtained. The distance between maxima in Fig. 4a is the wavelength. By measuring λ as a function of the frequency the dispersion $\omega = kc_s$ is identified. The turbulence consists of ion acoustic modes driven unstable by the relative drift between electrons and ions ($v_d \gg c_s$) in a plasma with $T_e > T_i$. Ion sound tubulence is likely to be responsible for the anomalous increase in the plasma resisitivity[12] and electron heating.[11]

The current sheet is also a source of magnetic fluctuations. Magnetic noise exists up to a few electron cyclotron harmonic frequencies ($\omega_{ce} \sim \omega_{pi}$). The cross-spectral function in this case is a tensor produced by correlating the three components of B for the reference and movable probe. In contrast to Fig. 4a, the cross-spectral function is a complex pattern,[13] due to the spatial interference of many wavenumbers at a given frequency. The three-dimensional wavenumber spectra at $f = 1$ MHz is shown in Fig. 4b for a single component ($B_{x1}B_{z2}$) of the Fourier transform of CSF tensor into k space. Each dot represents an observed random wave, and the size of the dot is proportional to the wave amplitude. All the modes lie within two surfaces that are the whistler wave dispersion

$$\left(\frac{kc}{\omega}\right)^2 \approx \frac{\omega_p^2}{\omega(\omega_c\cos\theta - \omega)} \tag{1}$$

for the maximum density and minimum magnetic field ($B_{y0} \simeq 10$ G) within the region probed.

All the possible modes are expected to fill the volume between these surfaces. Finally, a test of the random waves moving along the axial field showed the waves to be right circular polarized. In addition to these electromagnetic whistlers and ion sound waves, a second spectrum of electrostatic fluctuations is found near the electron plasma frequency ($f \simeq f_{pe} \lesssim 12$ GHz). These are Langmuir waves excited by the stream of high-energy electrons. They scatter the tail electrons and transfer some of the tail energy into the main body. By mode conversion on density gradients and by scattering off density fluctuations, Langmuir waves couple to electromagnetic waves that are observed outside the plasma with horn antennas.

Analysis of the particle-distribution functions and waves provides a better understanding of the causes for the anomalous fluid behavior. Figure 5a-c summarizes measurements of the electron distribution function in the current sheet in one-, two-, and three-dimensional velocity space, respectively. The electron flux versus energy observed along the $\pm y$ direction (Fig. 5a) reveals the presence of a tail of high-energy electrons streaming away from the cathode. These are electrons accelerated at the cathode sheath by a potential given by both the discharge supply voltage ($V \simeq 40$ volts) and the reconnection voltage ($V \simeq 70$ volts). Two-dimensional flux measurements yield the distribution function $f(v_y,v_z) = f(v_\parallel,v_\perp)$ (assuming isotropy in v_x,v_z) shown in Fig. 5b as a topographical map and a contour map. Finally, in three dimensions the distribution can be displayed as a set of nested surfaces of constant value $f(v_x,v_y,v_z) = $ a constant, one of which is shown in Fig. 5c. These data depict, with increasing resolution, the anisotropy of the electron distribution in the current sheet. The tail electrons are mainly accelerated by the reconnection electric field that, due to space charge effects, has been distributed nonuniformly along the separator. Such field localizations are likely to occur also at nonuniformities in density, conductivity, cross section, or magnetic fields. The consequence of localized reconnection fields is the production of runaway particles[14] that modify trans-

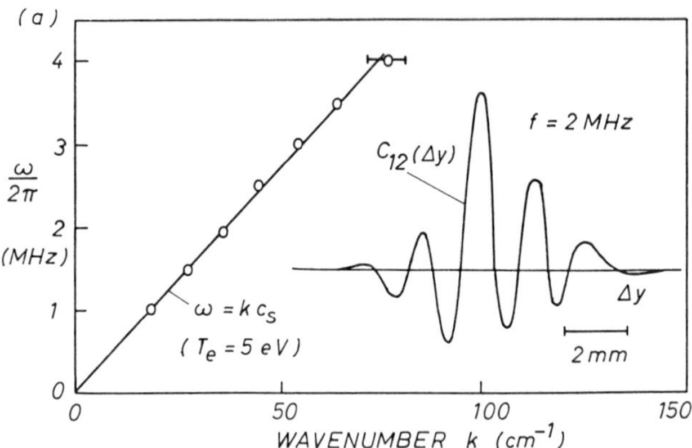

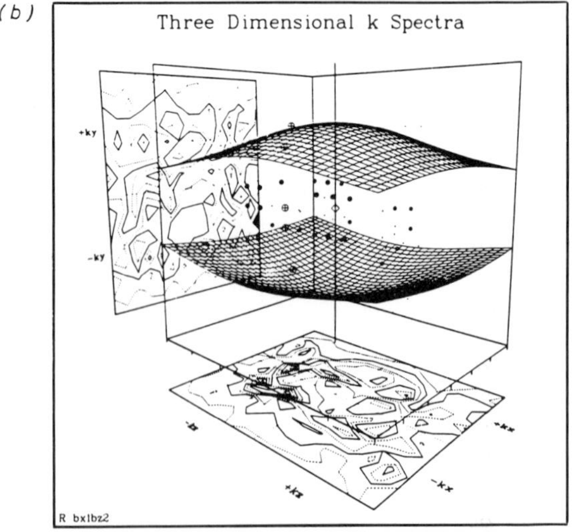

Figure 4—(a) Ion acoustic dispersion relation measured using cross-spectral function techniques. (b) Three-dimensional wavenumber spectrum of magnetic fluctuations. The experimentally observed modes (dots with radius proportional to mode amplitude) are bounded by whistler wave dispersion curves at $\omega = 1$ mHz.

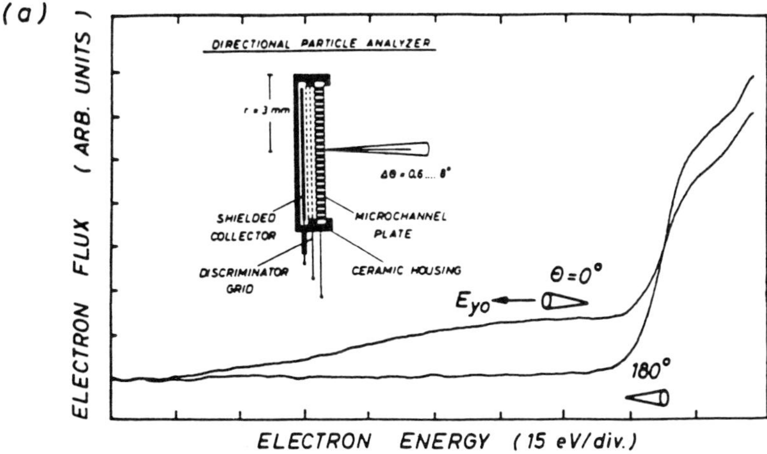

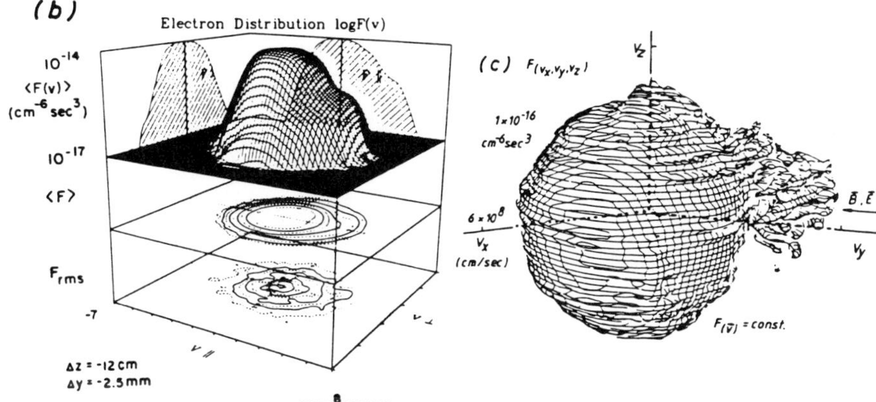

Figure 5—Characteristics of the electron distribution in the current sheet. (a) Electron flux versus energy as detected along the $\pm y$ direction with a directional velocity analyzer schematically shown in the insert. A large flux of energetic electrons opposite to E_y is observed. (b) Distribution function in two velocity dimensions, $\log f(v_y,v_z)$. Cuts through $\log f(v)$ show $F_\parallel = \log f(v_y,0)$, $F_\perp = \log f(0,v_z)$, and a contour plot $\log (v_y,v_z) = $ const. A second contour plot shows the rms fluctuations in the distribution function while the upper traces refer to ensemble average values. Note the tail in $f(v)$ and enhancements of fluctuations due to the anisotropy. (c) Distribution function in three-dimensional velocity space is displayed as surfaces of $f(v_x,v_y,v_z) = $ const. Shown in a surface $f(v)/f(0) = 2 \times 10^{-3}$, which cuts through the high-energy tail and shows its rich structure.

port coefficients (e.g., resistivity, heat flow), cause spatial nonuniformities (anisotropy), and can drive various kinetic microinstabilities.

The electron distribution function was analyzed[15] using an anisotropy test[16] that indicates that it is prone to a parallel whistler instability. The measured phase velocities of the Doppler shifted whistlers $\omega - \omega_c/k_\parallel$ falls into the range of the observed tail. In addition, the reduced parallel distribution function $g(v_\parallel) = \int f(E)dE_\perp$ exhibits a positive slope satisfying the criteria for Langmuir stabilities.

CURRENT SHEET DISRUPTIONS

While the previous observations described the properties of quasistationary reconnection, the present section deals with impulsive reconnection events that involve rapid changes in field topology and plasma properties. Two types of disruptions have been studied: spontaneous and controlled. Spontaneous disruptions arise when the current in the center of the neutral sheet is raised and are accomplished by biasing the central part of the end anode positively.[17] The cause for these current interruptions has been inferred from both the global circuit properties and the local plasma properties. They result in the production of double layers within the current sheet.

A second class of current disruption is caused by interruption of the entire plasma current. Such experiments can locally model the interruption of the cross-tail current. It is conjectured that the redirected currents may result in auroras at the earth's poles. In any event, the investigation of a current-sheet breakup addresses some fundamental questions. If the stored energy is embedded in a highly conducting plasma, at what rate can the energy be released during a disruption? How does the magnetic field topology change? The dynamics of unstable current systems is an open subject.

After establishing a well-formed current sheet, a magnetic switch is activated producing a thin slab of normal magnetic field B_z, preventing the electron flow between cathode and anode. The switch itself consists of a set of fine (0.02 mm) copper wires (Fig. 1c) placed midway between the oxide-coated cathode and grounded anode of the device. The current switch-off is performed rapidly ($\Delta t \simeq 3\ \mu s$) compared with the Alfvén transit time across the plasma ($t_A \gtrsim 30\ \mu s$) so as to distinguish subsequent magnetic field propagation and dissipation processes. The total plasma current may be disrupted with high reproducibility.

All three vector components of $\boldsymbol{B}$ are measured throughout the volume of the plasma between the wire array and the anode at 4000 spatial locations and 1024 time steps. The field measured at each position is an ensemble average over five pulses of the device. The volume measurement takes about a week (at a 1 Hz repetition rate) and is only possible because the source is extremely stable. The total magnetic field in this case is that due to the currents in plates at the plasma boundaries, the plasma current, and a constant axial field ($B_{y0} = 6$ G). There is also a stray magnetic field (B_{zs}) due to the switch, that drops off rapidly (over a few cm) in the y direction.

The transverse magnetic field energy density in a plane midway between the switch and anode is shown in Fig. 6. Before the disruption, a neutral sheet is observed near $z = 0$. Shortly after the disruption, $B_\perp^2/2\mu_0$ looks the same as it does in vacuum. When integrated over the volume, one finds that the stored magnetic energy disappears within $\Delta t = 5 \ldots 10$ μs, which is shorter than the propagation time of Alfvén waves to the boundaries.

By calculating the local current density $j = \nabla \times B/\mu_0$ and fitting field lines through this vector field, one can follow the space-time variation of the disrupted current flow as is shown in Fig. 7. Before the disruption, $t = -0.6$ μs, the current flows in a laminar sheet along the $-y$ direction. As the electron inflow at the left x-z plane is inhibited by the magnetic switch, the current begins to circulate within the plasma volume forming a spatially random pattern of small-scale current loops and filaments as well as decays. The magnetic fields associated with these currents cancel, giving rise to X-point topology for the transverse magnetic field. This process of cascading from large- to small-scale structures lowers the effective magnetic Reynolds number ($R_m \propto l$) and enhances the magnetic diffusion.

When the vertical magnetic field appears, the current

does not stop instantaneously. Its flow changes from a two-dimensional sheet to a fully three-dimensional inhomogeneous pattern. This may be seen in Plate IV-3, which shows current flux tubes before and after the disruption. The tubes are calculated by specifying a flux $|J_F|$ and start point. The local direction of j is calculated and then an area such that $|J_F| = \int j \cdot nda$. The calculation proceeds by storage of points on the boundary of the area and n advancing slightly along the local

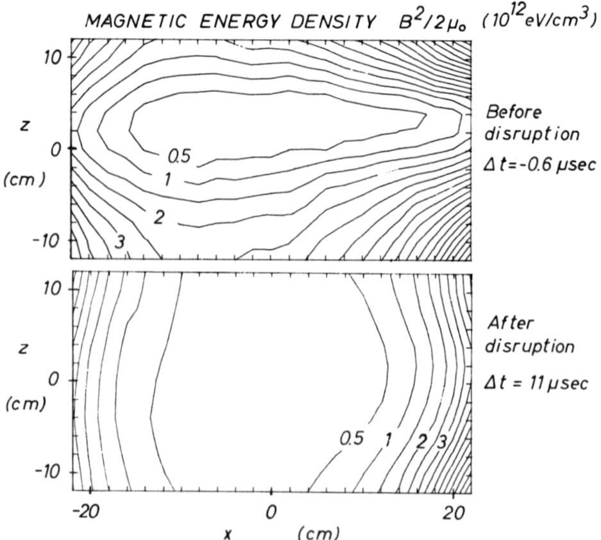

Figure 6—Magnetic energy density $B_\perp^2/2\mu_0$ versus (x,z) before and after the disruption. The magnetic energy associated with the plasma current is dissipated anomalously fast.

Figure 7—Field lines for the current density vector $j(x,y,z)$ at different times of the current disruption. The initially ($t = -0.6$ μs) laminar current sheet cascades during the disruption into small-scale current loops and filaments. Due to the triggered disruption, the spatially random pattern is highly repeatable from pulse to pulse.

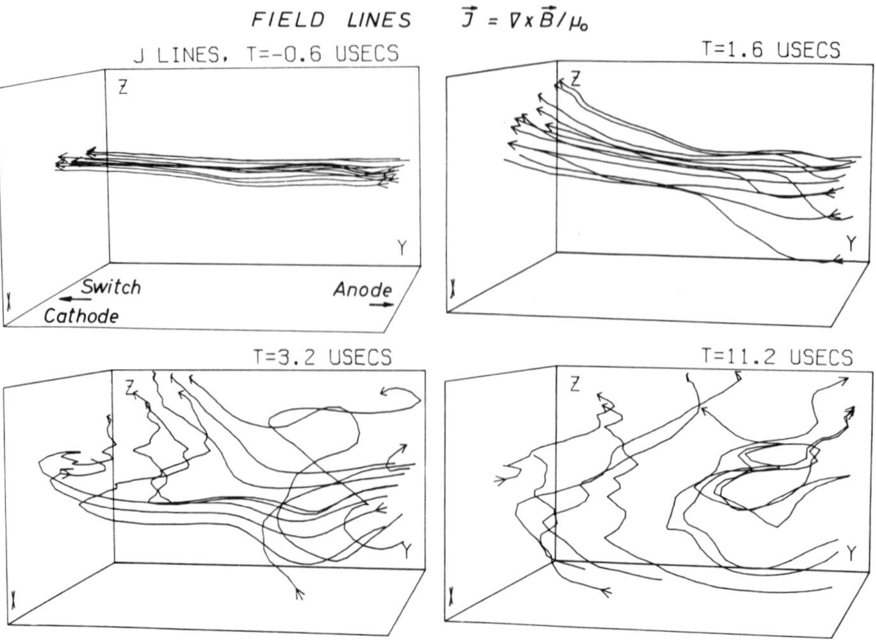

(A)

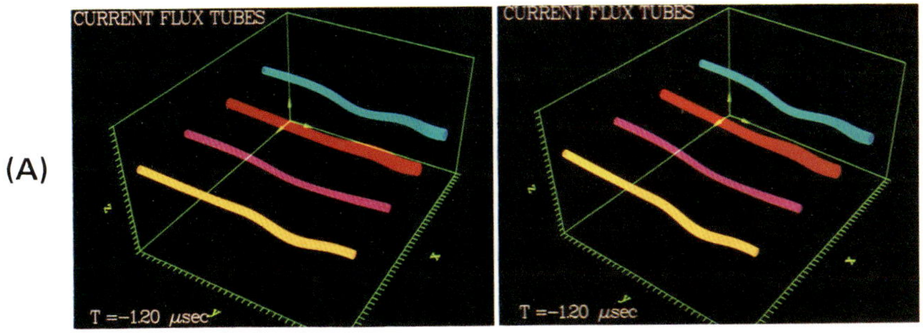

(B)

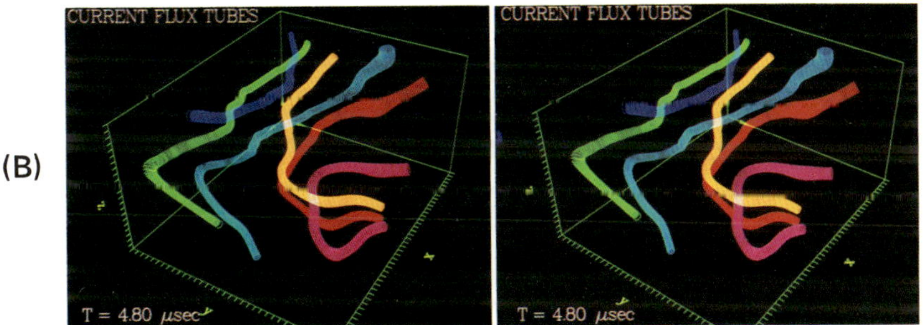

Plate IV·3—Stereo pairs showing current flux tubes: (A) before and (B) during the disruption. Initially the tubes are parallel; they then veer and weave around each other as they avoid the switch.

direction of *j* and repeating until a boundary is reached or 1000 areas are stored. The flux tubes are presented as three-dimensional stereo pairs, best seen with the viewer provided in this book. Before the disruption, the current tubes are a set of parallel cylinders within the neutral sheet. Afterward, the currents can criss-cross past each other (Plate IV-3B), implying that the flux associated with them can now link. This flux linkage gives rise to a change in the magnetic helicity.

HELICITY GENERATION

The magnetic helicity is a pseudoscalar quantity that may be identified with the linkage of magnetic field lines. It is defined as the volume integral

$$H = \int \mathbf{A} \cdot \mathbf{B} \, dV \qquad (2)$$

where A is the vector potential. For two untwisted magnetic flux tubes linked once $H = \pm 2\Phi_1\Phi_2$ where Φ is the flux in each tube and the sign depends upon the sense of linkage.[6] Equation 2 is a special case of the integral $\int z \cdot \nabla \times z \, dv$, which reflects the topology of $\nabla \times z$. As shown by Berger and Field,[18] the helicity can arise from twists and kinks within a single flux tube (internal helicity) and/or linkage and by knotting of separate flux tubes (external helicity). In ideal MHD, they are independently conserved. Taylor[19] has conjectured that the helicity is conserved in highly conducting plasmas, although reconnection can take place and the magnetic topology can change. This would be a consequence of a transfer between internal and external helicity.

In the current disruption, the ions are not magnetized (Fig. 7); they may be considered as nearly fixed ($V_{Ar} = 2 \times 10^5$ cm/s, $\Delta t = 10$ μs). The magnetized electrons that carry all the current may be considered a resistive fluid. When currents decay, the helicity does also,[20] according to

$$\frac{dH}{dt} = -2 \int \eta \mathbf{j} \cdot \mathbf{B} dv \qquad (3)$$

where η is the plasma resistivity. Conventional fluid theory is not applicable in this experiment because significant space-charge electric fields have been observed[2] and it is possible that a theory for a single electron fluid may make other predictions.

Various schemes to inject helicity have been proposed for fusion devices to maintain a given magnetic configuration. They are equivalent to adding flux to the system (or the imposition of an electric field) to keep the current flowing.

The integral for H (Eq. 2) is over all space and therefore includes the current systems that give rise to the vector potential and field

$$A(r) = \frac{\mu_0}{4\pi} \int \frac{j(r')}{|r - r'|} d^3r' \qquad (4a)$$

$$B = \nabla \times A \qquad (4b)$$

In this case, the current consists of at least two contributions. Current is forced to flow through the plates that bound the plasma column and to return through the chamber wall. This produces a vector potential A_{yp} (p = plate) along the y axis and the vacuum X-point magnetic field in any x-z plane. There is an additional constant axial field ($B_{y0} = 6$ G) imposed by solenoidal coils that surround the chamber.

The plasma currents arise self-consistently as A_{yp} changes slowly on the time scale of the disruption ($dt_{\text{plate}} \sim 80$ μs, $dt_{\text{disrup}} \sim 5$ μs). In addition, image currents are induced on the plates. Since the external circuit is not affected by the disruption, we evaluate the relative helicity[18]

$$H' = \int (A_{\text{pl}} + A_{\text{image}}) \cdot B_{\text{pl}} dv' \qquad (5)$$

A_{pl} is the vector potential, due to plasma currents, calculated from Eq. 4a.

The current is evaluated from the measured magnetic field, $j = (1/\mu_0) \nabla \times B$. B_{pl} is then obtained from $\nabla \times A_{\text{pl}}$. The image currents are obtained from the discontinuity of the tangential component of B_{pl} at the plates. To check the data evaluation, the vector potential A_{yp} was calculated and when added to $A_{\text{pl}} + A_{\text{image}}$, the field lines and vector magnetic field of the neutral sheet (Fig. 2b) could be reproduced. The volume integrated over is sandwiched between the plates on top and bottom ($|z| > 17$), the vacuum on the sides ($|x| > 35$), and the magnetic switch ($y = 0$) and anode ($y = 66$).

The temporal dependence of H' is shown in Fig. 8; a rapid change ($T \ll T_{\text{Alfvén}}$) in the relative helicity occurs at the time of the disruption. Subsequently, the helicity relaxes toward its initial value as the neutral sheet

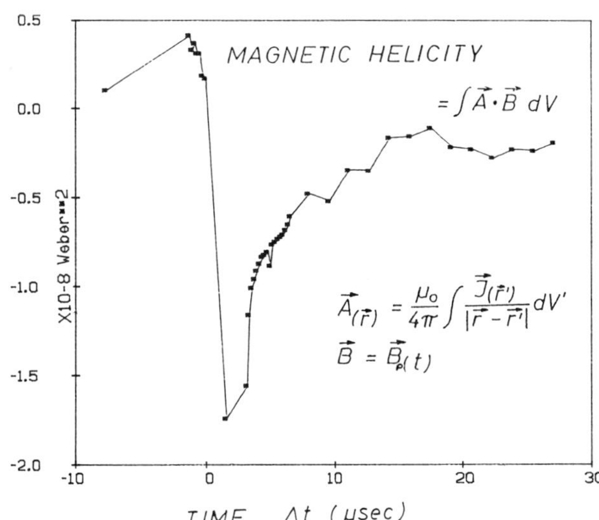

Figure 8—The magnetic helicity as a function of time. The disruption occurs at $t = 0$. The neutral sheet was formed at $t = -30$ μs.

reestablishes itself. The change in helicity is accompanied by changes in the field topology. Plate IV-4 shows two flux tubes evaluated from the magnetic field due to plasma current. The initial current (Fig. 7a) is predominant ly along the device axis and the flux encircles it. After the disruption, the geometry is so complicated that the linkage can only be seen in stereoscopic views. Plate IV-4B shows two flux tubes after the disruption. Although the tubes have nearly common endpoints, one of them (colored magenta) reverses in direction and loops through the other (red). Not all pairs of flux tubes link; in Plate IV-4C the pair simply spirals along in parallel.

A careful analysis of the plasma properties during the disruption is in progress. Although possible, a measurement of the plasma resistivity has not been made to see if Eq. 3 is verified. Nevertheless, some conjectures concerning the jump in helicity can be made. First of all, it cannot be attributed to injection from the switch. The field lines associated with the switch do not properly link with those surrounding the plasma current flux tubes (Fig. 9). The next (unlikely) possibility is that the change in relative helicity on the other side of the switch is equal and opposite to that in Fig. 8, thus conserving the total helicity. A contribution to the helicity could come from a change in wave activity caused by the current disruption. The helicity density carried by circularly polarized waves[18-21] is B_w^2/k, where B_w is the wave magnetic field and k the wavenumber. If the current disruption destroys the whistlers within the current sheet, a helicity transfer could occur, although these waves could only account for a fraction ($<5\%$) of it ($\bar{B}_w \sim 0.5$ G, $\langle\lambda\rangle \sim 10$ cm). The electromagnetic wave spectrum, however, extends to low frequencies[13] and wavelengths approaching the system size ($\lambda \sim 1$ m). These were not fully investigated; if polarized, they could carry considerable helicity. Finally, since kinetic effects are important here, it is conceivable that whatever instability causes the current filaments to snake about generates helicity as well.

SUMMARY AND CONCLUSIONS

The first three sections of this paper were a brief review of previously published work,[7] written to orient the reader to the geometry and plasma parameters in this reconnection experiment. Once a neutral sheet is established, a magnetic switch is used to disrupt the plasma current. The magnetic field topology rapidly changes and is accompanied by an abrupt jump in the helicity.

A review of these topological changes is summarized in Fig. 10, which is arranged as a matrix. The first row depicts field lines generated from the data and projected onto an x-z plane located 8 cm from the magnetic switch. The second row is contours of the axial component of the magnetic field B_y, and the last row synthesizes this with a view of some representative field lines in three space. The vacuum condition (column 1) is the superposition of an X-point (Fig. 10a) with a constant axial field (Fig. 10b), which results in the sheared topology of Fig. 10c. These fields are produced by currents in the plates bounding the plasma (Fig. 1a) and solenoidal coils on the device axis. The second two columns show the magnetic field, B_{pl}, due to plasma currents only. Before ($t = -7.6 \mu s$) the current disruption the transverse field lines are nested ellipses (Fig. 10d) and little axial magnetic field is generated from plasma currents. This is not the case after the disruption (column 3, $t = 4.8 \mu s$) where the transverse field strongly varies along y and considerable axial field is generated. The total measured field ($B_{\text{pl}} + B_{\text{vac}} + B_{\text{image}}$), shown in the final column, indicates that the neutral sheet (Fig. 2a) has collapsed into a distorted X-point. It is interesting that the axial field in this case (Fig. 10k) does not show the structure observed in Fig. 10h. This is because the axial magnetic fields generated by image currents in the plates nearly cancel those produced by plasma currents. One may speculate that full or partial interruptions of the cross-tail current could also generate "image current systems" in highly conducting plasma near the plasma sheet envelope. Finally, although the three-dimensional field configuration differs fom the vacuum case comparison of Fig. 10c and 10k is not enough since the plasma currents depend on field gradients.

Although helicity has been measured indirectly in a Spheromak experiment[22] and estimated within the solar wind from the spectrum of magnetic turbulence,[23] this is, to the authors' knowledge, the first in situ measurement from direct evaluation of A and B. The complete experimental database made it possible to show that the helicity generation was accompanied by linkage of magnetic flux tubes. The change in helicity occurs when current flux tubes intertwine and may be accompanied by a jump in resistivity (Eq. 3) or a large change in plasma turbulence. This would accompany the observed cascade of the laminar current sheet into small-scale current filaments and loops of high spatial turbulence.

It is apparent that a variety of complex interdependent processes has been observed in this laboratory experiment and it is germane to ask if these have analogs in the earth's magnetotail. An MHD description has been the standard tool for a global description of the magnetosphere and recent computer simulations (see articles by Walker and Ogino in this volume) have reproduced it. This laboratory plasma cannot be described by a single fluid MHD model. In the first place, the ion Larmor radius (Fig. 1b) in argon is of order of the transverse dimension between the plates on the plasma boundaries The magnetic Reynolds number ($R_m = \mu_0 \sigma L V_A$, where L = scale length, V_A = Alfvén velocity) is orders of magnitude lower than in space ($R_m \cong 10$ Ar, 25 He), mainly because of the enormous scale lengths up there. With this in mind, it is remarkable that the magnetic field and fluid (ion) flow patterns observed in the laboratory under quiescent reconnection conditions (Fig. 2a,c) are what is classically predicted.

These experiments, however, are not meant to model the magnetosphere, only the merging region. It is becoming evident from recent measurements[24] that microstructure exists within the tail. Spacecraft particle detectors can now measure distribution functions at

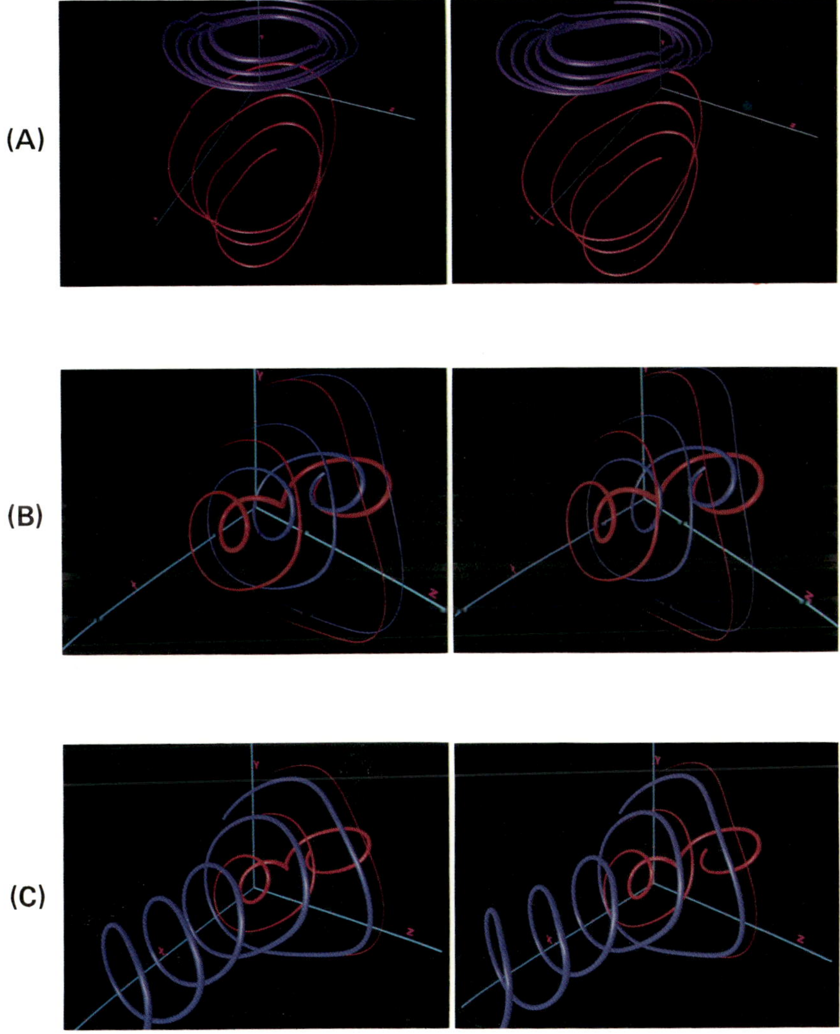

Plate IV-4—Stereoscopic pairs of magnetic flux tubes: (A) before; (B) 6.6 μs into the disruption, depicting tubes that link; and (C) two flux tubes at $t = 6.6$ μs that do not link. The origin defined by three arrows in the rear corner is at $x = -21$, $y = -16$, $z = -11$. The axis lengths are $\Delta x = 42$, $\Delta y = 32$, $\Delta z = 22$.

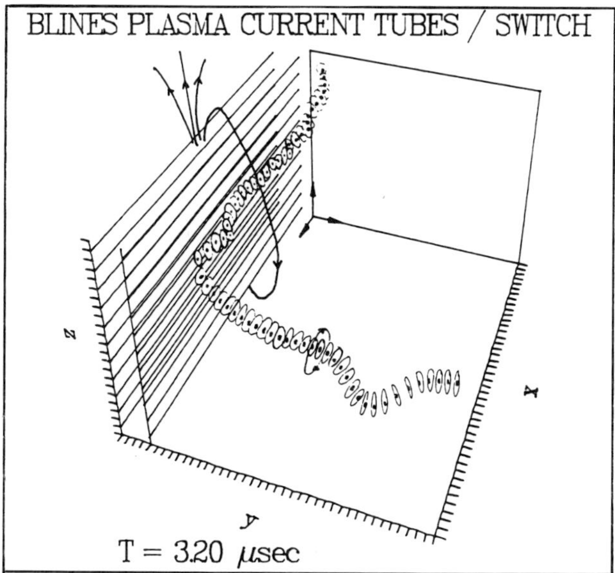

Figure 9—A typical measured current line surrounded by magnetic field near the switch. The stray magnetic field of the switch envelops the azimuthal field that surrounds the current line but does not link it.

type of activity in the tail exist over relatively small-scale lengths and the Reynolds number associated with them may not be large.

In this laboratory experiment, the neutral-sheet current is carried by electrons. A single fluid MHD model predicts otherwise for the tail and this has been the unproven belief of much of the space physics community until now. There is no reason, however, that electrons should not respond to E_y along a neutral line. Recent analysis of asymmetries in distribution functions of both species obtained by spacecraft[25] indicate this may be the case.

Several years ago, without the current accumulated wealth of satellite data, this laboratory experiment could only be superficially compared to what was then the model of the tail. It was assumed (perhaps with a sigh of relief) that the complexity observed in the laboratory would be ironed out in the highly conducting fluid in space. It is also true that as diagnostic techniques develop, structure previously averaged over always appears.

No claim is made that this laboratory experiment is a scaled-down version of the tail neutral sheet. In the first place the boundary conditions differ; secondly the spacecraft data are incomplete. However, a laboratory basic physics experiment can serve to stimulate fresh interpretation of signatures seen in space.

many angles in velocity space. The data are replete with ion conics, non-Maxwellian electron distribution functions with beam-like tails, and local heating. This, coupled with observations of large variations in B_y, cannot be described within an MHD framework. Regions of this

ACKNOWLEDGMENT—The authors gratefully acknowledge support from the U.S. National Science Foundation under grants NSF ATM 84-01322 and PHY 84-10495. We also appreciate L. Xu's help with the computer system.

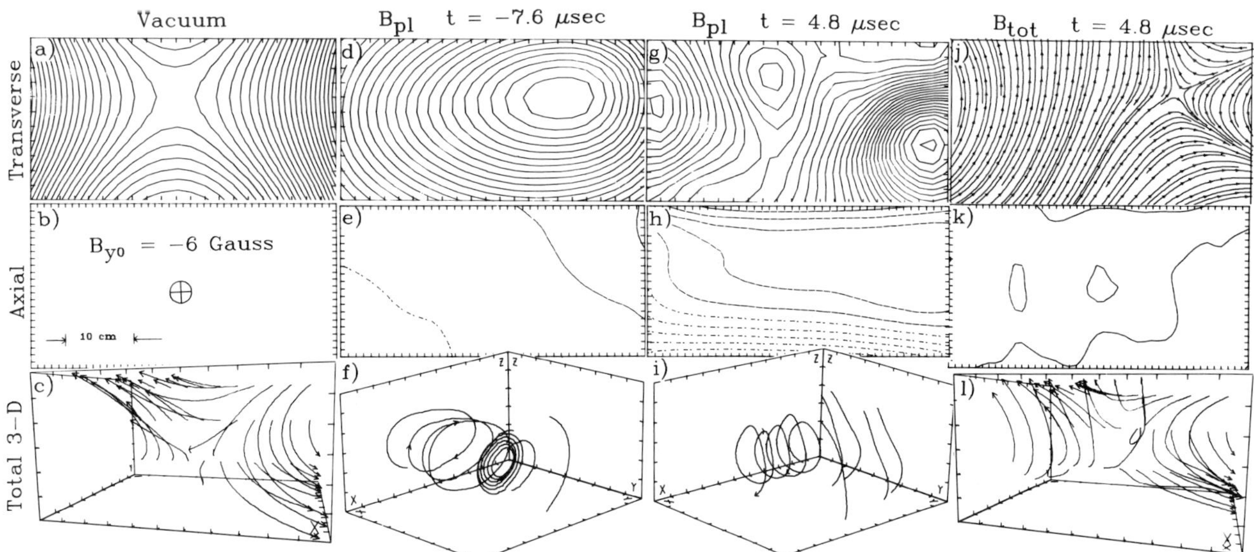

Figure 10—Matrix summary showing transverse field lines (row 1) and axial magnetic field (row 2) 8 cm from the magnetic switch as well three-dimensional field lines. The first column, (a) – (c), is the vacuum case, the second and third, (d) – (i), is the field due to plasma currents only, and the final column (j) – (l) is the total field. The axial fields in (e), (h), and (b) are 1 G/contour; dotted contours are into the page (+ y direction). In (b), the − 6 G field is suppressed. Distance scale for the first two rows is shown in (k). The tick marks in row three occur every 4 cm, (c) and (d) are views from the switch looking toward the anode, in (f) and (i) the switch is located 2 cm to the rear of x-z plane, and the anode is off to the right. The current-carrying plates are 2 cm above and below the x-z plane in all cases.

REFERENCES

[1] Furth, "Nonideal Magnetohydrodynamic Instabilities and Torodial Magnetic Confinement," *Phys. Fluids* **28**, 1595-1611 (1985).

[2] B. U. O. Sonnerup, "Magnetic Field Reconnection," in *Solar System Plasma Physics,* Vol. 3, North Holland Publishing (1979).

[3] H. A. Carmichael, "A Process for Solar Flares," in *AAS-NASA Symposium on Solar Flares,* W. N. Hess, ed. (1964).

[4] D. S. Spicer, "Magnetic Energy Storage and Conversion in the Solar Atmosphere," *Space Sci. Rev.* **31**, 351 (1981).

[5] M. A. Berger and G. B. Field, "The Topological Properties of Magnetic Helicity," *J. Fluid Mech.* **147**, 133-148 (1984).

[6] H. K. Moffatt, *Magnetic Field Generation Electrically Conducting Fluids,* Cambridge University Press (1978).

[7] R. L. Stenzel, W. Gekelman, and J. M. Urrutia, "Laboratory Experiment on Magnetic Reconnection and Turbulence," in *Proceedings of the Workshop on Magnetic Reconnection and Turbulence,* Cargese (Corsica-France) (July 7-13, 1985) (in press).

[8] R. L. Stenzel and W. Gekelman, "Magnetic Field Line Reconnection Experiments 1. Field Topologies," *J. Geophys. Res.* **86**, 649 (1981).

[9] J. W. Dungey, *Cosmic Electrodynamics,* Cambridge University Press, New York (1958).

[10] W. Gekelman, R. L. Stenzel, and N. Wild, "Magnetic Field Line Reconnection Experiments 3. Ion Acceleration, Flow, and Anomalous Scattering," *J. Geophys. Res.* **87**, 101 (1982).

[11] W. Gekelman and R. L. Stenzel, "Magnetic Field Line Reconnection Experiments 2. Plasma Parameters," *J. Geophys. Res.* **86**, 659 (1981).

[12] R. L. Stenzel, W. Gekelman, and N. Wild, "Magnetic Field Line Reconnection Experiments 4. Resistivity, Heating, and Energy Flow," *J. Geophys. Res.* **87**, 111 (1982).

[13] W. Gekelman and R. L. Stenzel, "Magnetic Field Line Reconnection Experiments 6. Magnetic Turbulence," *J. Geophys. Res.* **89**, 2715 (1984).

[14] R. L. Stenzel, W. Gekelman, and N. Wild, "Electron Distribution Functions in a Current Sheet," *Phys. Fluids* **26**, 1949 (1983).

[15] W. Gekelman and R. L. Stenzel, "Measurement and Instability Analysis of Three-Dimensional Anisotropic Electron Distribution Functions," *Phys. Rev. Lett.* **54**, 2414 (1985).

[16] C. F. Kennel and H. V. Wong, "Resonant Particle Instabilities in a Uniform Magnetic Field," *J. Plasma Phys.* **1**, 75 (1967).

[17] R. L. Stenzel, W. Gekelman, and N. Wild, "Magnetic Field Line Reconnection Experiments 5. Current Disruptions and Double Layers," *J. Geophys. Res.* **88**, 4793 (1983).

[18] M. A. Berger and G. B. Field, "The Topological Properties of Magnetic Helicity," *J. Fluid Mech.* **147**, 133-148 (1984).

[19] J. B. Taylor, "Relaxation of Torodial Plasma and Generation of Reverse Magnetic Fields," *Phys. Rev. Lett.* **33**, 1139 (1974).

[20] T. H. Jensen and M. S. Chu, "Current Drive Helicity Injection," *Phys. Fluids* **27**, 2881-2885 (1984).

[21] D. Montgomery and L. Turner, "Anisotropic Magnetohydrodynamic Turbulence in a String External Magnetic Field," *Phys. Fluids* **24**, 825-831 (1981).

[22] C. Barnes, et al., "Determination of the Conservation and Balance of Magnetic Helicity Between Source and Spheromak in the CTX Experiment," Los Alamos report LA-UR-85-2993 (1985).

[23] W. H. Matthaeus, M. L. Goldstein, and C. Smith, "Measurement of the Rugged Invariants of Magnetohydrodynamic Turbulence in the Solar Wind," *J. Geophys. Res.* **87**, 6011-6028 (1982).

[24] J. V. Bieber, E. C. Stone, E. W. Hones, Jr., D. V. Baker, S. J. Bame, and R. P. Lepping, "Microstructure of Magnetic Reconnection in Earth's Magnetotail," *J. Geophys. Res.* **89**, 6705-6717 (1984).

[25] J. D. Williams, personal communication, The Johns Hopkins University Applied Physics Laboratory.

HIGHLIGHTS ON KINETIC ASPECTS OF MAGNETOTAIL DYNAMICS

C. Y. Huang*

The importance of the boundary layers of the magnetotail has been established for some time. It is clear that transport of energy and momentum takes place within these spatially limited regions. These points were central to several of the papers in this session. However, there remains a number of outstanding problems that have not yet been addressed. Some examples are:

1. The spatial structure of the plasma-sheet boundary layer, particularly in the radial direction, is poorly understood. The ICE data set provides the first detailed samples of the distant tail boundaries. Combined with simultaneous near-earth measurements these could provide the first global view of the region.

2. The correlation of the boundary layer with geomagnetic activity has only been studied in terms of the spatial motion of the boundary during the onset and recovery phases of substorms. However, it is obvious that temporal changes do occur, e.g., there is energization of the streaming plasma that is evident upon plasma-sheet recovery. While it is always difficult to separate temporal from spatial variations, the proper understanding of this acceleration process is essential to any complete theory of magnetospheric substorms.

3. There is a natural tendency to focus on ion observations, especially during substorm events. However, the field-aligned currents, carried essentially by electrons, provide the connection to ground-based observations of aurora. A statistical study of the quiet-time structure of the tail currents has only recently been initiated. Temporal and spatial changes with increased activity are still unknown.

4. At present, the only theory that deals with boundary-layer formation at all times is current-sheet acceleration. There is no theoretical explanation for the generally observed source location at approximately 100 R_e from the earth. In addition, neither theory nor observation provides any information on the spatial extent or motion of the source during active times. (This point is related to point 2 above.)

5. The importance of direct particle entry of magnetosheath or lobe plasma via the low-latitude or plasma-sheet boundary layer into the central plasma sheet has only been partly investigated. This may be a significant source of plasma for maintenance of the central plasma sheet during extended periods of northward-directed interplanetary magnetic field.

6. The structure and particle characteristics of the low-latitude boundary layer in the tail, particularly at large radial distances, are virtually unknown. At large distances, it is to be expected that particle kinetics and viscous interactions could be important in magnetotail dynamics, e.g., generation of the plasma-sheet boundary layer. Currently, such theories are largely speculative.

Despite such questions (which remain outstanding) it seems clear that the observational results are far more sophisticated than current theoretical models that still favor fluid theory. As the dynamical behavior of the magnetotail is confined to relatively thin layers, it is questionable whether anything other than a fully kinetic approach will prove useful.

*Department of Physics and Astronomy, The University of Iowa, Iowa City, Iowa 52242.

KINETIC ASPECTS OF TAIL DYNAMICS:
THEORY AND SIMULATION

T. W. Speiser*

Kinetic theories relevant to the geomagnetic tail are reviewed. The topics discussed include kinetic instabilities, simulations, and current-sheet particle acceleration. Tearing mode and reconnection theories are emphasized. Kinetic treatment is appropriate for these topics since the tail plasma is collisionless. Fluid calculations are appropriate when stochastic processes dominate and for studies where long wavelengths are important. However, fluid treatments of tearing modes and reconnection require a finite resistivity in the diffusion region. Thus, although "anomalous resistivity" can be guessed or in some cases calculated, ideally the kinetic treatment is often to be preferred. Particle motion and acceleration in the current sheet can give rise to beam-like distributions in the plasma-sheet boundary layer. Studies of current-sheet particle motion have also been used as the basis for "kinetic" tail equilibrium models. Furthermore, quite recently current-sheet particle motion is used directly in Coroniti's explosive tail reconnection model. The "inertial conductivity" from the equilibrium models provides the "dissipation" necessary for reconnection.

INTRODUCTION

In the geomagnetic tail, there are several length scales of interest. The tail diameter is about 40 R_e (earth radii = 6.4×10^6 m), the length is greater than several hundred R_e, the plasma-sheet thickness varies from a fraction of an R_e to about 10 R_e, and the scale for variation along the length of the tail is also of order R_e to tens of R_e. Estimating the collisional mean free path, L_f, for the tail plasma sheet ($n \sim 1$ cm^{-3}, $\kappa T \sim$ 1 keV): $L_f \sim 10^{16}$ m (10^9 R_e). Thus, particle-particle collision processes are expected to play virtually no role in the geomagnetic tail. However, collective phenomena do exist, as evidenced by observations of waves,[1] and thus interaction lengths for scattering of particles by waves can often be defined giving some basis for the use of fluid theories. On the other hand, kinetic theories based on solutions to the Vlasov-Boltzmann equations or on individual particle orbits are valid even when fluid theories become inappropriate because strong scattering processes are absent. The kinetic theories are generally more complicated, however, and some approximations are usually necessary.

In this review we will briefly summarize past contributions on kinetic equilibrium and stability calculations and reconnection (second section), on simulations (third section), and on current-sheet particle accelerations (fourth section), relevant to the geomagnetic tail. Theories relating to tearing mode and reconnection are emphasized, and microinstabilities leading to "anomalous resistivity" are only briefly touched upon (second section).

In a sense, this whole field of endeavor was started by Dungey's[2,3] pioneering efforts, investigating neutral-point stability. Using fluid theory, Dungey showed that a magnetic neutral point, as sketched in Fig. 1, in a

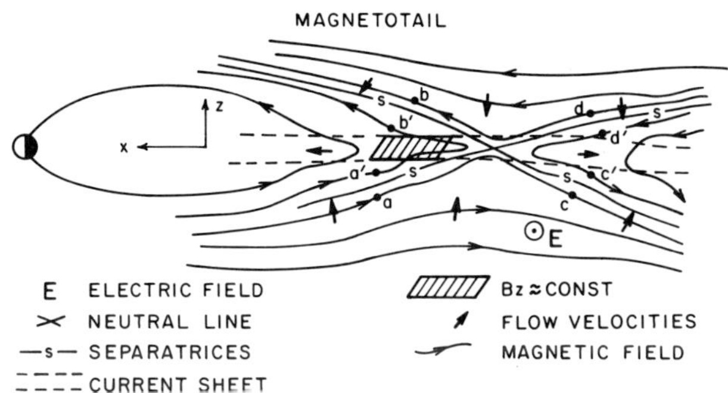

Figure 1—Sketch of magnetic field configuration in the geomagnetic tail. Lobe field lines like *a-c* or *b-d* cross the separatrices with point *a* moving to *a'*, *b* to *b'*, etc., and *a'-b'* joined together after the reconnection process.

magneto-fluid is unstable and with a small perturbation will quickly collapse, with the current density and inverse thickness going to infinity in a finite time. It was a fortuitous situation wherein the system was described by a finite set of equations with a finite set of unknowns. The basic approximations were neglect of pressure gradients and resistivity. Dungey argued that pressure gradients might change the time constants, but would not stop the "discharge," whereas a finite conductivity would stop the collapse to a finite thickness, with a finite current density. Under this situation, a finite electric field is left across the region (*y*-direction in Fig. 1), and plasma flows across the separatrices, labeled "s". As the plasma is frozen to the field in the external region, it is as if a field line, say *a-c* or *b-d* (Fig. 1), moves across the separatrices but this requires the field line to be broken as it passes through the neutral point, and to be reconnected or rejoined with its partner. Thus, for example, point *a* goes to *a'*, *b* goes to *b'*, and *a'-b'* become connected together. The direction of the flow is indicated on the figure by the double arrows. Dungey's paper also contained the first qualitative descrip-

*Department of Astrophysical, Planetary and Atmospheric Sciences, University of Colorado, Boulder, Colorado 80309-0391.
Also SEL, National Oceanic and Atmospheric Administration, Boulder, Colorado 80303.

tion of particle motion in an X-type field with electric field. So although Dungey's formulation was originally fluid, it was the trigger for the explosive release of publications, both fluid and kinetic, dealing with current-sheet stability.

KINETIC THEORY OF EQUILIBRIA AND INSTABILITIES RELEVANT TO THE MAGNETOTAIL

Kinetic Equilibria

In one of the first kinetic equilibrium studies of a current-sheet topology, Harris[4] solved the combined Vlasov-Maxwell equations. Since the Vlasov (or collisionless Boltzmann) equation is identical to the Liouville theorem ($Df/Dt = 0$) in phase space, a distribution function built up from constants of the motion for individual particle orbits will satisfy the Vlasov equation. Harris utilized this fact and started with a distribution function of the form

$$f_s = (M_s/2\pi\theta_s)^{3/2} N_s \exp[-(M_s/2\theta_s)$$

$$(\alpha_1^2 + (\alpha_2 - V_s)^2 + \alpha_3^2)] \qquad (1)$$

where s is an index for species; M_s, θ_s, and N_s are the species mass, thermal energy, and density; V_s is the bulk speed; and α_1, α_2, and α_3 are the constants of motion. (These are combinations of the total energy, and x and y are generalized (canonical) momenta. For this one-dimensional model, the dimension of variation is taken to be z, with e_z the direction perpendicular to the current sheet.) Using this distribution function along with the first and second moments to get the charge and current density, the Poisson equation and Ampere's law then give the coupled differential equations for the electrostatic (ϕ) and magnetic (A_y) potentials. One solution is for $\phi = 0$ and then $A_y \sim \log \cosh (z/h)$, where $h = cL_D/V$ is the scale for the variation. (c is the speed of light, L_D is the Debye length, and V is the streaming velocity in the y-direction.) The magnetic field thus varies as $\tanh (z/h)$ and the plasma density as $\cosh^{-2} (z/h)$, results which are widely quoted in the literature as well as used as a starting point for stability studies. Harris noted that his result showed that the magnetic pressure at infinity was just balanced by the plasma pressure at the center of the current sheet. In addition, using his solutions, it can be shown that the total pressure, gas plus magnetic, is constant everywhere. Finally, Harris showed that the fields could be found in other systems moving relative to his by a Lorentz transformation, and that his solution becomes the Bennett pinch solution in cylindrical coordinates.

Galeev[5] shows that for a weakly two-dimensional current sheet, the simplest generalization of the Harris sheet includes a weak normal magnetic field component, with the additional magnetic field tension being balanced by a weak pressure gradient directed along the current sheet.

In the following section, stability of the current sheet equilibria is discussed.

Stability and Tearing Modes

In the tearing mode instability, a current sheet "tears" or induces filamentation as parallel current elements attract. For the tearing mode, the classic MHD result is due to Furth, Killeen, and Rosenbluth.[6] There it was found that the growth rate, $\gamma \sim \eta_c^{3/5} k^{-2/5}$, where η_c is the resistivity and k the wavenumber, favoring growth at longer wavelengths. Furth[7] also discussed the collisionless tearing mode from the standpoint of the Vlasov equation. He started with the one-dimensional equilibrium solution, as discussed above, and considered perturbations of the form $f(z) \exp (\gamma t + ikx)$ where z is the dimension of variation perpendicular to the sheet, and x is the dimension along the sheet in the direction of the equilibrium field. (We changed Furth's notation to correspond to the usual solar-magnetospheric coordinates in the geomagnetic tail.) For flow velocities (V) large compared to the thermal speed (V_t), Furth estimates the collisionless growth rate as $\gamma \sim (4\pi k N_0 h e^2/Mc^2)^{1/2} V$, where N_0 is the maximum unperturbed plasma density, e and M the ion charge and mass, h the scale for the z-variation (as in the Harris sheet), and V the bulk velocity. Furth then notes that this result is equivalent to replacing the MHD collisional resistivity, η_c, by $\eta^* = \eta_c + \eta_i$, where η_i is an effective resistivity due to the ion mass and where $\eta_i \propto \gamma$, the collisionless growth rate. To quote Furth: "The extra term corresponds to the dj/dt term in Ohm's law. We can use all of the results of section A (MHD), simply replacing η by η^*." Thus, in the collisionless case, "current-carrier inertia" takes the place of resistivity in fluid calculations.

Coppi et al.,[8] using semi-intuitive arguments, analyze the stability of the Harris sheet. They find a growth rate of the order of $V_{the}/de (r_{Le}/h)^2$ where V_{the} is the electron thermal speed, r_{Le} is the mean electron Larmor radius, h, as before, is the sheet thickness, and $de \simeq (r_{Le}h)^{1/2}$ is the effective slab thickness centered on the neutral sheet, where macroscopic energy is transferred to resonant electrons. The result is for long wavelengths such that $k^2h^2 < 1$ and agrees with the more formal result of Laval et al.[9] Using values from IMP 1 of $2h = 600$ km, $B_0 = 16$ nT, $\theta_i \simeq 1$ keV, Coppi et al. find a growth time of order 5–15 s, for electron thermal energies from 1–0.01 keV. They also estimate that electrons could be energized to about 10 keV in times of order of the growth time. Note that these results are for a rather special case of thin current sheet and large ion drift velocity.

A very powerful method to investigate stability of a system in equilibrium utilizes a variational approach on a function δW where W is the system potential energy, and δW is its perturbation. δW_{min} is the minimum value and $\delta W_{min} \geq 0$ implies stability and $\delta W_{min} < 0$ implies instability (e.g., Refs. 5, 10–12). As δW represents perturbed system potential energy, and δK represents the perturbed kinetic energy and $\delta W + \delta K = 0$, if δW is negative, δK must be positive, i.e., the system kinetic energy grows and the system is unstable. If we can find the minimum of δW and show that it is positive, then the system is stable. Conversely, if the minimum is negative, then the system is unstable.

In many cases, the function δW may be separated into two parts: $\delta W = \delta F + \delta G$, where δF always has the same form:

$$\delta F = \frac{1}{2} \int \frac{(\nabla A_1)^2}{\mu_0} - A_1^2 \frac{d^2 P_0}{dA_0^2} \, d^2r , \qquad (2)$$

and δG depends on which model is being considered. Here, $P_0(A_0)$ is the equilibrium pressure, A_0 is the equilibrium y-component of the vector potential (considering field lines in the x-z plane), and A_1 is the first order perturbation. Schindler and Birn[12] show that for a large class of equilibria, $\delta G \geq 0$ for all A_1. So a sufficient criterion for stability is $\delta F_{min} \geq 0$.

Grad[13] showed that the minimum of the expression $F = \int [B^2/2\mu_0 - P] \, d^2r$ for ideal MHD, and under constant pressure variations, gives a sufficient criterion for instability. Schindler and Birn show that the first variation of this function gives rise to the equilibrium condition $-\nabla^2 A_0/\mu_0 = dP_0(A_0)/dA_0$ (magnetic force balanced by the pressure gradient), and the second variation yields Eq. 2 above. Schindler and Birn identify F with the free energy of the system. If $\delta F \geq 0$, the system is stable since all neighboring states have higher free energy than the equilibrium state. If $\delta F < 0$, then there are some (one?) neighboring states that have less free energy than the equilibrium and the system is potentially unstable, but the full δW, including δG, must be investigated. For various models, δG has the following forms:

Ideal MHD
(Bernstein et al.[14])

$$\delta G = \int \gamma P (\nabla \cdot (A_1 \nabla A_0 / |\nabla A_0|^2))^2 \, d^2r , \qquad (3)$$

Incompressible, resistive MHD,
with resistivity "frozen-in"
(Tasso[15])

$$\delta G = \int (d^2 P_0/dA_0^2) \, \bar{A}_1^2 \, d^2r \qquad (4)$$

Vlasov Theory
(Schindler,[11] Schindler et al.[16])

$$\delta G = \int |\partial \rho_0/\partial \phi_0| \{(\langle \psi_1 \rangle - \{\langle \psi_1 \rangle\})^2\} \, d^2r \qquad (5)$$

where $\partial \rho_0/\partial \phi_0 = \Sigma \, e^2 \int (\partial F_0/\partial H_0) \, d^3v$; ρ_0 and ϕ_0 are equilibrium density and electric potential, $F_0(H_0,P_0)$ is the particle distribution function for a given species as a function of the constants of the motion H_0 (the

Hamiltonian) and P_{y0} (the generalized momentum conjugate to the y-coordinate), $\partial F_0/\partial H_0$ is assumed to be negative; $e\psi_1$ is the perturbed Hamiltonian, $e\psi_1 = -(e/c)v_y A_1 + e\phi_1$; Σ sums over the particle species. The various averages have different meanings; see Schindler and Birn[12] for details.

Utilizing these concepts, Schindler and Birn show that for the magnetospheric tail, free energy is in fact available and that for the realistic case of a diverging tail $\delta F_{min} < 0$ always, while the tail would be stable if it converged (cross-sectional area reduces with increasing distance away from the earth). If $P(x,0)$ and B_x vary monotonically with x, then the stability condition can be simply written as follows: $B_z > 0$ for all z implies F-stability, whereas if B_z changes sign, then F-instability is implied. (B_z would change sign for example if $B_z > 0$ in the plasma sheet and yet with increasing z, the tail diverges, so after some z, B_z would become negative.)

In order to find a growth rate, the dynamic equations have to be studied as indicated above for the Harris sheet perturbation by Coppi et al.[8] For sufficiently weak B_z, where both electrons and ions are nonadiabatic in the neutral sheet, Schindler et al.[16] find the growth rate for the electron tearing mode of order $\gamma_e \sim V_{Te} \, a_e^{3/2} \, L_z^{-5/2}$, where V_{Te} is the electron thermal velocity.

For an intermediate B_z, where the electrons are magnetized but ions are still not adiabatic, one finds the growth rate (γ_i) for the ion tearing mode (Ref. 11) as:

$$\gamma_i = V_{Ti} a_i^{3/2} L_z^{-5/2} .$$

The adiabatic electrons can lead to a stabilizing effect;[17,18] however, according to the earlier theories, for a range of B_z values, there still exists a considerable range where the ion-tearing mode may take place. But more recently, Lembege and Pellat[19] find that "compressibility" of the adiabatic electron gas does indeed stabilize. Coroniti[20] estimates that instability begins for wavelengths greater than about 100 R_e, and this is so long compared to observed scales in the tail as to be unreasonable. On the other hand, Goldstein and Schindler[21] did obtain a tearing unstable mode for more reasonable wavelengths of order R_e, but the assumed tail plasma parameters were not very realistic. Lui and Krimigis[22] estimate tearing mode wavelengths in the range 2–17 R_e along the tail, on the basis of measurements of Birkeland current-sheet thicknesses in the tail. Swift[23] argues that electrical coupling of the plasma sheet to the ionosphere will discharge any electrostatic fields that tend to tie ions to magnetized electrons, thus negating the adiabatic electron stabilizing effect, but actual simulations including both ion and electron dynamics in a finite B_z model need to be done (also suggested by Goldstein and Schindler[21]).

Taking a different tack of investigating the effect on tearing of initially anisotropic distribution functions, Chen and Palmadesso[24] (see also Chen and Lee[25]) and Chen et al.[26] show that the tearing growth rates can be substantially enhanced for ion temperature anisotropies

$T_\perp/T_\parallel > 1$. Conversely, anisotropies with $T_\perp/T_\parallel < 1$ strongly stabilize the mode. Other potentially destabilizing conditions, such as electron pitch-angle scattering, are discussed in the excellent reviews by Galeev[5] and Coroniti.[20]

Finally, Baker et al.[27] study the role of heavy ions in the initiation of the tearing mode. Regions of enhanced ionospheric O^+, for example, can lead to locally enhanced tearing growth rates. These stability calculations are linear. We now turn to a few nonlinear results.

Explosive Tearing and Reconnection

Past the onset of instability, Galeev[5] (Galeev et al.[28]) estimates that the amplitude of the perturbation $b_1(t)$ increases explosively as $b_1(t) = b_1(0)/(1 - t/\tau_R)$, where τ_R is the characteristic time of an explosive growth. Note that $b_1(t)$ is proportional to the initial perturbation $b_1(0)$. This solution, therefore, is reminiscent of Dungey's[2] discharge, the starting point for ideas about reconnection. (See also Refs. 29 and 30.) For typical tail parameters, Galeev estimates $\tau_R \sim 100$ s, with highest possible energies of accelerated protons of order 1 MeV, and the acceleration time for these protons of ~ 100 s. Galeev also shows that the ion spectrum goes like a power law (as observed) and that observed inverse velocity dispersion can be produced by the model as well. (The lower energy ions are produced at an earlier time by the explosively growing electric field.)

Zelenyi et al.[31] use the above results of the explosively growing tearing mode and simulate particle dynamics in this model. They then compare the simulation with observations of proton spectra in the tail by Sarris et al.[32] The calculations are in reasonable agreement with the observations, but the observations were made during quiet to moderately disturbed times, while the model should of course apply to the beginning of a substorm.

Coroniti[20] has developed a model incorporating the nonlinear explosive phase plus saturation. In the final saturation phase, steady collisionless reconnection flow takes place at a rate comparable to the maximum Petschek rate. In Coroniti's model, "inertial conductivity" (see fourth section) is responsible for the explosive behavior. The particle lifetime in the system diminishes as the vector potential increases, and as the total current is fixed by the external solution, the electric field has to increase faster than linearly to compensate for the decreasing conductivity and current layer thickness. The reconnection time scale is found to be of order 5–10 minutes, in agreement with substorm onset times, and cross-tail potentials of order 1 MV are predicted during the explosive phase.

Anomalous Resistivity

It has often been thought that nature would always supply any necessary turbulent dissipation for processes like reconnection and tearing mode. A considerable amount of literature has developed investigating various instabilities that can lead to a saturated wave-particle interaction and thus a turbulent dissipation. For a review of observations and theories pertaining, see Refs. 20 and 33. Coroniti's basic conclusions are that although plasma-wave turbulence is often observed in various magnetospheric regions, either the strength of the anomalous dissipation is too weak or the description of the anomalous dissipation process is inappropriate. The basic problem is that often local processes just don't work, because hot magnetospheric plasmas "communicate" over large distances and are inherently global in nature (see also Mitchell[34]). These processes are important in many cases but will not be considered further here.

It is, however, interesting that Haerendel[35] reported that in the observations of the various ion clouds launched by the AMPTE-IRM satellite, absolutely no evidence has been found of any anomalous resistive or diffusive process.

KINETIC SIMULATIONS

Simulations are done to try and imitate a physical model with fewer approximations than the preceding, and with more of the microphysics included. However, simulations still have inherent limitations.

Simulations of the earth's magnetosphere and tail fall into several categories: MHD, kinetic, mixed kinetic and fluid, and laboratory simulations. Additionally, simulations may be local, dealing with one small part, or global in nature—attempts to mimic the entire solar wind-magnetosphere interaction. Here a few of the computer simulations that are mostly based on kinetics will be reviewed. For reviews of MHD models, see Refs. 36 and 37. For a review of laboratory simulations, see Refs. 38–40

An ideal kinetic simulation would contain full electron and ion dynamics, over a volume large enough to model a significant portion of the geomagnetic tail. Unfortunately, this is not possible at this time. Katanuma and Kamimura[41] have included complete electron and ion dynamics but in their treatment they are limited to scales of the order of several Debye lengths, i.e., approximately hundreds of meters—small compared with relevant lengths in the tail. Terasawa[42] presented results of a hybrid simulation where the electrons were treated as a fluid, but full ion dynamics were considered. Terasawa's results were basically consistent with the nonlinear theory of Galeev et al.[28] (GCA, Section 2), but electrostatic effects were not included. After a linear tearing mode phase, the nonlinear explosive phase is found, depending on the initial perturbation (as with GCA), but also depending on the system scale size. The explosive mode was found for the larger system ($L_x/\rho \sim 51$) but not for one-half as large. The nonlinear growth rate depended on the equilibrium drift velocity V_0 and varied as $V_0^{3/2}$, consistent with GCA, but a factor of 3 smaller in magnitude. The energetics were very interesting. Flow velocities were reduced inside the magnetic island in the explosive phase as compared to the linear phase, consistent also with a reversed electric field inside the island. Acceleration of particles by the rapidly growing electric field is significant in the vicinity of the X-line, but since the density is lower there as compared to the magnetic island where deceleration occurs, the deceleration effect dominates, considering the system as a whole. In a study in which both electrons and ions were treated kinetically, Le Boeuf et al.[43] essentially confirmed the explosive growth, but they assumed an ion-to-electron mass ratio of 10, which according to Lembege and Pellat[19] might have caused an important underestimation of the electron compressibility.

Swift[44] also used a particle code that included ion dynamics. He concluded that the dawn-dusk electric field results in plasma-sheet thinning and that the electric field along with a lobe particle population was necessary for the maintenance of the extended magnetotail. Birn and Schindler,[45] however, pointed out that Swift's system was not in equilibrium initially, and they questioned the addition of an initially constant E_y (see Fig. 1) everywhere. For Birn and Schindler's dynamic MHD simulation, only a small fraction of the lobe E_y penetrates into the interior of the plasma sheet. In an update of his first model, Swift[46] shows that the revisions allow two-dimensional simulations for more extended time periods and under more realistic conditions. He also finds that when the magnetic field starts to collapse, most of the energy goes into earthward streaming of plasma-sheet particles. The previous criticism of Birn and Schindler may still apply, but Swift notes that he obtains an equilibrium convection electric field proportional to $B_x B_z$ which is nearly the equilibrium condition obtained from particle orbits (Section 4 of Ref. 47; Hill[48] etc.). (Swift's expression, his equation 14, however, is not directly comparable to the particle orbit results in the next section, as it appears to be dimensionally incorrect.)

CURRENT-SHEET PARTICLE MOTION

Motion in Quasisteady Fields

Following Dungey's[2] original ideas concerning neutral-point acceleration, particle motion in current sheets has been studied by many authors (e.g., Parker,[49] Speiser,[50-53] Schindler,[54] Alexeev and Kropotkin,[55] Sonnerup,[56] Eastwood,[57,58] Pudovkin and Tsyganenko,[59] Cowley,[60-63] Jaeger and Speiser,[64] Swift,[65] Stern,[66] Wagner et al.,[67] Lyons and Speiser,[68] Speiser and Lyons,[69] Lyons[70]). In the absence of stochastic processes, Liouville's theorem says that the particle distribution function is constant along a particle orbit. Thus, in this approximation, particle orbit solutions are equivalent to solutions of the Vlasov equation. The simplest model of a current sheet involves a sheared magnetic field that changes direction by 180° on passing through a current sheet of finite width. Such a linear reversal approximates the tanh (z) dependency of the Harris[4] sheet discussed earlier. The field strength goes to zero everywhere at the sheet center. There may or may not be an electric field within the current sheet perpendicular to the field but tangential to the sheet. If there is an electric field, particles become trapped in the sheet, execute damped oscillations about the sheet, and become accelerated by the electric field.[50,51] In this configuration, plasma $E \times B$ drifts into the current sheet from both sides (e.g., from the top and bottom in Fig. 1), becomes accelerated in the interior (in the y-direction), and satisfies the definition of "reconnection" from the fluid point of view.[71-73] Such a simple model would not be too interesting if applied to the geomagnetic tail as all accelerated particles would get "shot out" the dawn (electrons) and dusk (ions) sides of the tail into the sheath region. However, the Alfvén[72] consistency condition ($\phi_A = B_{x0}^2/\mu_0 n_0 e$) allows the prediction of the overall voltage drop as a function of the field strength and plasma density, and was, therefore, one of the earliest predictions of a reconnection rate (more strictly speaking, annihilation rate for this simple model). This model may also be applicable in a limited region around a neutral point or neutral line.

In the next approximation, we now add a constant, weak, normal magnetic field, B_z, to the current sheet. The region of applicability is sketched as the shaded area in Fig. 1. The particle orbits now include a gyration about the weak normal field in addition to the oscillation about the sheet.[50-53,68,69] The particles, however, are turned by the gyromotion 90° to the accelerating electric field and when turned this much, they "pop out" of the sheet with small pitch angles. (The Lorentz force causing the oscillation about the sheet changes sign, thus ejecting the particles from the sheet.) The Alfvén consistency condition becomes generalized like the Eastwood[57] condition, which is now a function of the normal field ($\phi_E = B_{x0} B_z L/(2\mu_0 n_0 m)^{1/2}$, where B_{x0} is the external, tangential field component, B_z the normal field component, n_0 the external plasma density, L the current sheet width, and m the ion mass (see Ref. 57 for the connection between ϕ_E and ϕ_A). The Eastwood consistency condition is close to, but not quite, the prediction we would like to have of the reconnection rate, as the relevant value of B_z is unknown. Of course, we can take typical observed values for the geomagnetic tail and find the predicted electric field, for example. Values of the cross-tail potential drop come out to be of order 100 kV, i.e., about what is expected. In one sense, this model is simpler than the first model, for the electric field can be transformed away by moving with speed E_y/B_z toward the earth.[50] This moving frame, where $E = 0$, is sometimes referred to as the DeHoffman-Teller frame. The particle interaction with the current sheet can be likened to specular reflection off a moving mirror.[61] Particles initially moving away from the earth (into the mirror) gain twice the transformation velocity. Those moving toward the earth with the transformation velocity are not energized, i.e., do not interact with the current sheet. Current-sheet oscillation is expected when B_z is weak enough so that the usual adiabatic theory (μ = const) is violated. (When B_z is sufficiently strong, particles merely follow field lines through the sheet[57,66,67] but the moving mirror picture also works for the adiabatic case[74]). However, because of the new oscillatory character of the orbits about the current sheet, new adiabatic invariants come into play.[53,54,57] When B_z is weak enough, both electrons and ions undergo current sheet oscillation, and both contribute to the requisite cross-tail current. For larger B_z, such that the electrons remain adiabatic (μ = const) the ions undergo the current sheet oscillation, the ions would supply most of the cross-tail current, and the electron drift would be in the $E \times B_z$ direction, i.e., toward the earth (see Fig. 1). Speiser,[53] in fact, showed that for this simple, constant B_z mode, there is a j, E relationship even though this is single particle motion with no collisions. The average—say ion—drift is roughly the E_y/B_z transformation velocity, and the negative of that for the (nonadiabatic) electrons. Therefore, the net current density is just $j = ne \, \bar{V}_D$ where $\bar{V}_D$ is the average velocity difference between the ions and electrons, or $j \approx ne \, (E_y/B_z) \, e_y$, giving our j, E relation. Cowley[75] shows that for a one-dimensional current sheet, trapped particles, whether adiabatic or not, produce no net current. However, untrapped particles can give rise to a net current. For our use here, net currents can be produced if these accelerated particles do not return (or return to some other location where conditions have changed). The effective (or inertial) conductivity is therefore $\sigma_i \approx ne/B_z$ or $\sigma_i \approx ne^2/m\omega_g$, where ω_g is the gyrofrequency about the weak B_z. Thus, this gyrofrequency takes the place of a collision frequency in the conductivity expression. Energy dissipation is from the electric field into accelerated particles and not into stochastic, heat producing, processes (although these may be important at other stages). It is this "inertial" concept that Coroniti[20] uses for his explosive reconnection model (see above).

The Eastwood equilibrium condition has been generalized by Hill[48] to include the effects of initial streaming and pressure anisotropy. These effects will of course change the simple j, E relation above. Lyons and Speiser[47] showed how the Eastwood-Hill results (except for the anisotropy terms) arise from the results of single particle motion. The consistency in Hill's more general form is just the stress balance condition which reduces to the marginal firehose condition just outside the current sheet.

Lyons and Speiser[68] followed many ion orbits numerically in a model of the tail current sheet corresponding to this second model (E_y = const, B_z = const). For various initial pitch angles, particles can be ejected from the current sheet over a wide range of pitch angles, although there is some tendency for initially very low energy particles to come out field aligned, and otherwise for pitch angles to be preserved. (In the adiabatic (μ = const) limit, particles obviously preserve pitch angle, and Alexeev and Kropotkin[55] have shown that pitch angle is also preserved in the limit of a very thin current sheet or, equivalently, in the limit of very large initial particle energies.) For nominal fields, the energy increase is markedly higher for the particles ejected with small pitch angles than those ejected with large pitch angles. Assuming an initial "mantle" or "boundary layer" distribution, the calculated orbits allow a mapping to be made to the energized, streaming distribution on the edge of the plasma sheet. Comparison of this model-

eled distribution with observations resulted in good agreement. When B_z is weak enough, analytic solutions may be applied and initial distributions may be analytically mapped into energized, streaming distributions in the plasma-sheet boundary layer.[69]

For this (B_z = const) model, Cowley[62] shows how one can simply map an initial distribution into an accelerated, streaming distribution by reflecting the initial distribution about the plane $V_\parallel = V_F$, where V_F is the transformation velocity, E_y/B_z, with the direction toward the earth (Fig. 1).

Using a model slightly more complicated than the B_z = const model, Jaeger and Speiser[64] again followed ion orbits numerically. This model had a varying B_z, falling off with distance as observed and as would be expected, for example, with a neutral-line model. However, the rate of falloff was as imposed by a dipole field (r^{-3}) and therefore somewhat unrealistic. Again (using Liouville's theorem), initial distributions were mapped into accelerated streaming distributions. The initial distributions were from fluid models in the magnetosheath, incorporating reasonable solar wind compositional values. Final (post-acceleration) energies were nearly—but not exactly—equal energy per unit charge, i.e., which would be expected for various particles falling through the same potential drop. This was surprising as the previous B_z = const model predicts iso-velocity energization. These results, with those of Speiser[51] show that heavy ions should be found in streaming distributions in the plasma-sheet boundary layer inside the proton layer. Taking alpha particles as an example, the reason for this is that for the same energy, alpha particles will have a larger gyroradius about the weak normal field, B_z. Therefore, in a region where the proton gyroradius is about the tail width, proton beams will be formed toward the earth, but alpha particles will not get turned sufficiently and will thus be shot out the sides of the tail. If we now move the interaction region closer to the earth (where B_z is larger in this model), alpha particles will now be turned toward the earth and ejected along field lines that are necessarily inside (i.e., closer to the midplane) the energetic-proton streaming layer. Möbius et al.[76] have indeed found evidence of alpha particle layers within proton layers. However, it should be noted that any model (such as Sarris and Axford[77]) which has alpha particles moving more slowly than protons will tend to produce this type of separation via $E \times B$ drifts. One problem with the Jaeger and Speiser model was that the intensities were much larger than the observations. Lyons and Speiser used a more realistic initial distribution and comparisons with observations were more favorable.

As previously discussed, for the B_z = const model, B_z may be "weak" as seen by one species, yet "strong" as seen by another. For a tail field of 20 nT and B_z of 1 nT, ions with moderate energies, for example, essentially violate the condition μ = const and thus participate in "current-sheet oscillation." Electrons, on the other hand, would be expected to remain "adiabatic." Lyons[70] has followed many electron orbits in

282

the B_z = const model. He finds that most electrons incident on the current sheet become trapped in it, and they gain energy from the cross-tail electric field. Even for nominal parameters, as above, slight violations of the adiabatic condition are important for electrons. Their energy gain goes as $(B_z(x))^{2/3}$. Some small fraction of electrons within the current sheet may undergo pitch-angle scattering and thus end up in the atmospheric loss cone.

In the next section, results near neutral points or lines are summarized, i.e., near regions expected to form during tearing and reconnection.

Motion Near Neutral Points or Lines

There have been a few studies made of particle motion near magnetic neutral lines (see Fig. 1) or points (e.g., Refs. 78–80). As indicated in the previous section, numerical studies of particle orbits with a model with B_z varying did produce somewhat different results than the constant B_z model.[64] Aström categorized the types of electron trajectories in the vicinity of an X-type neutral line, and found orbits resembling trochoids, meandering, and figure-8 types. He also investigated the stability of orbit types if they were perturbed out of the initial plane. Rusbridge was interested in the change of a particle's magnetic moment on interaction with an X-type neutral line. He found that the maximum change was by a factor of about 5.75, and the key factor was the particle's phase angle at the point where it ceased to be adiabatic. Stern analyzed equations of motion near O-type neutral lines in some limiting cases and found that particles become decoupled from the field line motion and then could undergo a runaway acceleration mode along the neutral line. Furthermore, Stern concludes that the acceleration process is more efficient along O-type than X-type neutral lines, as there is a kind of stability there. However, the tearing mode models (e.g., Galeev[5]) predict a reversed electric field along the O-lines, so there could be a net deceleration there. Note also that Vasyliunas[81] shows that the O-line voltage is necessarily limited to the Alfvén voltage.

Martin[82] studied particle motion in an X-type magnetic neutral-line system with an electric field along the line. He shows that the system is basically nonintegrable, but can be studied qualitatively with a two-dimensional effective potential (see also Stern[80]). For a thin current sheet, the motion is relatively simple and particles can be significantly energized. For a thick current sheet, the motion can be chaotic (many transitions are made and the motion becomes randomized) for larger pitch-angle particles, and particles can lose energy. For the chaotic motion, Martin estimates an effective conductivity (minimum value) of about 10^{-4} mho/m. Martin also finds that all particles entering the neutral-line region become chaotic. Chen and Palmadesso[83] find a similar regime of stochastic orbits for a tail-like magnetic field, but they also find it in all three types of orbits—and these remain distinct in the absence of noise. The distribution will be strongly non-Maxwellian and may well have implications for the plasma dynamics.

Motion in Time-Dependent Fields

In the preceding section, it has been assumed that E_y $\approx$ const in space and time. Obviously, there are times such as substorms when the tail fields and plasmas suffer marked changes. In the reconnection sense, at the onset of reconnection, one would expect time changes in the fields to be large until saturation or quasisteady state is reached (if indeed that ever happens). (See, e.g., Refs. 11,28,84.) Even when the fields are varying in time and space, the current-sheet orbits just discussed may at times be valid if the variations are on a scale large compared to the particle scale lengths in the current sheet or if the variations are slow compared to the time spent by the particles in the current-sheet interaction. Nevertheless, we would not expect cross-tail potentials typically to exceed say ~ 200 kV and thus we would be hard pressed to explain 0.5 MeV particles with the previous current-sheet acceleration models. Note, however, that a high-energy tail in the particle population is generally observed and thus an energization mechanism may only be required to shift this high energy tail. Pellinen and Heikkila[85] follow particle orbits in such time-dependent fields, and find that the most efficient accelerating mechanism is a two-step process, with initial acceleration along a neutral line followed by betatron acceleration. Energies into the MeV range can possibly be achieved. Galeev[5] in fact shows that the explosive phase of the tearing instability in the magnetospheric tail also generates particle bursts with a power law spectrum, and for typical tail parameters, he estimates maximum proton energies of ~ 1 MeV, acceleration time ~ 10 s, and a power law of aproximately $-5 (J(E_p) \propto E_p^{-5})$, which compares favorably with observations. Inverse velocity dispersion can also be qualitatively explained, as the lower energy portion is generated earlier by the explosively growing electric field. The induction electric field is also generated near the X-line and directed from dawn to dusk, leading to harder protons (electrons) on the evening (morning) side of the tail. Electron acceleration would take place but with smaller intensities as compared with the ions since their acceleration region would be reduced in size.

SUMMARY

1. Collisionless linear tearing mode analysis[16] has progressed to the point that it looks like several unstable regimes exist. Electron magnetization stabilization in finite B_z fields can apparently be overcome by pitch-angle scattering and/or by electrical coupling to the ionosphere, but the final story is not yet in. Anisotropy with $T_\perp > T_\parallel$ leads to enhanced growth rates.

2. Galeev, Coroniti, Ashour-Abdalla (GCA) explosive nonlinear tearing has reasonable onset times and megavolt potentials, and has been verified by Terasawa's ion simulation and also by a fully kinetic[43] simulation. Some uncertainties still exist.

3. Coroniti's explosive model uniquely depends on particle inertia directly from the current-sheet dynamics. Reasonable onset times and megavolt potentials are deduced. Electron Hall and mirror dynamics are ignored.

4. Simple quasisteady current-sheet particle motion leads to consistency conditions between the fields and particles. Motion near neutral points or lines introduces a new chaotic regime. Few studies have been made in time-dependent fields.

ACKNOWLEDGMENT—This study was supported by grants ATM-8219436 (The Johns Hopkins University, APL Task G203) and ATM-8318203 from the National Science Foundation, and by grant NAGW497 (The Johns Hopkins University, APL Task G207) from the National Aeronautics and Space Administration. I would like to thank Drs. D. J. Williams and S. M. Krimigis for their hospitality at APL during the preparation of this manuscript. I also appreciate useful discussions with Drs. D. J. Williams, D. Mitchell, K. Schindler, R. Martin, P. Dusenbery, and L. Lyons.

REFERENCES

[1] D. A. Gurnett, L. A. Frank, and R. P. Lepping, "Plasma Waves in the Distant Magnetotail," *J. Geophys. Res.* **81**, 6059 (1976).

[2] J. W. Dungey, "Conditions for the Occurrence of Electrical Discharges in Astrophysical Systems," *Phil. Mag.* **44**, 725 (1953).

[3] J. W. Dungey, *Cosmic Electrodynamics* (1958).

[4] E. G. Harris, "On a Plasma Sheath Separating Regions of Oppositely Directed Magnetic Field," *Il Nuovo Cimento* **23**, 115 (1962).

[5] A. A. Galeev, "Magnetospheric Tail Dynamics," in *Magnetosopheric Plasma Physics*, A. Nishida, ed., *Developments in Earth and Planetary Sciences*, No. 04, D. Reidel, Boston (1982).

[6] H. P. Furth, J. Killeen, and M. N. Rosenbluth, "Finite Resistivity Instabilities of a Sheet Pinch," *Phys. Fluids* **6**, 459 (1963).

[7] H. P. Furth, "Instabilities Due to Finite Resistivity or Finite Current-Carrier Mass," in *Advanced Plasma Theory*, M. N. Rosenbluth, ed., Academic Press, New York, p. 159 (1964).

[8] B. Coppi, G. Laval, and R. Pellat, "Dynamics of the Geomagnetic Tail," *Phys. Rev. Lett.* **16**, 1207 (1966).

[9] G. R. Laval, R. Pellat, and M. Vuillemin, in *Proc. Second International Conference on Plasma Physics and Thermonuclear Fusion*, International Atomic Energy Agency, Vienna, V2, p. 259 (1965).

[10] K. Schindler, "A Self-Consistent Theory of the Tail of the Magnetosphere," in *Earth's Magnetospheric Processes*, M. McCormac, ed., D. Reidel, Dordrecht-Holland (1972).

[11] K. Schindler, "A Theory of the Substorm Mechanism," *J. Geophys. Res.* **79**, 2803 (1974).

[12] K. Schindler and J. Birn, "Magnetospheric Physics," *Phys. Rep.* **47**, 109 (1978).

[13] H. Grad, "Some New Variational Properties of Hydromagnetic Equilibria," *Phys. Fluids* **7**, 1283 (1964).

[14] I. B. Bernstein, E. A. Frieman, M. D. Kruskal, and R. M. Kulsrud, "An Energy Principle for Hydromagnetic Stability Problems," *Proc. R. Soc.* **A244**, 17 (1958).

[15] H. Tasso, "Energy Principle for Two-Dimensional Resistive Instabilities," *Plasma Phys.* **17**, 1131 (1975).

[16] K. Schindler, D. P. Pfirsch, and H. Wobig, "Stability of Two-Dimensional Collision-Free Plasmas," *Plasma Phys.* **15**, 1165 (1973).

[17] A. A. Galeev and L. M. Zelenyi, "Metastable States of Diffuse Neutral Sheet and the Substorm Explosive Phase," *JETP Lett.* **22**, 170 (1975).

[18] A. A. Galeev and L. M. Zelenyi, "Tearing Instabilities in Plasma Configurations," *Sov. Phys. JETP* **43**, 1113 (1976).

[19] B. Lembege and R. Pellat, "Stability of a Thick Two-Dimensional Quasi-Neutral Sheet," *Phys. Fluids* **25**, 1995 (1982).

[20] F. V. Coroniti, "Explosive Tail Reconnection: The Growth and Expansion Phases of Magnetospheric Substorms," *J. Geophys. Res.* **90**, 7427 (1985).

[21] H. Goldstein and K. Schindler, "Large-Scale, Collision-Free Instability of Two-Dimensional Plasma Sheets," *Phys. Rev. Lett.* **48**, 1468 (1982).

[22] A. T. Y. Lui and S. M. Krimigis, "Association Between Energetic Particle Bursts and Birkeland Currents in the Geomagnetic Tail," *J. Geophys. Res.* **89**, 10741 (1984).

[23] D. W. Swift, "Auroral Arcs and Tearing-Mode Instabilities in the Magnetotail," *Geophys. Res. Lett.* (submitted, 1985).

[24] J. Chen and P. J. Palmadesso, "Tearing Instability in an Anisotropic Neutral Sheet," *Phys. Fluids* **27**, 1198 (1984).

[25] J. Chen and Y. C. Lee, "Collisionless Tearing Instability in a Non-Maxwellian Neutral Sheet: An Integro-Differential Formulation," *Phys. Fluids* **28**, 2137 (1985).

[26] J. Chen, P. J. Palmadesso, J. A. Fedder, and J. G. Lyon, "Fast Collisionless Tearing in an Anisotropic Neutral Sheet," *Geophys. Res. Lett.* **11**, 12 (1984).

[27] D. N. Baker, E. W. Hones, Jr., D. T. Young, and J. Birn, "The Possible Role of Ionospheric Oxygen in the Initiation and Development of Plasma Sheet Instabilities," *Geophys. Res. Lett.* **9**, 1337 (1982).

[28] A. A. Galeev, F. V. Coroniti, and M. Ashour-Abdalla, "Explosive Tearing Mode Reconnection in the Magnetospheric Tail," *Geophys. Res. Lett.* **5**, 707 (1978).

[29] T. W. Forbes and T. W. Speiser, "Temporal Evolution of Magnetic Reconnexion in the Vicinity of a Magnetic Neutral Line," *J. Plasma Phys.* **21**, 107 (1979).

[30] T. G. Forbes, "Implosion of a Uniform Current Sheet in a Low-Beta Plasma," *J. Plasma Phys.* **27**, 491 (1982).

[31] L. M. Zelenyi, A. S. Lipatov, D. G. Lominadze, and A. L. Taktakishvili, "The Dynamics of the Energetic Proton Bursts in the Course of the Magnetic Field Topology Reconstruction in the Earth's Magnetotail," *Planet. Space Sci.* **32**, 313 (1984).

[32] E. T. Sarris, S. M. Krimigis, A. T. Y. Lui, K. L. Ackerson, and L. A. Frank, "Relationship Between Energetic Particle and Plasmas in the Distant Plasma Sheet," *Geophys. Res. Lett.* **8**, 349 (1981).

[33] F. V. Coroniti, "Space Plasma Turbulent Dissipation: Reality or Myth?" *Space Sci. Rev.* **42**, 399 (1985).

[34] D. Mitchell, "Kinetic Aspects of Tail Dynamics: Observations," this volume (1987).

[35] G. Haerendel, "Tail Explosion and Tail Formation," this volume (1987).

[36] F. V. Coroniti, "MHD Aspects of Magnetotail Dynamics," this volume (1987).

[37] J. Birn, "Computer Simulation of Reconnection in Planetary Magnetospheres," in *Unstable Current Systems and Plasma Instabilities in Astrophysics*, M. R. Kundu and G. D. Holman, eds., p. 167 (1985).

[38] R. L. Stenzel and W. Gekelman, "Laboratory Experiments on Magnetic Field Line Reconnection," *Proceedings of the 1982-84 MIT Symposia on Physics of Space Plasmas*, T. Chang, ed., MIT (1984).

[39] P. Baum and A. Bratenahl, "Magnetic Reconnection Experiments," in *Advances in Electronics and Electron Physics*, Academic Press, p. 54 (1980).

[40] I. M. Podgorny, "Active Experiments in Space, Laboratory Experiments, and Numerical Simulation," Report 11p-65I, Acad. of Sciences USSR Space Research Institute to IAGA Assembly, Edinburgh (1981).

[41] I. Katanuma and T. Kamimura, "Simulation Studies of the Collisionless Tearing Instabilities," *Phys. Fluids* **23**, 2500 (1980).

[42] T. Terasawa, "Numerical Study of Explosive Tearing Mode Instability in One-Component Plasmas," *J. Geophys. Res.* **86**, 9007 (1981).

[43] J. N. Le Boeuf, T. Tajima, and J. M. Dawson, "Dynamic Magnetic X-Points," *Phys. Fluids* **25**, 784 (1982).

[44] D. W. Swift, "Numerical Simulation of the Interaction of the Plasma Sheet with the Lobes of the Earth's Magnetotail," *J. Geophys. Res.* **87**, 2287 (1982).

[45] J. Birn and K. Schindler, "Computer Modeling of Magnetotail Convection," *J. Geophys. Res.* **90**, 3441 (1985).

[46] D. W. Swift, "A Two-Dimensional Simulation of the Interaction of the Plasma Sheet with the Lobes of the Earth's Magnetotail," *J. Geophys. Res.* **88**, 125 (1983).

[47] L. R. Lyons and T. W. Speiser, "Ohm's Law for a Current Sheet," *J. Geophys. Res.* **90**, 8543 (1985).

[48] Hill, "Magnetic Merging in a Collisionless Plasma," *J. Geophys. Res.* **80**, 4689 (1975).

[49] E. N. Parker, "Newtonian Development of the Dynamical Properties of Ionized Gases of Low Density," *Phys. Rev.* **107**, 924 (1957).

[50] T. W. Speiser, "Particle Trajectories in Model Current Sheets, 1, Analytical Solutions," *J. Geophys. Res.* **70**, 4219 (1965).

[51] T. W. Speiser, "Particle Trajectories in Model Current Sheets, 2," *J. Geophys. Res.* **72**, 3919 (1967).

[52] T. W. Speiser, "On the Uncoupling of Parallel and Perpendicular Motion in a Neutral Sheet," *J. Geophys. Res.* **73**, 1112 (1968).

[53] T. W. Speiser, "Conductivity without Collisions or Noise," *Planet. Space Sci.* **18**, 613 (1970).

[54] K. Schindler, "Adiabatic Orbits in Discontinuous Fields," *J. Math. Phys.* **6**, 313 (1965).

[55] I. I. Alexeev and A. P. Kropotkin, "Interaction of Energetic Particles with the Neutral Layer of the Magnetospheric Tail," *Geomag. Aeron.* **10**, 615 (1970).

[56] B. U. O. Sonnerup, "Adiabatic Particle Orbits in a Magnetic Null Sheet," *J. Geophys. Res.* **76**, 8211 (1971).

[57] J. W. Eastwood, "Consistency of Field and Particle Motion in the 'Speiser' Model of the Current Sheet," *Planet. Space Sci.* **20**, 1555 (1972).

[58] J. W. Eastwood, "Some Properties of the Current Sheet in the Geomagnetic Tail," *Planet. Space Sci.* **23**, 1 (1975).

[59] M. I. Pudovkin and N. A. Tsyganenko, "Particle Motions and Currents in the Neutral Sheet of the Magnetospheric Tail," *Planet. Space Sci.* **21**, 2027 (1973).

[60] S. W. H. Cowley, "A Self-Consistent Model of a Simple Magnetic Neutral Sheet System Surrounded by a Cold, Collisionless Plasma," *Cosmic Electrodynam.* **3**, 448 (1973).

[61] S. W. H. Cowley, "Plasma Populations in a Simple Open Model Magnetosphere," *Space Sci. Rev.* **26**, 217 (1980).

[62] S. W. H. Cowley, "The Causes of Convection in the Earth's Magnetosphere: A Review of the Developments During the IMS," *Rev. Geophys. Space Phys.* **20**, 531 (1982).

[63] S. W. H. Cowley, "The Distant Geomagnetic Tail in Theory and Observation," in *Proc. Chapman Conference on Magnetic Reconnection*, E. W. Hones, Jr., ed., American Geophysical Union, Washington, D.C. (1984).

[64] E. F. Jaeger and T. W. Speiser, "Energy and Pitch Angle Distributions for Auroral Ions Using the Current Sheet Model," *Astrophys. Space Sci.* **28**, 129 (1974).

[65] D. W. Swift, "The Effect of the Neutral Sheet on Magnetospheric Plasma," *J. Geophys. Res.* **82**, 1288 (1977).

[66] D. P. Stern, "Adiabatic Particle Motion in a Nearly Drift Free Magnetic Field: Application to the Geomagnetic Tail," Rep. X-602-77-89, Goddard Space Flight Center, Greenbelt, Md. (1977).

[67] J. S. Wagner, J. R. Kan, and S.-I. Akasofu, "Particle Dynamics in the Plasma Sheet," *J. Geophys. Res.* **84**, 891 (1979).

[68] L. R. Lyons and T. W. Speiser, "Evidence for Current Sheet Acceleration in the Geomagnetic Tail," *J. Geophys. Res.* **87**, 2276 (1982).

[69] T. W. Speiser and L. R. Lyons, "Comparison of Analytical Approximation for Particle Motion in a Current Sheet with Precise Numerical Calculations," *J. Geophys. Res.* **89**, 147 (1984).

[70] L. R. Lyons, "Electron Energization in the Geomagnetic Tail Current Sheet," *J. Geophys. Res.* **89**, 5479 (1984).

[71] J. W. Dungey, "Interplanetary Magnetic Field and Auroral Zones," *Phys. Rev. Lett.* **6**, 47 (1961).

[72] H. Alfvén, "Some Properties of Magnetospheric Neutral Surfaces," *J. Geophys. Res.* **73**, 4379 (1968).

[73] L. M. Vasyliunas, "Theoretical Models of Magnetic Field Merging, 1," *Rev. Geophys. Space Phys.* **13**, 303 (1975).

[74] S. C. Chapman and S. W. H. Cowley, "Acceleration of Lithium Test Ions in the Quiet-Time Geomagnetic Tail," *J. Geophys. Res.* **89**, 7357 (1984).

[75] S. W. H. Cowley, "A Note on the Motion of Charged Particles in One-Dimensional Magnetic Current Sheets," *Planet. Space Sci.* **26**, 539 (1978).

[76] E. Möbius, F. M. Ipavich, N. Scholer, G. Gloeckler, D. Hovestadt, and B. Klecker, "Observations of a Nonthermal Ion Layer at the Plasma Sheet Boundary During Substorm Recovery," *J. Geophys. Res.* **85**, 5143 (1980).

[77] E. T. Sarris and W. I. Axford, "Energetic Protons Near the Plasma Sheet Boundary," *Nature* **277**, 5696, 460 (1979).

[78] E. Aström, "Electron Orbits in Hyperbolic Magnetic Fields," *Tellus* **8**, 260 (1956).

[79] M. G. Rushbridge, "Non-Adiabatic Charged Particle Motion Near a Magnetic Field Zero Line," *Plasma Phys.* **13**, 977 (1971).

[80] D. P. Stern, "The Role of O-Type Neutral Lines in Magnetic Merging During Substorms and Solar Flares," *J. Geophys. Res.* **84**, 63 (1979).

[81] V. M. Vasyliunas, "Upper Limit on the Electric Field Along a Magnetic O Line," *J. Geophys. Res.* **85**, 4616 (1980).

[82] R. F. Martin, Jr., "The Effect of Plasma Sheet Thickness on Ion Acceleration Near a Magnetic Neutral Line," Geophysical Monograph no., American Geophysical Union, Washington, D.C. (in press, 1986).

[83] J. Chen and P. J. Palmadesso, "Chaos and Nonlinear Dynamics of Single-Particle Orbits in a Magnetotail-Like Magnetic Field," *J. Geophys. Res.* (1985).

[84] F. V. Coroniti, F. L. Scarf, L. A. Frank, and R. P. Lepping, "Microstructure of a Magnetotail Fireball," *Geophys. Res. Lett.* **4**, 219 (1977).

[85] R. J. Pellinen and W. J. Heikkila, "Energization of Charged Particles to High Energies by an Induced Substorm Electric Field Within the Magnetotail," *J. Geophys. Res.* **83**, 1544 (1978).

DISCUSSION

K. Schindler: Comment: In a sense the $E \cdot J$ in a Speiser-type field is energetically correct if forces (e.g., pressure gradients) are taken into account. The $E \cdot J$ term then indeed energizes the plasma. This, however, is not an irreversible process.

THE "CONVECTION SURGE" MECHANISM
OF ION ACCELERATION DURING SUBSTORMS

B. H. Mauk*

To explain the presence of the so-called "bounce-phase-bunched" ion distributions (eV range to ~20 keV) observed in the earth's geosynchronous magnetosphere in association with substorm expansion phases, Quinn and Southwood proposed the "convection surge" mechanism of ion energization. This mechanism is associated with a sudden earthward displacement of curved field lines resulting from the short-lived application of an intense, east-west electric field. A quantitative computer model of the mechanism has been constructed here. It is found that the mechanism easily generates the bounce-phase-bunched ion distributions in question. Additionally, the mechanism can generate dramatically field-aligned distributions. It is proposed that a key general characteristic of geosynchronous ion pitch angle distributions, field-aligned low energies, and field-perpendicular high energies is a signature of this energization mechanism. The presence of field-aligned ions cannot be presumed to be a signature of recent ionospheric extraction.

INTRODUCTION

Within the geosynchronous regions of the earth's magnetosphere, spatially bunched clusters of ions have been observed bouncing back and forth along field lines between the northern and southern hemispheres.[1] On the nightside, these so-called "bounce-phase-bunched" ion distributions are observed only at small pitch angles, involve ion energies from the eV range up to (occassionally) 20 keV, and are associated in time with the expansion phase of substorms. To explain the presence of these distributions, Quinn and Southwood[2] proposed the "convection surge" mechanism of ion energization. The convection surge is associated with a sudden earthward displacement of curved field lines resulting from the short-lived application of an intense, east-to-west electric field.

To confirm, illustrate, and quantify the claims made by Quinn and Southwood[2] a numerical computer model of the convection surge process has been constructed here. Incorporated in this model are the results of recent measurements of transient electric fields, and also the dipolarization of the field-line shapes. The computer model numerically integrates the equations of motion of ions that populate a single flux tube that goes through the convection surge or "dipolarization" transformation. This paper summarizes a more extensive presentation submitted elsewhere.[3]

MODELING THE CONVECTION SURGE PROCESS

The top panel of Fig. 1 shows the concept envisioned by Quinn and Southwood.[2] The figure shows field lines before (labeled $k = 2$) and after ($k = 0$) the "convection surge" or "dipolarization" transformation. The field lines shown on this top panel were drawn using a simple, analytic field-line model first presented, for other purposes, by Quinn and McIlwain.[1] The initial value of the free parameter of the model, $k = 2$, is reasonable given field inclination angle measurements in the $R = 6$ to 9 R_e region of the earth's magnetosphere. The final value of $k = 0$ corresponds to a purely dipolar shape. The transformation of k from 2 to 0 represents the dipolarization of the field-line shapes during the initiation of the expansion phase of substorms (e.g., Nagai[4]).

The analytical field-line model can only tell one the off-equatorial electric and magnetic parameters once the equatorial parameters are known. The time evolution of the equatorial parameters is determined by the properties of the imposed transient electric fields. Powerful transient electric fields pointing perpendicular to $\vec{B}$ have been measured within the near-geosynchronous, near-equatorial regions by Shepard et al.[5] and by Aggson et al.[6] In the nightside regions, and relatively close to the magnetic equator, the electric field transients generally point east to west, have magnitudes up to 15–30 mV/m, and last for periods of 1–2 minutes.

For the present study the transient, equatorial, east-to-west electric field E_0 is represented as a step function in time, with a constant magnitude of 20 mV/m lasting for a time period T of 1 minute (as viewed by the equatorial position of the evolving field line). Off-equatorial values of the electric field are calculated self-consistently with the field-line motions (i.e., $v = cE/B$). It is assumed that the free parameter k of the field-line model varies linearly with time between 2 and 0 during the 1-minute time period T during which the equatorial electric field transient is applied.

The initial equatorial parameters R_0 (radial position), B_0 (field strength), and β' [$\equiv (\partial B/\partial r)/B - 1/r$] are set, respectively, at 9 R_e, 40 nT, and $-0.28/R_e$ (where r is an independent radial position parameter). Note that at 9 R_e the parameter β' has been set so that the spatial scale length for magnetic field strength variations is twice the dipole value, corresponding to tail-like distortions. The variation of B_0 with time is determined by the frozen-in condition and conservation of magnetic flux.

*The Johns Hopkins University Applied Physics Laboratory, Laurel, Maryland 20707.

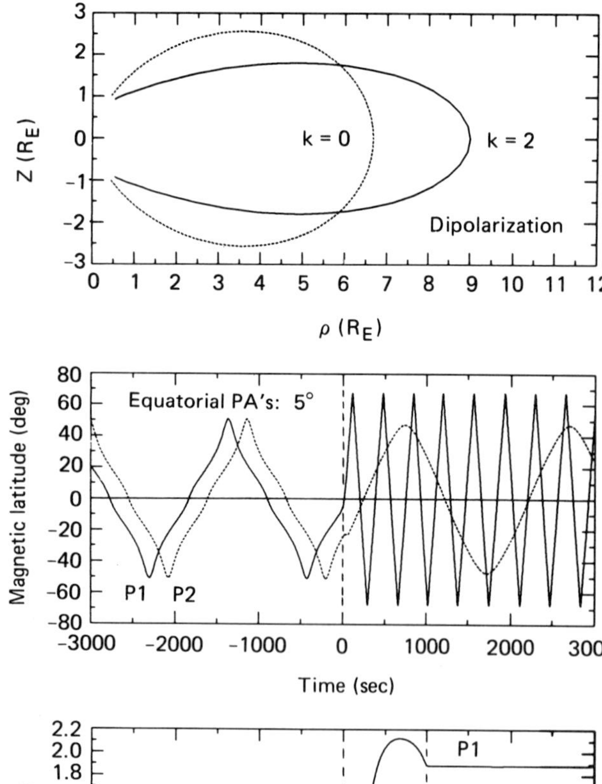

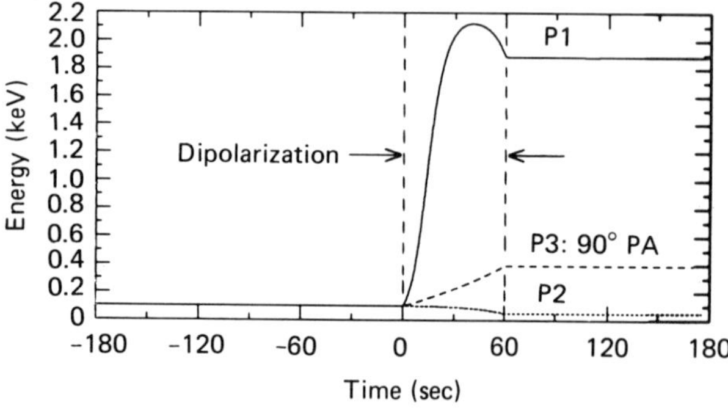

Figure 1—(Top) Field-line shapes before (labeled $k=2$) and after ($k=0$) "convection surge" or "dipolarization" modeled in this study. The lines were drawn with the analytic field line model of Quinn and McIlwain.[1] (Middle) Numerically determined trajectories of two different particles constrained to the flux tube diagrammed in the top panel. The dipolarization transformation occurs between 0 and 60 s near the middle of this panel. The two particles (P1 and P2) both start out with energies of 100 eV and equatorial pitch angles of 5°. (Bottom) On an expanded time scale, this panel shows the total energy of the two particles (P1, P2) shown in the middle panel, plus that of an equatorially mirroring particle (P3).

The medium compresses as the field line moves earthward due to the finite value of β' and the corresponding fact that $\nabla \cdot V_c \neq 0$ (where V_c is the convection speed). The value of R_0 is determined from B_0 by solving the equation $dR_0/dt = cE_0/B_0$. With solutions for R_0 and B_0, the expressions given by Quinn and McIlwain[1] can be used to determine R and B along the field line. Additional parameters calculated at all positions along the evolving field line are V_c, $\partial B/\partial s$, R_c, and $(\partial b/\partial t)_s$; which are, respectively, the perpendicular plasma (or field line) convection speed, the gradient of

the magnetic field strength B parallel to $\bar{B}$, the radius of curvature of the field line, and the time rate of change of the field-line pointing direction at constant positions along the field line.

The trajectories of particles confined to travel with the dipolarizing field line are calculated by numerically integrating the equations of motion. At each differential time step the perpendicular energization is calculated by conserving the first adiabatic invariant μ of each particle. The parallel motion is determined by solving the equation (e.g., Northrup[7]):

$$\frac{md\mathrm{v}_\parallel}{dt} = -\mu\,\frac{\partial B}{\partial s} + mV_c \cdot \frac{db}{dt},$$

where b is the unit vector in the direction of $\bar{B}$ at the particles' position, and where db/dt combines $\mathrm{v}_\parallel/R_c$ and $(\partial b/\partial t)_s$.

The middle panel of Fig. 1 shows the magnetic latitude versus time of two sample particles, P1 and P2, with initial energies of 100 eV and initial equatorial pitch angles of 5°. The only difference between the two particles is the time phasings of their bounce motions along the field line. The convection surge or dipolarization occurs briefly between 0 and 60 seconds near the very center of the panel. This middle panel shows dramatically that the response that the particles have to the dipolarization process depends critically on the phasings of the particles' bounce motions.

The bottom panel of Fig. 1 shows the particles' total energy displayed on an expanded time scale. The particle labeled P1 that on the middle panel was very close to the magnetic equator during the dipolarization has gained a factor of ~ 19 in energy. In contrast, the particle labeled P2 that was well away from the equator at the time of the dipolarization has actually lost energy. The bottom panel of Fig. 1 also shows the change in energy of an equatorially mirroring particle, labeled P3. This particle has gained a factor of ~ 4 in energy. The special particle with a small equatorial pitch angle and located close to the equator during the dipolarization has gained much more energy then has the equatorially mirroring particle, which means that the second adiabatic invariant (which would have yielded a factor of ~ 1.7 energization for P1 and P2) has been very substantially violated.

ION DISTRIBUTION RESPONSES TO THE DIPOLARIZATION

Figure 2 shows, in simulated spectrogram form, the response of an ensemble of particles to the dipolarization process. This figure was generated by loading the initial field line shown on Fig. 1 with particles such that at each position along the field line the ion distribution was on average Maxwellian in energy distribution, angularly isotropic out of the loss cone, and empty within the loss cone. The distribution was generated by weighting the output of random number generators. The trajec-

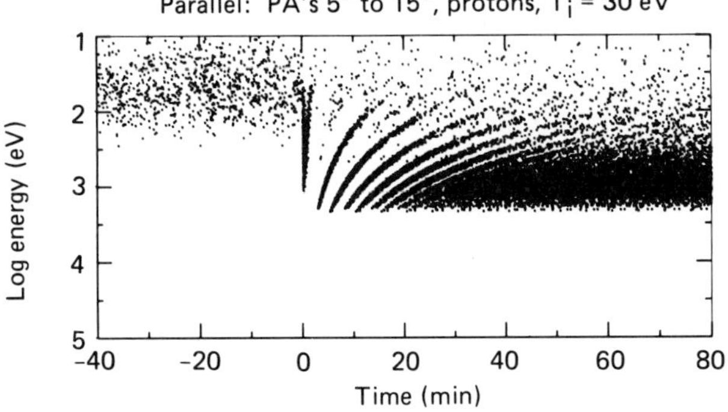

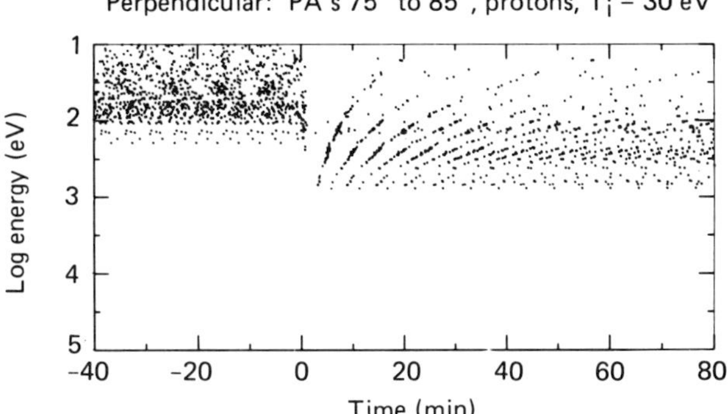

Figure 2—Simulated energy-time ion spectrograms generated by following the trajectories for an ensemble of particles constrained to the flux tube diagrammed on the top panel of Fig. 1. On average, the initial ensemble is Maxwellian as a function of energy at each position along the initial field line and angularly isotropic but with empty loss cones. Each point on the displays corresponds to the time and energy of each particle every time it crosses the magnetic equator. The top panel was generated by selecting only "parallel" particles, with pitch angles between 5° and 15°. The bottom panel shows only "perpendicular" particles, with pitch angles between 75° and 85°. The density of points is approximately proportional to the particles' differential energy flux (eV/cm^2-s-ster-eV). The initial distribution used here consisted of protons with a Maxwellian temperature of 30 eV.

tory of each of the resulting ≈ 10,000 particles was then followed. Every time a particle crossed the equatorial plane, its energy, pitch angle, and crossing time were recorded. Each point on Fig. 2 corresponds to one recorded crossing. The figure shows particle energy versus time for two selected pitch-angle windows. The top panel shows the parallel spectrogram, representing pitch angles between 5° and 15°, while the bottom panel shows the perpendicular spectrogram, with pitch angles between 75° and 85°. The density of points on the display is at each position approximately proportional to the differential energy flux (eV/cm^2-s-ster-eV) of the distribution. (The appearance of nonrandomness in the predipolarization, perpendicular spectrogram results from the fact that statistical fluctuations are propagated, given the procedure used.) The initial distribution consisted of protons with a Maxwellian temperature of 30 eV.

Of most immediate interest are the very striking and high-contrast dispersive streaks that appear in the top panel. The repetition period of the streaks at any one

energy corresponds to half the intrahemispherical bounce period of those particles. It is an obvious conclusion that these simulated streaks correspond to the dispersive streaks observed by Quinn and McIlwain.[1] The model calculations clearly support the conclusion that the observed bounce-phase-bunched ion distributions are generated by the convection surge or dipolarization process. Dispersive features are also observed in the simulated perpendicular spectrogram, but the flux values and overall contrast are lower, making these features difficult to observe.

Figure 3 shows a more quantitative view of the net effects on the distribution. This figure shows differential intensity, on a logarithmic scale, plotted versus a logarithmic energy scale. The initial distribution is shown as a solid line and the final distribution is shown as a short-dashed line for perpendicular particles and a long-dashed line for parallel particles. One sees from this figure that, for this case, the most dramatic response to the dipolarization was in the parallel direction, and that the "apparent" bulk energization in the parallel direction was a factor somewhere between 10 and 30. The dipolarization mechanism can energize a very substantial fraction of the whole distribution in the parallel direction, resulting in dramatically field-aligned distribution shapes.

Additional simulations have been run with different initial particle parameters. Using protons it is found that relatively high contrast bounce-phase-bunched distributions are generated using initial temperatures up to several keV; however, field-aligned final distributions are generated only for initial temperatures up to several (3–5) hundred eV. More dramatic effects are found using heavy ions. Using oxygen ions, field-aligned distributions (with peak intensities at up to 10–20 keV) can be gener-

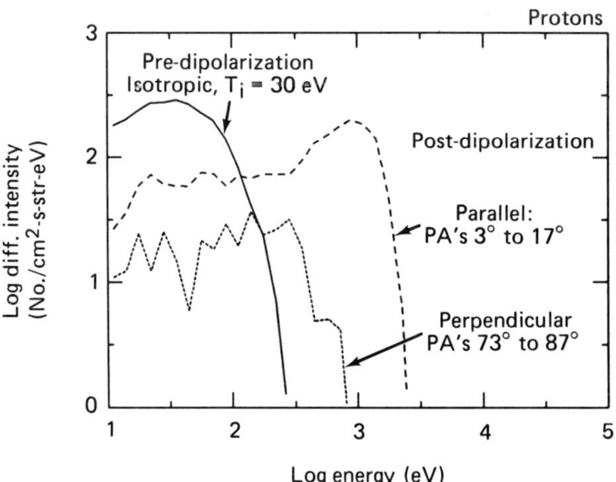

Figure 3—Predipolarization and postdipolarization ion distributions corresponding to the simulation run shown in Fig. 2 (protons with 30 eV initial temperature). The figure shows differential intensity plotted versus energy. The solid line corresponds to the initial (isotropic) distribution while the short-dashed line corresponds to the postdipolarization "perpendicular" distribution and the long-dashed line to the postdipolarization "parallel" distribution.

ated for initial temperatures as high as several (3–5) keV. At very low initial temperatures (~ 3 eV) very dramatic "bulk" parallel energization occurs (factors of 50–100), but the responses for different mass species are not as dramatically different as for the higher initial temperatures.

DISCUSSION

A very common feature of ion populations observed within the geosynchronous regions is the field-aligned character of the lower to intermediate energies (the eV to 1–10 keV ranges). In previous studies the field-aligned character of these ions has been taken as a signature of recent ionospheric extraction.[8-12] It is clear from the present study that this conclusion was premature. The observed characteristics are just what would be expected based on the convection surge or dipolarization mechanism. Certainly, upflowing ionospheric ions have been observed at low altitudes within the auroral regions.[13] The question that remains is where and in what form do those ions end up? It has, for instance, been argued (Lyons and Evans[14] and references therein) that the regions of the discrete aurora (where one might expect the ionospheric extraction of the sort observed by Shelley et al. to occur) corresponds (by magnetic mapping) to the outer edges of the plasma sheet population. In contrast, the geosynchronous injected plasmas are commonly associated (if in an energized form) with the inner edge of the plasma sheet population (Refs. 15,16) and with the diffuse auroral regions.[17] Hence, to associate the geosynchronous, sub-keV to keV range field-aligned ions with recent ionospheric extraction one must concern oneself with the subsequent transport of those ions from the outer to the disturbed inner edges of the plasma sheet population. And, it has been argued that that transport is of an extremely turbulent nature, to the extent that pitch-angle structure could be eliminated.[18]

Certainly one cannot exclude the possibility that the signature of ionospheric extraction persists within the low- to intermediate-energy geosynchronous particle populations. What one can say is that the success of the modeling discussed in this paper (using observed parameters) in terms of reproducing the commonly observed bounce-phase-bunched ion distributions, and the corresponding ease by which the model generates field-aligned ion distributions, means that the convection surge mechanism must play an important role in forming the observed pitch-angle distributions. This statement is supported additionally by the evidence[18,19] that all transport to the geosynchronous regions occurs in association with impulsive, spatially localized processes, presumably driven by the inductive electric fields discussed in the report.

ACKNOWLEDGMENT—We thank C.-I. Meng and D. G. Mitchell for helpful discussions. This work was supported by the Atmospheric Sciences Division, National Science Foundation grant ATM-8315041, and by the Air Force Office of Scientific Research grant 84-0049, and by NASA contract to The Johns Hopkins University Applied Physics Laboratory and Department of the Navy under task I2UOS1P of contract N00024-85-C-5301.

REFERENCES

[1] J. M. Quinn and C. E. McIlwain, "Bouncing Ion Clusters in the Earth's Magnetosphere," *J. Geophys. Res.* **84**, 7365 (1979).

[2] J. M. Quinn and D. J. Southwood, "Observations of Parallel Ion Energization in the Equatorial Region," *J. Geophys. Res.* **87**, 10536 (1982).

[3] B. H. Mauk, "Quantitative Modeling of the Convection Surge Mechanism of Ion Acceleration," *J. Geophys. Res.* **91**, 13423 (1986).

[4] T. Nagai, "Observed Magnetic Substorm Signatures at Synchronous Altitude," *J. Geophys. Res.* **87**, 4405 (1982).

[5] G. G. Shepard, R. Bostrom, H. Derblom, C.-G. Falthammar, R. Gendrin, K. Kaila, A. Korth, A. Pedersen, R. Pellinen, and G. Wrenn, "Plasma and Field Signatures of Poleward Propagating Auroral Precipitation Observed at the Foot of the GEOS-2 Field Line," *J. Geophys. Res.* **85**, 4587 (1980).

[6] T. L. Aggson, J. P. Heppner, and N. C. Maynard, "Observations of Large Magnetospheric Electric Fields During the Onset Phase of a Substorm," *J. Geophys. Res.* **88**, 3981 (1983).

[7] T. G. Northrup, *The Adiabatic Motion of Charged Particles*, Wiley Interscience, New York (1963).

[8] B. H. Mauk and C. E. McIlwain, "UCSD Auroral Particles Experiment," *IEEE Trans. Aerosp. Electron Syst.* **AES-11**, 1125 (1975).

[9] W. Lennartsson and D. L. Reasoner, "Low-Energy Plasma Observations at Geosynchronous Orbit," *J. Geophys. Res.* **83**, 2145 (1978).

[10] J. L. Horwitz, "Conical Distributions of Low-Energy Ion Fluxes at Synchronous Orbit," *J. Geophys. Res.* **85**, 2057 (1980).

[11] J. F. Fennell, D. R. Croley, Jr., and S. M. Kaye, "Low-Energy Ion Pitch Angle Distributions in Outer Magnetosphere, Ion Zipper Distributions," *J. Geophys. Res.* **86**, 3375 (1981).

[12] S. M. Kaye, E. G. Shelley, R. D. Sharp, and R. G. Johnson, "Ion Composition of Zipper Events," *J. Geophys. Res.* **86**, 3383 (1981).

[13] E. G. Shelley, R. D. Sharp, and R. G. Johnson, "Satellite Observations of an Ionospheric Acceleration Mechanism," *Geophys. Res. Lett.* **3**, 654 (1976).

[14] L. R. Lyons and D. S. Evans, "An Association Between Discrete Aurora and Energetic Particle Boundaries," *J. Geophys. Res.* **89**, 2395 (1984).

[15] C. E. McIlwain, "Substorm Injection Boundaries," in *Magnetospheric Physics*, B. M. McCormac, ed., D. Reidel, Hingham, Mass., p. 143 (1974).

[16] M. G. Kivelson, S. M. Kaye, and D. J. Southwood, "The Physics of Plasma Injection Events," in *Dynamics of the Magnetosphere*, S.-I. Akasofu, ed., D. Reidel, Hingham, Mass., pp. 385-405 (1980).

[17] C.-I. Meng, B. H. Mauk, and C. E. McIlwain, "Electron Precipitation of Evening Diffuse Aurora and its Conjugate Electron Fluxes Near the Magnetic Equator," *J. Geophys. Res.* **84**, 2545 (1979).

[18] J. B. Cladis and W. E. Francis, "The Polar Ionosphere as a Source of the Storm Time Ring Current," *J. Geophys. Res.* **90**, 3457 (1985).

[19] B. H. Mauk and C.-I. Meng, "Dynamical Injections as the Source of Near Geostationary Quiet Time Particle Spatial Boundaries," *J. Geophys. Res.* **88**, 10011 (1983).

[20] B. H. Mauk and C.-I. Meng, "Macroscopic Ion Acceleration Associated with the Formation of the Ring Current in the Earth's Magnetosphere," in *AGU Geophysical Monograph on Ion Acceleration*, T. Chang, ed., American Geophysical Union, Washington, D.C., p. 351 (1986).

MODEL FOR THE GEOMAGNETIC ION SPECTROMETER IN THE MAGNETOTAIL LOBES

J. L. Horwitz*

A two-dimensional kinetic "tail lobe ion spectrometer" model for the transport of ionospheric ions from the polar cleft ionosphere into the tail lobes is developed to simulate semiquantitatively the behavior of the observed O^+ ion streams in the magnetotail lobes. The consequences of the present model include (a) the increase in velocity of the O^+ streams away from the tail midplane and (b) the existence of "tongues" of O^+ density from the cleft ionosphere into the tail lobes whose distribution configuration depends on the convection electric field and the source thermal energies. These consequences are reasonably consistent with the trends in lobe streams observed by recent spacecraft.

INTRODUCTION

There have been several observations recently of tailward streaming O^+ in the magnetotail and lobe regions.[1-4] At the 15–23 R_e range of the ISEE 1 and 2 orbits, O^+ streams with tailward velocities ranging from 50–250 km/s (Ref. 4) and O^+ densities of $\sim 10^{-4}$ – 1 ions/cc (Refs. 3, 4) have been reported. The Candidi et al.[4] observations in particular show that the O^+ parallel velocities increase with distance away from the tail X-axis whereas the densities tend to decline from the axis, at least in the approximately upper half of the northern tail lobe. These results are qualitatively consistent with concepts of polar ionosphere supply and convection filtering of the O^+ in its transport through the tail lobes. Interestingly, the H^+ densities observed by Candidi et al.[4] tend to increase away from this X-axis, in contrast to the O^+ distribution. Candidi et al.[4] interpreted this result as indicating that the H^+ origin was different from that of the O^+, presumably the solar wind injection and formation of the mantle.

Recently, we have described a kinetic model for the investigation of ion transport in the magnetosphere.[5-8] The model has been used principally to simulate low-energy ion dynamics recently observed in the polar magnetosphere with the DE-1 satellite.[7-12]

The purpose of the present report is to extend the application of this kinetic model to investigate the transport of ionospheric ions into the magnetotail lobes. In general approach, the calculation of ion trajectories, distribution functions, and ion bulk parameters in the magnetosphere is described in Refs. 5 and 7 and will be summarized briefly here.

Ion trajectories are computed for transport from the cleft ionosphere to the tail in a two-dimensional plane appropriate to a noon-midnight meridional cross section of the magnetosphere. The trajectories now include the effect of parallel acceleration by convection through curved magnetic field regions (e.g., Cladis[13]). The evaluation of the distribution functions in the magnetosphere is based on Liouville's theorem and the ion trajec-

tory code. An array of energies and pitch angles specifies the phase space at a given location over which we wish to define the distribution function. For each point in this phase space we integrate the ion trajectory backward in time. If intersection with a specific cleft ionospheric source regions occurs, the distribution value is that of the source distribution at the mapped energy and pitch angle. The source distribution is taken to be the upgoing half of a bi-Maxwellian distribution with specified density and parallel and perpendicular energies. This source distribution is specified to exist within a given latitudinal range and is zero outside this latitudinal range, at an altitude of about 3000 km.

The distribution functions are calculated over an extended spatial array of grid points for the northern half of the polar and tail magnetosphere. Moments of the distribution functions are integrated to obtain the spatial distribution of various bulk parameters. The parameters are contoured in the two-dimensional magnetosphere to represent their distribution in this meridian.

The principal objective here is to illustrate the effects of various cleft ionosphere source energies and convection electric fields on the resulting distributions of such bulk parameters as ion density and parallel velocities. The primary results are obtained for O^+, and some of these results are compared to the tail lobe O^+ observations by Candidi et al.[4]

PARAMETRIC ILLUSTRATION OF IONOSPHERIC ION BULK PARAMETER DISTRIBUTIONS IN THE TAIL LOBES

In this section, we illustrate the two-dimensional distributions of ionospheric ion bulk parameters supplied by the cleft source to the tail lobes, chiefly for O^+ as a function of the source thermal energies and ionospheric convection electric fields. We consider that upgoing ions from the cleft topside ionosphere extend over a range of energies from a few eV into the keV range and that the complete distribution is highly variable; hence by investigating the effect of substantially different source thermal energies, we are actually examining in some sense the contribution of different parts of the energy spectrum

*Department of Physics, The University of Alabama in Huntsville, Huntsville, Alabama 35899.

of cleft source ions. For the purposes of convenient intercomparison, we maintain the source densities constant at 100 ions/cc at the source altitude and the source latitude range at 70°–80° on the dayside for most of the cases presented here.

Figure 1a and b shows first the O^+ densities in the tail magnetosphere for intermediate source energies and three convection electric fields. The representative thermal energies chosen are $kT_\perp = 100$ eV, $kT_\parallel = 50$ eV. We do not investigate here a sensitivity to variations in the temperature anisotropy ratio $T_\perp/T_\parallel$, since in the tail lobes the ion energies are almost entirely converted to parallel energy anyway.

Figure 1a illustrates then the O^+ densities for a distribution source for three ionospheric convection electric field values. In the 20 and 50 mV/m cases, ion densities of $\sim 10^{-2}$ O^+/cm^3 are observed out to near 20 R_e

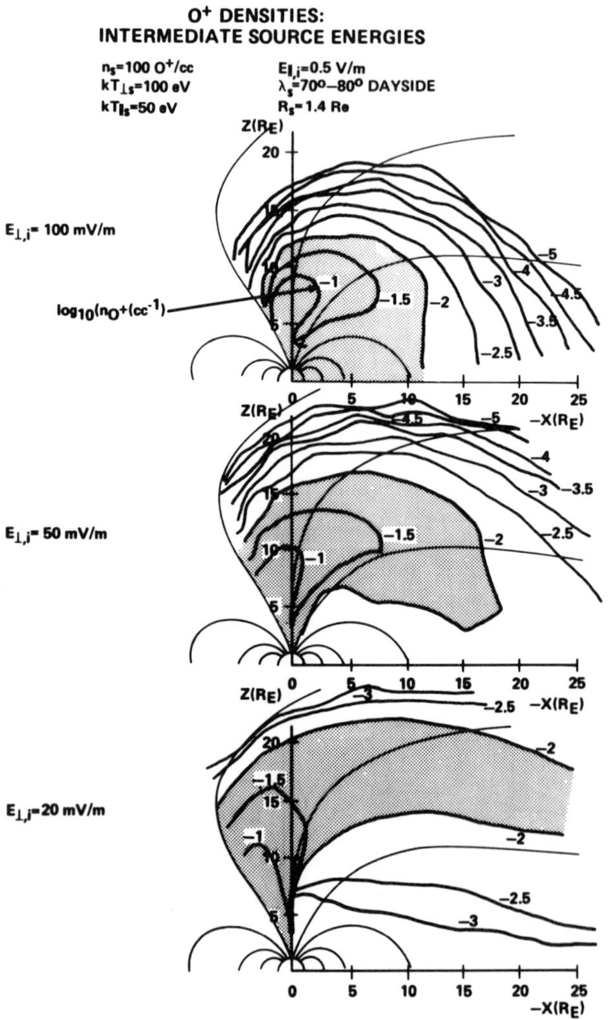

Figure 1—O^+ tail lobe concentrations depicted in the noon-midnight plane for intermediate source thermal energies at the polar cleft ionosphere for three different polar ionospheric ($E_{\perp,i}$) convection electric fields. The density "tongue" extending into the lobes is indicated for $n_{O^+} > 10^{-2}$ O^+/cc by the shaded region. This "tongue" moves away from the tail midplane with decreasing convection electric field.

geocentric distance in the tail. The effects of both the magnetic field divergence and convective transport on the ion densities can be seen; the magnetic field divergence is responsible for the major part of the approximately radial decline of the densities from the source. In each case there is a "tongue" configuration of density distribution that extends out from the cleft ionosphere into the tail lobes. This density tongue extends further down toward the tail midplane with increasing convection electric field. It is interesting that the observations of O^+ density distributions in the tail lobes by Candidi et al.[4] appear to display a peak at an intermediate distance off the tail X-axis. This would be consistent with the tongue feature seen in the present calculations (see discussion below). To some extent, the tongue aligns with the ion trajectory from the cleft for the ions with parallel energy of the order $kT_\perp$.

For the low source thermal energies of the order 5–10 eV (not shown), particularly for moderate-to-high convection electric fields, very little O^+ fills the tail lobes beyond about 20 R_e. Even presuming larger source densities and perhaps extended source longitudes in the longitudinally extended auroral oval, it does not appear that the ~ 10 eV cusp source can account for the population of O^+ observed in the lobes by Sharp et al.[3] and Candidi et al.[4] This is in contrast to the < 5 R_e polar cap magnetosphere that these low-energy O^+ ions tend to populate under at least moderate-to-high convection electric field conditions. The intermediate and high thermal energy source components should be more likely to supply the distant tail lobe O^+ populations observed by ISEE.

The parallel velocities in Fig. 2 are shown only for the one set of source energies for the three different convection electric field values. This is because, in the tail regions at least, the velocity filter is effective in selecting a narrow range of velocities for ions to have access to a particular region from the confined source. Hence, the bulk velocities are fairly insensitive to the source characteristics and only depend significantly on the convection electric field. The range of O^+ velocities in the geocentric range 15–23 R_e sampled by ISEE from Candidi et al.[4] is most consistent with the velocity distribution for $E_{\perp,i} = 50$ mV/m, which seems reasonable since this value is probably typical of the convection electric field in the ionospheric polar cap. In all cases, the ion velocities decline in moving away from the tail midplane to higher magnetic latitudes, in agreement also with the Candidi et al.[4] results, and as is consistent with the generally accepted topologies of the polar and tail magnetic and electric fields.

DISCUSSION

Some of these O^+ bulk parameter distributions may be compared to the distributions observed by Candidi et al.[4] In Figs. 3 and 4, we have compared the results for an intermediate thermal energy source (e.g., Fig. 1b) at $E_{\perp,i} = 50$ mV/m but with a source density of 10^3 O^+/cc (see below). Candidi et al.[4] plot some of their results versus the coordinate $S = (\text{sign } Y)(Z^2 + Y^2)^{1/2}$

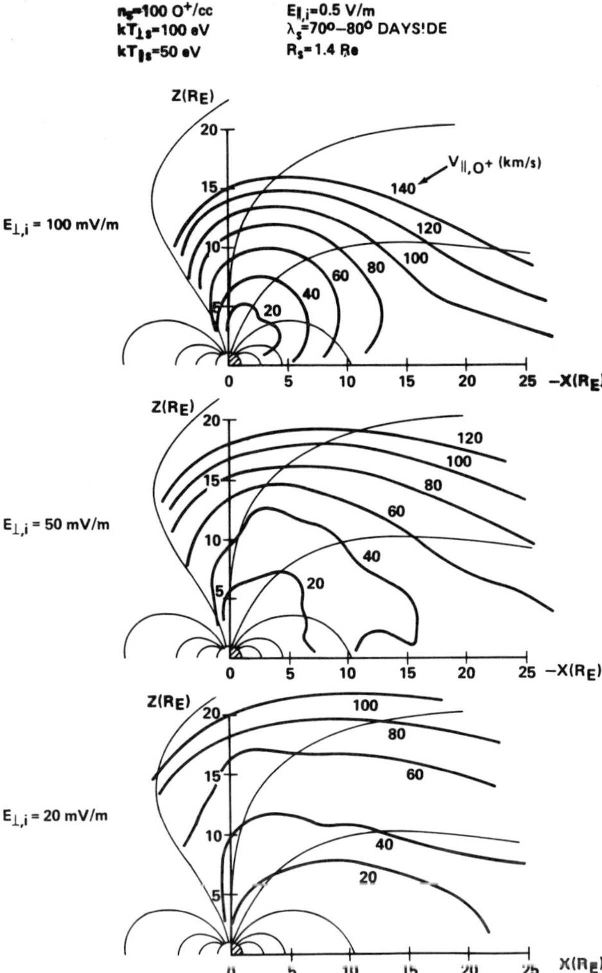

Figure 2—O$^+$ outward parallel velocities for three different convection electric fields. Although the intermediate source energies cases are presented here, the velocities are rather insensitive to the source energies in the tail lobe regions of interest here, owing to the velocity filter effect.

of the ISEE sampling location from the GSM X-axis in the tail, where "sign Y" refers to the sign of the Y-coordinate. In order to compare results from our present two-dimensional model, we have used the Z distance from the tail midplane in our model as analogous to the S coordinate. Candidi et al.[4] used data from a range of geocentric distances 15–23 R_e in their scatter plots of parameters versus this coordinate S. We observe in the model calculations significant variations versus X and Z, so perhaps their data could be more usefully organized versus both X and S. In any case, in order to compare with their results we used model calculations along a locus in which Z and X varied in concert, as would be the case for an ISEE 2 orbit. Table 1 shows the set of points defining this locus of spatial locations for which we use model calculations. The specific locations of the discrete values in this table correspond to grid points for which we obtained model bulk parameters.

Table 1—R, λ, and S coordinates in comparison with results of Candidi et al.[4]

R (R_e)	λ	S (or Z) (R_e)
13.8	15	3.5
13.8	25	5.8
16.6	25	7.0
16.6	35	9.6
19.4	35	11.1
19.4	35	13.7
22.2	45	15.7
22.2	55	18.2

Figure 3 shows the scatter plot of Candidi et al.[4] for the O$^+$ bulk velocities versus S (their Fig. 6). Superimposed on these data is the model calculation for 50 mV/m. As noted before, the bulk velocities are relatively insensitive to the source thermal energy characteristics, and depend mainly on the value of the convection electric field. The general agreement between their data and our model calculations for the velocities supports the concept that the velocity filter, which combines effects of $E \times B/B^2$ drift and field-aligned flow, is the principal governing influence on the bulk flow velocities. It would be of interest to have the available data organized in S,X coordinates in some fashion to determine the X variations as well. Figure 2 indicates that the X variations of the parallel velocity should be significant particularly at average to high convection electric fields. For instance, at very low latitudes near the tail midplane for

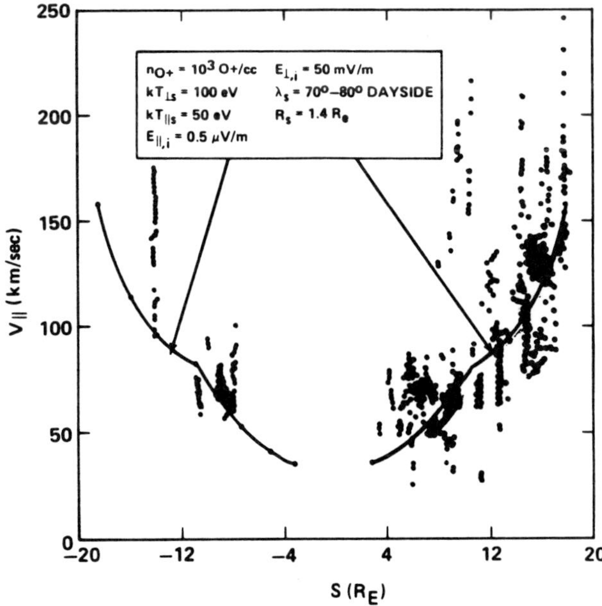

Figure 3—Comparison of the distribution of O$^+$ parallel velocities versus S = (sign Y) $(Y^2 + Z^2)^{1/2}$ from the X-axis in the tail as observed by Candidi et al.[4] versus model calculations of $V_{\parallel}$ for a polar convection electric field of 50 mV/m, using the tail locations indicated in Table 1.

the $E_{\perp,i} = 100$ mV/m case in Fig. 2, the velocity gradient is almost entirely radial, as the convection would be dominating the inflow of O^+ toward the midplane.

The corresponding O^+ density scatter plot of Candidi et al.[4] (their Fig. 7) is shown in Fig. 4. In order to obtain tail densities of the order 10^{-1} O^+/cc, the rather large source density of 10^3 O^+/cc at 3000 km altitude in the cleft was used for the model comparison. The upward flux from such an ion distribution is of the order 10^9 ions/cm²-s in the ionosphere. This is in the range of cleft upward ion fluxes reported by Lockwood et al.,[14] but these were for ions of ~10 eV energies. Klumpar[15] reported differential fluxes at 1400 km altitude in the cleft of 30–300 eV ions whose upper flux range might be consistent with 10^9 ions/cm²-s. Possibly a significant fraction of the low-energy ion fluxes observed by Lockwood et al.[14] might be energized into the 50–100 eV range at slightly higher altitudes in the cleft; however, this is speculative at the present time. Another parameter that affects the tail lobe densities is the horizontal distance over which the ion source extends at 3000 km altitude. For resulting tail-lobe densities and fluxes, increases in this source extent are approximately equivalent to proportionate increases in the source number density. For the cleft runs presented here, we used a source latitudinal width of 10° which is about 1000 km in horizontal extent. Though presented in a two-dimensional model, we tend to regard this range as actually contributed by a combination of latitudinal and longitudinal extent for flux tubes convecting partly along a source region in the dayside auroral oval before entering the polar cap region through the throat convection region. In a rough sense, then, we could use a source density of 10^2 O^+/cc and a source horizontal extent of

10^4 km to be equivalent to a source density of 10^3 O^+/cc and 10° latitudinal width. This would correspond to a source extending along most of the auroral oval for a flux tube convecting along it.

Although we are somewhat uncertain about the source parameters to be used in the density comparison here, we should note that the O^+ densities observed by Sharp et al.[3] are peaked in occurrence frequency near ~10^{-2} O^+/cc and are broadly spread in occurrence down to ~10^{-4}, which is near the detection threshold of the instrument used by Sharp et al.[3] These densities are generally consistent with those resulting from our presented model results with a 10^2 O^+/cc and 10° latitudinal width (e.g., Fig. 1b, middle panel).

Nevertheless, a higher density source is required to fit the general pattern presented by Candidi et al.[4] One interesting aspect of this figure is that despite the uncertainty in the source density, there is apparently a peak in the density distribution near $S = 10$ R_e off the tail X-axis which is broadly reproduced in the calculations shown in the model curve (solid line). This density peak off the X-axis may be associated with the tongue feature observed in the density contour plots presented here.

ACKNOWLEDGMENT – This research was supported in part by NASA contract NAS8-33982 and NSF grant ATM-8503102.

REFERENCES

[1] D. A. Hardy, J. W. Freeman, and H. K. Hills, "Double-Peaked Ion Spectra in the Lobe Plasma: Evidence for Massive Ions?" *J. Geophys, Res.* **82**, 5529 (1977).

[2] L. A. Frank, K. L. Ackerson, and M. Yeager, "Observations of Atomic Oxygen (O^+) in the Earth's Magnetotail," *J. Geophys. Res.* **82**, 129 (1977).

[3] R. D. Sharp, D. L. Carr. W. K. Peterson, and E. G. Shelley, "Ion Streams in the Magnetotail," *J. Geophys. Res.* **86**, 4639 (1981).

[4] M. Candidi, S. Orsini, and V. Formisano, "The Properties of Ionospheric O^+ Ions as Observed in the Magnetotail Boundary Layer and Northern Plasma Lobe," *J. Geophys. Res.* **87**, 9097 (1982).

[5] J. L. Horwitz, "Features of Ion Trajectories in the Polar Magnetosphere," *Geophys. Res. Lett.* **11**, 1111 (1984).

[6] J. L. Horwitz and M. Lockwood, "The Cleft Ion Fountain: A Two-Dimensional Kinetic Model," *J. Geophys. Res.* **90**, 9749 (1985).

[7] J. L. Horwitz, M. Lockwood, J. H. Waite, Jr., T. E. Moore, C. R. Chappell, and M. O. Chandler, "Transport of Accelerated Low-Energy Ions in the Polar Magnetosphere," in *Monograph on Ion Acceleration in the Magnetosphere and Ionosphere* (submitted, 1985).

[8] J. L. Horwitz, J. H. Waite, Jr., and T. E. Moore, "Supersonic Ion Outflows in the Polar Magnetosphere Via the Geomagnetic Spectrometer," *Geophys. Res. Lett.* **12**, 757 (1985).

[9] M. Lockwood, M. O. Chandler, J. L. Horwitz, J. H. Waite, Jr., T. E. Moore, and C. R. Chappell, "The Cleft Ion Fountain," *J. Geophys. Res.* **90**, 9736 (1985).

[10] M. Lockwood, T. E. Moore, J. H. Waite. Jr., C. R. Chappell, J. L. Horwitz, and R. A. Heelis, "The Geomagnetic Mass Spectrometer-Mass and Energy Dispersions of Ionospheric Ion Flows into the Magnetosphere," *Nature,* **316**, 612 (1985).

[11] T. E. Moore, C. R. Chappell, M. Lockwood, and J. H. Waite, Jr., "Suprathermal Ion Signatures of Auroral Acceleration Processes," *J. Geophys. Res.* **90**, 1611 (1985).

[12] J. H. Waite, Jr., T. Nagai, J. F. E. Johnson, C. R. Chappell, J. L. Burch, T. L. Killeen, P. B. Hays, G. R. Carignan, W. K. Peterson, and E. G. Shelley, "Escape of Suprathermal O^+ Ions in the Polar Cap," *J. Geophys. Res.* **90**, 1619 (1985).

[13] J. B. Cladis, "Parallel Acceleration and Transport of Ions from Polar Ionosphere to Plasmasheet," *Geophys. Res. Lett.* (in press, 1986).

[14] M. Lockwood, J. H. Waite, Jr., T. E. Moore. J. F. E. Johnson, and C. R. Chappell, "A New Source of Superthermal O^+ Ions Near the Dayside Polar Cap Boundary," *J. Geophys. Res.* **90**, 4099 (1985).

[15] D. M. Klumpar, "Transversely Accelerated Ions: An Ionospheric Source of Hot Magnetospheric Ions," *J. Geophys. Res.* **84**, 4229 (1979).

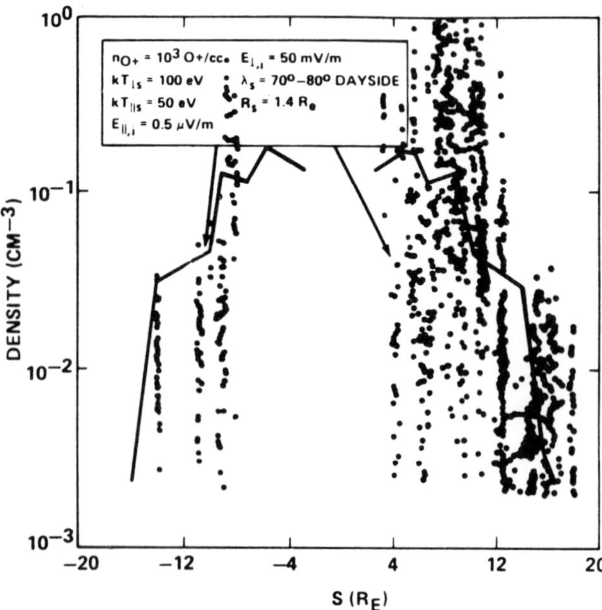

Figure 4—Comparison of the O^+ density distributions in the tail versus S from Candidi et al.[4] with model calculations using the indicated source parameters and the locations of Table 1.

GENERATION OF BROADBAND NOISE IN THE MAGNETOTAIL

P. B. Dusenbery[*][†] and L. R. Lyons[‡]

We have evaluated the generation of electrostatic noise in the geomagnetic tail by ion beams. We assume a stationary plasma-sheet electron distribution and streaming-ion distributions. Both warm ion streams, as observed within the plasma-sheet boundary layer, and cold ion streams, as expected from upward flowing ionospheric ions, are considered. Warm ion streams by themselves are found to be stable, whereas a cold ion stream by itself is unstable to the beam acoustic mode. However, wave growth is increased if both cold and warm streams are simultaneously present. These results suggest that the interaction between the warm and cold ion streams is responsible for the peak in electrostatic wave intensities observed within the plasma-sheet boundary layer. For cold and warm ions streaming in the same direction, we find wave growth peaks for wave normal angles $\theta = 0°$ and wave frequencies ~ 0.1 times the electron plasma frequency. However, for antiparallel streaming cold and warm ions, wave growth peaks near $\theta \approx 90°$ and wave frequencies are an order of magnitude smaller. Including counterstreaming warm ions in addition to a cold ion stream results in wave growth that is a superposition of that for the above two cases.

INTRODUCTION

Intense electrostatic waves are frequently observed in the plasma-sheet region of the earth's magnetotail in the frequency range 10 Hz to several kHz with maximum intensities near 10–50 Hz. The first report of these waves was by Scarf et al.[1] using IMP 7 data.

Subsequently, Gurnett et al.[2] made more detailed measurements from the wave experiments on the IMP 8 satellite. In addition to the broad frequency range, Gurnett et al.[2] found an average rms electric field amplitude of about 1 mV/m and was able to deduce that at times the waves propagated within about $\pm 20°$ from perpendicular to the magnetic field. The highest frequency of occurrence of the electrostatic noise was found by Gurnett et al.[2] in the region near the plasma-sheet boundary layer when anisotropic fluxes of ions streaming either earthward or anti-earthward were present.

Earthward and anti-earthward streaming ions of keV energies have been observed within the plasma-sheet boundary layer using IMP 8 measurements (Frank et al.,[3] DeCoster and Frank[4]) and ISEE 1 measurements (Williams,[5] Lui et al.[6]). These particles likely result from energization in the tail current sheet as proposed by Lyons and Speiser.[7] They are dominantly protons with streaming speeds of 500–1500 km/s, temperatures of 0.1–1 keV, and densities < 1 cm^{-3}. Their thermal energy is approximately equal to the beam energy.

Cold ion streams of ionospheric origin have been observed in the lobes as well as in the plasma sheet (Sharp et al.[8]). The composition of the cold ions is usually H$^+$ or O$^+$. Typical number densities are < 0.1 cm^{-3}, temperatures of 50 eV or less, and streaming speeds 10–1000 km/s. Figure 1 is a schematic of the noon-midnight meridian cross section of the earth's magnetosphere emphasizing the suggested importance of both ion beams within the plasma-sheet boundary layer.

Plate IV-5 is from Grabbe and Eastman[9] illustrating the association between ion beams and broadband waves. Plate IV-5A is the ISEE 1 LEPEDEA particle data for day 85 of 1978 from 0000–0600 UT. The ion beam was measured near 0325 UT, coincident with the intensification of the measured electrostatic noise by the SFR wave experiment on ISEE 1 (Plate IV-5B). Prior to and following 0325 UT, ISEE 1 was in the lobe and central plasma sheet, respectively, where the noise was observed to be much weaker (T. Eastman, personal communication, 1985). The smooth ion distribution in Plate IV-5A, part c near the origin in velocity space is not real and represents an artifical one-count level in the particle detector. In fact, it is very rare for stationary ~ 100 eV ion distributions to be measured when ISEE 1 is either in the lobes or in the plasma-sheet boundary layer (C. Huang, personal communication, 1985).

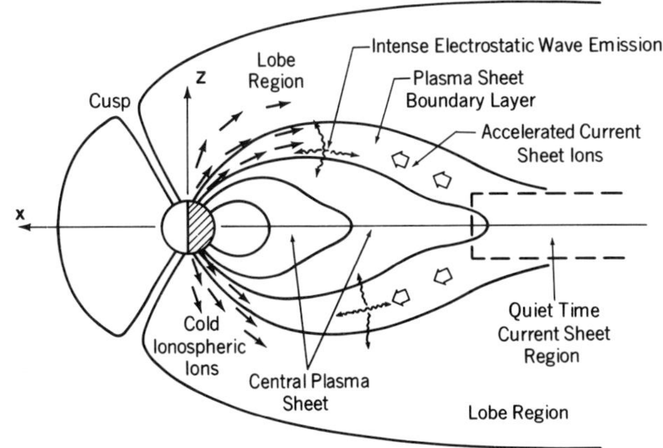

Figure 1—A noon-midnight meridian cross section of the earth's magnetosphere illustrating the regions of importance to the generation of broadband electrostatic noise.

*Department of Astrophysical, Planetary, and Atmospheric Sciences, University of Colorado, Boulder, Colorado 80309-0391.
†Also at Space Environment Lab R/E/SE, National Oceanic and Atmospheric Administration, 325 Broadway, Boulder, Colorado 80303.
‡The Aerospace Corporation, Space Sciences Laboratory, Mail Station M2-260, P. O. Box 92957, Los Angeles, California 90009.

Because of the strong correlation between ion beams and broadband noise,[2] it is reasonable to investigate the role played by ion-beam instabilities in explaining the generation of broadband waves in the plasma-sheet region. Grabbe and Eastman[9] modeled the plasma-sheet particle populations by streaming keV ions, cold stationary ions (~ 100 eV), and stationary electrons of several hundred eV. They found that broadband waves were excited only when the ion beams were very cold (thermal energy $<1\%$ of the beam energy). Neither such ion beam thermal energies nor stationary ion distributions are typical of warm plasma-sheet boundary-layer ions so that the Grabbe and Eastman[9] study does not explain the association of the intense broadband noise with the warm boundary-layer ion beams. In addition, Grabbe and Eastman[9] calculated that the growth rate of the excited waves decreases with increasing wave normal angle, which cannot explain wave intensities peaking at wave normal angles near $70°$. However, Omidi[10] showed that if the ions are properly treated as unmagnetized, then the distribution model assumed by the Grabbe and Eastman study leads to growth rates that peak near wave normal angles of $\sim 70°$.

Assuming that there are two streaming-ion distributions, a warm current-sheet population, and a cold ionospheric population, Dusenbery and Lyons[11] investigated the role number density, relative ion drift, and current-sheet ion temperature had in destabilizing the fast and slow beam acoustic mode. Their model predicted that the wave intensity of broadband noise should peak in the plasma-sheet boundary layer due to the strong interaction between the warm and cold ion beams. Observations of less intense electrostatic waves in the lobes and plasma sheet were likely a result of cold ionospheric beams in the absence of warm ion beams, which resulted in smaller growth rates.

A limitation of our previous study was its restriction on wave normal angle (parallel propagating waves were only considered) and its choice of distribution model (ion beams drifting in the same direction). The primary goal of the present study is to investigate several different distributions for the streaming ions and to evaluate the frequency and growth-rate characteristics of the beam acoustic instability as a function of wavenumber and wave normal angle.

THEORY AND RESULTS

In this paper, we assume both ions and electrons are unmagnetized. For the magnetic field in the tail (<50 γ), the wave frequencies we consider are much greater than the ion gyrofrequency (≤ 0.8 Hz) and the wavenumbers k are much greater than ρ_i^{-1} where ρ_i is the ion gyroradius ~ 300 km for 300 eV H^+ ions.

The complex zeroes of the longitudinal dielectric function, ϵ (Ichimaru[12]), give the dispersive properties for waves that are purely electrostatic. From Dusenbery,[13] the solution of ϵ for unmagnetized ions and electrons is given by

$$\epsilon(k,\omega') = 1 + \sum_\alpha \frac{1}{k^2\lambda_{D\alpha}^2} \, W(x_\alpha) \tag{1}$$

where

$$x_\alpha = \frac{2^{1/2}}{kV_\alpha} \, (\omega' - k_\parallel v_\alpha)$$

$$V_\alpha = (2T_\alpha/m_\alpha)^{1/2}$$

$$\lambda_{D\alpha} = (T_\alpha/4\pi\bar{n}_\alpha e^2)^{1/2}$$

$$\omega' = \omega + i\gamma$$

and W is the W function defined by Ichimaru,[12] which is closely related to the plasma dispersion function of Fried and Conte.[14] Parallel ($\parallel$) and perpendicular ($\perp$) refer to directions with respect to a constant, external, plasma-sheet magnetic field B_0 aligned along the x axis. The subscript α denotes species. The parameter v_α is the streaming speed along B_0, V_α is the thermal speed, $\lambda_{D\alpha}$ is the Debye length, and $\bar{n}_\alpha$ is the number density.

An analytic expression for the beam acoustic wave frequency was derived by Dusenbery[13] valid for any wave normal angle, θ, given by

$$\omega = k \left[v_c \cos\theta \pm C_s \sqrt{\frac{\bar{n}_c/\bar{n}_e + 3T_c/T_e}{1 + k^2\lambda_{De}^2}} \right] \tag{2}$$

where the subscript c refers to the cold ion beams, $C_s = (T_e/m_c)^{1/2}$ is the ion acoustic speed, and $\pm$ refers to the two branches of the beam acoustic mode. Dusenbery and Lyons[11] showed that the two branches of the beam acoustic mode could be characterized by positive or negative wave energy (W_E) when looked at in the electron rest frame and found that for a single cold ion stream, only the slow mode ($W_E < 0$) was growing. Dusenbery[13] derived an analytic expression for the beam acoustic growth rate (γ) valid for $T_c \ll T_e$ and $k\lambda_{De} < 1$. Assuming a single cold ion stream ($\bar{n}_e = \bar{n}_c$), then γ for the slow mode is given by

$$\gamma = \frac{\sqrt{\pi}}{2^{3/2}} \, \sqrt{m_e/m_c} \, [1 + 3T_c/T_e]^{3/2}$$

$$\times \left(kv_c \cos\theta - C_s \sqrt{1 + 3T_c/T_e} \right) \tag{3}$$

As long as $|v_c| > C_s$, Eq. 3 implies that γ should be maximum at $\theta = 180°$ (for $v_c < 0$ or cold ions streaming anti-earthward) and equal to zero at $\theta = \theta_0$ where

$$\cos\theta_0 = C_s \sqrt{1 + 3T_c/T_e} \, / \, v_c \tag{4}$$

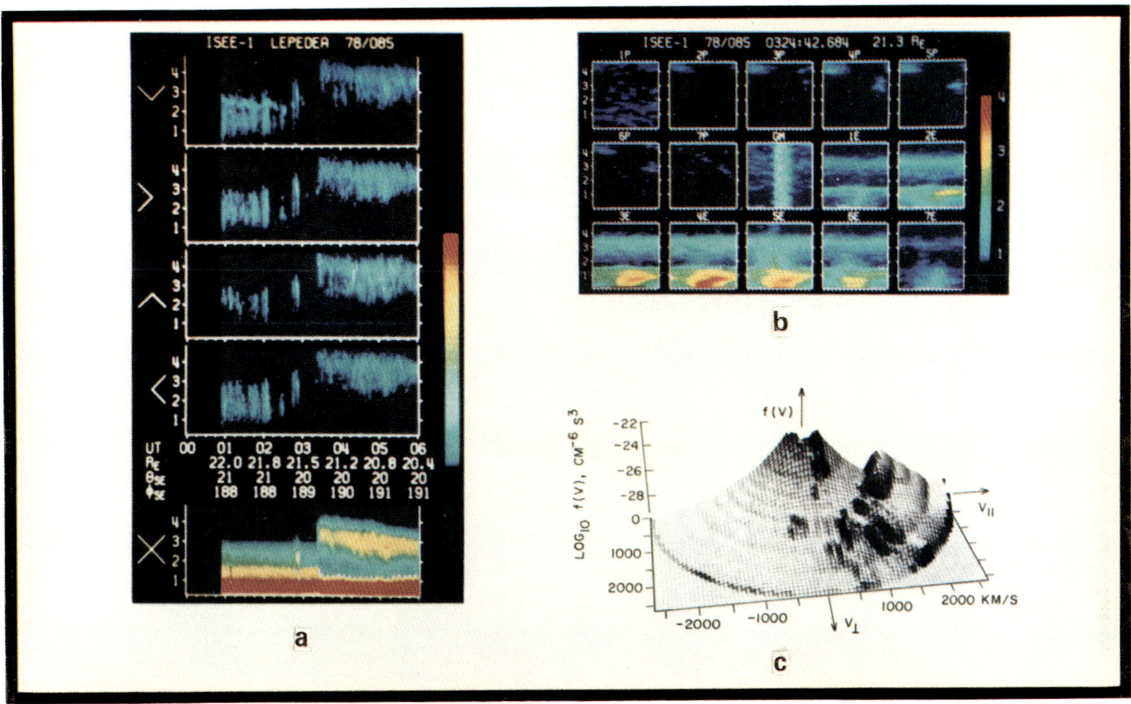

Plate IV-5—(top) ISEE 1 LEPEDEA particle data for day 85 of 1978. (a) Energy versus time spectrogram from 0000 to 0600 UT. (b) Energy versus ϕ near 0326 UT showing a single beam near $\phi = 0$. (c) Perspective plot of the three-dimensional ion distribution for 0324:42 to 0326:50 UT when ISEE 1 was in the plasma-sheet boundary layer. The smooth distribution centered at the origin of velocity space is artificial. (bottom) ISEE 1 SFR wave data for day 85 of 1978. Notice that the wave intensity increases near 0324 UT when ISEE 1 enters the boundary layer (courtesy of C. Huang, University of Iowa).

Figure 2 shows γ versus θ in panel (a) and γ/ω_{pe} versus $k\lambda_{De}$ in panel (b) for the slow mode with parameters typical of the plasma-sheet region for a single cold ion stream. Using these values, we find from Eq. 4 that $\theta_0 \simeq 111°$, which compares favorably with Fig. 2a for the θ at which $\gamma = 0$. Notice from Fig. 2b that the growth rate peaks at $k\lambda_{De} \lesssim 1$ with $(\gamma/\omega_{pe})_{max} \approx 4 \times 10^{-4}$. Such growth is due solely to free energy in the resonant electrons. Dusenbery and Lyons[11] have shown, however, that the beam acoustic growth rate can be significantly enhanced by resonant interactions with a warm streaming current sheet ion distribution. This is the topic we consider next.

While single cold ion streams predominate in the lobe region,[8] particle observations in the plasma-sheet boundary layer[5,15] indicate that one of the following three conditions usually prevails:

1. Parallel streaming—warm and cold ion beams drifting in the same direction.
2. Antiparallel streaming—warm and cold ion beams drifting in opposite directions.
3. Counterstreaming—warm and cold ion beams drifting opposite to a warm ion beam.

It is our intent in this paper to determine if any model dependence exists in the behavior of the calculated wave frequency and growth rate of the beam acoustic mode for the models considered above.

Omidi[10] noted that in order to calculate obliquely propagating beam acoustic modes under plasma-sheet conditions when the ions were assumed to be unmagnetized, the ion distributions had to be projected onto the phase velocity vector, parallel to the wave vector k. This projection implies that hot ions with speed $v = \omega/k$ are in Landau resonance with beam acoustic waves, satisfying Eq. 2. If there are significant numbers of such ions and if they have sufficient free energy, then a strong ion-ion resonant interaction occurs. Figure 3 illustrates the difference between parallel and antiparallel streaming

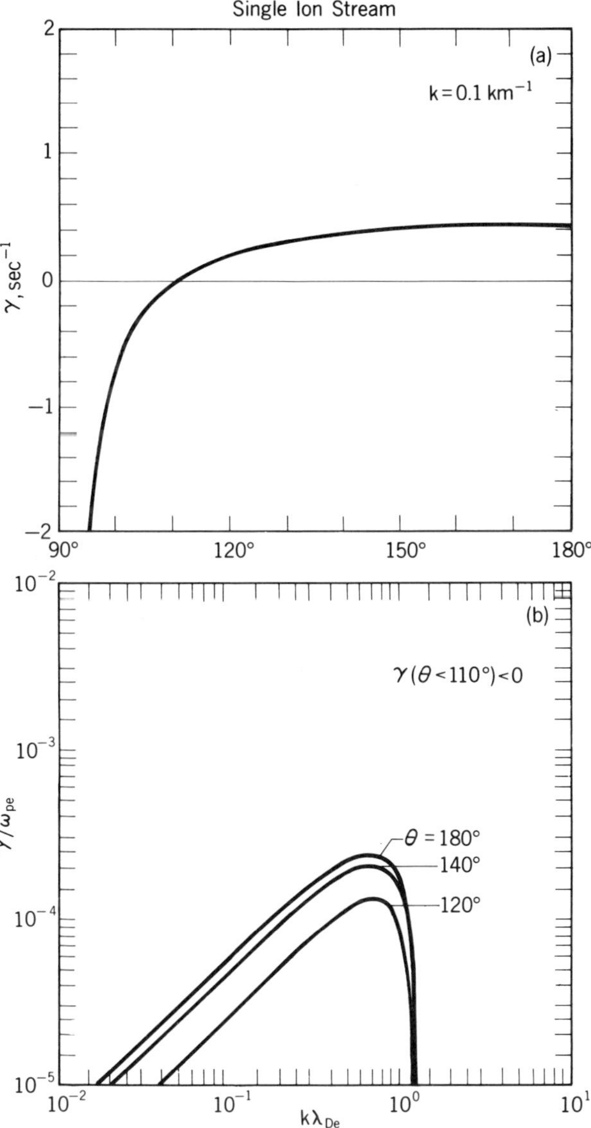

Figure 2—Growth rate, γ, versus wave normal angle (a) and versus normalized wave number (b) for a single ion stream. $T_e = 300$ eV, $T_i = 10$ eV, $\bar{n}_e = \bar{n}_i = 1$ cm^{-3}, and $v_i = -500$ km/s.

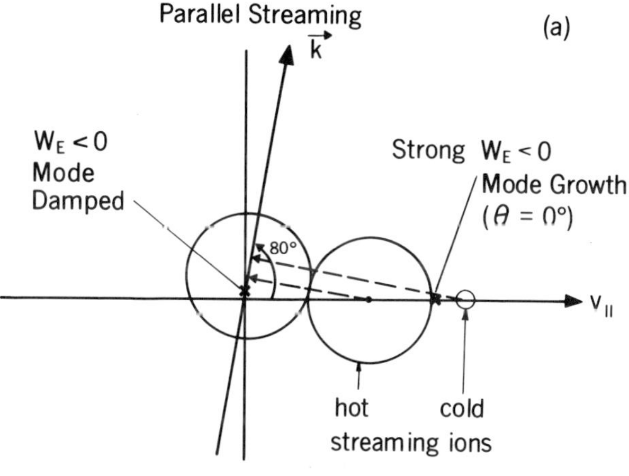

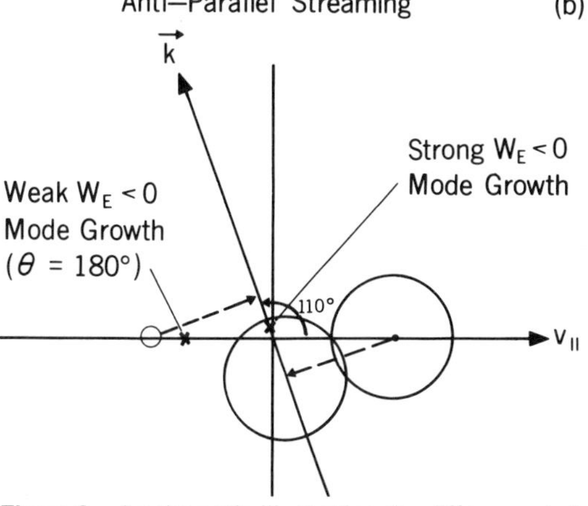

Figure 3—A schematic illustrating the difference between parallel and antiparallel streaming of cold and hot ion distributions in generating the beam acoustic mode for different θs. An obliquely propagating beam acoustic wave is in resonance with streaming ions that have an effective drift speed given by projecting the drift velocity onto the wave vector, k. Wave growth should therefore maximize at $\theta = 0°$ in (a) and $\theta \approx 110°$ in (b).

models for generating the slow-beam acoustic mode for different θs. The x identifies where Landau resonance occurs in velocity space for the θs considered.

Recall from Dusenbery and Lyons[11] that the slow mode grows (is damped) when resonant particles diffuse to higher (lower) energies in velocity space and such growth (damping) depends upon the velocity gradient in the distribution function at resonance. Thus, Fig. 3a predicts that the growth rate for the parallel streaming model should peak at $\theta = 0°$ and decrease with increasing θ. On the other hand, Fig. 3b predicts the opposite behavior for the growth rate where a strong ion-ion interaction now occurs for a θ much closer to 90° because the projected distributions are now much closer together in velocity space.

Figures 4 and 5 are plots of wave frequency and growth rate versus θ and versus wavenumber, respectively, for the parallel streaming model. Notice in Fig. 4 that the growth rate peaks at $\theta = 0°$ as expected from Fig. 3a and that the wave frequency is maximum there. From Fig. 5, it is seen that the frequency at peak growth ($k\lambda_{De} \lesssim 1$) occurs near 0.1 f_{pe}, where f_{pe} is the electron plasma frequency. The frequency and growth-rate behavior for the antiparallel streaming model is quite different. Figures 6 and 7 are for the same parameters used in the two previous figures except that the drift velocity of the cold ion beam has been reversed. From Fig. 6b, the growth rate is quite small for waves propagating along the magnetic field and peaks at a θ close to 90° as expected from Fig. 3b. Notice, however, that the wave frequency is small near where the growth rate is

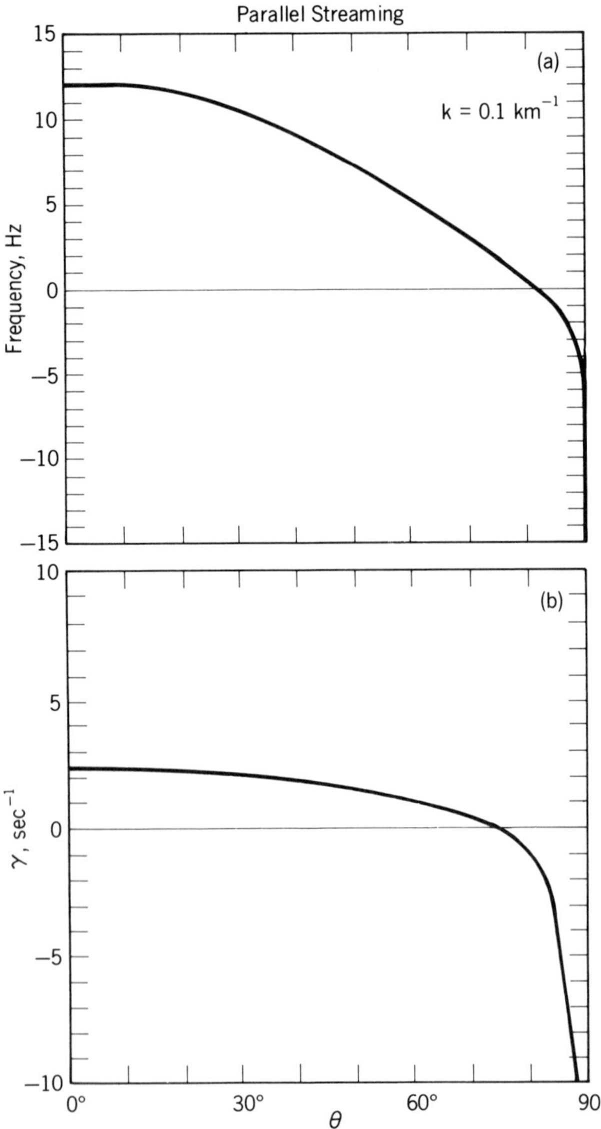

Figure 4—Frequency (a) and wave growth (b) dependence on wave normal angle, θ, for the parallel-streaming case described in Fig. 3. Wave growth maximizes at $\theta = 0°$. Plasma parameters are $T_e = T_h = 300$ eV, $T_c = 10$ eV; $\bar{n}_c = \bar{n}_h = \bar{n}_e/2$ and $\bar{n}_e = 1$ cm⁻³; $v_c = 900$ km/s and $v_h = 500$ km/s. The subscript h refers to the warm or hot plasma-sheet ion distributions.

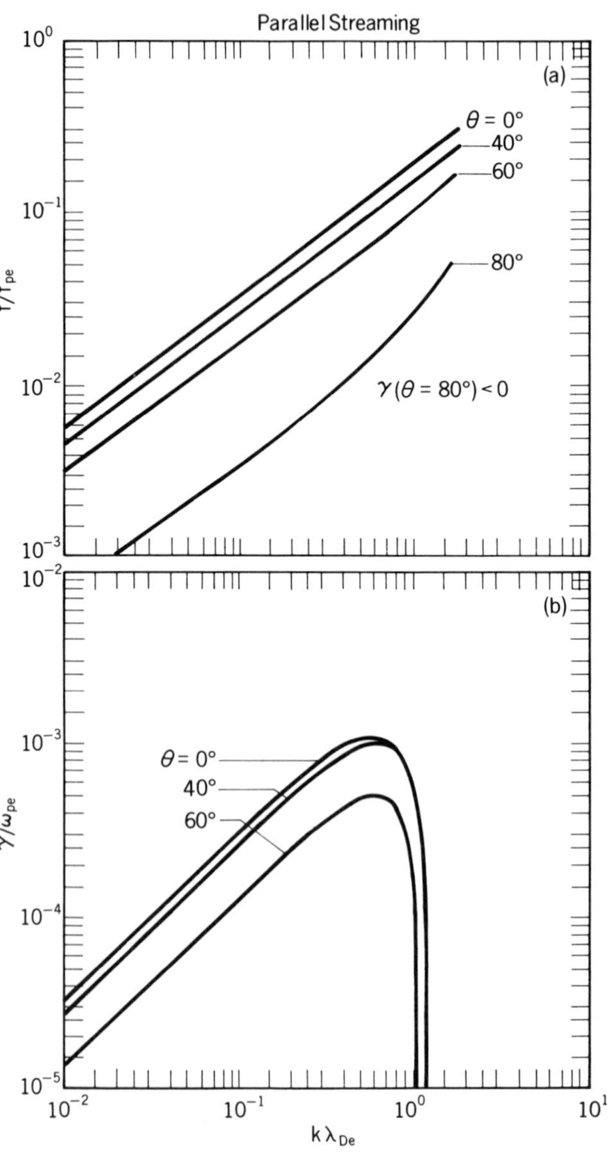

Figure 5—Dispersion plots of normalized wave frequency (a) and normalized wave growth (b) versus normalized wavenumber for various wave normal angles. The plasma parameters are defined in Fig. 4.

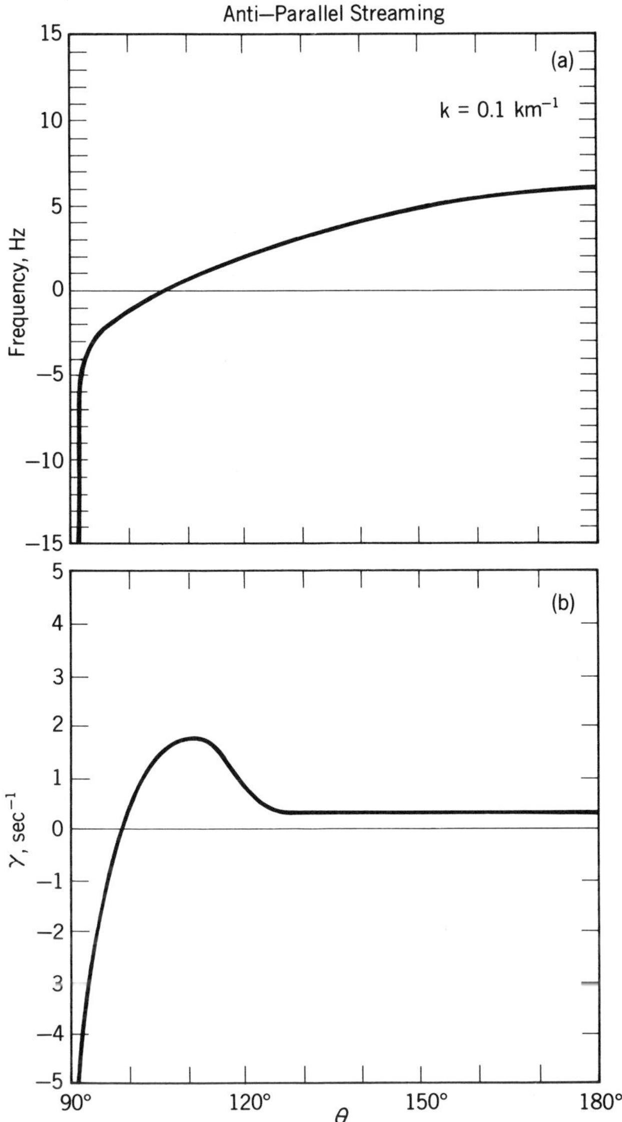

Figure 6—Similar to Fig. 4 except that $v_c = -900$ km/s. This is an example of the antiparallel-streaming case described in Fig. 3. Wave growth maximizes at $\theta \approx 110°$.

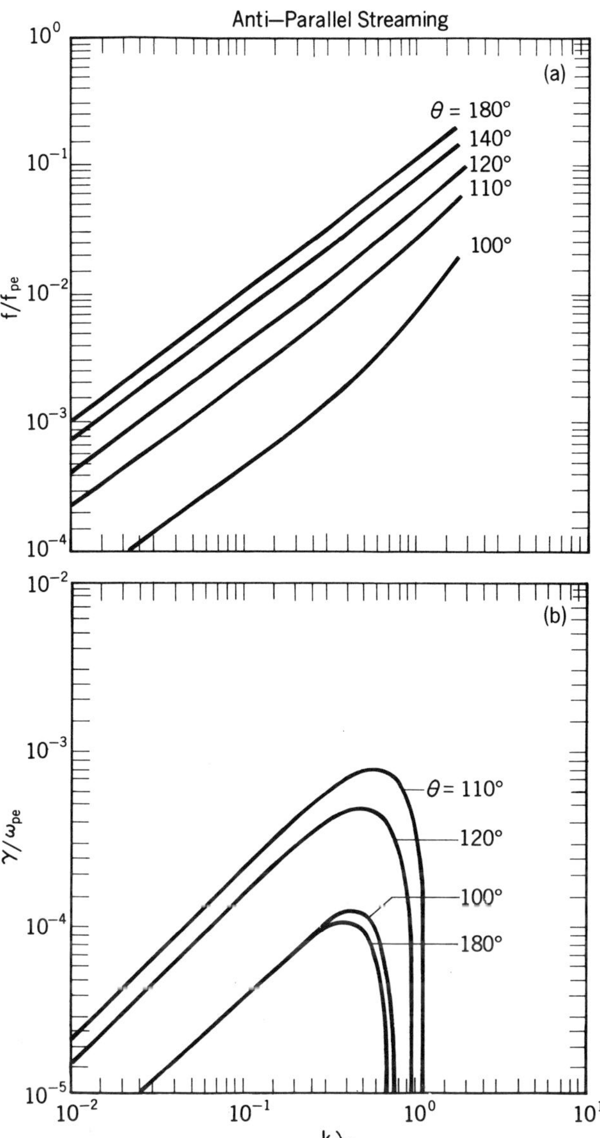

Figure 7—Normalized wave frequency (a) and normalized growth rate (b) versus normalized wavenumber for various wave normal angles for the antiparallel-streaming case considered in Fig. 6. Notice that wave growth increases to a maximum value at $\theta \approx 110°$ and then decreases as expected from Fig. 6b.

maximum. In fact, the frequency at peak growth (see Fig. 7) is now $\sim 0.01\, f_{pe}$, nearly a magnitude smaller than the corresponding frequency for the parallel streaming model. The frequency at peak growth for the antiparallel streaming model can be reduced further by reducing the drift velocity of the cold ions. Comparing Figs. 5 and 7 to Fig. 2, it is seen that the addition of a warm ion distribution significantly enhances the growth of the beam acoustic instability. The final model we will consider is the counterstreaming model.

Including counterstreaming warm ion distributions in addition to a cold ion stream with $|v_c| > |v_h|$ results in growth rate behavior that is a superposition of the parallel and antiparallel streaming models. This behavior is shown in Fig. 8a, a plot of growth rate versus θ for $k = 0.1$ km^{-1}. The growth rate is maximum at $\theta =$

180° due to a parallel streaming interaction but the growth rate does not monotonically decrease as θ approaches 90°. Instead, an increase in γ is evident near $\theta \approx 100°$–110° due to an antiparallel streaming interaction. The reason the growth rate peak at $\theta \sim 100°$ is not as large as in Fig. 6b is due to damping from the $v_h < 0$ ion distribution. In addition, the growth rate of the beam acoustic mode decreases with decreasing number density.[11] This accounts for the reduction in the overall growth rate for the counterstreaming model relative to the two previous models, because the same total ion density (1 cm^{-3}) must now be apportioned over three distributions instead of two.

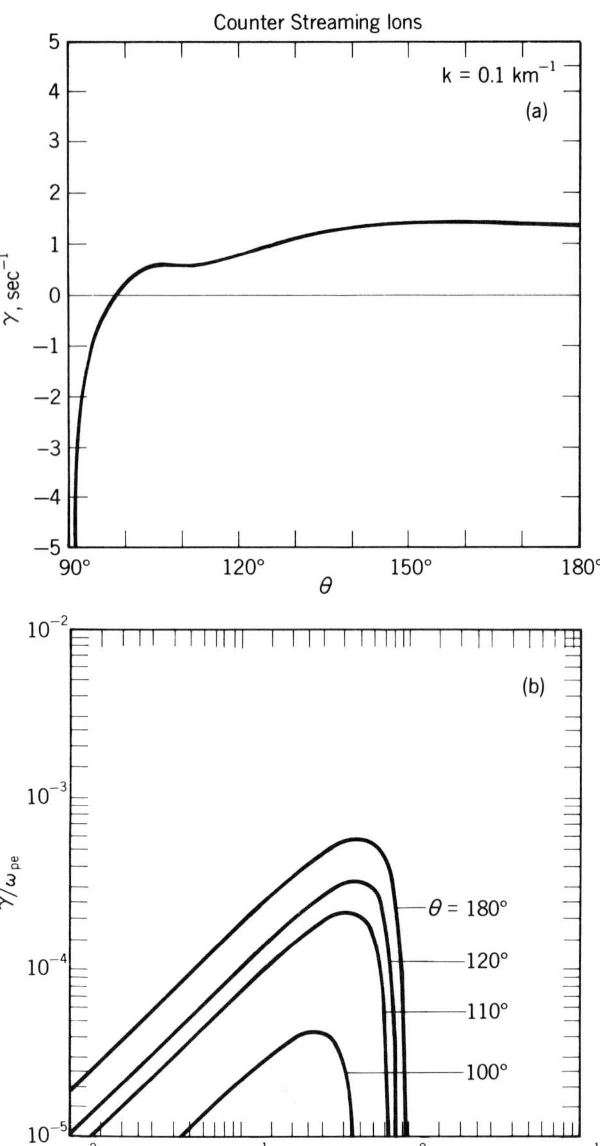

Figure 8—Growth rate versus θ for a given k and normalized growth rate versus normalized wavenumber for various θs for a counterstreaming ion model. Plasma parameters are similar to those in Fig. 4 except that $v_c = -900$ km/s and $v_h = \pm 500$ km/s; $\bar{n}_c = 0.4$ cm^{-3} and each hot ion distribution has $\bar{n}_h = 0.3$ cm^{-3}.

More details concerning the ion beam acoustic mode may be found in Ref. 13 where analytic expressions for the wave frequency and growth rate are derived; i.e., group velocities and amplification lengths for the beam acoustic modes are calculated.

CONCLUSIONS

We have investigated the generation of electrostatic noise in the plasma-sheet region of the geomagnetic tail. The wave intensity of the noise is observed to peak in the plasma-sheet boundary layer but weaker wave events are frequently observed in the tail lobes and central plasma sheet. Previous studies found that if only warm boundary-layer ion beams are present, then no broadband waves are excited. However, if cold ion beams (of ionospheric origin) are present, then a growing beam acoustic mode referred to as the slow mode can be excited. If both warm and cold ion beams are present, then both the slow and fast beam acoustic modes may be unstable with growth rates significantly greater than those for a single cold ion beam.

This study focused on the wave frequency and growth rate characteristics of the beam acoustic mode for various particle distribution models. The models were:

1. A single cold ion beam.
2. Parallel-streaming cold and warm ion beams.
3. Antiparallel-streaming cold and warm ion beams.
4. Counterstreaming warm ion beams and a cold ion beam.

Details for each model may be found in the second section. Single and counterstreaming warm ion beams are stable to the beam acoustic mode in the absence of a cold ion beam. In the rest frame of the plasma-sheet electrons, only streaming ions—no stationary ions at the origin in velocity space—were assumed to be present. To date, ISEE 1 particle observations show this assumption to be reasonable for the plasma-sheet boundary layer. The single cold ion stream model adequately accounts for the weak wave emissions observed in the lobe. It is argued here and in Ref. 11 that the peak wave intensities observed in the plasma-sheet boundary layer are a result of cold-warm ion beam interactions.

It has been argued elsewhere in the literature (e.g., Grabbe,[16] Omidi[10]) that the peak growth rate of the beam acoustic mode must occur near a wave normal angle close to 90°. Our study finds that this is not always the case, depending upon the combination of ion beams present. In particular, it was found that the growth rate did indeed peak near 90° for the antiparallel-streaming model but that the growth rate peaked at 0° for the parallel-streaming model. In addition to the differences in the behavior of the growth rate for these models, the wave frequency also behaves differently. The wave frequency at peak growth rate was much larger for the parallel-streaming model than for the antiparallel model. Thus, instances when earthward (warm) and antiearthward (cold) ion distributions are present imply that the wave intensity should be observed to peak at low frequencies (10–50 Hz) with wave normal angles close to 90°. On the other hand, whenever cold and warm parallel streaming ion distributions are present, the wave intensity is predicted to shift to higher frequencies (100–500 Hz) and peak for wave normal angles close to 0° (parallel propagating). Including counterstreaming warm ions in addition to a cold ion stream results in wave growth that is a superposition of that for the above two models.

In summary, we feel that ion beam instabilities offer a reasonable explanation for the generation of broadband electrostatic noise and adequately account for the observations of such noise not only in the plasma-sheet boundary layer, but in the lobes and central plasma-sheet regions as well. It must be emphasized that the distribution of wave energy in k-space may not always be predictable from a linear growth-rate analysis. Knowing the quasilinear and nonlinear evolution of the instability can only give what the final distribution will be. In this study, we implicitly assume that linear theory is adequate but recognize the need to investigate such nonlinear characteristics of the instability as particle diffusion and wave saturation. The wave frequency and growth rate behavior of the noise were found to be sensitive to the choice of particle distribution functions. Thus, the observed noise spectrum and simultaneous particle observations should provide a definitive test for the theory.

ACKNOWLEDGMENT—We would like to thank T. E. Eastman and C. Y. Huang for their many helpful comments and for the use of the ISEE 1 wave and particle data from the University of Iowa. This research was supported in part by the Atmospheric Research Section of the National Science Foundation grant ATM-8204647, by NASA Solar Terrestrial Theory Program contract NASW-3839 with The Aerospace Corporation, and by the Space Division of U.S. Air Force Systems Command under contract F0470185-C-086.

REFERENCES

[1] F. Scarf, L. Frank, K. Ackerson, and R. Lepping, "Plasma Wave Turbulence at Distant Crossings of the Plasma Sheet Boundaries and Neutral Sheet," *Geophys. Res. Lett.* **1**, 189 (1974).

[2] D. A. Gurnett, L. A. Frank, and R. P. Lepping, "Plasma Waves in the Distant Magnetotail," *J. Geophys. Res.* **81**, 6059 (1976).

[3] L. A. Frank, K. L. Ackerson, and R. P. Lepping, "On Hot Tenuous Plasma, Fireballs, and Boundary Layers in the Earth's Magnetotail," *J. Geophys. Res.* **81**, 5859 (1976).

[4] R. J. DeCoster and L. A. Frank, "Observations Pertaining to the Dynamics of the Plasma Sheet," *J. Geophys. Res.* **84**, 5099 (1979).

[5] D. J. Williams, "Energetic Ion Beams at the Edge of the Plasma Sheet: ISEE 1 Observations Plus a Simple Explanatory Model," *J. Geophys. Res.* **86**, 5507 (1981).

[6] A. T. Y. Lui, D. J. Williams, T. E. Eastman, and L. A. Frank, "Observations of Ion Streaming During Substorms," *J. Geophys. Res.* **88**, 7753 (1983).

[7] L. R. Lyons and T. W. Speiser, "Evidence for Current-Sheet Acceleration in the Geomagnetic Tail," *J. Geophys. Res.* **87**, 2276 (1982).

[8] R. D. Sharp, D. L. Carr, W. K. Peterson, and E. G. Shelley, "Ion Streams in the Magnetotail," *J. Geophys. Res.* **86**, 4639 (1981).

[9] C. L. Grabbe and T. E. Eastman, "Generation of Broadband Electrostatic Noise by Ion Beam Instabilities in the Magnetotail," *J. Geophys. Res.* **89**, 3865 (1984).

[10] N. Omidi, "Broadband Electrostatic Noise Produced by Ion Beams in the Earth's Magnetotail," *J. Geophys. Res.* **90**, 12,330 (1985).

[11] P. B. Dusenbery and L. R. Lyons, "The Generation of Electrostatic Noise in the Plasma Sheet Boundary Layer," *J. Geophys. Res.* **90**, 10,935 (1985).

[12] S. Ichimaru, *Basic Principles of Plasma Physics*, W. A. Benjamin, Reading, Mass. (1973).

[13] P. B. Dusenbery, "Generation of Broadband Noise in the Magnetotail by the Beam Acoustic Instability" (preprint, 1986).

[14] B. D. Fried and S. D. Conte, *The Plasma Dispersion Function*, Academic Press, Orlando, Fla. (1961).

[15] T. E. Eastman, L. A. Frank, W. K. Peterson, and W. Lennertsson, "The Plasma Sheet Boundary Layer," *J. Geophys. Res.* **89**, 1553 (1984).

[16] C. L. Grabbe, "New Results in the Generation of Broadband Electrostatic Waves in the Magnetotail," *Geophys. Res. Lett.* **12**, 483 (1985).

ON THE GENERATION OF BROADBAND ELECTROSTATIC NOISE

M. Ashour-Abdalla* and H. Okuda[†]

In order to understand the mechanisms generating broadband electrostatic noise (BEN) and the effects of BEN on particles as well as the nonlinear saturation level in the geomagnetic tail region, linear theory analysis and particle simulation studies were carried out. The plasma-sheet particle population was modeled by streaming ion beams as well as by warm ions and electrons. When the beam temperature, T_b, is much smaller than that of the warm plasma sheet electrons, T_e, both beam-ion acoustic and ion-ion two-stream instabilities grow. The frequency of the ion acoustic instability is given by $\omega = \pm(k(n_b/n_e)^{1/2} c_s - k_\parallel U) \approx \pm k_\parallel U \leq \omega_{pe}$ where $c_s = (T_e/m_i)^{1/2}$ is the ion sound speed and U is the ion beam speed relative to the warm plasma-sheet electrons. When the beam ion temperature is comparable to the temperature of the warm plasma-sheet particles, electrostatic ion cyclotron instabilities become unstable, giving rise to low-frequency noise at $\omega \leq n\Omega_i$ where Ω_i is the ion gyrofrequency. In the presence of a two-temperature electron plasma distribution and a warm ion beam, electron acoustic waves are found to be unstable. Simulation results confirm the presence of broadband electrostatic noise extending from $\omega = 0$ to $\omega = k \cdot U \leq \omega_{pe}$, with an amplitude of $E \sim$ mV/m, which is in reasonable agreement with satellite data.

INTRODUCTION

Intense broadband electrostatic noise (BEN) has been observed in the boundary layer region of the earth's magnetosphere.[1,2] Observations showed the noise to be present over a broad range of frequencies extending from about 10 Hz–1 kHz with intensities ranging from about 50 μV/m–5 mV/m. By examining particle distributions in IMP 8 data it was found that the broadband electrostatic noise was associated with anisotropic ions streaming either toward or away from the sun. The properties of ion flows in the magnetotail were studied extensively by DeCoster and Frank,[3] Frank et al.,[4] Frank and Ackerson,[5] and Eastman et al.[6] It was found that anisotropies in the ion velocity distribution were particularly prevalent near the boundary layer of the plasma sheet. Further studies of energetic ions in the plasma-sheet boundary layer established that anisotropies correspond to well-defined ion beams, and that these ion beams form a characteristic signature of the plasma-sheet boundary layer.[6,7]

There have been several theoretical research efforts made to further our understanding of BEN excitation. To this end Grabbe and Eastman,[7] Dusenbery and Lyons,[8] and Omidi[9] presented an extensive study of ion beam instabilities in the geomagnetic tail. These authors concentrated on excitation mechanisms of broadband electrostatic noise by linear ion acoustic instabilities, whereas Akimoto and Omidi[10] discussed growth of BEN due to ion-ion instabilities.

The above theoretical attempts were limited to linear theory, which was certainly the correct way to approach the problem initially. Yet BEN is one of the most intense forms of electrostatic noise observed in the earth's

magnetosphere: electrostatic waves have an average rms electric field amplitude of ~1 mV/m. With such amplitudes as these it is certainly conceivable that nonlinear effects will be important. These large-amplitude waves should certainly affect electron and ion distributions. In order to understand the effects of the waves on particle distributions as well as to obtain wave spectra directly comparable with satellite data, we initiated a simulation study. In this study we focused our interest on BEN excited in the geomagnetic tail. As such we assumed the free energy source to be ion beams, and we computed the excited wave spectrum as well as the particle distribution consistent with such a wave spectrum. In the next section we review the linear theory of electrostatic instabilities, and in the section following, we discuss our simulation results; our results are summarized in the concluding section.

LINEAR THEORY

We modeled the plasma-sheet particle population, using warm ions and electrons as well as counter-streaming ion beams. The model velocity distribution for the background plasma sheet ions and electrons can be written as

$$f_{ps}^{i,e}(v) = \frac{n_{ps}^{i,e}}{(2\pi T_{i,e}/m_{i,e})^{3/2}} \exp\left(-\frac{m_{i,e}v^2}{2T_{i,e}}\right) \quad (1)$$

while the velocity distribution for the beam ions is taken to be

$$f_b^i(v) = \frac{n_b^i}{(2\pi T_b/m_i)^{3/2}}$$

$$\left[\exp\left(-\frac{m_i(v-U)^2}{2T_b}\right)\right.$$

$$\left.+ \exp\left(-\frac{m_i(v+U)^2}{2T_b}\right)\right] \quad (2)$$

*Department of Physics and Institute of Geophysics and Planetary Physics, University of California, Los Angeles, California 90024.
[†]Princeton University, Plasma Physics Laboratory, Princeton, New Jersey 08544.

where U is the drift velocity along the magnetic field. The electrostatic dispersion relation for this case has been studied by a number of authors: Grabbe and Eastman,[7] Dusenbery and Lyons,[8] Omidi,[9] Stringer,[11] Forslund and Shonk,[12] Weibel,[13] Perkins,[14] Ashour-Abdalla and Okuda[15].

When unstable waves propagate obliquely with respect to the magnetic field B_0, the dispersion relation can be written as

$$1 = \frac{1}{2k^2\lambda_i^2} Z'\left(\frac{\omega}{\sqrt{2}kv_i}\right) + \frac{1}{2k^2\lambda_e^2}$$

$$Z'\left(\frac{\omega}{\sqrt{2}k_\parallel v_e}\right) - \frac{k_\perp^2}{k^2}\frac{\omega_{pe}^2}{\Omega_e^2}$$

$$+ \frac{1}{2k^2\lambda_b^2}\left[Z'\left(\frac{\omega-k_\parallel U}{\sqrt{2}kv_b}\right)\right.$$

$$\left. + Z'\left(\frac{\omega+k_\parallel U}{\sqrt{2}kv_b}\right)\right] \tag{3}$$

where the ions are assumed to be unmagnetized (an assumption valid for $\omega \gg \Omega_i$ and $k_\perp \rho_i \gg 1$), and the electrons are assumed to be strongly magnetized (a valid assumption for $\omega \ll \Omega_i$ and $k_\perp \rho_e \ll 1$).

When the drift speed of the ions, U, is less than the electron thermal speed of the plasma-sheet particles ($U < v_e$), it is well known that two kinds of electrostatic waves become unstable, for $T_e \gg T_b$ (Ref. 11). When $U > c_s = (T_e/m_i)^{1/2}$, beam-ion acoustic waves become unstable from electron inverse Landau damping. The frequency of the beam-ion acoustic waves in the beam-ion frame is given by $\omega = k(n_b/n_e)^{1/2}c_s$ and is Doppler-shifted to $\omega = \pm(k(n_b/n_e)^{1/2}c_s - k_\parallel U)$ in the electron frame where the $\pm$ sign corresponds to beam drifting at $\pm U$.

When the drift velocity of the ion beam is less than the ion acoustic speed ($U \leq c_s$) but more than the beam-ion thermal velocity ($U > v_b$), ion-ion two-stream instabilities can become unstable if $T_e \gg T_b$ (Refs. 11 and 12). The threshold of this instability is approximately $U > 1.3\, v_i$. Since the distribution of the ions is symmetric with respect to $v_\parallel = 0$, the ion-ion modes between two ion beams are purely growing in the frame of the plasma sheet while the modes between beam ions and the plasma-sheet ions are oscillatory in the same frame of reference.

If, on the other hand, the beam temperature is comparable to the electron temperature, we would not expect either the ion acoustic or the ion-ion two-stream instabilities to grow so long as the condition $U < v_e$ is satisfied. In this case, electrostatic ion cyclotron waves can become unstable and propagate nearly perpendicular to the magnetic field. The dispersion relation in this case is given by

$$1 = \frac{1}{2k^2\lambda_e^2} Z'\left(\frac{\omega}{\sqrt{2}k_\parallel v_e}\right) + \frac{1}{2k^2\lambda_i^2}$$

$$\left[1 + \sum_n \Gamma_n \frac{\omega}{\sqrt{2}k_\parallel v_i} Z\left(\frac{\omega-n\Omega_i}{\sqrt{2}k_\parallel v_i}\right)\right]$$

$$+ \frac{1}{2k^2\lambda_b^2}\left[1 + \sum_n \Gamma_n\right.$$

$$\left. \frac{\omega-k_\parallel U}{\sqrt{2}k_\parallel v_b} Z\left(\frac{\omega-k_\parallel U-n\Omega_i}{\sqrt{2}k_\parallel v_b}\right)\right]$$

$$+ \frac{1}{2k^2\lambda_b^2}\left[1 + \sum_n \Gamma_n\right.$$

$$\left. \frac{\omega+k_\parallel U}{\sqrt{2}k_\parallel v_b} Z\left(\frac{\omega-k_\parallel U-n\Omega_i}{\sqrt{2}k_\parallel v_b}\right)\right] \tag{4}$$

where $\Gamma_n = I_n(\mu)\exp(-\mu)$, $\mu = k_\perp^2 v_i^2/\Omega_i^2$ and I_n is the modified Bessel function of order n (Refs. 13 and 16).

Now if we assume the presence of cold and hot electrons and thermal ions as well as a single ion beam drifting along the magnetic field, the dispersion relation for parallel propagation may be written as

$$1 = \frac{1}{2k^2\lambda_i^2} Z'\left(\frac{\omega}{\sqrt{2}kv_i}\right) + \frac{1}{2k^2\lambda_b^2} Z'\left(\frac{\omega-kU}{\sqrt{2}kv_b}\right)$$

$$+ \frac{1}{2k^2\lambda_h^2} Z'\left(\frac{\omega}{\sqrt{2}k^2\lambda_h^2}\right)$$

$$+ \frac{1}{2k^2\lambda_c^2} Z'\left(\frac{\omega}{\sqrt{2}kv_c}\right) \tag{5}$$

where the subscripts, h, c, i, and b represent hot electrons, cold electrons, thermal ions, and beam ions, respectively.

Both the ions and electrons at the plasma sheet are assumed to be isotropic Maxwellian distributions with temperatures T_{eh} and T_i, respectively. The ion beam is assumed to be a drifting Maxwellian with temperature T_b and drift $+U$, whereas the cold electrons are modeled as isotropic Maxwellians with a temperature T_{ec}.

It is well known that electron acoustic waves propagate at a phase speed satisfying $v_c < \omega/k < v_h$, so that little damping exists even in the absence of ion beams.[17] In the presence of beam ions, an electron acoustic mode may be destabilized if the electron Landau damping is exceeded by inverse ion Landau damping from beam ions.

Assuming $v_c \ll \omega/k < v_h$, $\omega/kv_i \gg 1$, and $(\omega - kU)/kv_b \lesssim 1$, we find from Eq. 5

$$1 = \frac{\omega_{pi}^2}{\omega^2} + \frac{1}{2k^2\lambda_b^2}\left(-2 - 2i\sqrt{\pi}\,\frac{\omega - kU}{\sqrt{2}kv_b}\right)$$

$$+ \frac{1}{2k^2\lambda_h^2}\left(-2 - 2i\sqrt{\pi}\,\frac{\omega}{\sqrt{2}kv_h}\right)$$

$$+ \frac{1}{2k^2\lambda_c^2}\,\frac{2k^2v_c^2}{\omega^2}$$

$$= -\left(\frac{1}{k^2\lambda_b^2} + \frac{1}{k^2\lambda_h^2}\right) + \frac{\omega_{pc}^2}{\omega^2} - i\left(\frac{\pi}{2}\right)^{1/2}$$

$$\left(\frac{1}{k^2\lambda_b^2}\,\frac{\omega - kU}{kv_b} + \frac{1}{k^2\lambda_h^2}\,\frac{\omega}{kv_h}\right) \tag{6}$$

Solving Eq. 6 for $\omega = \omega_r + i\gamma$, we find

$$\omega_r = \frac{\omega_{pc}}{(1/k^2\lambda_b^2 + 1/k^2\lambda_h^2)^{1/2}}$$

$$= k\left(\frac{n_c}{n_h}\right)^{1/2} v_h \Big/ \left(1 + \frac{T_{eh}}{T_{bi}}\,\frac{n_b}{n_h}\right)^{1/2} \tag{7}$$

and

$$\frac{\gamma}{\omega_r} = \frac{1}{2}\,\frac{\omega_r^2}{\omega_{pc}^2}\left(\frac{\pi}{2}\right)^{1/2}\left[\frac{kU}{k^2\lambda_b^2 kv_b} - \frac{\omega_r}{k^2\lambda_b^2}\right.$$

$$\left.\left(\frac{1}{kv_b} + \frac{1}{k^2\lambda_h^2/\lambda_b^2}\,\frac{1}{kv_h}\right)\right]. \tag{8}$$

Figure 1 summarizes the four different velocity distributions that are considered: (a) when $T_e \sim T_i \gg T_b$ and $c_s < U < v_e$, beam-ion acoustic waves become unstable; (b) for $T_e \gtrsim T_i \gg T_b$ and $v_i < U < c_s$, we expect ion-ion two-stream instability growth; (c) for $T_e \sim T_i \sim T_b$ and $v_i < U < v_e$, electrostatic ion cyclotron harmonic waves can be excited propagating nearly perpendicular to B_0; (d) for $T_{eh} \simeq T_i \sim T_b$ where a cold electron component is present with $T_{oc} \ll T_{eh}$ and $v_{eh}(n_c/n_h)^{1/2} < U < v_{eh}$, electron acoustic instabilities grow.

RESULTS OF SIMULATION

In order to study the nonlinear behavior of BEN and its associated particle distribution, one- and two-dimensional elecrostatic particle simulation codes have been used. Plasma-sheet ions and electrons as well as streaming ion beams propagating along the magnetic field are included in the simulations. The codes utilize the standard finite-size particles in a spatial grid with the fast Fourier transform techniques.[18] Periodic boundary conditions are used throughout the simulations reported here that are appropriate to study a plasma free from boundary effects. Typical simulation parameters are the following. Mass ratio $m_i/m_e = 100, 400$; 64×64 grid

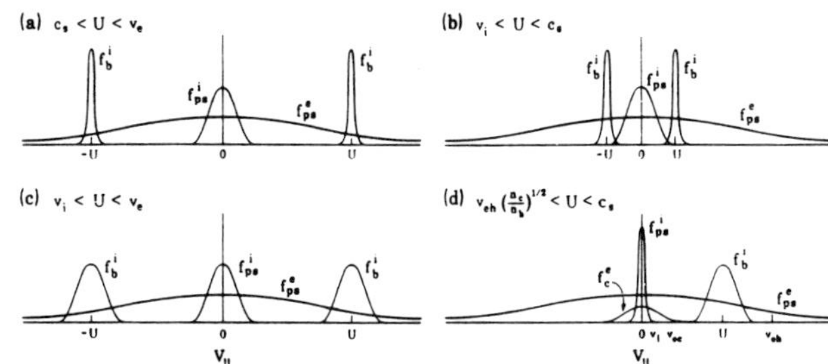

Figure 1—A schematic of the four different conditions considered in the text: (a) cold ionospheric counterstreaming ion beams (f^i_b) streaming through warm plasma-sheet ions (f^i_{ps}) and electrons (f^e_{ps}). Here $T_e \sim T_i \gg T_b$ and $c_s < U < v_e$ so that the dominant electrostatic instability is the ion acoustic waves on the beam ions; (b) $T_e > T_i \gg T_b$ and $v_i < U < c_s$ so that the ion-ion two-stream instability is the dominant mode; (c) $T_e \simeq T_i \simeq T_b$ and $v_i < U < v_e$ so that electrostatic ion cyclotron waves are unstable modes; (d) $T_{eh} = T_i = T_b$ and $v_{eh}(n_{ec}/n_{eh})^{1/2} < U < c_s$ in the presence of a cold electron component so that electron acoustic waves can be unstable.

(two-dimensional) and 256 grid (one-dimensional); grid size $\sim$ Debye length; and the number of particles per grid is $9 \sim 16$ (two-dimensional) and $40 \sim 80$ (one-dimensional).

The first example (case I) considered corresponds to a situation where counterstreaming, cold ionospheric ion beams interact with warm ions and electrons at the plasma sheet. Both ions and electrons at the plasma sheet are considered to have isotropic Maxwellian distributions with $T_e = T_i$ while ionospheric ion beams are drifting Maxwellians along a magnetic field with a temperature of T_b and a drift speed of $\pm U$. $T_i/T_b = 25$, $U/v_i = 20$, and $n_b^+ = n_b^- = n_i = n_e/3$ are used where $n_b^\pm$, n_i, and n_e are ionospheric beam density, plasma sheet ion density, and plasma sheet electron density, respectively. The above choice of simulation parameters corresponds to, for example, the case where cold ionospheric ions at 1 eV are accelerated to a drift speed of 10 keV by the double-layer electric field and collide with a 25 eV plasma at the plasma-sheet boundary layer. These parameters are not too different from the spacecraft observations except for the beam ion temperature; the temperature observed by ISEE 1 is considerably higher. Later, we will discuss possible causes of the heating of the ionospheric beam ions in the magnetotail.

Figure 2 shows (a) the electron and (b) the ion velocity distributions along the magnetic field at $t = 0$ (initial) and $t = 450\,\omega_{pe}^{-1}$ (final). Figure 2a clearly shows that the electron distribution develops plateaus on both sides of the positive and negative velocity space, a phenomenon caused by ion acoustic waves propagating in both directions. The electron kinetic energy increases by a factor of about three.[19] The heating of ions remains small as shown in Fig. 2b in which both beams suffered little slowing. Warm ions of the plasma sheet remain nonresonant since their thermal spread is much

Velocity Distributions

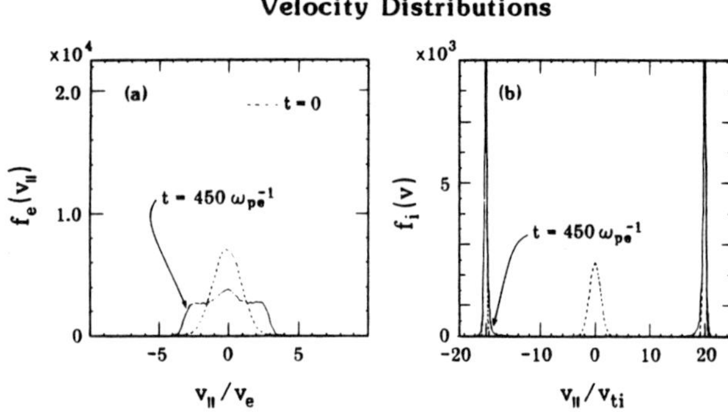

Figure 2—(a) Electron and (b) ion velocity distributions along the magnetic field at $t = 0$ and $t = 450\,\omega_{pe}^{-1}$. Note the formation of a plateau on the electrons while little change occurs in the ion distribution.

Time History of Fourier Mode   **Power Spectrum**

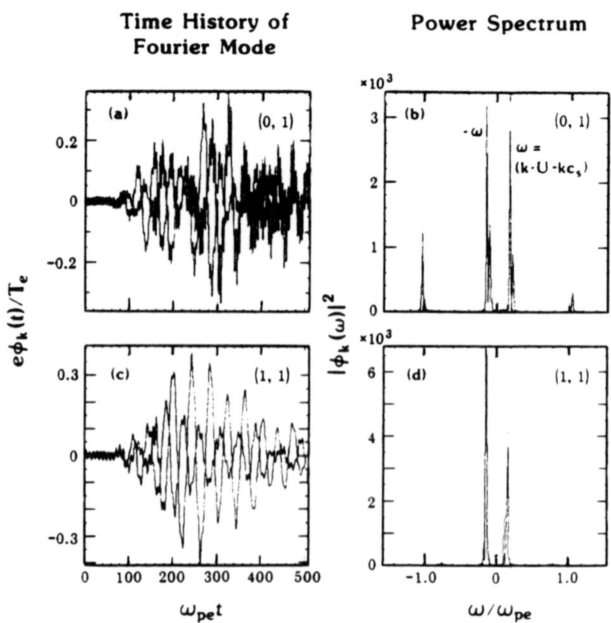

Figure 3—The Fourier mode (0,1): (a) the time history; (b) the power spectrum. Note that $e\phi_k/T_e$ reaches ~0.3 and slowly decays thereafter. Four peaks of the power spectrum correspond to the ion acoustic waves on the beam ions ($\omega \simeq \pm 0.18\,\omega_{pe}$) and the electron plasma waves ($\omega = \pm\omega_{pe}$). The Fourier mode (1,1): (c) the time history; (d) the power spectrum. Note that in panel d the plasma waves are no longer unstable, so that the spectrum peaks only near the ion acoustic frequency of the beam ions.

smaller than the cold beam drift speed, and no appreciable changes are observed. It is clear that the nonlinear saturation of the ion acoustic waves is caused by a flattening of the electron distribution.[20]

The frequency spectrum of the electrostatic potential for different Fourier modes is shown in Fig. 3. In the upper panel, we see the time history and the power spectrum for mode (0,1) which propagates in the direction of the magnetic field, whereas the lower panel shows the same plots for mode (1,1) where $k_x = k_y = 0.1\,\lambda_e^{-1}$. The time history of the electrostatic potential for a parallel propagating mode shows that it reaches a maximum at $e\phi_k(t)/T_e \sim 0.35$ before saturating at $e\phi_k/T_e \sim 0.2$. The corresponding electric field intensity for the plasma-sheet boundary layer is estimated to be $E = 7$ mV/m for $k = 0.1\,\lambda_e$, $\lambda_e = 0.003$ m^{-1} for $n_e = 1$ cm^{-3} and $T_e = 25$ eV. This value of the electric field intensity obtained from the simulation is of the same order of magnitude as that of the satellite observations. Note the presence of several frequencies, shown in the power spectrum in Fig. 3b. There are two large peaks near $\omega \simeq \pm 0.2\,\omega_{pe}$ and two smaller peaks near $\omega = \pm\omega_{pe}$. The large peaks (at lower frequencies) are ion acoustic waves of beam ions observed in a stationary frame; the frequencies are given by $\omega = \pm(k(n_b/n_e)^{1/2}c_s - k_\parallel U)$. Note that $c_s/U = 0.1/20 = 0.005 \ll 1$; thus, the frequencies are essentially given by the Doppler shift $\omega \simeq \pm k_\parallel U$. Since the initial drift speed is larger than the electron thermal speed, electron plasma waves are also unstable. Such waves correspond to the two peaks above the electron plasma frequency.

Obliquely propagating waves are also excited to large amplitudes as is shown in the lower panel of Fig. 3. For this mode, two low-frequency peaks corresponding to the ion acoustic waves remain at about the same frequency, $\omega = \pm(k_\parallel U - k(n_b/n_e)^{1/2}c_s) = \pm\mathbf{k}\cdot\mathbf{U}$ since $U \gg c_s$. Note that near the electron plasma waves the peaks with frequency $\omega = \pm(k_\parallel/k)\omega_{pe} \sim 0.7\,\omega_{pe}$ remain very small as they are not unstable for oblique propagation.

The power spectra for individual Fourier modes (Fig. 3) show the presence of coherent peaks near the Doppler-shifted ion acoustic frequency of beam ions. However, if we sum up all the wavelengths of the simulation modes we have a much broader distribution with many peaks (Fig. 4). This is reminiscent of the broadband electrostatic noise observed on ISEE 1 which extends all the way to ω_{pe}. The reason we have coherent peaks persisting at lower frequencies is that the simulation plasma is very small, only $64\,\lambda_e \times 64\,\lambda_e$, which limits to a small number of the available modes. If the simulation plasma is larger, for example $256\,\lambda_e \times 256\,\lambda_e$, the available number of unstable modes increases by a factor of 16, and we would expect the integrated frequency spectrum given here to be broadband noise without any significant peaks. It is clear that the plasma volume sampled by a satellite antenna is much larger than the size of any simulation plasmas.

As stated earlier, the thermal spread of the beam ions found in ISEE 1 data shows $T_b \leq 100$ eV, which is much larger than the ionospheric temperature used in the preceding simulation. One possibility is that instabilities other than the ion acoustic instability may be operative in the plasma-sheet boundary layer and may be responsible for the heating of the ionospheric plasma. We shall now study one such example, the ion-ion two-stream instability.

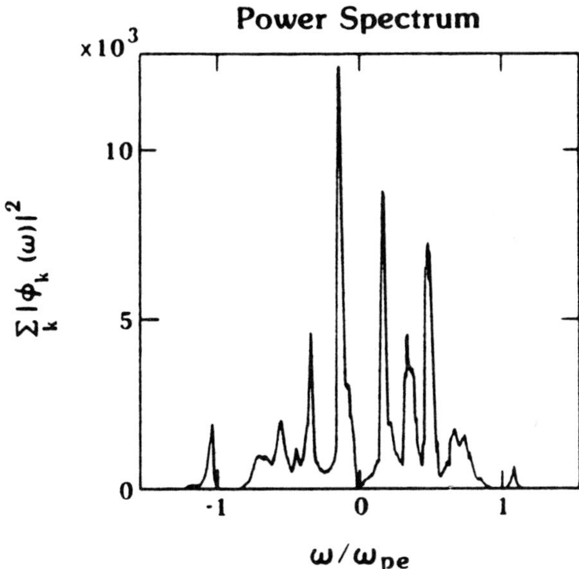

Power Spectrum

Figure 4—Summation of the power spectrum over k-space, which corresponds to a single point measurement. While a significant broadening of the spectrum is obtained, there are several coherent peaks that persist.

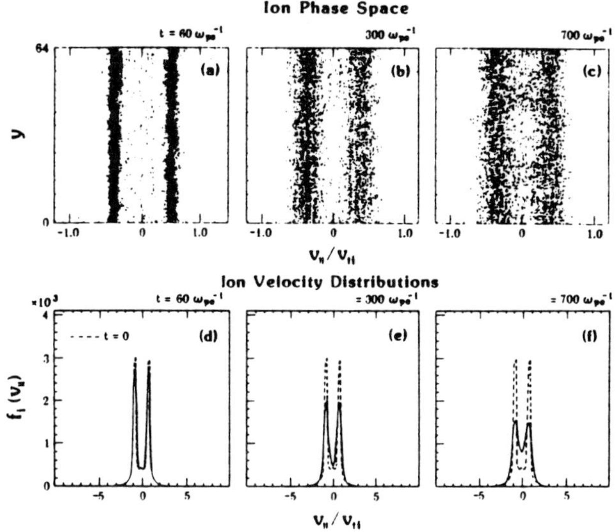

Figure 5—The panels (a), (b), and (c) show, respectively, the ion phase along the magnetic field at $t = 60 \ \omega_{pe}^{-1}$, 300 ω_{pe}^{-1}, and 700 ω_{pe}^{-1} for a two-dimensional ion-ion streaming instability. The panels (d), (e), and (f) show, respectively, the ion velocity distribution along the magnetic field at $t = 60 \ \omega_{pe}^{-1}$, 300 ω_{pe}^{-1}, and 700 ω_{pe}^{-1}. Note the rapid diffusion of beam and plasma sheet ions associated with the ion-ion two-stream instability.

We now turn to the second case (case II). In this case the beam velocity is less than the ion sound speed, $U < c_s$, so that the ion acoustic waves are stable. The ion-ion two-stream instability is expected to grow as long as $U > v_i$. We have taken $T_e = T_i = 100 \ T_b$, $U = 8 \ v_b = 0.8 \ c_s$. Figure 5 shows the time development of the ion phase space (upper panel) and the ion velocity distributions (lower panel) in the presence of the ion-ion instability at times $\omega_{pe}t = 60$, 300, and 700.

In Fig. 5a, cold counterstreaming ion beams are well-defined in a thermal background of plasma-sheet ions. As the ion-ion instability grows, cold beam ions start to diffuse, and finally spread out so that the relative drift speed approaches the beam-ion thermal speed. The velocity distribution for the ions clearly indicates a strong velocity space diffusion of cold ions. Well-defined peaks occur at $\omega_{pe}t = 60$ (Fig. 5d), spread out (Fig. 5e), and finally reach a steady state (Fig. 5f). At $t = 700 \ \omega_{pe}^{-1}$, the two cold ion beams have largely merged so that the relative drift speed is now comparable to their thermal speed. When this happens, the ion-ion instability must be stabilized. While we have not shown it here, the electron parallel velocity distribution does not change at all during the course of this run, confirming that the ion acoustic waves are stable. It also confirms that the electrons contribute to the ion-ion mode only as a neutralizing background consistent with linear theory.[11] The amplitude of the instability is found smaller in this case compared with the previous case, $e\phi_k/T_e \lesssim 0.05$. The electric field, however, is not small since $E = k\phi$ and k is large here, $k \gtrsim \lambda_e^{-1}$.

Now we turn to the case of warm beam ions (case III). The simulation parameters are $T_e = T_i = T_b$ with $m_i/m_e = 100$, $\Omega_e/\omega_{pe} = 2$, $U/v_i = 8$, and $B_0 =$

$(B_x,B_y,0)$ and $B_x/B_0 = 0.1$. In this case, neither the ion acoustic instability nor the ion-ion two-stream instability is expected to grow. Indeed simulation results indicate an absence of those instabilities in this case. Instead, ion cyclotron waves, driven by the relative streaming between the beam ions and the beam-ion plasma sheet, are expected to become unstable. Such electrostatic ion cyclotron waves propagate nearly perpendicular to the magnetic field at frequencies $\omega = 0$, Ω_i, $2\Omega_i$, ... (Ref. 16).

The power spectrum for mode (0,3) propagates nearly perpendicular to the magnetic field with $k_\parallel/k = 0.1$ and $k_\perp \rho_i = 1.5$ and is shown in Fig. 6. We observe peaks near $\omega = -\Omega_i$ and Ω_i that correspond to ion cyclotron waves. Smaller peaks are seen near $\omega = 0$, $2\Omega_i$, and $-2\Omega_i$. Comparing the power spectrum of the waves in this case, we find that the ion acoustic waves given by Figs. 3 and 4 have a wave energy that is a magnitude larger than the ion cyclotron waves. While the ion-beam-driven electrostatic ion cyclotron waves grow even in an isothermal plasma where neither ion acoustic nor ion-ion instabilities grow, ion cyclotron waves can propagate only nearly perpendicular with respect to magnetic fields. Ion acoustic and ion-ion instabilities can grow at nearly all the angles of propagation. In this sense, the ion cyclotron waves may contribute to BEN in the low frequency $\omega \lesssim \Omega_i$ and near perpendicular propagation $k_\parallel/k \ll 1$ in the frequency and wavenumber domain.

Now we turn to the last case (case IV). This case is still warm beam, $T_e = T_i = T_b$, but we now add a population of thermal electrons with temperatures $\sim$ eV

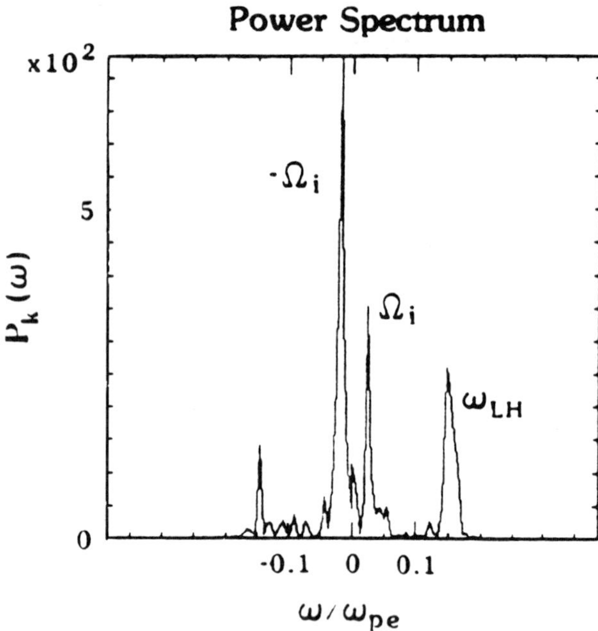

Figure 6—Power spectrum of Fourier mode (0,3), which propagates nearly perpendicular to the magnetic field. In addition to the peaks, $\omega = -\Omega_i, 0, \Omega_i$, which corresponds to the electrostatic ion cyclotron wave. We note the presence of two peaks near $\omega = \pm 0.15\,\omega_{pe}$, which corresponds to the thermal lower hybrid noise.

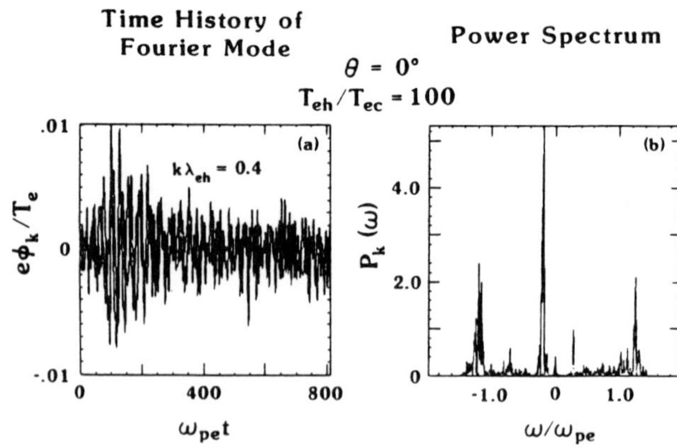

Figure 7—(a) Time history of the electrostatic potential of the second mode $k\lambda_{eh} = 0.4$ and (b) the corresponding power spectrum.

(Frank, personal communication). Thus we consider a four-component plasma, hot plasma-sheet electrons, cold electrons with density ratio $n_{ec}/n_{eh} = 0.2$, and temperature $T_{ec}/T_{eh} = 100$, as well as warm plasma sheet ions and an ion beam.

If we take an ion beam with a drift speed U that is less than the hot electron thermal speed but greater than the phase velocity of the electron acoustic wave, then the electron acoustic waves should grow.

One-dimensional electrostatic code was used in this example in which exact dynamics of both electrons and ions are followed in time $m_i/m_e = 400$ and $L = 256\lambda_e$ are used.

The time history of the elecrostatic potential and the frequency spectrum of the most unstable mode in the system, $k\lambda_{eh} = 0.4$, are shown in Fig. 7. The exponentially growing waves reach $e\phi_k/T_e \sim 0.01$ at $\omega_{pe}t = 100$ and saturate thereafter. The corresponding power spectrum clearly indicates the presence of the peak near $\omega \simeq 0.2\,\omega_{pe}$, which agrees with the electron acoustic frequency (Eq. 7). The peaks near $\pm\omega_{pe}$ are the electron plasma oscillations. Note that the saturation level of $e\phi/T_e$ is smaller in this case than in the previous instabilities shown in Fig. 3. However, the wavelengths of the unstable modes are shorter in this case resulting in a relatively large electric field. For $T_{eh} = 25$ eV and $k\lambda_{eh} = 0.4$, we find $E \lesssim 1$ mV/m for $e\phi/T_e \lesssim 0.01$. While the amplitude of the instability after saturation is as shown in Fig. 7a, the saturation level stays constant in the presence of constant flow of beam ions.[21]

The nonlinear effects of the electron acoustic waves are shown in Fig. 8. Here we plot the time history of (a) the ion drift velocity, (b) the ion kinetic energy, (c) the cold electron temperature, and (d) the hot electron temperature. All these quantities are normalized to their initial values. We note that (a) the ion drift velocity, (b) the ion kinetic energy, and (d) the hot electron temperature hardly change during the course of the simulation. However the cold electron temperature (c) increases rapidly in time for $\omega_{pe}t \lesssim 200$ followed by a slower rate of increase thereafter. Near the saturation of the instability, $T_{eh}/T_{ec} = 4$, which is marginally stable for the excitation of the electron acoustic waves for $n_{ec}/n_{eh} = 0.25$. Clearly the heating of cold electrons to temperatures comparable to that of the hot electrons will stabilize the instability, since the electron acoustic waves would be heavily damped in the absence of cold electrons.[22]

Since the electron acoustic instability can grow even in the presence of warm ion beams where ion acoustic and ion-ion instabilities cannot, it is a potential candidate for BEN, especially when no cold ionospheric ion beams exist. It is clear that the electron acoustic waves can propagate at various angles with respect to the magnetic field so that broadband noise can be generated as a result of this mode. This is in contrast to the ion cyclotron waves, which are rather coherent modes. The amplitude of the electric field associated with the electron acoustic wave is in reasonable agreement with space observations.

DISCUSSION

Assuming the presence of streaming ion beams in the plasma sheet, electrostatic instabilities with $\omega \lesssim \omega_{pe}$, $\rho_i^{-1} \lesssim k_\perp \ll \rho_e^{-1} \ll 1$ are studied analytically and by using numerical simulations. It is shown that when the beam temperature is much smaller than the plasma-sheet temperature, both ion acoustic instabilities and ion-ion two-stream instabilities grow to large amplitudes. The amplitude of ion acoustic waves is found to reach ~ 7 mV/m, which is larger than those observed by space-

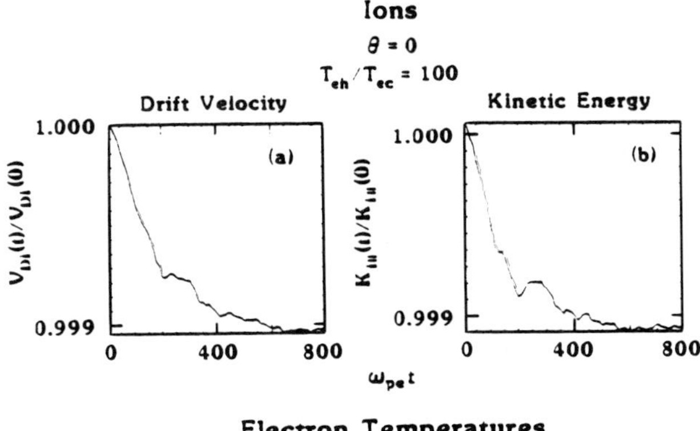

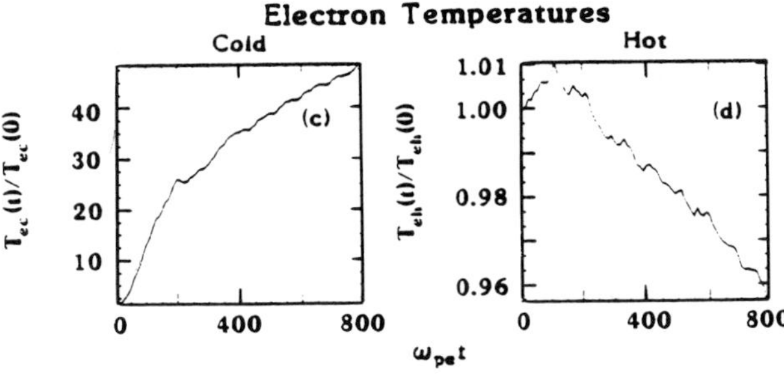

Figure 8—(a) Time history of the beam ion drift velocity, (b) ion kinetic energy, (c) temperature of the cold electrons, and (d) temperature of the hot electrons. All quantities are normalized to their initial values.

craft due to the choice of cold, energetic beam ions. Even though the power spectrum of each Fourier mode indicates the presence of sharp peaks, summation over the Fourier modes, which corresponds to a single point observation such as is done by a satellite, reveals a broader spectrum with frequencies extending from $\omega \simeq 0$ to the local electron plasma frequency. While the ion acoustic instability is responsible for heating the plasma-sheet electrons, the ion-ion instability heats the beam and plasma-sheet ions.

When the beam temperature is comparable to that of the plasma-sheet ions, electron acoustic and ion cyclotron waves grow. The electron acoustic wave heats the cold electrons while the ion cyclotron instability heats the beam ions and the plasma-sheet ions in a perpendicular direction. The saturation amplitude of the electron acoustic wave is found to be ≤ 1 mV/m (this value is close to amplitudes found from space observations).

In conclusion, ion acoustic and ion-ion instabilities may be the cause of the BEN when the cold ion beam is present while the electron acoustic modes may be responsible for the BEN when the beam ions are warm.

Those instabilities generate an electric field whose amplitude is consistent with space observations.

ACKNOWLEDGMENT—We would like to thank B. Buti and N. Omidi for useful comments and discussions. This work was supported by NASA STTP grant NAGW-78, Air Force contract F19628-85-K-0027, National Science Foundation grants, ATM85-12512 and ATM85-13215, and Department of Energy contract DE-AC02-76-CHO-3673.

REFERENCES

[1]F. Scarf, L. A. Frank, K. L. Ackerson, and R. P. Lepping, "Plasma Wave Turbulence at Distant Crossings of the Plasma Sheet Boundaries and Neutral Sheet," *Geophys. Res. Lett.* **1**, 189 (1974).

[2]D. A. Gurnett, L. A. Frank, and R. P. Lepping, "Plasma Waves in the Distant Magnetotail," *J. Geophys. Res.* **81**, 6059 (1976).

[3]R. J. DeCoster and L. A. Frank, "Observations Pertaining to the Dynamics of the Plasma Sheet," *J. Geophys. Res.* **84**, 5099 (1979).

[4]L. A. Frank, K. L. Ackerson, R. J. DeCoster, and B. G. Burek, "Three-Dimensional Plasma Measurements Within the Earth's Magnetosphere," *Space Sci. Rev.* **22**, 739 (1978).

[5]L. A. Frank and K. L. Ackerson, "Several Recent Findings Concerning Dynamics of the Earth's Magnetotail," *Space Sci. Rev.* **23**, 375 (1979).

[6]T. E. Eastman, L. A. Frank, and W. Peterson, "The Plasma Sheet Boundary Layer," *J. Geophys. Res.* **89**, 1553 (1984).

[7]C. L. Grabbe and T. E. Eastman, "Generation of Broadband Electrostatic Noise by Ion Beam Instabilities in the Magnetotail," *J. Geophys. Res.* **89**, 3865 (1984).

[8]P. B. Dusenbery and L. R. Lyons, "The Generation of Electrostatic Noise in the Plasma Sheet Boundary Layer," *J. Geophys. Res.* **11**, 10935 (1985).

[9]N. Omidi, "Broadband Electrostatic Noise Produced by Ion Beams in the Earth's Magnetotail," *J. Geophys. Res.* **12**, 12330 (1985).

[10]K. Akimoto and N. Omidi, "A Study of the Generation of Broadband Electrostatic Noise by an Ion Beam in the Magnetotail," *Geophys. Res. Lett.* (submitted, 1985).

[11]T. E. Stringer, "Electrostatic Instabilities in Current-Carrying and Counter-Streaming Plasmas," *Plasma Phys.* **6**, 267 (1964).

[12]D. W. Forslund and C. R. Shonk, "Numerical Simulation of Electrostatic Counterstreaming Instabilities in Ion Beams," *Phys. Rev. Lett.* **25**, 281 (1970).

[13]E. S. Weibel, "Ion Cyclotron Instability," *Phys. Fluids* **13**, 3003 (1970).

[14]F. W. Perkins, "Ion-Streaming Instabilities: Electromagnetic and Electrostatic," *Phys. Fluids* **19**, 1012 (1976).

[15]M. Ashour-Abdalla and H. Okuda, "Electron Acoustic Instabilities in the Geomagnetic Tail," *Geophys. Res. Lett.* (submitted, 1986).

[16]H. Okuda and K.-I. Nishikawa, "Ion-Beam-Driven Electrostatic Hydrogen Cyclotron Waves on Auroral Field Lines," *J. Geophys. Res.* **89**, 1023 (1984).

[17]K. Watanabe and T. Taniuti, "Electron Acoustic Mode in a Two Temperature Electron Plasma," *J. Phys. Soc. Japan* **43**, 1819 (1977).

[18]H. Okuda, "Introduction to Numerical Simulations of Plasmas and the Application to Electrostatic Waves in Space," in *Numerical Simulations of Space Plasmas*, H. Matsumoto and T. Sato, eds., Terra Scientific Pub., Tokyo, p. 1 (1985).

[19]M. Ashour-Abdalla and H. Okuda, "Theory and Simulation of Broadband Electrostatic Noise in the Magnetotail," *J. Geophys. Res.* (submitted, 1985).

[20]D. Biskamp and R. Chudura, "Computer Simulation of Anomalous DC Resistivity," *Phys. Rev. Lett.* **27**, 1553 (1971).

[21]H. Okuda and M. Ashour-Abdalla, "Acceleration of Hydrogen Ions in the Auroral Zone," *J. Geophys. Res.* **88**, 899 (1983).

[22]P. S. Gary and R. Tokar, "The Electron Acoustic Mode," *Phys. Fluids* **28**, 2439 (1985).

SIMULATION OF ELECTROSTATIC TURBULENCE IN THE PLASMA-SHEET BOUNDARY LAYER WITH ELECTRON CURRENTS AND ION BEAMS

K.-I. Nishikawa,* L. A. Frank,* T. E. Eastman,[†] and C. Y. Huang*

Plasma data from ISEE 1 show the presence of electron currents as well as energetic ion beams in the plasma-sheet boundary layer. Broadband electrostatic noise and low-frequency electromagnetic bursts are detected in the plasma-sheet boundary layer, especially in the presence of strong ion flows, currents, and sharp gradients in the fluxes of few-keV electrons and protons.

A linear theory analysis and particle simulations have been performed to investigate electrostatic turbulence driven by a cold electron beam and/or ion beams with the temperature anisotropy. The linear theory indicates that the growth rates of the ion acoustic waves driven by the cold ion beams become larger in the presence of the cold electrons. On the other hand, the effect of the temperature anisotropy ($T_\perp > T_\parallel$) of ion beams reduces the growth rates of ion acoustic instability. The simulation results show that the cold electron beam excites ion acoustic waves with the real frequency $\omega \approx k_\parallel c_s$ as predicted by the linear theory. The simulation results also show that the counterstreaming ion beams as well as the counterstreaming of the cold electron beam and the ion beam excite ion acoustic waves with the Doppler-shifted real frequency $\omega \approx \pm k_\parallel (c_s - V_{i2\parallel})$. The wave spectra of the electric fields at some points for simulations show turbulence generated by growing waves. The frequency of these spectra ranges from Ω_i to ω_{pe}, which is in qualitative agreement with the satellite data.

INTRODUCTION

Broadband electrostatic noise was discovered in the magnetotail by Scarf et al.[1] and studied in detail by Gurnett et al.[2] They found waves in the plasma-sheet boundary layer with frequencies from about 10 Hz up to several kHz. This electrostatic noise is also observed along these field lines at low altitudes over the discrete aurora.[3] These plasma waves, however, are relatively absent in the central plasma sheet, including the current sheet centered on the neutral sheet.[4] Gurnett et al.[2] concluded that the upper cutoff frequency might be near the electron gyrofrequency (about 1 kHz), although they noted that, in some cases, it appeared to go above that frequency. More recent studies have shown that it extends up to the electron plasma frequency near 10 kHz (Grabbe and Eastman[5]) and possibly higher (Parks et al.[6]).

Broadband electrostatic noise is detected in the plasma-sheet boundary layer especially in the presence of strong ion flows and field-aligned currents. Many characteristics of the ion flows in the magnetotail have been investigated by DeCoster and Frank,[7] Frank and Ackerson,[8] Williams,[9] and Eastman et al.[10] Field-aligned currents are also observed in the boundary layer of the plasma sheet.[11] Detailed descriptions of plasmas in the plasma sheet and the boundary regions are presented by Eastman et al.[10] and Frank.[12]

Ashour-Abdalla and Thorne[13] proposed an electrostatic instability at the $n + \frac{1}{2}$ ion cyclotron harmonics for the auroral broadband electrostatic noise. The instability is quenched for electron temperatures $T_e <$

10 eV. The lower-hybrid drift instability in the magnetotail was proposed by Huba et al.[14] This instability produces a spectrum with frequencies ranging up to the lower-hybrid frequency. A theoretical study was made of ion-beam instabilities in the plasma-sheet boundary layer with very cold beams by Grabbe and Eastman.[5] Their linear theory predicts that a spectrum of growing waves that propagate nearly parallel to the magnetic field can be driven for frequencies from 0.001 ω_{pe} up to ω_{pe}, the electron plasma frequency, with a spectral peak typically near 0.01 ω_{pe} or lower.

Other research has been done on the theory of the generation of broadband electrostatic noise using ion-streaming instabilities similar to that of Grabbe and Eastman[5] (see Refs. 15–21). Recently it was shown that growth rates rise rapidly at propagation angles near 70° and this mode is identified as the ion-ion instability.[19,20] Also, Grabbe[15-17] has shown that a second component of cold electrons can generate higher frequencies in the spectrum close to the plasma frequency. Dusenbery and Lyons[18] and Ashour-Abdalla and Okuda[20] have considered the presence of more than one ion beam in their studies. They have shown that in the presence of both cold and warm ion beams the growth rates are enhanced.[18] Also, Ashour-Abdalla and Okuda[22] also performed simulations of ion acoustic instabilities and ion-ion two-stream instabilities, assuming the presence of counterstreaming ion beams in the plasma sheet and of electrostatic instabilities with $\omega \lesssim \omega_{pe}$, Ω_e, and $k_\perp \rho_e \ll 1$. Ashour-Abdalla and Okuda[22] have shown that electron acoustic waves in a two-temperature electron plasma can be driven unstable in the presence of beam ions ($T_{ib} = T_{eh}$).

In this paper, we investigate electrostatic turbulence driven by field-aligned currents[11] and/or ion beams,[10,23] both theoretically and by particle simulations. However, we restrict ourselves to investigating some general fea-

*Department of Physics and Astronomy, The University of Iowa, Iowa City, Iowa 52242.
[†]Space Plasma Physics Branch, NASA Headquarters, Washington, D.C. 20546.

tures of available observations such as cold electron beams and the temperature anisotropy of beam ions. With realistic plasma parameters based on observations in the plasma-sheet boundary layer ($\Omega_e/\omega_{pe} \approx 1/3$–$1/6$ at $R \approx 20$ R$_e$), we use the full kinetic equations for both electrons and ions. In the next section, we present a theoretical analysis of instabilities driven by a cold electron beam and/or ion beams with a temperature anisotropy, along with the simulation model and results and in the section following, we present concluding remarks and discussion.

SIMULATION MODEL AND RESULTS

We simulate plasmas in the plasma-sheet boundary layer at ~ 12–20 R$_e$ where comprehensive measurements are available. We consider both magnetized electrons and ions with a cold electron beam and ion beams. The observational data provide the temperature anisotropy for the warm and cold electrons and for the ion beams (see, e.g., Ref. 12). The general dispersion relation in the electrostatic limit for the plasmas is given by Miyamoto.[24]

$$\epsilon_{\perp\perp} k_\perp^2 + 2\epsilon_{\perp\parallel} k_\perp k_\parallel + \epsilon_{\parallel\parallel} k_\parallel^2 = 0$$

where

$$\epsilon_{\perp\perp} = 1 + \sum_j \frac{\omega_{pj}^2}{\omega^2} \sum_n [\zeta_0^j Z(\zeta_n^j)$$
$$- (1 - 1/\lambda_T^j)(1 + \zeta_n^j Z(\zeta_n^j))] \frac{n^2 I_n^j}{b_j} e^{-b_j} ,$$

$$\epsilon_{\perp\parallel} = - \sum_j \frac{\omega_{pj}^2}{\omega^2} \sum_n [\zeta_0^j Z(\zeta_n^j) - (1 - 1/\lambda_T^j)(1$$
$$+ \zeta_n^j Z(\zeta_n^j))](2\lambda_T^j)^{1/2} \eta_n^j \frac{n I_n^j}{\alpha_j} e^{-b_j} ,$$

$$\epsilon_{\parallel\parallel} = 1 + \sum_j \frac{\omega_{pj}^2}{\omega^2} \sum_n [\zeta_0^j Z(\zeta_n^j)$$
$$- (1 - 1/\lambda_T^j)(1 + \zeta_n^j Z(\zeta_n^j))]$$
$$(2\lambda_T^j)(\eta_n^j)^2 I_n^j e^{-b_j} + \sum_j \frac{\omega_{pj}^2}{\omega^2} 2\lambda_T^j (\eta_0^j)^2 ,$$

$$\zeta_n^j = \frac{\omega - k_\parallel V_j + n\Omega_j}{k_\parallel (2T_{\parallel j}/m_j)^{1/2}} ,$$

$$\eta_n^j = \frac{\omega + n\Omega_j}{k_\parallel (2T_{\parallel j}/m_j)^{1/2}} ,$$

$$\lambda_T^j = \frac{T_{j\parallel}}{T_{j\perp}} ,$$

$$\alpha_j = k_\perp V_{Tj\perp} / \sqrt{2}\ \Omega_j ,$$

$$b_j = \alpha_j^2 ,$$

$$V_{Tj\perp} = (2T_{j\perp}/m_j)^{1/2} ,$$

$$\Omega_j = -q_j B_0/m_j .$$

Here, $Z(\zeta_n^j)$ is the plasma dispersion function, and ω_{pj} and I_n^j are j-th species plasma frequency and n-th modified Bessel function, respectively.

The dispersion relation for a cold electron beam is investigated here for a simulation study. We consider a plasma consisting of warm electrons, cold electron beams, and ions immersed in an external homogeneous magnetic field B_0. The external magnetic field B_0 is taken in the x-z plane ($B_{0x} = 0.985\ B_0$, $B_{0z} = 0.172\ B_0$). We use the two-dimensional electrostatic code for which the simulation plane lies in the x-y plane. The parameters used for the simulation are the following: $\Omega_e/\omega_{pe} = 1/3$, $m_i/m_e = 100$, $T_{eH} = T_i$, $T_{eC} = 0.1\ T_{eH}$, $\bar{n}_{eC} = \bar{n}_{eH} = 0.5\bar{n}_i$, $V_{eC} = 2.5\ v_{eCt} = 0.79\ v_{eHt}$ ($v_{eCt} = (2T_{eC}/m_e)^{1/2}$ and $v_{eHt} = (2T_{eH}/m_e)^{1/2}$). The real frequency and growth rate of the ion acoustic instability driven by the cold electron beams are shown in Fig. 1a as a function of $k_\parallel \lambda_{e\parallel}$ for the plasma described above. This mode propagates along the x-direction with different mode numbers. Therefore, the value $k_\perp \lambda_{e\parallel}$ is proportional to $0.175\ k_\parallel \lambda_{e\parallel}$ in this case.

On the other hand, ion beams are also a candidate for generation of broadband electrostatic noise in the plasma-sheet boundary layer. Using observations to gain a realistic example, we simulate a plasma consisting of hot and cold electrons ($\bar{n}_{eC} = \bar{n}_{eH}$, $T_{eC} = 0.1\ T_{eH}$, $v_{eHt} = (2T_{eH}/m_e)^{1/2}$), and counterstreaming ion beams with an anisotropic temperature ($\bar{n}_i^+ = \bar{n}_i^-$, $V_{i\parallel}^+ = -V_{i\parallel}^- = 0.7\ v_{eHt}$, $T_{i\perp} = T_{eH}$, $T_{i\parallel} = 0.1\ T_{i\perp}$). The parameters $\Omega_e/\omega_{pe} = 1/3$, $M_i/m_e = 100$, and $B_{0x}/B_0 = 0.985$ are kept the same for all the simulations reported here.

The real frequency and growth rates of the ion acoustic instabilities for three different cases are shown in Fig. 1b as a function of $k_\parallel \lambda_{e\parallel}$ for the comparison. The real frequency marked by B is shown only for the plasma described above because of very small differences among other cases. The growth rate marked by B is the case for the plasma described above. The growth rate is larger for cold ion beams with the isotropic temperature ($T_{i\parallel} = T_{i\perp}$) as shown by the line A. On the other hand, in the case for no cold electrons, the growth rate is smaller as shown by the line C. This is consistent with the results presented by Grabbe.[15-17]

We have also investigated the case for counterstreaming of the cold electron beams ($\bar{n}_{eC} = \bar{n}_{eH}$, $T_{eC} = 0.1\ T_{eH}$, $V_{eC} = -0.25\ V_{eHt}$) and the single ion beam with

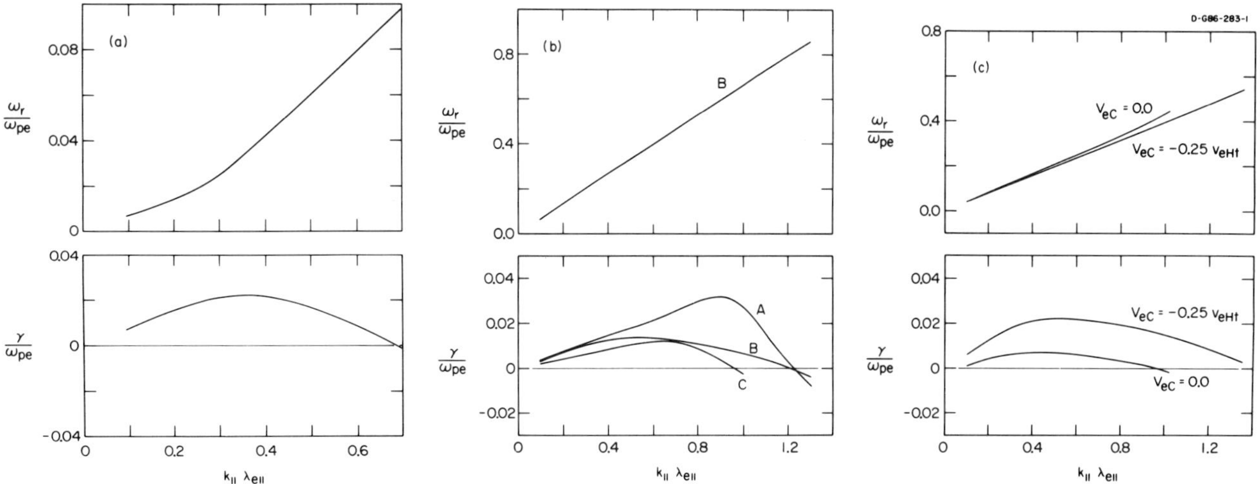

Figure 1—The real frequency (upper) and the growth rate (lower) are shown as a function of $k_\parallel \lambda_{e\parallel}$ ($\Omega_e/\omega_{pe} = \frac{1}{3}$, $m_i/m_e = 100$).
(a) For the case of the cold electron beam ($\bar{n}_{eC} = \bar{n}_{eH} = 0.5\bar{n}_i$, $T_{eC}/T_{eH} = 0.1$, $T_i = T_{eH}$, $V_{eC} = 0.79\ v_{eHt}$).
(b) For the case for the counterstreaming ion beams ($\bar{n}_i^- = \bar{n}_i^+$, $V_{i\parallel}^+ = -V_{i\parallel}^- = 0.7\ v_{eHt}$).
    A $\bar{n}_{eC} = \bar{n}_{eH}$, $T_{eC} = 0.1\ T_{eH}$, $T_{i\parallel} = T_{i\perp} = 0.1\ T_{eH}$
    B $\bar{n}_{eC} = \bar{n}_{eH}$, $T_{eC} = 0.1\ T_{eH}$, $T_{i\perp} = T_{eH}$, $T_{i\parallel} = 0.1\ T_{i\perp}$
    C $\bar{n}_{eC} = 0$, $T_{i\perp} = T_{eH}$, $T_{i\parallel} = 0.1\ T_{i\perp}$
(c) For the case of the cold electron beam ($\bar{n}_{eC} = \bar{n}_{eH}$, $T_{eC} = 0.1\ T_{eH}$, $V_{eC} = -0.25\ v_{eHt}$) and the anisotropic ion beam ($V_{i\parallel} = 0.45\ v_{eHt}$, $T_{i\perp} = T_{eH}$, $T_{i\parallel} = 0.1\ T_{i\perp}$). The growth rate for the counterstreaming case is larger than that for the no-cold electron beam.

the anisotropic temperature ($V_i = 0.45\ v_{eHt}$, $T_{i\perp} = T_{eH}$, $T_{i\parallel} = 0.1\ T_{i\perp}$). The growth rate with the cold electron beam ($V_{eC} = -0.25\ V_{eHt}$) is larger than that without the beam ($V_{eC} = 0$) as shown in Fig. 1c.

We have performed the particle simulation for several cases. A two-dimensional electrostatic code is used that retains the full dynamics of both electrons and ions in three-dimensional velocity space. At first we investigate the ion acoustic instability driven by the cold electron beams as shown in Fig. 1a. We use a system length $L_x = L_y = 64\Delta$ where Δ is the grid space, which is equal to the electron Debye length λ_e and $\bar{n}_e \lambda_e^2 = 9$. In this case, the uniform recycling model is used to model electron flow along the field lines.[25,26] At each time step ($\omega_{pe}\Delta t = 0.06$), 0.6% of the cold electrons are randomly replaced by cold electrons with distribution of their initial temperatures and drift velocities. Because of the uniform recycling model, decreases of the cold electron drift are small and increases of the cold electron parallel temperature are also small. The simulation is also performed without the recycling of cold electrons. As expected, even though the instability grows, it saturates at small amplitude due to decreases of the cold electron drift and increases of the cold electron parallel temperature. Both simulations show that the cold electron beam excites ion acoustic waves with real frequency $\omega \approx k_\parallel c_s$ as predicted by the linear theory. In these cases, the frequencies of excited waves range only up to 0.08 ω_{pe}. Therefore, the cold electron beam is not likely responsible for broadband electrostatic noise by itself.

Then we have performed the simulation for the plasma consisting of the counterstreaming ion beams with the temperature anisotropy ($\bar{n}_i^+ = \bar{n}_i^-$, $V_i^+ = -V_i^- =$

0.7 v_{eHt}, $T_{i\perp} = T_{eH}$, $T_{i\parallel} = 0.1\ T_{i\perp}$) in the presence of the cold electrons ($\bar{n}_{eC} = \bar{n}_{eH}$, $T_{eC} = 0.1\ T_{eH}$, $v_{eHt} = (2T_{eH}/m_e)^{1/2}$). In this case, to reduce the thermal noise, we use the code with $L_x = L_y = 32\Delta$ and $\bar{n}_e\lambda_e^2 = 36$, by which the maximum available memory is used at the NCAR computer center. These parameters are kept the same for the remainder of the simulations reported here.

Growing ion acoustic waves for two modes are found, for example, in the time evolution of the perturbations of electrostatic potential as shown in Fig. 2a and c. The real frequencies of these modes are slightly smaller than those predicted by the theory as shown in Fig. 1b. Ashour-Abdalla and Okuda[20] also found this instability in their simulations with isotropic ion beams ($T_{i\parallel} = T_{i\perp}$). It is found that the observed frequencies increase with parallel wavenumber. Some obliquely propagating waves are also excited.

The excited instability is saturated due to the plateau formations on both sides of the positive and negative velocity space in the cold electron parallel distribution. At the same time the cold electrons are heated in the parallel and perpendicular velocity distribution. The heating of counterstreaming ions remains small and the slowing down of beams is also small.

The plasma with the counterstreaming of cold electron beams ($\bar{n}_{eC} = \bar{n}_{eH}$, $T_{eC} = 0.1\ T_{eH}$, $V_{eC} = -0.25\ v_{eHt} = -0.79\ V_{eCt}$, $V_{eCt} = (2T_{eC}/m_e)^{1/2}$) and the single ion beam with a temperature anisotropy ($T_{i\perp} = T_{eH}$, $T_{i\parallel} = 0.1\ T_{i\perp}$, $V_{i\parallel} = 0.45\ v_{eHt} = 4.5\ V_{i\perp t}$, $V_{i\perp t} = (2T_{i\perp}/m_i)^{1/2}$) is investigated also by the particle simulation. In this case, we performed the two simulations with and without the recycling of the cold electrons

315

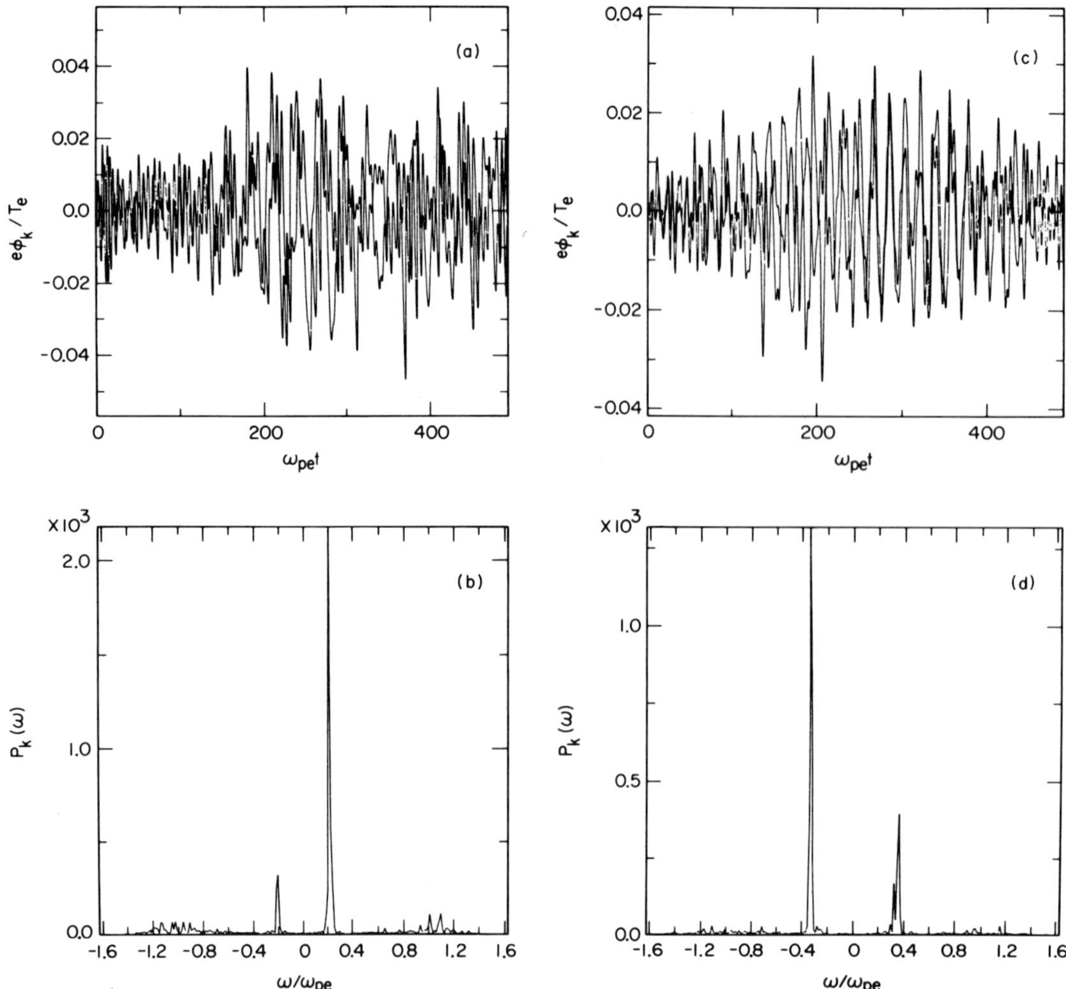

Figure 2—The time evolutions (a, c) and their power spectra (b, d) of perturbations of electrostatic potential for two modes in the case of the counterstreaming ion beams (shown by line B in Fig. 1b).

$$\text{(a), (b)} \quad (0,2) \quad (k_\parallel \lambda_{e\parallel} = 0.985 \times 2\pi \times 2/32 = 0.386,$$
$$k_\perp \lambda_{e\parallel} = 0.172 \times 2\pi \times 2/32 = 0.067)$$
$$\text{(c), (d)} \quad (0,3) \quad (k_\parallel \lambda_{e\parallel} = 0.985 \times 2\pi \times 3/32 = 0.580,$$
$$k_\perp \lambda_{e\parallel} = 0.172 \times 2\pi \times 3/32 = 0.101)$$

(0.3% of the electrons are replaced by the cold electrons randomly at each step). As expected, the simulations show that the instability for the case with the recycling of the cold electrons grows larger than that for the case without it. The parallel velocity distribution of the cold electron beam and the ion beams are shown in Fig. 3a and b. As shown in Fig. 3a, the cold electrons are heated at the positive side; however, no flattening of distribution occurs due to the recycling of cold electrons. In the case without the recycling of the cold electrons, the plateau is formed at the positive side of the cold electron parallel velocity distribution that causes saturation of the instability. The sufficient slowing down of the beam ions is shown in Fig. 3b.

Figure 4 shows time evolutions in perturbations of the electrostatic potential and their spectra for the (0,2) and (0,3) modes. The time evolution shows that the electrostatic potential reaches its maximum at $|e\phi_k/T_e| \approx$

0.088 for the (0,2) mode. The real frequencies of these modes are slightly smaller than those predicted by the theory as shown in Fig. 1c.

ISEE wave data indicate broadband electrostatic noise extending up to ω_{pe} or more. The wavelength of waves is not resolved because the electric fields (E_x, E_y) are diagnosed at sample points in the simulation region. The time evolutions of the x and y components of the electric field at $x = 16$ and $y = 16$ for the two cases is shown in Fig. 5a and c. The power spectra of these electric fields as shown in Fig. 5b and d resemble those of turbulence. Due to the smaller simulation region ($L_x = L_y = 32\Delta$), based on line B in Fig. 5b, only five or six modes become unstable for the quasiparallel modes. For example, if we use the system size $L_x = L_y = 128\Delta$, 20 or 21 modes would become unstable and at lower frequencies we would expect that with longer simulations the spectrum would be broadband noise without any sig-

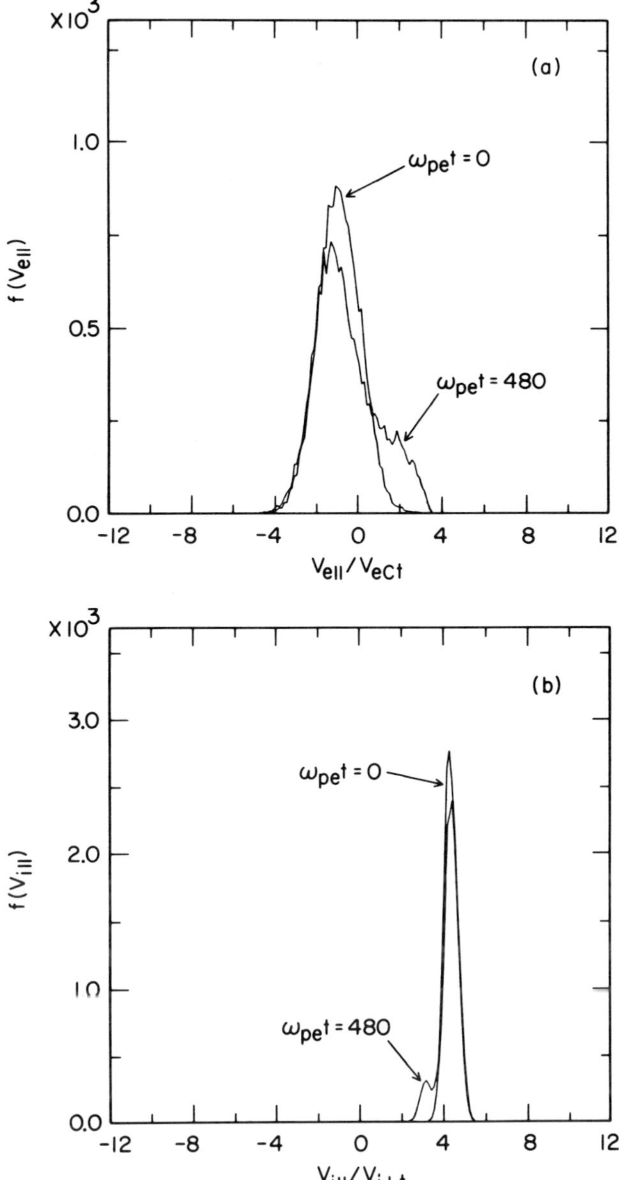

Figure 3—The parallel velocity distribution: (a) the cold electron beam and (b) the ion beam at $\omega_{pe}t = 0$ and 480.

nificant peaks. However, in order to compare this spectrum obtained by the simulation with the observed broadband electrostatic noise, we need further development of the simulations with respect to the observations. We will report these results in a separate paper.

DISCUSSION

We have studied the electrostatic turbulence driven by a cold electron beam and/or ion beams in order to understand the processes generating broadband electrostatic noise analytically and by means of numerical simulations. Linear theory indicates that the presence of cold electrons increases the growth rates of ion acoustic waves driven by the ion beams. On the other hand, the growth rates are reduced by increasing temperature anisotropy ($T_\perp > T_\parallel$) of ion beams.

Several simulations have been performed to confirm the linear theory. For the first case, we have investigated the plasma with a cold electron beam ($\bar{n}_{eC} = \bar{n}_{eH} = 0.5\,\bar{n}_i$, $T_{eC} = 0.1\,T_{eH}$, $V_{eC} = 0.79\,v_{eHt}$) in background warm electrons and ions ($T_{eH} = T_i$). In this case, the uniform recycling model is used in order to model the cold electron flow along the field lines.[25,26] The growing ion acoustic waves are observed with real frequency $\omega \approx k_\parallel c_s$ as predicted by the linear theory.

The second simulation shows that the counterstreaming ion beams with the temperature anisotropy ($T_{i\parallel} = 0.1\,T_{i\perp}$) excite various electrostatic electron and ion waves in the presence of the cold electrons. The spectra of density perturbations of electrostatic potential show the ion acoustic waves with the Doppler-shifted real frequency $\omega \approx \pm k_\parallel (c_s - V_{i2\parallel})$.

The third simulation shows that the counterstreaming of the cold electron beam and the single ion beam with the temperature anisotropy also excites ion acoustic waves with the Doppler-shifted real frequency $\omega \approx -k_\parallel (c_s - V_i)$. The overall frequency range is obtained by observing the electric field at some points in the simulation region for the second and third cases. As shown in Fig. 5c, the electric fields reach their maxima at $|E_{x,y}| \approx 6$ mV/m for the plasma with $\bar{n}_e = 1.0$ cm^{-3}, and $T_e = 200$ eV. The power spectra of the electric field exhibit turbulence that is qualitatively similar to that measured in the plasma-sheet boundary layer. Further simulations and detailed comparisons with observed plasma velocity distributions and wave power spectra are required for further progress in the interpretation of broadband electrostatic noise, which will be reported in the separate paper.

ACKNOWLEDGMENT—We would like to thank Prof. M. Ashour-Abdalla for useful comments and discussions. This research was supported by the National Aeronautics and Space Administration under contract NAS5-28700 and the National Science Foundation, grant ATM-8519528. The simulations were performed at NCAR with support from NSF.

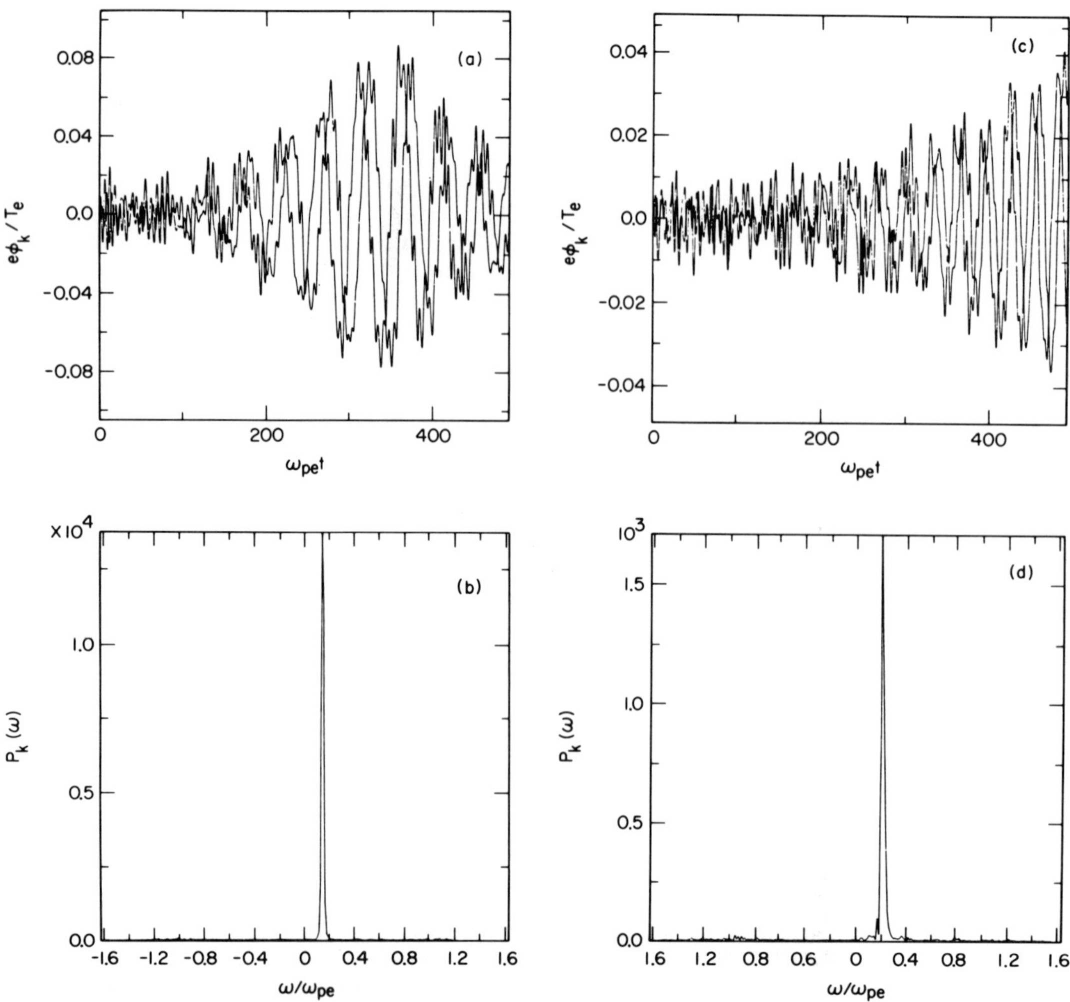

Figure 4—The time evolutions (a, c) and their power spectra (b, d) of perturbations of electrostatic potential for two modes in the case of the counterstreaming of the cold electron beam and the ion beam with the temperature anisotropy applying the recycling of the cold electron beam (shown in Fig. 1c).

$$\begin{aligned}
&\text{(a), (b)} \quad (0,2) \quad (k_\parallel \lambda_{e\parallel} = 0.985 \times 2\pi \times 2/32 = 0.386, \\
&\qquad\qquad\qquad\quad k_\perp \lambda_{e\parallel} = 0.172 \times 2\pi \times 2/32 = 0.067) \\
&\text{(c), (d)} \quad (0,3) \quad (k_\parallel \lambda_{e\parallel} = 0.985 \times 2\pi \times 3/32 = 0.580, \\
&\qquad\qquad\qquad\quad k_\perp \lambda_{e\parallel} = 0.172 \times 2\pi \times 3/32 = 0.101)
\end{aligned}$$

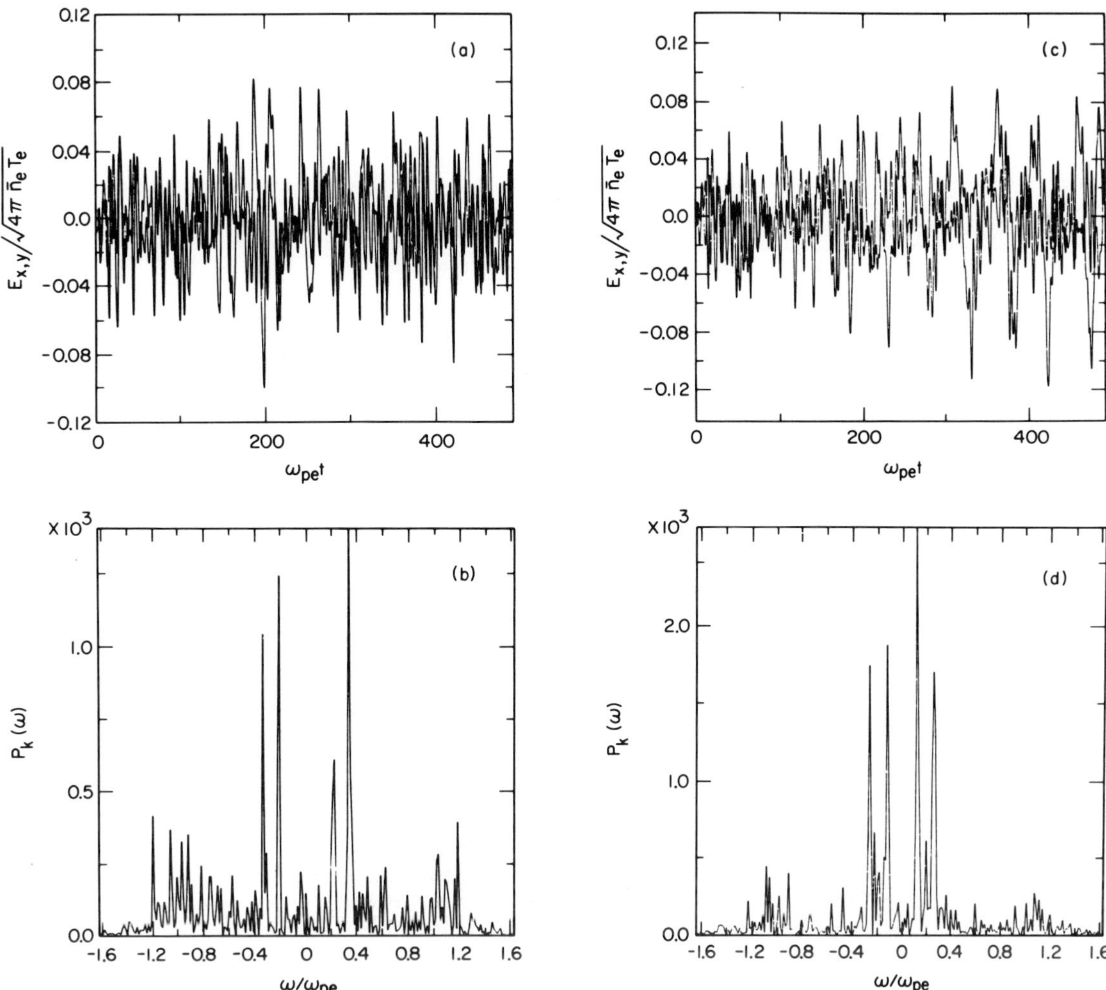

Figure 5—The time evolution (a, c) and their power spectra (b, d) of the x and y components of the electric field sampled at $x = 16$ and $y = 16$.

(a), (b)　The case of the counterstreaming ion beam (the same as Fig. 2).

(c), (d)　The case of the cold electron beam and the anisotropic ion beam with the temperature anisotropy applying the recycling of the cold electron beam (the same as Fig. 3).

REFERENCES

[1] F. L. Scarf, L. A. Frank, K. L. Ackerson, and R. P. Lepping, "Plasma Wave Turbulence at Distant Crossings of the Plasma Sheet Boundaries and the Neutral Sheet," *Geophys. Res. Lett.* **1**, 189 (1974).

[2] D. Gurnett, L. A. Frank, and R. Lepping, "Plasma Waves in the Distant Magnetotail," *J. Geophys. Res.* **81**, 6059 (1976).

[3] D. A. Gurnett and L. A. Frank, "A Region of Intense Plasma Wave Turbulence on Auroral Field Lines," *J. Geophys. Res.* **82**, 1031 (1977).

[4] R. R. Anderson, "Plasma Waves in Planetary Magnetospheres," *Rev. Geophys. Space Phys.* **21**, 674 (1983).

[5] C. L. Grabbe and T. E. Eastman, "Generation of Broadband Electrostatic Noise by Ion Beam Instabilities in the Magnetotail," *J. Geophys. Res.* **89**, 3865 (1984).

[6] G. K. Parks et al., "Particle and Field Characteristics of the High-Latitude Plasma Sheet Boundary Layer," *J. Geophys. Res.* **89**, 8885 (1984).

[7] R. J. DeCoster and L. A. Frank, "Observations Pertaining to the Dynamics of the Plasma Sheet," *J. Geophys. Res.* **84**, 5099 (1979).

[8] L. A. Frank and K. L. Ackerson, "Several Recent Findings Concerning the Dynamics of the Earth's Magnetotail," *Space Sci. Rev.* **23**, 375 (1979).

[9] D. J. Williams, "Energetic Ion Beams at the Edge of the Plasma Sheet: ISEE 1 Observations Plus a Simple Explanatory Model," *J. Geophys. Res.* **86**, 5507 (1981).

[10] T. E. Eastman, L. A. Frank, and W. K. Peterson, "The Plasma Sheet Boundary Layer," *J. Geophys. Res.* **89**, 1553 (1984).

[11] L. A. Frank, R. L. McPherron, R. J. DeCoster, B. G. Burek, K. L. Ackerson, and C. T. Russell, "Field-Aligned Currents in the Earth's Magnetotail," *J. Geophys. Res.* **86**, 6817 (1981).

[12] L. A. Frank, "Plasmas in the Earth's Magnetotail," in *Space Plasma Simulations*, M. Ashour-Abdalla and D. A. Dutton, eds., D. Reidel Pub. Co., Holland, p. 211 (1985).

[13] M. Ashour-Abdalla and R. M. Thorne, "Toward a Unified View of Diffuse Auroral Precipitation," *J. Geophys. Res.* **83**, 4755 (1978).

[14] J. Huba, N. Gladd, and K. Papadopoulos, "Lower-Hybrid-Drift Wave Turbulence in the Distant Magnetotail," *J. Geophys. Res.* **83**, 5217 (1978).

[15] C. L. Grabbe, "Generation of Broadband Electrostatic Noise in the Magnetotail for Two-Component Electron Populations," *Proc. Int. Conf. Plasma Phys.*, Lausanne, Switzerland (1984).

[16] C. L. Grabbe, "New Results on the Generation of Broadband Electrostatic Waves in the Magnetotail," *Geophys. Res. Lett.* **12**, 483 (1985).

[17] C. L. Grabbe, "Numerical and Analytical Study of the Spectrum of Broadband Electrostatic Noise in the Magnetotail," *J. Geophys. Res.* **91** (in press, 1986).

[18] P. B. Dusenbury and L. R. Lyons, "The Generation of Electrostatic Noise in the Plasma Sheet Boundary Layer," *J. Geophys. Res.* **90**, 10939 (1985).

[19] N. Omidi, "Broadband Electrostatic Noise Produced by Ion Beams in the Earth's Magnetotail," *J. Geophys. Res.* **90**, 12330 (1985).

[20] M. Ashour-Abdalla and H. Okuda, "Theory and Simulations of Broadband Electrostatic Noise in the Geomagnetic Tail," *J. Geophys. Res.* **91**, 6833 (1986).

[21] K. Akimoto and N. Omidi, "A Study of the Generation of Broadband Electrostatic Noise by an Ion Beam in the Magnetotail," *Geophys. Res. Lett.* **3**, 97 (1986).

[22] M. Ashour-Abdalla and H. Okuda, "Electron Acoustic Instabilities in the Geomagnetic Tail," *Geophys. Res. Lett.* **13**, 366 (1986).

[23] T. E. Eastman and L. A. Frank, "Boundary Layers of the Earth's Outer Magnetosphere," in *Magnetic Reconnection in Space and Laboratory Plasmas*, Geophysical Monograph No. 30, American Geophysical Union, Washington, D.C., p. 249 (1984).

[24] K. Miyamoto, *Plasma Physics for Nuclear Fusion*, The MIT Press, Cambridge, Mass. (1980).

[25] H. Okuda and M. Ashour-Abdalla, "Acceleration of Hydrogen Ions and Conic Formation Along Auroral Field Lines," *J. Geophys. Res.* **88**, 899 (1983).

[26] K.-I. Nishikawa and H. Okuda, "Heating of Light Ions in the Presence of a Large-Amplitude Heavy Ion Cyclotron Wave," *J. Geophys. Res.* **90**, 2921 (1985).

DISCUSSION

S. Peter Gary: Comment: The electrostatic approximation that you have used is probably valid in the short wavelength ion acoustic region, but is not necessarily valid for long wavelength ion cyclotron waves that propagate nearly perpendicular to the magnetic field. Finite effects may stabilize the ion cyclotron instability and should be considered in future research.

S. Peter Gary: Dr. Abdalla's simulations show relatively low amplitudes for the ion cyclotron mode, whereas your work shows relatively large amplitude of this instability. Can you account for this difference?

K. I. Nishikawa: The results shown at the conference showed the very weak growth of the ion acoustic waves due to the small ion drift velocity with the warmer ion beam with parallel temperature. The simulation results in the paper show the larger amplitude of the ion acoustic waves than that of the ion cyclotron waves driven by the ion beam with the larger drift velocity and the colder parallel temperature. Beam-driven instabilities in finite-B plasmas will be investigated in near future.

NON-MAXWELLIAN FREE-ENERGY GENERATION IN THE MAGNETOTAIL DUE TO CHAOTIC PARTICLE MOTION

J. Chen and P. J. Palmadesso*

It has previously been shown that the charged-particle motion in a magnetotail-like magnetic field is generally nonintegrable and that the particle orbits can be classified into three distinct types: the bounded integrable orbits, unbounded stochastic orbits, and unbounded transient orbits. The three distinct types of orbits occupy disjoint regions of the phase space. As a result, the different regions of the phase space exhibit qualitatively different response to external influences, resulting in "differential memory" of single-particle distributions. In this paper, we consider the current distribution in the vicinity of the equatorial plane and show that the stochastic and transient orbits carry current near $Z = 0$ while the integrable orbits do not. We then consider a scenario whereby differential memory can produce non-Maxwellian features in plasma distribution functions. It is noted that a certain type of distribution functions possess the free energy necessary to drive the non-Maxwellian tearing mode in neutral sheets. Physical implications for magnetotail dynamics are considered.

INTRODUCTION

The motion of charged particles in the earth's magnetotail has been under extensive investigation for the past two decades in order to fully understand the properties of the magnetotail plasmas from both the experimental and theoretical points of view. A number of specialized aspects of the single-particle motion have been analyzed using various methods including approximate analytical methods[1-6] and numerical methods.[7-13]

The magnetotail field may be modeled, in its simplest form, by a neutral sheet magnetic profile $B_0(z)x$ with $B_0(z = 0) = 0$ and a superimposed component $B_n z$ normal to the plane of the neutral sheet (the so-called "quasineutral sheet" geometry). We use the standard magnetotail coordinate system with x in the earthward direction. In the above works, the fundamental question regarding the integrability of the particle motion was not addressed. However, it has recently been shown[14] that the magnetotail-like system is intrinsically nonintegrable due to the presence of the normal component B_n. Subsequently, Chen and Palmadesso,[15] have identified various phase space structures in detail. More specifically, using the Poincaré surface-of-section method, they have shown that particle motions in the magnetotail-like configuration can be classified into three distinct types of orbits occupying disjoint regions of the phase space—the bounded integrable orbits, unbounded stochastic orbits, and unbounded transient orbits.

In studying individual orbits, a number of researchers have noted "randomness" in certain orbits. For example, Swift[10] noted that some orbits appear to randomize in the equatorial plane after a few crossings. Wagner et al.[11] found that certain orbits exhibit sensitive dependence on initial conditions. Gray and Lee[12] noted that the magnetic moments for some particles exhibit randomness across the equatorial plane. We now understand that this randomness is a manifestation of the intrinsic nonintegrable nature of the system. The work of Chen and Palmadesso[14,15] provides a systematic and unified understanding of the nature and properties of the particle motion throughout the entire phase space. Previously, Kim and Cary[16] studied stochastic orbits in an O-type neutral point. More recently, Martin[17] showed that particle motion near an X-point is chaotic. Also, it has long been known[18] that motion in the dipole field is not integrable.

The integrability of the particle motion is not purely of mathematical interest. The existence of distinct classes of orbits has profound effects on the plasma particle distribution and on the response of the magnetotail plasmas to external influences. In this regard, the concept of "differential memory" was introduced by Ref. 15 to describe the property that disjoint regions of the phase space occupied by distinct types of orbits cannot communicate with each other so that different regions respond to external influences on different time scales. In particular, different regions of the phase space retain the memory of the existing particle distribution for different lengths of time. This implies that a plasma distribution function has a natural tendency to develop non-Maxwellian features in response to changes in physical conditions. In turn, the distribution and character of the charged-particle orbits can have a significant impact on other dynamical properties of the magnetotail. One important example is the collisionless tearing-mode instability[19-21] which has long been thought to play an important role in magnetic field reconnection.[22,23] It has recently been shown that the growth rates of the collisionless tearing instability can be enhanced by up to a few orders of magnitude due to the presence of temperature anisotropy or other non-Maxwellian features in the particle distribution.[24-26] Thus, the particle dynamics plays a fundamental role in influencing the magnetotail dynamics. A fundamental question, then, is whether the magnetotail can sustain free energy carrying non-Maxwellian features that can support large-scale instabilities

*Geophysical and Plasma Dynamics Branch, Plasma Physics Division, Naval Research Laboratory, Washington, D.C. 20375-5000.

of potential relevance to magnetotail dynamics, such as the non-Maxwellian collisionless tearing mode.

In this paper, we will examine the way the magnetic field topology of a quasineutral sheet can generate non-Maxwellian features in plasma distribution functions and the attendant free energy that can drive certain plasma instabilities. In the next section, we review the particle dynamics using the Poincaré surface-of-section method. In the section after next and the final section, current distributions and the generation of non-Maxwellian free energy in the magnetotail plasmas are discussed. We will refer primarily to the ion motion. The modifications needed for electrons are trivial.

SINGLE-PARTICLE DYNAMICS

In order to model the motion of charged particles in the magnetotail, we consider a magnetic field given by $B = B_0 f(z)x + B_n z$ where B_0 is the asymptotic field in the x direction, B_n is the uniform normal field, and $B_0 f(z)$ is a neutral sheet profile such that $f(-z) = -f(z)$. We will primarily use the Harris configuration

$$f(z) = \tanh(z/\delta) \tag{1}$$

where δ is the characteristic scale length of the magnetic field. The treatment is also applicable to other quasineutral sheet configurations and another example, $f(z) = z/\delta$, is discussed in Ref. 15.

We choose the gauge such that the vector potential is $A_y(x,z) = -B_0 F(z) + B_n x$, where $dF(z)/dz = f(z)$. The single-particle motion is described by the equation of motion

$$m\frac{dv}{dt} = \frac{q}{c} v \times B \tag{2}$$

This vector equation possesses three exact constants of motion: the Hamiltonian $H = mv^2/2$ where $v^2 = v_x^2 + v_y^2 + v_z^2$, the canonical momentum $P_y = mv_y + (q/c)A_y(x,z)$, and a constant $C_x = m(v_x - \Omega_n y)$ associated with the x-motion. Here, we use $\Omega_n \equiv qB_n/mc$ and $\Omega_0 \equiv qB_0/mc$ for each species. If we take the Poisson brackets of the constants of motion, we find $[H, P_y] = 0$ and $[H, C_x] = 0$. However, for C_x and P_y, we find

$$[C_x, P_y] = -m\Omega_n \tag{3}$$

so that C_x and P_y are not in involution. This indicates that the particle orbits may be stochastic. As a general remark, a Hamiltonian system with N degrees of freedom is integrable if and only if there exist N constants of motion that are in involution. The existence of such global constants of motion means that one can find a canonical transformation to a frame in which the N coordinates are cyclic. For more detailed properties of Eq. 2, see Ref. 15.

In describing the orbits, the following normalization will be used: $b_n = B_n/B_0$, $X = (x - P_y/m\Omega_n)/b_n\delta$, $Y = (y + C_x/m\Omega_n)/b_n\delta$, $Z = z/b_n\delta$, $\tau = \Omega_n t$, and

$H = H/(mb_n^2\Omega_n^2\delta^2)$. A useful technique for displaying the long-time properties of the particle motion is to use the Poincaré surface-of-section method (see, for example, Ref. 27). For our purpose, a surface-of-section plot at $Z = 0$ for a given value of H is constructed by following an orbit by numerical integration and plotting each point, X and $\dot{X} = dX/d\tau$, where the orbit crosses the equatorial plane. Figure 1a, reproduced from Ref. 15, shows such a plot for $H = 500$ and $b_n = 0.1$, evaluated at $Z = 0$. All kinematically allowable orbits are confined within the circle of radius $(2H)^{1/2}$.

The orbits in the region marked A are bounded and integrable. There exists an additional invariant in this localized region of the phase space. This means that the motion of an integrable orbit is constrained to a two-dimensional invariant surface whose cross section through $Z = 0$ is a closed curve. However, this additional invariant is not a global isolating constant and is not expressible in closed form in terms of elementary functions. Note also that integrable orbits are not necessarily adiabatic.

Figure 1 also demonstrates existence of a large stochastic region, marked B. The stochastic region is disjoint from the integrable region A; there is no orbit that can connect the two regions. The orbits in region B are stochastic in the sense that they are sensitively dependent on initial conditions, with nearby orbits diverging rapidly with time. In fact, a large number of orbits originating in a small area anywhere in region B can cover region B. For $z \gg \delta$, the motion is regular. We note that Fig. 3 of Ref. 11 corresponds to a stochastic orbit.

The nonintegrable orbits may be thought of as forming two flux tubes, which are mirror images of each other about the $Z = 0$ plane, originating from and escaping to infinity. The regions C1 through C5 are the $Z = 0$ cross sections of the flux tubes as they thread the equatorial plane. All orbits at infinity that can reach $Z = 0$ are mapped into region C1. The orbits then successively cross regions C2 through C5. These regions, C1 through C5, have interesting substructures as shown in Fig. 1b. A subset of the orbits entering C1, those crossing S1 and T1, enter the stochastic region B after crossing S5 or T5. That is, region B can only be accessed from region C5. After entering region B, a stochastic orbit will intersect region B repeatedly and eventually escape to infinity after crossing one of the cross-hatched regions. Stochastic orbits typically make finite numbers of crossings before escaping to infinity. However, we have found orbits crossing $Z = 0$ a few thousand times and it appears that the crossing number has no upper bound. The remainder of orbits in C1 escape to infinity after crossing C5 or C4 just above T4. The latter orbits are referred to as transient orbits by Ref. 15. These transient orbits appear to be the type studied by Speiser.[7] Only orbits of this type are shown in Fig. 1b. In Fig. 2, we show a surface-of-section plot for $H = 500$ and $b_n = 0.15$. Note that all the basic features are similar to those in Fig. 1a ($b_n = 0.1$). However, with a stronger B_n component, a larger fraction of the phase space (i.e., the regions corresponding to C1 through C5) is occupied by

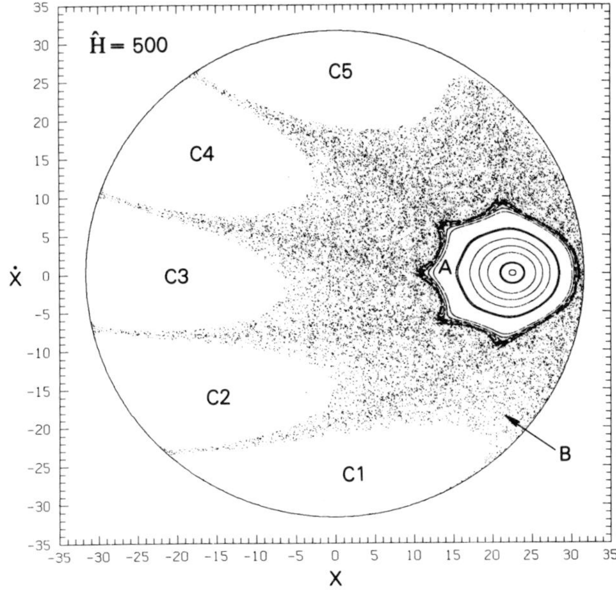

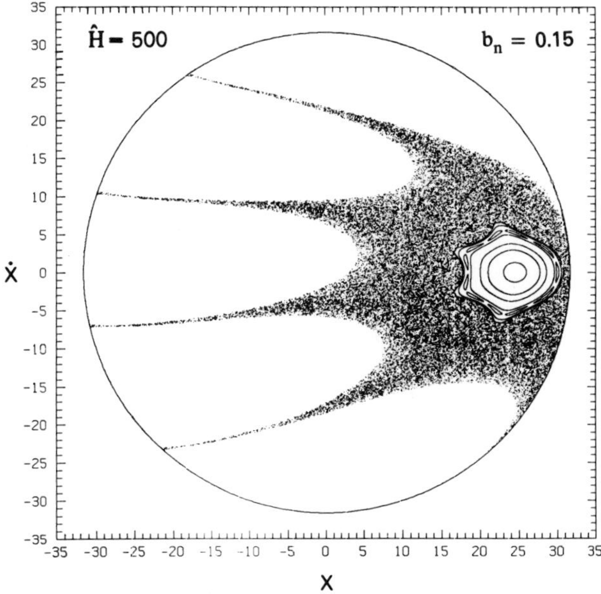

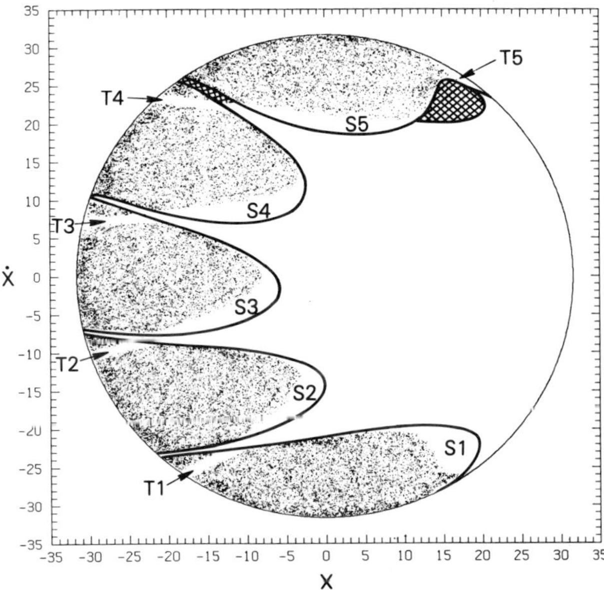

Figure 1—(From Ref. 15) Surface-of-section plots for the Harris-type field with $Z = 0$, $b_n = 0.1$, and $H = 500$. (a) Representative integrable orbits in region A and stochastic orbits in region B; 60,000 crossing points. (b) Transient orbits, showing the substructures in regions C1 through C5; 42,000 crossing points.

Figure 2—Surface-of-section plot. The Harris-type field with $b_n = 0.15$ and $H = 500$. The structures are similar to those in Fig. 1.

tion of the phase-space structures for both cases can be found in Ref. 15.

CURRENT DISTRIBUTIONS

The surface-of-section plots provide another important piece of information. It is clear that the integrable (regular) orbits in region A carry no net current. On the other hand, the nonintegrable orbits, i.e., the transient and stochastic orbits, do carry currents because they come from and escape to infinity. For example, for $H = 500$ in the Harris case, all nonintegrable orbits (ions) enter the equatorial plane through region C1 (Fig. 1a) and escape from C5 and the nearby cross-hatched regions. This means that regardless of the actual path for different orbits, there is a net drift in the $+Y$ direction in the vicinity of the equatorial plane ($|Z| \lesssim \delta$). The electron motion is, of course, the opposite so that the electron and ion currents are additive. For the parabolic case in which essentially all nonintegrable orbits are eventually reflected back, the orbits still set up a net drift near the equatorial plane. In this case, the return drift is established away from the $Z = 0$ plane with zero total drift integrated over all Z. Thus, for both the Harris and parabolic case, the nonintegrable orbits form a net current in the dawn-to-dusk direction near the equatorial plane. Note that Stern and Palmadesso[4] showed that particles that are trapped between mirror points exhibit no net drift across the equatorial plane. Subsequently, Cowley[5] and Pellat and Schmidt[6] generalized the proof. From our analysis, we conclude that the proof is exactly applicable to the integrable orbits. The transient and stochastic orbits in the Harris case do carry a net drift because the motion is unbounded in the z direction. In the parabolic field, the proof is still applica-

the flux tube structure and the transient orbits therein. This is because the field lines are "straighter" for larger b_n so that the orbits can translate in the z direction more easily. (In the limit $b_n \to \infty$, all orbits move in the z direction freely.) Similar structures exist in the surfaces of section for different values of H. As the value of H is reduced, large-scale integrable regions become more complicated. For $b_n = 0.1$, large integrable regions vanish for $H \lesssim 6.2$ and essentially all orbits are of the stochastic or the transient types. For the parabolic case, similar features exist. A more detailed descrip-

323

ble but there is a net drift near the $Z = 0$ plane as described above.

GENERATION OF FREE ENERGY

A prominent feature of the particle motion in the quasineutral sheet geometry is the existence of disjoint regions in the phase space near $Z = 0$, each region consisting of orbits of distinct nature: (a) bounded integrable orbits, (b) unbounded stochastic orbits, and (c) unbounded transient orbits. In effect, there exist boundary surfaces between the disjoint regions, which do not allow randomization of information or particle energy because of the phenomenon of differential memory.[28] If noise fields are introduced to the magnetic field, the boundary surfaces will break up and allow "diffusion" of orbits. The time scale for this process depends on the strength of the noise field and is slow for low levels of noise. The basic disjointness of the regions and differential memory are expected to be insensitive to low-level noises.

The concept of differential memory has a number of physically important implications. Suppose the system contains a population of charged particles in thermal equilibrium. There must also be a steady supply of particles in the distant regions to maintain equilibrium. If the parameters of the distant plasma distribution are changed, then the fact that the orbits are divided into distinct types belonging to disjoint regions of the phase space results in a highly non-Maxwellian distribution because of differential memory. The resulting non-Maxwellian distributions may be written in the form $F(H,P_y)$ for the nonintegrable regions and $F(H,P_y,g)$ for the integrable regions where g is the additional invariant of motion. It is the existence of boundary surfaces that can establish the free energy to be tapped by suitable plasma instabilities.

Note that care must be used in observational interpretations. For example, for 1 keV protons in a magnetotail-like field with $B_0 = 20\gamma$, $B_n = 2\gamma$ ($b_n = 0.1$) and $\delta = 1$ R_e, we have $H \simeq 7$. For $\delta \simeq 0.1$ R_e, we have $H \simeq 600$. As shown in Ref. 15, the characteristic time between two successive crossings for transient orbits is given by $t_1 = 2^{1/4}\pi(\Omega_n^{-1}H^{-1/4})$ and such particles make $N \simeq (H/2)^{1/4} + 1$ crossings so that they spend a total time of approximately $T \equiv \pi\Omega_n^{-1}$ in the vicinity of the equatorial plane. For typical transient orbits with $H = 500$ (Fig. 1b) and $\Omega_n \simeq 0.2$ s^{-1}, we obtain $T \simeq 15$ s. For stochastic orbits, if we take, say, 20 crossings for randomization, then our numerical calculations show that the time needed is roughly 5 minutes. Thus, the transient orbits can traverse the system in about one minute and the stochastic orbits can be randomized in a matter of several minutes. In addition, in a steady-state physical system with noises, the entire phase space can be randomized across the boundary surfaces in a sufficiently long time. Thus, the quiet-time magnetotail should be essentially isotropic. If a satellite observation of the particle distribution is made, then the measurement is likely to reveal essentially isotropic distributions. This is consistent with the observed near-isotropy.[29] The above

scenario for the apparent isotropy is to be contrasted with, for example, the work of Nötzel et al.,[30] which uses anisotropy-driven instabilities to explain isotropy. However, following (unspecified) changes in the physical conditions, the distribution function can maintain non-Maxwellian features for a certain amount of time, determined by time scales of diffusion between the different disjoint regions of the phase space due to noise fields as described above. During this time, the non-Maxwellian tearing-mode[28] may be important. Indeed, this instability itself would also tend to isotropize the distribution function.

In the context of the earth's magnetotail, it is believed that the collisionless tearing instability may play an important role in reconnection processes. It has recently been shown[24,26] that the growth rate of the collisionless tearing mode in a non-Maxwellian neutral sheet can be substantially greater, by a few orders of magnitude, than in the Maxwellian case. The general form of the distribution functions that can provide the necessary free energy is $F(H,P_y,C)$ where C is an independent constant of motion. Thus, if the magnetotail, which is initially in thermal equilibrium, is subjected to changes in external conditions, e.g., the solar wind, pressure distribution, magnetopause, etc., then the magnetic topology can develop appropriate plasma distributions to drive the non-Maxwellian tearing mode. An important point is that the free energy for this instability is the particle energy in addition to the conventional tearing-mode free energy. That is, changes in external conditions can generate by the process of differential memory an additional source of free energy in the form of non-Maxwellian distributions.

We add, however, that the above discussion does not take into account the influences of the normal magnetic field on the instability itself under certain circumstances.[31-33] This point warrents further consideration and will be addressed in a future paper. Finally, this paper is based on single-particle motion. In order to determine more quantitatively the nature and associated time scales of these processes, it is necessary to follow an ensemble of particles. This work is currently under way.

ACKNOWLEDGMENT—This research was supported by the National Aeronautics and Space Administration and the Office of Naval Research.

REFERENCES

[1] T. W. Speiser, "Particle Trajectories in Model Current Sheets, 1. Analytical Solutions," *J. Geophys. Res.* **70**, 4219 (1965).

[2] I. I. Alekseyev and A. P. Kropotkin, "Interaction of Energetic Particles with the Neutral Layer of the Tail of the Magnetosphere," *Geomagn. Aeron.* **10**, 615 (1970).

[3] B. U. O. Sonnerup, "Adiabatic Particle Orbits in a Magnetic Null Sheet," *J. Geophys. Res.* **76**, 8211 (1971).

[4] D. Stern and P. Palmadesso, "Drift-Free Magnetic Geometries in Adiabatic Motion," *J. Geophys. Res.* **80**, 4244 (1975).

[5] S. W. H. Cowley, "A Note on the Motion of Charged Particles in One-Dimensional Magnetic Current Sheets," *Planet. Space Sci.* **26**, 539 (1978).

[6] R. Pellat and G. Schmidt, "Absence of Particle Drift in Magnetic Fields of Translational Symmetry," *Phys. Fluids* **22**, 381 (1979).

[7] T. W. Speiser, "Particle Trajectories in Model Current Sheets, 2. Applications to Auroras Using Geomagnetic Tail Model," *J. Geophys. Res.* **72**, 3919 (1967).

[8] S. W. H. Cowley, "The Adiabatic Flow Model of a Neutral Sheet," *Cosmic Electrodynam.* **2**, 90 (1971).

[9] J. W. Eastwood, "Consistency of Fields and Particle Motion in the 'Speiser' Model of the Current Sheet," *Planet. Space Sci.* **20**, 1555 (1972).

[10] D. Swift, "The Effect of the Neutral Sheet on Magnetospheric Plasma," *J. Geophys. Res.* **82**, 1288 (1977).

[11] J. S. Wagner, J. R. Kan, and S.-I. Akasofu, "Particle Dynamics in the Plasma Sheet," *J. Geophys. Res.* **84**, 891 (1979).

[12] P. Gray and L. C. Lee, "Particle Pitch Angle Diffusion Due to Nonadiabatic Effects in the Plasma Sheet," *J. Geophys. Res.* **87**, 7445 (1982).

[13] T. W. Speiser and L. R. Lyons, "Comparison of an Analytical Approximation for Particle Motion in a Current Sheet with Precise Numerical Calculations," *J. Geophys. Res.* **89**, 147 (1984).

[14] J. Chen and P. J. Palmadesso, "Chaos and Nonlinear Dynamics of Single-Particle Orbits in a Quasi-Neutral Sheet with Normal Field," *Eos* **65**, 1065 (1984).

[15] J. Chen and P. J. Palmadesso, "Chaos and Nonlinear Dynamics of Single-Particle Orbits in a Magnetotail-like Magnetic Field," *J. Geophys. Res.* **91**, 1499 (1986).

[16] J.-S. Kim and J. R. Cary, "Charged Particle Motion Near a Linear Magnetic Null," *Phys. Fluids* **26**, 2167 (1983).

[17] R. F. Martin, "Chaotic Particle Dynamics Near a Two-Dimensional Magnetic Neutral Point with Application to the Geomagnetic Tail," *J. Geophys. Res.* **91**, 11985 (1986).

[18] A. J. Dragt and J. M. Finn, "Insolubility of Trapped Particle Motion in a Magnetic Dipole Field," *J. Geophys. Res.* **81**, 2327 (1976).

[19] H. P. Furth, "The 'Mirror Instability' for Finite Particle Gyroradius," *Nucl. Fusion Suppl., Pt. 1,* 169 (1962).

[20] D. Pfirsch, "Mikroinstabilitä ten vom Spiegeltyp in inhomogenen Plasmen (Mirror Type Instabilities in Inhomogeneous Plasmas)," *Z. Naturforsch* **17a**, 861 (1962).

[21] G. Laval, R. Pellat, and M. Vuillemin, "Instabilities Electromagnetiques des Plasmas sans Collisions," in *Plasma Physics and Controlled Nuclear Fusion Research,* Vol. 2, International Atomic Energy Agency (1966).

[22] B. Coppi, G. Laval, and R. Pellat, "Dynamics of the Geomagnetic Tail," *Phys. Rev. Lett.* **16**, 1207 (1966).

[23] K. Schindler, in *Proceedings of the Seventh International Conference on Phenomena in Ionized Gases,* Vol. II. Gradevinska Knjiga, Beograd, Yugoslavia, p. 736 (1966).

[24] J. Chen and P. Palmadesso, "Tearing Instability in an Anisotropic Neutral Sheet," *Phys. Fluids* **27**, 1198 (1984).

[25] J. Chen, P. Palmadesso, J. A. Fedder, and J. G. Lyon, "Fast Collisionless Tearing in an Anisotropic Neutral Sheet," *Geophys. Res. Lett.* **11**, 12 (1984).

[26] J. Chen and Y. C. Lee, "Collisionless Tearing Instability in a Non-Maxwellian Neutral Sheet: An Integro-Differential Formulation," *Phys. Fluids* **28**, 2137 (1985).

[27] A. J. Lichtenberg and M. A. Lieberman, *Regular and Stochastic Motion, Springer-Verlag,* New York (1983).

[28] J. Chen, P. J. Palmadesso, and Y. C. Lee, "Magnetic Reconnection in a Non-Maxwellian Neutral Sheet" (this volume, 1986).

[29] G. S. Stiles, E. W. Hones, S. J. Bame, and J. R. Asbridge, "Plasma Sheet Pressure Anisotropies," *J. Geophys. Res.* **83**, 3166 (1978).

[30] A. Nötzel, K. Schindler, and J. Birn, *J. Geophys. Res.* **90**, 8293 (1985).

[31] A. A. Galeev and L. M. Zelenyi, "Tearing Instability in Plasma Configuration," *Sov. Phys. JETP,* 1113 (1976).

[32] B. Lembège and R. Pellat, "Stability of a Thick Two-Dimensional Quasineutral Sheet," *Phys. Fluids* **25**, 1995 (1982).

[33] F. V. Coroniti, "On the Tearing Mode in Quasi-Neutral Sheets," *J. Geophys. Res.* **85**, 6719 (1980).

DISCUSSION

T. Armstrong: How robust are the regions of chaotic and organized motion to changes in the magnetic field?

P. J. Palmadesso: Two kinds of changes can be considered. First, we can ask how sensitive our conclusions are to the choice of magnetic field topology. In general, the kinds of behavior we see are typical of Hamiltonian systems (i.e., partition of phase space into chaotic/ergodic and regular/integrable subregions). While a different magnetotail-like choice of field-line shape would change the shape of the various regions, the qualitative behavior we have seen should persist. A second kind of change in the magnetic field that should be considered is a random temporal change of small amplitude (noise field). The likely consequence of this is to break the K.A.M. surfaces (roughly, the 2D surfaces in the 3D phase space that intersect the reference place to form the simple closed curves in the regular/integrable region) into island chains, on to distort these surfaces, depending on the nature of the noise field. The result is that information diffuses into scale, long compared to the rate of information flow within the chaotic region. The concept of "differential memory" remains valid unless the noise is very strong.

MAGNETIC RECONNECTION IN A NON-MAXWELLIAN NEUTRAL SHEET

J. Chen,* P. J. Palmadesso,* and Y. C. Lee[†]

The collisionless tearing-mode instability is believed to play an important role in magnetospheric reconnection processes. It has been found that the linear growth rate for a non-Maxwellian neutral sheet can be enhanced by a few orders of magnitude in comparison with the conventional isotropic case and that the short wavelength perturbations ($k\delta \gg 1$) are strongly favored, where δ is the characteristic half-width of the neutral sheet. For magnetotail parameters, it is found that the linear e-folding time can be reduced to a matter of minutes or less. The resulting small scale islands are expected to coalesce rapidly. The additional free energy is the particle energy in the form of non-Maxwellian distribution. Differential equation analysis and full integro-differential analysis using exact unperturbed orbits are discussed. Physically, if non-Maxwellian features are generated in the plasma distribution by changes in external conditions such as the solar wind conditions, then the magnetotail may become strongly unstable, resulting in rapid development of small-scale islands, coalescence, and reconnection. This process may manifest itself as "triggering" of substorms following such changes.

INTRODUCTION

The collisionless tearing mode[1-3] has attracted considerable attention in connection with the earth's magnetotail.[4,5] It was suggested[4] as a possible mechanism for magnetic field reconnection in the magnetosphere. The tearing mode may also be relevant to the dayside magnetopause (e.g., Refs. 6 and 7). The basic idea is that the magnetic field energy stored in the magnetosphere as a result of interaction with the solar wind is released in other forms of energy via reconnection processes. The collisionless tearing mode is a possible instability that allows necessary changes in the magnetic topology in the absence of resistivity.

The geometry most often used to model the magnetotail is the neutral-sheet geometry. The standard analysis (e.g., Ref. 3) uses a Maxwellian plasma distribution function. However, in a collisionless plasma such as the magnetotail, the motion of particles parallel to the magnetic field is decoupled from the perpendicular motion and non-Maxwellian features may persist. Laval and Pellat[8] showed that the collisionless tearing mode ($k_\parallel B_0$) can be strongly modified by weak electron temperature anisotropy, $|T_{e\perp}/T_{e\parallel} - 1| < \rho_e/\delta \ll 1$, where k is the wave vector, B_0 is the equilibrium magnetic field, ρ_e is the typical electron Larmor radius, δ is the characteristic half-width of the neutral sheet, and $T_{e\perp}$ and $T_{e\parallel}$ are the electron temperatures associated with the particle motion perpendicular and parallel to B_0, respectively. Subsequently, Forslund[9] obtained an approximate dispersion relation for weakly anisotropic electrons. The work showed a substantial enhancement of the growth rate for $T_{e\perp} > T_{e\parallel}$, consistent with Laval and Pellat.[8]

In the work of Forslund,[9] the effects of axis-crossing ion orbits extending beyond the electron inner region of the order of $(\rho_e \delta)^{1/2}$ were taken to be negligible (the conventional "two-region" approximation). Recently, Chen and Palmadesso[10] have shown the existence of an ion intermediate region with a thickness of the order of $(\rho_i \delta)^{1/2} \gg (\rho_e \delta)^{1/2}$ where the axis-crossing ion orbits make a major contribution. Inclusion of the ion intermediate region (the "three-region" approximation) shows that a given degree of ion anisotropy ($T_{i\perp}/T_{i\parallel} > 1$) can increase the growth rate by nearly one order of magnitude over the two-region approximation results. Furthermore, in the two-region approximation, because the ion anisotropy effects are neglected, a neutral sheet with $1 - T_{e\perp}/T_{e\parallel} \approx \rho_e/\delta \ll 1$ is completely stable unless unrealistically high degrees of ion anisotropy are included. However, the three-region treatment shows that only modest ion anisotropy is needed to destabilize such a neutral sheet. This is of particular interest since any significant anisotropy in the electron population would tend to be isotropized on very fast time scales by the tearing instability itself or by other instabilities (see, for example, Ref. 11). The time scale for the ions is longer, of the order of minutes, making it more relevant for magnetotail reconnection processes.

The three-region approximation also shows that a more adequate treatment of systems containing anisotropic ions requires an accurate treatment of large ion orbits. In the above works, the complicated orbit integrals were approximated. The inability to treat orbits exactly leads to a number of limitations such as the constant-ψ approximation (ψ, the perturbed vector potential, is constant in the inner region) and weak anisotropy. In addition, if anisotropic ion effects dominate in the ion intermediate region while the electrons still dominate in the inner region, then the orbits for both species with disparate scale sizes must be treated accurately. Holdren[12] carried out full integro-differential

*Geophysical and Plasma Dynamics Branch, Plasma Physics Division, Naval Research Laboratory, Washington, D.C. 20375-5000.
[†]Laboratory of Plasma and Fusion Studies, University of Maryland, College Park, Maryland 20742.

calculations of the tearing mode in a (relativistic) neutral sheet. In this work, the orbits were calculated numerically and an iterative method was used with discrete variables. As a result, the numerics required were substantial. Recently, Chen and Lee[13] developed an integro-differential method that treats *all* the equilibrium orbits exactly and analytically for both species. This method makes it possible to remove the above limitations and they provide an accurate treatment of highly non-Maxwellian neutral-sheet plasmas with minimal numerical calculation. Indeed, it was found that the eigenmode structure is highly localized at the null plane and that the constant-ψ approximation is not valid for the non-Maxwellian case.

In this paper, we incorporate the recent results into a model of magnetic reconnection in the magnetotail. We will first review the theoretical results described above pertaining to the non-Maxwellian collisionless tearing-mode instability in a neutral sheet. In this paper, we will not directly consider the effects of the magnetic field component normal to the equatorial plane.

THE NON-MAXWELLIAN TEARING MODE (NMTM)

Consider a neutral sheet whose magnetic field profile is given by $B_0(z) = B_x(z)\,x$ such that $B_x(-z) = -B_x(z)$. The current density is given by $J(z) = J_0(z)y = (c/4\pi)\nabla \times B_0$ with $J_0(-z) = J_0(z)$. The equations of motion admit three independent constants of motion: $H_\perp = (\tfrac{1}{2})mv_\perp^2$, $P_y = mv_y + (q/c)A_y(z)$, and $H_\parallel = (\tfrac{1}{2})mv_\parallel^2$, where $v_\perp^2 = v_y^2 + v_z^2$ and $v_\parallel = v_x$. Here $A(z) = A_y(z)y$ is the equilibrium vector potential, and the equilibrium electric field is set equal to zero. The tearing instability can be described by perturbations of the form $\psi(x,z,t) = \psi(z)\exp(ikx - i\omega t)$ where the wave vector $k = kx$ is parallel to the equilibrium magnetic field. Although not necessary, we will consider perturbations with $|\omega| \ll \omega_{ci}$ and neglect perturbed scalar potential. We assume charge neutrality to first order.

We consider a class of equilibria described by distribution functions of the type $F_j = F_j(H_\perp,P_y,H_\parallel)$ where j is the species index. Using the standard method of characteristics, the first order Vlasov distribution function f_j for the species j can be written as

$$f_j = \frac{q_j}{c}\,\frac{\partial F_j}{\partial P_{yj}}\,\psi + i\omega q_j\,\frac{\partial F_j}{\partial H_{\perp j}}\,S_j$$
$$- ikq_j\left(\frac{\partial F_j}{\partial H_{\parallel j}} - \frac{\partial F_j}{\partial H_{\parallel j}}\right)v_x S_j \quad (1)$$

where ψ is the perturbed vector potential $\psi = A_{1y}$ and S_j is the orbit integral given by

$$S_y = -\frac{1}{c}\int_\infty^t dt'\,v_y'\,\psi \quad (2)$$

The integration is carried out along equilibrium orbits. The perturbed vector potential then satisfies the equation

$$\frac{d^2\psi}{dz^2} - k^2\psi + \frac{4\pi}{c}\,J_{1y}(z) = 0 \quad (3)$$

where

$$J_{1y} = \sum_j q_j \int d^3v\,v_y\,f_j$$

As a general remark, the full solution of the first-order problem can be obtained by solving the integro-differential equation (Eq. 3). In the work of Forslund,[9] an approximate dispersion relation for weakly anisotropic (bi-Maxwellian) electrons was obtained using velocity moments to express the orbit integral. Essentially the same dispersion relation was obtained by Chen and Palmadesso[10] by matching the inner $|z| \le (2\rho_e\delta)^{1/2}$ and outer $|z| > (2\rho_e\delta)^{1/2}$ solutions which are both analytically accessible. In this work, the straight-line orbit approximation[4] was used for the axis-crossing orbits in the electron inner region for both species. Both approaches, however, neglect the fact that the axis-crossing ion orbits extend far beyond the electron region. By including the axis-crossing ion orbits, Chen and Palmadesso[10] showed the existence of an ion intermediate region, $(2\rho_e\delta)^{1/2} < |z| < (2\rho_i\delta)^{1/2}$, in which the anisotropic ion contribution dominates. By matching the solutions in the three regions, they found that, for a given degree of anisotropy, the growth rate based on the three-region approximation can exceed, by nearly one order of magnitude, the growth rates obtained by the two-region approximation. Figure 1 summarizes these results, showing the growth rates for several values of $T_{i\perp}/T_{i\parallel}$ with isotropic electrons. In particular, Curve e ($T_{i\perp}/T_{i\parallel} = 1.15$), which is obtained using the two-region approximation, is much lower than Curve d, which is obtained by the three-region approximation for the same value of $T_{i\perp}/T_{i\parallel}$. Note that the maximum-$\gamma$ wavelengths are considerably shorter for the anisotropic case with $T_{i\perp}/T_{i\parallel} > 1$ than for the isotropic case.

In the above works, various approximations were used to treat the orbit integrals such as the straight-line orbit approximation, constant-ψ approximation, and expansion in terms of velocity moments. These simplifications constitute severe limitations on the applicability of the analyses to physically interesting systems. Chen and Lee[13] developed an integro-differential equation method that uses all the equilibrium orbits for both species exactly and analytically. They showed that the dispersion relation for $|\omega| < \omega_{ci}$ has the general form

$$\frac{\bar{\gamma}}{\bar{k}} = \left(\frac{\rho_e}{\delta}\right)D \quad (4)$$

where $\bar{\gamma} = \gamma/\omega_{ci}$ with $\omega_{ci} = eB_0/m_i cB_0$ and $\bar{k} = k\delta$. Here, D is a universal function of $T_{e\perp}/T_{e\parallel}$, $T_{i\perp}/T_{i\parallel}$,

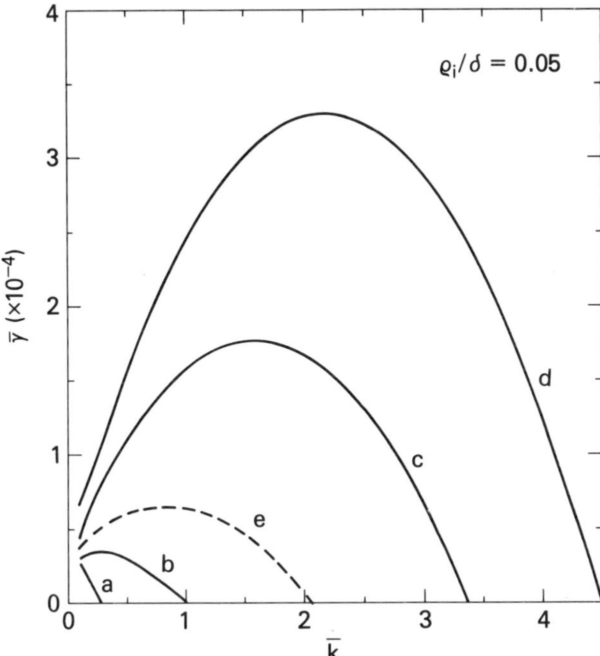

Figure 1—Normalized growth rates $\bar{\gamma}$ using the three-region approximation. The values of $T_{i\perp}/T_{i\parallel}$ are a, 0.9; b, 1.0; c, 1.1; and d, 1.15. Curve e is the result of the two-region approximation for $T_{i\perp}/T_{i\parallel} = 1.15$.

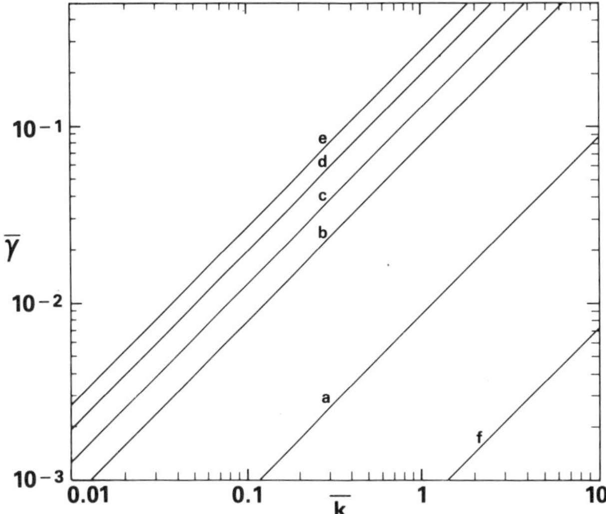

Figure 2—(From Ref. 13) The dispersion relation for a highly non-Maxwellian neutral sheet (Eq. 4). The values of $T_{e\perp}/T_{e\parallel}$ are a, 1.0; b, 1.25; c, 1.5; d, 2.0; e, 3.0; and f, 3.0, for the second branch to become unstable. Here, $T_{i\perp}/T_{i\parallel} = 1$, $T_{e\perp}/T_{i\perp} = 0.5$, and $\rho_i/\delta = 0.02$.

and $T_{e\perp}/T_{i\perp}$. Using the Galerkin method, they solved the integro-differential equation (Eq. 3) and obtained the universal function D and the eigenmode structure. Figure 2, reproduced from Chen and Lee,[13] shows the dispersion relation for several values of $T_{e\perp}/T_{e\parallel}$. The preferred wavelength is seen to be reduced significantly although the neglect of the scalar potential did not allow the accurate determination of the value of k for the maximum growth rates. In this work, both species are allowed to be highly non-Maxwellian. As a result, the instability is dominated by the electrons. A non-Maxwellian ion population with isotropic electrons is expected to have similar behavior but with smaller enhancement of the growth rate. The numerical simulation of Ambrosiano et al.[14] closely supports the behavior described above for the bi-Maxwellian case.

As an order-of-magnitude example, we estimate the linear e-folding times for relatively thick ($\rho_i/\delta \ll 1$) non-Maxwellian neutral sheets. For example, with $\rho_i/\delta = 0.05$ and $B_0 = 20\gamma$, Fig. 1 shows $\gamma^{-1} \simeq 27$ min for $T_{i\perp}/T_{i\parallel} = 1.15$ while $\gamma^{-1} \simeq 4.5$ hours for the isotropic case. For $\rho_i/\delta = 0.1$ and $T_{i\perp}/T_{i\parallel} = 1.5$, we find $\gamma^{-1} \simeq 4$ min (see Ref. 15). For thinner neutral sheets, the fastest e-folding time decreases roughly as $(\delta/\rho_e)^{5/2}$. For the highly non-Maxwellian case with $\rho_i/\delta = 0.02$, Fig. 2 shows that γ/ω_{ci} can range from 10^{-3} to 10^{-1}, corresponding to e-folding times of $\gamma^{-1} \simeq 8$ min to $\gamma^{-1} \simeq 0.1$ min, respectively. In the above estimates, we have used $T_{i\perp}/T_{e\perp} = 2$. Thus, a non-Maxwellian neutral sheet can exhibit e-folding times ranging from a small fraction of a minute to tens of minutes depending on the distribution functions. This is to be contrasted with the isotropic e-folding times of hours for comparable thicknesses.

MAGNETIC RECONNECTION IN A NON-MAXWELLIAN PLASMA

Equation 1 shows that the type of non-Maxwellian features must be representable by distribution functions of the form $F = F(H_\perp, P_y, C)$, since the first two terms are the usual tearing terms. Here, C is a constant of motion independent of $H_\perp$ and P_y. In the above discussion, we have used the case with $C = H_\parallel$. Note that it is possible to construct a non-Maxwellian distribution of the form $F(H_\perp, P_y)$, etc. However, this type of distributions can only give rise to the conventional tearing mode. The NMTM can rearrange the internal particle energy as well as the magnetic energy on a fast time scale (possibly as high as $\gamma \approx \omega_{ci}$). The instability should lead to the formation of small-scale islands. The islands then should coalesce rapidly,[16] which is a well documented process.[17–20] This scenario is also borne out by the anisotropic neutral-sheet simulation of Ambrosiano et al.[14] In addition, their results show that the conversion of magnetic energy to kinetic energy takes place faster and more efficiently. We note that the above process may play a role in "driven reconnection" since the driving forces may preferentially increase the perpendicular temperature. The magnitude of anisotropy and hence the growth rate depend on the strength of the forcing function.

We now consider a possible role the non-Maxwellian tearing mode may play in a physical system such as the magnetotail. Consider a Maxwellian equilibrium neutral sheet. Suppose the particle distribution function undergoes some large-scale changes due to changes in the external conditions. This leads to a rearrangement of particle energy. If the system admits three independent invariants of motion as described in the preceding paragraphs, then the free energy associated with the non-Maxwellian features can drive the NMTM. This occurs along with the usual tearing mode. For typical magnetotail parameters, the linear *e*-folding time can be reduced to a matter of minutes or less depending on the non-Maxwellian features. Subsequently, small-scale ($k\delta \gg 1$) islands are formed which then coalesce rapidly. The entire process may manifest itself as a triggering of reconnection following changes in physical conditions. It may also appear to be explosive.[21-24]

The above process suggests that the coupling of the solar wind to the internal particle energy of the magnetotail is important. The non-Maxwellian distribution that can be generated by external changes depends on the strength of the changes and can drive the instability accordingly. In the context of a substorm, the substorm may appear to be triggered or directly driven in response to some changes in the solar wind conditions with the NMTM providing a mechanism for the "solar wind control." The time delay is determined by the fast non-Maxwellian tearing time scales and the subsequent coalescence of islands. Note that there has been considerable effort to classify substorms in two basic categories: storage-unloading of energy and direct-drive by the solar wind.[25] The above mechanism contains attributes of both types. In this regard, the differential memory of particle distributions suggested by Chen and Palmadesso[26] provides a natural means for the generation of necessary non-Maxwellian distributions following changes in external conditions.

Finally, it has been suggested[27,28] that the magnetic field component normal to the equatorial plane (not treated here) modifies the tearing mode properties significantly if the electrons are magnetized, possibly stabilizing the ion tearing mode.[29] However, Coroniti[30] has pointed out that pitch angle scattering can effectively demagnetize the electrons so that the tearing mode can still be unstable. Recently, Chen and Palmadesso[10] have shown that the normal component renders the single-particle motion in a quasineutral sheet nonintegrable so that the electrons may be only partially magnetized. In addition, it has been suggested by Goldstein and Schindler[31] that the ionospheric influences may contribute to the destabilization of the (conventional) tearing mode. Thus, there are strong indications that the non-Maxwellian tearing mode is important for magnetotail dynamics.

ACKNOWLEDGMENT—This research was supported by the National Aeronautics and Space Administration and the Office of Naval Research.

REFERENCES

[1] H. P. Furth, "The Mirror Instability for Finite Particles Gyroradius," *Nucl. Fusion Suppl. Pt. 1*, 169 (1962).

[2] D. Pfirsch, Mikroinstabilitäten vom Spiegeltyp in inhomogenen Plasmen (Mirror Type Instabilities in Inhomogeneous Plasmas), *Z. Naturforsch* **17a**, 861 (1962).

[3] G. Laval, R. Pellat, and M. Vuillemin, "Instabilities Electromagnetiques des Plasmas sans Collisions," in *Plasma Physics and Controlled Nuclear Fusion Research*, Vol II, International Atomic Energy Agency, Vienna, p. 259 (1966).

[4] B. Coppi, G. Laval, and R. Pellat, "Dynamics of the Geomagnetic Tail," *Phys. Rev. Lett.* **16**, 1207 (1966).

[5] K. Schindler, in *Proceedings of the Seventh International Conference on Phenomena in Ionized Gases*, Vol II, Gradevinska Knjiga, Beogard, Yugoslavia, p. 736 (1966).

[6] J. B. Greenly and B. U. O. Sonnerup, "Tearing Modes at the Magnetopause," *J. Geophys. Res.* **86**, 1305 (1981).

[7] K. B. Quest and F. V. Coroniti, "Tearing at the Dayside Magnetopause," *J. Geophys. Res.* **86**, 3280 (1981).

[8] G. Laval and R. Pellat, "Stability of the Plane Neutral Sheet for Oblique Propagation and Anisotropic Temperature," in *Proceedings of the ESRIN Study Group*, Frascati, Italy (European Space Research Organization, Neuilly-sur-Seine, France), p. 5 (1968).

[9] D. W. Forslund, Ph.D thesis, Princeton University (1968).

[10] J. Chen and P. Palmadesso, "Tearing Instability in an Anisotropic Neutral Sheet," *Phys. Fluids* **27**, 1198 (1984).

[11] B. Coppi and M. N. Rosenbluth, "Model for the Earth's Magnetic Tail," in *Proceedings of the ESRIN Study Group*, Frascati, Italy (European Space Research Organization, Neuilly-sur-Seine, France), p. 1 (1968).

[12] J. P. Holdren, Ph.D thesis, Stanford University (1970).

[13] J. Chen and Y. C. Lee, "Collisionless Tearing Instability in a Non-Maxwellian Neutral Sheet: An Integro-Differential Formulation," *Phys. Fluids* **28**, 2137 (1985).

[14] J. Ambrosiano, L. C. Lee, and Z. F. Fu, "Simulation of the Collisionless Tearing Instability in an Anisotropic Neutral Sheet," *J. Geophys. Res.* **91**, 113 (1986).

[15] J. Chen, P. Palmadesso, J. A. Fedder, and J. G. Lyon, "Fast Collisionless Tearing in an Anisotropic Neutral Sheet," *Geophys. Res. Lett.* **11**, 12 (1984).

[16] J. M. Finn and P. K. Kaw, "Coalescence Instability of Magnetic Islands," *Phys. Fluids* **20**, 72 (1977).

[17] P. L. Pritchet and C. C. Wu, "Coalescence of Magnetic Islands," *Phys. Fluids* **22**, 2140 (1979).

[18] A. Bondeson, "Linear Analysis of the Coalescence Instability," *Phys. Fluids* **26**, 1275 (1983).

[19] J. N. Leboeuf, T. Tajima, and J. M. Dawson, "Dynamic Magnetic X Points," *Phys. Fluids* **25**, 784 (1982).

[20] A. Bhallacharjee, F. Brunel, and T. Tajima, "Magnetic Reconnection Driven by the Coalescence Instability," *Phys. Fluids* **26**, 3332 (1983).

[21] A. A. Galeev, F. V. Coroniti, and M. Ashour-Abdalla, "Explosive Tearing Mode Reconnection in the Magnetospheric Tail," *Geophys. Res. Lett.* **5**, 707 (1978).

[22] Teresawa (1981).

[23] F. V. Coroniti, "Explosive Tail Reconnection: The Growth and Expansion Phases of Magnetospheric Substorms," *J. Geophys. Res.* **90**, 7427 (1985).

[24] T. Tajima and J.-I. Sakai, "Explosive Coalescence of Magnetic Islands and Explosive Particle Acceleration" (preprint, 1986).

[25] S.-I. Akasofu, "Energy Coupling Between the Solar Wind and the Magnetosphere," *Space Sci. Rev.* **28**, 121 (1981).

[26] J. Chen and P. J. Palmadesso, "Non-Maxwellian Free-Energy Generation in the Magnetotail Due to Chaotic Particle Motion" (this volume, 1986).

[27] A. A. Galeev and L. M. Zelenyi, "Tearing Instability in Plasma Configurations," *Sov. Phys. JETP* **43**, 1113 (1976).

[28] B. Lembege and R. Pellat, "Stability of a Thick Two-Dimensional Quasineutral Sheet," *Phys. Fluids* **25**, 1995 (1982).

[29] K. Schindler, "A Theory of the Substorm Mechanism," *J. Geophys. Res.* **79**, 2803 (1974).

[30] F. V. Coroniti, "On the Tearing Modes in Quasineutral Sheets," *J. Geophys. Res.* **85**, 6719 (1980).

[31] H. Goldstein and K. Schindler, "On the Role of the Ionosphere in Substorms: Generation of Field-Aligned Currents," *J. Geophys. Res.* **83**, 2574 (1978).

A PARTICLE SIMULATION OF FLUX TRANSFER EVENTS AT THE EARTH'S MAGNETOPAUSE

D. Q. Ding, L. C. Lee, and Z. F. Fu*

The flux transfer events observed by ISEE satellites indicated that the dayside magnetic reconnection geometry is different from that in Dungey's single X-line model. The multiple X-line reconnection process proposed by Lee and Fu seems to explain the flux transfer events observed at the dayside magnetopause. We have studied the multiple X-line reconnection by computer simulations based on a two-dimensional magnetostatic particle code. A driven boundary condition is applied at the two sides of the simulation domain with an incoming plasma flow while a free boundary condition is used at the other two sides with an outflow. Repeated formation and convection of magnetic islands are observed in the simulation. Magnetic islands with various sizes are observed. Particles are found to be heated during the reconnection process. The convection speed of the island may exceed the Alfvén speed. The recurrence time of magnetic islands is found to be 5–15 min. Our results are consistent with satellite observations of flux transfer events. Superthermal ions with energy of 1–50 keV are found to be present during the multiple X-line reconnection process.

INTRODUCTION

The open magnetosphere model was first proposed by Dungey in 1961. In this open model, the interplanetary magnetic fields (IMFs) and geomagnetic fields reconnect along a single neutral line (X-line) near the subsolar region of the dayside magnetopause. However, the flux transfer events observed by ISEE satellites[1] have suggested that the dayside magnetic reconnection geometry is different from that in Dungey's quasisteady-state model.[2] Localized open magnetic flux tubes with a cross section of 1 R_e^2 were observed to occur repeatedly every 5–15 minutes. Magnetic flux tubes with smaller cross sections are also observed .[3,4] Recent observations of FTEs[3,5] suggested that magnetic field lines inside the flux tubes are highly twisted, indicating the presence of field-aligned currents.

Recently, Lee and Fu[6] proposed a multiple X-line reconnection model for the dayside magnetopause, that can explain the occurrence and most of the observed properties of FTEs. The multiple X-line reconnection can be considered as a magnetic reconnection process between two approaching magnetized plasmas, which is modified by the development of tearing instability. The multiple X-line reconnection is not a steady process. Repeated formation and convection of magnetic islands (magnetic flux tubes) are expected to occur at the magnetopause. A schematic sketch of the magnetic field lines at the magnetopause, based on the multiple X-line reconnection model, is reproduced as Fig. 1 from the paper by Lee and Fu.[6] It is a snapshot of the dynamical magnetopause, in which three X-lines are assumed to exist. Figure 1a is a three-dimensional perspective view of the magnetopause and of the reconnected magnetic field lines, while Fig. 1b is the projection of the field lines in the noon-midnight meridian plane. It is seen that two magnetic flux tubes containing twisted field lines are formed as a natural consequence of the multiple X-line reconnection. The flux tubes northward and southward of the central X-lines will be convected, respectively, to the north and south cusp regions by the magnetosheath plasma flow and the tension force of the reconnected flux tubes. The sense of field twisting as shown in Fig. 1a is consistent with the ISEE observations.[6].

Recently, Fu and Lee[7] carried out an MHD simulation of the multiple X-line reconnection process, in which repeated formation and convection of magnetic islands are observed. However, in the MHD simulation, the value of the anomalous resistivity has to be arbitrarily assumed, which may not be realistic in the collisionless magnetospheric plasma. Simulations of collisionless tearing or reconnection have also been reported by Brunel et al.,[8] Leboeuf et al.,[9] and Ambrosiano et al.[10] However, some periodic boundary conditions are used in their simulations and hence repeated formation and convection of magnetic islands are not observed.

In this paper we study the multiple X-line reconnection process based on a particle simulation code and apply the results to the FTEs at the dayside magnetopause. It is found that magnetic flux tubes can indeed be formed intermittently and repeatedly, as observed by ISEE satellites.

SIMULATION MODEL

The simulation in this study was carried out by using a two-dimensional magnetoinductive particle code, which is described in a separate paper. This simulation code is similar to that used by Ambrosiano et al.,[10] in which the electron currents and electrostatic effects are ignored and only the ion current contributions are included. The electron effects on reconnection may be important (e.g., Refs. 11, 12) and will be examined in a future study.

The simulation is carried out in the x-z plane, in which the current sheet is parallel to the $x = 0$ plane. The initial plasma and magnetic field configurations consist of

*Geophysical Institute and Department of Physics, University of Alaska, Fairbanks, Alaska 99775-0800.

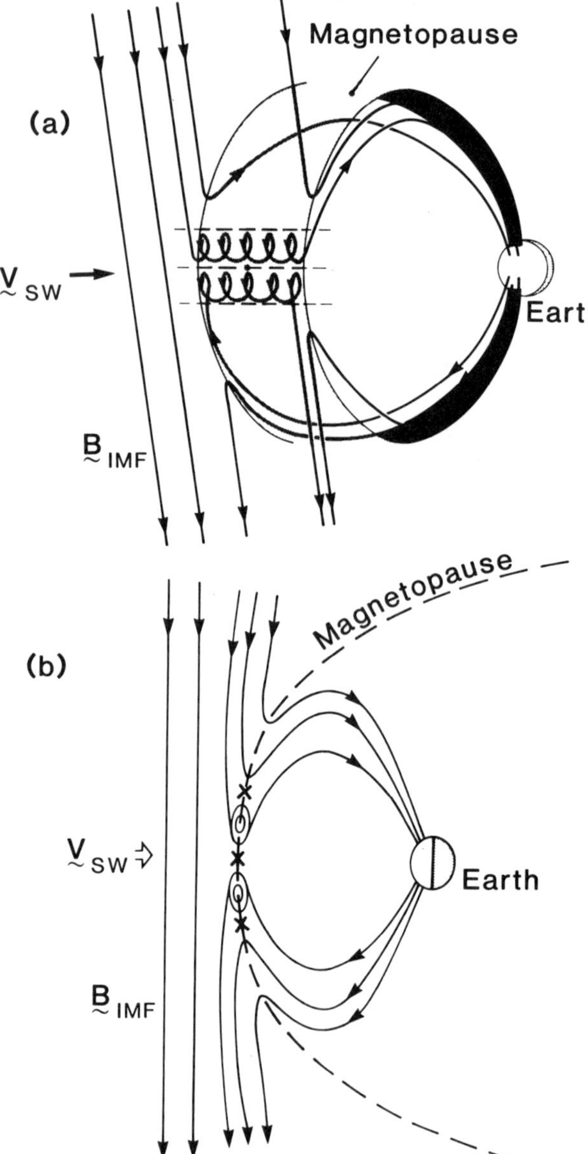

Figure 1—(a) A perspective view of the open magnetic flux tubes and the regular open field lines in the multiple *X*-line model. (b) The projection of flux tubes in the noon-midnight meridian plane (from Ref. 6).

a uniform background plasma that is added to a Harris[13] current sheet. The magnetic field $B(x,z,0)$ and plasma number density $n(x,z,0)$ at $t = 0$ are given by

$$B(x,z,0) = B_0 \tanh(x/\lambda)\,\hat{z} \qquad (1)$$

$$n(x,z,0) = N_c \operatorname{sech}^2(x/\lambda) + N_b , \qquad (2)$$

where B_0 is the magnitude of the magnetic field outside the current sheet, λ is the thickness, N_c is the particle number density associated with the current sheet, and N_b is the number density associated with the background species. The velocity distribution of the background particles is a Maxwellian distribution with a

thermal speed V_{th}, while the current sheet particles (N_c) have a drift-Maxwellian distribution with a thermal speed V_{th} and a drift velocity V_d in the y-direction.

Let the ion thermal energy be $T_i = m_i V_{th}^2/2$, the ion gyrofrequency be $\Omega = eB_0/m_i c$, and the gyroradius be $\rho = V_{th}/\Omega$, where e is the magnitude of an electron charge, m_i is the ion mass, and c is the speed of light. Let the Alfvén speed be $V_A = B_0 (4\pi N_b m_i)^{-1/2}$, which can be related to the thermal speed by $V_A = V_{th}\beta^{-0.5}$, where $\beta \equiv 8\pi N_b T_i/B_0^2$. The drift speed V_d can be related to the thickness of the current sheet λ by $V_d/V_{th} = \rho/\lambda$. The grid size is taken to be Δ. The dimension of the simulation domain is $L_x \times (2L_z)$, where $L_x = 32\Delta$ and $L_z = 64\Delta$.

A symmetric boundary condition is imposed in the $x = 0$ plane. At $x = L_x$, a driven boundary condition is imposed, in which background particles are injected into the simulation domain with a flux density of $N_b V_1$, where V_1 is the incoming drift speed of the background particles. Let $A(x,z,t)$ be the vector potential, which is related to the magnetic field by $B = \nabla \times (A\hat{y})$. At $x = L_x$ we have $A(x = L_x, z, t) = A(x = L_x, z, 0) - V_1 B_0 t$ which is equivalent to imposing at the boundary a convection electric field $E_1 = V_1 B_0/cy$. A free boundary condition is imposed at $z = \pm L_z$. The particles are free to escape from the buffer zones which are located outside the boundaries at $z = \pm L_z$. The boundary condition for the vector potential at $z = \pm L_z$ is chosen to be $\partial A/\partial z = 0$. It is noted that the "free" boundary condition allows the outward connection of magnetic islands formed in the simulation region even when E_y is applied uniformly at the boundary.

SIMULATION OF FLUX TRANSFER EVENTS

We have simulated several cases in which the inflow speed $V_1 = 0.1, 0.2, 0.3,$ and $0.42\ V_A$. An example of the simulation is shown in Fig. 2a and b, in which the magnetic field lines, positions of particles, and the flow patterns at various times are plotted. The parameters used are $R_0 = V_1/V_A = 0.42$, $V_d = 0.25\ V_{th}$, $\lambda = 4\Delta$, and $\beta = 1.0$. The initial number of particles is 25,000, of which 5000 particles belong to the current-sheet species and the remaining 20,000 are the background particles. Figure 2a shows that magnetic islands (flux tubes) with various sizes are repeatedly formed and convected out of the boundaries. The coalescence of two magnetic islands is observed, leading to the formation of a larger island, such as magnetic island A shown in Fig. 2a. Magnetic islands can grow to a large size before they are convected out of the simulation domain. Following the formation and convection of a large island, several small islands tend to be formed. The length (S_z) and width (S_x) of the large magnetic islands are, respectively, $S_z \simeq 60\rho$ and $S_x \simeq 30\rho$. The recurrence time τ of magnetic islands on each side of the simulation domain is found to be $\tau \simeq 200$–$300\Omega^{-1}$.

Figure 2b shows the distribution of particles and the flow pattern in the x-z plane. The particle density inside the magnetic islands is high. We also find that the

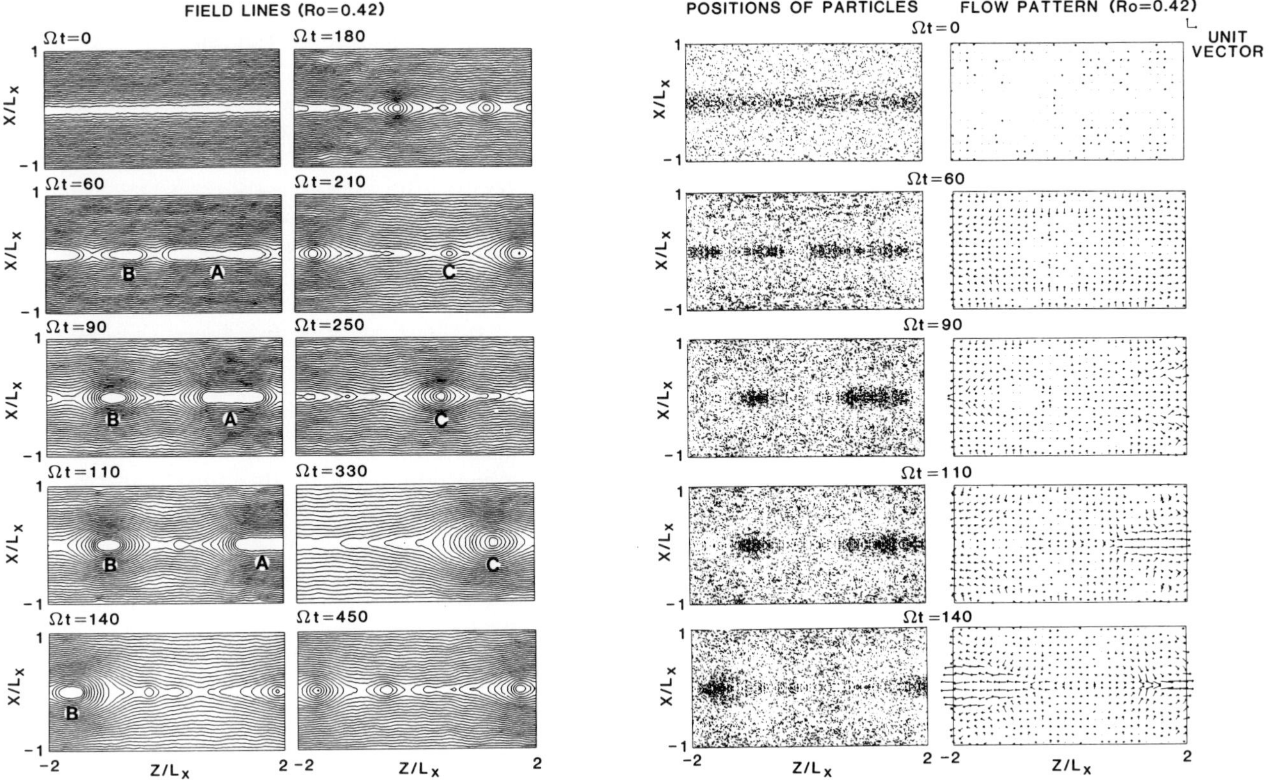

Figure 2—(a) Magnetic field lines at various simulation times Ωt. the imposed reconnection rate at $x = L_x$ is $R_0 = 0.42$. (b) The corresponding positions of particles and flow patterns at various times. The unit vectors correspond to the ion thermal speed V_{th}.

removal of a large magnetic island usually leads to a drastic decrease of the total particle number in the simulation domain. From the flow patterns, it is found that high-speed plasma flows ($\gtrsim V_A$) are associated with the convection of large magnetic islands.

Figure 3 shows the positions of the centers of three magnetic flux tubes A, B, and C as a function of time t. It is noted that the convection speed at the earlier stage is small and may increase to Alfvén speed at the later stage. For example, the convection speed of flux tube A is $\sim 0.3 V_A$ at $\Omega t = 70$ and is $\sim 1.2 V_A$ at $\Omega t = 110$.

The particle kinetic energy, magnetic energy, and the total energy are plotted in Fig. 4 as a function of time. It is seen that (a) the kinetic energy and magnetic energy seem to vary together in time, and (b) the increase (decrease) of kinetic and magnetic energy is associated with the formation (ejection) of the magnetic islands.

From the diagnosis of the particle thermal energy (not shown) it is found that the particle thermal energy increases from T_i at the injection boundary ($x = L_x$) to 2–$3T_i$ at the center of the current sheet or magnetic islands during the phases of enhanced reconnection. The increase of thermal energy is mainly associated with the presence of superthermal particles. Figure 5 shows the energy distribution functions of ions at $\Omega T = 0$ and 150. It can be seen that many superthermal ions, with energy up to $30T_i$, are present at $\Omega t = 150$. We also find that the number of superthermal ions is greatly reduced

at $\Omega t = 200$ when the reconnection rate is small. Note that this result cannot be obtained from MHD simulations.

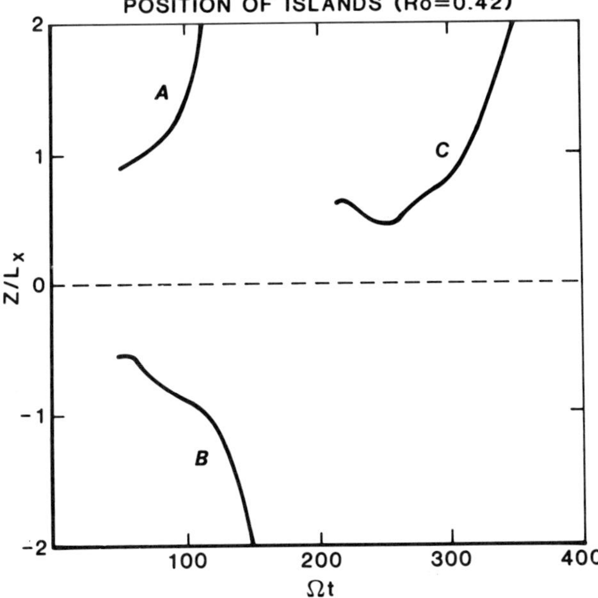

Figure 3—The central positions of three large flux tubes A, B, and C as a function of time Ωt.

333

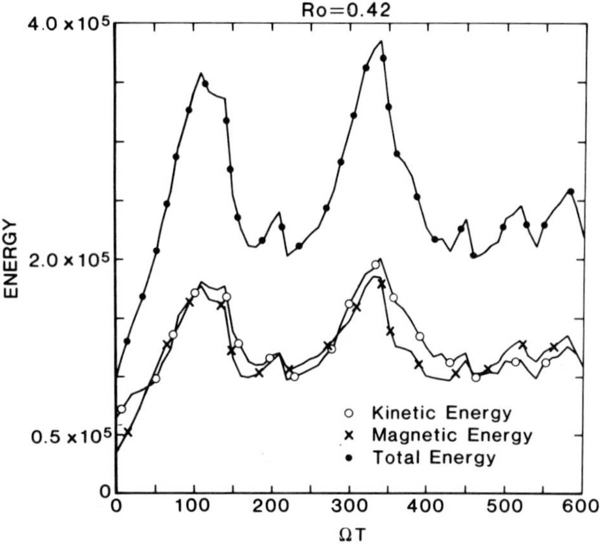

Figure 4—The particle kinetic energy, magnetic energy, and the total energy in the simulation domain as a function of time.

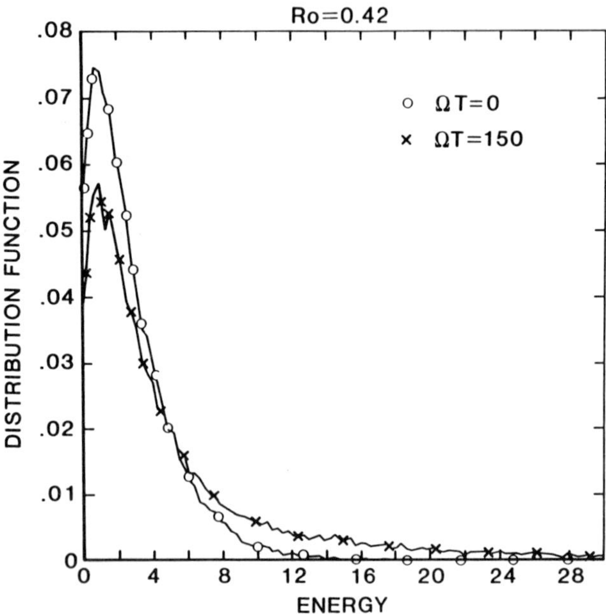

Figure 5—The normalized particle distributions as a function of particle energy at $\Omega t = 0$ and 150. The energy is in units of $T_i/2$.

COMPARISON WITH OBSERVATIONS

The simulation results can be compared with the satellite observations of flux transfer events at the dayside magnetopause. The typical plasma and field parameters observed at the dayside magnetosheath are $B = 20\text{–}30$ nT, $n = 10\text{–}20$ cm^{-3}, and $T_i = 0.2\text{–}1$ keV. Therefore,

we have $\Omega = 2\text{–}3$ s^{-1}, $V_A \simeq 200\text{–}300$ km/s, $V_{th} \simeq 200\text{–}400$ km/s, and $\rho \simeq 70\text{–}150$ km. The sizes of the large magnetic flux tubes are $S_x \simeq 30\rho \simeq 0.35\text{–}0.75$ R$_e$ and $S_z \simeq 60\rho \simeq 0.7\text{–}1.5$ R$_e$, which are consistent with observations.[1,12] It is noted that the size of flux tubes is independent of the initial current thickness since reconnection occurs after the current sheet is pinched by the driving force at the boundary.

The recurrence time of flux transfer events for the case in Fig. 2 is $\tau \simeq 200\text{–}300\Omega^{-1} \simeq 2\text{–}3$ min. However, we have also simulated cases with the imposed rate $R_0 = V_1/V_A = 0.1, 0.2$, and 0.3. It is found that the recurrence time τ is approximately inversely proportional to R_0 (cf, Ref. 14). Therefore, if $R_0 \simeq 0.1$ at the dayside magnetopause, we have $\tau \simeq 8\text{–}12$ min, which is consistent with observations.

The convection speed of flux tubes can reach the Alfvén speed ($V_A \simeq 200\text{–}300$ km/s) as shown in Fig. 3. ISEE observation also indicates that the convection speed of FTEs is on the order of 150–300 km/s.

In conclusion, our particle-code simulation of the multiple X-line reconnection process can reproduce some of the observed features of FTEs. The results provide further support to the multiple X-line reconnection model of the dayside magnetopause proposed by Lee and Fu.[6] Furthermore, superthermal particles are found to be present during the reconnection process.

ACKNOWLEDGMENT—This work is supported by National Science Foundation grant ATM85-21115 and Department of Energy grant DE-AT06-76R0005 to the University of Alaska.

REFERENCES

[1] C. T. Russell and R. C. Elphic, "ISEE Observations of Flux Transfer Events at the Dayside Magnetopause," *Geophys. Res. Lett.* **6**, 33 (1979).

[2] J. W. Dungey, "Interplanetary Field and the Auroral Zones," *Phys. Rev. Lett.* **6**, 47 (1961).

[3] M. A. Saunders, C. T. Russell, and N. Sckopke, "Flux Transfer Events: Scale Size and Interior Structure," *Geophys. Res. Lett.* **11**, 131 (1984).

[4] R. P. Rijnbeck, S. W. H. Cowley, D. J. Southwood, and C. T. Russell, "A Survey of Dayside Flux Transfer Events Observed by ISEE-1 and 2 Magnetometers," *J. Geophys. Res.* **89**, 786 (1984).

[5] G. Paschmann, G. Haerendel, I. Papamastorakis, N. Sckopke, S. J. Bame, J. T. Gosling, and C. T. Russell, "Plasma and Magnetic Characteristics of Magnetic Flux Transfer Events," *J. Geophys. Res.* **87**, 2159 (1982).

[6] L. C. Lee and Z. F. Fu, "A Theory of Magnetic Flux Transfer at the Earth's Magnetopause," *Geophys. Res. Lett.* **12**, 105 (1985).

[7] Z. F. Fu and L. C. Lee, "Simulation of Multiple X-line Reconnection at the Dayside Magnetopause," *Geophys. Res. Lett.* **12**, 291 (1985).

[8] F. Brunel, T. Tajima, and J. M. Dawson, "Fast Magnetic Reconnection Processes," *Phys. Rev. Lett.* **49**, 323 (1982).

[9] J. N. Leboeuf, T. Tajima, and J. M. Dawson, "Dynamic Magnetic X Points," *Phys. Fluids* **25**, 784 (1982).

[10] J. J. Ambrosiano, L. C. Lee, and Z. F. Fu, "Simulation of the Collisionless Tearing Instability in an Anisotropic Neutral Sheet," *J. Geophys. Res.* **91**, 113 (1986).

[11] B. Lembege and R. Pellat, "Stability of a Thick Two-Dimensional Quasi-Neutral Sheet," *Phys. Fluids* **22**, 1995 (1982).

[12] D. W. Swift, "Numerical Simulations of Tearing Mode Instabilities," *J. Geophys. Res.* **91**, 219 (1986).

[13] E. G. Harris, "On a Plasma Sheath Separating Regions of Oppositely Directed Magnetic Field," *Nuovo Cimento* **23**, 15 (1962).

[14] L. C. Lee, Z. F. Fu, and S.-I. Akasofu, "A Simulation of Forced Reconnection Processes and Magnetospheric Substorms and Storms," *J. Geophys. Res.* **90**, 10896 (1985).

V. ACTIVE DIAGNOSIS OF THE GEOMAGNETOTAIL

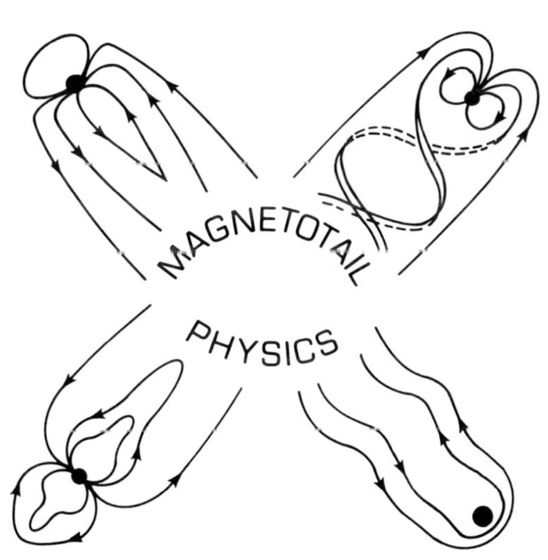

TAIL EXPLORATION AND TAIL FORMATION
WITH ARTIFICIAL PLASMA CLOUDS

G. Haerendel*

Only very few plasma injection experiments were carried out in the earth's magnetotail. The first of them, in 1969, led to the first high-altitude determination of substorm associated field-aligned currents. Four barium and lithium plasma injections in the plasma sheet in 1985 were not successful with respect to the desired study of plasma transport in the tail by means of tracer ions. However, useful (but not yet analyzed) data bearing on the tail dynamics were provided by in-situ measurements and optical coverage of the injected barium clouds.

The second part of the paper deals with the formation processes of comet-like tails in barium releases in the solar wind. The momentum coupling between the two plasmas is investigated from an MHD and a one-particle point of view, respectively. The compression of the interplanetary field in the artificial comet's head can be well understood. The formation of a tail is seen essentially as a consequence of momentum balance leading to ion injection from the rear side of the plasma cloud in the downstream direction. Other processes contributing to formation and dynamics of the tail are ion drag along the magnetic field due to the expansion of a heated electron component (toothpaste effect) and separations of magnetic substructures from the comet head (disconnection events). The lifetime of the artificial comet phenomena is ultimately limited by ion extraction from head and tail by the external electric field. All of these processes have their counterparts in natural comets and similar systems like the Venusian ionosphere.

INTRODUCTION

The subject of this conference is the investigation of magnetospheric tails. The task given to this author was to report about exploration of the earth's magnetic tail by means of plasma cloud injections. Such injections have taken place in very low number, one in 1969 and four in 1985. Whereas the results of the first have been reported by Haerendel and Lüst[1] and Haerendel et al.,[2] there is not yet much to report on the four injections in 1985 because of the complexity of the data reduction. For this reason, the focus of this paper was directed to another topic, the formation of visible magnetic tails in barium plasma injection experiments in the solar wind (artificial comet experiments). Although of completely different nature than the object of this conference, they are of considerable interest in an era of intensive cometary research (Giacobini-Zinner and Halley fly-bys). But also the interaction of the solar wind with the ionosphere of planet Venus has several features in common with the behavior of the artificial comets created during the AMPTE mission.

TAIL EXPLORATION

Table 1 summarizes the characteristics of the altogether five releases of barium and lithium plasmas in the magnetic tail of the earth. Of course, numerous barium and other releases have been made at the ionospheric ends of field lines extending into the tail's plasma sheet or lobes. But, although their results bear on the tail dynamics (e.g., the plasma convection), they will not be reviewed in this context. We will strictly adhere to releases in the tail proper, i.e., to experiments above about 10 R_e distance.

*Max-Planck-Institut für Physik and Astrophysik, Institut für extraterrestrische Physik, 8046 Garching b. München, W. Germany.

The first of these plasma injections, from ESA's HEOS I satellite, was made in the near-earth part of the northern tail lobe, the other four inside the plasma sheet. Subsequently, a short summary of the published results of the 1969 release will be given. It will be followed by some very preliminary statements on findings from the 1985 releases.

All barium plasma injections in the weak magnetic fields of the outer magnetosphere and tail have one common property—the time constant, τ_0, of momentum coupling to the ambient plasma is longer than the observation time. Therefore, it is not easily possible to deduce the strength of the ambient electric field from the observed displacements of the plasma cloud, in contrast to the situation in the ionosphere. This time constant, first derived by Scholer,[3] is given by the ratio of the flux-tube content of artificially injected plasma and the rate by which ambient mass gets involved in the momentum coupling along the affected flux tube

$$\tau_0 = \frac{M_{\text{ion}}}{2A_c \, \rho_a \, v_A} \tag{1}$$

Here A_c is the cross section of the flux tube loaded by the mass, M_{ion}, of injected ions, ρ_a is the ambient mass density, and v_A the ambient Alfvén speed. The factor of 2 in the denominator reflects the two Alfvén wings. The expression applies to a plasma cloud extending as an infinite sheet in the direction of motion. In the realistic situation of a finite cloud, the volume to which the momentum is being coupled exceeds that defined by the cross section of the cloud. So, A_c should be multiplied by a factor of 2 (Haerendel[4]).

When inserting the parameters of the three barium releases of Table 1, one obtains values of τ_0 of several

Table 1—Characteristics and ambient parameters of five plasma injection experiments in the earth's magnetotail. τ_0 is the time constant of momentum coupling.

Date	r (R_e)	Element	M_{ion} (kg)	B (nT)	n_a (cm^{-3})	R_c (km)	τ_0 (s)
March 18, 1969	12.4	Ba	0.2	49	$\leqslant 0.1$	40	1.8×10^4
March 21, 1985	12.1	Ba	2.0	9	0.9	280	6.7×10^3
April 11, 1985	18.7	Li	0.4*	11	0.2	30†	5.5×10^2
April 23, 1985	12.1	Li	0.4*	40	1	30†	7×10^1
May 13, 1985	14.4	Ba	2.0	17	0.2	140	3×10^4

*Refers to total released mass.
†Refers to central, initially diamagnetic part of cloud.

hours. This is substantially longer than the optical tracking time in all three cases.

There are two further complications. The first stems from the nonuniform density of the injected plasma. The coupling constant τ_0 is a variable quantity that reaches its highest values in the center of the loaded flux tube. As a consequence, the displacement of the cloud follows a spatially variable acceleration rate given by

$$\frac{dv_c}{dt} = \frac{v_a - v_c}{\tau_0} \qquad (2)$$

(index a stands for ambient and c for cloud). A severe distortion is typically the result. The other complication arises from the large gyroradius. Whenever the characteristic length defined by ambient flow speed (measured from the frame of plasma injection) and gyrofrequency, Ω_i, of the injected ion

$$l_{\text{inj}} = \frac{v_a}{\Omega_i} \qquad (3)$$

is of the order of, or even exceeds, the transverse radius of the cloud, a simple treatment like that underlying Eqs. 1 and 2 is no longer possible. Ion extraction or pick-up effects come into play that introduce new features into the gross behavior of the visible cloud. The relevance of such effects, even for the tail injections, is easily assessed when realizing that with a very moderate ambient convection speed of ~ 10 km/s in a field of 10 nT, l_{inj} exceeds 1000 km, whereas typical transverse cloud dimensions are of the order of a few hundred kilometers.

A last general remark concerning the momentum exchange with the ambient plasma is in order. Although τ_0 exceeds, typically, by far the travel time of an Alfvén wave to the ionospheric boundaries of the system, no manifestations of reflection effects have been observed. All clouds behaved as if embedded in an infinite medium without walls. The Alfvén waves emitted by the injected plasma appear to be damped over scales shorter than the dimensions of the ambient medium (Haerendel[5]).

The Barium Injection Experiment of March 18, 1969

Although optical tracing of the motion of plasma clouds in the tail has limited diagnostic value, there is another property that may provide significant answers. It is the orientation of the long axes of such clouds or their substructures. Naturally, they are aligned with the ambient magnetic field and reflect changes of the magnetic vector. Of course, such information can be readily obtained from on-board magnetometers; however, plasma clouds "paint" individual flux tubes and do not rush through them as a satellite does. The presence of the seeded plasma changes the orientation of the flux tubes very little, as long as the ambient speed is well sub-Alfvénic.

This property was used when analyzing the observed changes of the long axis of the cloud injected at 12.4 R_e during a substorm on March 18, 1969 (Haerendel et al.[2]). The location of this plasma cloud was in the northern tail lobe adjacent to the plasma sheet at 04:30 MLT (Fig. 1). Comparison of the cloud's orientation with the readings of the magnetometer aboard HEOS 1 and ground magnetograms (Fig. 2) on the one hand proved the response of the cloud's alignment to changes

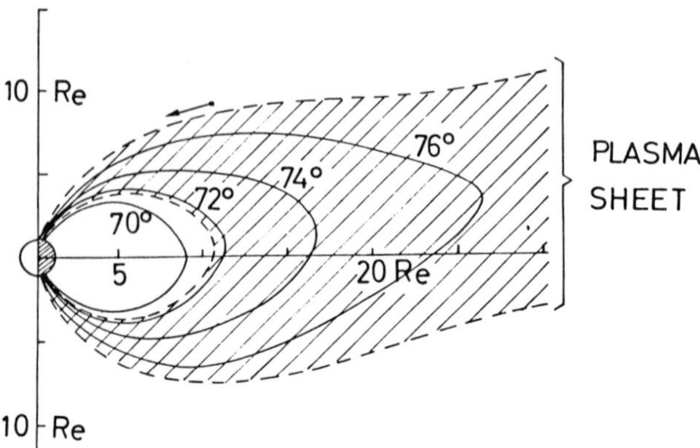

Figure 1—Position of the barium plasma cloud of March 18, 1969 in the meridional plane at ~04:30 LT (Ref. 1).

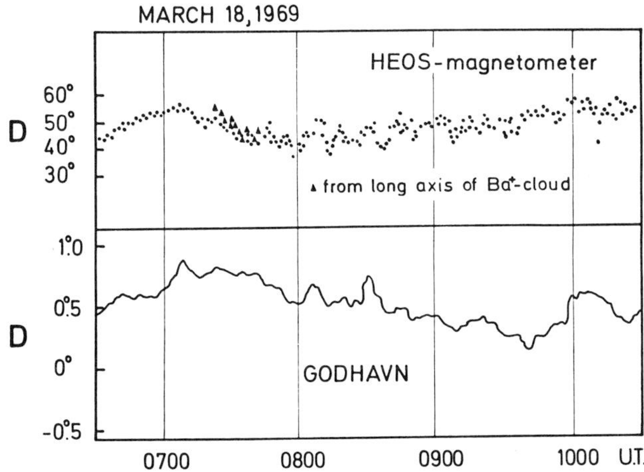

Figure 2—Declination of magnetic field as observed on HEOS I and in Godhavn together with data (triangles) deduced from the orientation of the long axis of the barium cloud; March 18, 1969 (Ref. 2).

of B and, on the other hand, led to the discovery of the existence of field-aligned currents on the morning side of the plasma sheet. Their total strength was determined to be 0.5×10^6 A, directed inward. This is fully compatible with present day's knowledge of the overall pattern of field-aligned currents and of the net inward current on the morning side of a substorm region (Baumjohann et al.[6]).

Except for this modest contribution of the exploration of the large-scale current flow between tail and ionosphere, little more has been learned about the dynamics of the tail from this experiment.

AMPTE Plasma Injections in the Earth's Magnetotail

The first goal of the AMPTE mission was the tracing of magnetospheric transport processes by means of artificially injected tracer ions. After two lithium releases in front of the bow shock, four releases of barium and lithium (two each) were performed in the tail in the spring of 1985. Each of the latter occurred within one hour of local midnight. No tracer ions were detected by the collaborating AMPTE spacecraft CCE (Charge Composition Explorer), which was situated inside 9 R_e at near-equatorial latitudes and at somewhat later local time (McEntire et al.[7]). The reasons for this failure are not easily assessed, but may be closely related to the strong variability of the convection electric field (Cladis and Francis[8]).

In the present context, we are interested in diagnostic information derived from the visible plasma clouds. Only the two barium releases of March 21 and May 13, 1985 can serve this purpose; the lithium plasma is not visible. At the time of writing, hardly any results are available that would shed some light on the tail dynamics. The magnetic conditions were quite different between March 21 and May 13, in that the latter was much more perturbed and the average convection speed substantially

higher. Both clouds developed some field aligned fine structure that showed interesting particular motions. The visible lifetime of both clouds was about ½ hour. What led to the final fairly abrupt brightness decrease was not clear in the first experiment; in the second it was obviously caused by a substantial increase of the external electric field. The development of the cloud of May 13, 1986 is shown in Plate V-1. It was subject to severe shear motions and sudden changes of shape and brightness near 05:50 UT, when the plasma flow velocity reached values above 200 km/s. v_p is shown in Fig. 3 together with the magnetic field as measured on the Ion Release Module (IRM) (courtesy H. Lühr, G. Paschmann, and W. Baumjohann). Note that while the magnetic field is essentially directed in the antisolar direction, the flow was toward dawn, i.e., it represented a transverse convection with E_0 pointing in $+z$-direction.

In order to understand the response of the plasma cloud to the external conditions, it must be realized that the injection parameter (Eq. 3) exceeded 10,000 km around 05:50 UT, whereas the cloud extended no more than a few thousand kilometers. So, ion extraction by the enhanced field was most likely responsible for its rapid shearing and disappearance. Quite clearly, it is not possible to interpret the behavior of the cloud simply in terms of convective motions. The prime goal of the further data analysis will be to understand the development of the barium plasma in the light of the changing external parameters as measured on the IRM. This work is presently going on. Only thereafter may it be possible to gain additional insights into the dynamics of the tail.

TAIL FORMATION

Observations

The AMPTE project offered two opportunities for plasma injections in the solar wind, the first of their kind (Valenzuela et al.[9]). The external conditions of the two were significantly different in that one release was in front and the other behind the earth's bow shock. The Alfvénic Mach numbers were 6 and 2.3 (with respect to the flow component perpendicular to B). The sequence of events was very similar in both cases. The first minute was dominated by the expansion of the neutral gas cloud and its ionization. The total mass of free barium atoms was about 2 kg. Because of the high initial pressure, a magnetic cavity was formed. After the plasma cloud had reached a radius of about 80–90 km, pressure balance was reached with the ambient and compressed magnetic field. At this time the barium density had dropped to a value of about 3×10^3 cm^{-3}. During the following half minute or so, the compressed field penetrated into the barium cloud while the field lines outside the cloud were swept downstream and draped around the main body of the barium plasma in a way reminiscent of real comets. The penetration of the field proceeded like a snowplow, i.e., it swept up the ions and caused a temporary density enhancement by a factor of 3–5 (Lühr et al.,[10] Haerendel et al.,[11] Haerendel,[12] Gurnett et al.[13]). Tail formation began with the

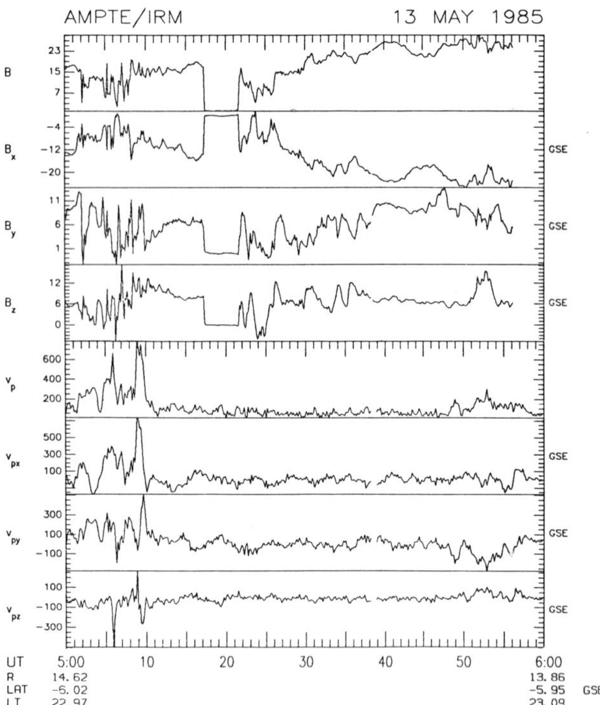

Figure 3—Magnetic and velocity vectors as measured on the Ion Release Module of AMPTE for one hour around the barium injection experiment of May 13, 1985 (courtesy of H. Lühr, G. Paschmann, and W. Baumjohann).

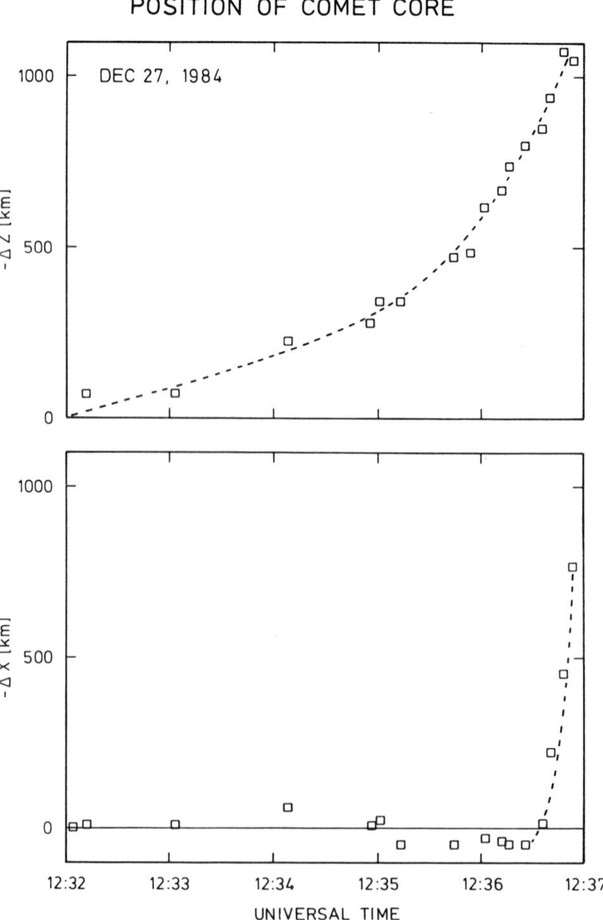

Figure 4—Displacement of the head of the artificial comet of December 27, 1984, with respect to the point of injection. The solar wind flows approximately in $-x$-direction; the interplanetary field is nearly aligned with the $+z$-axis.

penetration and draping of the field during the second minute after injection.

With regard to formation and decay of the tail, the following further observations are of significance. The comet head consisted of a substructure of magnetic arches that could be well resolved in the second experiment. In both cases, such substructures were observed to separate from the comet head and move downstream along the main tail, which pointed essentially, though not precisely, in the direction of solar-wind flow. The total lifetime of both comets was about five minutes.

The most surprising finding was that while a clear distinction between head and tail could be made, the head did not at all move in the direction of solar-wind flow, but rather at right angles to it. Figure 4 shows the path of the center of the visibly identifiable comet head with respect to the release point for the December 27, 1984 release. Only after 4.5 min, when the head was essentially drained of its plasma content, did it start to move in the expected direction (downstream) whereby it quickly dissolved and disappeared in the background light. The interpretation given by Haerendel et al.[11] is that the initial absence of a downstream motion of the comet head is due to the fact that the total momentum received from the solar wind, via the stresses of the trapped and draped magnetic field, is transferred to the ions injected from the rear side of the cloud into the tail. Their recoil keeps the comet head near the point of injection. Momentum balance and tail formation are thus intimately related.

A similar interpretation is given of the lateral displacement. It is seen as a consequence of the extraction of ions by the ambient electric field, which is modified in strength and direction by the plasma cloud (note that l_{inj} (Eq. 3) is about three orders of magnitude larger than the cloud's radius). The inertial force of the extracted ions is balanced by magnetic stresses. The compressed magnetic field, on the other hand, exerts a corresponding force on the visible bulk of the plasma in head and tail and moves it into the direction opposite to the external field, $E_0 = -v_{sw} \times B$. Hence, we see a lateral recoil effect in response to the ion extraction by the ambient electric field. A deflection of solar-wind ions in the same direction adds to the momentum balance and has to be taken into account.

Cheng[14] and Papadopoulos and Lui[15] have also dealt with the cause of the lateral displacement. Besides the recoil effect discussed here, they emphasize the direct interaction with the solar wind ions. A critical assessment of these approaches vis-à-vis the ion extraction and recoil effects is not possible in this context and will be published elsewhere.

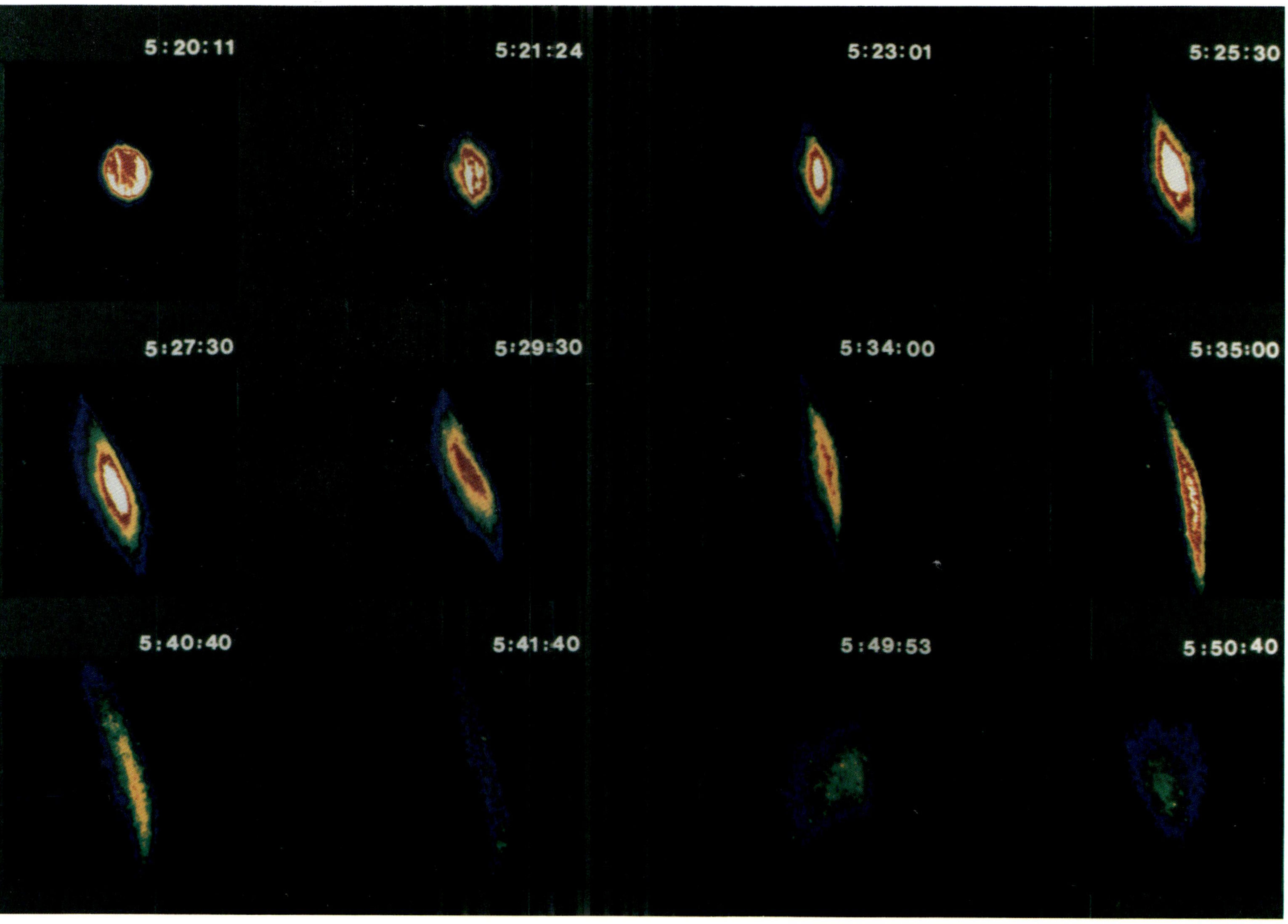

Plate V-1—Sequence of low-light-level TV images of the barium cloud released in the tail at 14.4 R_e on May 13, 1985 (false colors). The cloud is strongly sheared and rapidly dissolved around 05:50 UT.

Ion injection or extraction from the main cloud into the tail and out of head and tail into the environment are therefore key features of the solar-wind/comet interaction. They will be the main subject of the remainder of this paper. The extraction of ions from head and tail into the direction of E_0 is nowhere better displayed than in Plate V-2 taken at $4^m\ 25^s$ after the injection on December 27, 1984 from an aircraft over the South Pacific (Valenzuela et al.[9]). Background radiation has been subtracted and sensitivity variations of the SID-TV camera and other instrumental effects have been carefully corrected (courtesy of H. Höfner). The asymmetry at lower brightness levels exhibits the ion motion. By coincidence, E_0 is very nearly within the plane of projection, pointing upward, nearly perpendicular to the direction of the tail. The ever-increasing speed of the ions picked up by E_0 leads to a quick drop of their density and hence of the Ba II emissions. The ion extraction process is also indicated in the sketch of Fig. 6b.

Balance of Forces

MHD Approach. We consider the balance of forces acting on a flux tube filled with barium plasma of density ρ_c over a length l_c. The solar wind couples its force to the cloud via magnetic normal and shear stresses of the draped field as shown in Fig. 5. The pressure force is neglected in view of the low thermal speed of the injected plasma.

$$\rho_c l_c \frac{dv_x}{dt} = l_c \frac{\partial}{\partial x}\left(\frac{B_y^2}{8\pi}\right) + \frac{2B_x B_y}{4\pi} \qquad (4)$$

is a very simplified way of expressing the force balance. The magnetic shear stresses contain the jump, $2B_x$, of the field component in the flow direction. It is related to the unperturbed field, B_0, by Walén's relation (see Fig. 5)

$$\frac{B_x}{B_0} = \frac{v_{sw} - v_x}{v_A} \qquad (5)$$

where v_A is the Alfvén velocity.

The perpendicular field, B_y, is compressed inside and near the cloud. We define a compression factor, κ, by

$$\kappa = \frac{B_y}{B_0} \qquad (6)$$

We know from the observations that the visible head of the artificial comet is not subject to acceleration in the flow (x) direction for several minutes. Hence, $\dot{v}_x \approx 0$ over a large volume; the magnetic normal and shear stresses balance each other to a great extent. Introducing the transverse scale length, $l_\perp$, of the magnetic pressure gradient we find from Eq. 5

$$\frac{l_c}{l_\perp}\kappa^2 = \frac{v_{sw} - v_x}{v_A}\,4\kappa \qquad (7)$$

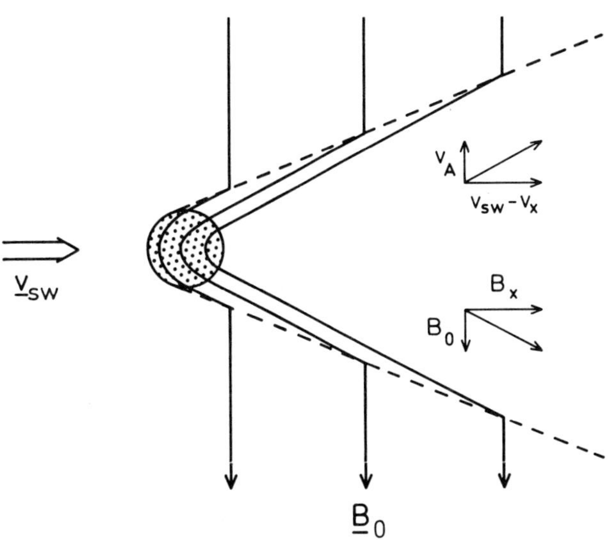

Figure 5—Sketch of Alfvén wings of the magnetic field trapped in a plasma cloud in a super Alfvénic flow. Demonstration of Walén's relation.

With $v_x \ll v_{sw}$, $M_A = v_{sw}/v_A$, and the estimate $l_c/l_\perp \approx 2$, we find

$$\kappa \approx 2M_A \qquad (8)$$

for the compression factor. In comparison with the observed field, this must be regarded as the maximum possible compression.

Naturally, $v_x \approx 0$ cannot hold everywhere. It breaks down on the rear side of the cloud, where magnetic normal and shear stresses are pointing in the same direction, i.e., downstream. Here, we can add the two magnetic stress components and write

$$\rho_c l_c \frac{v_x}{\tau_{acc}} = 8\kappa\,\frac{v_{sw} - v_x}{v_A}\,\frac{B_0^2}{8\pi}$$

In order to obtain a definite value for the characteristic acceleration time, τ_{acc}, we set $v_x = 0.5\,v_{sw}$. This leads to

$$\tau_{acc} = \frac{\pi \rho_c\,l_c\,v_A}{\kappa\,B_0^2} = \frac{\tau_0}{2\kappa} \qquad (9)$$

where τ_0 is the momentum coupling time derived by Scholer[3] (Eq. 1). So, Eq. 9 is a generalization of τ_0 for the case of high Alfvénic Mach number.

Table 2 contains a comparison of expected and measured quantities of κ and the displacement, Δx, of the cloud after four minutes. According to Eq. 2, Δx is given by

$$\Delta x = \frac{v_{sw}}{2\tau_{acc}}\,(\Delta t)^2 \qquad (10)$$

343

Table 2—Comparison of theoretical and observed magnetic compressions (κ) and displacements (Δx) of two plasma clouds injected in the solar wind.

Experiment	$M_{A\perp}$	κ_{max}	τ_{acc} (s)	Δx (km)
Dec. 27, 1984				
Observed	6	12	–	<100
Theory	–	12	5700	2700
July 18, 1985				
Observed	2.3	4.5	–	<100
Theory	–	4.6	4100	2000

Although we know that the magnetic forces were essentially applied to the ions injected from the rear side into the tail and not to the bulk of the plasma, we use the central density for the estimate of ρ_c for the estimate of τ_{acc}. Δx then tells us what displacements should be seen, had the forces been applied to the cloud as a whole.

The outcome of this comparison is that the simple MHD consideration is quite appropriate for an assessment of the field compression, whereas the net force acting on the bulk plasma is about two orders of magnitude smaller than expected in a simple slingshot model because of a balancing recoil force exerted by the ions injected into the tail.

Ion Injection Into the Tail. The injection of ions is as sketched in Fig. 6a, b. Only the ions on the rear flank of the cloud feel the force exerted by the solar wind. In the MHD picture, this force would be the sum of magnetic normal and shear stresses. The single ion, however, is subject mainly to the electric force. (We will see that the Lorentz force is negligible.) Pressure effects will be ignored for a while as in the previous MHD consideration.

We use an extremely simple approach in describing the momentum balance on the rear side and the injection of ions into the tail. An electric field pointing into the tail ($-x$-direction) accelerates the ions

$$\frac{dv_{ix}}{dt} = \frac{e}{m_i} E_x \tag{11}$$

The electrons are magnetized and perform a perpendicular drift (see Fig. 6b)

$$V_{e,z} = c\,\frac{E_x}{E_y} \tag{12}$$

This drift constitutes a "Hall" current

$$j_z = -en\,v_{ez} = -enc\,\frac{E_x}{B_y}$$

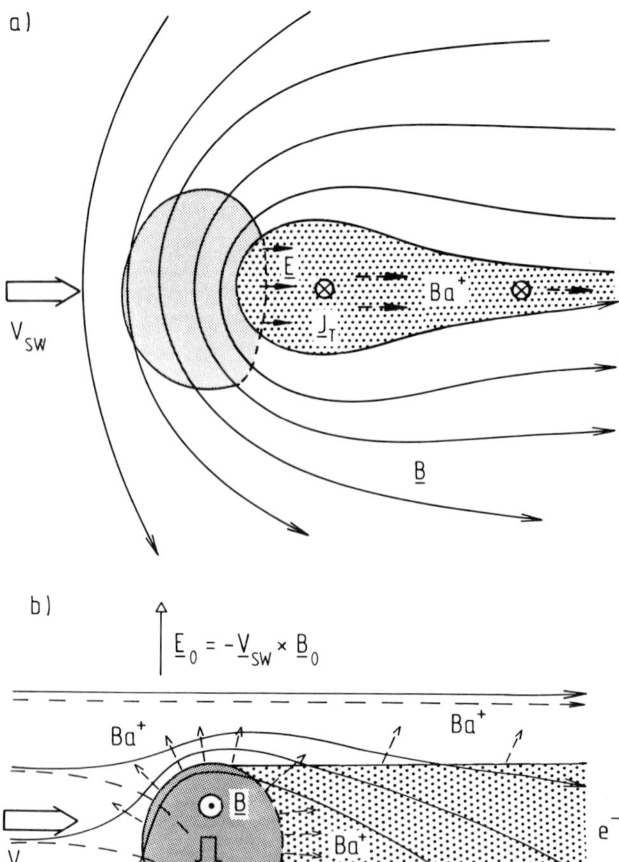

Figure 6—(a) Principle of tail formation by ion injection in the plane of the magnetic field; (b) trajectories of solar wind ions (H⁺) and electrons and of extracted barium ions in a plane $\perp B_0$ through the center of plasma cloud and tail current sheet (Ref. 12).

that is shielding the magnetic field of the comet head from the tail. Since magnetic normal and shear stresses are approximately equal on the rear side, we can estimate the total electric force on the tail axis by

$$en\,E_x = \frac{1}{c}\,B_y\,j_z \approx 2\,\frac{\partial}{\partial x}\left(\frac{B_y^2}{8\pi}\right) \tag{13}$$

Equations 11 and 13 express the balance of forces exerted by the solar wind and ultimately applied via E_x to the injected ions. In a stationary situation (i.e., injection time of ions is short compared with lifetime of comet), the relations yield

$$m_i n\,v_{ix}\,\frac{\partial v_{ix}}{\partial x} = \frac{\partial}{\partial x}\left(\frac{B_y^2}{4\pi}\right) \tag{14}$$

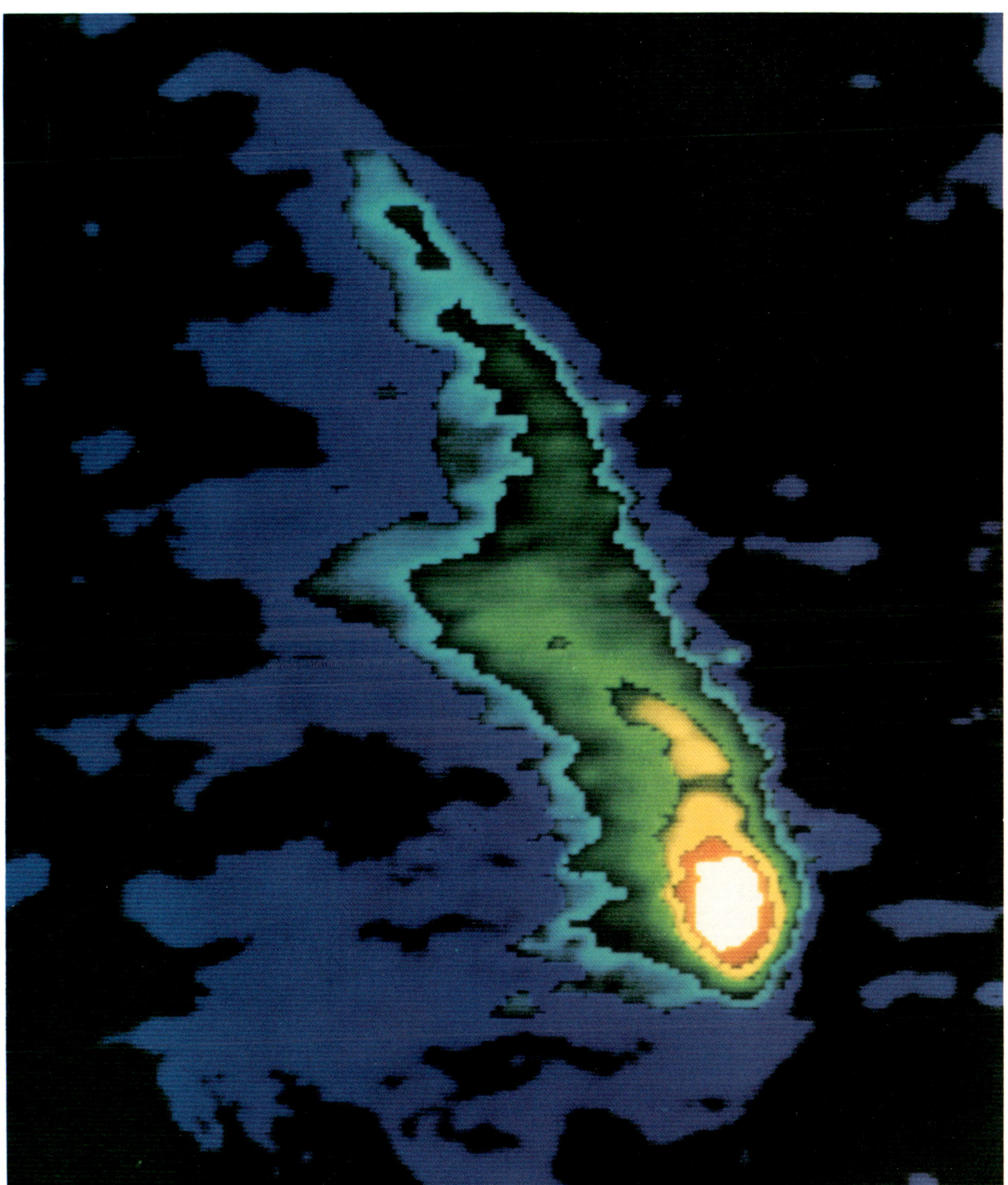

Plate V-2—False color image of the artificial comet taken by a low-light-level TV camera from an aircraft on December 27, 1984 (Ref. 9). The image has been corrected for distortions, sensitivity variations, and sky and instrumental background by H. Hofner (MPE).

Since the ion flux, $F_{ix} = n\, v_{ix}$, can be taken as approximately constant, we can write

$$\frac{\partial}{\partial x}\left(m_i\, F_i\, v_{ix} - \frac{B_y^2}{4\pi}\right) \approx 0 \qquad (15)$$

B_y drops from maximum compression to a small value, B_T, on the tail axis, and v_{ix} grows from zero to a final value, which depends on the density, n_T, in the tail current sheet

$$v_{ix} \rightarrow v_{iT} \approx \frac{B_{y,\max}}{\sqrt{4\pi m_i n_T}} \qquad (16)$$

$B_{y,\max} = \kappa\, B_0$. The density on the tail axis must be of the order of the unperturbed ambient density, since (as shown in Fig. 6b) the electrons, which neutralize the ion current, are supplied from the flanks out of the (somewhat perturbed) solar-wind flow. Hence $n_T \sim n_a$ must be a reasonable estimate. This leads to

$$v_{iT} \cong \frac{2}{\sqrt{\mu_i}} \cdot v_{Aa} \qquad (17a)$$

or, since $\kappa = 2\, M_A$ (Eq. 8),

$$v_{iT} = \frac{2}{\sqrt{\mu_i}}\, v_{sw} = 0.17\, v_{sw} \qquad (17b)$$

for barium ions, where μ_i is the atomic weight. The simplicity of the geometric model as well as the neglect of the pressure force do not alter this estimate significantly.

Whereas it is easy to see that the gyroradius of an ion in the tail current sheet is large ($> 10{,}000$ km) compared with its width (~ 100 km), this is not obvious at the rear flank, where the acceleration maximizes. The injection scale length, l_{inj}, defined analogous to Eq. 3, can be estimated from Eq. 13 to be

$$l_{\mathrm{inj}} = c\, \frac{E_x}{B_y\, \Omega_i} = \left(\frac{c}{\omega_{pi}}\right)^2 \frac{1}{L} \qquad (18)$$

where L is the gradient length of the magnetic pressure. The latter is less than 100 km. The c/ω_{pi} for barium is the same magnitude for $n = 10^3$ cm^{-3}, i.e., close to the density maximum. This means that except for the very beginning of the rear side acceleration, the ions will essentially follow the electric force with negligible bending of their path by the Lorentz force. Our use of Eq. 11 as a starting point is entirely justified.

The neglect of the pressure forces in the stress balance perpendicular to B is not trivial. The gradient of the electron pressure on the rear side generates an electric current opposite to that constituted by the Hall drift.

This means that inclusion of the electron pressure would even enhance the accelerating field for a given magnetic field profile. The above estimated injection speed may, therefore, even be on the low side. However, a much more careful investigation is needed to assess this contribution. The role of p_e on the dynamics parallel to B will be discussed below.

Equation 17 then predicts ion injection speeds of 90 and 50 km/s, respectively, into the tail for the two experiments, in good agreement with preliminary evaluations of the observed motions of density markers in the tail.

In summary, we see that the main force balance between solar wind and plasma cloud has two consequences: a compression of the field, estimated by κ in Eq. 8, and an injection of ions into the tail, estimated by v_{iT} in Eq. 17. These two equations, therefore, represent the essence of the tail formation process of the artificial comet. The lateral force balance can be analyzed in a similar fashion, and the observed recoil of the comet head can be satisfactorily explained. This will be done in a different paper.

The Tail Current System

Nearly all our information on the physics of tail formation has to be derived from the optical data. In situ measurements were made in the head of the comet by the IRM and at its dawn flank (outside the main cloud) by the United Kingdom Subsatellite (in the December 27, 1984 experiment only) (Lühr et al.,[10] Haerendel et al.,[11] Rodgers et al.[16]). Because of the recoil of the comet head due to ion extraction by the ambient electric field (see Fig. 4), the IRM made a transition from the center to the exterior in the direction of E_0 (Haerendel et al.[11]). This allows verification and detailed determination of the ion extraction, but sheds little light on the tail formation, with two exceptions. The compression of the magnetic field in the head is a consequence of overall momentum balance and thus of the total force acting on the tail (ion injection). Secondly, part of the electron component found inside the head made its entry through the Alfvén wings, i.e., through the tail field. From the local determination of p_e, we get a good indication of the pressure force acting on the tail.

Enhancements of both density and temperature of the directly measured electron component (20 eV–30 keV) were found in the comet head. They maximized during the compression phase (Haerendel et al.,[11] Bryant et al.[17]). The pressure reached a magnitude somewhat below $B^2/8\pi$ of the compressed field. The majority of electrons, however, had energies below the lower threshold of measurement. Their contribution to the total pressure cannot be assessed reliably. It may be approximately equal to, but not greater than, that of the hot electrons. The latter reached average energies of several hundred electron volts during the compression phase.

With this knowledge and some guesses of the modification of the external electric field near the comet head, it is possible to derive a picture of the overall electric current system. First, there is a current encircling the comet head (in the plane transverse to the unperturbed field, B_0). It confines the trapped and compressed magnetic field. To a large extent it is constituted by the $E \times B$ drift of cloud and solar-wind electrons. The pressure gradient drift corresponds to a current in the opposite direction, since high density and magnetic field are correlated. The electric field on the upper flank of the comet head (Fig. 6b) is the modified external field. On the rear flank, it is the field causing the ion injection. Below the density maximum, but well inside the comet head, there may be a downward pointing electric field that is instrumental in achieving the recoil motion of the ions that are not fully magnetized. At the lower border, finally, E turns back into the direction and magnitude of the unperturbed external field, E_0. In addition, as shown in Fig. 6b, a current component is constituted by the deflected solar-wind ions (Haerendel et al.[11]). They contribute to, but do not fully carry, the shielding current on the lower side of the comet head.

The tail current system consists of two parts. The central sheet current is carried by electrons (partially solar wind, partially cloud electrons) that transit from the upper to the lower side (in the sense of E_0). Close to the head, this current is quite strong and due to the $E \times B$ drift driven by the ion injection field. Further downstream, it is probably just the pressure gradient current of the ambient electrons needed for neutralization of the injected barium ions (see Fig. 7). The transverse current on the rear side of the comet head, which is carried by the electrons left behind by the injected ions, has a finite divergence owing to the density decrease along the electron drift path. This constitutes a sink of field-aligned current from the tail lobes, as sketched in Fig. 7. This current largely balances the ion injection current into the central-tail plasma sheet. To a smaller extent, it continues the current of solar-wind ions that concentrate on the underside of head and tail (see Fig. 6b). Eventually, i.e., much further downstream, all these currents must close.

The effect of the ion injection and field-aligned currents on the tail field is to bend the field lines downward with increasing distance. The total ion injection current is of the order of

$$I_{\text{tail}} = e \, n_a \, v_{iT} \cdot \pi \, R_c^2$$

For the December 27, 1984 parameters, it is about 2 kA. If we adopt a width of $2 \, R_c$, we find a jump of the perpendicular component of B of $\Delta B_z \cong$ 10–15 nT. Whereas the total tail field adjacent to the central plasma sheet should be of the order of ~ 50 nT (maximum field in the comet head 120 nT !), the perpendicular component ($\frac{1}{2} \Delta B_z$) is about one order of magnitude lower. Of course, these estimates are not very reliable. The resulting aberration of the tail field would thus be of the order of 5°–10°.

Another current component must be present in the tail as a result of the ion extraction from the upper side. This current is not included in the sketch of Fig. 7. It would lead to field bending in the same sense.

Further Tail Formation Processes

The ion injection into the tail as a result of the force balance with the solar wind is not the only process that leads to the visible appearance and development of the tail. Two more processes are of importance, (a) ion drag along field lines by the expansion of heated electrons and (b) occasional disconnections of whole flux tubes from the comet head. The first process is easily estimated. Since the driving force is the electron pressure gradient that acts on the ions via an ambipolar electric field and is balanced by the ion inertia, the outflow speed must be the ion-acoustic speed

$$v_{i,\parallel} = \sqrt{\frac{kT_e}{m_i}} = c_{si}$$

It may not be possible to assign a unique temperature to the electrons, since they are a mixture of cold photo and hot solar-wind electrons. So kT_e has to be replaced by the mean electron energy. Taking the electron pressure as derived from the measurements of the three-dimensional plasma instrument (Paschmann and Carlson, personal communication), we deduce outflow speeds of the order of 10 km/s at the rear flank. The evaluation and calibration of the images of both artificial comets is still going on; no reliable assessment of field-aligned movements can yet be given. However, speeds of a few tens of kilometers per second parallel to B are consistent with preliminary assessments.

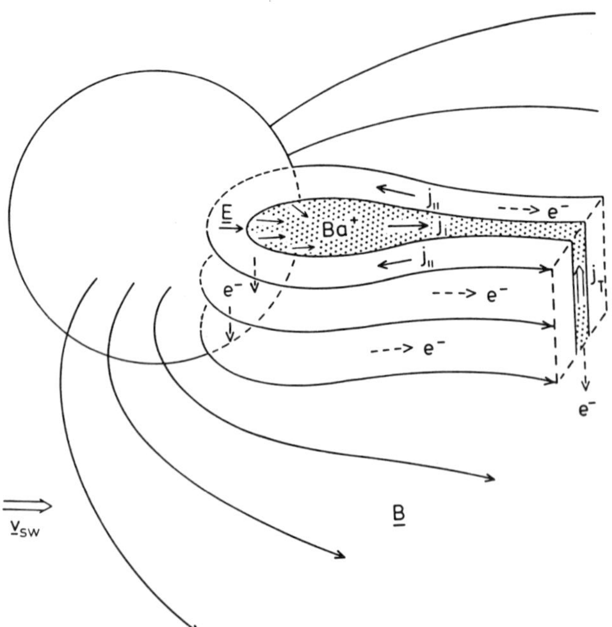

Figure 7—Three-dimensional view of the ion and electron currents flowing in the central regions of the tail of an artificial comet. Only a few field lines of the draped magnetic field are shown.

In both artificial comet experiments, events have been observed that appear to be separations of flux tubes filled with barium plasma. They move downstream along the tail axis and slowly dissolve. Plate V-3 gives one example from the July 18, 1985 experiment. It is tempting to relate these events to so-called disconnection events observed in real comets, although their origin has been tentatively ascribed to reversals of the interplanetary field and reconnection (Niedner and Brandt[18]). In our experiment some changes in direction but no drastic reversals of B were observed. It looks as if the repulsive force that keeps the comet head as a whole near the point of injection can become reduced in individual magnetic structures whereby they start to move downstream and separate from the head plasma.

All effects together, ion injection, parallel ion drag (toothpaste effect), and disconnection events, produce the tail and its dynamical features. In addition, ion extraction by E_0 operates and limits the tail's life, i.e., its visible length. The loss of plasma from head to tail, just by ion injection, is an important factor in limiting the overall existence of the visible phenomenon. An equivalent loss time is defined by

$$\tau_{\text{tail}} = \frac{N_{\text{ion}}}{n_a \; v_{iT} \; \pi \; R_c^2}$$

It is of the order of 600 s ($N_{\text{ion}} \approx 9 \times 10^{24}$). If one adds the other processes, in particular direct ion extraction from the head, one can readily account for the artificial comet's lifetime of only 5 min.

CONCLUSIONS

Whereas only small contributions to the study of the earth's magnetotail have been made so far by plasma releases in the tail proper, a related phenomenon could be extensively investigated, i.e., the creation of a comet-like tail by temporary trapping of interplanetary magnetic field in barium plasma clouds produced in the solar wind. The dominant processes leading to tail formation are ion injection from the rear of the plasma cloud, ion drag along field lines by the expanding hot electron component, and separation of entire flux tubes from the comet head. The first process has been identified as direct expression of the force balance with the magnetic stresses exerted by the solar wind. As a consequence, the comet head is found not to move downstream, but is kept near the point of injection by the recoil of the ions injected into the tail. Head and tail perform, however, a lateral movement that is again interpreted as a recoil effect, in this case due to ions extracted by the solar-wind electric field (i.e., transverse to B and v_{sw}).

All these processes seem to act in some way also in natural environments. The artificial comet experiments may help in their elucidation. For instance, Ness and Donn[19] interpret cometary tail rays as magnetic neutral or current sheets. They suggest a different ion injection process. The here-described injection by large electric fields near the comet head, where the magnetic stresses are applied, may well be the dominant process, at least at the origin of the central tail ray of a comet. Another frequently discussed process is the toothpaste effect that proceeds with ion-acoustic speed and may dominate the growth of lateral rays. Major disconnection events in real comets, as identified by Niedner and Brandt,[18] may actually be caused by interplanetary field reversals and rupture of flux tubes by reconnection. However, a simple dragging of flux tubes through and finally out of the coma should occur continuously and is probably responsible for the wavy structures seen to populate the central parts of type I comet tails. The nonstraight appearance of these structures seems to signal the relief of magnetic stresses because of untrapping of the flux tubes from the coma where these stresses were balanced before by drag forces.

Finally, ion extraction by the solar-wind electric field should be a nonnegligible process also in real comets. It has not yet been identified there, but in the environment of planet Venus (Luhmann et al.[20]). Rather than in a recoil motion, this ion extraction manifests itself in an asymmetry of the strength of the draped magnetic field. This is quite understandable, as the lateral magnetic pressure force must balance the inertial force of the extracted ions. These are only a few remarks as to how the study of the artificial comets and of natural phenomena may support each other.

ACKNOWLEDGMENT—I thank Dr. W. Baumjohann (MPE), H. Höfner (MPE), Dr. H. Lülu (TU Braunschweig) and Dr. G. Paschmann (MPE) for making some of the processed data available to me prior to publication in a different context. The AMPTE mission was supported by the German Federal Ministry for Research and Technology (BMFT) through grant 01 0M 010-ZA/WRK 275. Mission preparation and control was in the hands of the German Satellite Operations Center of the DFVLR at Oberpfaffenhofen.

REFERENCES

[1] G. Haerendel and R. Lüst, "Electric Fields in the Ionosphere and Magnetosphere," in *Particles and Fields in the Magnetosphere*, B. M. McCormac, ed., pp. 213-228 (1970).

[2] G. Haerendel, P. C. Hedgecock, and S.-I. Akasofu, "Evidence for Magnetic Field-Aligned Currents During the Substorms of March 18, 1969." *J. Geophys. Res.* **76**, 2382-2395 (1971).

[3] M. Scholer, "On the Motion of Artificial Ion Clouds in the Magnetosphere," *Planet. Space Sci.* **18**, 977-1004 (1970).

[4] G. Haerendel, "Alfvén's Critical Velocity Effect Tested in Space," *Z. Naturforsch.* **37a**, 728-735 (1982).

[5] G. Haerendel, "Plasma Confinement and Interaction Experiments," in *Proc. Int. Symp. Active Experiments in Space*, ESA SP-195, pp. 337-340 (1983).

[6] W. Baumjohann, R. J. Pellinen, H. J. Opgenoorth, and E. Nielsen, "Joint Two-Dimensional Observations of Ground Magnetic and Ionospheric Electric Fields Associated with Auroral Zone Current, 4. Current Systems Associated with Local Auroral Breakups," *Planet. Space Sci.* **29**, 431-447 (1981).

[7] R. W. McEntire, S. M. Krimigis, G. Haerendel, G. Gloeckler, E. G. Shelley, J. B. Cladis, and B. H. Mauk, "The AMPTE Magnetotail Tracer Release Experiments" (abstract), 26th COSPAR Symp., Toulouse (1986).

[8] J. B. Cladis and W. E. Francis, "On the Transport of Ions Released in the Magnetotail by the AMPTE-IRM Satellite" (abstract), 26th COSPAR Symp., Toulouse (1986).

[9] A. Valenzuela, G. Haerendel, H. Föppl, F. Melzner, H. Neuss, E. Rieger, J. Stöcker, O. Bauer, H. Höfner, and J. Loidl, "The AMPTE Artificial Comet Experiments," *Nature* **320**, 700-703 (1986).

[10] H. Lühr, D. J. Southwood, N. Klöcker, M. W. Dunlop, W. A. C. Mier-Jedrzejowicz, R. P. Rijnbeek, M. Six, B. Häusler, and M. Acuña, "In Situ Magnetic Field Observations of the AMPTE Artificial Comet," *Nature* **320**, 708-711 (1986).

[11] G. Haerendel, G. Paschmann, W. Baumjohann, and C. W. Carlson, "Dynamics of the AMPTE Artificial Comet," *Nature* **320**, 720-723 (1986).

[12] G. Haerendel, "Künstliche Kometen," *Phys. Bl.* **42**(5), 134-137 (1986).

[13] D. A. Gurnett, R. R. Anderson, B. Häusler, G. Haerendel, O. H. Bauer, R. A. Treumann, H. C. Koons, R. H. Holzworth, and H. Lühr, "Plasma Waves Associated with the AMPTE Artificial Comet," *Geophys. Res. Lett.* **12**, 851-854 (1986).

[14] A. F. Cheng, "Transverse Deflection and Dissipation of Small Plasma Beams and Clouds in Magnetized Media," *J. Geophys. Res.* (submitted, 1986).

[15] K. Papadopoulos and A. T. Y. Lui, 'On the Initial Motion of Artificial Comets in the AMPTE Releases," *Geophys. Res. Lett.* (submitted, 1986).

[16] D. J. Rodgers, A. J. Coates, A. D. Johnstone, M. F. Smith, D. A. Bryant, D. S. Hall, and C. P. Chaloner, "UKS Plasma Measurements Near the AMPTE Artificial Comet," *Nature* **320**, 712-716 (1986).

[17] D. A. Bryant, C. P. Chaloner, D. S. Hall, and D. R. Lepine, "Electron Heating in the Solar Wind: Results from AMPTE," *J. British Interplanet. Soc.* **39**, 199-206 (1986).

[18] M. B. Niedner, Jr., and J. C. Brandt, "Interplanetary Gas, XXIII, Plasma Tail Disconnection Events in Comets: Evidence for Magnetic Field Line Reconnection at Interplanetary Sector Boundaries?" *Astrophys. J.* **223**, 655-670 (1978).

[19] N. F. Ness and B. Donn, "Concerning a New Theory of Type I Comet Tails," in *Nature et Origine des Comètes*, 13th Liège Symp., Cointe-Sclessin, pp. 343-362 (1966).

[20] J. G. Luhmann, C. T. Russell, J. R. Spreiter, and S. S. Stahara, "Evidence for Mass-Loading of the Venus Magnetosheath," *Adv. Space Res.* **5**, 307-311 (1985).

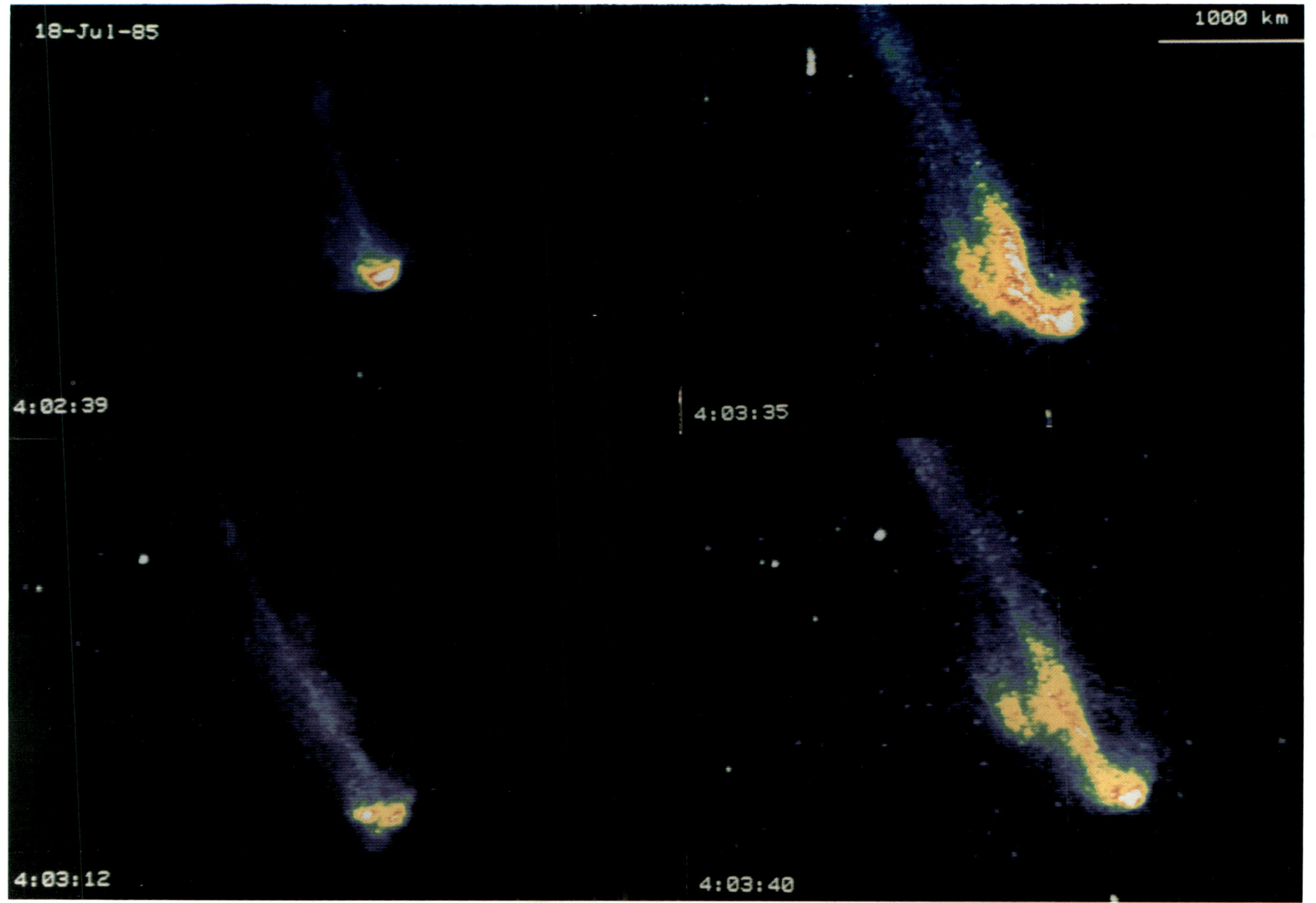

Plate V-3—Development of the artificial comet (barium clouc) of July 18, 1985 as observed with a low-light-level TV camera (false colors). The formation of a cusp on the rear side, of magnetic arches in the head, and the separation of magnetic structure should be noted.

THE MARCH 21, 1985 AMPTE BARIUM MAGNETOTAIL RELEASE: IMAGING AND DOPPLER IMAGING OBSERVATIONS FROM THE NASA AIRBORNE OBSERVATORY

D. Rees*

The Ion Release Module (IRM) of the Active Magnetospheric Particle Tracer Explorers (AMPTE) made a barium thermite release at a geocentric distance of 70,000 km (within the earth's magnetotail) at 0921 UT on March 21, 1985. This release, which produced a yield of 2–3 kg of barium vapor, was observed from the NASA Convair 990. The rapid initial expansion and photoionization of the barium vapor created a shell of ions about 600 km in diameter by 0924 UT. This shell partly collapsed by 0927 UT, forming a strong central striation (core). The initial shell and later the core were surrounded by a diffuse, rapidly expanding, ion halo. The diffusion coefficient of the halo ions was estimated to be about 2×10^9 m^2/s. This halo was observable for 40 min, although the core became indistinct by 0956 UT (+ 35 min). The core drifted slightly northward relative to the motion of the IRM until 0946 UT (+ 25 min). By 0950 UT, as a result of the rapid loss of barium ions from the core into the halo, the central intensity of the core had dropped below about 20% of its initial value. A large southward and smaller westward drift of the core was then observed (up to 7 km/s perpendicular component). The diffuse halo appeared to drift eastward and northward between 0929 and 0936 UT, and later moved generally southward between 0936 and 0946 UT. Several distinct striations formed within the halo, on the eastward side between 10 and 20 min after the release and later, on the west side, thus roughly correlating with the direction of drift of the halo relative to the core. The UCL Doppler Imaging System showed that the core initially moved away from the earth, following the track of the IRM. Between 0931 and 0941 UT, the core showed a shearing motion with the northward part of the cloud (away from the earth) having a small velocity away from the earth, while the south part of the cloud moved earthward. Up to 0941 UT, line-of-sight velocities within the core were of the order of 2–4 km/s. In the last DIS image at 0947 UT, there was a 10 km/s earthward velocity component, possibly related to the subsequent southward acceleration of the core after 0950 UT. The ion core and halo faded rapidly after 0940 UT. There was an associated sharp decrease of the total inventory of barium ions within the instrument field of view (8000 km). If the radial barium-ion velocity distribution was a monotonically increasing function of distance from the core, a total energy input of some 3.75×10^9 J, or about 5×10^{-8} W/m^2 would have been required. An alternative loss mechanism, which seems unlikely, is the further photoionization of Ba II to Ba III.

INTRODUCTION

At 0921 UT on March 21, 1985, the Ion Release Module (IRM) of the three-satellite AMPTE project (Krimigis et al.[1]) released a barium cloud (yield 2–3 kg of barium ions) at a geocentric distance of 70,000 km within the earth's geomagnetic tail. The release occurred south of the geomagnetic equator, with a viewing geometry such that the undisturbed terrestrial magnetic field was directed approximately 10° from the line of sight (LOS) (Haerendel[2]). Further background information on AMPTE, the March 21 release, and the in-situ magnetic, electric field, and plasma measurements made by the IRM and attempted by the Charge Composition Explorer (CCE) spacecraft can be found elsewhere.

THE OBSERVATIONS

The observations to be discussed in this paper were obtained from two optical systems on board the NASA airborne observatory, Gallileo II, flying from Moffett Field (NASA, Ames Research Center) California, that was located approximately 40.4° N, 123.6° W at the time of the release. The instruments were a very sensitive Imaging Photon Detector imaging system (McWhirter et al.[3]) and a Doppler Imaging System for mapping LOS ion velocities (Rees et al.[4]). Both these optical systems recorded the 455.4 nm resonance line emission of sunlit barium ions through narrow-band interference filters. Other optical observations were made from ground-based locations and from an Argentine Boeing 707 aircraft, flying from Tahiti.[2]

Initial Phase

Immediately after the release, the barium ions formed by photoionization of the evaporated neutral barium atoms expanded at a velocity of the order of 1.5 km/s, forming a spherical shell. This shell had a maximum diameter of about 600 km, some 200 s after the release (Haerendel,[2] Bernhardt[5]). The shell, produced by initial equilibrium between the expanding barium plasma and the magnetospheric magnetic field and plasma, contained a diamagnetic cavity. By 0927, the ion shell had partly collapsed to form an intense central field-aligned ion core and the terrestrial magnetic field penetrated to the central part of the ion cloud (Luhr[6]).

A faint, diffuse halo of barium ions expanded rapidly from the shell. The halo continued its expansion after the collapse of the ion shell, reaching a diameter of 3000 km by 0929 UT (Plate V-4A). By 0939 UT, the halo extended beyond the field of view (FOV) of the imager (8000 km).

*Department of Physics and Astronomy, University College London, Gower Street, London WC1E 6BT, United Kingdom.

Motion of the Core of the Ion Cloud

The motion of the ion cloud core, relative to that of the IRM, is shown in Fig. 1. Until the collapse of the shell at 0927 UT, it was not possible to separate the motion of the ion cloud from that of the IRM. Later, the ion cloud appeared to drift slightly northward of the IRM (until 0939 UT). After this time there was a rapid southward acceleration to a maximum apparent perpendicular velocity of about 7 km/s at the time of the last discernible image of the core (0955 UT).

Expansion of the Diffuse Halo

The surface brightness of the halo (Plate V-4A, B) was the order of 2% of that of the ion cloud core until about 0937 UT. Mean radial expansion velocities as high as 8 km/s are required to populate the outer detectable contours of the halo. Also the halo may have extended beyond the detection threshold of the imaging system. The halo expansion was nearly isotropic, perpendicular to the earth's magnetic field (noting that the LOS was within 10° of the magnetic field vector at the barium release). The loss of barium ions from the core into this halo continued throughout the period of observations. From the measured rate of radial expansion of the halo until 0937 UT, the estimated diffusion coefficient was about 2×10^9 m^2/s. This value is two orders of magnitude larger than the Bohm diffusion coefficient for thermal (say 1.5 km/s) barium ions (Haerendel[2]).

Individual barium ions gyrate within the terrestrial magnetic field, once outside the initial shell. Their mean velocity therefore had to be significantly larger than the 8 km/s radially outward diffusion velocity of individual ions. An estimate of the mean velocity of the halo

ions would be about $2 \times \pi \times 8$ km/s, or 50 km/s. The gyroradius of 1.5 km/s barium ions in the observed terrestrial field of 8 nT was about 260 km, with a gyro period of about 200 s (Haerendel[2]). At 50 km/s, a barium ion has a gyroradius (8 nT field) of about 3000 km, a value that is comparable with the observed size of the halo.

It is possible that a very small fraction of the initially produced ions had initial velocities of the order of 50 km/s (high-energy tail). However, to explain the continuous loss of barium ions from the core to the halo, and the subsequent rapid expansion of the halo, all the barium ions initially created by photoionization following the thermite release had eventually to reach velocities of the order of 50 km/s. Within 40 min (2000 s) a total of at least 3.75×10^9 J of energy was coupled into the ion cloud. A range of low-frequency electrostatic and electromagnetic waves are endemic within the magnetosphere (Heppner[7]). Assuming a source dimension of 6000 km, comparable with the instrument FOV, the mean energy density extracted from the background wave field by the barium ions would be about 5×10^{-8} W/m^2. The mode and possible efficiency of this energy coupling has not yet been fully treated.

Initially, the halo was almost symmetrical, with a slightly enhanced extension in the field-aligned (north/south) direction until 0929 UT. Between 0929 and 0939 UT, the halo appeared to have a generally eastward drift relative to the core. After 0941 UT, the motion of the halo changed to southward. Later, the halo extended beyond the 8000 km FOV of the imager, and it was impossible to determine the halo boundaries or motion. Figure 2 shows the outermost detectable contours of the halo, relative to the core of the cloud, between 0929 and 0944 UT.

Formation of Striations Within the Halo

Within 6 min of the release, several large-scale striations formed within the halo (contrasting with the small-scale striations reported on the initial shell boundary by Bernhardt and Haerendel). These striations were confined to the westward and northwestward parts of the halo, as shown in Plate V-4B (0939 UT). Between 0939 and 0941 UT (Plate V-4C), a strong striation formed within the eastward extension of the halo, while those striations in the westward halo extension weakened rapidly. This change of the basic patterns of striations appeared to coincide with the onset of the general southward drift of the halo. After 0944, when the halo extended beyond the limits of the FOV, weak striated structures were observed until the end of observations (1001 UT).

Depletion of the Core of the Ion Cloud

The temporal variations of the peak surface brightness of the core and of the total integrated intensity of the core and halo of the barium cloud are shown in Fig. 3. Both quantities are relatively constant between 0925 and 0937 UT and then decrease very rapidly after 0940 UT. When the residual mass of Ba II in the core fell

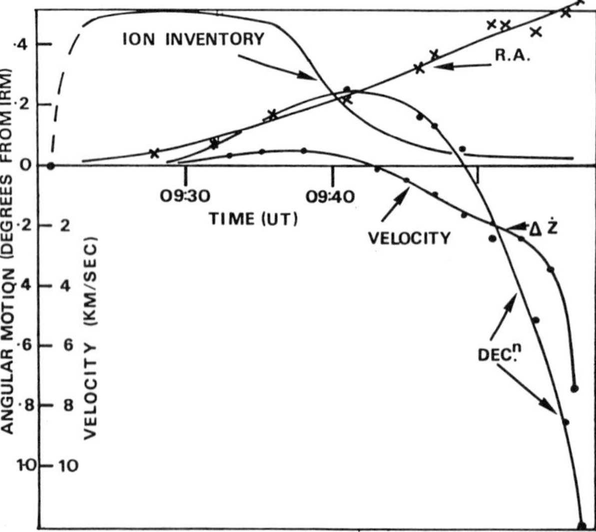

Figure 1—Apparent motion of the ion core relative to the track of the IRM between 0921 and 0957 UT. The right ascension increases continuously, while the drift in declination is initially northward, prior to the rapid southward drift after 0945, when the total ion inventory falls rapidly. The projected southward core velocity is also shown ($\Delta \dot{Z}$).

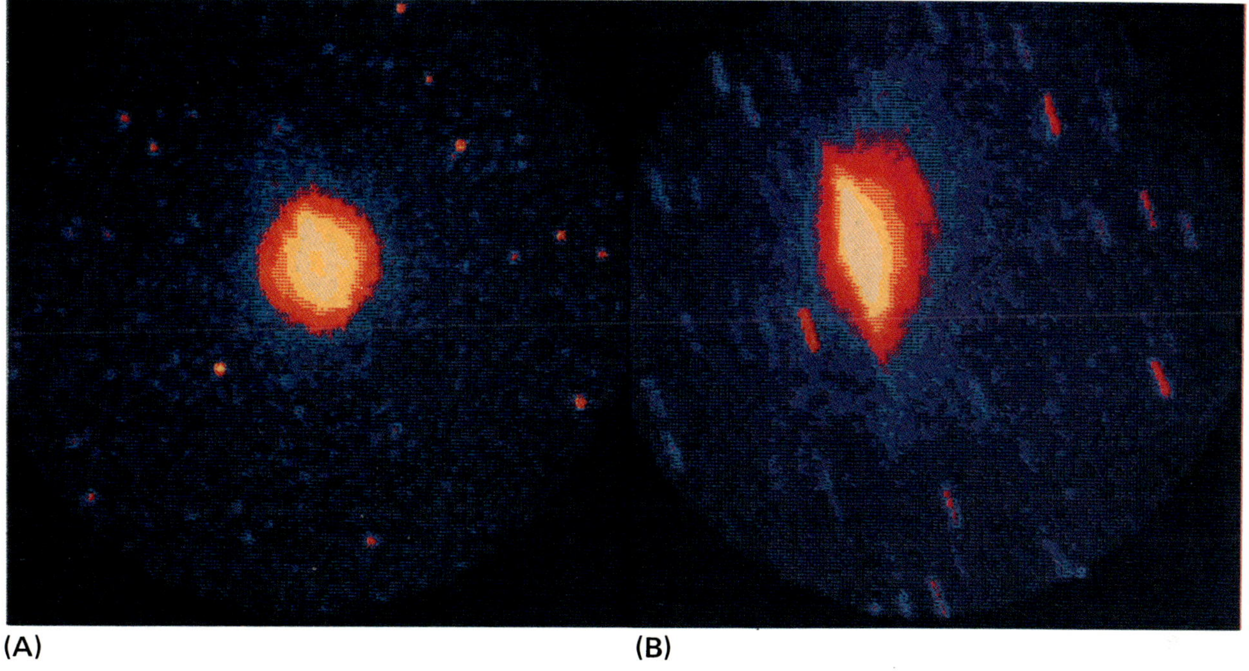

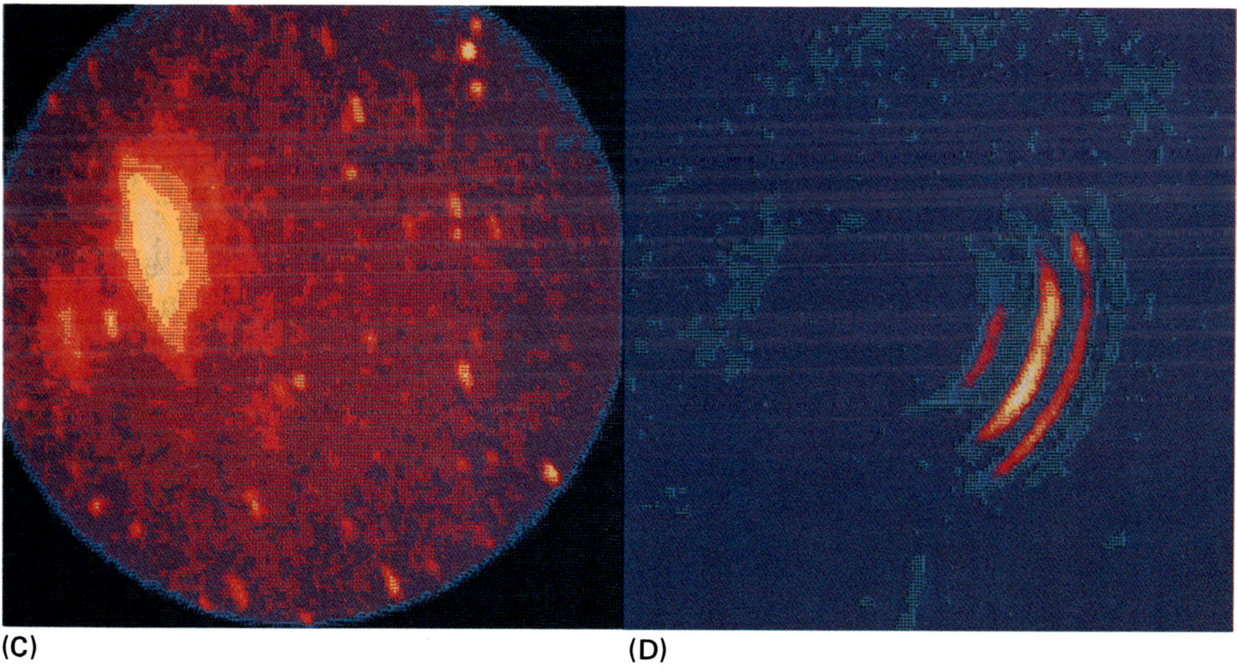

Plate V-4—(A) Image of the magnetotail release of March 21, 1985 at 0929 UT. The image shows the central striation that formed when the initial 600 km diameter shell collapsed. It also shows the early symmetric expansion of the diffuse halo to a diameter of about 3000 km. (B) The ion cloud at 0939 UT. The core is still distinct but significantly weaker than at 0929 UT. The diffuse halo now extends beyond the edges of the FOV of the imager (8000 km diameter). (C) The ion cloud at 0941 UT. Whereas until 0939 UT the striations appeared on the westward side of the core, in this image a strong striation can be seen on the east side of the core. (D) Image of the barium cloud at 0932 UT obtained by the DIS. The fringes are obtained from the core of the cloud and are correlated with those from a calibration lamp to obtain the LOS velocity distribution over the cloud.

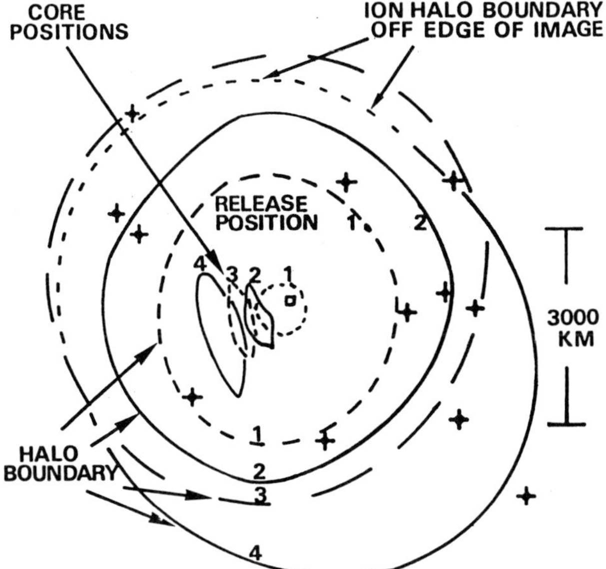

Figure 2—The apparent boundary of the halo and the core of the ion cloud at 0924 (1), 0929 (2), 0935 (3), and 0944 (4) (time in UT). Bright stars are marked. The boundary is somewhat subjective, limited by the threshold of the imager. There is a major change of overall appearance about 0942 UT, as the initial halo drift (northward and westward) changes to a southward and eastward direction. The northeast limit of the halo of (4) is limited by the edge of the instrument FOV.

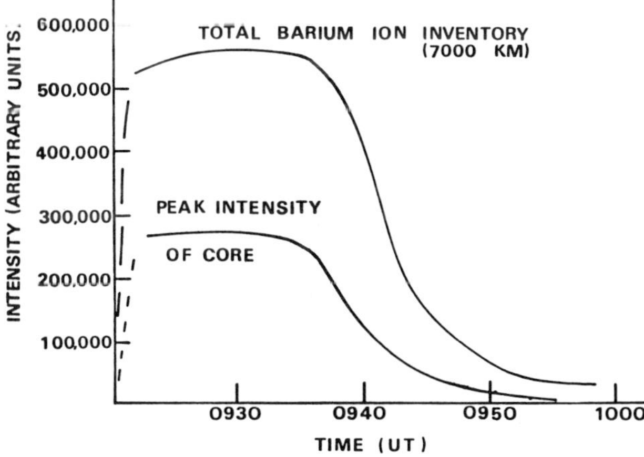

Figure 3—Variation of the total intensity of the core and halo and the peak intensity of the core in the period 0924 to 1000 UT. The total integrated intensity of the barium ions within the 7000 km diameter FOV of the imager is also shown.

below about 20% of the mass of initially-created ions (0950 UT), the core accelerated rapidly southward.

Several models of ion expansion within the halo have been tested against the observed intensity distributions and variations between 0925 and 0956 UT. A model where the ion velocity within the halo is a monotonically increasing function of radial distance from the core fits the radial and temporal variations acceptably. However, this places a high demand on the source of energi-

zation of the halo ions (at least 3.75×10^9 J total, 5×10^{-8} W/m^2). An alternative loss mechanism might be the destruction of Ba II ions by further photoionization with a time-constant of decay of Ba II to Ba III (ionization potential 10 eV) of about 30 min. As far as is known, however, no emissions corresponding to Ba III resonance fluorescence have ever been observed in conjunction with ionospheric or magnetospheric barium releases.

OBSERVATIONS BY THE DOPPLER IMAGING SYSTEM

Doppler Imaging Observations of the Core—Initial Velocity

An image of the barium ion cloud through the Fabry-Perot etalon of the DIS is shown in Plate V-4D. The Fabry-Perot fringes (concentric annulae) formed by the monochromatic Ba II resonance-fluorescence emission at 455.5 nm are illuminated by the core of the barium cloud. These sharp fringe patterns obtained from the magnetotail release contrasted sharply with the lack of such patterns in the images obtained by the same instrument during the December 27, 1984 solar wind "comet" release (Rees et al.[8]). The Doppler analysis is conducted by comparing the observed fringe locations (fringe radius) with the fringes from a calibration lamp (providing the "rest" frame of reference). Many individual estimates of LOS velocity can be made over the surface of the barium cloud. Each fringe can be azimuthally sectioned, limited by the optical resolution and the number of gathered photons (Rees et al.[4]).

At 0932 UT, the Doppler shift of the central fringes of the DIS (Plate V-5A) indicated a mass motion of the core of about 2 km/s, away from the earth, consistent with the initial IRM/earth LOS velocity. There was also a velocity shear, with away velocities (4 km/s) at the north edge of the core, and very small LOS velocities at the south edge.

Doppler Imaging Observations of the Core—Velocity Shears

Between 0929 and 0944 UT, the DIS observations provided a consistent pattern of LOS velocity distributions. There was a persistent north-south shear, of the order of 2 km/s, with the LOS velocity of the south part being more earthward than that of the north part. As time progressed (Plate V-5B, 0936 UT; Plate V-5C, 0940 UT), the initial mean LOS component of about 2 km/s, directed away from the earth, changed into a mean earthward component of about 2 km/s (Plate V-5C, 0940 UT).

In the last of the useful DIS images (Plate V-5D, 0947 UT) the south boundary of the core had an earthward LOS velocity of 5 km/s. This may be linked to the rapid (lateral) southward acceleration observed after 0950 UT by the imager. At the south edge of the core, the DIS showed an earthward LOS velocity exceeding 10 km/s. This LOS velocity component (nearly the field-aligned component) was consistent with the lateral velocities of the entire core after 0950 UT. That would imply that the late-time velocity may have had a strong earthward field-aligned component.

Doppler Imaging Observations of the Ions of the Halo

As in the case of the December 27, 1984 comet release,[8] the Fabry-Perot fringes in the halo surrounding the core are of low contrast. This indicates a large LOS velocity distribution of the halo ions. The range of toward and away velocities should have exceeded 10 km/s. Such a velocity range would have been consistent with the velocities of barium ions diffusing to populate the extreme regions of the diffuse and rapidly expanding halo. These DIS observations do not themselves define the nature of the process by which the halo ions were energized, but indicate that the radial ion acceleration started close to the core boundary.

SUMMARY

Within 200 s of the release, a diffuse ion halo was observed, which expanded rapidly from the core of the ion cloud. A "diffusion coefficient" of about 2×10^9 m^2/s would characterize the halo expansion. This halo was observable for 40 min, although the core became indistinct 35 min after the release. Bohm diffusion of thermal (1000 K) ions cannot explain the rapid radial diffusion.

The continuous loss of plasma from the core into the halo is accompanied by a radially-outward acceleration. The mechanism that produces the acceleration may be the coupling of energy from the low frequency spectrum of electrostatic and electromagnetic waves endemic within the magnetosphere. As a result, the halo plasma is dispersed through very large regions ($\gg 1$ R_e) of the magnetosphere, rather than reaching a larger quasistable cloud diameter. This mechanism is a fundamental cause of the eventual dispersal and disappearance of all artificial magnetospheric ion clouds.

The surface brightness of the core of the total halo ion inventory within the 8000 km diameter FOV decreased very rapidly between 20 and 40 min after the release. A monotonically increasing radial velocity of halo ions with increasing distance from the core might explain the observations, requiring an energy input of about 5×10^{-8} W/m^2. Photoionization of Ba II with a time constant of the order of 30 min (to Ba III) could be a contributing loss mechanism, but seems unlikely in view of the 10 eV ionization potential of Ba II.

During the first 25 min after the release, the core of the cloud drifted slightly northward relative to the motion of the IRM. After the central intensity of the core had dropped below about 20% of its initial value (0950 UT), a large southward and smaller eastward drift was observed (up to 7 km/s perpendicular component). The core velocities after 0950 UT were of a magnitude similar to those of the in-situ magnetospheric plasmas observed by the IRM (Möbius[9]).

The diffuse halo appeared to drift westward and northward during the period 8–15 min after the release, and later moved generally eastward and southward between 15 and 25 min after the release. Distinct striations formed within the halo, on the westward side between 10 and 20 min after the release and later on the east side. These observations need to be carefully compared with the in-situ plasma drift observations.

The Doppler Imaging System showed that the ions initially moved away from the earth, following the IRM. Between 10 and 20 min after the release, the core of the cloud showed a north/south shearing motion with the northward part of the cloud (away from the earth) having a small velocity away from the earth, while the south part of the cloud moved earthward. Up to 20 min after release, the LOS velocities within the core region were of the order of 2–4 km/s. At 0947 UT, the last useful DIS record, there was a suggestion of a larger earthward velocity component at the southward boundary of the core and halo, immediately before the core accelerated southward and westward (0950 UT).

The early-time drift of the core of the ion cloud did not reflect the ambient magnetospheric ion flows, due to the inertia of the cloud. However, the initial drift of the halo, northward and westward, should have represented average flow conditions. The DIS velocity maps indicated that as the residual mass of the core decreased, by about 0937 UT, the core ions became more responsive to ambient flow conditions. The dramatic change of relative drift of the halo about 0940 UT, accompanied by major changes in the patterns of striations, indicated an abrupt change of ambient magnetospheric flow. The core of the cloud reflected this change of ambient plasma flow after about 0944 UT, by which time its mass and inertia had been reduced below about 20% of the initial values.

The halo of the barium cloud produced low contrast fringes in the DIS. This indicated a relatively high LOS velocity distribution throughout the halo, which is consistent with its rapid expansion. Such observations provide boundary conditions for the mechanism which energized the continuous expansion of the ion halos of magnetospheric (and perhaps also the solar wind) releases.

A vaporized mass of barium of about 200–400 g would be easily tracked by the present instruments. Multiple small releases, rather than single large releases, would be a more effective way of diagnosing magnetospheric plasma drifts.

ACKNOWLEDGMENT—Many scientists and engineers within the United States, West German, and United Kingdom communities involved with the AMPTE project played critical roles in the planning, coordination, and execution of the AMPTE chemical release experiments. Dr. Gerhard Haerendel, Dr. Tom Krimigis, and Dr. Mario Acuña, in particular, facilitated the complex arrangements and information flow essential for the success of the entire project. The airborne observations were possible through the technical and engineering assistance of the Medium Altitude Missions Branch of NASA Ames Research Center, under the direction of Bob Cameron. The University College London involvement was sponsored and partly funded by the United Kingdom Science and Engineering Research Council. It is a pleasure to acknowledge several valuable discussions with Drs. James P. Heppner and Tom Hallinan.

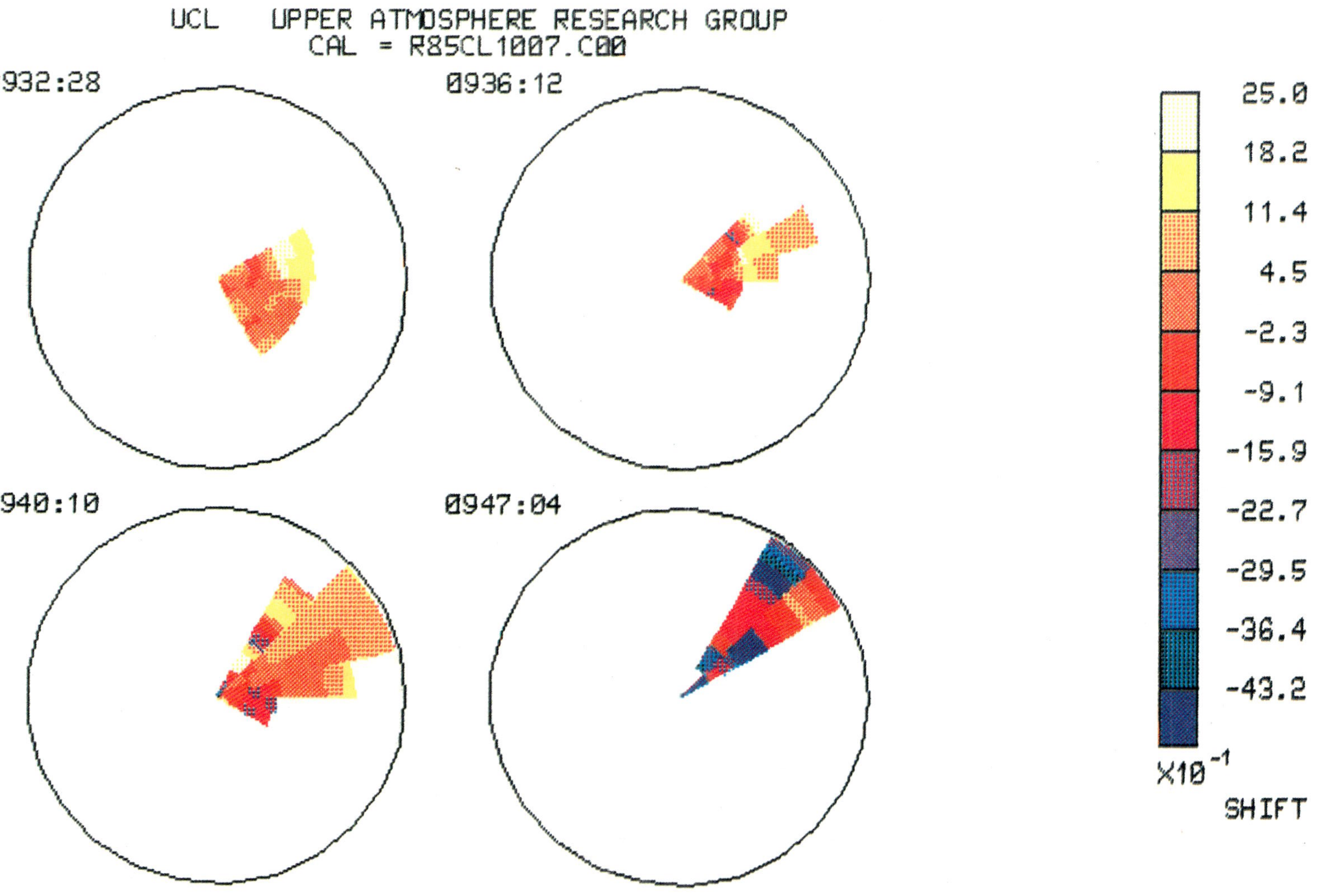

Plate V·5—(A) The pattern of LOS velocities obtained from the DIS at 0931 UT. At this time, the ion cloud core is moving away from the earth with a mean velocity of about 2 km/s with the highest away velocities at the northern side. (B) At 0936 UT, the LOS velocity distribution shows a shear from "away" velocities at the north boundary of the core (2 km/s) to earthward velocities of similar magnitude at the south boundary. (C) At 0940 UT, there is an increasing trend of earthward velocities with a mean earthward velocity of about 1 km/s. (D) In this last useful DIS velocity map, the south boundary of the core shows an earthward LOS velocity component of 5 km/s. In a small region at the south boundary of the core of the cloud, there is an earthward component of more than 10 km/s.

REFERENCES

[1] S. M. Krimigis, G. Haerendel, R. W. McEntire, G. Paschmann, and D. A. Bryant, "The Active Magnetospheric Particle Tracer Explorers Program," *Eos* **63**(45), 843 (1982).

[2] G. Haerendel, "Optical Data from the AMPTE Geomagnetic Tail Releases," in *Minutes of the AMPTE Joint Science Working Group, 18-20 June 1985*, D. A. Bryant and R. Blake, eds., Rutherford Appleton Laboratory (1985).

[3] I. McWhirter, D. Rees, and A. H. Greenaway, "Miniature Imaging Photon Detectors. III. An Assessment of the Performance of the Resistive Anode IPD," *J. Phys. E: Sci. Inst.* **15**, 145 (1982).

[4] D. Rees, A. H. Greenaway, R. Gordon, I. McWhirter, P. J. Charlton, and A. Steen, "The Doppler Imaging System: Initial Results on the Auroral Thermosphere," *Planet. Space Sci.*, 273-285 (1984).

[5] P. A. Bernhardt, personal communication (1985).

[6] H. Luhr, "IRM Magnetic Field Observations During the AMPTE Geomagnetic Tail Releases," in *Minutes of the AMPTE Joint Science Working Group, 18-20 June 1985*, D. A. Bryant and R. Blake, eds., Rutherford Appleton Laboratory (1985).

[7] J. P. Heppner (1986).

[8] D. Rees, T. J. Hallinan, H. C. Stenbaek-Nielsen, M. Mendillo, and J. Baumgardner, "Optical Observations of the AMPTE Artificial Comet from the Northern Hemisphere Stations," *Nature* **320**, 704 (1986).

[9] E. Möbius, "Observations During the AMPTE Magnetotail Releases on the IRM," in *Minutes of the AMPTE Joint Science Working Group, 18-20 June 1985*, D. A. Bryant and R. Blake, eds., Rutherford Appleton Laboratory (1985).

HIGHLIGHTS ON "ACTIVE DIAGNOSIS OF THE GEOMAGNETOTAIL"

S. M. Krimigis*

The session on "Active Diagnosis" during the Chapman Conference on Magnetotail Physics consisted principally of preliminary results from the recent AMPTE releases of lithium and barium ions in the magnetotail. In addition, there was discussion of tail formation processes in the artificial comet releases of December 27, 1984 and July 18, 1985. The speakers presented much observational evidence that was unusual and unexpected, in the context on prerelease expectations. I will not delve into the specific details of the data and the interpretations, but rather will note a couple of points that are truly remarkable and challenging to both our current understanding and our view on how to proceed in future investigations.

First, G. Haerendel[1] stated that (and I paraphrase) in the eight releases that were made, not once was there any evidence that anomalous resistivity processes were operating as diagnosed by the waves measured in situ, where amplitudes were lower than expected by several orders of magnitude. I believe this to be a truly remarkable finding. I did not hear anyone in the audience challenge that statement, despite the fact that anomalous resistivity is such a central concept in many of the numerical simulations that are done in magnetospheric physics, including many of the papers presented at this meeting. The obvious question is: Will these results cause us to stop and think about what we are doing, or will we conveniently ignore the evidence and continue to run our codes without regard to relevance to reality?

Second, R. W. McEntire et al.,[2] in discussing the tracer results, showed that the released ions did not convect to that part of the magnetosphere where they were expected to appear. Of course, we cannot conclude that they did not convect to the inner magnetosphere; however, the modeling that was done gave reasonable expectation that, at least in some of the cases, the ions should have been observed within the range of energies and sensitivities of the instrumentation on the Charge Composition Explorer spacecraft. So, what does that mean in terms of the steady-state convective mass transfer from the outer magnetosphere/magnetotail into the inner magnetosphere? It would have been far better, of course, if some ions had been observed; then we would have been able to tweak the models and see how we could fit the observations to the predictions. I don't think that we can just ignore the fact that we came within factors of two or so in intensity, according to the models, where we should have seen the ions—and didn't! We must now seriously think about the validity of our convection

models, especially the fact that all of them are steady state and require a magnetosphere in a constant state over a few hours time, when in fact most of the observations that we have show that there is no such thing as a steady-state magnetosphere for periods longer than a few tens of minutes. Ideally, what one would like to have is a complete specification of the electric field throughout the nightside magnetosphere and its time evolution for periods comparable to the transit time of the ions from the point of release in the tail to synchronous altitudes where they could be observed. It is clear that we are a long way from such a goal.

Third, the microphysics of the releases produced several surprises in both the magnetotail and the solar wind. Bernhardt et al.[3] pointed out that the March 21 barium release grew to substantially larger size than was expected by the balance between the dynamic pressure of the expanding cloud and the magnetic field and plasma pressures. Yet many of our concepts and calculations in space plasma physics rely heavily on such balance to explain phenomena such as the standoff distance of the bow shock or effects of the ring current. Another surprise was the observation by Gurnett et al.[4] of a narrowband electrostatic emission near the barium-ion plasma frequency in both the magnetotail and solar-wind releases. They interpret this observation as evidence for an ion acoustic mode, although no such mode was predicted for these releases.

Finally, the tail formation in the two artificial comets exhibited features that were far from predictions prior to release, as pointed out by Haerendel.[1] The tail did not form uniformly downstream, the "nucleus" was deflected to the side instead of downwind, there was significant distortion of the cavity, etc.; most of these phenomena were not predicted in advance. Of course, Haerendel has provided an elaborate model that appears to account, post-facto, for several of the observations. Nevertheless, there does not seem to be unanimity on the interpretations, as we all heard Papadopoulos say that his explanations are entirely different from Haerendel's. Here is a case where we thought that we would conduct a fairly straightforward experiment where we would inject cold plasma into a hot plasma and with our knowledge of plasma physics, we would be able to reliably predict the outcome—and didn't!

This brings us back to the question, "Where do we go from here?" I am not as optimistic as George Siscoe that launching 10^4 spacecraft, to separate time from spatial variations, is going to give us the answer. (I'm sure it was meant as a joke!) It wouldn't work in any case, since we would probably end up with 10^4 prin-

---

*The Johns Hopkins University Applied Physics Laboratory, Laurel, Maryland 20707.

cipal investigators each wanting to analyze his/her data in isolation from all the others! In a more serious vein, I believe that the artificial injection of plasmas in the vicinity of the earth has already proved to be a tool for gaining new knowledge about the physics of the natural environment and should be of great value in our search for physical understanding of many of the processes in space plasma physics.

ACKNOWLEDGMENT—This work has been supported by NASA under Task I of Contract N00024-85-C-5301.

REFERENCES

[1] G. Haerendel, "Tail Exploration and Tail Formation," in *Proc. Chapman Conf. on Magnetotail Physics, Oct. 28-31, 1985*, The Johns Hopkins University Applied Physics Laboratory, p. 29 (1985).

[2] R. W. McEntire, J. B. Cladis, S. M. Krimigis, G. Haerendel, G. Gloeckler, E. G. Shelley, and B. H. Mauk, "Tracer Aspects of AMPTE Tail Releases," in *Proc. Chapman Conf. on Magnetotail Physics, Oct. 28-31, 1985*, The Johns Hopkins University Applied Physics Laboratory, p. 34 (1985).

[3] P. A. Bernhardt, R. A. Roussel-Dupre, M. B. Pongratz, and J. Zinn, "Images of the Quiet and Disturbed Magnetotail During AMPTE," in *Proc. Chapman Conf. on Magnetotail Physics, Oct. 28-31, 1985*, The Johns Hopkins University Applied Physics Laboratory, p. 30 (1985).

[4] D. A. Gurnett, R. R. Anderson, O. H. Bauer, G. Haerendel, R. A. Treumann, H. C. Koons, and R. Holzworth, "Ion Plasma Waves Associated with the AMPTE Barium Releases," in *Proc. Chapman Conf. on Magnetotail Physics, Oct. 28-31, 1985*, The Johns Hopkins University Applied Physics Laboratory, p. 33 (1985).

VI. MAGNETOTAILS OF CELESTIAL OBJECTS

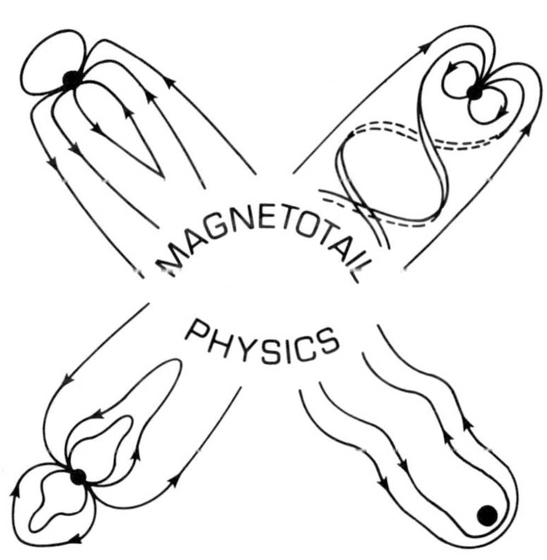

COMETARY PLASMA TAILS AND THE COMA SOURCE REGION

W.-H. Ip*

Several recent theoretical ideas are discussed that concern comet/solar-wind interaction, which is essential to the eventual formation of cometary ion tails. With emphasis on the magnetospheric aspects, processes relevant to the generation and transport of cometary pickup ions, the wave-particle interactions, and the configuration and dynamics of the cometary magnetic tails are explored. The importance of the spacecraft missions to comet Giacobini-Zinner and comet Halley are outlined to clarify these issues. Some comparisons of the magnetotail structures of different planetary plasma environments with similar solar-wind/neutral atmosphere interactions are made in the case of Venus and Titan.

INTRODUCTION

To a great extent, the modern research on cometary plasma physics started only after the recent International Cometary Explorer (ICE) flyby of the tail of comet Giacobini-Zinner (G-Z). The initial experimental reports immediately after the encounter indicated that a very intense flux of energetic ions and an extremely high level of plasma-wave activities were observed already at great distance ($\sim 10^6$ km) from the cometary nucleus.[1-3] The energetic ions originating from ionization of the cometary neutrals in the expanding coma of very large dimension would initially have a ring-like velocity distribution that is unstable against plasma instabilities. Similar problems have been investigated in the case of solar-wind interactions with the atmosphere of Mercury,[4,5] with the atmosphere of Venus (Curtis[6]), and then with the artificially released barium and lithium clouds in the AMPTE mission.[7-9] The latter, which was meant to simulate artificial comets[10] indeed found many interesting analogies with the actual comet/solar-wind interaction observed at comet G-Z. This is particularly true with respect to plasma turbulences and electron heating reflecting the dissipation and transfer of the free energies of the cometary pickup ions via a number of plasma processes.

The ICE spacecraft traversed the tail of comet G-Z at a downstream distance of about 8000 km. In macroscopic terms, it can be said that the spacecraft went through the very center of the coma and not the proper ion tail as identified in photographic observations (Fig. 1). It is therefore reassuring that the magnetic field measurements showed the formation of a magnetic tail near the center completed with a current sheet. The finding that the tail field could reach a magnitude of 50 γ or higher only makes the whole problem even more interesting.[11] The large degree of magnetic field fluctuations observed in the outer coma is also noteworthy.

While the unique dataset of the observations of the G-Z encounter would take years to analyze, so that the complexities and richness of the cometary plasma pro-

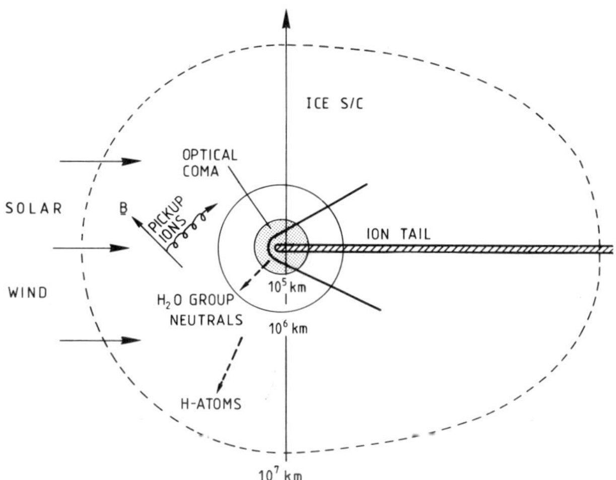

Figure 1—A schematic view of the streaming of the solar-wind plasma into the cometary neutral coma.

cesses could be fully unveiled, it is tempting to assemble some of the preliminary information available now and to compare it with the theoretical ideas and interpretations of cometary plasma phenomena proposed prior to the G-Z observations. The same thing may be said about observations of other planetary plasma systems connected with plasma-neutral interactions. Topics of interest include: (a) plasma dynamics and energetics, (b) particle-wave interactions, and (c) magnetic tail configuration and dynamics.

From these brief descriptions it will be clear that while several of the basic processes (i.e., generation of pickup ions and of plasma turbulences) have been anticipated one way or the other, the first in-situ measurement by the ICE spacecraft nevertheless took us by surprise in many of these novel features. The flyby observations by the armada of six spacecraft (Vega 1 and 2, Sakigake, Suisei, Giotto, and ICE) at comet Halley have added much more information on the basic nature of comet/solar-wind interaction. It is, however, the tail passage by the ICE spacecraft at comet G-Z that is most relevant to the study of cometary ion tail physics.

*Max-Planck-Institut für Aeronomie, D-3411 Katlenburg-Lindau, Federal Republic of Germany.

By hindsight, a first look at the characteristic features of cometary ion tails was gained, long before the onslaught of the space missions to comets Giacobini-Zinner and Halley. The solar-wind interaction with the upper atmosphere of Venus, for example, has been explored in great detail by space probes from the Soviet Venera program and the U.S. Pioneer Venus mission. Reviews of the corresponding solar-wind/atmosphere interaction process can be found in Refs. 12 through 17. In the following we will only highlight some new aspects in connection with the comparative views of the solar-wind interaction with the neutral atmosphere of planets (and satellites) and with comets.

Figure 2 illustrates the magnetic field configuration in the vicinity of Venus' ionospheric tail (cf. Ref. 17). It basically shows how the interplanetary magnetic field may be "draped" around the planetary ionosphere—as was predicted by Alfvén's[18] model for the cometary magnetic field—and that the magnetospheric tail is bound by a current boundary layer. (Interestingly enough, a similar boundary layer was seen at comet G-Z with a cross-sectional radius of about 4500 km.[19]

The Voyager 1 flyby observations of Titan in its wake region have shown that the tail field of Titan is also characterized by the formation of two lobes of opposite magnetic polarities separated by a current sheet.[20,21] Interestingly enough, the plasma plumes of Titan of exospheric origin might have been seen by the Voyager plasma experiment[22] making it a very intriguing analog of the cometary ion tail but in a planetary magnetospheric environment. The Voyager 1 observations at Titan have also shown that the detailed configuration and dynamics of the magnetic tail could be significantly influenced by the process of exospheric ion pickup.[20,21] It is partly for this reason that our consideration of the cometary ion tails will start with a discussion of the plasma dynamics and the wave-particle interaction involved. In fact, in the case of comets, the upstream processes are even more closely related to the formation and variability of the ion tails, just like the study of solar wind in which case the final answers to all the plasma dynamics and ion composition eventually have to be sought at the source regions near the solar corona; the first step of cometary ion tail dynamics has to do with the coma region.

PLASMA DYNAMICS
Generation of Pickup Ions

An example of direct observations of He^+ ions picked up by the solar wind from the interstellar wind is given in Fig. 3a. This energy spectrum of the M/Q = 4 channel of the SULEICA spectrometer on the AMPTE-IRM spacecraft shows a prominent shoulder with a sharp cutoff at 23 keV/e, which corresponds to four times the bulk energy of the high-speed solar wind.[23] The SULEICA measurements also showed that pitch-angle scattering has led to randomization of the velocity vectors of these pickup ions in the solar-wind frame while the gyration energy is still conserved. In the case of the intersteller ions implanted in the solar system, the sweeping effect of the solar wind will cause continuous adiabatic deceleration of these He^+ ions as they move outward.[24] As for cometary pickup ions (see Fig. 3b), unlike the intersteller He^+ ions that could be treated as test particles, their bulk flow and thermal motion in the vicinity of the comet will be determined by the pressure gradient, the plasma turbulence, and the magnetic field configuration resulting from their corresponding kinetic effect. Some of these effects will be mentioned in the following. But meanwhile, we would first like to review briefly the physical nature of the source region of the ion tail, namely, the neutral coma.

The general assumption about the chemical composition of the volatile ice of cometary nuclei is that H_2O is the dominant species followed by CO_2 and CO.[26] The abundances of CO and CO_2 relative to that of H_2O vary from comet to comet. In Fig. 4a the number density distributions of a coma with an assumed composition of $[H_2O] : [CO_2] : [CO] = 1 : 0.1 : 0.1$ are shown with gas production rates corresponding to comets Halley and Giacobini-Zinner at 1 AU solar distance. The effects of photoionization and photodissipation are to produce rapid cutoffs of the number densities beyond $r \gtrsim 10^6$ km. The curve of the H_2O group is made up of the sum of H_2O and its photodissociation fragments, OH and O. The actual abundances of CO and CO_2 could be considerably different from what are depicted in Fig. 4, however.

The cometary neutrals such as H_2O and CO_2 have an expansion velocity on the order of 1 km/s. Other fragments from photodissociation could gain large random speeds at production. For example, the photodissociation of H_2O into OH and H would supply a flux of H-atoms with an excess speed of 8 km/s, and the sub-

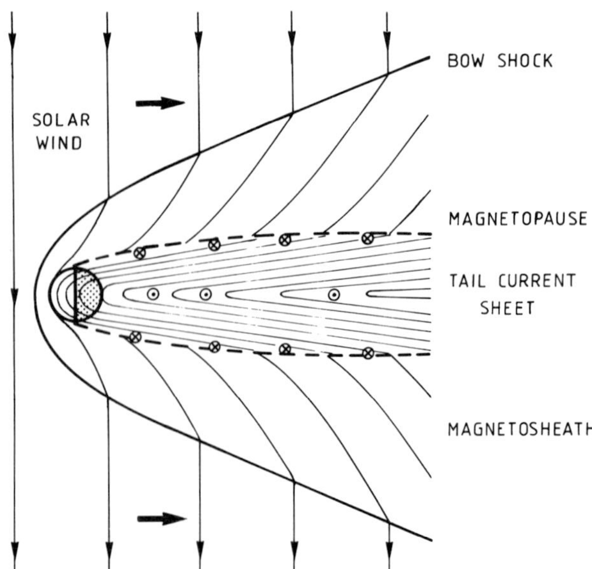

Figure 2—Two-dimensional schematic model illustrating the cross-tail current system in the Venus tail and the average magnetic field structure therein.

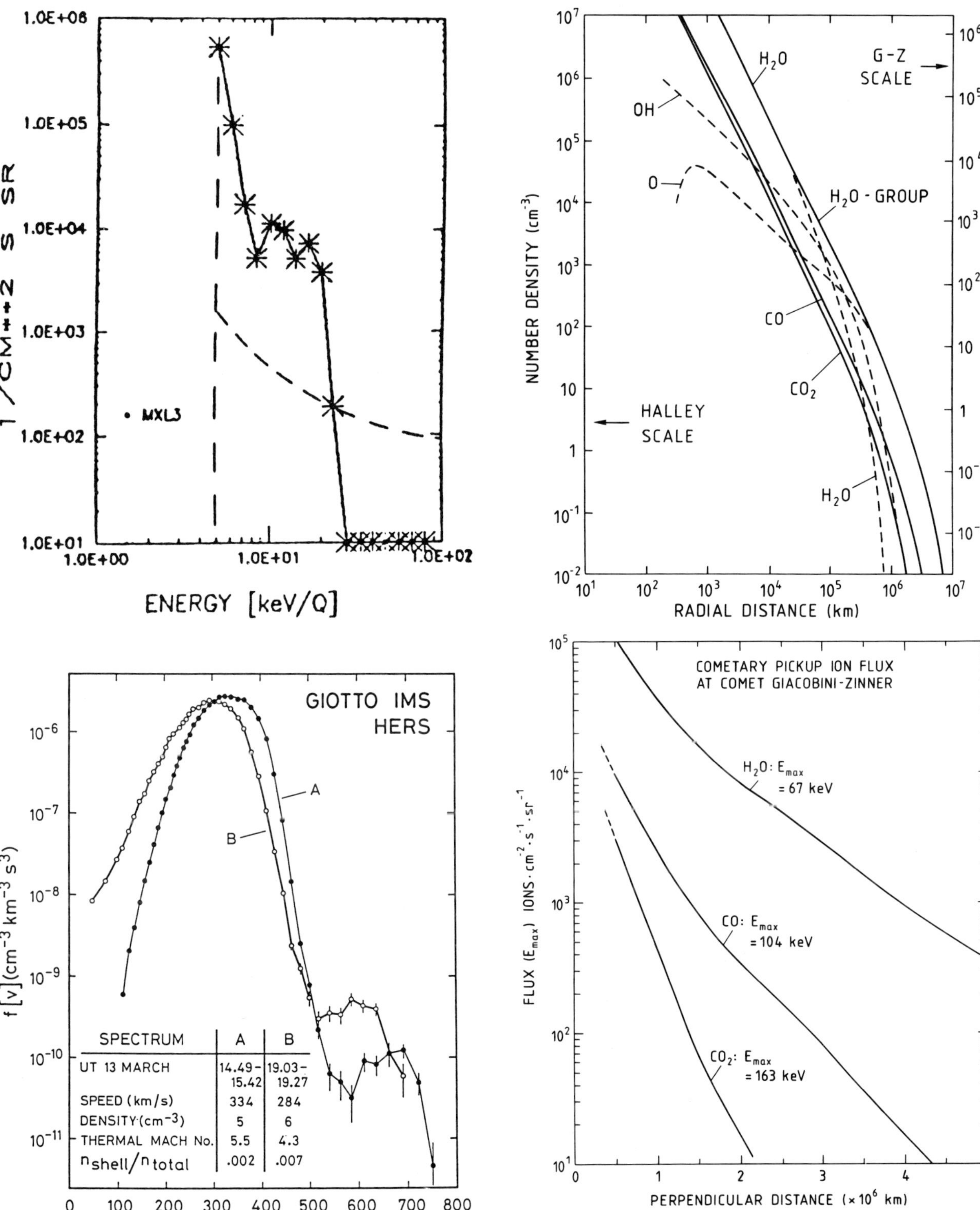

Figure 3—(a) An energy spectrum of the intersteller He$^+$ ions picked up in the solar wind as observed by the SULEICA spectrometer on the AMPTE-IRM spacecraft.[23] (b) The velocity distributions, parallel to the solar-wind velocity vector, of solar-wind protons and hydrogen ions of cometary origin for two intervals before the Giotto spacecraft crossed the cometary bow shock. Data from the Giotto Ion Mass Spectrometer experiment.[25]

Figure 4—(a) A model of the density distributions of the H_2O group, CO_2, and CO. The relative abundances are assumed to be [H_2O] : [CO_2] : [CO] = 1 : 0.1 : 0.1. (b) Variations of the pickup H_2O^+, CO^+, and CO_2^+ ion fluxes along the trajectory of the ICE spacecraft through the tail of comet G-Z. The energies are the maximum values for $B^\perp V_{sw}$ and are measured in the spacecraft frame (i.e., $E = 2\,m_i\,V_{sw}^2$, where m_i is the ion mass). In this model calculation, the relative production rates are taken to be Q[H_2O] : Q[CO] : Q[CO_2] = 1.0 : 0.1 : 0.1.

sequent dissociation of OH into O and H would produce H atoms with an excess speed of 20 km/s.[27] The neutral coma of the heavier species is therefore enveloped by an even more extensive halo of hydrogen atoms as indicated by UV observations of their Lyman-α emissions in many comets.[28,29] The ionization and pickup of the H^+ ions by the solar wind in this extended hydrogen cloud (diameter $\sim 10^7$ km) would be a precursor of the mass loading effect as the solar-wind plasma streams further inward. Note that the photodissociation of the CO molecules could also produce a flux of fast-moving atomic (C and O) fragments making plasma effects associated with cometary ion pickup processes observable at large distances from the nucleus (cf. Ref. 3).

At great distances perpendicular to the comet-Sun line, the solar-wind velocity maintains its original speed and the stream lines for the solar-wind plasma can be approximated as a straight line through the cometary coma. If the collective effect in assimilating the pickup ions into the solar-wind flow is not significant, the cometary ions with large gyroradii will execute $E \times B$ guiding center drift (Fig. 1). The fluxes of the H_2O^+, CO_2^+, and CO^+ ions along the trajectory of the ICE spacecraft so estimated for the special case when the solar-wind velocity is perpendicular to the IMF are given in Fig. 4b. It is seen that for the H_2O-group ions, the particle instruments measuring energetic ions with energies ~ 70 keV (in the spacecraft frame) should see a flux of 3×10^3 ions/cm^2 sr at 10^6 km away from the comet if the pickup ions are isotropized in their pitch-angle distribution. The actual flux as a function of pickup energy should depend on the orientation of the interplanetary field and the solar-wind condition in general. For example, an increase of the solar-wind speed from 400 km/s to 700 km/s could lead to the appearance of H_2O^+ ions at maximum energy in the 200 keV channel (and 400 keV for the CO_2^+ ions) whereas in the low-speed solar-wind regions with $V_{sw} = 300$ km/s, the H_2O^+ pickup ions would be seen at the 40 keV channel (and 90 keV for the CO_2^+ ions). In the spacecraft frame, if α is the angle between the interplanetary field and the solar-wind flow direction, the three components of the particle velocity will be

$$V_x = V_{sw} (1 + \sin^2 \alpha \cos \phi - \cos^2 \alpha)$$

$$V_y = V_{sw} \sin \alpha \sin \phi \qquad (1)$$

$$V_z = -V_{sw} \cos \alpha \sin \alpha (1 + \cos \phi)$$

where $\hat{y} \parallel V_{sw}$, and ϕ is the phase angle of the gyration motion; the maximum relative velocity of the pickup ions is

$$V_{max} = 2V_{sw} \sin \alpha \qquad (2)$$

and the maximum energy is

$$E_{max} = 2 M V_{sw}^2 \sin^2 \alpha . \qquad (3)$$

A sudden change of the local magnetic field direction therefore would shift the energy spectra of the newborn pickup ions. Such a complexity must be taken into consideration in the data analysis. Indeed, the ICE measurements have shown a very good correlation between the energetic ion flux and the angle α.[17] See Fig. 5.

At a distance of a few times 10^5 km, perpendicular to the comet-Sun axis, the mass-loading effect is insufficient to cause a noticeable slowdown of the solar-wind flow in the case of comet G-Z (but not for comet Halley). Thus in the energy spectra of the pickup ions there may be only a small dispersion from the maximum energy as given by Eq. 3 if the solar-wind condition remains steady. But as soon as the spacecraft reaches within a perpendicular distance of 10^5 km, the flow velocity will change substantially; the pickup ions swept along a certain stream tube would be characterized by a range of gyration energies. The cometary ions accreted by the solar wind at great distance from the comet should have the peak energy; in the region where the plasma flow is significantly reduced, increasing amounts of cometary ions of lower energies will be added. Utilizing the plotted results from the MHD cometary flow model by J.A. Fedder et al.[32] (which was kindly provided by M.B. Niedner) simulating the flow field and magnetic field configuration at comet G-Z, we could fold in the neutral coma model in Fig. 3 to estimate the fluxes of different components of the pickup ions in the inner coma. The spatial evolutions of the velocity distributions (in the solar-wind frame) for the H_2O^+ ions along a stream tube are illustrated in Fig. 6. Note that for cometary ions in stream lines far away from the comet their distribution functions should be nearly monoenergetic as long as the flow speed is not changed much. However, for the stream tube with $r_\perp (\infty) = 2.2 \times 10^4$ km,

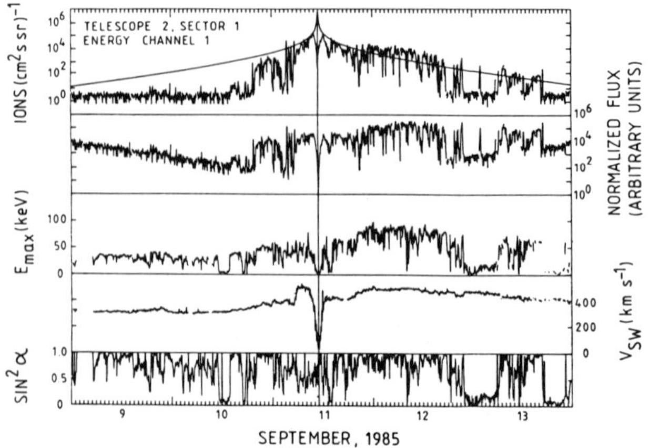

Figure 5—Fluxes of energetic ions measured in the coma of comet G-Z using the energetic particle experiment on the ICE spacecraft.[31] The energy range for the channel displayed is E1: 65–95 keV. From top to bottom: observed ion flux and predicted flux, flux normalized to the predicted flux, E_{max}, the solar-wind velocity, and the angle between the solar-wind velocity vector and the magnetic-field direction for the 5-day period commencing at 00:00 UT on September 9, 1985.

the flow velocity will be reduced significantly as the comet is being approached; and as a result, the H_2O^+ ions with energies ~ 1 keV will be assimilated and become the main bulk of the cometary plasma in the central part of the ion tail.

Since our main interest here is on the tail flow, the stream lines as defined by the velocity field in the MHD model have to be followed from large upstream distances to the tailward side. Numerical computations dealing with the MHD of solar-wind interaction will be required as a guidance. On the other hand, if we are aiming at the solar-wind inflow along the comet-Sun line, the treatment could be simplified greatly as was first carried out by Wallis.[33]

In his analytical model, Wallis showed that, if the pickup ions do not suffer from thermalization and that the magnetic moments of the cometary ions could be assumed to be invariant, an adiabatic heating effect will be introduced into the pickup ion population as the plasma flow slows down and the magnetic field strength increases. The extension of the velocity distribution beyond the initial value of the pickup velocities in Fig. 6 reflects this possible effect. However, for CO_2^+ ions with a gy-

ration energy of 44 keV, the corresponding gyroradius will be $r_g \sim 1.8 \times 10^4$ km, which is comparable to the characteristic dimension of the magnetic field variation in the case of $r_\perp \sim 2.2 \times 10^4$ km; the motion of individual particles thus could not be approximated as just following the bulk flow as assumed in the MHD calculation or being described by a combination of the electric drift, magnetic gradient, and curvature drift.[34] For these populations of energetic ions with origin at large upstream distances, kinetic effect will be important in determining their motion in the large gyroradius limit.

Cometary Shocks

As originally suggested by Wallis[35] and later confirmed by numerical calculations,[32,36,37] the shock jump should take place at the location with $M \sim 2$. For a gas production rate of about 2.5×10^{28} H_2O/s at 1 AU, the bow shock of comet G-Z should form at a subsolar distance of 5×10^4 km. The subsonic flow behaves like an incompressible flow with thermal pressure, and remains constant until reaching the region where charge exchange recombination will become important in getting rid of the hot ions picked up upstream.[35,38] The inflow of the cometary plasma will be eventually stagnated by frictional interaction with the neutral gas. The nominal stagnation distance can be estimated to be on the order of 200 km for comet G-Z.

In a fluid description, the function of a collisionless shock is to allow the supersonic flow to accommodate the obstacle, namely, the mass-loading effect of the cometary coma.[39] The main issue is therefore the position, strength, and shape of the cometary shocks. Even though the cometary ions enter as sources of mass and thermal energy, they are assumed to be efficiently assimilated and thermalized by the solar-wind plasma. This approach does not deal with the situation in which the pickup ions do not couple strongly with the solar-wind protons, and the local kinetic effect (i.e., dissipation process) at the shock front is not addressed at all.

To understand the coupling between the solar-wind plasma and the heavy cometary ions, a two-fluid description must be given to the whole interaction process. Hybrid simulation methods as developed in the investigations of the earth's collisionless bow shock are therefore very useful tools. Initial reports on such a numerical approach have been presented by Galeev and Lipatov[37] and Omidi et al.[40] While the one-dimensional simulation codes treat the electromagnetic fields and the motion of the particle ions and electron fluid self-consistently, the location of interest is limited to a small interval (about a few 10^4 km) including the transition zone where the shock forms and the immediate upstream and downstream regions. The initial velocity distributions of the solar-wind proton and heavy cometary ions and the number density of the pickup ions must be assumed. For instance, in the study of Omidi et al.,[40] cases with different velocity distributions of the heavy ions (varying from ring-beam to Maxwellian distribution) and different Mach numbers have to be considered.

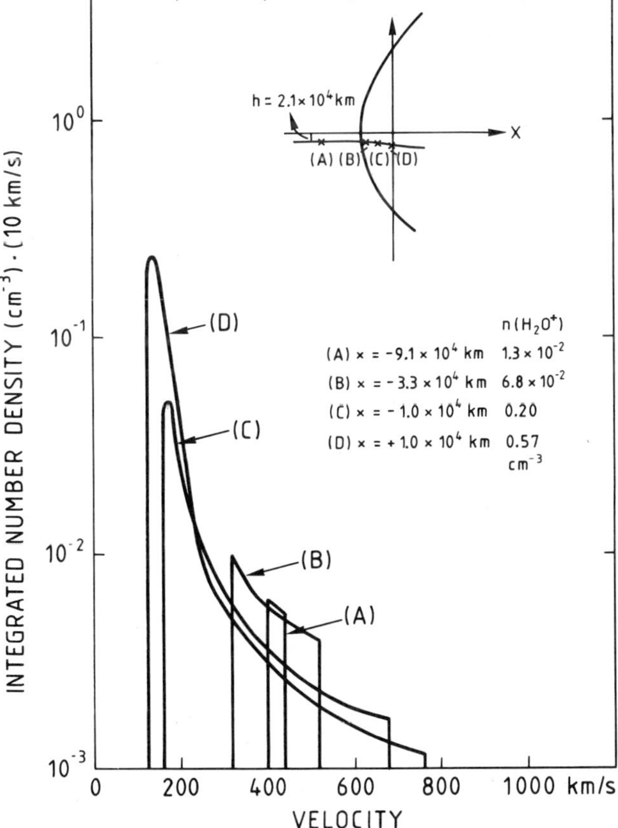

Figure 6—The number density distributions as a function of gyration speed in the solar-wind frame. Variations of the density distributions along a stream line are computed for a vertical distance $r_\perp = 2 \times 10^4$ km at large upstream distance from the comet. The calculations were performed for the case of comet G-Z.

While observations at comet G-Z and comet Halley should shed light on which case study is the most appropriate, the numerical calculations by Galeev and Lipatov[37] and Omidi et al.[40] have all reached the conclusion that the motion of the heavy ions could have a very significant effect on the cometary bow shock. The basic reason is that the gyration of the energetic heavy ions plays a very important role in the microphysics at the shock front. One example of pickup ions (O^+) with a ring-beam distribution and the supersonic solar-wind flow with a high Mach number ($M = 8$) is shown in Fig. 7 in which it is seen that a long foot is created in front of the shock. The large gyroradii ($r_g \sim 5 \times 10^3$ km) of the O^+ ions allows repeated crossing of the shock front as indicated by the enhancement of the O^+ number density in front of and behind the shock. As for the solar-wind protons, the shock is characterized by the existence of an extended foreshock and then an overshoot behind the shock. It was suggested that the heavy ions play a role similar to that of the reflected ions in the formation of the earth's bow shock.

In the above simulation calculation, the O^+ number density is assumed to be 3% of that of the solar-wind protons. With this relative abundance of cometary ions, the plasma flow should have been slowed down appreciably and the corresponding Mach number might be reduced due to heating (at least in the MHD framework); the situation of a weak shock might arise.[32,35,36] From their simulation computations for the cases of low Mach numbers, Omidi et al.[40] found that the shocks could be transitory due to interaction of the heavy ions with the shock. The characteristic of bunching of the heavy ions near the shock front due to their gyration motion persists, however.

As discussed above, the whole issue of cometary shocks is extremely complicated. While different approaches have led to clarifications of various aspects of the problem in different locations and length scales, no consistent picture has been developed yet. Even though,

strictly speaking, the ICE spacecraft to comet G-Z and the multispacecraft missions to comet Halley explored only the flank sides of the cometary shocks, it is nevertheless interesting to compare observations with theoretical results (for example, the formation of a layer of energetic cometary ions near the shock jump would be one feature to look for).

According to Sagdeev et al.,[41] the presence of the extended foreshock region in the cometary shock simulation experiments yields the impression that the situation may be equivalent to the diffuse cosmic-ray shock transition in which the energetic cosmic-ray particles carry most of the thermal pressure and the plasma flow carries most of the mass (see Ref. 42). However, unlike the cosmic-ray shocks, the foreshock of a cometary shock is mediated by the kinetic motion of the energetic heavy ions and not by any diffusion process as required in the cosmic-ray shocks. In any event, the possibility of a diffusive-shock type acceleration in the foreshock region of a comet has been suggested by Amata and Formisano.[43] Their argument, which is quite interesting, is that the creation of the pickup ions with a ring-beam velocity distribution should lead to a number of plasma instabilities that, in turn, would generate a high level of large-amplitude waves. With the pickup ions as seed particles and the waves as scattering centers, Fermi acceleration by means of the slowdown of the solar-wind flow could take place. From a scaling of the coupling between the hydromagnetic wave excitation and ion acceleration upstream of the earth's bow shock, Amata and Formisano[43] proposed that the pickup ions could be accelerated to energies of 500 keV or higher at comet Halley. At the same time, stochastic acceleration of the cometary pickup ions could be effective. A simple consideration of the second order Fermi acceleration process by Ip and Axford[44] indicated that the mass-dependent energization effect could be more effective in producing H^+ ions with energies easily exceeding 70 keV and H_2O^+ ions exceeding 300 keV in the coma of comet Halley. It should also be noted that the existence of very heavy molecules with mass ~ 100–200 AMU in the cometary coma could allow the direct acceleration of energetic ions to energies of more than 500 keV via pickup process alone (W. I. Axford, personal communication, 1985). Composition analysis at this energy range would thus be important to separate different acceleration mechanisms.

One consequence of wave scattering is that the initial ring-beam distribution will be randomized by pitch-angle diffusion. As a result, there will be field-aligned streaming of the pickup ions. If unimpeded, there could be substantial loss of the pickup ions from the source region. On the other hand, the super-Alfvénic streaming of the energetic cometary ions along the field line should excite ion cyclotron waves;[41] the scattering effect of the waves might modify the motion of energetic ions in turn such that a diffusion process (and hence stochastic acceleration) is more appropriate.[45,46] The injection of cometary ions to the tail region is thus potentially regulated by the hydromagnetic waves excited in the coma.

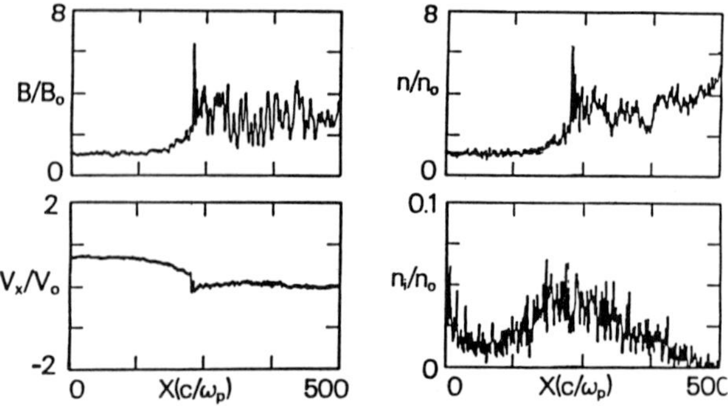

Figure 7—The result of a simulation for the effect of a ring-beam O^+ distribution on the structure of cometary bow shock: B/B_0, V_x/V_0, n/n_0, and n_i/n_0 are the normalized values of the magnetic field, solar-wind proton velocity, solar-wind proton number density, and oxygen ion-number density, respectively. The initial Mach number is 8 and $(n_i/n_0)_0$ = 0.03. From Ref. 40.

In addition, other types of wave-particle interactions could have direct effect on the production of cometary ions and the configuration of the plasma tail. These are discussed in the next section.

PARTICLE-WAVE INTERACTIONS

Almost all types of plasma instabilities have been found in the plasma environment of comet G-Z.[3] What is happening may be guessed by looking at an analogous process, namely, the expansion of an emerging solar magnetic flux loop into the turbulent solar atmosphere. Figure 8 is an adaptation of a scenario developed for the interpretation of solar type I noise storms in which the formation of a weak shock is able to generate unstable cross-field electrostatic instabilities by the perpendicular currents at the shock front.[47] The lower hybrid waves so excited will accelerate electrons stochastically,[48] leading to further generation of upper hybrid waves and other wave emissions in the wake of the shock. (The generation of cometary kilometric radiation (CKR) at comet Halley as discovered by the Sakigake spacecraft[49] could be of a similar cause.)

From a phenomenological point of view, this seems to fit into the preliminary picture of the plasma environment of comet G-Z as observed by the ICE spacecraft. The most important difference, of course, is that it is the production of cometary ions that provides the free-energy source for most of the observed plasma turbulences, and the initial ring-beam distribution of the pickup ions is sufficient to drive the lower hybrid emission and ion acoustic waves. The interaction of the electrostatic waves with the electron gas is to produce a population of suprathermal electrons carrying the heat flux.

Under normal solar-wind conditions with a number density of $n_e = 10$ cm^{-3} and an electron temperature of 10^5 K, the corresponding electron impact ionization time of the H_2O molecules is 2×10^7 s whereas at a raised temperature of $3 \times 10^5 - 5 \times 10^5$ K, the ionization time will be reduced to $(2-3) \times 10^6$ s, which is comparable to the ionization time from photoionization and solar-wind proton charge exchange (this also makes multiple ionization possible). Thus, if plasma turbulent heating of the solar-wind electrons occurs in sheet-like (or burst-like) structures with a width $\leq 10^4$ km, enhanced ionization would result in the formation of plasma inhomogeneities. The required energy transfer rate (from the gyrating pickup ions to the electron gas) of about 10% is compatible with the theoretical value derived by Formisano et al.[50] for low ionization limit. Such a nonuniform ionization process in cometary comas has been suggested to be a plausible cause for the formation of narrow ion rays often observed in cometary plasma tails.[15] From plasma measurements by the Vega spacecraft at comet Halley, Galeev et al.[51] have identified several episodes of enhanced ionization in the inner coma of comet Halley.

Figure 9 provides a composite view of the energetic particle intensity variation[2] and the plasma-wave activity[3] in the coma of comet G-Z over a distance of 2×10^6 km from the center. The extensive nature of the coma interaction is truly impressive. At smaller time resolution, as afforded by the particle experiment of Hynds et al.,[1] bursts of energetic particles over a duration of 10–20 min can be seen in the outer region (i.e., for perpendicular distances $>5 \times 10^5$ km for an inbound pass). These particle bursts are each separated by approximately one hour and, in fact, represent plasma structures of high thermal pressure (since the pickup ions contribute significantly to the plasma pressure). As mentioned before, the modulation of the intensity and energy spectrum of cometary pickup ions upstream of the coma may be determined by the direction of the IMF as well as by the solar-wind velocity. One possible consequence of such IMF modulation effect would be the generation of optical ion rays often seen in pairs. The ICE electron measurements[52]—showing that large-amplitude fluctuations of electron density with a peak-to-peak variation of 5–40 cm^{-3} could occur in the coma within 10^5 km of the center—might also be relevant to the issue of ion-ray generation. One example of symmetrical ion-ray formation observed at comet G-Z in its return in 1959 is shown in Fig. 10. But no one-to-one correspondence between the electron density variations and optically observable plasma inhomogeneities in the ion tail has been reported.

As mentioned above, the large anisotropy characterized by the ring-beam distribution of the newborn cometary ions would necessarily generate strong wave activities via ion cyclotron resonance.[41,53,54] As discussed in the case of the earth's magnetosphere, different ion species in a collisionless plasma could be coupled via the ICWs;[55] ion heating (similar to heating of He$^+$ ions in the earth's magnetosphere) may be possible. In this con-

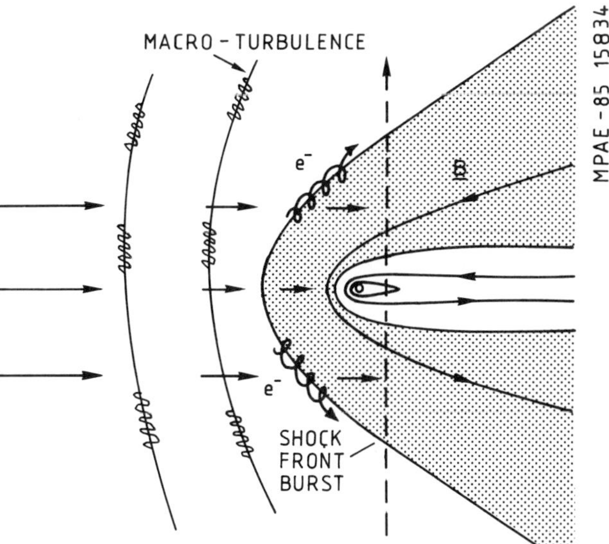

Figure 8—A schematic sketch of plasma-wave activities induced by solar-wind interaction with the cometary pickup ions, before and behind the weak shock front. An analogous view has been given for the interaction of a system of emerging magnetic flux with the turbulent solar atmosphere.

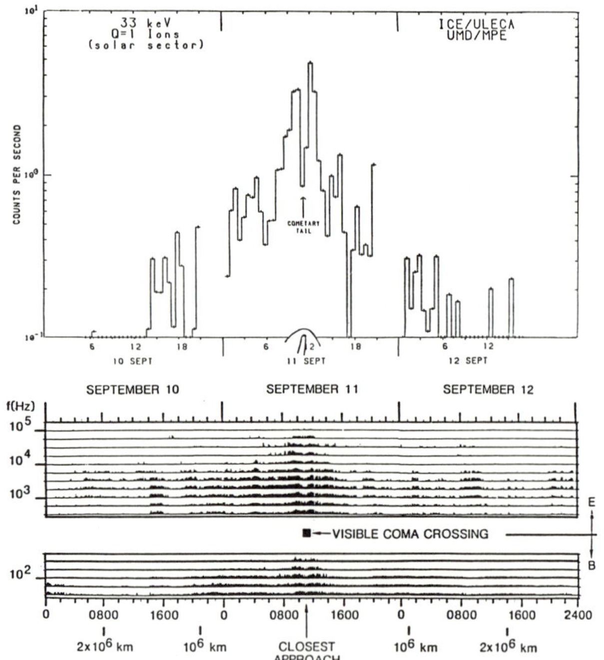

Figure 9—A comparison of the energetic ion intensity and the plasma wave activity as observed by the ICE spacecraft in the vicinity of comet G-Z. Particle data are from Ipavich et al.[2] and plasma wave data are from Scarf et al.[3]

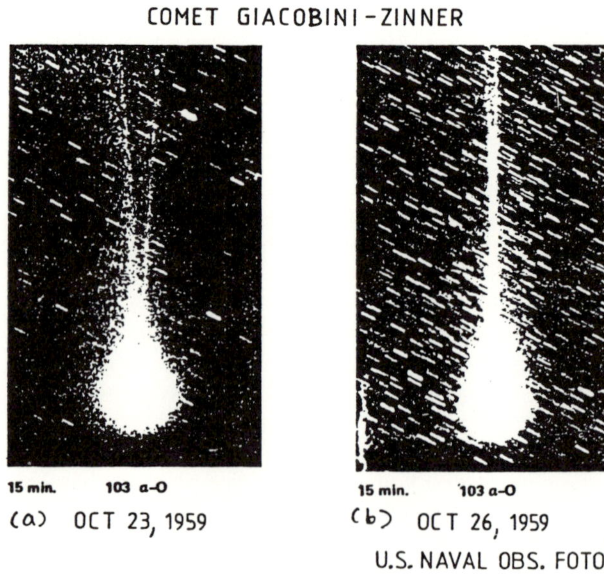

COMET GIACOBINI-ZINNER

Figure 10—Examples of the ion tail morphologies of comet G-Z observed in its 1959 return: (a) the formation of a pair of symmetric ion rays; (b) the appearance of one single central tail. (U.S. Naval Observatory photographs).

nection, it should be mentioned that in the solar wind there appears to be a strong coupling between the protons and the alpha particles by the fact that the bulk speed of the alpha particles tends to exceed that of protons by approximately the Alfvén speed.[56] This effect has led to the idea that the acceleration of minor ions in the solar wind is somehow related to gyroresonance with ICWs.[57] How about the cometary plasma? From the PLASMAG measurements at comet Halley, Gringauz et al.[58] reported the presence of a "chemical" boundary at a distance of about 1.6×10^5 km separating the external solar-wind-driven plasma from the internal plasma flow of cometary ions. The motion of the proton population and the heavy ions appears to be decoupled in the sense that the protons have a larger flow speed. One possibility is that wave-particle interaction might play a role in maintaining the observed velocity difference in addition to the neutral frictional effect as suggested by Ref. 58.

COMETARY PLASMA TAILS

In addition to the formation of pairs of well-defined symmetric ion rays with widths $\approx 10^3$–10^4 km as shown in Fig. 9, the cometary ion tails sometimes are characterized by the presence of bundles of numerous filaments of even narrower widths. The rich variety of the ion tail configurations should mean that the morphologies of cometary plasma flows basically reflect the time variations of the interplanetary conditions and observational geometries. From this point of view, it is interesting to compare the fluctuations in the electron density observed in the coma of comet G-Z[52] with the ion rays known to ground-based observers long since. The ICE measurements showed that large-amplitude variation in the electron density started to occur in a distance of about 10^5 km from the closest encounter. The plasma inhomogeneities lasted from about one to several minutes and hence should have a spatial scale of a few thousand km, and the peak-to-peak density variation is from 5–30/cm^3. Since the locations of these plasma density fluctuations should not be directly connected to the cometary ionosphere, considering the magnetic field topology in the mass-loaded cometary plasma flow, the physical condition at large distances upstream of the comet might be more appropriately the factor controlling the inhomogeneous structures. The bursty nature of the flux of the cometary pickup ions observed at an inbound distance of 10^6 km may testify to this effect. This is because the thermal pressure in the solar-wind accretion flow is dictated by the pickup ions; and for neighboring flux tubes, one filled with 20 keV H_2O^+ ions and the other much less so, the convection of these flux tubes through the bow shock could eventually lead to the generation of density enhancement and rarefaction as a result of the pressure equilibrium condition.

Not to be forgotten, of course, is the earlier suggestion that electrons accelerated at a strong cometary shock should cause enhanced impact ionization and thus the ray structures along the magnetic field lines.[39,59,60] While the ICE observations at comet G-Z have shown that the supersonic-flow/subsonic-flow transition could not be readily described as a strong shock, but the intense plasma wave activities in the turbulent subsonic flow region are nevertheless indicative of an electron heating process,[3,61] and as discussed above, additional ionization effects from electron impact is thus possible.

A different approach has been attempted by Schmidt and Wegmann[36] in the question of ion-ray formation. These authors have simulated the MHD response of the cometary ion tail to solar-wind tangential discontinuities defined by a sudden 90° change in the direction of the IMF and found that narrow streamers could be produced this way. Furthermore, as discussed above, the orientation of the IMF is critical to the kinetic energy of the pickup ions as well as to the growth rate of magnetic field turbulence and hence to the process of collective pickup; cometary ion rays could thus have an upstream origin. One good example may be the plasma structures observed in the coma of the CO^+-rich comet, Morehouse 1908III, in which the weak continuum and neutral emissions were not able to mask the strong brightness profile of the CO^+ ions. According to several authors who have studied the observational materials in detail,[62-64] ion structures in the form of parabolic envelopes first formed at a projected distance of 1–1.5 $\times$ 10^5 km from the optical center (Fig. 11). Immediately after their rapid formation, the envelopes would always start to shrink toward the center with a speed of about 10 km/s. After 3 to 4 hours the gradually sharpened structure of the plasma envelope would slow down to a much lower speed. And as it recedes to a distance of 5 $\times$ 10^4 km it would intermingle with the coma emission and become unrecognizable. At one time, three or more parabolic envelopes were observed to be in different stages of contraction or collapse. In the composite picture of the formative sequence of receding envelopes in Fig. 11 as depicted by Lüst,[63] the envelopes can be seen to lengthen into ray-like symmetric structures at the final stage. A further comparison of the structure of the tail rays and envelopes is illustrated in Fig. 12. In this sketch, the tail rays are all convergent toward a point just ahead of the optical center. This effect certainly produces the impression that the contracting envelopes are the precursors of folding ion rays and that both plasma structures are related to processes upstream of the comet head.

From investigation of the photographic plates of comet Morehouse 1908III, Wurm[64] and Wurm and Rahe[65] had also made the interesting point that the inner coma with a radius of 1–2 $\times$ 10^3 km is the site of strongest CO^+ emission with the ions concentrated in several sunward-pointing jets or streamers. They suggested that the ion rays in the tail could actually connect to these ionospheric streamers. A more appropriate explanation appears to be that the plasma envelopes and rays are determined by the solar-wind inflow through the cometary coma whereas the jets in the inner coma are related to the flow pattern of the expanding ionospheric outflow. A projection effect could then produce an overlapping of these two classes of ion structures. In any event, the study of the inner coma CO^+ emission of comet Morehouse 1908III and other comets has led Wurm[64] to the opinion that, besides photoionization and solar-wind charge exchange, there must be an ad-

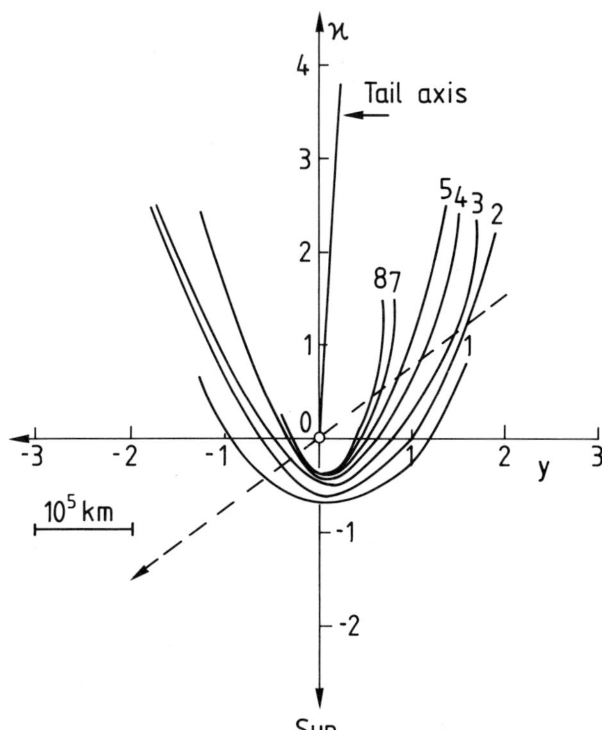

Figure 11—The development of a very well-defined envelope observed on October 21-22, 1908 in comet Morehouse 1908III. The time interval between (1) and (8) is 5.25 hr. The orbital direction of the comet is indicated by the dashed line. (From Ref. 63).

ditional ionization effect (the so-called internal ionization source) operating in the inner coma to produce rapid ionization (the CO^+ emission therein was observed to have a very short-time scale of variation, $\sim 10^3$ s). It is interesting to note that short-term outbursts of CO_2^+ emission in the coma of comet Halley had been detected during the time interval of spacecraft flyby observations in March 1986 by the IUE satellite.[66]

If the suggestion of an internal ionization source is followed up, we can see two possibilities in generating the required ionization effects. First, of special interest to us here, there are auroral activities possibly resulting from precipitation of energetic particles.[67,68] As in the case of a planetary magnetosphere, this process required discharge of the current system in the ion tail or the ion coma. As for steady-state acceleration in the tail current sheet region, the maximum energy that might be derived can be written as $\Phi = V_A B_t L_t$, where V_A is the Alfvén speed in the ion tail, B_t the tail field, and L_t the width of the ion tail. As a first approximation, the ICE observation at comet G-Z gives $B_t \sim 60 \gamma$, $n_i \sim 100$ cm^{-3}, and $L_t \sim 3 \times 10^4$ km; thus, $\Phi \sim 6$ keV. To produce an ionization time scale of 10^3 s in the inner coma, the flux of the keV-electrons required would be of the order of $F \sim 10^{11}$ cm^{-2} and a corresponding energetic electron number density of 100 cm^{-3}. This is certainly too high a value for the keV-electron number density. On the other hand, it should

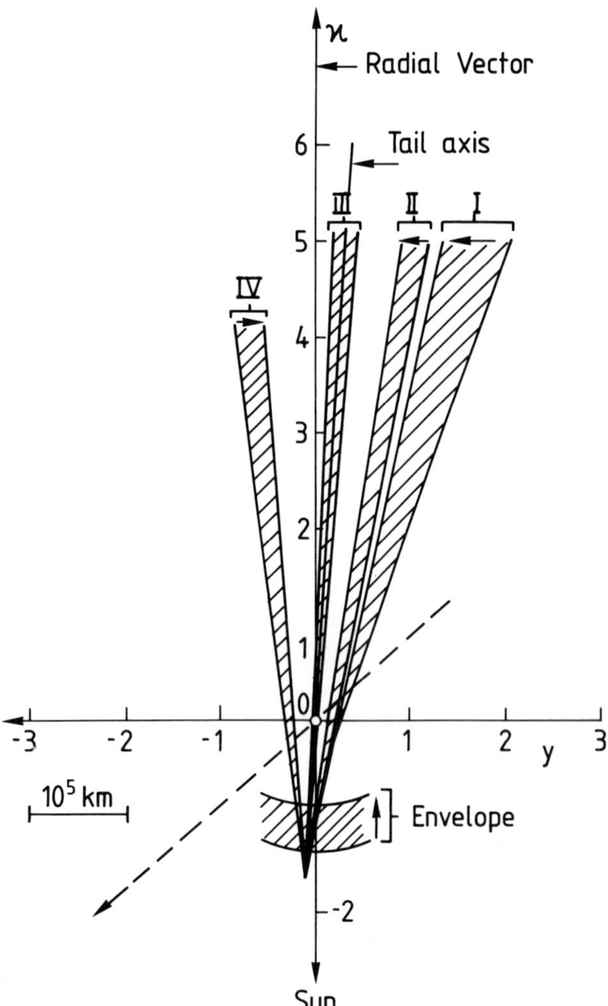

Figure 12—Movements of four ion rays and an envelope in comet Morehouse 1908III as summarized from photographs taken on October 3-4. The arrows indicate the direction of motions of the structures, and the displacements are represented by the shaded regions. (From Ref. 63).

be noted that with a number density of just 1 cm^{-3}, the electron impact ionization rate would be 10^{-5}/s, which is considerably faster than the photoionization rate ($\sim 10^{-6}$/s) of the H_2O molecules. Also, even if current sheet acceleration is not particularly efficient, the field-aligned streaming of hot electrons energized in the outer coma could cause significant ionization and dissociation effects as the magnetic field lines thread through the inner coma. The detection of keV-electrons in the coma of comet Halley by Gringauz et al.[69] is therefore particularly of interest.

One extreme situation of electron heating via wave-particle interaction in an ionizing system of counter-streaming neutral gas and plasma is the critical velocity ionization mechanism.[50,70-72] The fundamental point is that when the streaming velocity of a neutral cloud relative to a magnetized plasma (field-perpendicular speed is the relevant parameter) exceeds the critical value

$$V_c = \left(\frac{2\,e\,\Phi_i}{m} \right)^{1/2} / \eta \qquad (4)$$

where Φ_i is the ionization potential of the neutral atoms, m its mass, and η the efficiency of energy transfer from the cometary ions to the ionizing electrons, anomalous ionization effect from electron impact would occur. Such critical velocity ionization effect has been seen in laboratory experiments and space active experiments (see review by Newell[73]). The very high level of plasma wave turbulence and relatively significant electron heating effect observed by the ICE spacecraft at comet G-Z may be related to this effect even though the resulting electron impact ionization rate in the outer coma is still lower than the combined photoionization and solar-wind charge-exchange ionization rates.

In a recent study, Galeev and Lipatov[37] have compared the ionization time scale, the transit time scale, and the Townsend condition for avalanche ionization in the inner coma of a comet. They found that anomalous ionization effect is not probable in a cometary ionosphere unless in a very special situation. The ICE observations had not been able to clarify this point due to the tail pass of the spacecraft trajectory. Galeev et al.[51] have investigated several events of plasma density enhancement in the inner coma of comet Halley and found that they may be interpreted in terms of non-stationary critical ionization effects.[74] The overall effect of these ionization events could be evaluated to be small, however.[75]

In discussing the physical origin of the ion rays, we have tentatively divided a cometary ion coma into two parts: the outer one containing the mass-loaded solar-wind plasma and the inner one containing only ionospheric plasma from the central region. These two regions are separated by a contact surface. It is within this context that we delineated the source regions of the ion rays as of external origin (i.e., plasma inhomogeneities convected downstream by the solar-wind flow) or of internal origin (i.e., from enhanced ionization effect in the ionosphere). In reality, there might be no clear separation between these two plasma flows because of a number of instabilities at the boundary surface.[76,77] In other words, the magnetized plasma from upstream could infiltrate inside the ionosphere from time to time. An intermediate scenario for ion-ray formation would be plausible. For instance, Wolff et al.[78] have recently considered the scenario of how infiltration of magnetic flux tubes inside the cometary ionosphere via the Kelvin-Helmholtz instability could lead to the formation of pairs of flux tubes hung on the ionosphere, channeling the dense ionospheric plasma outward. This scenario, in part, was inspired by the observations of plasma irregularities in Venus' ionosphere by the Pioneer Venus spacecraft, in which case the Kelvin-Helmholtz instability was suggested to be the mechanism responsible for the twisted ionospheric flux ropes.[79] However, one major difference here is that the plasma source region for comets is not limited to the ionosphere, with a radius of no more than about 100 km in the case of comet G-Z (see below). By virtue of atmospheric expansion, the whole coma of very extensive nature could supply cometary ions to the plasma flow. In other words, even without invoking the Kelvin-Helmholtz instability at the "ionopause" cold cometary plasma from photoionization will be funneled along the magnetic field lines as they get caught up at the inner coma.

Thus far, only ion tail formation during relatively quiescent conditions of cometary plasma tails has been considered. These small-scale structures may be related to variations in the microscopic structures of the solar wind and IMF. For large-scale structures in the solar-wind, dynamical disturbances in the ion tails will be correspondingly more dramatic. One of these concerns the passage of interplanetary shocks or corotating interaction regions (CIR). Both phenomena are characterized by enhanced solar-wind dynamic pressure that might be conducive to the abrupt increase of cometary ion production via increase in the charge-exchange ionization or electron impact ionization rates.[54] With the sudden introduction of ionized matter, large-scale plasma structures should form in the ion tails. Beushausen and Jockers[80] had investigated this effect using a one-dimensional time-dependent model and found that an ion cloud with density enhancement by a factor about 3 can be formed during a CIR passage through a comet coma.

A combination of the dynamic pressure front and rapid variation of the solar-wind flow direction could also lead to the disturbances of the ion tails in the form of helical or kinky features. Rapid turning of one segment of an ion tail had also been observed in several cases. According to the so-called wind-sock model of cometary ion tails,[81] such dynamical behavior is consistent with the sudden change of the solar-wind velocity vector in direction perpendicular to the radial direction and its propagation through the ion tails. However, the very fast adjustment of the ion tail orientation to the new solar-wind flow condition must mean that there is a very strong momentum coupling between the solar wind and the cometary plasma, perhaps, via certain instability effects. If a boundary of sharp velocity gradient exists at the ion tail, then the Kelvin-Helmholtz instability would be a viable candidate.[82] On the other hand, the sideway interaction of the solar-wind flow with the ion tail as expected from the large perpendicular component of the solar wind at the CIR could also cause the occurrence of the Rayleigh-Taylor instability.[83] Taking the example of the rapid turning event of comet Bradfield 1979X in which the change in velocity is $\Delta V_\perp \sim 100$ km/s over a time interval $\Delta t \sim 2$ hr[84,85] we may approximate the Rayleigh-Taylor growth time as

$$t_{R\text{-}T} \sim \left(\frac{L\ \Delta t}{2\pi\ \Delta V_\perp} \right)^{1/2} \tag{5}$$

where L is the wavelength of the disturbances, for $L \sim 10^4$ km, $t_{R\text{-}T} \sim 350$ s. Thus, small-scale irregularities at

the ion tail boundary could be generated to facilitate solar-wind momentum coupling.

Sometimes there are indications that more complex phenomena were happening at disturbed ion tails. As illustrated in Fig. 13 the ion tail of comet Kohoutek (1973XII) was observed to develop helical structures as well as a kink in the middle of the tail; these disturbances were associated with the passage of a CIR.[86-88] One feature of particular interest is the appearance of filamentary plasma structures emanating from the "hinge" point defining the kink. Since the source region of these so-called side rays is too far away from the central coma to be supplied by photoionization of the cometary neutrals, the ray formation then must originate from leakage of cometary plasma already contained in the main tail. If magnetic X- and O-lines could form in the distant ion tails under disturbed solar-wind conditions[54,89] a way of channeling the cometary ions to the solar wind may exist. But much of the detail remains to be worked out before the cometary ion tail reconnection can be considered as a probable explanation for the side rays.

Since the cometary magnetic field may be interpreted as the result of a draping of the interplanetary field,[18] with the magnetic field morphology being describable in terms of an extensive current system (if need be) it is reasonable to expect that there should be a reversal of the magnetic field polarities in the ion tail after the crossing of an IMF sector boundary. Niedner and Brandt[87] and Niedner et al.[90] had applied this idea to an interpretation of the observed disruption events of ion tails from comet heads. In order to fit some of the observational features, Niedner et al.[90] developed a scenario in which the oppositely pointing magnetic fields are supposed to be reconnected at the frontside of the ion coma such that the magnetic flux accumulated in the previous IMF sector will be eroded away until a new magnetic tail with polarity orientation consistent with the present sector is formed (see Fig. 14). Such peeling of the magnetic flux, according to Niedner et al.,[83] may be related to the detachment of cometary plasma clouds from comet heads, hence the ion tail disconnection. However, it should be noted that the magnetometer experiment on board the Sakigake spacecraft detected multiple crossings of a nearly horizontal heliospheric neutral sheet during March 10–12, 1986 in which no apparent disconnection event was found in comet Halley's ion tail.[91] Instead, a bending of the ion tail was usually observed during the crossing of the IMF sector boundary.

As an alternative to the frontside reconnection process, the possibility of reconnection in the tailward side has been suggested.[15,16,92] Russell et al.[92] pointed out the difficulty with the reconnection process in the near-tail region, i.e., the presence of a stabilizing magnetic field perpendicular to the current sheet and the low value of the Alfvén speed; and they argued that the reconnection process may be triggered anyhow by several different processes such as interplanetary shocks. The underlying idea is that the density rarefaction produced in the ion tail during comet interaction with an IMF sector structure, say, could lead to a rapid reconnection pro-

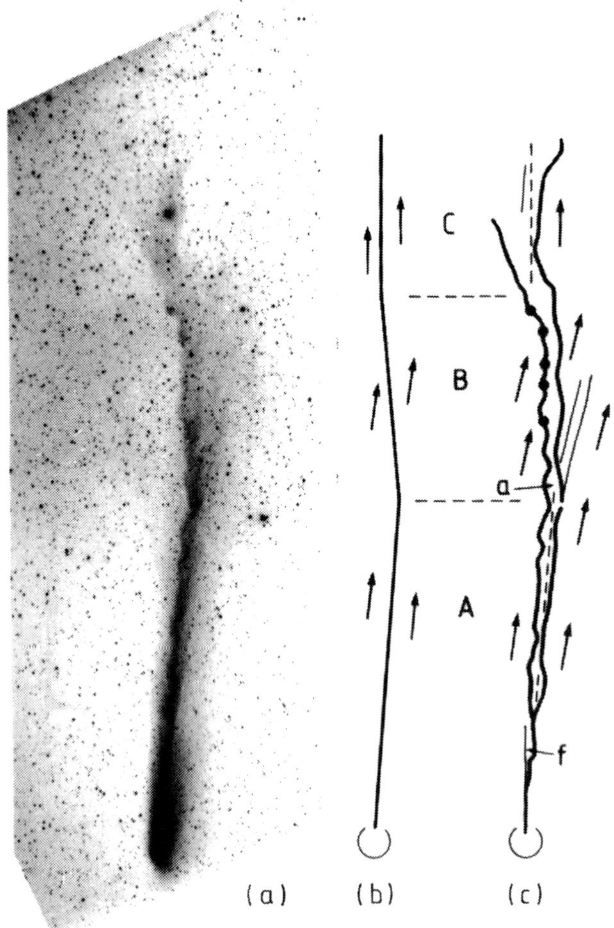

(a) (b) (c)

Figure 13—(a) An ion tail disturbance of comet Kohoutek 1973XII photographed on January 20, 1974 at 01:56 UT. (b) and (c) Delineation of the gross structure and relation with the solar-wind flow. This picture shows the formation of a kink (segment B) and large-scale turbulence along the ion tail. One additional feature is the parallel side rays emanating from one side of the ion tail (most conspicuous from the kink) like a curtain. (Courtesy of K. Jockers).

cess with a faster Alfvén speed at localized regions.[32] In addition to these two types of large-scale changes in the magnetic field morphology of cometary ion tails, there is a third possibility, namely, that of the enhanced ionization effect of the CIR which might also contribute to the production of a large plasma cloud moving away from the comet head.[54,80]

Published works on MHD simulations of comet-solar wind interaction by Schmidt and Wegmann[36] and Fedder et al.[32] have shown that the distribution of the cometary plasma would evolve from an initial configuration of cylindrical symmetry to a thin sheet bound by two lobes of opposite magnetic polarities. One major success of the ICE mission is that the spacecraft managed to cross the center of the ion tail of comet G-Z so that a number of the interesting physical phenomena taking place in the cometary ionosphere can be studied on a firmer ground. Figure 15a compares some of the ob-

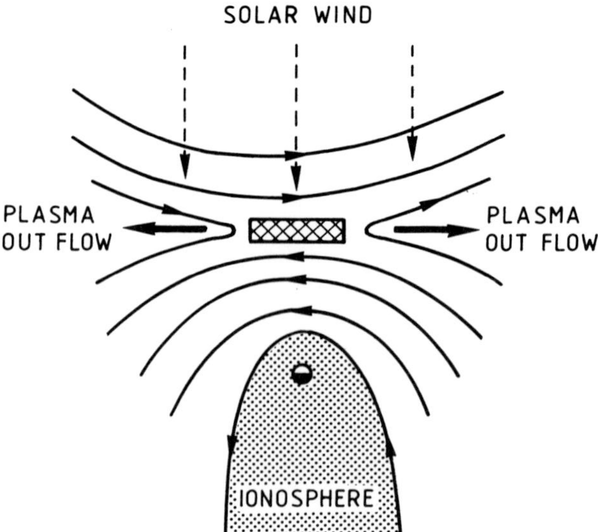

Figure 14—A scenario of reconnection of the interplanetary magnetic field in the frontside coma of a comet that (according to Niedner and Brandt[86]) is responsible for the ion tail disruption event observed at many comets.

served features of the plasma environment of the inner region within 2×10^4 km to the point of closest approach[11,52,93] with the numerical model by Fedder et al. (J. Fedder, personal communication, 1985).

To be noticed first is that the central current sheet with a thickness of 1000–2000 km should contain the bulk of the cold plasma from the ionosphere with an electron temperature of $T_e \sim 1$ eV (see Fig. 15b). The low temperature of the electrons (presumably the ions also) is the result of very efficient cooling by the H_2O molecules in the inner coma.[94,95] Because of the draping of the magnetic field by the ionosphere, there is a magnetic connection to the ionospheric region where the electron cooling effect is important even when the spacecraft was at a perpendicular distance of about 2×10^4 km. It is perhaps for this reason that the cold electron population was observed in a region much more extensive than the current sheet.

The cometary plasma with a peak electron number density of about $n_e \sim 600$ cm^{-3} (Ref. 93) is accelerated tailward by the $J \times B$ force provided by the magnetic tension. As discussed earlier, the penetration of the magnetic field into the current sheet may be the result of the Rayleigh-Taylor instability triggered, once again, by the magnetic tension of the magnetic field piled up at the ionopause.[76] More recently, the Kelvin-Helmholtz instability has also been invoked as a means of allowing the entry of magnetic flux tubes into the ionosphere.[77] However, for comet G-Z during the ICE encounter, the so-called ionopause might be nonexistent or quite small for the following reason.

Suppose the solar-wind plasma is kept outside the ionospheric region by a contact surface (or ionopause). The plasma density will be determined by the equilibrium between photoionization and electron dissociative recom-

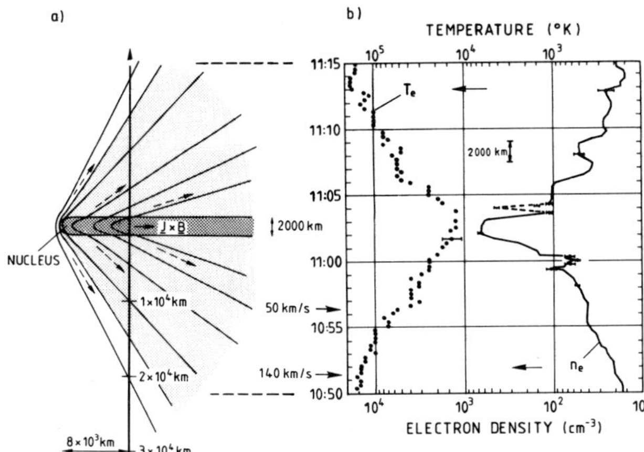

Figure 15—(a) A probable configuration of the magnetic field in the near vicinity of the ion tail of comet G-Z during the ICE spacecraft encounter. Both the ICE magnetometer measurements[11] and MHD model calculations (J. A. Fedder, personal communication) lead to similar patterns. The cold ionospheric plasma is subject to the $J \times B$ acceleration in the tailward direction. (b) Plasma density and temperature across the comet tail of comet G-Z (resolution: 18 s or 380 km). (From Ref. 93).

bination of the molecular ions (i.e., H_2O^+, H_3O^+, etc.). We have then

$$\alpha\, n_i\, n_e \;=\; \frac{n}{t_i} \tag{6}$$

where n, n_i, n_e are the number densities of neutrals, ions, and electrons, respectively; t_i is the photoionization time and α is the electron dissociative recombination coefficient. With the total gas production rate written as Q, the neutral expansion velocity V_n and the radial distance from the cometary nucleus r, we have

$$n \;=\; \frac{Q}{4\pi\, V_n\, r^2} \tag{7}$$

for spherically symmetric expansion, and thus

$$n_i \;=\; \left(\frac{Q}{4\pi\, V_n\, \alpha\, t_i\, r^2}\right)^{1/2} \tag{8}$$

If it is assumed that the r-dependence of the cometary ion number density can be described by Eq. 8 throughout the ion coma until reaching the boundary with the magnetized plasma convected from upstream regions, the equilibrium between the magnetic curvature force and the ion-neutral frictional force can be expressed as[54]

$$\frac{B^2}{4\pi r} \;\sim\; (n_i\, m_i\, V_n) \cdot (n\, V_n\, \sigma) \tag{9}$$

where m_i is the mass of the cometary ions and σ the ion-neutral momentum-transfer cross section.

The radial distance of the stagnation point therefore can be written as

$$r_s \;\sim\; (4\pi\, m_i\sigma/B^2)^{1/2} \cdot (Q/4\pi)^{1/4} \cdot (V_n/\alpha\, t_i)^{1/4} \tag{10}$$

Substituting $B \sim 60\,\gamma$, $m_i = 18\,m_p$, $\sigma \sim 3 \times 10^{-15}$ cm^2, $Q \sim 2.3 \times 10^{28}$ H_2O/s, $V_n \sim 1$ km/s, $\alpha \sim 10^{-6}$ cm^3/s, and $t_i \sim 10^6$ s, we find $r_s \sim 260$ km for comet G-Z (and $r_s \sim 3100$ km for comet Halley if $Q_{\text{Halley}} \sim 6 \times 10^{29}$ H_2O/s).

From the above consideration we may conclude that comet G-Z could only maintain a very small ionosphere (if it existed at all) and it may not be appropriate to treat the various boundary effects in the similar manner as practiced in the ionopause of Venus. On the other hand, if the ionization rate is effectively much faster than the photoionization rate, a sizable ionosphere would form and much of the discussions on the instability of the ionopause would be applicable.

In the issue of ion-neutral coupling in the ion coma, Biermann[96] introduced the concept of collisional coupling radius. The idea is that for a cometary ion moving toward the nucleus, it will suffer on the average one collision with the expanding neutral gas at an axial distance along the comet-Sun line defined as

$$\int_{\infty}^{r_c} n\, \sigma\, dr \;=\; 1 \tag{11}$$

or

$$r_c \;=\; \frac{Q\, \sigma}{4\pi\, V_n} \tag{12}$$

Hence $r_c \sim 600$ km for comet G-Z and about 30,000 km for comet Halley if $\sigma \sim 3 \times 10^{15}$ cm^2.

The important thing to remember is that, as the flow of pickup ions moves to a distance of a few times r_c from the nucleus, the collisional effect with the neutral gas will cause rapid charge-exchange recombination loss and collisional cooling. Both processes should result in strong cooling of the cometary plasma in this region.[97] One surprising result from the in-situ measurements at comet Halley is to discover that the dimension of the stagnant cometary plasma-flow region ($r_\perp \sim 1.5 \times 10^5$ km) is significantly larger and the boundary (i.e., the cometopause as defined by Gringauz et al.[58]) and collision pause (as by Neugebauer et al.[98]) is relatively sharp. It is most likely that the main bulk of the ion tails observed in optical emissions, for example, from CO^+ and H_2O^+ ions, is contained within the cometocollisional pause.

For the energetic pickup ions with large gyroradii ($r_g \sim 10,000$ km for 200 keV H_2O^+ ions with $B \sim 20\,\gamma$), they would feel the charge-exchange effect at large upstream distances already. Galeev et al.[38] considered the corresponding cooling effect to be of importance in reducing the thermal pressure of the loaded solar-wind flow so that it would be reaccelerated toward the comet head instead of being stagnated. Wallis and Dryer[99] noted such a cooling process in the inner part of the cometary coma and suggested that it could lead to a convergence of the solar-wind flow toward the central tail axis; the reexpansion of the plasma flow in this fashion should then cause the shock transition at the flank sides to decay away. The observations at comet Halley appeared not to have confirmed both theoretical scenarios. The depletion of the energetic pickup ions in the central part of the ion tail of comet G-Z nevertheless has to be related to the charge-exchange loss process. And the resulting diamagnetic effect leads to the maintenance of a magnetic field strength of about 60 γ in the near-tail region traversed by the ICE spacecraft. In the distant tail, because of flaring geometry, the tail field should reach a value of about 10 γ as estimated by Ershkovich.[82]

EPILOG

This report was initially prepared between the ICE encounter on September 11, 1985 with comet G-Z and the eve of the encounter with comet Halley by the Vega probe on March 6, 1986. Between these two landmarks of cometary research, there were the Uranus encounter by the Voyager 2 spacecraft on January 24 and the Challenger explosion on January 28. A very eventful period indeed, and it simply highlights the tremendous effort that is required to bring space physics forward, step by step. This consideration also demonstrates how important the data and information from the G-Z and Halley observations will be in the next few years in stimulating research activities and new ideas about cometary physics and space plasma physics in general. In a longer time scale, we could look forward to a comet rendezvous mission to provide us a far more complete picture of the dynamical behavior and very complicated chemical processes in the cometary coma and ion tail. As for solar-wind and magnetospheric interactions with planetary and satellite atmospheres, the Galileo mission to Jupiter will be fundamental in telling us about the electrodynamical interaction at Io. At the same time, there is a project by the name of Cassini, which is under study for a new mission to Saturn and Titan. It would be the next step for in-depth exploration of Titan's atmospheric interaction and magnetic tail process. After Voyager 2's encounter with Neptune and Triton in 1989 (then we shall know whether or not Neptune follows the Busse similarity law for dynamo action), the Pluto-Charon system and the asteroids will be the only planetary objects for whose plasma environments we still have no information. As far as the asteroids are concerned, the gap may perhaps be filled by performing flyby or rendezvous missions to these small bodies. From this point of view, a Mercury orbiter mission will also be important since we now know Mercury has very strong sodium emission[100] and may even have a quite extensive sodium exosphere and sodium tail—not unlike comets. To return to Venus, we could envision a multispacecraft program like those of ISEE 1 and 2 so that the many fine-scale plasma structures in the ionosphere and the magnetotail could be quantitatively investigated. With all these things remaining to be done, magnetotail physics as discussed in this conference has really just begun.

ACKNOWLEDGMENT—I would like to thank my colleagues, W. I. Axford, P. W. Daly, and K. Jockers, at the Max-Planck-Institut für Aeronomie, and the Giotto IMS team with H. Balsiger as the P.I., for useful discussions, F. L. Scarf for giving me the opportunity to participate in the exciting moment of the ICE encounter with comet Giacobini-Zinner, M. B. Niedner and J. A. Fedder for supplying numerical data of an MHD model of comet Giacobini-Zinner, and last but not the least, A. T. Y. Lui for patience (and courage) for seeing this manuscript through.

REFERENCES

[1] R. J. Hynds, S. W. H. Cowley, T. R. Sanderson, J. J. Van Rooijen, and K.-P. Wenzel, "Observations of Energetic Ions from Comet Giacobini-Zinner," *Science* **232**, 361-365 (1986).

[2] F. M. Ipavich, A. B. Galvin, G. Gloeckler, D. Hovestadt, B. Klecker, and M. Scholer, "Observations of Ionized Heavy Particles at the Comet Giacobini-Zinner," *Science* **232**, 366-369 (1986).

[3] F. L. Scarf, F. V. Coroniti, C. F. Kennel, D. A. Gurnett, W.-H. Ip, and E. J. Smith, "Initial Report on Plasma Wave Observations at Comet Gioacobini-Zinner," *Science* **232**, 377-381 (1986).

[4] C. S. Wu and R. C. Davidson, "Electromagnetic Instabilities Produced by Neutral Particle Ionization in Interplanetary Space," *J. Geophys. Res.* **77**, 5399-5406 (1972).

[5] C. S. Wu and R. E. Hartle, "Further Remarks on Plasma Instabilities Produced by Ions Born in the Solar Wind," *J. Geophys. Res.* **79**, 283-285 (1974).

[6] S. A. Curtis, "Solar Wind Pick-Up of Ionized Venus Exosphere Atoms," *J. Geophys. Res.* **86**, 4715-4270 (1981).

[7] D. Winske, C. S. Wu, Y. Y. Li, Z. Z. Mou, and S. Y. Guo, "Coupling of Newborn Ions to the Solar Wind by Electromagnetic Instabilities and their Interaction with the Bow Shock," *J. Geophys. Res.* **90**(A3), 2713-2726 (1985).

[8] D. A. Gurnett, T. Z. Ma, R. R. Anderson, O. H. Bauer, G. Haerendel, B. Häusler, G. Paschmann, R. A. Treumann, H. C. Koons, R. Holzworth, and L. Lühr, "Analysis and Interpretation of the Shock-Like Electrostatic Noise Observed During the AMPTE Solar Wind Lithium Releases," *J. Geophys. Res.* **91**, 1301-1310 (1986).

[9] A. T. Y. Lui, C. C. Goodrich, A. Mankofsky, and K. Papadopoulos, "Early Time Interaction of Lithium Ions with the Solar Wind in the AMPTE Mission," *J. Geophys. Res.* **91**(A2), 1333-1338 (1986).

[10] S. M. Krimigis, G. Haerendel, R. W. McEntire, G. Paschmann, and D. A. Bryant, "The Active Magnetospheric Particle Tracer Explorers Program," *Eos Trans. AGU,* **63**, 843 (1982).

[11] E. J. Smith, B. T. Tsurutani, J. A. Slavin, D. E. Jones, G. L. Siscoe, and D. A. Mendis, "ICE Encounter with Giacobini-Zinner: Magnetic Field Observations," *Science* **232**, 382-385 (1986).

[12] C. T. Russell and O. Vaisberg, "The Interaction of the Solar Wind with Venus," in *Venus,* D. M. Hunten et al., eds., Univ. Arizona Press, pp. 873-940 (1983).

[13] P. A. Cloutier, T. F. Tascione, R. E. Daniell, Jr., H. A. Taylor, and R. S. Wolff, "Physics of the Interaction of the Solar Wind in the Ionosphere of Venus: Flow/Field Models," in *Venus,* D. M. Hunten et al., eds., Univ. Arizona Press, pp. 941-979 (1983).

[14] K. I. Gringuaz, "The Bow Shock and the Magnetosphere of Venus According to Measurements from Venera 9 and 10 Orbiters," in *Venus,* D. M. Hunten et al., eds., Univ. Arizona Press, pp. 980-993 (1983).

[15] W.-H. Ip, "Cometary Plasma Physics: Large-Scale Interaction," in *Advances in Space Plasma Physics,* B. Buti, ed., World Scientific Pub. Co., Singapore, pp. 1-21 (1985).

[16] W.-H. Ip, "Solar Wind Neutral Atmospheres," in *Proc. Workshop on Future Missions in Solar, Heliospheric and Space Plasma Physics,* ESA SP-235, ESTEC, Noordwijk, pp. 65-82 (1985).

[17] M. A. Saunders, C. T. Russell, and J. G. Luhmann, "Interactions with Planetary Ionospheres and Atmospheres: A Review," in *Proc. on Comparative Study of Magnetospheric Systems,* La Londe des Maures, France (in press, 1985).

[18] H. Alfvén, "On the Theory of Comet Tails," *Tellus* **9**, 92-96 (1957).

[19] J. A. Slavin, E. J. Smith, B. T. Tsurutani, G. L. Siscoe, D. E. Jones, and D. A. Mendis, "Giacobini-Zinner Magnetotail: ICE Magnetic Field Observations," *Geophys. Res. Lett.* **13**, 283-286 (1986).

[20] D. A. Gurnett, F. L. Scarf, and W. S. Kirth, "The Structure of Titan's Wake from Plasma Wave Observations," *J. Geophys. Res.* **87**, 1395-1403 (1982).

[21] F. M. Neubauer, D. A. Gurnett, J. D. Scudder, and R. E. Hartle, "Titan's Magnetospheric Interaction," in *Saturn,* T. Gehrels and M. S. Matthews, eds., Univ. Arizona Press, pp. 760-787 (1984).

[22] A. Eviatar, G. L. Siscoe, J. D. Scudder, E. C. Sittler, Jr., and J. D. Sullivan, "The Plumes of Titan," *J. Geophys. Res.* **87**, 8091 (1982).

23 D. Hovestadt, E. Möbius, B. Klecker, M. Scholer, G. Gloeckler, and F. M. Ipavich, "Observation of Pick-Up Ions in the Solar Wind: Evidence for the Source of the Anomalous Cosmic Ray Component?" in *Proc. 19th Int. Cosmic Ray Conf., La Jolla, Calif.,* Vol. 5, pp. 176-179 (1985).

24 V. M. Vasyliunas and G. L. Siscoe, "On the Flux and the Energy Spectrum of Interstellar Ions in the Solar System," *J. Geophys. Res.* **81**, 1247-1252 (1978).

25 H. Balsiger, K. Altwegg, F. Bühler, J. Geiss, A. G. Ghielmetti, B. E. Goldstein, R. Goldstein, W. T. Huntress, W.-H. Ip, A. J. Lazarus, A. Meier, M. Neugebauer, U. Rettenmund, H. Rosenbauer, R. Schwenn, R. D. Sharp, E. G. Shelley, E. Ungstrup, and D. T. Young, "Ion Composition and Dynamics at Comet Halley," *Nature* **321**, 330-334 (1986).

26 A. H. Delsemme, "Chemical Composition of Cometary Nuclei," in *Comets,* L. L. Wilkening, ed., Univ. Arizona Press, pp. 85-130 (1982).

27 H. U. Keller, "Wasserstoff als Dissoziationsprodukt in Kometen," *Mitt. Astron. Gesell.* **30**, 143-148 (1971).

28 H. U. Keller, "The Interpretations of Ultraviolet Observations of Comets," *Space Sci. Rev.,* **18**, 641-684 (1976).

29 P. D. Feldman, "Ultraviolet Spectroscopy of Comae," in *Comets,* L. L. Wilkening, ed., Univ. Arizona Press, pp. 461-479 (1982).

30 K.-P. Wenzel, T. R. Sanderson, I. G. Richardson, S. W. H. Cowley, R. J. Hynds, S. J. Bame, R. D. Zwickl, E. J. Smith, and B. T. Tsurutani, "In-Situ Observations of Cometary Pick-Up Ions $\geq$ 0.2 AU Upstream of Comet Halley: ICE Observations," *Geophys. Res. Lett.* (in press, 1986).

31 T. R. Sanderson, K.-P. Wenzel, P. Daly, S. W. H. Cowley, R. J. Hynds, E. J. Smith, S. J. Bame, and R. D. Zwickl, "The Interaction of Heavy Ions from Comet P/Giacobini-Zinner with the Solar Wind," *Geophys. Res. Lett.* **13**, 411-414 (1986).

32 J. A. Fedder, S. H. Brecht, and J. G. Lyon, "MHD Simulation of a Comet" *Icarus* (in press, 1985).

33 M. K. Wallis, "Shock-Free Deceleration of the Solar Wind?" *Nature* **233**, 23-25 (1971).

34 M. K. Wallis and A. D. Johnstone, "Implanted Ions and the Draped Cometary Field," in *Cometary Exploration, I,* T. I. Gombosi, ed., Central Res. Inst. for Physics, Hungarian Acad. Sciences, Budapest, pp. 307-311 (1982).

35 M. K. Wallis, "Weakly-Shocked Flows of the Solar Wind Plasma Through Atmospheres of Comets and Planets," *Planet. Space Sci.* **21**, 1647-1660 (1973).

36 H. U. Schmidt and Wegmann, "Plasma Flow and Magnetic Fields in Comets," in *Comets,* L. L. Wilkening, ed., Univ. Arizona Press, pp. 538-560 (1982).

37 A. A. Galeev and A. S. Lipatov, "Plasma Processes in Cometary Atmospheres," *Adv. Space Res.* **4**(9), 229-237 (1984).

38 A. A. Galeev, T. E. Cravens, and T. I. Gombosi, "Solar Wind Stagnation Near Comets," *Astrophys. J.* **289**, 807-819 (1985).

39 W. I. Axford, "The Interaction of Solar Wind with Comets," *Planet. Space Sci.* **12**, 719-720 (1964).

40 N. Omidi, D. Winske, and C. S. Wu, "The Effects of Heavy Ions on the Formation and Structure of Cometary Bow Shocks," *Icarus* **66**, 165-180 (1986).

41 R. Z. Sagdeev, V. D. Shapiro, V. I. Shevchenko, and K. Szegö, "MHD Turbulences in Solar Wind-Comet Interaction Region," *Geophys. Res. Lett.* **13**, 85-88 (1986).

42 W. I. Axford, "Acceleration of Cosmic Rays by Shock Waves," in *Proc. 17th Int. Cosmic Ray Conf., Paris 1981,* Vol. 12, Service de Docum. du CEN, Saclay, pp. 155-203 (1982).

43 E. Amata and V. Formisano, "Energization of Positive Ions in the Cometary Foreshock Region," *Planet. Space Sci.* **33**, 1243-1250 (1985).

44 W.-H. Ip and W. I. Axford, "The Acceleration of Particles in the Vicinity of Comets," *Planet. Space Sci.* (in press, 1986).

45 R. Kulsrud and W. P. Pearce, "The Effect of Wave-Particle Interactions on the Propagation of Cosmic Rays," *Astrophys. J.* **156**, 445 (1969).

46 D. G. Wentzel, "Hydromagnetic Waves Excited by Slowly Streaming Cosmic Rays," *Astrophys. J.* **152**, 987 (1969).

47 D. S. Spicer, A. O. Benz, and J. D. Huba, "Solar Type I Noise Storms and Newly Emerging Magnetic Flux," *Astron. Astrophys.* **105**, 221-228 (1981).

48 C. S. Wu, J. D. Gaffey, Jr., and B. Liberman, "Statistical Acceleration of Electrons by Lower-Hybrid Turbulence," *J. Plasma Phys.* **25**, 391-401 (1981).

49 H. Oya, A. Morioka, W. Miyake, E. J. Smith, and B. T. Tsurutani, "Discovery of Cometary Kilometric Radiations and Plasma Waves at Comet Halley," *Nature* **321**, 307-310 (1986).

50 V. Formisano, A. A. Galeev, and R. Z. Sagdeev, "The Role of the Critical Ionization Velocity Phenomena in the Production of Inner Coma Cometary Plasma," *Planet. Space Sci.* **30**, 491 (1982).

51 A. A. Galeev, et al. (1986).

52 S. J. Bame, R. C. Anderson, J. R. Asbridge, D. N. Baker, W. C. Feldman, S. A. Fuselier, J. T. Gosling, D. J. McComas, M. F. Thomsen, D. T. Young, and R. D. Zwickl, "Comet Giacobini-Zinner: A Plasma Description," *Science* **232**, 336-361 (1986).

53 C. F. Kennel and H. E. Petschek, "Limit on Stably Trapped Particle Fluxes," *J. Geophys. Res.* **71**, 1-28 (1966).

54 W.-H. Ip and W. I. Axford, "Theories of Physical Processes in the Cometary Comae and Ion Tails," in *Comets,* L. L. Wilkening, ed., Univ. Arizona Press, pp. 588-634 (1982).

55 A. Roux, S. Perraut, J. L. Rauch, C. De Villedary, G. Kremser, A. Korth, and D. T. Young, "Wave-Particle Interactions Near $\Omega_{He}+$ Observed on Board Geos 1 and 2. 2. Generation of Ion Cyclotron Waves and Heating of He$^+$ Ions," *J. Geophys. Res.* **87**(A10), 8174-8190 (1982).

56 M. Neugebauer, "Observations of Solar Wind Helium," *Fundam. Cosmic Phys.* **7**, 131-199 (1981).

57 J. F. McKenzie, W.-H. Ip, and W. I. Axford, "The Acceleration of Minor Ion Species in the Solar Wind," *Astrophys. Space Sci.* **64**, 183-211 (1978).

58 K. I. Gringauz, T. I. Gombosi, M. Tatrallyay, M. I. Verigin, A. P. Remizov, A. K. Richter, I. Apathy, I. Szemerey, A. V. Dyachkov, O. V. Balakina, and A. F. Nagy, *Geophys. Res. Lett.* (in press, 1986).

59 D. B. Beard, "The Theory of Type I Comet Tails," *Planet. Space Sci.* **14**, 303-311 (1966).

60 D. B. Beard, "Cometary Tails," *Astrophys. J.* **245**, 743-752 (1981).

61 S. Klimov, S. Savin, Ya. Aleksevich, G. Avanesova, V. Balebanov, M. Baiikhin, A. Galeev, B. Gribov, M. Nozdrachev, V. Smirnov, A. Sokolov, O. Vaisberg, P. Oberc, Z. Krawczyk, S. Grzedzielski, J. Juchniewicz, K. Nowak, D. Orlowski, B. Parfianovich, S. Woźniak, Z. Zbyszynski, Ya. Voita, and P. Triska, "Extremely-Low-Frequency Plasma Waves in the Environment of Comet Halley," *Nature* **321**, 292-293 (1986).

62 A. S. Eddington, "The Envelope of Comet Morehouse (1908c)," *Mon. Not. R. Astron. Soc.* **70**, 442-458 (1910).

63 Rh. Lüst, "Bewegung von Strukturen in der Koma und im Schweif des Kometen Morehouse," *Z. Astrophys.* **65**, 236-250 (1967).

64 K. Wurm, "Structure and Kinematics of Cometary Type I Tails," *Icarus,* **8** 287-300 (1968).

65 K. Wurm and J. Rahe, "Type I Tail Structures of Comets Within the Inner Coma Region," *Icarus* **11**, 408-412 (1969).

66 P. D. Feldman, M. F. A'Hearn, M. C. Festou, L. A. McFadden, H. A. Weaver, and T. N. Woods, "Is CO_2 Responsible for the Outbursts of Comet Halley?" *Nature* (submitted, 1986).

67 W.-H. Ip and D. A. Mendis, "The Generation of Magnetic Fields and Electric Currents in Cometary Plasma Tails," *Icarus* **29**, 147-151 (1976).

68 W.-H. Ip, "Currents in the Cometary Atmosphere," *Planet. Space Sci.* **27**, 121-125 (1979).

69 K. I. Gringauz, T. I. Gombosi, A. P. Remizov, I. Apathy, I. Szemerey, M. I. Verigin, L. I. Denchikova, A. V. Dyachkov, E. Keppler, I. N. Klimenko, A. K. Richter, A. J. Somogyi, K. Szegö, S. Szendrö, M. Tátrallyay, A. Varga, and G. A. Vladimirova, "First In Situ Plasma and Neutral Gas Measurements at Comet Halley," *Nature* **321**, 282-285 (1986).

70 H. Alfvén, *On the Origin of the Solar System,* Oxford University Press, London (1954).

71 M. A. Raadu, "The Role of Electrostatic Instabilities in the Critical Ionization Velocity Mechanism," *Astrophys. Space Sci.* **55**, 125 (1978).

72 K. Papadopoulos, "On the Physics of the Critical Ionization Velocity Phenomena," in *Advances in Space Plasma Physics,* B. Buti, ed., World Scientific Pub. Co., Singapore, pp. 33-58 (1985).

73 P. T. Newell, "Review of the Critical Ionization Velocity Effects in Space," *Rev. Geophys.* **23**, 93-103 (1985).

74 G. Haerendel, "Plasma Flow and Critical Velocity Ionization in Cometary Comae," *Geophys. Res. Lett.* **13**, 255-258 (1986).

75 W.-H. Ip, "A Preliminary Consideration of the Electron Impact Ionization Effect in Cometary Comas," *Adv. Space Res.* (in press, 1986).

76 W.-H. Ip and D. A. Mendis, "The Flute Instability as the Trigger Mechanism for Disruption of Cometary Plasma Tails," *Astrophys. J.* **223**, 671 (1978).

77 A. I. Ershkovich and D. A. Mendis, "On the Penetration of the Solar Wind into the Cometary Ionosphere," *Astrophys. J.* **269**, 743 (1983).

78 R. S. Wolff, G. L. Siscoe, D. G. Sibeck, and M. Neugebauer, "Cometary Rays: Magnetically Channeled Outflow," *Geophys. Res. Lett.* **12**, 749-752 (1985).

79 R. S. Wolff, B. E. Goldstein, and C. M. Yeates, "The Onset and Development of Kelvin-Helmholtz Instability at the Venus Ionosphere," *J. Geophys. Res.* **85**, 7697-7707 (1980).

80 R. Beushausen and K. Jockers, "One Dimensional, Time-Dependent Models of the Interaction of the Solar Wind with a Comet," in *Asteroids, Comets, Meteors,* C.-I. Lagerkvist and H. Rickman, eds., Universitetet Reprocentralen HSC, Uppsala, Sweden, pp. 317-326 (1983).

[81] J. C. Brandt, "Observations and Dynamics of Plasma Tails," in *Comets,* L. L. Wilkening, ed., Univ. Arizona Press, pp. 519-537 (1982).

[82] A. I. Ershkovich, "Kelvin-Helmholtz Instability in Type-I Comet Tails and Associated Phenomena," *Space Sci. Rev.* **25,** 3-34 (1980).

[83] K. Jockers, "The Ion Tail of Comet Kohoutek 1973XII During 17 Days of Solar Wind Gusts," *Astron. Astrophys. Suppl.,* Ser. 62, 791-838 (1985).

[84] J. C. Brandt, J. D. Hawley, and M. B. Niedner, "A Very Rapid Turning of the Plasma-Tail Axis of Comet Bradfield 1979l on 1980 February 6," *Astrophys. J.* **241,** L51-L54 (1980).

[85] J.-F. Le Borgne, "Comet Bradfield 1979X Event on 1980 February 6: Correlation with an Interplanetary Solar Wind Disturbance," in *The Need for Coordinated Ground-Based Observations of Halley's Comet,* P. Veron, M. Festou, and K. Kjär, eds., European Southern Observatory Press, pp. 217-226 (1982).

[86] M. B. Niedner and J. C. Brandt, "Interplanetary Gas. XXIII. Plasma Tail Disconnection Events in Comets. Evidence for Magnetic Field Line Reconnection at Interplanetary Sector Boundaries?" *Astrophys. J.* **223,** 655-670 (1978).

[87] M. B. Niedner and J. C. Brandt, "Interplanetary Gas. XXIV. Are Cometary Plasma Tail Disconnections Caused by Sector Boundary Crossings or by Encounters with High-Speed Streams?" *Astrophys. J.* **234,** 723-732 (1978).

[88] K. Jockers, "Plasma Dynamics in the Tail of Comet Kohoutek (1973XII)," *Icarus* **47,** 397-411 (1981).

[89] P. J. Morrison and D. A. Mendis, "On the Fine Structure of Cometary Plasma Tails," *Astrophys. J.* **226,** 350-354 (1978).

[90] M. B. Niedner, J. A. Ionson, and J. C. Brandt, "Interplanetary Gas. XXVI. On the Reconnection of Magnetic Fields in Cometary Ionospheres at Interplanetary Sector Boundary Crossings," *Astrophys. J.* **245,** 1159-1169 (1981).

[91] T. Saito, K. Yumoto, K. Hirao, T. Nakagawa, and K. Saito, "Interaction Between Comet Halley and the Interplanetary Magnetic Field Observed by Sakigake," *Nature* **321,** 303-307 (1986).

[92] C. T. Russell, M. A. Saunders, J. L. Phillips, and J. A. Fedder, "Near-Tail Reconnections as the Cause of Cometary Tail Disconnections," *J. Geophys. Res.* **91**(A2), 1417-1423 (1986).

[93] N. Meyer-Vernet, P. Couturier, S. Hoang, C. Perche, J.-L. Steinberg, J. Fainberg, and C. Meetre, "Plasma Diagnosis from the Thermal Noise, and Limits on Dust Flux or Mass in Comet Giacobini-Zinner," *Science* **232,** 370-374 (1986).

[94] M. Shimizu, "The Structure of Cometary Atmospheres—I: Temperature Distribution," *Astrophys. Space Sci.* **40,** 149 (1976).

[95] M. L. Marconi and D. A. Mendis, "The Photochemical Heating of the Cometary Atmosphere," *Astrophys. J.* **260,** 386 (1982).

[96] L. Biermann, "Interaction of a Comet with the Solar Wind," in *Solar Wind Three,* C. T. Russell, ed., UCLA-IGPP Publication, pp. 396-414 (1974).

[97] M. K. Wallis and R. S. B. Ong, "Strongly-Cooled Ionizing Plasma Flows with Application to Venus," *Planet. Space Sci.* **23,** 713-721 (1975).

[98] M. Neugebauer, J. A. Slavin, and W.-H. Ip, "A Plasma Model for Comet Kopff," JPL-PD-699-10, Vol. XII, pp. VI.1-VI.24 (1985).

[99] M. K. Wallis and M. Dryer, "Decay of the Cometary Bow Shock," *Nature* **318,** 646-647 (1986).

[100] A. Potter and T. Morgan, "Discovery of Sodium in the Atmosphere of Mercury," *Science* **229,** 651-653 (1985).

MAGNETIC FIELD AND PARTICLE PRESSURE IN THE PLASMA SHEET OF JUPITER

L. J. Lanzerotti and C. G. Maclennan,* J. N. Broughton[†] and D. Venkatesan,[††] and R. P. Lepping[§]

The results of an analysis of the energetic particle and magnetic field data acquired by the Voyager 2 spacecraft at distances of ~40–70 R_J on the nightside of the planet are reported. As in a previous study of similar data at distances $\gtrsim 80$ R_J, we find the energy densities of ions (primarily protons) to be sufficient to provide the diamagnetic depressions measured in the magnetic field intensity as the spacecraft made successive encounters with the nightside plasma sheet. There is some evidence that the percent contribution of the protons to the energy balance decreases with increasing distance from the planet over this radial interval, although this conclusion is dependent upon the assumption that the proton and heavier ion (oxygen) energy spectra are similar.

INTRODUCTION

Large "oscillations" observed in the particle fluxes and the magnetic field intensity as a spacecraft traverses the Jovian magnetosphere on the dusk and night sides are among the most evident spatial/temporal variations measured. Following the Pioneer encounter with Jupiter's magnetosphere, there was considerable discussion in the literature as to the energy and composition of the ion and/or electron populations that comprised the plasma sheet and produced the observed magnetic field depressions. Walker et al.[1] inferred from magnetic field and energetic (0.5–1.8 MeV) proton data from Pioneer 10 that the Jovian current sheet was imbedded in a high β plasma in which "the high energy (>60 keV) ions" are the dominant constituent. Goertz et al.[2] concluded from a similar analysis of Pioneer 10 data that the energy range of the plasma sheet ions was 0.1–10 keV. Van Allen[3] concluded from an analysis of Pioneer 10 electron ($E > 60$ keV) data that these particles played no substantial role in the production of the diamagnetic depression at the center of the plasma sheet.

An analysis of Voyager 2 ion data at distances of ~80, 100, and 120 R_J was made by Lanzerotti et al.[4] These authors concluded that the configuration of the Jovian plasma sheet at these distances was "determined by ions (protons and heavier nuclei) of energies $\gtrsim 30$ keV." They also stated that the bulk direction of motion of the plasma is predominantly across the local magnetic field in a direction consistent with that expected from corotation of the planetary magnetic field.

Here we report the analysis of six sequential Jovian plasma-sheet crossings made by the Voyager 2 spacecraft in the interval ~40–70 R_J. We determine in a more quantitative fashion than that done in Ref. 4 the plasma/magnetic field pressure balances and the relative energy abundances of protons and heavier ions contributing to the plasma-sheet diamagnetic depressions.

The particle and magnetic field measurements used in this analysis were acquired by the Low Energy Charged Particle (LECP) instrument and in the magnetometer (MAG) instrument on the Voyager 2 spacecraft. The LECP instrument was described by Krimigis et al.[5] and consists of two double-ended solid-state detector telescopes, each optimized for certain specialized measurements in planetary magnetospheres and in interplanetary space. A stepping platform can rotate the LECP instrument through eight discrete sectors spanning 360° over intervals of time determined by the scientific objectives, and is commandable from the ground. Sector 8 is occulted by a sun shade that also provides a background measurement. Used principally in this paper are data from the Low Energy Magnetosphere Particle Analyzer (LEMPA), which measures ions (proton energies ~30 keV–3.5 MeV) and electrons. The Voyager 2 energy channels of interest for this study are listed in Table 1. The Low Energy Particle Telescope (LEPT) was designed as a composition instrument and is capable of identifying the major ion species at energies $\gtrsim 60$ keV/nucleon. The magnetometer instrument has been described by Behannon et al.[6] and consists of a low-noise flux-gate magnetometer and associated data processing and calibration electronics.

*AT&T Bell Laboratories, Murray Hill, New Jersey 07974.
[†]University of Calgary, Calgary, Alberta, Canada T2N 1N4.
[‡]Also at The Johns Hopkins University Applied Physics Laboratory, Laurel, Maryland 20707.
[§]NASA Goddard Space Flight Center, Greenbelt, Maryland 20771.

Table 1—Voyager 2 energy channels.

Energy Channel	H (keV)	O (keV)
PL01	28–43	66–108
PL02	48–80	93–150
PL03	80–137	144–220
PL04	137–215	220–340
PL05	215–540	340–677
PL06	540–990	677–1156
PL07	990–2140	1156–2320
PL08	2140–3500	2320–4200

The six plasma-sheet crossings that are analyzed occurred when the Voyager 2 spacecraft was oriented in its "Arcturus Reference Frame." In this reference frame the LECP instrument orientation was such that the particle measurements were made in a plane approximately perpendicular to the Jovian plasma sheet. In this reference frame corrections to the data for corotational effects are not appreciable and were omitted.

OBSERVATIONS

The Jupiter encounter of Voyager 2 as evidenced by the proton and ion fluxes measured by the LECP instrument is plotted in Fig. 1. Two ion channels and one proton (0.52–1.5 MeV) channel are shown. The lowest energy ion fluxes (43–80 keV) were measured by the LEMPA detector system, whereas the protons and the medium Z ions were measured by the LEPT system. Clearly seen in the data are the large oscillations in the fluxes following closest approach (CA). These oscillations are produced by the Jovian plasma sheet passing the spacecraft as the sheet corotates with the planet. Shown at the bottom of the figure are the reference orientation directions of the spacecraft. Of particular interest here are the plasma-sheet crossings occurring during the time of the Arcturus (ARC) orientation, which occurs during days 193–195, 1979, when the spacecraft was between ~40 and 70 R_J of the planet.

The plasma-sheet crossings for the low-energy ions (43–80 keV) during the Arcturus reference interval are shown in more detail in the bottom panel of Fig. 2. The top panel plots the total magnetic field intensity during the same time period. The six specific sets of analyzed plasma-sheet crossings are numbered A–F in the upper panel. There is considerable structure in the plasma-sheet crossings, giving rise to the variations observed in the particle fluxes and the magnetic field intensity. The magnetic field components and intensity are sampled more frequently than are the particle fluxes. Because of the orientation of the spacecraft, scan-averaged (over 48-s or 192-s intervals depending upon the location in the trajectory) particle fluxes are used in the analysis. Therefore, magnetic field data acquired at the time occurring at the center of the individual particle detector scans are used for the comparisons and calculations.

Two examples of ion spectra taken on day 193 and day 194 in the plasma sheet are plotted (assuming all proton fluxes) in the upper panels of Fig. 3. Both of the spectra are similarly shaped, tending to become somewhat flattened at the lowest energies measured by the

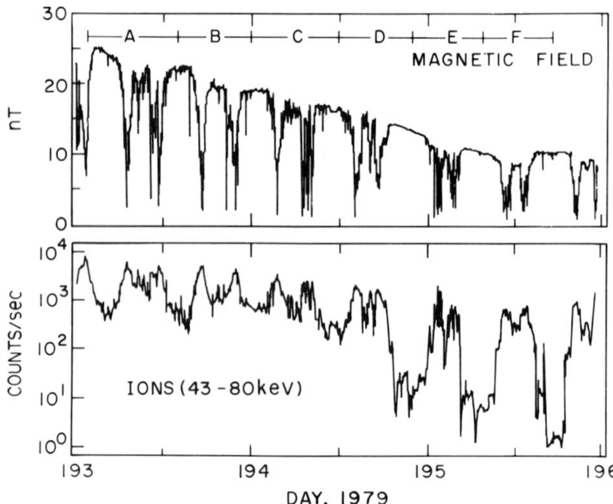

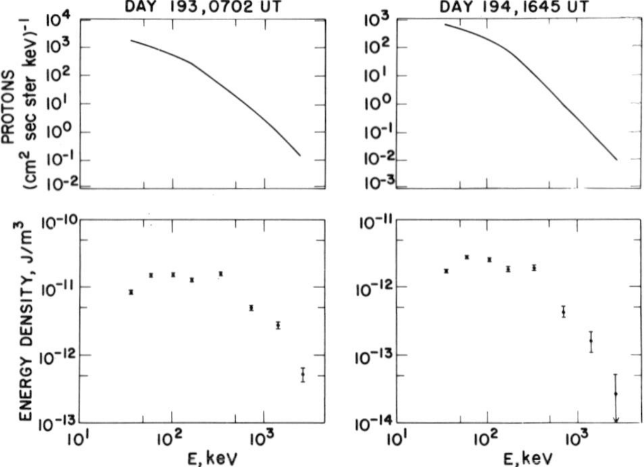

Figure 2—Magnetic-field and particle (ion) count rates for the plasma-sheet crossings during the Arcturus reference-frame interval. The specific crossings analyzed are designated A–F.

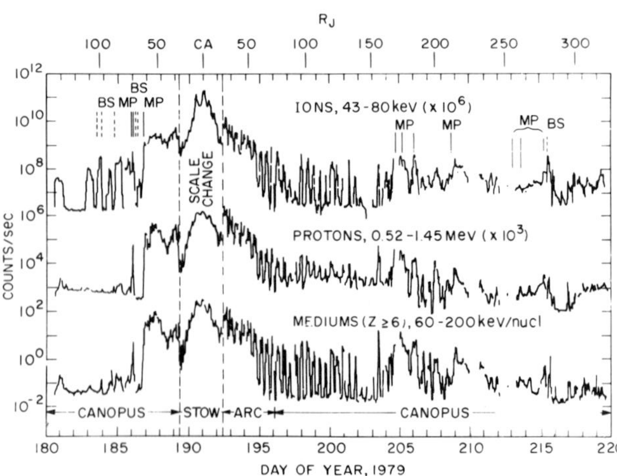

Figure 1—Particle count rates versus time during the Voyager 2 encounter with the magnetosphere of Jupiter. MP and BS signify crossings of the planet's magnetopause and bow shock, respectively. The periods of orientation of the spacecraft in the Canopus and the Arcturus (ARC) frames of reference are indicated at the bottom of the plot. During the interval around closest approach (CA) to the planet, the LECP instrument was held in a stowed (STOW) configuration, not scanning spatially the particle distributions as during the remainder of the encounter period.

Figure 3—Sample ion spectra (plotted assuming all protons; top panels) measured during two of the plasma-sheet crossings of Fig. 2. The lower two panels plot the energy densities contributed by each of the proton channels.

instrument. The particle fluxes become less intense with increasing radial distance, accounting for the overall decrease in the spectral intensity on day 194 compared to day 193. The lower panels plot the energy densities (assuming all protons) contributed by each of the energy channels. The total energy density at a given time is the sum of the individual points. Obviously, contributions to the total are expected from particles with energies lower than those measured by LECP.

The magnetic field and particle data from two of the plasma-sheet crossings (crossings A and D of Fig. 2) are plotted in Fig. 4. The magnetic field data plotted are the values measured at the same time as the sample at the center of each detector scan interval. These data are plotted in terms of the magnetic field energy density (or, equivalently for the units used, pressure), $B^2/8\pi$. The particle data have been plotted in terms of the energy densities of the ions, where each energy density ϵ_i has been computed by integrating over the spectrum of the measured particles. That is

$$\epsilon_i = (4\pi) \sum_{n=1}^{5} \Delta E_n \frac{1}{7} \left\{ \sum_{k=1}^{7} J_{nk} \right\} \frac{E_n}{v_n} \tag{1}$$

where J_{nk} is the ion flux in the k-th angular sector of the n-th energy channel, ΔE_n is the energy channel width, E_n is the logarithmic center energy of the channel, and v_n is the ion velocity corresponding to this energy. Obviously, this procedure (including only ions with energies between ~ 30 and 600 keV) results in underestimates of the total ϵ_i in the plasma sheet. Note also that for a Maxwellian plasma distribution the ϵ_i computed should be reduced by $\frac{2}{3}$ to compute the β of the plasma; i.e., $\beta = 16 \pi E_i/3B^2$, where B is the magnetic field intensity. In the center panel the plasma energies assume the ions are all protons. Such a representation provides a lower limit on the plasma-sheet particle pressure. A similar plot can be made under the assumption that the measured ions are another species such as oxygen that would be produced from the decomposition of the sulfur dioxide associated with the volcanism of Io. This would basically give an upper limit to the particle pressure.

The magnetic field data from the two upper panels are replotted in the two lower panels. For each of the six crossings (A–F) indicated in Fig. 2, an exponential (straight line on a semilog plot as shown here) best fit was made across the magnetic field energy densities. These best fits are shown as the dotted lines for the two cases shown in Fig. 4. The fundamental idea then is to determine the composition of the hot plasma particles that would produce the observed magnetic field depressions in the plasma sheet. That is, the fraction of hot protons is determined in a least-squares sense by determining the relation

$$\epsilon_{mod} - \epsilon_B = f_p \epsilon_p + (1 - f_p) \epsilon_o \tag{2}$$

where ϵ_{mod} is the "model" magnetic field energy density (i.e., the dotted lines in the lower panels of Fig. 4) and ϵ_o is the hot-plasma energy density assuming all oxygen ions. This implicitly assumes that the spectra of the two species are similar at each measurement time. The results of the best fit for f_p in determining ϵ_{mod} at each measurement point is shown by the dashed line close to the dotted "model" line in each lower panel. As can be seen for the two cases shown, the fits are rather good.

The values of the f_p for each of the six plasma sheet crossings are shown in Fig. 5, plotted at the center radial distance from Jupiter of each crossing, A–F. Each of the uncertainties shown are one standard deviation in the determinations. It is clear that hot plasma ions in a plasma sheet consisting primarily of protons can produce the diamagnetic depressions in the sheet magnetic field, particularly at distances ~ 40–50 R$_J$. If the LECP instrument could measure to energies as low as ~ 20 keV, it is conceivable that all of the observed diamagnetic depression could be produced by a hot proton plasma.

The model-dependent results of Fig. 5 suggest that there is a decrease in the abundance of protons contributing to the static pressure of the plasma sheet beyond ~ 50 R$_J$. That is, in order to balance the diamagnetic depres-

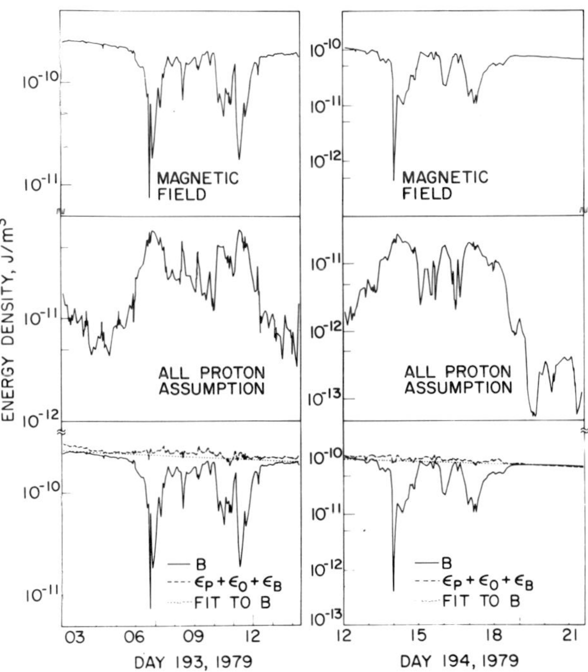

Figure 4—Top panels: Magnetic field intensity plotted in terms of energy densities for two of the plasma-sheet crossings identified in Fig. 2. Center panels: Energy densities of the ions during the plasma-sheet crossings, plotted as all protons. Lower panels: Magnetic field energy densities from the top panel (solid line) and a linear fit to the field data assuming no plasma sheet (dotted line). The irregular broken line is the result of a fit of the sum of the magnetic field energy densities plus the ion energy densities to the linear interpolation across the plasma sheet assuming an admixture of protons and oxygen ions in the plasma sheet.

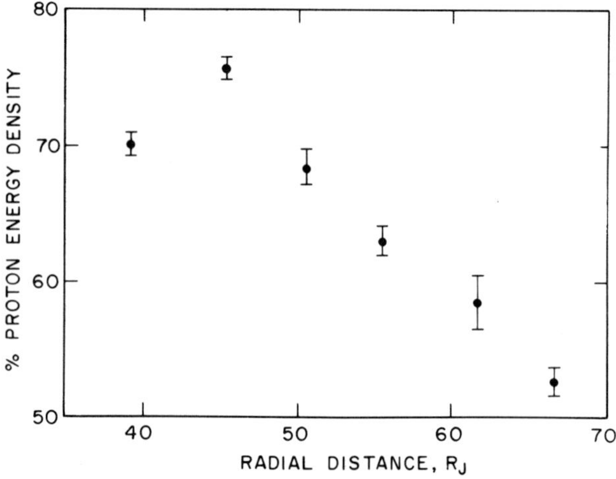

Figure 5— Percent proton energy densities deduced in the plasma sheet, resulting from the analyses as a function of radial distance from the planet. The values are plotted at the distances corresponding to the center of each of the crossings.

sions, the plasma sheet must become more abundant in low-energy oxygen ions with increasing distance in the region ~50–70 R$_J$. There is no noticeable, sudden change in the deduced composition in association with the change in thickness of the plasma sheet beyond ~60 R$_J$ (Figs. 1 and 2).

DISCUSSION

The diamagnetic depressions in the plasma sheet on the night side of Jupiter can be produced by a hot plasma consisting primarily of protons with energies ≥ 30 keV. The composition of the hot ions may change somewhat beyond ~50 R$_J$, decreasing from ~70% to ~55% proton energy abundance, although this conclusion is model dependent (also see below). These results are similar to those found by Lanzerotti et al.[4] for the plasma sheet at distances of ~80, 100, and 120 R$_J$. In that work, the proton energy abundance was found to be between 50% and 75% and no precise fitting procedure was used.

The results presented here are not affected by the particle and magnetic field stress balance considerations of Mauk et al.[7] This is because the centrifugal stresses and the pressure stresses are for different particle populations in the rotating magnetosphere plasma sheet. The centrifugal stresses at Jupiter appear to be produced by the colder plasma[8,9] whereas the hot plasma (energies ≥ 30 keV) appears to dominate the pressure balances as discussed here (see also Ref. 10).

A major assumption in the analysis that led to the composition conclusion is that the shapes of the proton and oxygen spectra that contribute to the energy densities (pressure balance) change similarly with radial distance from the planet. If this is not the case, then the conclusion would be altered. Indeed, the decrease in the

percent proton energy density contribution shown in Fig. 5 could partially be an artifact of the relative changes in the ion spectra.

If the ion differential energy spectra are taken to be represented as power laws (even though in actuality they begin to turn over at the lower energies; see Fig. 3 and Ref. 10) for the *i*-th ion,

$$\frac{dj_i}{dE} = A_i E^{-\gamma i} \tag{3}$$

then the energy density of the *i*-th ion ϵ_i can be written as

$$\epsilon_i = 4\pi A_i \left(\frac{m_i}{2}\right)^{1/2} \frac{E_l^{-\gamma i + 3/2}}{\gamma i - 3/2} \tag{4}$$

for $\gamma i \neq 3/2$, where m_i is the mass of the *i*-ion and E_l is the lowest energy measured (assumed, for simplicity, to be the same for all ions). The ratio of the plasma energy densities of two species (protons and oxygen) is then, apart from a constant factor,

$$\frac{\epsilon_p}{\epsilon_o} = \frac{\gamma_0 - 3/2}{\gamma_p - 3/2} E_l^{-\gamma_p + \gamma_o} \tag{5}$$

The ratios of the energy densities will vary if $\gamma_o = g\gamma_p$ and $g \neq 1$.

The composition of nonthermal ($E > 600$ keV/nucleon) ions at Jupiter was measured by the LEPT instrument on Voyager 2 and reported in Ref. 11. In the region ~40–60 R$_J$, γ_p is approximately constant at 3.5 and γ_o varies from ~5 at ~40 R$_J$ to ~3.5 at ~60 R$_J$. Assume that the proton and oxygen spectra can be extrapolated below 600 keV/nucleon with the same power-law exponents; then, with $E_l = 30$ keV, the ratio of Eq. 5 will change from ~100 to ~1 as g changes from ~1.4 to 1. Thus, as the oxygen spectrum becomes harder (smaller γ_o) with increasing radial distance, the ratio of Eq. 5 will tend to become smaller; i.e., less proton energy contribution relative to the oxygen energy contribution with increasing radial distances. This is the trend of the composition results seen in Fig. 5. However, caution must be exercised in that the spectral extrapolations by more than a decade in energy below ~600 keV/nucleon are likely to be uncertain.

Nevertheless, it is plausible that changes in the oxygen energy spectrum relative to the proton energy spectrum could produce some or all of the composition changes suggested in Fig. 5. Other considerations enter, of course, including differences in the E_l with species, changes in the A_i, and nonpower-law spectra at the lowest energies.

In summary, the hot ions ($E \geq 20$ keV) in the Jovian magnetotail can produce the diamagnetic depressions in the plasma-sheet magnetic field intensities between ~40 and ~120 R$_J$ (invoking as well the initial results of Lanzerotti et al.[4]). There may be an increase in the relative oxygen (i.e., "heavy" ion) abundance in the plasma sheet with radial distance from ~40 to ~70 R_J. A portion of this deduced composition change may arise from changes in the oxygen energy spectra relative to the proton energy spectra.

REFERENCES

[1] R. J. Walker, M. G. Kivelson, and A. W. Schardt, "High β Plasma in the Dynamic Jovian Plasma Sheet," *Geophys. Res. Lett.* **5**, 5799 (1978).

[2] C. K. Goertz, A. W. Schardt, J. A. Van Allen, and J. L. Parish, "Plasma in the Jovian Current Sheet," *Geophys. Res. Lett.* **6**, 495 (1979).

[3] J. A. Van Allen, "Energetic Electrons in Jupiter's Dawn Magnetodisk," *Geophys. Res. Lett.* **6**, 309 (1979).

[4] L. J. Lanzerotti, C. G. Maclennan, S. M. Krimigis, T. P. Armstrong, K. Behannon, and N. F. Ness, *Geophys. Res. Lett.* **7**, 817 (1980).

[5] S. M. Krimigis, T. P. Armstrong, W. J. Axford, C. O. Bostrom, C. Y. Fan, G. Gloeckler, and L. J. Lanzerotti, "The Low Energy Charged Particle (LECP) Experiment on the Voyager Spacecraft," *Space Sci. Rev.* **21**, 329 (1977).

[6] K. W. Behannon, M. H. Acuña, L. F. Burlaga, R. L. Lepping, N. F. Ness, and F. M. Neubauer, "Magnetic Field Experiment for Voyager 1 and 2," *Space Sci. Rev.* **21**, 235 (1977).

[7] B. H. Mauk, S. M. Krimigis, and R. P. Lepping, "Particle and Field Stress Balance within a Planetary Magnetosphere," *J. Geophys. Res.* **90**, 8253 (1985).

[8] A. J. Lazarus and R. L. McNutt, Jr., "Low-Energy Plasma Ion Observations in Saturn's Magnetosphere," *J. Geophys. Res.* **88**, 8831 (1983).

[9] R. L. McNutt, Jr., "Force Balance in the Outer Planet Magnetosphere," in *Proc. 1982-4 Symp. Phys. Space Plasmas, SPI Conf. Proc. Reprint Ser., 5*, H. S. Bridge et al., eds., Scientific Publishers, Cambridge, Mass. (1984).

[10] S. M. Krimigis, J. F. Carbary, E. P. Keath, C. O. Bostrom, W. I. Axford, G. Gloeckler, L. J. Lanzerotti, and T. P. Armstrong, "Characteristics of Hot Plasma in the Jovian Magnetosphere: Results from the Voyager Spacecraft," *J. Geophys. Res.* **86**, 8227 (1981).

[11] D. C. Hamilton, G. Gloeckler, S. M. Krimigis, and L. J. Lanzerotti, "Composition of Non-Thermal Ions in the Jovian Magnetosphere," *J. Geophys. Res.* **86**, 8301 (1981).

THE AVERAGE CONFIGURATION OF THE INDUCED VENUS MAGNETOTAIL

D. J. McComas,* H. E. Spence, and C. T. Russell[†]

We discuss the interaction of the solar-wind flow with Venus and the morphology of magnetic field line draping in the Venus magnetotail. In particular, we describe the importance of the interplanetary magnetic field (IMF) X-component in controlling the configuration of field draping in this induced magnetotail, and using the results of a recently developed technique, we examine the average magnetic configuration of this magnetotail. The derived $J \times B$ forces must balance the average, steady-state acceleration of, and pressure gradients in, the tail plasma. From this relation the average tail-plasma velocity, lobe and current-sheet densities, and average ion temperature have been derived. In this study we extend these results in three areas. First, we examine the apparent angle of the magnetic field draping across the Venus magnetotail and show that this angle is consistent with IMF angle at Venus modified by the slowing of the flow in the magnetosheath and coordinate rotations used in the study. Next we make a connection between the derived consistent plasma flow speed and density, and the observational energy/charge range and sensitivity of the Pioneer Venus Orbiter (PVO) plasma analyzer, and demonstrate that if the tail is principally composed of O^+, the bulk of the plasma should not be observable much of the time that the PVO is within the tail. Finally, we examine the importance of solar-wind slowing upstream of the obstacle and its implications for the temperature of pick-up planetary ions, compare the derived ion temperatures with their theoretical maximum values, and discuss the implications of this process for comets and AMPTE-type releases.

Certain results of our previous analysis of the Venus magnetotail[1] are extended in this presentation. We begin with a summary of the basic interaction of the solar wind with Venus and certain of the results of our previous analysis that are needed to put the present work into the proper context. The basic interaction of the solar wind with Venus is shown schematically in Fig. 1. The solar wind flows radially outward from the sun carrying the IMF embedded in it. This plasma is then slowed and deflected and the embedded magnetic field is compressed as it flows through the Venus bow shock. Downstream from the shock in the magnetosheath region, the plasma and field vary with position in a complicated fashion as shown in the gas dynamic model of Spreiter and Stahara.[2] The portions of the magnetosheath flow that intercept the mass loading and conducting obstacles of Venus' extended atmosphere and ionosphere are slowed and diverted around the planet, while the portions of the flow distant from the planet continue to move at the shocked solar wind speed. Since the magnetic field links these two regions, the field on either side of the tail becomes draped back into roughly pointing in the sunward and antisunward directions and forms the draped "lobes" of the Venus magnetotail.

The slowing of the plasma, as it encounters the Venus obstacle, causes the component of the field perpendicular to the upstream velocity vector to drape about the planet as shown qualitatively in Fig. 1. This process was first suggested by Alfvén[3] to describe the formation of comet tails. In the draping process, however, the component of the field parallel to the upstream flow, or X-component, also plays a fundamental role in determining the actual magnetic topology of the Venus magnetotail. As Fig. 1 indicates, the X-component of the field maps into the magnetotail, preferentially into the lobe on the left side of the figure for a normal Parker spiral IMF configuration and into the right side for an antispiral oriented field. This causes the quantities of magnetic flux contained in the two lobes, at any given distance behind the planet, to be quite different. Since the lobes are low-beta regions, as indicated by our previous analysis, the current sheet must move toward the side of lesser flux content in order to achieve a state of pressure-balanced equilibrium within the tail.

The IMF spiral angle (defined as the angle between the magnetic field and the radial direction, in the x-y plane of a Venus-centered solar orbital coordinate system), however, typically is highly variable on a time scale of a few minutes. As a consequence of this variability, the current sheet in the Venus tail flaps back and forth due to the varying quantities of magnetic flux and, therefore, magnetic pressures in the two tail lobes. This phenomenon, the solar-wind flow direction, has made it impossible to determine the average magnetic field draping in the Venus magnetotail, until recently. A new analysis technique[1] has been developed, which statistically removes the effects of current-sheet and tail flapping by measuring locations with respect to the internal tail structures themselves. From these internal measurements a coordinate system was constructed in which Y^* is defined to be parallel to the Y axis and measures the distances from the center of the current sheet. In this coordinate system the average properties of the tail are readily determined. The actual procedure for carrying this out is rather complex and has been described in detail in our previous paper.

*Los Alamos National Laboratory, Los Alamos, New Mexico 87545.
[†]Institute of Geophysics and Planetary Physics, University of California, Los Angeles, California.

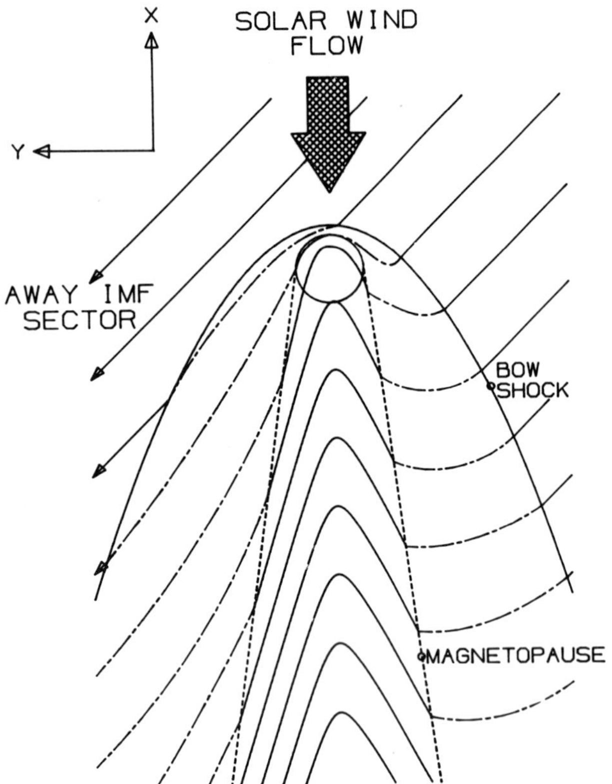

Figure 1—Schematic diagram of the interaction of the solar wind with Venus. The solar wind is slowed and the magnetic field is deflected in the transition through the bow shock and magnetosheath region (dashed to indicate the uncertainty of the exact configuration). Above the planet, along the near-stagnation streamlines, the flow is appreciably slowed and, consequently, the field becomes draped in the Venus magnetotail. The X-component of the external field maps into the magnetotail causing the quantities of magnetic flux in the two lobes to be quite different at a given distance down the tail. Consequently, the current sheet moves toward the side of lesser flux content and, as the upstream spiral angle varies, causes the current sheet to flap back and forth within the tail.

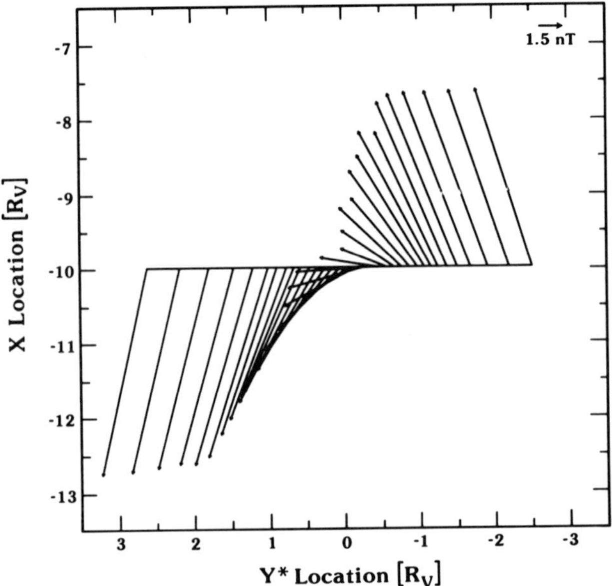

Figure 2—Magnetic field vectors in the draped Venus magnetotail showing the variations of the draping angle and field magnitude as a function of location across the tail. The data are plotted at $X = -10$ R$_V$, which is a typical downtail distance for the data in our dataset. The derived variation is an average over X and Z based on 9423 tail data points collected over the first 10 Venus years that PVO was in orbit. There is clearly a smooth variation from one lobe, through the current sheet, and across the other lobe.

In this constructed coordinate system the approximate average properties of the magnetic field variations in the Venus magnetotail have been examined for the first time. Figure 2 shows the average magnetic field variation across the tail in this new coordinate system, shown at a typical X distance (-10 R$_V$) for the data studied. This is the true average magnetic configuration of the tail in the range from -8 to -12 R$_V$ since it was derived from averaging the entire dataset within this X range over observed values of Z ($-1.6 < Z < 1.6$ R$_V$). It is interesting to note that the center of the current sheet is offset toward the $-Y^*$ side of the tail by ~ 0.4 R$_V$. This displacement of the current sheet is qualitatively consistent with the effect of mapping the X component of the field into the tail as described above, since a normal spiral configuration causes an enhancement of flux on the $+Y^*$ side of the tail and a consequent displacement of the current sheet toward the $-Y^*$ side. Quantitatively, this displacement causes all of the

magnetic flux on the $-Y^*$ side of the tail at $X \simeq -11$ R$_V$ to cross the $+Y^*$ side at this distance within $Y^* < 1.7$ R$_V$. For $1.7 < Y^* < 2.6$ R$_V$, the average lobe field angle is $\sim 72°$ from the Y^* axis toward the $-X$ axis, and therefore, the field lines map an average of ~ 2.8 R$_V$ further back on the $+Y^*$ side than on the $-Y^*$ side. This difference in downtail mapping distances, combined with the average 5.1 R$_V$ width of the tail at ~ -10 R$_V$, gives an apparent spiral angle of the field of $\sim 60°$ connecting across the tail.

This value is appreciably larger than the nominal IMF spiral angle of $\sim 35°$ expected at the orbit of Venus. Two important effects probably combine to principally account for this difference. The first is a purely geometric effect which was suggested by John Phillips (personal communication, 1985); the original coordinate system was chosen so that the magnetic field was entirely in the x-y plane since the Venus obstacle is cylindrically symmetric and only ordered by the draping magnetic field direction. Therefore, the cone angle of the field (angle from the X-axis) is mapped into the tail to determine the "apparent spiral angle," while the IMF spiral angle is defined as the projection of this vector into the solar ecliptic x-y plane. Therefore the apparent spiral angle is always greater than or equal to the IMF spiral angle.

The other major effect is simply based on the bending of the magnetic field behind the shock in the magnetosheath. In the magnetosheath, at the downtail distances studied here, the flow speed is slower than it is in the upstream solar wind,[2] and therefore the spiral angle is larger. For both of these reasons, the draped magnetic field lines connect across the magnetotail at angles that are larger than the upstream IMF spiral angle.

From the average field draping pattern shown in Fig. 2, and from the derived variation of the component of the field crossing the tail, as a function of distance down the tail, our previous analysis determined certain internally consistent plasma properties of the Venus magnetotail. Continuity of the tangential electric field across the magnetopause boundary was used to determine the average tailward velocity, which was shown to be given by

$$\langle V_X(X) \rangle_Y = -1540/(11.53 + 0.68X) \qquad (1)$$

where V_X is in km/s, X is in R_V, the $\langle \rangle_Y$ brackets indicate that this velocity has been averaged over the entire width of the tail (valid for $-8 < X < -12$), and the upstream solar-wind velocity is assumed to be -440 km/s. This result indicates that throughout most of the range the magnetotail plasma falls behind the surrounding sheath plasma, and the tail field becomes increasingly kinked. Beyond a critical distance ($\sim -11.8\ R_V$) Eq. 1 indicates that the plasma velocity is greater than the outside solar-wind speed. The plasma in the tail is accelerated by the Maxwell stresses in the kinked field configuration, and the field beyond this distance restraightens toward the initial IMF configuration.

From the variations of the magnetic field as a function of location within the Venus magnetotail, we derived the current density distributions in the tail and $J \times B$ magnetic forces. By assuming that the average tail configuration represents a state of equilibrium and using the MHD momentum equation and derived $J \times B$ forces we determined the average plasma properties in the Venus magnetotail. The results of this previous analysis are summarized in Table 1. The average lobe and current-sheet densities, from -8 to $-12\ R_V$ down the tail, are ~ 0.07 and ~ 0.9 amu/cm^3 which, if the tail is composed entirely of planetary O^+, corresponds to ~ 0.005 and $\sim 0.06\ O^+/cm^3$, respectively. The average temperature of the tail plasma is between $\sim 6 \times 10^6$ K,

if the tail is composed entirely of protons and $\sim 9 \times 10^7$ K if it is entirely O^+.

These average plasma properties have interesting implications for the detection of pick-up ions with the PVO plasma analyzer.[4] As suggested by Slavin et al.,[5] O^+ ions with bulk flow speeds in excess of about 300 km/s will have an E/Q of greater than 8 keV/e, and will not be measurable with the plasma analyzer. From Eq. 1, and the distribution of tail data as a function of X, we calculate that the bulk tailward plasma flow speed is > 300 km/s about 85% of the time that the PVO spacecraft is within the Venus magnetotail. This occurs because the PVO orbit is rather eccentric and thus PVO spends the longest fraction of time in the tail near apoapsis of $\sim 12\ R_V$, where the average flow speed is relatively large. Even though the finite plasma temperature broadens the velocity distribution of the particles and theoretically should make it possible to observe the low energy tail of the O^+ distribution, the low O^+ density makes this difficult in practice. Saunders and Russell[6] indicate that none of the PVO plasma instruments have sufficient sensitivity to resolve a plasma with a density of less than ~ 1 cm^{-3}. Since the average current sheet densities given in the table range from $\sim 0.9/cm^3$ (H^+) to $\sim 0.06/cm^3$ (O^+), the instrument sensitivity should be insufficient to detect the average tail plasma density. Due to both of these effects, our analysis indicates that the bulk of the tail plasma distributions should be unobservable with the PVO plasma analyzer much of the time that the spacecraft is within the tail current sheet and virtually all of the time that it is within the tail lobes. Mihalov and Barnes[7] and Slavin et al.[5] indicate that the magnetotail plasma is, indeed, often unobservable within the energy range and sensitivity of the PVO plasma analyzer.

A field reversing current sheet self-consistently forms to separate the oppositely directed magnetic fields in the tail lobes as described above and shown qualitatively in Fig. 1. The portions of the field lines that populate the tail current sheet are mass loaded as they approach the stagnation region above the dayside ionopause. Those portions of the field lines that penetrate the deepest are most slowed and spend the longest time in the environs of the extended Venus atmosphere. Thus, the majority of the mass is picked up by these near stagnation portions of the field that eventually slip around the planet and populate the tail current sheet. Since the majority of this newly ionized material is added in regions where the flow has already been slowed, the average pick-up ion gains only a fraction of the perpendicular thermal velocity available in the pick-up process by the upstream solar-wind flow. As a consequence, the ions of planetary origin in the Venus current sheet are cooler and have smaller gyroradii than would be naively expected if they were picked up by the unimpeded solar wind.

If the Venus magnetotail is composed entirely of planetary O^+, the derived average temperature of the plasma (see table, results from Ref. 1) is $\sim 9 \times 10^7$ K, while for a tail plasma composed entirely of protons, this temperature is $\sim 6 \times 10^6$ K. These values are very

Table 1—Averge consistent plasma properties.

	p^+	O^+
ρ_{lobe}	$\sim 1.2 \times 10^{-22}$ kg/m^3	
n_{lobe}	$\sim .07$ cm^{-3}	$\sim .005$ cm^{-3}
ρ_{CS}	$\sim 1.6 \times 10^{-21}$ kg/m^3	
n_{CS}	~ 0.9 cm^{-3}	$\sim .06$ cm^{-3}
$T_{isothermal}$	$\sim 6 \times 10^6$ K	$\sim 9 \times 10^7$ K
	$\beta_{lobe} \simeq .08$	$\beta_{CS} \simeq 12$

approximate due to the numerous assumptions in their derivation, and the actual ion composition and temperature probably lie approximately somewhere between the two extremes. Upstream of the planet, in the near stagnation region, the magnetic field is nearly perpendicular to the magnetosheath flow so that virtually all of the relative velocity between newly created ions and the bulk flow is turned into perpendicular thermal velocity. For an average upstream solar-wind speed of 440 km/s, an oxygen ion picked up by the unimpeded solar wind would gain perpendicular thermal energy equivalent to a temperature of $\sim 1.9 \times 10^8$ K. The average pick-up ion in the Venus tail has, therefore, been ionized in a region where the bulk flow speed is less than half of that upstream in the unimpeded solar-wind flow.

The importance of mass loading is probably much greater in the interaction between the solar wind and comets or AMPTE-type releases. At these bodies there is no "hard" conductive obstacle such as the gravitationally bound Venus ionosphere to help divert the flow. In the regions antisunward of such obstacles, plasma temperatures should, therefore, be appreciably cooler and ion gyroradii appreciably smaller than for single particles picked up by the unimpeded solar-wind flow.

We have summarized and extended certain salient features of our previous analysis. These extensions occur in three separate areas. The first is the explanation of the $\sim 60°$ spiral angle of the magnetic field connecting across the Venus magnetotail. In this paper we showed that although the IMF spiral angle at Venus is only $\sim 35°$, the larger value across the magnetotail can be accounted for by slowing of the flow in the magnetosheath region and the coordinate transformations used to average together data with varying upstream magnetic configurations. The second is that we demonstrated that the derived average consistent plasma flow speed and density are outside the observational E/Q range and sensitivity of the PVO plasma analyzer. Therefore, the plasma should not be observable much of the time that the PVO is within the tail current sheet, and virtually never when the PVO is within the tail lobes, as appears to be the case. Finally, we showed that magnetosheath slowing upstream of the Venus obstacle has the effect of reducing the temperature and gyroradii of the picked up planetary ions compared to single ions picked up by the unimpeded solar-wind flow. Such lessons learned from the Venus interaction with the solar wind may have valuable implications for the study of other space physics phenomena.

ACKNOWLEDGMENT—We gratefully acknowledge valuable discussions with J. G. Luhmann, J. L. Phillips, J. A. Slavin, and J. R. Spreiter and are particularly grateful to J. T. Gosling for reading this manuscript and providing numerous useful suggestions. Work done at the IGPP, UCLA, was supported under NASA contract NAS2-9491, while work done at Los Alamos was supported by Los Alamos National Laboratory under the auspices of the United States Department of Energy.

REFERENCES

[1] D. J. McComas, H. E. Spence, C. T. Russell, and M. A. Saunders, "The Average Magnetic Field Draping and Consistent Plasma Properties of the Venus Magnetotail," *J. Geophys. Res.* (in press, 1986).

[2] J. R. Spreiter and S. S. Stahara, "Solar Wind Flow Past Venus: Theory and Comparisons," *J. Geophys. Res.* **85**, 7715 (1980).

[3] H. Alfvén, "On the Theory of Comet Tail," *Tellus* **9**, 92 (1957).

[4] D. S. Intriligator, J. H. Wolfe, and J. D. Mihalov, "The Pioneer Venus Orbiter Plasma Analyzer Experiment," *IEEE Trans. Geosci. Remote Sensing,* **GE-19**(1), 39 (1980).

[5] J. A. Slavin, E. J. Smith, and D. S. Intriligator, "A Comparative Study of Distant Magnetotail Structure at Venus and Earth," *Geophys. Res. Lett.* **11**, 1074 (1984).

[6] M. A. Saunders and C. T. Russell, "Average Dimension and Magnetic Structure of the Distant Venus Magnetotial," *J. Geophys. Res.* (in press, 1986).

[7] J. D. Mihalov and A. Barnes, "The Distant Interplanetary Wake of Venus: Plasma Observations of Pioneer Venus," *J. Geophys. Res.* **87**, 9045 (1982).

DYNAMIC SUBSTORM INJECTIONS:
SIMILAR MAGNETOSPHERIC PHENOMENA AT EARTH AND MERCURY

S. P. Christon,* J. Feynman,[†] and J. A. Slavin[†]

In Earth's magnetotail at geosynchronous orbit, single-point observations of earthward-propagating dynamic substorm injections near local midnight are characterized by the sudden appearance of hot plasma electrons associated with a simultaneous increase in the magnetic field strength and a rapid northward rotation of the field to a more dipolar orientation. During the first encounter of the Mariner 10 spacecraft with the planet Mercury, similar, simultaneous hot electron plasma and magnetic field signatures accompanied two portions of one of the four major energetic particle enhancements known historically as the A, B, C, and D events, which were observed in the magnetic equatorial region of Mercury's nightside magnetosphere and dawn magnetosheath. Following suggestions that the transient magnetic field and particle disturbances observed in Mercury's magnetotail are very similar to substorm related phenomena in Earth's magnetotail, we have reexamined the correlations between energetic electrons (>35 keV and >175 keV), plasma electrons (13–688 eV), and magnetic fields during the Mercury 1 energetic particle events and compare them to several well-documented substorm injections at Earth. It is found that the B and B′ events possess the same characteristics as single-point observations of terrestrial dynamic injections. In addition, several previously unnoticed correlations between the energetic electrons, plasma electrons, and magnetic fields at Mercury have been discovered and are documented herein.

INTRODUCTION

On its pioneering flight into the inner solar system during the years 1973 to 1975, Mariner 10 encountered the planet Mercury three times at a heliocentric distance of ~ 0.5 AU. Mariner 10's trajectory during the first and third Mercury encounters, commonly called Mercury 1 and Mercury 3, were close passes ($\lesssim 1.3$ R$_M$ (Mercury radii), 1 R$_M$ = 2439 km) through the nightside equatorial and high-latitude regions, respectively (see articles in *Science* **185**, pp. 141-183, 1974). Mercury 2 was a relatively distant (~ 20 R$_M$) sunward pass that allowed some investigation of upstream plasma and magnetic fields.[1] The observed interaction of Mercury with the solar wind during the first and third encounters determined that the planet had a small intrinsic dipole-like magnetic field[2] about a factor of 5×10^{-4} the strength of Earth's.

The presence of Mercury's magnetic field as an obstacle to solar-wind flow creates both a bow shock in the solar wind and a magnetospheric cavity, much like that found at Earth, in which the intrinsic planetary magnetic field and magnetic fields due to the magnetospheric current systems dominate and exclude the solar wind and interplanetary magnetic field (IMF).[3,4] Dynamically, Mercury is expected to respond to directional variations in the IMF much like Earth, since its dipole moment also points in the general direction of the south celestial pole. Magnetic field and electron plasma observations in Mercury's magnetosphere strongly suggest that its interaction with the solar wind results in a near planet ($\lesssim 2$ R$_M$) environment morphologically similar to an appropriately scaled region in Earth's magnetosphere ($\lesssim 13$ R$_e$).[1,2,5] The recent discovery of sodium (undetectable by Mariner 10 instruments) as a major atmospheric constituent at Mercury[6] introduces the tantalizing possibility that there may yet be a tenuous ionosphere at Mercury. Theoretical analyses have indicated that the spatial and temporal scales for magnetospheric phenomena at Mercury are smaller and shorter by factors of ~ 7 and ~ 50, respectively, than at Earth.[5] Mean energies of plasma electrons are nearly the same in the two magnetospheres, 100–300 eV in the cool plasma region and >1 keV in the hot plasma region. The electron number density in Mercury's magnetosphere is consistent with the terrestrial value scaled up by about a factor of 5, the ratio of solar-wind number densities at the two planets.[1] At Mercury 1 a number of energetic particle enhancements (the A, B, and C events) were observed within the magnetosphere and another (the D event) in the dawn magnetosheath.[7] No other energetic particle enhancements were observed in the vicinity of Mercury. Since Mercury occupies a larger volume of its magnetosphere than Earth, permanently trapped particle radiation belts are not expected and were not reported.[7]

The transitory nature of the B and C energetic particle enhancements and the turbulent magnetic fields observed in the hot plasma region at Mercury 1 have kindled interest in the substorm analogy between Earth and Mercury.[1,2,5,7,8] Eraker and Simpson[8] have suggested that strong similarities exist between the B and C events at Mercury and the substorm injections at Earth. They have also argued further that the B, C, and D events are sequences of particle bursts with repetition rates of 5 to 6 s and decay times of ~ 1.5 s/decade. We test the substorm injection analogy by first describing single-point observations of earthward propagating dynamic injections in the near-Earth magnetotail, and then

---

*Physics Department, California Institute of Technology, Pasadena, California 91125.
[†]Jet Propulsion Laboratory, California Institute of Technology, Pasadena, California 91109.

inspecting the energetic particle events at Mercury 1 using the electron flux and magnetic field description of dynamic injections at Earth as a pattern. Only the B and B′ events at Mercury appear to fit clearly the pattern of terrestrial dynamic injections (see Refs. 9, 10 for more detailed discussions of "dynamic" injections).

DYNAMIC INJECTIONS AT EARTH

Near geosynchronous orbit, a number of well-defined dynamic injections have been found to be propagating earthward with velocities of 10–100 km/s.[9] The events shown in Fig. 1 highlight the important characteristics of single-point observations of dynamic injections. (GOES 3 is 20 min east of, and measures magnetic fields that well represent those at, ATS 6.[9]) At the onset of the events, the electron flux increases suddenly at higher energies (≥ 1 keV) and decreases at lower energies (≤ 70 eV). These flux variations are accompanied by clear magnetic field perturbations in which B increases by a factor of 2–3, and there is a rapid reorientation of the field direction to a more dipolar configuration with $\textbf{\textit{BZ}}$ increasing by about the same factor as and dominating $\textbf{B}$. Note that we use orthogonal cartesian magnetic field coordinate systems in which $\textbf{\textit{BZ}}$ increases antiparallel to the planetary magnetic dipole, $\textbf{\textit{BX}}$ is toward the sun and $\textbf{\textit{BY}}$ is toward dusk. At Earth this system is geographic, that is, radial, north-south, and east-west,[9] and at slowly rotating Mercury, the system is the Mercury Orbital (MO) coordinate system (see e.g., Ref. 11).

Each of the panels in Fig. 2 depicts schematically the near-planet magnetic field and motion of the injection disturbance during subsequent phases of the substorm process (see e.g., Ref. 12). In the upper schematic of each panel we view the planet looking toward the sun from slightly above the planetary orbital plane in the dusk magnetosheath region. In the lower schematic, we view the planet from dusk in the planetary orbital plane. Dashed equatorial rays in the upper schematic are for reference. Small circles along the planet-sun line in the lower schematic represent the cross-tail current system. The heavy solid arrow through the planet represents the intrinsic magnetic dipole moment. The heavy open arrows represent $\textbf{\textit{BZ}}_{\text{IMF}}$, the component of the IMF along the planetary dipole axis. Solid (dashed) magnetic field lines are the present (previous) positions assuming only meridional motion.

The general phases of the substorm process illustrated are: (a) quiet conditions with $\textbf{\textit{BZ}}_{\text{IMF}}$ northward; (b) southward turning of $\textbf{\textit{BZ}}_{\text{IMF}}$ results in flux erosion on the dayside and inflation of the nightside magnetotail;[13]

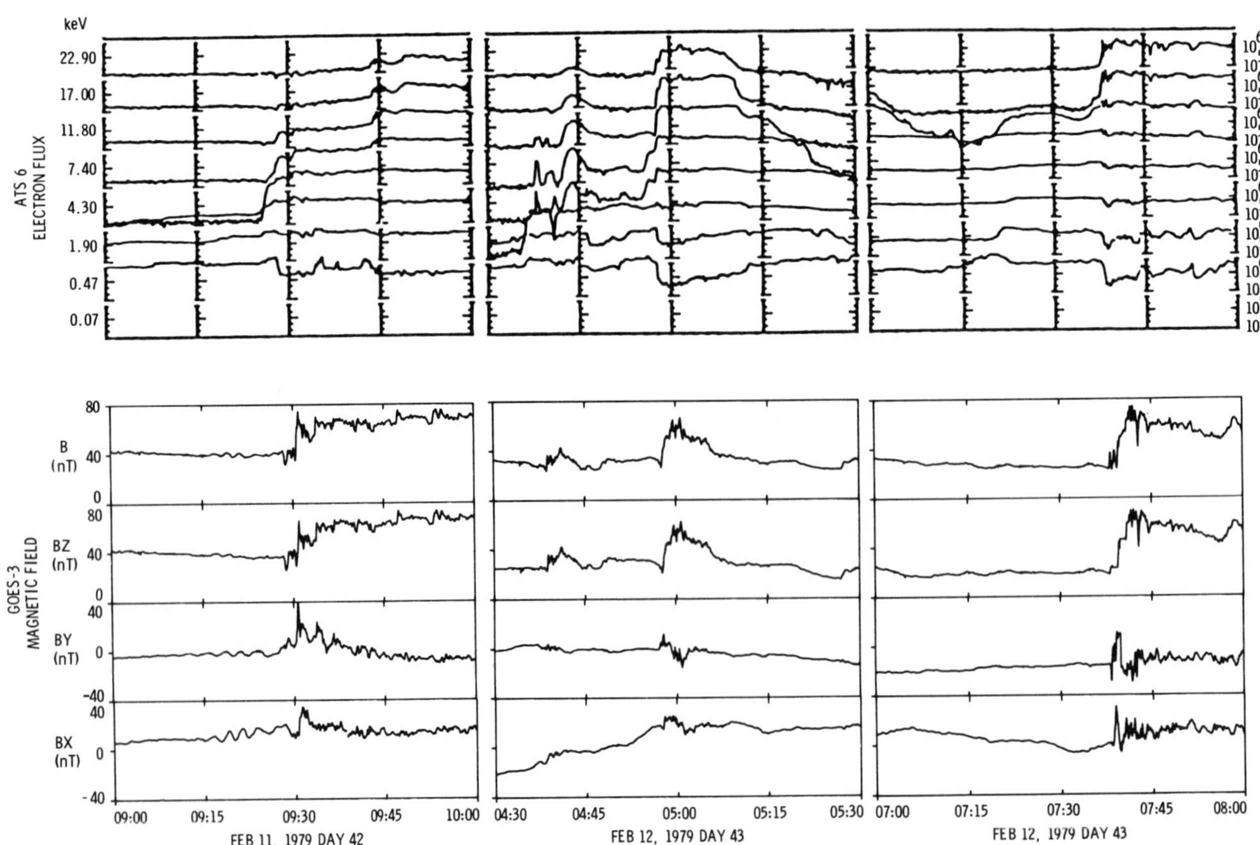

Figure 1—Electron flux and magnetic field measurements from three of the dynamic injections studied by Moore et al.[9] Times of the events are: 0930 day 42, 1979; 0458 and 0740 day 43, 1979. ATS 6 electron flux measurements are adapted from Fig. 8 of Moore et al.[9] GOES 3 magnetic field measurements are plotted so that $\textbf{B}$ and $\textbf{\textit{BZ}}$ can be closely compared to the electron flux measurements.

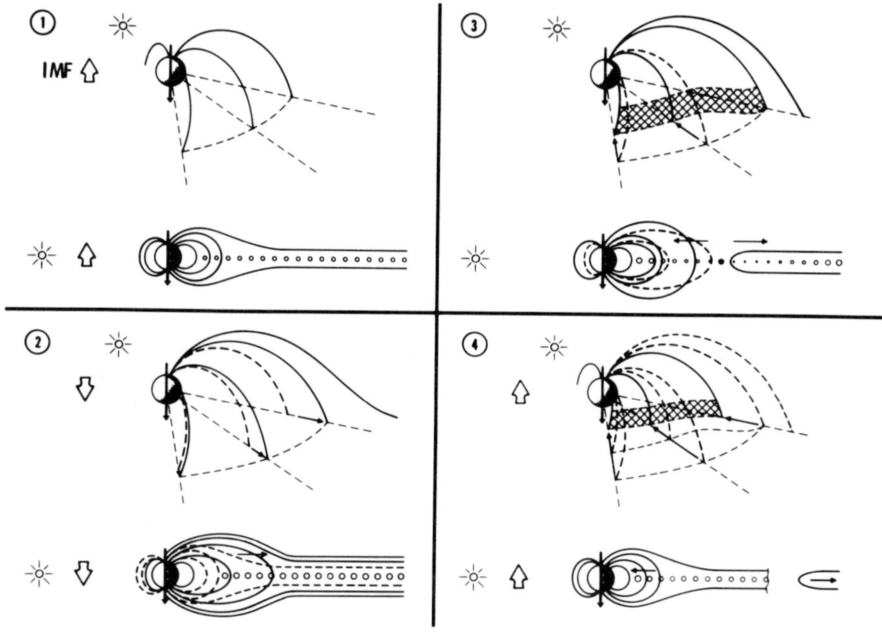

Figure 2—Schematic drawings of near planet field line positions during the substorm injection process. Details of the drawing are discussed in the text: (1) quiet conditions, (2) dayside erosion and nightside inflation, (3) substorm onset, and (4) post-substorm reconfiguration.

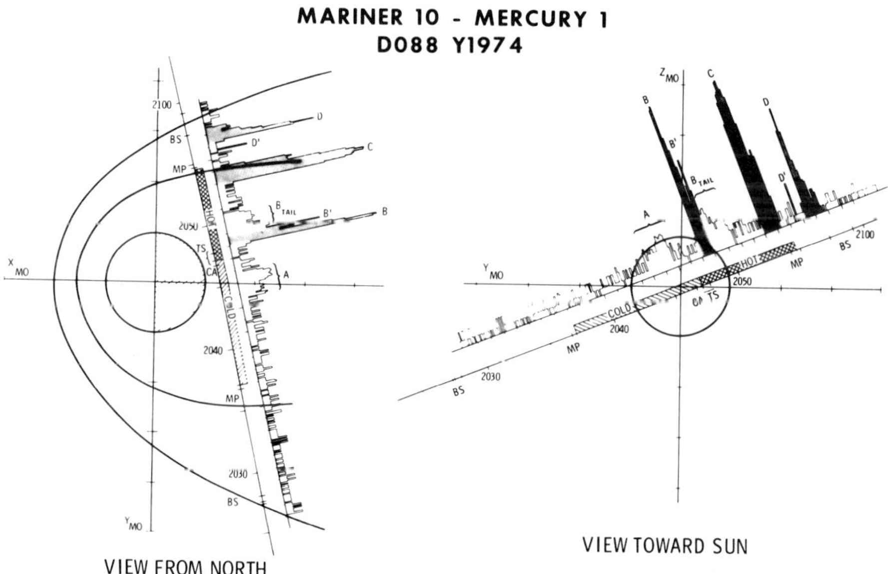

Figure 3—Mariner 10 trajectory plot in Mercury Orbital (MO) coordinates showing bow shock (BS), magnetopause (MP), and tail sheet (TS) crossings as well as closest approach (CA). Six-second averages of the ID1 counting rate from the charged particle experiment show the major energetic electron flux events A, B, B′, B_{TAIL}, C, D′, and D discussed in the text. The measured electron fluxes have energies >175 keV except during the shaded portions of the events when a considerable contribution to the D1 counting rate was most probably from intense fluxes of 35–175 keV electrons, energies normally below the instrumental electronic threhold. In this mode of operation the measured counting rate is not necessarily a direct measure of the electron flux.

(c) some substorm intialization event, such as disruption of an azimuthally restricted segment of the cross-tail current (represented by the asterisk), leads to the propagation of a dynamic injection toward the planet (see e.g., Ref. 14 and references therein); (d) the near-planet tail field is reconfigured and the plasma sheet is reforming as tail fields and plasmas relax into a quiescent state.

TRANSIENT PHENOMENA AT MERCURY

The Mariner 10 trajectory at Mercury 1 is shown in Fig. 3 with a sketch suggesting bow shock (BS) and magnetopause (MP) surfaces. Observed bow shock, magnetopause,[3,4] and tail sheet (TS)[11] crossings and closest approach (CA) are shown. The twofold advantages of the Mercury Orbital coordinate system are (a) north-south symmetry may be present because the planet's orbital velocity vector is in the XY_{MO} plane, and (b) estimates of Mercury's magnetic dipole moment place it within ~15° of Z_{MO},[15] making this a convenient dipole-aligned coordinate system. The spacecraft entered the magnetosphere in the plasma sheet and remained in the sheet until the outbound magnetopause crossing. Plasma electron characteristics suggested two major divisions—the cold and the hot plasma sheet regions.[1] The tail sheet crossing from 2047:00 to 2047:40[11] separates these general regions.

395

Six-second averages of the IDl counting rate showing the four major energetic electron flux enhancements at Mercury 1,[7,16] are plotted in Fig. 3 on a timeline offset from the Mariner 10 trajectory. Flux enhancement B is complex and has been separated into three portions B, B′, and B_{TAIL}. Note that the >175 keV electron flux enhancements A and B_{TAIL} appear on either side of the tail sheet crossing and that the >35 keV flux enhancements B, B′, and C appear in the hot plasma-sheet region. Shaded portions of the counting rates indicate times when this counting rate was most probably responding to intense fluxes of 35–175 keV electrons.[16-18] At all other times the IDl counting rate was responding to >175 keV electrons. Some controversy has surrounded interpretations of the energetic particle measurements at Mercury 1.[16,18] Because of this, the procedure used for identifying and separating response modes of the IDl counting rate is outlined in the appendix.

The upper five panels of Fig. 4 contain all electron flux information gathered during Mercury 1 at the highest time resolution—6.0 s samples for energies below 700 eV and 0.6 s averages for energies above 35 keV. This is the first time that all of the particle data have been presented together. The top panel clearly shows the increasing frequency of the >175 keV electrons that make up the A and B_{TAIL} events. At this temporal resolution, each readout of the instrument is shown. The B, B′, C, and D events clearly display different characteristics than the A or B_{TAIL} enhancements. Note that the particular nomenclature "B_{TAIL}" was chosen for the >175 keV electron flux portion of the B enhancement in order to clearly separate it from the B and B′ events.

Measurements from the fifteen energy channels of the plasma instrument, equally spaced on a logarithmic scale, are separated into three groups of five adjacent channels for display in Fig. 4. The transition from cold to hot plasma-sheet regions appears to occur soon after the A event and before closest approach at 2046:38, rather than at 2047 as noted by Ogilvie et al.[1] Intensities at lower plasma energies (the lowest of which often represent the electron plasma number density[3] decrease suddenly after the end of the A event. Note that the density increases again during the tail sheet crossing. The increasing difference between electron intensities at the highest and lowest plasma energies indicates that the spacecraft encounters progressively hotter electron plasma as it travels dawnward from inbound to outbound magnetopause crossings. This result is consistent with being either a temporal variation such as substorm heating (e.g., Ref. 5) or a spatial variation such as the dawn-dusk electron temperature asymmetry observed in Earth's plasma sheet.[19] A period of cooler plasma observation during the A event, represented by a significant decrease of higher energy plasma electron intensities by a factor of >10, suggests an association of the >175 keV electrons with plasma flow variations.[5] This decrease at electron energies of a few hundred eV is consistent with the plasma electron flux variations at the same energies in the terrestrial events (Fig. 1). However, during the A event,

the magnetic field remains predominately tail-like[8] (i.e., dominated by *BX*), and the A event does not therefore closely resemble the dynamic injections at Earth. Interpretation of the A event is beyond the scope of this paper. We note, however, that no previous discussion details of the plasma and field variations during the A event.

High resolution magnetic field data (0.6 s averages) appear in the lower four panels of Fig. 4. The magnetic field model of Whang,[11] incorporating a two-dimensional cross-tail current sheet and an image dipole to estimate fields due to magnetopause currents, is also plotted (connected dots). Shaded periods in the *BZ* panel are when *BZ* is more positive (more dipolar) than expected by this model. Note that these periods of more northerly *BZ* generally coincide with periods of enhanced electron flux above either the >35 keV or >175 keV thresholds, and sometimes both. Although *BZ* is more northerly than Whang's model fit during virtually all periods of enhanced particle flux, *BZ* is directly correlated with *B* only after the tail sheet crossing. A similar comparison of magnetic field observations with the offset, tilted dipole model of Ness et al.[4] has the observed *BZ* approaching the model fit in the A event and more positive than the fit in the B and C events. However, this early model neglected any external terms and is not complex enough to fully model the magnetospheric magnetic field.[2] Temporal variations of *B* are not correlated with *BZ* in the A event, and, although there is a flux decrease at electron energies of a few hundred eV as in the dynamic injections at Earth, the A event does not have the clear magnetic field signature of the terrestrial dynamic injections. Therefore, the A event does not appear to be a dynamic injection.

Comparison of energetic electron counting rates and magnetic field observations in Fig. 4 reveals that nearly simultaneous, rapid increases of the >35 keV electrons, *B*, and *BZ* apparent at the onset of the B and B′ events is not evident in either the C or D events. In the C event, the >35 keV electron flux intially increases as *B* and *BZ* are decreasing. The most intense C event flux increases are seen when *B* and *BZ* are at local minima. (This can be seen easily in Fig. 14 of Ref. 8). Therefore, the B and C events do not appear to be similar in all major aspects as suggested by Eraker and Simpson.[8] Baker et al.[20] have suggested that the C event may represent the dawnward signature of drifting, post-injection electron clouds.[9] The D event, occurring in the magnetosheath on the planetward side of the bow shock, is possibly a more dawnward signature of an injection event, with the injected electrons drifting through the magnetopause and being lost. Simpson et al.[7] found the D event electron flux intensity variations were strongly correlated with oscillatory changes in the magnetic field, possibly implying magnetic channeling from the magnetopause to the point of observation. Electron flux maxima in the D event occurred when BZ_{IMF} was southward, suggesting escape through the dawn magnetopause possibly via flux transfer events (Russell and Walker[21]). We now compare the high resolution data

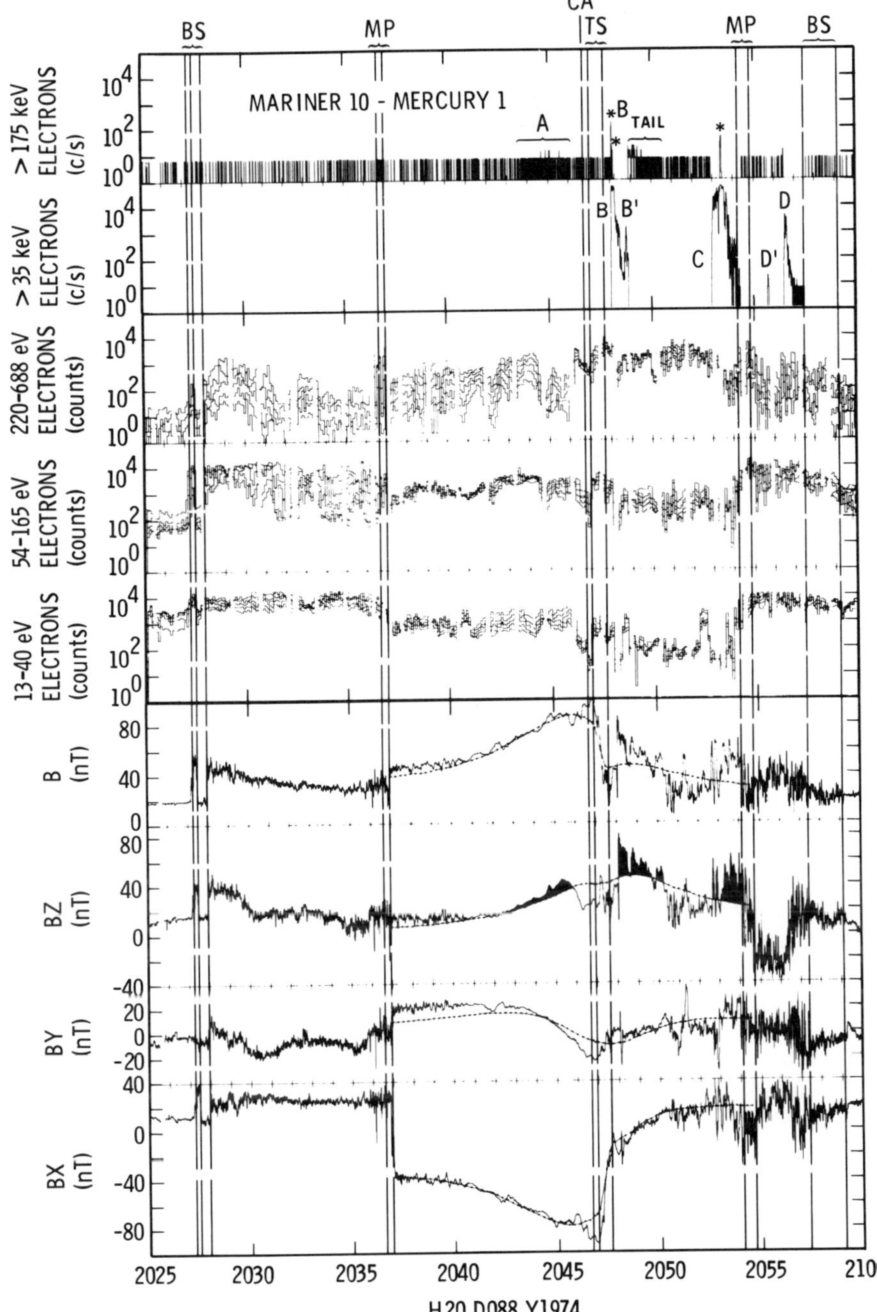

Figure 4—Simultaneous energetic electron, plasma electron, and magnetic field data observed during Mercury 1. Every readout from the energetic electron (0.6 s) detectors and every readout from the electron plasma (6.0 s) detectors, as well as high resolution magnetic field data (0.6 s) are plotted. Asterisks above peaks in the >175 keV electron counting rate denote measurements from the L12 coincidence counting rate of LET (Christon et al.[16]). Dotted curves in the magnetic field panels are the model fit by Whang.[11] Darkened portions in the BZ panel highlight observed BZ increases above model fit during periods of energetic electron flux enhancements.

from the Mercury 1 B and B′ events with the terrestrial events shown in Fig. 1.

THE B AND B′ EVENTS

Detailed comparison of the >35 keV electron flux (0.6 s averages) and magnetic field data (0.04 s samples) during the B and B′ events is shown in Fig. 5. Rigorous identification of associations between the energetic electrons and the magnetic field magnitude and direction is difficult because of the coarser temporal resolution of the electron flux measurement. However, after the first 12 s of the B event, that is, after 2048:15, some positive

correlation between the energetic electron flux, B, and BZ is evident throughout a number of local maxima (dashed lines) and local minima. This correlation of 3–6 s variations in the energetic electron flux and magnetic field argues for the presence of small-scale MHD compressional fluctuations or waves. If the energetic electron flux at energies >35 keV is dominated by ~35 keV electrons, a factor of 10^4 variation in the IDI counting rate may represent only a factor of ~12–50 variation in the flux of energetic electrons.[16] The bursty appearance during these events may represent relatively small variations in the plasma density and/or temperature. We

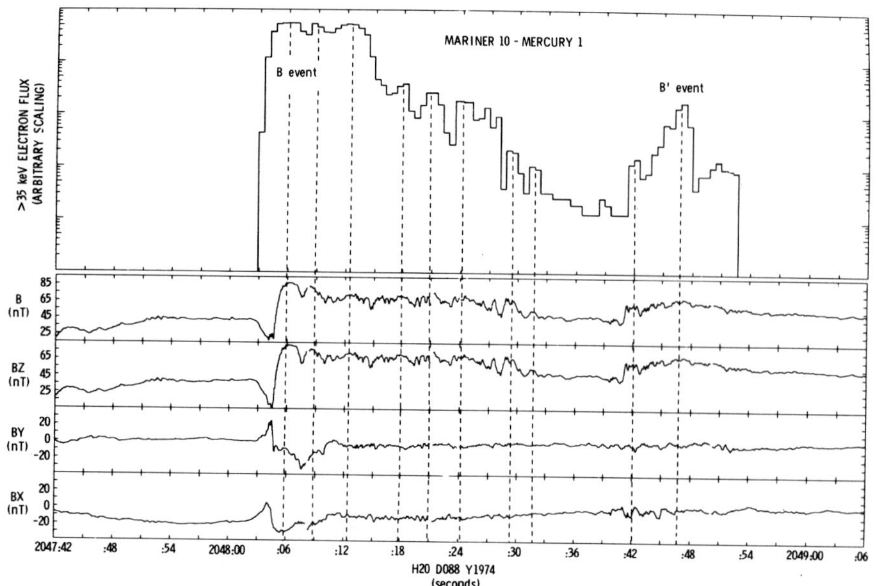

Figure 5—Detailed comparison of >35 keV electron flux (0.6 s resolution) and magnetic field (0.04 s resolution) data during the B and B′ events. Dashed lines lead from increases in the electron flux to corresponding increases in **B** and **BZ**. Note that except near the onsets of the two events, there is little corresponding variation in **BX** or **BY**.

note that this is a radically different interpretation of these quasiperiodic variations than that put forth by Eraker and Simpson,[8] who suggest that bursts of newly accelerated particles may be observed every 3–6 s.

Prior to the onset of the B event the field varies smoothly for ~15 s. The onset of the B event is identified by a sudden increase in the >35 keV electron flux and a simultaneous wave-like disturbance in the field[7,8] in which the average value of **B**, dominated by **BZ**, increases by a factor of ~1.5, roughly consistent with the size of the field increases associated with dynamic injections at Earth. Except for the wave-like variations during the first six seconds of the B event, **BX** and **BY** appear relatively unaffected, with **BX** steadily increasing and **BY** constant. Fluctuations in the field are present at times when hot plasma appears and are not present during the lowest flux period before the B′ event. The enhanced value of **B** decreases so that by 2048:33, **B** and **BZ** are only slightly higher than the pre-B event values. The >35 keV electron flux also decreases and the onset of the B′ event is again identified by a sudden increase in the >35 keV electron flux[7,16] and a simultaneous disturbance in the field in which **B**, again dominated by **BZ**, increases by a factor of ~1.5 times the average pre-onset value. The pattern of correlated energetic electron and field variations in the B event are repeated in the B′ event, although on a smaller scale. After the B′ event the field magnitude has increased by ~20 nT from its pre-B event value and the field direction has become more northerly by ~25°—both significant increases. So the dipolarization of the field appears to increase with each subsequent injection in a steplike process as at the Earth during multiple onset substorms (see e.g., Refs. 12 and 22).

Time series of the plasma electron differential energy spectra, j, in electrons/(cm²-s-sr-keV), are shown in Fig. 6 for three consecutive minutes including the B and B′ events. Scanning horizontally one can compare both densities and characteristic energies separated by 1 min intervals. The plasma in the B event is hotter than during the preceding minute in the tail sheet crossing when the mean electron energy appears to be in the tens to hundreds of eV range. Ogilvie et al.[1,3] argued that the mean energy in the hot-plasma region lies above 688 eV, the upper limit of the plasma instrument. During the 6 s period starting at 2048:08 (not plotted), the counting rates in the intermediate channels decreased to background levels leaving only low rates in the energy channels below 20 eV and above 389 eV.[3] Hot plasma is

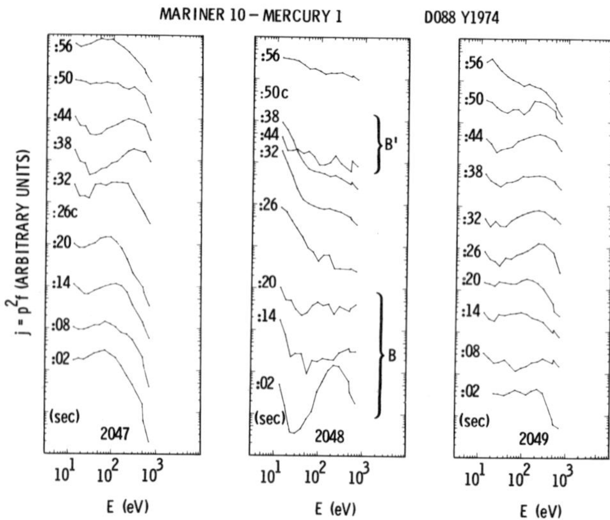

Figure 6—Six-second samples of electron plasma differential energy spectra from 2047:02 until 2049:56. Periods during the B and B′ events are indicated by curly brackets. Successive time periods are offset by the same vertical separation in each panel so that flux levels at successive 1 min intervals may be compared by scanning horizontally.

398

present from the onset of the B event until 2048:26 when a "cooler" plasma electron energy spectrum is evident until the onset of the B' event at $\geq$ 2048:42 when hotter plasma reappears. The cooler plasma is observed during the period of reduced magnetic field variance noted above. Intensity of the lowest energy plasma electrons decreases at the onset of both the B and B' events, in a manner similar to terrestrial substorm injections (see Fig. 7 in Ref. 9).

Additional evidence points to the B-B' event combination as being two separate injections similar to multiple onset substorms at Earth, rather than subsequent passes through a flapping plasma-magnetic field structure. The clear magnetic perturbation accompanying the onset of both the B and B' events is consistent with those of dynamic injections as described by Moore et al.[9] and Feynman.[10] The short ($\leq$ 1 s) periods of constant increase in **BZ** at $\sim$ 2048:05 and 2048:41 are similar to the field variations at the leading edges of the dynamic injections in Fig. 1. Minimum variance analyses for these two short periods results in minimum variance directions generally along the X_{MO} axis. There is little rotation in the plane of the current sheet, consistent with a planar surface passing the spacecraft on its way toward the planet. The dynamic injections at Earth were observed to be consistent with inward motion by timing arrival differences at two spacecraft. One might expect propagating compressional waves to be steeper at Mercury than at Earth because of the lower Alfvén speed at Mercury implied by a generally higher density[1] and similar magnetic field magnitudes in the region of the observations (see Figs. 1 and 5). The shock-like character of the B event onset argues for this interpretation. Note that Eraker and Simpson[8] first deduced the presence of dawn-dusk current sheets, or magnetic waves, at the onsets of both the B and C events by investigating field variations encompassing the onsets of these two events and referring to analogous interpretations of field variations near dynamic injections at Earth. However, as we mentioned earlier, we see no clear evidence for nearly simultaneous increases in electron flux, **B**, and **BZ** at the onset of the C event.

CONCLUSIONS

In conclusion, the following new points relevant to the comparison of magnetospheric dynamics at Mercury and Earth have been revealed:

1. We find that observations during the B and B' events at Mercury 1 are similar to single spacecraft observations of several well-documented earthward propagating dynamic injections by direct comparison, confirming the analogy suggested by Eraker and Simpson.[8] The B and B' events represent two separate, consecutive dynamic injections, perhaps corresponding to multiple onset substorms at Earth.

2. The C event does not appear to be the same type of feature as the B event, contrary to the suggestion of Eraker and Simpson.[8] It may represent the near-dawn signature of the injection process, but it is not of the same dynamical character as the B and B' events.

3. In the B event, a positive correlation exists between > 35 keV electron flux increases and increases in **B** and **BZ** throughout the first 30 s of the event. This supports the interpretation that the disturbances manifest in the B and B' events are compressional waves and may help in further understanding of the nature of these fluctuations. We do not feel that these fluctuations represent sequences of repeated neutral line formations and subsequent electron acceleration on 6 s time scales as suggested by Eraker and Simpson.[8]

4. There is a marked correlation between the presence of hotter plasma electrons and the > 35 keV electrons in the B and B' events. The B and B' events appear to be separated by $\sim$ 18 s of much cooler plasma electrons.

5. In the A event, the > 175 keV electron flux increase occurs at a time when the magnetic field is not dominated by **BZ**, although the highest energy plasma electron flux decreases signficantly; hence the A event is not obviously a dynamic injection.

6. A general correlation between the A, B, B', B_{TAIL}, and C particle events and increases in **BZ** exists. Increases in **BZ** in the near-planet region at Earth often represent restructuring of the magnetospheric current systems where the inflated tail field relaxes to a lower tail flux. This corroborative evidence argues in support of the similarity of the Mercury and Earth magnetospheres.

APPENDIX

Identification and separation of > 35 keV and > 175 keV electron fluxes in the IDl counting rate is based upon analyses of four counting rates and simultaneously observed IDl energy deposit data from the charged particle experiment at Mercury 1 by Christon and Tuzzolino (unpublished reports, 1975; 1976; 1977) using counting rate and energy deposit information from laboratory calibrations of the single counting (> 175 keV) and pileup (35–119 keV) electron responses of the University of Chicago charged particle telescopes on Mariner 10 (Christon et al.[16]). The procedure was as follows: (a) Energy deposit and average counting rate information for a specified time block of data (event or period between events) was initially interpreted as resulting from single counting electrons with energies > 175 keV having a power-law differential energy spectrum. A chi-squared minimization calculation using the prelaunch single counting electron calibrations resulted in best-fit values for the intensity and energy spectral index. (b) These values were then used as input for a calculation of the pileup and singles contributions to the various counting rates assuming now that the power-law differential energy spectrum extended down to electron energies of $\sim$ 35 keV. The calculated pileup contributions to the average IDl counting rate during the B, B',

C, D′, and D events were about 75%, 100%, 30%, 100%, and 100%, respectively. In all other time periods IDl had negligible pileup contributions. Note that the few periods of >175 keV electron flux identified by asterisks, appearing during the most intense >35 keV electron flux periods, are obtained from the L12 counting rate when IDl counting rate was $>50 \times 10^3$ c/s.[16] (c) The B and C events were scrutinized further by examining the dependence of energy deposited in the Dl detector on the observed IDl counting rate. In the laboratory calibrations and the computer simulations based upon them, the energy deposited increased linearly with the observed counting rate, relatively independent of the energy of the incident subthreshold electrons. The data observed at Mercury 1 for both the B and C events followed this pattern, so both were judged to be significantly contaminated by subthreshold energy electrons. Although the temporal coverage of fluxes above either energy threshold is discontinuous and spectral information is indeterminate for the >35 keV flux periods due to the nature of the instrument response, important information can be extracted from these data. In fact, it is a fortuitous circumstance that the ID1 counting rate did register a response to >35 keV electrons as well as the intended >175 keV electrons. Without this chance occurrence, precious information about Mercury's magnetosphere would have been missed. Budgetary considerations in Mariner 10's planning stages resulted in the absence of a medium-energy particle detector spanning the energy gap between the plasma and energetic particle detectors.[8]

ACKNOWLEDGMENT—We thank N. F. Ness, K. W. Ogilvie, J. A. Simpson, and the NSSDC for providing us with high-resolution data from the Mariner 10 Mercury encounters. The assistance of J. E. Wolf with certain computational aspects of this study is gratefully acknowledged. We thank E. C. Stone for use of computer facilities at the California Institute of Technology and R. E. Mewaldt for a critical reading of this paper. The research carried out at the Jet Propulsion Laboratory of the California Institute of Technology was performed under contract to the National Aeronautics and Space Administration.

REFERENCES

1. K. W. Ogilvie, J. D. Scudder, V. M. Vasyliunas, R. E. Hartle, and G. L. Siscoe, "Observations at the Planet Mercury by the Plasma Electron Experiment: Mariner 10," *J. Geophys. Res.* **82**, 1807 (1977).
2. N. F. Ness, "The Magnetosphere of Mercury," in *Solar System Plasma Physics*, Volume II, C. F. Kennel, L. J. Lanzerotti, and E. N. Parker, eds., North-Holland, pp. 183-206 (1979).
3. K. W. Ogilvie, J. D. Scudder, R. E. Hartle, G. L. Siscoe, H. S. Bridge, A. J. Lazarus, J. R. Asbridge, S. J. Bame, and C. M. Yeates, "Observations at Mercury Encounter by the Plasma Science Experiment on Mariner 10," *Science* **185**, 143 (1974).
4. N. F. Ness, K. W. Behannon, R. P. Lepping, Y. C. Whang, and K. H. Schatten, "Magnetic Field Observations Near Mercury: Preliminary Results from Mariner 10," *Science* **185**, 150 (1974).
5. G. L. Siscoe, N. F. Ness, and C. M. Yeates, "Substorms on Mercury?" *J. Geophys. Res.* **80**, 4359 (1975).
6. A. Potter and T. Morgan, "Discovery of Sodium in the Atmosphere of Mercury," *Science* **229**, 651 (1985).
7. J. A. Simpson, J. H. Eraker, J. E. Lamport, and P. H. Walpole, "Electrons and Protons Accelerated in Mercury's Magnetosphere," *Science* **185**, 160 (1974).
8. J. H. Eraker and J. A. Simpson, "Acceleration of Charged Particles in Mercury's Magnetosphere," *J. Geophys. Res.* **91**, 9973 (1986).
9. T. E. Moore, R. L. Arnoldy, J. Feynman, and D. A. Hardy, "Propagating Substorm Injection Fronts," *J. Geophys. Res.* **86**, 6713 (1981).
10. J. Feynman, A Comment on Electron Injections (this volume, 1986).
11. Y. C. Whang, "Model Magnetosphere of Mercury," *Phys. Earth Planet. Interiors* **20**, 218 (1979).
12. R. L. McPherron, "Magnetospheric Substorms," *Rev. Geophys. Space Phys.* **17**, 657 (1979).
13. J. A. Slavin and R. E. Holzer, "The Effect of Erosion on the Solar Wind Stand-Off Distance at Mercury," *J. Geophys. Res.* **84**, 2076 (1979).
14. D. N. Baker, S.-I. Akasofu, W. Baumjohann, J. W. Bieber, D. H. Fairfield, E. W. Hones, Jr., B. Mauk, and R. L. McPherron, "Substorms in the Magnetosphere," Chapter 8 in *Solar Terrestrial Physics: Present and Future*, NASA Reference Publication 1120 (1984).
15. R. P. Lepping, N. F. Ness, and K. W. Behannon, "Summary of Mariner 10 Magnetic Field and Trajectory Data for Mercury I and III Encounters," NASA Technical Memorandum 80600 (1979).
16. S. P. Christon, S. F. Daly, J. H. Eraker, M. A. Perkins, J. A. Simpson, and A. J. Tuzzolino, "Electron Calibration of Instrumentation for Low Energy, High Intensity Particle Measurements at Mercury," *J. Geophys. Res.* **84**, 4277 (1979).
17. S. P. Christon and J. A. Simpson, "Relativistic Electrons in Neutral Sheet Regions at Mercury, Earth, and Jupiter," (unpublished manuscript), invited paper to *IAGA/IAMAP Joint Assembly*, Seattle, Wash. (1977).
18. T. P. Armstrong, L. J. Lanzerotti, and S. M. Kirmigis, "Comment on 'Electron Calibration of Instrumentation for Low Energy, High Intensity Particle Measurements at Mercury' by Christon, Daly, Eraker, Perkins, Simpson, and Tuzzolino," *J. Geophys. Res.* **84**, 4468 (1979).
19. S. J. Bame, J. R. Asbridge, H. E. Felthause, E. W. Hones, and I. B. Strong, "Characteristics of the Plasma Sheet in Earth's Magnetotail," *J. Geophys. Res.* **72**, 113 (1967).
20. D. N. Baker, J. A. Simpson, and J. H. Eraker, "A Model of Impulsive Acceleration and Transport of Energetic Particles in Mercury's Magnetosphere," *J. Geophys. Res.* **91**, 8742 (1986).
21. C. T. Russell and R. J. Walker, "Flux Transfer Events at Mercury," *J. Geophys. Res.* **90**, 11067 (1985).
22. R. L. McPherron, "Magnetic Variations During Substorms," in *Dynamics of the Magnetosphere*, S.-I. Akasofu, ed., D. Reidel, Dordrecht, p. 631 (1980).

VII. DIALOGS ON CONTROVERSIAL TOPICS

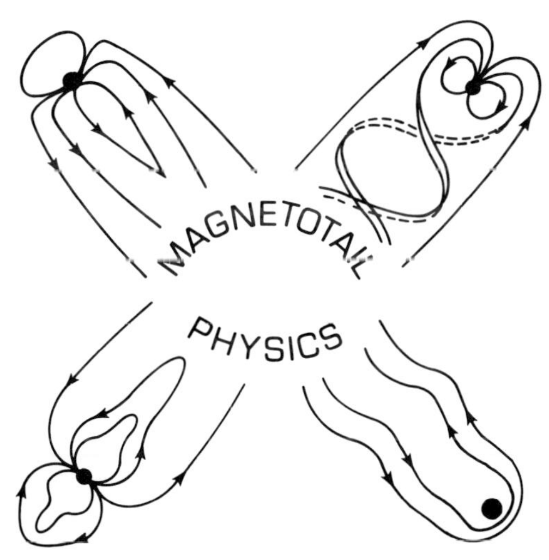

DIALOG ON INJECTION-BOUNDARY VERSUS ALFVÉN-LAYER MODELS

PANEL MEMBERS: M. G. Kivelson, J. Feynman, B. H. Mauk, and R. A. Wolf

INTRODUCTION BY PANEL CHAIR: M. G. KIVELSON

Immediately following the onset of an auroral substorm, the flux of charged particles increases markedly at many locations in the near-earth magnetotail. This type of flux enhancement is referred to as an injection. The controversial issue is the description in space and time of the source of the observed particles.

Some disagreements arise because of ambiguity in terminology, a situation that can be avoided by agreeing on definitions. The observed injections, for example, may be operationally defined as the increase of particle flux in a detector of finite bandwidth at a specific spatial location.[1] The observations, though extensive, come overwhelmingly from satellites near geosynchronous orbit, although measurements at other locations in space (e.g., Moore et al.;[2] Feynman et al.[3]) and ground-based measurements of precipitation and other associated phenomena have also been incorporated into studies. It follows that interpretations of the process rely on data spatially aliased by the limited locus of space observations and the ambiguity of magnetosphere-ionosphere coupling during disturbed times. Further confusion may arise from examination of datasets that cover only a narrow portion of phase space.

The conflicting models are similar in structure. The injection boundary model was introduced by McIlwain[4] who, in the words of Mauk and Meng,[5] "proposed that particles are energized and/or transported very quickly (on time scales faster than ordinary convection times) so that all energies of the distribution fill at one time only in an extended region which lies tailward of a sharply defined, nightside, spiral-like 'injection boundary'." In this picture, the initial injection is followed by ordinary quasistationary convection. Subsequent work[5] showed that a double spiral boundary with additional free parameters markedly improved fits to the geostationary orbit data. These models are purely phenomenological and do not seek to identify the mechanism that impulsively energizes and/or transports the particles.

The Alfvén-layer model (Kivelson et al.;[1] Kaye and Kivelson,[6] and references therein) emphasizes a mechanism for creating spatial boundaries and explores the consequences in a simplified quasistationary model (dipole field geometry, one or two parameter cross-tail field, and no loss mechanisms). The underlying concept is that of an Alfvén layer—a separatrix or surface not penetrated by particles drifting inward from the tail while conserving the first two adiabatic invariants. In applying the concept to observations, it is more relevant to identify regions of space at which particles of given energy may be found following inward drift. The boundaries of such regions have been termed steady-state demarcation boundaries (SSDB). The Alfvén-layer model suggests that many (but not all) injections can be modeled by assuming that prior to a substorm, SSDBs consistent with a small cross-tail field (E_0) exist in the near-tail regions. (The emphasis on the near-tail region is necessary because losses, neglected in this model, may become important when particles move to great distances from their source.) At substorm onset, an enhanced convection electric field (E_1), where $E_1 \gg E_0$, sets the preexisting boundaries into motion, producing time-dependent convection boundaries (TDCB). This model is not applicable when $\partial B/\partial t$ is large nor does it explain dispersionless injection. Proponents of the model have focused on behavior within an hour or two of substorm onset when the assumptions of no further time variations and limited particle losses are realistic.

To account for the frequent observation of dispersionless injection, Moore et al.[2] have modified the model in two ways. First, they note that SSDBs are modified by losses produced by energy-dependent pitch-angle scattering—an effect that reduces the spatial dispersion of the inner boundaries of particles of different energies. Second, they also point out that if inward motion is considered from the view of MHD wave motion, the front steepens as it moves inward toward regions of lower phase velocity. Both of these features would decrease energy dispersion of the arrival of particles at a geostationary satellite.

Neither of the models fully accounts for all important processes, e.g., the injection-boundary model has proved useful for organizing the data but leaves the process producing the injected particles unexplained. The Alfvén layer model must certainly be modified to explain dispersionless events. Even modified, it has never been applied systematically to a large dataset with a view to determining the extent to which it can account for the observations over time scales consistent with the assumptions of the model. Coupling with the ionosphere is found to be critical in more complete and complex treatments, such as the Rice Convection Model (see Wolf's remarks, which follow) but these models have not been applied to large datasets from geostationary satellites.

None of the models faces the problem of injected MeV particles that require inductive acceleration (see evidence in Ref. 7). Baker[8] argues that they are associated with a close-in neutral line (though that structure does not seem necessary to the author). He discusses a cloud rather than an advancing front.

Many questions remain unanswered. Some that may stimulate thought close this introduction:

1. Could the inductive field of a collapsing tail create a front that approximates the double-spiral injection boundary of Mauk and Meng[5]? Is this the compressional wave of Moore et al.[2]? Can it explain MeV injection?

2. To what degree are we misled by ambiguities in observations? For example:
   a. Do steady-state convection boundaries ever exist and can we distinguish them from moving boundaries?
   b. Do we distinguish properly between fronts advancing inward from the tail and fronts expanding poleward from the equatorial plane?
   c. Do we map correctly between inward propagating fronts in the tail and poleward moving ionospheric disturbances?

3. Are we misled by oversimplified theoretical models? For example:
   a. What is the consequence of strong shielding skewing as in the electric field $E = - \nabla [AR^4 \sin(\phi + \pi/4)]$ proposed by Hultqvist et al.[9]?
   b. What are the consequences of realistic ionospheric boundary conditions? For example, Maltseev[10] shows that nonuniform conductivity diverts magnetospheric flow parallel to the mapped surface of the conductivity discontinuity and similar results emerge from the Rice Convection Model.
   c. Does the nondipolar field of the real magnetosphere significantly modify the SSDB and the predictions based on it?

COMMENTS FROM PANEL MEMBERS

J. Feynman

In order to test competing models of particle injections we need a reliable, accurate, and complete description of observed substorm injection phenomena. The effort to produce such a description has been hampered by two factors: first, until recently most studies used a single geosynchronous satellite and, second, many particle detectors had only a limited energy range. Studies using more than one satellite have begun to give a wider perspective on injection events. In addition, the perspective has been further widened by the near geosynchronous satellite SCATHA, which covered the distance range from 5.3 R_e to 7.9 R_e and carried a magnetometer and several particle detectors covering a wide energy range. SCATHA data permitted three different types of electron injections to be clearly distinguished and the positions at which they occurred in the magnetosphere to be mapped. The defining charactersitics of these three event types—plasma-sheet entry, dynamic injection, and return from dropout—were very briefly described during the dialog session reported on here. More complete definitions of the event types and descriptions of the regions of the near-earth magnetotail at which these events took place have been included in this volume in

a brief report entitled "Distinguishing Among Electron Injection Types." These event types can be difficult or impossible to distinguish in datasets in which the particle detectors have limited energy ranges. Confusion among them can lead, and I believe has led, to a muddied description of substorm events seen by geosynchronous spacecraft.

The three types of events easily distinguished in SCATHA data do not exhaust the event types occurring during a substorm, and the development of a "second generation" phenomenological description using reasonably clearly defined event types and describing the positions at which each type occurs is needed in order to definitively test the current conflicting substorm injection models.

B. H. Mauk

It is without question that global, quasi-curl-free convection plays an important role in the transport of particles from the near magnetotail regions to the middle (<8 R_e or near geosynchronous) regions of the earth's magnetosphere. The question that remains is whether that form of transport to the middle regions ever exists in isolation from the highly transient, spatially localized transport processes driven by inductive electric fields. My answer is no. Due to space constraints the reader is referred to Ref. 11 for more complete referenced support for the arguments given here.

A fundamental prediction of the global convection picture is that during relatively undisturbed time periods, the plasmas within the magnetotail will fill out a quasi-time-stationary, energy-dispersed spatial pattern within the middle regions. As a result of this so-called "Alfvén layers" structure, a satellite should observe dispersed particle energy cutoffs that correspond to the energy-dispersed spatial pattern. A corollary to the Alfvén-layer picture is that some dynamic particle injections can be understood as the temporal redistribution of the spatial pattern of particles resulting from a sudden change in the solar-wind/magnetopause global electric-field dynamo.

The problem with the evidence presented in favor of the Alfvén-layer picture is that it has always been highly selective, not only in the events that are chosen for comparison with theory but in the portions of any one event that are chosen. In the regions between dusk and midnight, the model predicts that a geosynchronous satellite will pass from a void region, the so-called "forbidden zone," into a region where low-to-moderate energy plasmas are present. In the regions between noon and dusk, the satellite should pass from a region where the plasmas are present into the void or forbidden region. Of crucial importance is the sense of the energy dispersion that is observed.

It is the particle signatures in the dusk-to-midnight regions that are commonly used to support the Alfvén-layer picture. The problem is that only a portion of that signature is used to support the model. In the classic work of Kivelson et al.,[1] the authors show that the electron portion of an observed pattern is reproduced quite nicely by calculations based on the steady Alfvén-layer model. What the authors do not point out (nor have others who have used this feature to support the model) is that the ion portion of the observed dispersion pattern is dramatically inconsistent with the Alfvén-layer model ions have penetrated inside their forbidden zones.[11] (These include very energetic ions—E $\geq$ 20 keV—whose convection orbits are simply nested as a function of magnetic moment. See the comments of M. G. Kivelson.) When the full ion-electron feature is considered, the Alfvén-layer model is not supported. Even more problematical is the signature observed in the noon-to-dusk quadrant. Here it has been shown that the observed energy dispersion in both electrons and ions has a sense that is exactly inverted from the sense that is expected based on the Alfvén-layer model. This inverted dispersion is regularly observed from very quiet to very active time periods. Also, it is argued that neither the introduction of particle losses nor time dependences to the Alfvén model can explain the inverted sense of the observed energy dispersion.[11]

It is without question that at least at times the transport of particles to the middle regions is associated with transient, localized, inductively driven processes that are distinct from the global, quasi-curl-free (even time-dependent) processes discussed above. Intense, transient inductive electric fields have been observed within these regions in association with particle injections. Also, it has been shown that the so-called "bounce-phase-bunched" ions and the parallel ion acceleration observed commonly in association with local injections are signatures of this inductive electric field. Additionally, plasma injections are often observed occurring in an energy and species dispersionless manner. Finally, on energy-time spectrograms, sharp time-dispersed boundaries in energy often appear between freshly injected populations and apparently undisturbed preexisting populations.

The so-called "injection boundary model"[4] is a phenomenological model that was developed to accommodate the above described signatures. It is hypothesized that particles are energized and/or transported very quickly (on time scales faster than ordinary convection times) so that all energies and species of the particle distributions fill, at one instant only, an extended region tailward of a sharply defined, nightside boundary. Earthward of the boundary, the preexisting populations are left comparatively undisturbed. Finally, portions of the shape of the injection boundary, most notably in the midnight-to-dawn quadrant, are very distinct from the shape of the commonly drawn Alfvén-layers.

Key to the present discussion is the fact that it has been shown that the use of the injection boundary model (which clearly must be associated with transient, localized processes), when used in conjunction with subsequent global, quasi-curl-free convection, explains essentially all of the commonly occurring particle-energy dispersion signatures observed within the middle- or near-geosynchronous regions. As has been argued above, not even one complete ion-electron signature has been explained within these regions by the use of the global, quasi-curl-free convection processes alone. We conclude that while global, quasi-curl-free convection has an important role to play, the transport to the middle regions always happens in association with the transient, localized, inductively driven processes. Note that magnetic signatures will not be observed in association with highly dispersed particle signatures because the satellite is not located at the site of the inductively driven injection.

The relation between the global and localized processes is probably as follows. During relatively quiet times the "forbidden zone" (if it exists at these times) lies well outside of the middle regions. During enhanced periods of activity that zone is driven to, or is formed at, the middle regions, and in the absence of other processes coming into play the ion and electron signatures would reflect the character of the quasi-time-stationary convection patterns. However, in one-to-one correlation with the activity that drives the forbidden zone earthward is the initiation of the impulsive, localized, inductively driven processes. These impulsive processes dictate the character of the particle signatures, and yet in between the impulses, the global convection patterns serve to determine the subsequent evolutions of the newly transported and energized plasmas.

M. G. Kivelson

Without doubt, the Alfvén-layer model and modifications thereof provide, at best, an incomplete explanation of particle injection in the middle magnetosphere. However, the ideas of SSDBs and TDCBs organize many observations and they are, therefore, extremely useful. Their most natural application is to cases in which particles are injected in the absence of significant magnetic perturbations. As such cases are found not infrequently in the ATS 5 and 6 data, I argue that convection in the absence of a significant inductive electric field can and does play a role in the middle magnetosphere.

Whether one considers injection with or without associated magnetic activity, the evidence for Alfvén layers is clearest in electron distributions and low-energy ion distributions because the Alfvén boundaries for these particles are highly organized; that is, they are nested. The nesting is such that higher-energy electron boundaries are farthest from the earth with low-energy electrons and low-energy ions successively closer to earth. Ion boundaries for ions greater than 5 keV are not all simply connected. This means that regions accessible to preexisting ion distributions and regions accessible to newly injected distributions are strongly intermingled in the vicinity of geostationary orbit. Analytic solutions are extremely sensitive to inaccurately modeled parameters (e.g., electric and magnetic field models). Further uncertainty arises because of the presence of convection drift orbits on which ions must pass near flow stagnation points before reaching a geostationary spacecraft. Near stagnation points, the flow is slowed and temporal fluctuations lead to strong losses. Consequently, a meaningful calculation of ion distributions must take into account more than conservative convective flow. The ions must be traced over their orbits for times of many hours, and loss processes must be considered. Thus, I argue that the clearest signature of processes important on the substorm time scale is carried by the electrons and the lowest energy ions, which are unaffected by the complexities described above.

I agree that a focus on electrons and low-energy ions is selective, but I suggest that this is a useful selectivity that eliminates the complexity associated with multiple unrelated processes. Lack of selectivity can be misleading. I believe that some of the features of the most recent injection models, especially the deep penetration of the boundary near midnight, accurately reflect the role of inductive electric fields in enhancing convection in localized regions in the vicinity of the substorm current wedge. On the other hand, I believe that the attempt to force a simple convection model to fit the distribution of ions ≥ 5 keV without taking loss processes into account may introduce extraneous and nonphysical features into the model.

Finally, as an example of the way in which the ideas of steady and time-varying demarcation boundaries are useful in organizing electron injection events, I turn to Feynman's contribution to this conference and note that in the three substorms illustrated in Fig. 2 of Ref. 12, the enhancements follow the pattern predicted from an Alfvén-layer model. Prior to the first substorm, the spacecraft begins to penetrate low-energy SSDBs but only up to 0.7 keV. The first substorm leads to enhancements for energies less than 52 keV but not above. In the convection boundary picture, the TDCBs for energies ≤ 9 keV have passed across the spacecraft and the 23–52 keV boundaries are partially penetrated. In the second substorm, no further enhancement occurs for energies ≤ 9 keV because the spacecraft is already inside the TDCBs for these energies. Enhancements in the channels up to 70 keV represent penetration of additional TDCBs in this event. A slow decay of fluxes above 50 keV occurs before the third substorm and the particle signature of that final event is weak, appearing mainly in the highest energy electron channels for which the flux decayed between substorms 2 and 3. As we would expect for an Alfvén-layer model, the signature in ions shows no relation to the substorm.

The Alfvén-layer model thus explains the systematic shift of the particle energy channels that do or do not respond to substorm onsets. I believe it would be difficult to interpret these observations in terms of the injection boundary model. This suggests that the Alfvén-layer model encapsulates the approximate structure and dynamical response of particle fluxes in the middle magnetosphere.

R. A. Wolf

I would like to discuss some elements of what I think is the essential physics of substorm-associated ring-current injection.

The substorm is, in my view, essentially a mechanism for releasing energy from certain plasma-sheet flux tubes, which, having been convected earthward from the far tail and adiabatically compressed, need to release ener-gy before they can take the almost-dipolar form of inner-plasma-sheet flux tubes. (This general view has been discussed by Erickson, Coroniti, and Fairfield in papers given at this conference.) After this excess energy has been released, apparently in the form of a plasmoid flowing down the tail, the remaining closed flux tubes collapse to a more dipolar form and contribute to the injection of the ring current.

Plasma-sheet particles are injected into the ring current by electric and magnetic fields, and I would like to emphasize the former in this brief discussion. I would identify four major contributors to the substorm electric field that exist in addition to the large-scale shielded convection field and corotation, and strongly affect injection of the ring current:

1. The induction electric field associated with the collapse of midnight-region magnetic field lines. This electric field is westward in the region of the collapse and can be large for a short time interval.
2. A westward potential electric field that is generated across the region of closed flux tubes that have released energy and have collapsed. This ionospheric response to the interruption of the tail current (caused by ejection of the plasmoid) is most conveniently described quantitatively in terms of a slightly modified form of Vasyliunas' equations,[13] namely,

$$J_{\|i}/B_i = [2(\int ds/B)^{5/3} B^2]^{-1} B$$

$$\cdot \nabla(\int ds/B) \times \nabla[p(\int ds/B)^{5/3}] \quad (1)$$

$$\nabla_i \cdot (\Sigma \cdot E') = J_{\|i} \sin I \quad (2)$$

where $J_{\|i}$ is the density of magnetic field-aligned current down into the northern ionosphere, B_i is the magnetic field there, I is the magnetic dip angle, p is plasma pressure, and $\Sigma \cdot E'$ represents a field-line integral of the product of the ionospheric-conductivity tensor and the electric field in the rest frame of the neutrals. The equations imply a downward (upward) current on the dawn (dusk) side of the collapsed region and a westward contribution to the electric field there.
3. An eastward potential electric field that is generated across the earthward edge of the collapsed region, owing to the inner edge of the plasma sheet being "dented earthwards" there. The equations imply an upward (downward) Birkeland current at the dawn (dusk) end of the dent, and an eastward electric field perturbation across the collapsed region.
4. A disturbance in the electric field caused by enhanced electron precipitation and ionospheric conductivity in the collapsed region.

Computer simulations carried out with the Rice Convection Model indicate that sources 1–3 play important roles in substorm-associated ring-current injection, and it is clear that source 4, though not included in the simulations yet, is also quantitatively significant. All of these electric fields are highly time dependent in a substorm. Alfvén-layer models, based on the assumption that electric fields are constant for about a drift period, and injection-boundary models, typically based on constancy of the electric field after substorm onset, may be useful approaches to organizing particle data; however, they should be used with caution and should not necessarily be accepted as realistic representations of reality. It is unlikely that the real magnetosphere is simpler than the computer simulations, which involve quite complex and dynamic electric fields.

SUMMARY REMARKS: M. G. KIVELSON

The issues of controversy seem to have been sharpened by the panel discussion, though areas of increasing agreement are also evident. All of us are now emphasizing the role of induction electric fields in creating the substorm-associated injection boundaries.

The issue of whether SSCBs exist at quiet times has been challenged by Mauk and treated with detachment by Wolf, although as Alfvén layer enthusiasts, both Kivelson and Feynman have found considerable support for their views in specific datasets. Mauk has examined more data and interpreted them with a multiparameter model; his challengers have looked at data at a larger range of distances down the tail. Mauk believes the Alfvén-layer model cannot explain quiet time ion signatures. Kivelson argues that ion signatures are highly sensitive to departures from idealized field models and, as well, are critically modified by diffusive losses through pitch-angle scattering, so she expects that their behavior will be poorly predicted by a two-parameter idealized model. It is important to continue examining the signatures and, following the lead of Wolf, Maltseev, etc., to include ionospheric coupling. As well, new studies must include simultaneous data from locations away from geostationary orbit to test the injection boundary model at locations removed from the region for which it was developed.

Controversy about details of quiet-time behavior will undoubtedly continue. Resolving these disagreements seems less important than developing a quantitative "first principles" description of the dynamical changes in particle-phase space density associated with substorm onset.

ACKNOWLEDGMENT—B. H. Mauk acknowledges support from the Atmospheric Sciences Division, National Science Foundation grant ATM-8315041 and the Air Force Office of Scientific Research grant 84-0049. M. G. Kivelson acknowledges support from the National Aeronautics and Space Administration grant NGL-05-007-004.

REFERENCES

[1] M. G. Kivelson, S. M. Kaye, and D. J. Southwood, "The Physics of Plasma Injection Events," in *Dynamics of the Magnetosphere*, S.-I. Akasofu, ed., D. Reidel, Hingham, Mass., pp. 385-405 (1980).

[2] T. E. Moore, R. L. Arnoldy, J. Feynman, and D. A. Hardy, "Propagating Substorm Injection Fronts," *J. Geophys. Res.* **86**, 6713 (1981).

[3] J. Feynman, D. A. Hardy, and E. G. Mullen, "The 40-keV Electron Durable Trapping Region," *J. Geophys. Res.* **89**, 1517 (1984).

[4] C. E. McIlwain, "Substorm Injection Boundaries," in *Magnetospheric Physics*, B. M. McCormac, ed., D. Reidel, Hingham, Mass., p. 143 (1974).

[5] B. H. Mauk and C.-I. Meng, "Characterization of Geostationary Particle Signatures Based on the "Injection Boundary" Model," *J. Geophys. Res.* **86**, 3055 (1983).

[6] S. M. Kaye and M. G. Kivelson, "Time-Dependent Convection Electric Fields and Plasma Injection," *J. Geophys. Res.* **84**, 4183 (1979).

[7] M. G. Kivelson, "July 29, 1977 Magnetospheric Studies: Impulsive Waves, Global Dynamics and Geomagnetic Indices," *J. Geophys. Res.* **87**, 5981 (1982).

[8] D. N. Baker, "Particle and Field Signatures of Substorms in the Near Magnetotail," in *Magnetic Reconnection in Space and Laboratory Plasmas*, E. Hones, ed. (1984).

[9] B. Hultqvist, H. Borg, L.-A. Holmgren, H. Reme, A. Bahnsen, M. Jespersen, and G. Kremser, "Quiet-Time Convection Electric Field Properties Derived from keV Electron Measurements at the Inner Edge of the Plasma Sheet by Means of Geos-2," *Planet. Space Sci.* **30**, 261 (1982).

[10] Yu. P. Maltseev, "Effect of Finite Parallel Conductivity on the Magnetospheric Convection," *Planet. Space Sci.* **33**, 493 (1985).

[11] B. H. Mauk and C.-I. Meng, "Macroscopic Ion Acceleration Associated with the Formation of the Ring Current in the Earth's Magnetosphere," in *Ion Acceleration in the Magnetosphere and Ionosphere*, T. Chang, ed., Geophysical Monograph 38, American Geophysical Union, Washington, D.C., p. 351 (1986).

[12] J. Feynman, "Distinguishing Among Electron Injection Types," this issue (1986).

[13] Vasyliunas "Mathematical Models of Magnetospheric Convection and Its Coupling to the Ionosphere," in *Particles and Fields in the Magnetosphere*, B. M. McCormac, ed., D. Reidel, Hingham, Mass., p. 60 (1970).

DIALOG ON THE RELATIVE ROLES OF
RECONNECTION AND THE "VISCOUS" INTERACTION
IN PROVIDING SOLAR-WIND ENERGY TO THE MAGNETOSPHERE

PANEL MEMBERS: G. Rostoker, D. N. Baker, J. Lemaire, and V. Vasyliunas

INTRODUCTION BY PANEL CHAIR:
G. ROSTOKER

It has long been known that the solar wind must supply energy at the rate of $\sim 10^{10}$ to 10^{12} J/s to the magnetosphere in order to account for energy dissipated in the auroral oval through particle collisional processes and the energy required for ring current formation. In 1961, two methods were independently proposed by which the energy could enter the magnetosphere. Axford and Hines[1] contended that viscous processes at the earth's magnetopause could lead to energization of the magnetized plasma in the interior of the magnetosphere. Their primary interest was in studying the implications of the ensuing convective motion of the plasma and, as a consequence, they did not explore the possible physical mechanisms through which such a viscous behaviour might arise at that time. In the same year, Dungey[2] suggested that the interconnection of the IMF with the earth's magnetic field could permit a reconfiguration of the magnetosphere with the development of a long magnetic tail. One crucial prediction of Dungey's theory was that the entry of solar wind energy would be modulated by the magnitude and direction of the component of the IMF parallel (antiparallel) to the earth's magnetic field lines at the dayside magnetopause. This type of dependency was confirmed by the late 1960s (Fairfield and Cahill;[3] Rostoker and Fälthammar[4]) and, coupled with the strong theoretical base developed for the reconnection hypothesis (Petschek;[5] Yeh and Axford[6]), further consideration of the viscous interaction was set aside. Despite the theoretical study of diffusive processes by Eviator and Wolf[7] and observation evidence by Freeman et al.[8] suggesting that viscous processes might be of importance, the last word for a decade seems to have gone to Axford[9] who stated:

> "However, although there are a few observations suggesting that the viscous effect could be important (e.g., Freeman et al., 1968), it seems unlikely that we would be so unlucky to have two competing mechanisms of equal significance. For the present, we will abandon the simple form of viscosity (i.e., momentum transfer without field-line reconnection) and concentrate on field-line reconnection as a means of exerting a shear stress on the body of the magnetosphere."

By the early 1970s, aided by the application of plausible physical processes to a comprehensive phenomenological model of reconnection processes by Coroniti and Kennel,[10] the Dungey hypothesis dominated thinking regarding the nature of the solar terrestrial interaction.

In the mid-1970s, a new set of observations came to light that led to a revival of the viscous interaction as a process that might have to be reckoned with. The discovery of the low-latitude boundary layer (LLBL) by Eastman et al.[11] provided evidence for a plasma population at low GSM latitudes just inside the magnetopause whose estimated particle number flux of $\sim 10^{26}$ to 10^{27}/s was adequate to populate the magnetotail and whose momentum flux ($v \simeq 150$ km/s) was large enough to make it of consequence in terms of the energetics of the solar terrestrial interaction. Subsequent studies by Lundin and Dubinen[12] and by Eastman et al.[13] provided more support for the view that this viscous interaction could not be ignored and the observations of "viscous cells" at ionospheric levels reported by Reiff[14] make it clear that the viscous interaction influences ionospheric processes.

The question then arises as to how significant a role viscous processes play. While Cowley[15] discounts their importance totally, based on the results of Wygant et al.[16] and Mozer,[17] several theoreticians have put forward mechanisms that suggest a more significant role for viscous processes. Two types of mechanisms under consideration are gradient drift entry[18] and impulsive "plasmoid" entry for which different mechanisms have been proposed by Heikkila[19] and Lemaire.[20] One of the most promising theoretical approaches is that of Miura[21] who considers the Kelvin-Helmholtz instability on the magnetopause as a mechanism for anomalous momentum transport that can provide up to 30 kV across the magnetotail.

At present, it seems probable that magnetic field-line reconnection is the dominant means by which solar-wind energy enters the magnetosphere (after which it is stored in the magnetotail and ultimately unloaded to power substorm expansive phase activity). The question is, how important is the role of the viscous interaction that seems to be responsible for the LLBL. A specific question that might be asked is how much of the (slowly varying) directly driven system[22] activity can be accounted for by a viscous interaction.

COMMENTS FROM PANEL MEMBERS

D. N. Baker

It is probably a fair assessment of nature to say that if a given physical process can occur, it will. I certainly feel this way about "viscous" interaction processes in solar-wind/magnetosphere coupling. However, the specific charge to this panel was to determine the relative importance of the two modes of interaction. The following are my impressions based on experimental results in the distant ($r > 100$ R$_e$) magnetotail.

I feel that the competition between viscous and reconnection processes in magnetotail formation is difficult to assess and quantify in a very precise way. From my own observational perspective, however, it appears that reconnection processes are of overwhelming importance in the distant magnetotail, with the tail owing its very existence and basic structure primarily to dayside reconnection and its consequences.

As shown by Gosling et al.,[23,24] the ISEE 3 data reveal broad plasma boundary layers immediately inside the distant tail magnetopause. As pointed out in those papers, the presence of such boundary layers at great tail distances shows very strongly that plasma from the magnetosheath often crosses the magnetopause locally along much of the tail length, implying that the tail is "open." This in turn implies that reconnection has taken place between solar wind and magnetospheric field lines to produce those effects. In examining asymmetries between the distant lobes (north and south), Gosling et al.[24] often found density differences of factors of 3–10. The density effects show dawn-dusk asymmetries that depend on the IMF (e.g., high density in the north-dawnside lobe for $+B_y$ etc.) that also strongly support a "reconnection" interpretation of the distant tail character.

These impressions have been further reinforced by the examination of higher energy electrons that serve as a kind of "test particle" population.[25,26] It is found that electrons in the 50–500 eV energy range commonly exhibit strong, field-aligned anisotropies in the tail lobes. Because of large tail motions, the ISEE 3 data provide extensive sampling of both the north and south lobes in rapid succession. The bidirectional fluxes are found to occur predominantly in the lobe directly connected to the sunward IMF in the open magnetosphere model (north lobe for away sectors and south lobe for toward sectors). Electron anisotropy and magnetic field data are presented that show the transition from unidirectional (sheath) electron populations to bidirectional (lobe) populations. Taken together, the evidence suggests that the bidirectional electrons that are observed in the distant tail are closely related to the polar rain electrons observed previously at lower altitudes. Furthermore, these data provide strong evidence that the distant tail is comprised largely of open magnetic field lines in strong distinction to some recently advanced models[27] that postulate a much larger role for viscous tail formation processes.

Finally, strong and, indeed, accelerating plasma flow observed as one goes deeper down the tail suggests an important role for magnetic reconnection at a distant nightside neutral line.[28,29] Viscous processes may play some role in the low-latitude boundary layers adjacent to the distant-tail magnetopause,[30] but for the vast majority of the distant-tail volume it seems probable that reconnection processes have been the important agents of the tail structure and development.

J. Lemaire

The question addressed in the following paragraph is, "How much can we rely on quasistationary (i.e., quasisteady state) models to describe the interaction between the solar wind and the earth's magnetic field?" An equivalent question is, "How much can we rely on quasistationary convection flow patterns, quasistationary electric equipotential contours and current systems deduced from the steady-state reconnection or viscous-like interaction models?"

Most of the time the answer to these questions will be, "Not very much!" Indeed, most of the time the solar wind is a very inhomogeneous medium. The solar-wind plasma impinging on the geomagnetic field is almost always nonuniform over distances smaller than the dimensions of the magnetosphere.

To convince oneself that the supersonic solar-wind plasma is generally not uniform over the surface of the magnetopause, it suffices to examine high-resolution magnetograms of the IMF. From these high-resolution magnetograms it can be seen that it is sometimes hard to find periods of time larger than half a minute during which the three components of the solar-wind magnetic field are strictly constant.

Even in the so-called quiet solar wind, when the plasma bulk velocity (V) is low, it is difficult to identify extended periods of time, Δt, when high-resolution IMF magnetograms are perfectly flat, i.e., when the solar wind was uniform over a distance Δx ($= V\Delta t$) exceeding 20 R$_e$ or the length of the magnetotail. When the magnetic-field components are averaged over 15 seconds (or over 1 hour of time) these small-scale variations and nonuniformities are washed out and are easily overlooked.

The same can be said for solar-wind plasma observations that are generally sampled at a rate much lower than spacecraft spin frequency. Therefore, these low time resolution plasma measurements were inadequate to identify small-scale plasma and field irregularities in the solar wind. Nevertheless, these plasma irregularities are present in the solar wind; they even became steep and compressed when they transit through the bow shock before they hit the magnetopause surface.

As a consequence, except for very rare periods of time when the solar wind is strictly constant for more than several minutes, the interaction between the solar-wind plasma and the geomagnetic field cannot be described in the framework of any steady-state theory. Most of the time steady-state reconnection or viscous-like interaction models are invalid representations of the actual magnetosphere because they are based on the simplifying assumption that the external boundary conditions are time independent. The rapidly changing boundary conditions at the surface of the magnetosphere induce time-dependent electric and magnetic fields inside the magnetosphere. The effects of these rapidly changing electromagnetic fields on magnetospheric and ionospheric plasma are commonly ignored. These electromagnetic field effects cannot be understood nor modeled in terms of slowly varying electrostatic or magnetostatic field distributions; they are electrodynamic effects generated by AC electric fields that cannot be understood in the framework of DC electric-field theory.

There are a number of conflicting observations that can only be explained with time-dependent models—the most obvious example of such electromagnetic fields being the irregular short-time-scale fluctuations that are almost always present in magnetospheric electric and magnetic field observations. These ultra-low-frequency field fluctuations are superimposed on the larger scale variations usually studied in the steady approximations.

It cannot even be argued that the magnetosphere could be described in terms of a composite model that would have attributes partly of both the reconnection (open) and the viscous-like (closed) steady-state models. Indeed, in such a composite steady-state magnetosphere model any of the anticipated electrodynamic effects (e.g., the observed rapid field fluctuations) expected from AC boundary conditions are still missing and overlooked.

The rapidly changing boundary conditions at the surface of the magnetopause induce transient magnetic fields and electric field perturbations inside the magnetosphere. This leads to transient and patchy opening (or interconnection) of magnetic field lines as in stationary reconnection pictures. But this leads also to antisunward drag of magnetospheric plasma along the flanks of the plasma boundary layer as in the viscous-like interaction models. This means that time-dependent interaction models are not only imposed on us by the nonuniform nature of the solar-wind flow but will be able to reconcile otherwise conflicting observations.

Therefore, instead of asking if reconnection is more likely than viscous-like interaction or vice versa (assuming we all would agree on a nonequivocal and undisputed definition of these two words), we should first ask the question, "Is the solar-wind interaction with the magnetosphere a steady-state or a dynamical one;" "Or can we always interpret our observations in terms of steady-state magnetospheric models with closed electric equipotential contours and stationary magnetic field line distributions?" The answer is, "No, most of the time we cannot."

Therefore, it is speculative to draw definite conclusions in favor of one or another of the many existing reconnection or viscous-like interaction models without verifying beforehand (for instance, from high-resolution IMF magnetograms) that the solar-wind field and plasma impinging on the magnetosphere have been uniform and stationary for a long enough period of time preceding the observations used to draw these conclusions. Otherwise, these conclusions may most probably be meaningless.

G. Rostoker

It is quite apparent, at this time, that at least two distinctive processes operate at the magnetopause so that energy and/or momentum penetrates into the magnetosphere from the solar wind. One of these processes involves the creation of a class of magnetic field lines with one foot on the earth and one foot on the sun. This

process is modulated by the magnitude and direction of the IMF and is moderately well predicted by a theory whose physical process is termed reconnection. Such a process must take place, if one is to accept Faraday's law (cf. Ref. 31). The evidence for reconnection in a magnetospheric context seems indisputable.

On the other hand, observations of regions adjacent to the magnetospheric equatorial plane and just inside the magnetopause reveal the presence of antiearthward flowing plasma on closed field lines whose existence is not accounted for by reconnection theory.[11] The existence of this low latitude boundary layer (LLBL) has been attributed to some type of viscous interaction at the magnetopause and the estimated electric fields (~ 1–3 mV/m) and total particle fluxes ($\sim 5 \times 10^{26}$/s) are sufficiently large to warrant attention. Physical mechanisms (e.g., anomalous momentum transport as enunciated by Miura[21]) do exist to account for the observed particle populations of the boundary layers (both LLBL and plasma mantle). The real question then lies in the relative contributions of reconnection and "viscous" processes for varying levels of magnetospheric activity.

In this respect, it has been traditional to ignore the possible effects of "viscous" processes since observations (e.g., see Refs. 15 and 15) seem to indicate a total contribution of no more than ~ 20 kV to the cross-polar cap (CPC) potential drop due to nonreconnection processes. This conclusion is based on the observation that during episodes of northward IMF, a CPC potential drop of ~ 20 kV exists independent of the magnitude of the IMF B_z component. The increase in the CPC potential drop with increasing magnitude of southward IMF B_z is attributed solely to reconnection. While such a modulation is well predicted by reconnection theory, what is not clear at this time is whether or not viscous processes are also modulated by the IMF B_z component. If such a modulation occurs, the "viscous" interaction might, in fact, mimic the reconnection process in terms of magnetospheric response. For example, in the Miura[21] concept of anomalous momentum transport involving the Kelvin-Helmholtz instability, it seems inconceivable that magnetic shear across the velocity shear zone would not alter the efficiency of the momentum transport. It is incumbent on the reconnection supporters to demonstrate that viscous processes either do not depend on IMF orientation or that the dependence does not follow the observations. Only if this "rule out" procedure is followed will it be demonstrated unequivocally that the reconnection process is dominant at all times in the solar terrestrial interaction.

V. M. Vasyliunas

In my view, the so-called "viscous" processes do not play any significant role in magnetotail formation. I arrive at this conclusion by an extension of the global stress balance arguments due primarily to George Siscoe (see Ref. 32 and references therein). Calculate F_x, the solar magnetospheric x component of the total force on the magnetotail, by integrating the stress tensor over a closed surface surrounding the magnetotail, a surface that I choose to extend to just outside the magnetopause and its boundary layers on the flanks (whereas in most earlier treatments the surface had been chosen just inside). F_x, which by Siscoe's well-known argument must vanish, is then given by

$$F_x = -(B_T^2/8\pi)A_T(1 - \delta) + S\Delta V_x = 0 \quad (1)$$

where the first term is the net stress of the highly stretched-out magnetic field B_T over the cross-sectional area A_T of the near-earth face of the magnetotail, reduced by plasma pressure effects δ, and the last term is the inertial stress of the total mass flow S of plasma through the magnetotail and the region of its interaction with the solar wind, the plasma undergoing an average change of velocity ΔV_x. The term δ arises primarily from pressure on the flanks of the flaring magnetotail and to a lesser extent from the plasma sheet; I estimate $\delta \approx 0.7$ from conventional models.

Equation 1 states that about ⅓ of the magnetic stress of the magnetotail must be balanced by the loss of momentum of plasma flow interacting with it. An interior stress balance is of course possible with pressure gradients (e.g., Ref. 33), but the magnetotail is ultimately formed as the result of solar-wind flow—otherwise it would not be aligned with it!—and must be maintained by an exterior stress balance (the integration surface to compute F_x was chosen precisely to tap these inertial stresses).

To test whether a proposed process is adequate to maintain the magnetotail, I estimate S, calculate from Eq. 1 the implied ΔV_x, and ask whether its value is reasonable. In all cases, S turns out to be large compared to the net input of solar-wind plasma to the magnetosphere. Hence, most of the plasma in S must be assumed to flow out again, which in turn implies that ΔV_x must be small compared to the average inflow speed V:

$$\Delta V_x/V \ll 1 \quad (2)$$

For a reconnection process, S is given by plasma inflow along open magnetic field lines. In an obvious notation,

$$S = \int dA\ \rho V_n = \int dA\ \rho B_n/(4\pi\rho)^{1/2} \quad (3)$$

and the total open magnetic flux is

$$\int dA\ B_n = B_T A_T \quad (4)$$

hence, from Eqs. 1, 3, and 4,

$$\Delta V_x/V \approx [(1 - \delta)/\sqrt{2}\ ][B_T^2/8\pi\rho V^2]^{1/2} \quad (5)$$

where ρ and V are average values in the inflow region comparable in this case to the solar-wind values ρ_{SW} and V_{SW}. From the observed values,

$$B_T^2/8\pi\rho_{SW}^2 \approx 0.1$$

and thus the reconnection process can maintain the magnetotail provided that

$$\Delta V_x/V \approx 0.07$$

which is consistent with Eq. 2.

For a viscous process, on the other hand, the mass flow S is provided by the magnetopause boundary layers, with $h \simeq 1\ R_e$ the layer thickness and $R_T \simeq 20\ R_e$, the tail radius at the near-earth edge of the magnetotail,

$$S = \int dA\ \rho V_n \approx 2\pi\ R_T\ h\ \rho V \quad (6)$$

and, from Eqs. 1 and 6,

$$\Delta V_x/V = [(1 - \delta)/2]\ [B_T^2/8\pi\rho V^2]\ [R_t/h] \quad (7)$$

In this case $\rho < \rho_{SW}$ and $V \leq V_{SW}/2$, approximately. Then the condition that the magnetopause boundary layer flow be able to maintain the magnetotail is

$$\Delta V_x/V \gtrsim 1.2$$

which is markedly inconsistent with Eq. 2.

It appears that the magnetospheric boundary layers (in contrast to the inflow implied by reconnection) simply do not carry enough momentum to build up the magnetic stress of the magnetotail, unless one supposes that their flow is completely stopped or reversed. But then S would represent a net mass input into the magnetosphere, and the value given by Eq. 6 implies

$$S/\rho_{SW} V_{SW} A_T \approx 0.02$$

which is too large by an order of magnitude. I conclude that, whatever other role "viscous" processes might have in the magnetosphere, if they lead to the boundary layers that are observed, they do not contribute significantly to the formation of the magnetotail.

SUMMARY: G. ROSTOKER

The views of the panel members were, as could be expected, rather diverse. Axford and Vasyliunas used analytical expressions based on the physics of the plasma processes involved to argue that "viscous" processes could not account for the energy flow required to power storms (Axford) or for the balance between the magnetic stress of the magnetotail and the loss of momentum of plasma flow interacting with the tail (Vasyliunas). Baker supported this viewpoint based on observations of the deep tail particle and field behavior obtained using the ISEE 3 satellite. On the other hand, Lundin[12] argued that the very existence of the low-latitude boundary layer provided one with a source of particles that was quite adequate to supply the plasma sheet and potential drops whose implications for high-latitude convection and electric current flow are non-trivial.

Lemaire took a somewhat different viewpoint, arguing that any steady-state assumptions made in assessing either reconnection or "viscous" processes are invalid because the solar wind represents a rapidly time-varying environment to which the magnetosphere is continuously exposed. (This concern was voiced at an earlier Chapman Conference on Solar-Wind/Magnetosphere Coupling by G. Rostoker (see Ref. 34) who noted that the IMF B_z component fluctuates on a characteristic time scale far shorter than the impulse response time of the magnetosphere.) Lemaire's viewpoint is shared by Heikkila[19] and there is evidently a question (which deserves to be addressed further) as to how valid it is to apply steady-state analytical formalisms to a situation where time-varying conditions are the norm.

In this dialog, while the existence of both "viscous" and reconnection effects was acknowledged, the question of the relative roles of the two processes was really not answered. Axford's demand that

$$\left(\frac{\rho v v}{\delta}\right)(v\,A) \sim 10^{19} \text{ ergs/s}$$

(where ($\rho v v/\delta$) is the viscous stress, v the magnetosheath velocity, and A the area of the surface across which the interaction takes place) and his contention that the "viscous" effects could not achieve this only proved that, by themselves, "viscous" effects were inadequate. In addition, the demand that 10^{19} ergs/s be supplied amounts to storm conditions where it appears certain that reconnection processes play an important role. The reader is tempted to ask how easily a demand of 10^{17}–10^{18} ergs/s (typical of average activity levels) could be met with "viscous" processes playing a more significant role proportionately. Vasyliunas' claim that the stress balance in the tail rules out a consequential role for "viscous" effects was a very instructive exercise in which he argued that the total mass flow $\Delta M/\Delta t$ through the magnetotail, calculated from

$$\frac{\Delta M}{\Delta t} = \frac{1}{3}\left(\frac{B_T^2}{8\pi}\right)\frac{A_T}{\Delta v_x}$$

is large compared to the net input of solar-wind plasma to the magnetosphere. He concludes that very little solar-wind flow energy is transferred to the magnetosphere (viz., $\Delta v_x/V_{sw} \ll 1$) and thus reconnection must be the dominant contributor to the global stress balance. What the balance might be between reconnection and "viscous" effects was not really spelled out quantitatively particularly as regard to how large a tail could be supported by "viscous" effects alone. One problem here is the varying estimates of $\Delta M/\Delta t$ stemming from early LLBL studies that seem to range from 10^{26}–10^{27}/s (Eastman, private communication). (For example, Heikkila often quotes particle fluxes of $\sim 10^{27}$/s for the LLBL contribution. For a change in boundary-layer velocity from ~ 200 km/s at the magnetopause to zero at the LLBL/CPS interface and $R_T = 15$ R_e with $B_T = 10$ nT, the stress balance equation of Vasyliunas is satisfied.) It will be incumbent on LLBL researchers to provide the community with a more thorough study of the mass fluxes in this important region of space if we are to be able to define accurately their relative roles in powering magnetospheric activity. Perhaps Baker best summarized the situation in noting that "... if a given physical process can occur, it will." The existence of the LLBL tells us that a process other than reconnection is at work. Establishing how that process is modulated by the interplanetary particle and field environment will go a long way in quantitatively assessing its impact on magnetospheric activity levels.

REFERENCES

[1] W. I. Axford and C. O. Hines, "A Unifying Theory of High-Latitude Geophysical Phenomena and Geomagnetic Storms," *Can. J. Phys.* **39**, 1443 (1961).

[2] J. W. Dungey, "Interplanetary Magnetic Field and the Auroral Zones," *Phys. Rev. Lett.* **6**, 47 (1961).

[3] D. H. Fairfield and L. J. Cahill, "Transition Region Magnetic Field and Polar Magnetic Disturbances," *J. Geophys. Res.* **71**, 155 (1966).

[4] G. Rostoker and C.-G. Fälthammar, "Relationship Between Changes in the Interplanetary Magnetic Field and Variations in the Magnetic Field at the Earth's Surface," *J. Geophys. Res.* **72**, 5853 (1967).

[5] H. E. Petschek, "Magnetic Field Annihilation," in *AAS-NASA Symp. Phys. Solar Flares*, W. N. Hess, ed., NASA SP-50, p. 425 (1964).

[6] T. Yeh and W. I. Axford, "On the Re-Connexion of Magnetic Field Lines in Conducting Fluids," *J. Plasma Phys.* **4**, 207 (1970).

[7] A. Eviator and R. A. Wolf, "Transfer Processes in the Magnetopause," *J. Geophys. Res.* **73**, 5561 (1968).

[8] J. W. Freeman, C. S. Warren, and J. J. MacGuire, "Plasma-Flow Densities at the Magnetopause, January 13 and 14, 1967," *J. Geophys. Res.* **73**, 5719 (1968).

[9] W. I. Axford, "Magnetospheric Convection," *Rev. Geophys.* **7**, 421 (1969).

[10] F. V. Coroniti and C. F. Kennel, "Changes in Magnetospheric Configuration During the Substorm Growth Phase," *J. Geophys. Res.* **77**, 3361 (1972).

[11] T. E. Eastman, E. W. Hones, Jr., S. J. Bame, and J. R. Asbridge, "The Magnetospheric Boundary Layer: Site of Plasma, Momentum and Energy Transfer from the Magnetosheath into the Magnetosphere," *Geophys. Res. Lett.* **3**, 685 (1976).

[12] R. Lundin and E. Dubinen, "Solar Wind Energy Transfer Regions Inside the Dayside Magnetopause. I. Evidence for Magnetosheath Plasma Penetration," *Planet. Space Sci* **32**, 745 (1984).

[13] T. E. Eastman, L. A. Frank, and C. Y. Huang, "The Boundary Layers as the Primary Transport Regions of the Earth's Magnetotail," *J. Geophys. Res.* **90**, 9541 (1985).

[14] P. H. Reiff, "Sunward Convection in Both Polar Caps," *J. Geophys. Res.* **87**, 5976 (1982).

[15] S. W. H. Cowley, "Solar Wind Control of Magnetospheric Convection," in *Proc. Conf. Achievements of the IMS*, ESA SP-217, p. 483 (1984).

[16] J. R. Wygant, R. B. Torbert, and F. S. Mozer, "Comparison of S3-3 Polar Cap Potential Drops with the Interplanetary Magnetic Field and Models of Magnetopause Reconnection," *J. Geophys. Res.* **88**, 5727 (1983).

[17] F. S. Mozer, "Electric Field Evidence on the Viscous Interaction at the Magnetopause," *Geophys. Res. Lett.* **11**, 135 (1984).

[18] W. F. Olson and K. A. Pfitzer, "Magnetosphere Responses to the Gradient Drift Entry of Solar Wind Plasma," *J. Geophys. Res.* **90**, 10823 (1985).

[19] W. J. Heikkila, "Impulsive Plasma Transport Through the Magnetopause," *Geophys. Res. Lett.* **9**, 159 (1982).

[20] J. Lemaire, "Plasmoid Motion Across a Tangential Discontinuity (with Application to the Magnetopause)," *J. Plasma Phys.* **33**, 425 (1985).

[21] A. Miura, "Anomalous Transport by Magnetodydrodynamic Kelvin-Helmholtz Instabilities in the Solar-Wind/Magnetosphere Interaction," *J. Geophys. Res.* **89**, 801 (1984).

[22] S.-I. Akasofu, "Energy Coupling Between the Solar Wind and the Magnetosphere," *Space Sci. Rev.* **28**, 121 (1981).

[23] J. T. Gosling, D. N. Baker, S. J. Bame, E. W. Hones, Jr., D. J. McComas, R. D. Zwickl, J. A. Slavin, E. J. Smith, and B. T. Tsurutani, "Plasma Entry into the Distant Tail Lobes: ISEE 3," *Geophys. Res. Lett.* **11**, 1078 (1984).

[24] J. T. Gosling, D. N. Baker, S. J. Bame, W. C. Feldman, R. D. Zwickl, and E. J. Smith, "North-South and Dawn-Dusk Plasma Asymmetries in the Distant Tail Lobes: ISEE 3," *J. Geophys. Res.* **90**, 6354 (1985).

[25] D. N. Baker, S. J. Bame, W. C. Feldman, J. T. Gosling, R. D. Zwickl, J. A. Slavin, and E. J. Smith, "Strong Electron Bidirectional Anisotropies in the Distant Tail: ISEE-3 Observations of Polar Rain," *J. Geophys. Res.* **91** (in press, 1985).

[26] D. N. Baker, S. J. Bame, W. C. Feldman, J. T. Gosling, R. D. Zwickl, J. A. Slavin, and E. J. Smith, "Bidirectional Electron Anisotropies in the Distant Tail," in *Magnetotail Physics*, Geophysical Monograph American Geophysical Union, Washington, D.C. (submitted, 1986).

[27] W. J. Heikkila, "Magnetospheric Topology of Fields and Currents," in *Magnetospheric Currents*, Geophysical Monograph No. 28, American Geophysical Union, Washington, D.C., p. 208 (1983).

[28] R. D. Zwickl, D. N. Baker, S. J. Bame, W. C. Feldman, J. T. Gosling, E. W. Hones, Jr., D. J. McComas, and E. J. Smith, "Evolution of the Earth's Distant Magnetotail: ISEE 3 Electron Plasma Results," *J. Geophys. Res.* **89**, 11007 (1984).

[29] J. A. Slavin, E. J. Smith, D. G. Sibeck, D. N. Baker, R. D. Zwickl, and S.-I. Akasofu, "An ISEE 3 Study of Average and Substorm Conditions in the Distant Magnetotail," *J. Geophys. Res.* (in press, 1985).

[30] W. J. Heikkila, J. A. Slavin, E. J. Smith, D. N. Baker, and R. D. Zwickl, "Neutral Sheet Crossings by ISEE 3 in the Distant Magnetotail," in *Proc. Comparative Study of Magnetospheric Systems*, La Londe-Les-Maures, France, 9-13 September 1985 (in press, 1986).

[31] S. W. H. Cowley, "Evidence for the Occurrence and Importance of Reconnection Between the Earth's Magnetic Field and the Interplanetary Magnetic Field," in *Magnetic Reconnection in Space and Laboratory Plasmas*, E. W. Hones, Jr., ed., American Geophysical Union, Washington, D.C., pp. 375-378 (1984).

[32] R. L. Carovillano and G. L. Siscoe, "Energy and Momentum Theorems in Magnetospheric Processes," *Rev. Geophys. Space Phys.* **32**, 289-353 (1973).

[33] F. J. Rich, V. M. Vasyliunas, and R. A. Wolf, "On the Balance of Stresses in the Plasma Sheet," *J. Geophys. Res.* **77**, 4670-4676 (1972).

[34] R. A. Wolf, S.-I. Akasofu, S. W. H. Cowley, R. L. McPherron, G. Rostoker, G. L. Siscoe, and B. U. O. Sonnerup, "Coupling Between the Solar Wind and the Earth's Magnetosphere: Summary Comments," in *Proc. AGU-Chapman Conference on Solar-Wind/Magnetosphere Coupling*, Los Angeles, Calif (in press, 1986).

DIALOG ON THE PHENOMENOLOGICAL MODEL OF SUBSTORMS IN THE MAGNETOTAIL

PANEL MEMBERS: K Schindler,[*] T. E. Eastman,[†] W. J. Heikkila,[‡] L. C. Lee,[§] R. P. Lepping,[‖]
L. R. Lyons,[#] R. L. McPherron,[**] and J. A. Slavin[††]

INTRODUCTION BY PANEL CHAIR: K. SCHINDLER

A substantial part of current work in magnetotail physics is devoted to the understanding of magnetospheric substorms. It is the aim of this panel discussion to assess present knowledge and to identify possible points of controversy and important areas for future work in this field.

A natural basis for discussions on substorm phenomena is provided by the "near-earth neutral line model." This model has been described extensively in the literature (e.g., Refs. 1–4).

The following list of "cornerstones" of the near-earth neutral line model is an attempt to characterize its major elements (in this combination, first conceived by Schindler[5]):

1. Mass and energy transfer from the dayside to the tail,
2. Energy storage in the tail (loading or growth phase),
3. Near-earth time-dependent reconnection leading to a newly-formed neutral line (or several neutral lines),
4. Formation and tailward acceleration of plasmoid(s),
5. Energy flow into the inner magnetosphere,
6. Particle acceleration,
7. Recovery.

Figure 1 shows the evolution of the magnetic field topology within the neutral-line model.[2]

These global processes are accompanied by a large class of additional small-scale effects or phenomena localized in space, energy, or time. It will be a long process to identify even the major consequences that the cornerstones (1–7) imply, and it will be interesting to see how this discussion will contribute to that process. Another important topic concerns the limitations of the model, because their assessment should also be a continuous task. As a representative example, I mention the recent results on two time scales of the solar-wind/magnetosphere coupling obtained by Bargatze et al.[6]

Although the neutral-line model seems to be able to organize a large volume of observational data and theoretical results, it must be realized that its appeal is based on consistency tests only. It seems that this is a general limitation of models,[7] which becomes particularly evident for systems of very large numbers of degrees of freedom. An acceptable level of confidence has to be built up through a sufficiently large number of successfully passed consistency tests. Consistency is most effectively tested by establishing decisions between mutually exclusive alternatives (falsification). An example was the exclusion of models predicting earthward bulk flow in the entire tail by confronting them with the dominance of tailward flow at sufficiently large radial distances.

On the basis of these arguments it is important for the panel (and the audience) to try to identify possible alternatives to the neutral-line model. Tentatively, we could define alternative models in terms of definite conflicts with cornerstones 1–7. Furthermore, it is of particular interest to discuss whether there exist important additional aspects within the model that have not yet received appropriate attention.

COMMENTS FROM PANEL MEMBERS
T. E. Eastman

Although we now understand the earth's magnetosphere to be open, both with respect to field connection and (limited) particle transfer, the specific neutral-line model appropriate to both quiet and active geomagnetic conditions has been difficult to identify uniquely due to the lack of global measurements. In particular, the near-earth neutral-line model was introduced before our present means for global auroral imaging from DE and the Viking spacecraft and before our present understanding of the dynamical importance of the magnetospheric boundary layers. These boundary layers have been a major problem for reconnection. Yet, as Alfred North Whitehead noted, "A clash of doctrines is not a disaster—it is an opportunity." We now recognize that the magnetospheric boundary layers are a crucial topic for magnetospheric physics and, further, that they are dynamically important for the entire system. For example, the field-aligned currents that are so closely associated with the magnetospheric boundary layers also link

[*] Ruhr-Universität Bochum, Theoretische Physik IV, 4630 Bochum 1, Federal Republic of Germany.

[†] Space Plasma Physics EES, NASA Headquarters, Washington, DC 20546 (previously at the Department of Physics and Astronomy, University of Iowa, Iowa City, Iowa 52242).

[‡] Danish Space Research Institute, Lundtoftevej 7, DK-2800 Lyngby, Denmark; permanent address: University of Texas at Dallas, Richardson, Texas 75083-0688.

[§] Geophysical Institute, University of Alsaka, Fairbanks, Alaska 99775-0800.

[‖] Goddard Space Flight Center, Greenbelt, Maryland 20771.

[#] Space Sciences Laboratory M2-260, The Aerospace Corporation, P. O. Box 92957, Los Angeles, California 90009.

[**] Institute of Geophysics and Planetary Physics, University of California Los Angeles, Los Angeles, California 90024.

[††] Jet Propulsion Laboratory, 4800 Oak Grove Drive, Pasadena, California 91109.

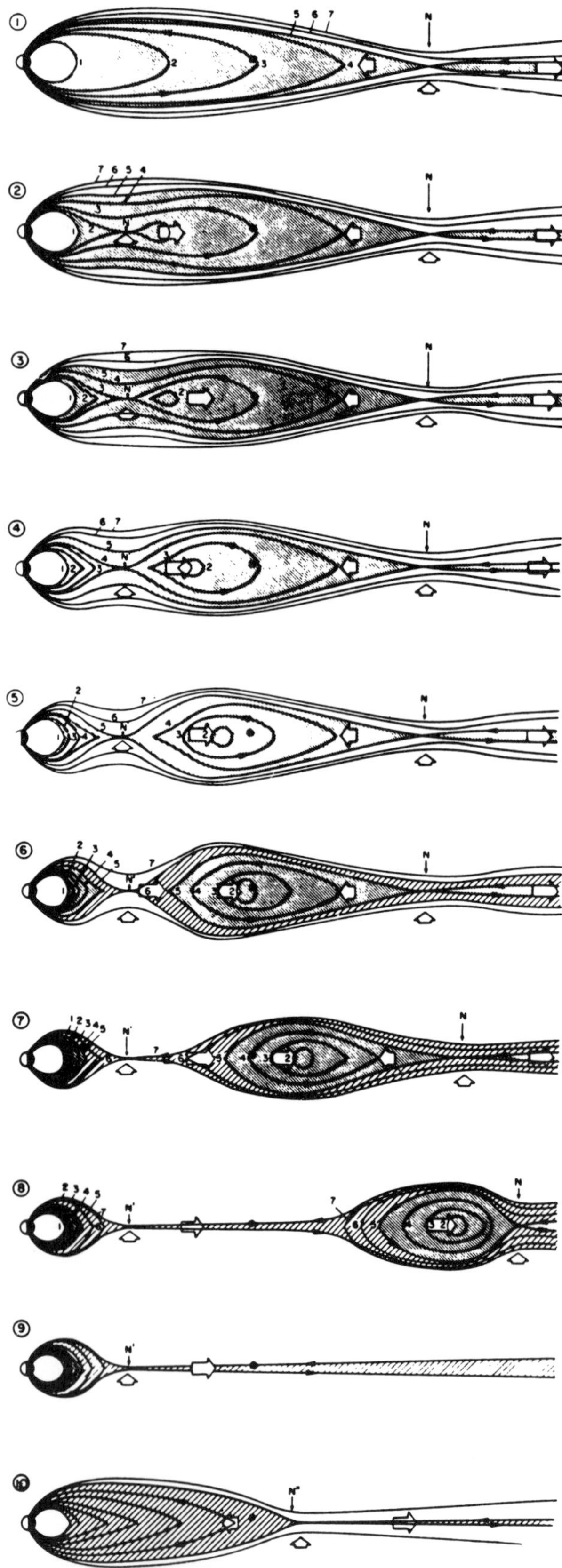

Figure 1—Development of the plasma sheet in the noon-midnight meridian plane during a substorm in the near-earth neutral line mode.

together the global system by providing the basic iono-sphere-magnetosphere connection. The magnetospheric boundary layers are also important for providing a unique testing ground for advancing our understanding of how to better link micro- and macro-plasma physical models.

The near-earth neutral-line model is often represented as leading to a plasma-sheet boundary layer as a secondary effect. There is a growing consensus that this model must incorporate the boundary layers as an integral, dynamically important element, or we need to more vigorously pursue alternative models for magnetospheric substorms. As an example of such alternative models, a boundary layer model for magnetospheric substorms has recently been formulated by Rostoker and Eastman,[8] based in part on the CDAW 7 workshop held in Edmonton in August 1984. This is a particular example of a model emphasizing boundary-layer dynamics and we expect more such models to be developed in coming years. Essential components of a boundary-layer-dynamics model are provided in Fig. 2. On the basis of these essential features of a viable substorm model capable of explaining currently available observations, some participants at the Edmonton workshop also formulated sets of predictions as shown in Fig. 3 for the two main competing substorm models.

In summary, reconnection is important for establishing the basic topology of the magnetosphere and for setting boundary conditions—the real issue is whether it plays any further role of dynamical significance. The approach of boundary-layer-dynamics models is that the basic engines are dynamo processes, current-sheet acceleration, acceleration by finite $E_\parallel$, or pressure gradients, and that reconnection primarily serves to set the stage for magnetospheric dynamics.

ACKNOWLEDGMENT—Primary contributors to the model components and list of predictions given in Figs. 1 and 2 are T. Eastman, L. Frank, C. Huang, A. T. Y. Lui, L. Lyons, D. Mitchell, G. Parks, and G. Rostoker.

W. J. Heikkila

It is certain that a low-latitude magnetospheric boundary layer (LLBL) exists just inside the magnetopause, essentially at all times, with solar-wind plasma flowing tailward, probably on closed magnetic field lines.[9,10]

In the boundary-layer model of the magnetotail, the plasma-sheet particles are supplied by the LLBL in a diffusive process[9,11] and not by the plasma mantle as in reconnection models.[12] Once inside the plasma sheet, after they cross the potential extremum, these particles are convected earthward and energized.[13] With sufficient energization they become nonadiabatic; in other words, the magnetic field loses the ability to control this energized plasma, perhaps through an ion-tearing instability.[5] It can now escape, and it will do so in the direction of least magnetic pressure, i.e., tailward (see Fig. 4). A plasmoid is formed, as fully confirmed by the ISEE 3 observations.[14]

416

1. The magnetotail plasma sheet cannot be adequately treated as a single entity. Various plasma regimes of the magnetotail, including the plasma sheet boundary layer, are present at all times although they may be subject to significant spatial and temporal variations. The plasma sheet always contains both the "central plasma sheet" (CPS) and the "plasma sheet boundary layer" (PSBL).

2. The CPS at topside ionospheric levels and the CPS in the magnetotail are morphologically one and the same region as mapped along magnetic field lines. Similarly, the "boundary plasma sheet" (BPS) at topside ionospheric levels and the PSBL in the magnetotail are topologically one and the same region as mapped along magnetic field lines.

3. The observed plasma flows are primarily field-aligned and they provide the primary means of plasma and momentum transport. The convective flow component is often small and not easily separated in the boundary layer regions where high-speed flow occurs. At low altitude, convection reversal occurs embedded in the BPS; at high altitude, this reversal boundary maps out through the PSBL and eventually through the ecliptic plane where the reversal is between tailward and sunward plasma flow. Such flow reversal probably occurs within the outer plasma sheet where source plasma (with low temperatures and spectra related to the magnetospheric boundary layer) is being turned around and accelerated earthward (via current-sheet acceleration, field-aligned potentials, etc.). Ion beams in the PSBL, created by such acceleration, act as a source for the hot CPS.

4. Field-aligned currents in the plasma sheet boundary layers are major contributors to locally measured magnetic fields within the magnetotail. Inclined field-aligned current sheets, often involving superimposed surface waves, can result in systematic field deflections, including $-B_z$.

5. The magnetotail plasma population has substantial contributions from both the ionosphere and the magnetosheath. The boundary layers are the major transport regions for bringing these populations ultimately into the central plasma sheet. Different sources and propagation effects result in the frequent observation of multi-component velocity distributions in the PSBL.

6. Substorm initiation occurs within the PSBL and not solely within the CPS. The discrete aurora maps to the high-altitude PSBL. This result is confirmed by the fact that the auroral breakup associated with substorm expansive phase occurs poleward of the driven system elctrojets which, at times, extend into the BPS region. The location of substorm initiation is not specified in detail by the BLD model.

7. Substorms result in substantially higher energies for ion beams observed in the PSBL and thus to higher energy plasma sheet distributions which are fed by that source.
   Substorm processes do not invoke dramatic topological changes but primarily changes in the locus and efficacy of particle acceleration in association with changing patterns in field-aligned current distributions.

Figure 2—Essential components of the boundary layer dynamics model.

Boundary Layer Dynamics Model	Near-Earth Neutral Line Model
• Acceleration processes in the magnetotail may occur over relatively large spatial regions and are associated with nonadiabatic motion or field-aligned potentials.	Acceleration is spatially localized and associated with X-type reconnection regions.
• High-speed flows both tailward and earthward, are always associated with boundary layers with $\bar{v}_\parallel > \bar{v}_\perp$ for the ion beams.	High-speed tailward ion flows in plasma sheet tailward of the neutral line.
• Plasma Sheet Boundary Layer (PSBL) always present.	The plasma sheet boundary layer often is not present at dropout near onset [currently not included in the neutral line model].
• Ion beam energies systematically increase after substorm onset on a long time scale ($\tau > 5-10$ min.).	Ion beam energies increase at substorm onset on a short time scale ($\tau <$ a few minutes).
• High-speed tailward flows occur in the boundary layers - both positive and negative B_z occur in conjunction with reduced number density and temperature.	Some cases of high-speed tailward flow in plasmoids should occur with high densities (typical of the central plasma sheet) and negative B_z.
• Earthward high-speed flow can sometimes occur in conjunction with large negative θ_B.	Only tailward high-speed flow occurs in conjunction with large negative θ_B.
• Auroral arcs (quiet and breakup) are associated with the ionospheric BPS and the tail PSBL. Pre-existing quiet arcs map to the PSBL in the magnetotail.	There is no association of specific regions in the tail plasma sheet with pre-existing quiet arcs.
• Driven-system electrojets should remain through onset.	Driven-system electrojets should be significantly disrupted upon plasmoid formation and ejection.
• Substorm initiation occurs along field lines containing pre-existing PSBL.	Substorm initiation occurs along field lines containing pre-existing CPS.

Figure 3—Predictions of two substorm models.

417

The formation of the plasmoid involves many basic processes, in particular the production of localized inductive electric fields as explained by Pellinen and Heikkila.[15] There is one major difference between our model and that proposed by Hones et al.:[14] in their model the escape is made possible by the assumption that the plasmoid is engulfed by open field lines, being in effect pulled by magnetic tension, while in ours the plasmoid has to face the earthward tension of closed field lines. In the latter case, there must be an outward force on the plasmoid; in fluid terms this means a $J \times B$ force due to a current J within the plasmoid. The current will be partly a circulating current within the magnetic island of the plasmoid, but also a net current from dusk to dawn,[16] which is continued by field-aligned currents to the polar ionosphere; there the closure current is an isolated Pedersen current within the poleward moving auroral forms, which has been observed by Iijima and Potemra.[17] This is the usual substorm current diversion of the cross-tail current into the ionosphere.

In our model the plasmoid continues its motion until it reaches the distant magnetopause, all the while maintaining a footprint in the polar cap, traveling poleward. The recovery phase auroral forms fade away at very high latitudes, 75°–80°; we have proposed that this is due to the plasmoid losing its magnetic connection to the ionosphere[18] at the distant X-line and magnetopause.

This boundary-layer model is a combination of an open magnetosphere and its associated magnetic merging, and a viscous interaction process acting through the boundary layer. The latter is more important as a steady-state feature while time-dependent magnetic merging in three dimensions is involved in the creation of the plasmoid. It is doubtful whether two-dimensional steady-state theories are applicable to such a process,[19] in large part by the assumption that curl $E = 0$.[20]

ACKNOWLEDGMENT—This work was supported by NASA grant NAG5-572 and by the Danish Board of Space Activities.

L. C. Lee

I would like to make two brief remarks: one on the reconnection process in the magnetotail and the other on the energy transfer process from the solar wind to the polar ionosphere.

Reconnection in the Magnetotail. The magnetic reconnection in the magentotail during magnetospheric substorms and storms is basically a driven process. A complete sequence of the event, going from the gradual pileup of magnetic flux to the formation and the antiearthward convection of the new X-line and to the ejection of plasmoids from the antiearthward end of the magnetotail, requires the presence of the driving force. Furthermore, based on the MHD simulation study by Lee et al.,[21] it is found that under a constant driving force the magnetic fields in the magnetotail tend to reconnect impulsively and the formation of X-line and plasmoids occurs intermittently and repeatedly every 2–4 h.

Energy Transfer Process from the Solar Wind to the Polar Ionosphere. Figure 5 displays the geomagnetic AE index and the latitudinal angle of the IMF during the storm of October 30, 1978. As shown in the figure, the power ϵ was fairly constant for about 12 h during the storm; the third panel shows the dissipation rate U_T of the magnetosphere, which was fairly constant for the period when ϵ was almost constant. The fourth panel shows the AE index, namely the superposed magnetic records from the AE stations; the bottom panel shows the DST index.

We can see clearly that the AE index showed several impulsive changes during the storms at ~1130, ~1400, ~1830, and ~2130 UT, respectively. The impulsive changes of the AE index during the storm were also associated with auroral substorm activity. The recurrence time of substorms is about 2–4 h. These features are superposed on an enhanced level of the AE index. We suggest that such impulsive changes of the AE index and the associated auroral substorms are manifestations of the repeated occurrence of plasmoids observed in the simulation study.

The enhanced level of the AE index is suggested to be due to the direct energy transfer from the solar wind through the near-earth magnetopause to the polar atmosphere. In fact, the polar cap potential drop ϕ_{pc} is controlled directly by the dayside reconnection process, which in turn is controlled by the solar-wind velocity and IMF B_z. An increase of ϕ_{pc} will lead to an enhancement of electric fields in the auroral oval region through the electric-field leakage in the conducting ionosphere, which in turn leads to the observed enhanced electrojets.

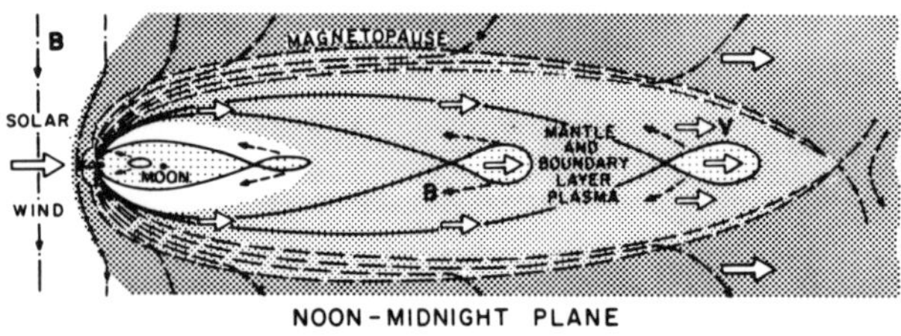

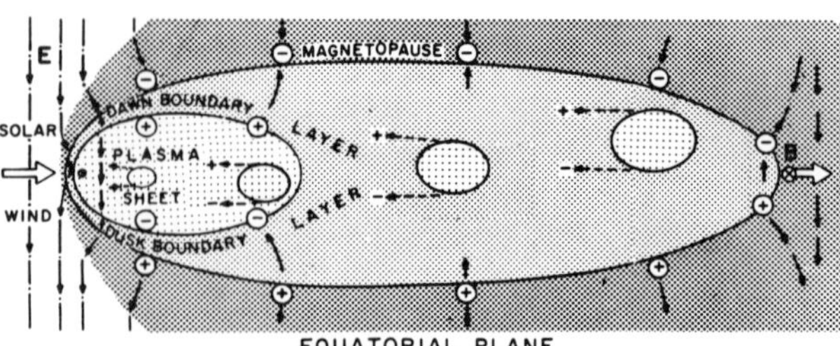

Figure 4—A cut of the magnetotail in the noon-midnight as well as the equatorial planes. In this model, most of the distant magnetotail is composed of the combined dawn and dusk boundary layers. After it is formed in the near-earth magnetotail within the plasma sheet, a plasmoid must burrow through a long region of closed field lines; during this time it must have an outward force that will be provided by a current loop back to the ionosphere. The result is recovery phase auroral forms poleward of the auroral oval (from Ref. 18).

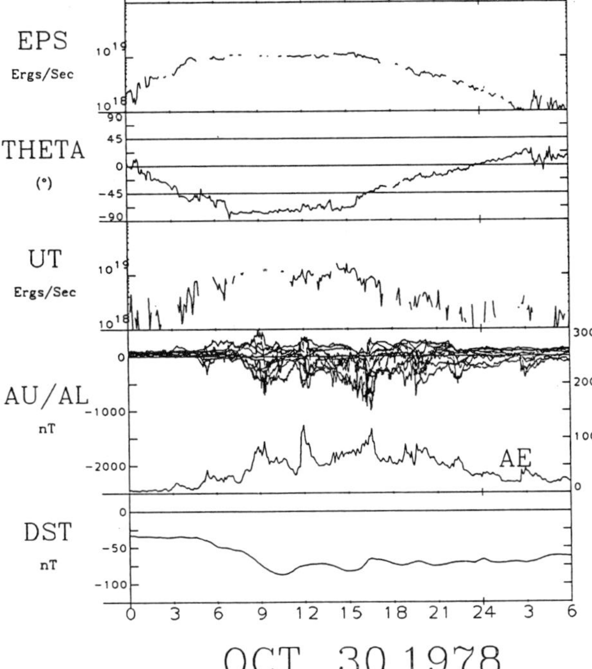

Figure 5—From the top, the solar wind-magnetosphere dynamo power, the IMF polar angle θ in the y-z plane, the total output rate U_T of the magnetosphere, and the two geomagnetic indices AU/AL and DST for the period on October 30, 1978.

Thus, manifestations of distant magnetotail phenomena associated with substorms in the polar upper atmosphere could often be overwhelmed by such direct effects; this may particularly be the case when the upper atmosphere phenomena are monitored by the auroral electrojet index.

R. P. Lepping

I wish to call attention to three important issues concerning aspects of the phenomenological model of substorms in the magnetotail starting with Fig. 6, a sketch by Hones.[22] The sketch shows a commonly held view of a dynamic solar-wind/magnetosphere plasma system where solid lines are magnetic field (**B**) lines and heavy white arrows indicate plasma bulk flow. Such models demand that the component B_n of **B** normal to the magnetopause (MP) (dashed line) be finite, as shown; the picture is tacitly presented as satisfactorily two dimensional; and magnetic merging is believed to occur in or near the plasma sheet, region C in the figure. My three points are: (a) Why is the tail MP usually observed to be either a tangential discontinuity (where $B_n = 0$) or a rotational discontinuity with a B_n so small that it cannot be distinguished from a tangential discontinuity? This seems to be true at Mercury, Earth, Jupiter, and Saturn. If true, the reconnection picture is questionable. (b) The two-dimensional model is inadequate. Near the tail merging region, B_y (normal to Fig. 6) can be quite large; it is not unusual for $|B_y| \sim |B_x|$. (c) What plasma regime in the tail participates in the merging pro-

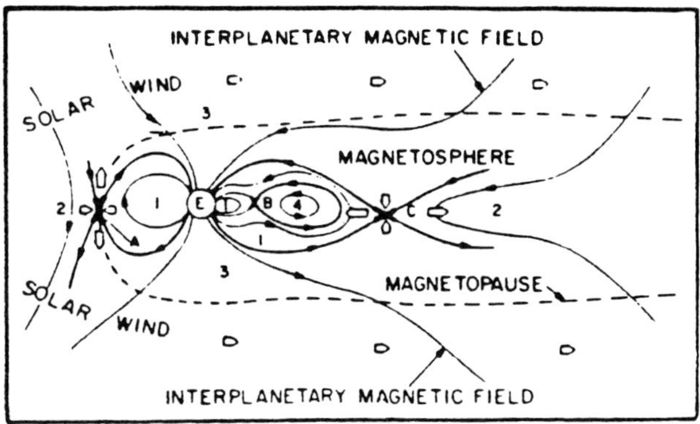

Figure 6—(see text, Lepping)

cess, plasma-sheet plasma or plasma-sheet boundary-layer plasma? I now address points (a) and (b). Point (c) is a challenge left to others.

For many good reasons the occurrence of magnetic merging at the MP (and hence finite B_n there) is becoming an accepted fact. For example, using ISEE 3 electron plasma data Gosling et al.[23] present convincing evidence that plasma does enter through the "distant" tail MP at open field regions ($B_n \neq 0$) and then $E \times B$ drifts toward the tail axis. Hence, finite B_n fields must exist at the earth's tail-MP, but they are apparently small compared to $|B|$, as magnetic field studies have shown. Therefore, only through such an analysis as that of Gosling et al. and/or the use of local detailed plasma distributions at the MP can the nature of the boundary be determined unambiguously.

Concerning point (b) I now discuss a magnetotail event studied by Bieber et al.[24] using IMP 8 data. Detailed (0.32 s) **B** data for the event are displayed in Fig. 7. The event occurred on April 18, 1974 during a substorm; onset is at about 1041 UT shown by the downward arrow in the figure. The spacecraft was very close to the nominal tail axis at about 34 R_e tailward from earth. Electron heating occurred at 1047–1049 UT near southward **B** ($B_z < 0$). For about 4 min starting at substorm onset, the B_y component is dominant; negative B_y is shown shaded. At 1044 UT $|B_y| \approx 2|B_x|$ and $|B_z| \approx 0$, which means that the field was $\approx 65°$ away from its usual direction, and its strength was over 20 nT at the time. How common are such strong B_y fields during substorms, and what are their implications? Their frequency of occurrence is not known, but there are enough examples, including some from ISEE 3 **B** data,[25] to convince us that the two-dimensional picture is inadequate, at least on the several-minute time scale. It deserves further study.

L. R. Lyons

The plasma-sheet boundary layer (PSBL) represents a transition between the central plasma sheet and the magnetotail lobes.[26-28] It contains high-speed flows with speeds of hundreds of kilometers per second, and the

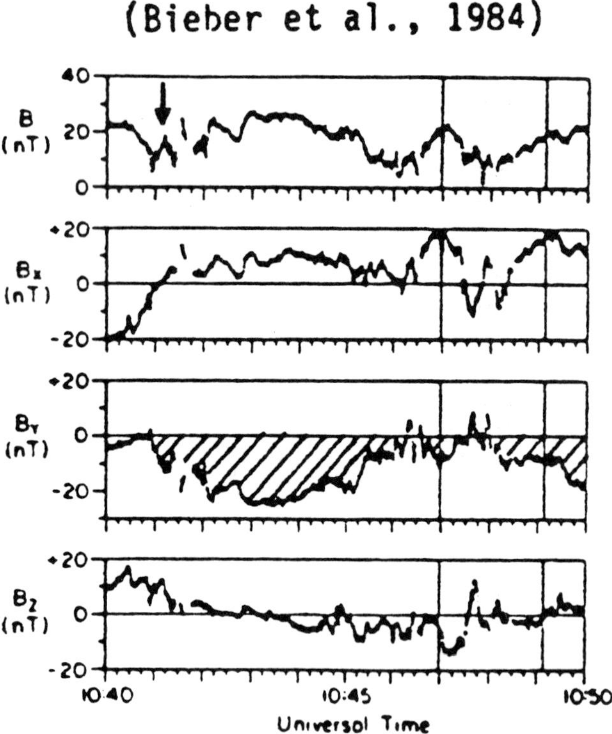

Figure 7—(see text, Lepping)

flows are nearly magnetic field aligned. The PSBL is a dynamic region with earthward and tailward ion beams, field-aligned currents, and enhanced plasma wave activity. The tailward beams can be explained as a direct result of the mirroring of earthward beams. The field-aligned currents have densities of 1–100×10^{-9} A/m^2,[29-33] which are of the magnitude expected for auroral currents extending into the tail. Outflowing ionospheric ions are also observed in the boundary layer. The currents and outflowing ions suggest that discrete auroras map to the plasma-sheet boundary layer.

Particle acceleration associated with the PSBL can be explained by well-known processes. Nothing else is necessary. Field-aligned electric fields and processes that form conics can account for ionospheric ions in the PSBL. Interactions with the tail current sheet can account for the earthward streaming ions.[34] Flows peaking at high energies are time dependent and can be simply explained by temporal variations in the current sheet source.[35]

The ion population within the PSBL contains particles that fill the loss cone. This allows the PSBL to be mapped to the auroral ionosphere, and the precipitating ions expected from the PSBL have been observed by low-altitude satellites. In addition auroral arcs, identified as regions of structured electron precipitation with precipitating energy fluxes $\gtrsim 10$ ergs/cm^2-s, have been observed in association with the PSBL ion precipitation.[36] The observed arcs occur at boundaries of the PSBL ion population, and the PSBL extends equatorward of all arcs.

Based on the high- and low-altitude satellite observations, I believe it is fair to conclude that discrete auroral arcs generally are on the same field lines as the plasma-sheet boundary layer, and in fact are generally on the outer part of the boundary layer.

Auroral observers often state that preexisting quiet auroral arcs break up at the initiation of a substorm. Certainly, a discrete aurora does not appear well equatorward of all preexisting discrete auroral arcs when a substorm begins. Therefore, substorms must be initiated on field lines containing the preexisting PSBL and thus on field lines that contain the processes that form the boundary layer. This implies that formation of the PSBL and initiation of substorms are spatially related processes, and it is reasonable to conclude that substorms are initiated along field lines containing the PSBL.

A commonly discussed model of substorm initiation involves the formation of a near-earth neutral line within the central plasma sheet. At present, the association of substorm initiation with the PSBL is not incorporated in the model. In order for this model to be valid, it must be modified to accommodate the PSBL association. Modification of the model must include either neutral line formation on those field lines that contain the processes responsible for the preexisting boundary layer, or neutral line formation on field lines equatorward of all preexisting discrete auroral arcs. The latter option requires that the neutral line does not directly cause an auroral breakup at the time of its formation. Presuming the boundary layer is formed by particle interactions with the tail current sheet, the first option implies that the region of non-guiding-center particle motion associated with the current sheet extends equatorward to the region where the neutral line forms.

Based on the evidence that the PSBL is formed by ions ejected from the tail current sheet, I conclude that substorm initiation is associated with the current sheet. I believe this association may be a fundamental aspect of substorms that should be included in any model of substorm initiation.

R. L. McPherron

My purpose is to suggest that the dominant response mode of the magnetosphere during intervals of magnetic disturbance depends on the waveform of B_z in the IMF. In this case either the driven or the unloading model can be a correct description of substorms, depending on conditions in the solar wind.

In the unloading model of the substorm it is assumed that the magnetosphere is open to the solar wind.[37,38] During southward IMF the solar wind transports flux from the dayside to the nightside and back in a closed convection cycle in which the field lines pass through two neutral lines, one on the dayside and one in the distant plasma sheet. During northward IMF there is no transport from day to night; instead, the solar wind circulates open magnetic field lines of the tail lobe in closed circuits entirely confined to the lobes.[39] In this case there is a neutral line present in the distant plasma sheet, but it is inactive and does not reconnect the lobe field.

420

The expansion phase of an isolated substorm begins when a new neutral line forms in a localized region near the earth. Initially, only closed field lines of the plasma sheet reconnect. Eventually, reconnection severs the last closed field line connected to the distant neutral line and a plasmoid is ejected from the tail. Subsequently the near-earth neutral line reconnects the lobe field until, at the beginning of the recovery phase, it moves tailward and replaces the preexisting distant neutral line.

The driven model does not explicitly describe the interaction with the solar wind beyond noting that it is a dynamo.[40] Presumably this implies that the earth's magnetic field is connected to the solar wind. Since the model states that energy storage and release are not important, it must be true that flux transport from front to back is balanced; otherwise, energy would accumulate. This implies the presence of two neutral lines, just as in the unloading model, except that the distant line is continuously active during southward IMF, returning flux as required to maintain equilibrium.

To reconcile the two substorm models, and to explain the bimodal nature of magnetic activity, we postulate the following. When the IMF turns southward, following a prolonged interval of northward IMF, the distant neutral line is not able to return the flux necessary to maintain equilibrium immediately. Consequently, energy accumulates and the configuration of the magnetosphere is progressively distorted until a new neutral line forms close to the earth. This neutral line returns flux as required. If the IMF remains southward, this neutral line remains near the earth and balances flux transfer. A convection bay ensues, and the magnetic activity can be characterized as driven. On the other hand, if the IMF turns northward, the neutral line moves outward to its more distant location. A subsequent southward turning of the IMF requires the formation of another near-earth neutral line and plasmoid. Such situations would produce a sequence of unloading events.

ACKNOWLEDGMENT—This work was supported at UCLA by grants from the National Science Foundation, ATM 83-18200; the Office of Naval Research, N00014-84-C-0158; the National Aeronautics and Space Administration, NGL-05-007-004; and the Institute of Geophysics and Planetary Physics at Los Alamos National Laboratory.

K. Schindler

My first comment deals with results derived from analytical theory and from computer modeling on the formation of near-earth neutral lines and plasmoids. It follows from both analytical and numerical studies[41-44] that a tangential electric field at the high- latitude magnetopause causes a time-dependent evolution of the tail with properties characteristic of the loading phase of the neutral line model. During this phase, energy is transferred from the magnetosheath to the magnetotail (presumably by both Poynting and kinetic energy fluxes). For the internal evolution of the tail in the loading phase nonideal processes are unimportant. As the tail is compressed, the plasma sheet becomes thinner and the tail current density increases. This development is associated with the growth of nonideal effects that can be caused by nonadiabatic ions or collective fluctuations. When the nonideal behavior (e.g., measured by the strength of the nonideal terms in Ohm's law) has become sufficiently pronounced, it causes the formation of new neutral lines. Figure 8 illustrates this effect for four different approaches: (a) a global resistive MHD computation, by Lyon et al;[45] (b) a local resistive computation, by Birn and Hones;[46] (c) laminar Vlasov theory, by Goldstein and Schindler;[47] and (d) a model based on a statistical mechanics description that excludes resistive forces along the direction of the main current, but allowing for viscous dissipation, by Kiessling et al.[48] In all cases neutral lines form due to the presence of nonideal effects. Plasmoids are clearly visible in cases (b), (c), and (d), but also in case (a) a plasmoid seems to have formed. (The authors refer to the inner field structure near the tailward boundary as the "earthward half of a large magnetic island.")

The only known way to avoid neutral line formation in an otherwise realistic tail configuration is to impose the validity of ideal Ohm's law.

Nonlinear analytical and numerical results[49,50] have confirmed that the tail configuration possesses a sufficient amount of free energy to feed a large-scale dynamic process such as postulated by the neutral line model. Explosive nonlinear growth based on ion inertia effects was found by Coroniti.[3] I emphasize that the shape of the dynamic mode is prescribed a priori in none of the approaches of Fig. 8. The systems themselves select the most unstable forms of dynamic evolution.

An important task for future work is to clarify what nonideal effects are responsible for the neutral line formation in the earth's magnetotail. In addition to the processes mentioned above, stochastic particle orbits and stochastic magnetic fields might also be relevant.

In my second comment I argue that large plasma bulk-flow velocities as observed at the boundary of the plasma sheet is a generic property of magnetic reconnection. Large parallel flows occur quite generally on field lines that connect to a region of weak magnetic field in the presence of an electric field perpendicular to $\boldsymbol{B}$. This effect is particularly pronounced in reconnection geometries with strongly localized nonideal and inertia effects. In a typical example, evaluated for the earth's magnetotail, bulk-flow speeds of the order of several hundreds of kilometers per second appear near the separatrix, i.e., near the plasma sheet boundary.[51] The effect has been shown to be not very sensitive to the MHD approximation. Depending upon the kinetic processes involved, the distribution function may well consist of a number of separate "beams" or "jets."

J. A. Slavin

The ISEE 3 observations taken in the distant tail constitute a new and important source of information on the configuration of the magnetosphere and the substorm process. In particular, the ISEE 3 observations have generally supported the "open" model of the magnetosphere.[52-56] However, an important feature of the

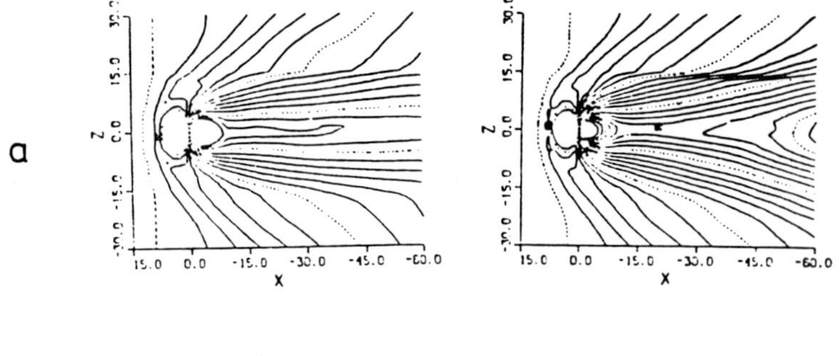

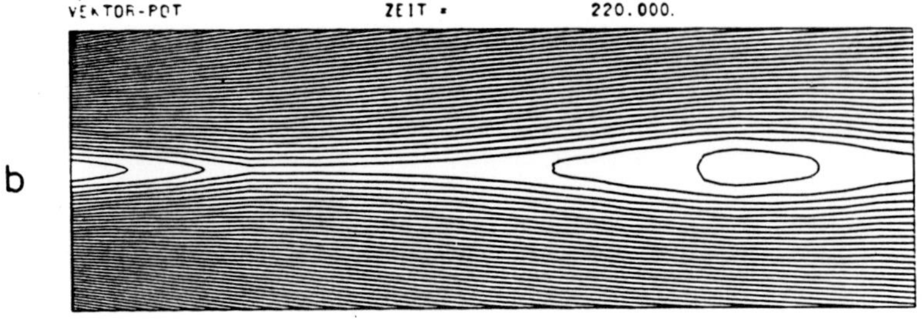

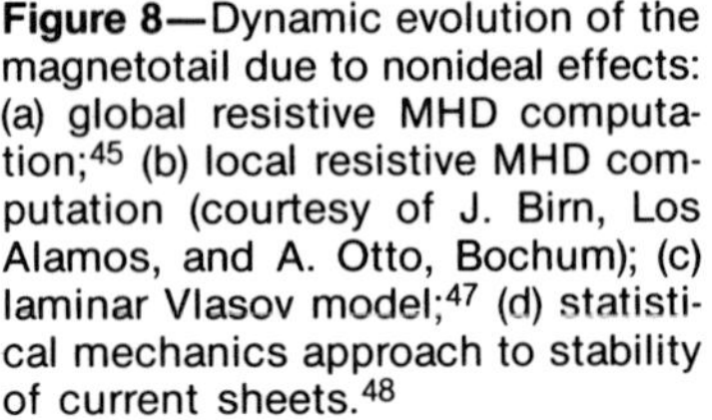

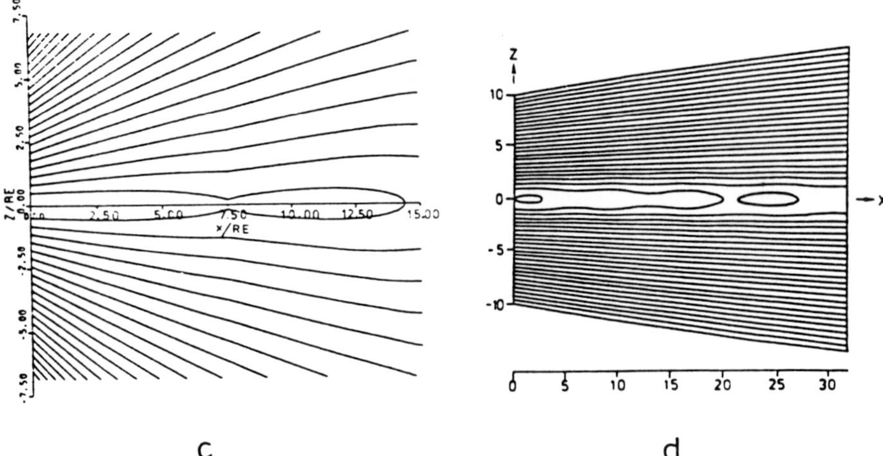

Figure 8—Dynamic evolution of the magnetotail due to nonideal effects: (a) global resistive MHD computation;[45] (b) local resistive MHD computation (courtesy of J. Birn, Los Alamos, and A. Otto, Bochum); (c) laminar Vlasov model;[47] (d) statistical mechanics approach to stability of current sheets.[48]

open model is the existence of a single, quasistationary neutral line in the distant tail during nonsubstorm conditions. Earthward of the neutral line the plasma-sheet flow is predicted to be sunward and threaded by closed (northward) field lines. Tailward of the neutral line the flow should be antisunward and contain disconnected (i.e., southward) field lines. ISEE 3 has determined that the flow in the central plasma sheet is indeed nearly always antisunward at $|X| > 100$ R$_e$.[57-59] However, when substorm activity is low, the plasma sheet contains northward magnetic fields out to the ISEE 3 apogee, ~240 R$_e$.[60] It is only when intervals containing substorm activity are considered that the plasma sheet field has a net southward component, and then only in the central portion of the tail.[56] The source of the stress responsible for the antisunward flow carrying closed field lines tailward during quiet times has not been determined. However, it is most probably the flux of tailward moving mantle and low-latitude boundary layer ions that are being added continuously to the plasma sheet.[61,62] The potential drop across the magnetosphere due to the processes appears to be several tens of kilovolts in agreement with high-latitude ionospheric convection studies.[63] This result is important for the reconnection models of substorms. If the distant neutral line is continuously "blown" downstream, then it may not play a significant role in the substorm process as is assumed by many phenomenological models.[37]

The other topic to be discussed concerns the substorm models that predict the formation of large magnetic islands termed plasmoids.[1,5] Substorm reconnection begins at $|X| = 15$–30 R$_e$ and initially only involves closed plasma-sheet field lines. This new X-type neutral line and the preexisting one in the distant tail create closed magnetic loops that coalesce to form the plasmoid. The Maxwell stress of reconnected lobe field lines and plasma pressure gradients combine to accelerate the plasmoid down the tail. The reconstruction of this picture from the near-earth observations[1] is nontrivial and has been the subject of much controversy.[64] At $|X| >$ 200 R_e, however, the plasmoids have had time to fully evolve and be accelerated up to speeds of 500–1000 km/s making it much easier to separate spatial and temporal behavior. The ISEE 3 observations argue quite convincingly that 50–100 R_e long plasmoids moving down the tail are a relatively common occurrence.[14,65,66] Furthermore, if plasmoids are large three-dimensional structures, as theory predicts (e.g., Ref. 21), then they should produce a region of compression in the lobes that accompanies the plasmoid as it moves down the tail. These traveling compression regions, or TCRs, have also been observed by ISEE 3.[67] Finally, it was suggested during the panel discussion that O^+ from the ionosphere should have been detected in the plasmoids if they did actually originate in the near tail. Unfortunately, the ISEE 3 solar-wind composition experiment was designed to measure the multiple ionized species flowing out of the sun's corona, and it cannot detect O^+ moving at the 500–1000 km/s speeds that are typical of plasmoids in the distant magnetotail magnetotail.[68]

In summary, the ISEE 3 observations in the distant magnetotail are in good agreement with many of the predictions of the reconnection model of substorms, including the formation and rapid antisunward motion of plasmoids. Additional studies of the distant tail observations should be directed toward understanding the predominance of tailward flow in the distant plasma sheet during "quiet" times and the relationship between plasmoid numbers and properties (i.e., length, speed, energy content,...) and the characteristics of individual substorms.

SUMMARY REMARKS: K. SCHINDLER

As expected (see the introduction), the near-earth neutral line model largely served as a frame of reference for the discussion. Besides a number of critical statements (described further below), the larger part of the contributors explicitly or implicitly assumed general validity of the model, commenting on particular advantages, on the relative weight of the various processes, and on necessary additions or modifications. A group of contributions (by D. N. Baker, E. W. Hones, Jr., R. L. McPherron, J. A. Slavin, and others) emphasized that most of the data related to magnetospheric activity, as collected in the solar wind, on earth, in the ionosphere, and in various regions of the magnetosphere, are consistently organized by the neutral-line model. In particular, several speakers addressed traveling plasmoids in the distant tail, which constitute one of the cornerstones of the model. In several theoretical and simulation studies, neutral-line formation is a dominant feature (Schindler).

The discussion, however, did not exhaust itself in an appraisal of the neutral-line model. A good deal of the statements emphasized its limitations, necessary extensions, and desirable future activities. Subjects of that category were the bimodal nature of the interaction between the solar wind and the magnetosphere (L. C. Lee, R. L. McPherron), the role of the B_y-component in tail dynamics (R. P. Lepping), the necessity to include parallel flow at the boundary of the plasma sheet (including suggestions concerning its origin) (L. R. Lyons and K. Schindler), and the need for further studies of the plasmoid acceleration process (W. J. Heikkila). It was also emphasized by several speakers (e.g., L. R. Lyons) that the association between the site of reconnection in the magnetotail with the location of discrete auroral arcs must be clarified in future investigation.

It seems to me that the common denominator of the discussion as sketched so far was the understanding that, in spite of clearly visible gaps and deficiency, the neutral-line model with its empirical and theoretical aspects is able to outline a rough but coherent picture describing substorm physics in terms of cause and effect.

However, criticism regarding basic cornerstones of the neutral-line model was also raised in some of the discussion, mostly in connection with boundary-layer aspects (T. E. Eastman, G. K. Parks, G. Rostoker). A thorough comparison between the boundary-layer ideas and the neutral-line model, although suggested (T. E. Eastman), did not become a main element of the discussion. My own interpretation of this fact is that such a comparison is extremely difficult at present because of large differences in emphasis. The boundary-layer picture emphasized aspects of geometry, magnetic connection between different regions, and the structure of plasma distribution functions including associated transport phenomena, while the neutral line model is formulated in terms of a sequence of processes (see the introduction) implying cause-and-effect relationships.

A definite conflict with the neutral-line model (at least with the version suggested in the introduction) is implied in a statement that emphasized boundary layer aspects but also explicitly disputed the presence of newly formed near-earth neutral lines and of plasmoids (L. A. Frank). In that picture a single neutral line rapidly changes its radial position. Further examination of neutral-line kinematics seems to be an interesting task for future work.

It should be added that it was also commented that the difference between a moving neutral line and a newly-formed neutral line may not be really fundamental.

Vivid discussions, many more hand-raisings than could be accommodated, and a concentrated audience was symptomatic of a group dealing with a particularly exciting topic in current research.

REFERENCES

[1] E. W. Hones, Jr., "Substorm Processes in the Magnetotail: Comments on 'On Hot Tenous Plasmas, Fireballs and Boundary Layers in the Earth's Magnetotail' by L. A. Frank, K. L. Ackerson, and R. P. Lepping," *J. Geophys. Res.* **82**, 5633 (1977).

[2] E. W. Hones, Jr., "Plasma Flow in the Magnetotail and Its Implications for Substorm Theories," in *Dynamics of the Magnetosphere*, S.-I. Akasofu, ed., D. Reidel Pub. Co., p. 545 (1979).

[3] F. V. Coroniti, "Explosive Tail Reconnection: The Growth and Expansion Phases of Magnetospheric Substorms," *J. Geophys. Res.* **90**, 1946 (1985).

[4] K. Schindler and J. Birn, "Magnetospheric Physics," *Phys. Rep.* **47**, 2 (1978).

[5] K. Schindler, "A Theory of the Substorm Mechanism," *J. Geophys. Res.* **79**, 2803 (1974).

[6] L. F. Bargatze, D. N. Baker, R. L. McPherron, and E. W. Hones, Jr., "Magnetospheric Impulse Response for Many Levels of Geomagnetic Activity," *J. Geophys. Res.* **90**, 6387 (1985).

[7] K. Popper, "The Logic of Scientific Discovery," Hutchinson, London (1959).

[8] G. Rostoker and T. E. Eastman, "A Boundary Layer Model of Magnetospheric Storms," *J. Geophys. Res.* (submitted, 1986).

[9] T. E. Eastman, B. Popielawska, and L. A. Frank, "Three-Dimensional Plasma Observations Near the Outer Magnetospheric Boundary," *J. Geophys. Res.* **90**, 9519 (1985).

[10] D. J. Williams, T. E. Eastman, L. A. Frank, and D. G. Mitchell, "Energetic Particle Observations in the Low-Latitude Boundary Layer," *J. Geophys. Res.* **90**, 5097 (1985).

[11] W. J. Heikkila, "Exit of Boundary Layer Plasma from the Distant Magnetotail," *Geophys. Res. Lett.* **10**, 218 (1983).

[12] S. W. H. Cowley, "Plasma Populations in a Simple Open Model Magnetosphere," *Space Sci. Rev.* **26**, 217 (1980).

[13] W. I. Axford, "Magnetospheric Convection," *Rev. Geophys.* **7**, 421 (1969).

[14] E. W. Hones, Jr., D. N. Baker, S. J. Bame, W. C. Feldman, J. T. Gosling, D. J. McComas, R. D. Zwickl, J. A. Slavin, E. J. Smith, and B. T. Tsurutani, "Structure of the Magnetotail at 220 R_e and Its Response to Geomagnetic Activity," *Geophys. Res. Lett.* **11**, 5 (1984).

[15] R. J. Pellinen and W. J. Heikkila, "Inductive Electric Fields in the Magnetotail and Their Relation to Auroral and Substorm Phenomena," *Space Sci. Rev.* **37**, 1 (1984).

[16] J. R. Walker and T. Sato, "Externally Driven Magnetic Reconnection," in *Magnetic Reconnection in Space and Laboratory Physics*, E. W. Hones, Jr., ed., Geophysical Monograph No. 30, American Geophysical Union, Washington, D.C., p. 272 (1984).

[17] T. Iijima and T. A. Potemra, "Large-Scale Characteristics of Field-Aligned Currents Associated with Substorms," *J. Geophys. Res.* **83**, 599 (1978).

[18] W. J. Heikkila and J.-P. Treilhou, *Proceedings of Results of the Arcad 3 Project*, p. 281, Cepadues-Editions, Toulouse, France (1985).

[19] W. J. Heikkila, "Criticism of Reconnection Models of the Magnetosphere," *Planet. Space Sci.* **26**, 121 (1978).

[20] W. J. Heikkila, "The Electromagnetic Field for an Open Magnetosphere," in *Magnetic Reconnection in Space and Laboratory Plasmas*, E. W. Hones, Jr., ed., Geophysical Monograph No. 30, American Geophysical Union, Washington, D.C., p. 39 (1984).

[21] L. C. Lee, Z. F. Fu, and S.-I. Akasofu, "A Simulation Study of Forced Reconnection Processes and Magnetospheric Storms and Substorms," *J. Geophys. Res.* **90**, 10896 (1985).

[22] E. W. Hones, Jr., *Eos*, **65**, 124 (1984).

[23] J. T. Gosling and R. P. Lepping, "North-South and Dawn-Dusk Plasma Asymmetries in the Distant Tail Lobes: ISEE-3," *J. Geophys. Res.* **90**, 6354 (1985).

[24] J. W. Bieber, E. C. Stone, E. W. Hones, Jr., D. N. Baker, S. J. Bame, and R. P. Lepping, *J. Geophys. Res.* **89**, 6705 (1984).

[25] B. Tsurutani, personal communication, 1985.

[26] A. T. Y. Lui, E. W. Hones, Jr., F. Yasuhara, S.-I. Akasofu, and S. J. Bame, "Magnetotail Plasma Flow During Plasma Sheet Expansions: Vela 5 and IMP 6 Observations," *J. Geophys. Res.* **82**, 1235 (1977).

[27] R. J. DeCoster and L. A. Frank, "Observations Pertaining to the Dynamics of the Plasma Sheet," *J. Geophys. Res.* **84**, 5099 (1979).

[28] T. E. Eastman, L. A. Frank, W. K. Peterson, and W. Lennartsson, "The Plasma Sheet Boundary Layer," *J. Geophys. Res.* **89**, 1553 (1984).

[29] M. P. Aubry, M. G. Kivelson, R. L. McPherron, C. T. Russell, and D. S. Colburn, "Outer Magnetosphere Near Midnight at Quiet and Disturbed Times," *J. Geophys. Res.* **77**, 5487 (1972).

[30] D. H. Fairfield, "Magnetic Field Signatures of Substorms on High-Latitude Field Lines in the Nighttime Magnetosphere," *J. Geophys. Res.* **78**, 1553 (1973).

[31] M. Sugiura, "Identification of the Polar Cap Boundary and the Auroral Belt in the High Latitude Magnetosphere: A Model for Field-Aligned Currents," *J. Geophys. Res.* **80**, 2057 (1975).

[32] L. A. Frank, R. L. McPherron, R. J. DeCoster, B. G. Burek, K. L. Ackerson, and C. T. Russell, "Field-Aligned Currents in the Earth's Magnetotail," *J. Geophys. Res.* **86**, 687 (1981).

[33] C. Y. Huang, L. A. Frank, and T. E. Eastman, "High-Altitude Observations of an Intense Inverted-V Event," *J. Geophys. Res.* **89**, 7423 (1984).

[34] L. R. Lyons and T. W. Speiser, "Evidence for Current-Sheet Acceleration in the Geomagnetic Tail," *J. Geophys. Res.* **87**, 2276 (1982).

[35] D. J. Williams, "Energetic Ion Beams at the Edge of the Plasma Sheet, ISEE 1 Observations Plus a Simple Explanatory Model," *J. Geophys. Res.* **86**, 5507 (1981).

[36] L. R. Lyons and D. S. Evans, "An Association Between Discrete Aurora and Energetic Particle Boundaries," *J. Geophys. Res.* **89**, 2395 (1984).

[37] R. L. McPherron, "Magnetospheric Substorms," *Rev. Geophys. Space Phys.* **17**, 657 (1979).

[38] E. W. Hones, Jr., "Transient Phenomena in the Magnetotail and Their Relation to Substorms," *Space Sci. Rev.* **23**, 393 (1979).

[39] K. Maezawa, "Magnetospheric Convection Induced by the Positive and Negative Components of the Interplanetary Magnetic Field: Quantitative Analysis Using Polar Cap Magnetic Records," *J. Geophys. Res.* **81**, 2289 (1976).

[40] S.-I. Akasofu, "Energy Coupling Between the Solar Wind and the Magnetosphere," *Space Sci. Rev.* **28**, 121 (1981).

[41] G. M. Erickson and R. A. Wolf, "Is Steady Convection Possible in the Earth's Magnetotail?" *Geophys. Res. Lett.* **7**, 897 (1980).

[42] K. Schindler and J. Birn, "Self-Consistent Theory of Time-Dependent Convection in the Earth's Magnetotail," *J. Geophys. Res.* **87**, 2263 (1982).

[43] J. Birn and K. Schindler, "Self-Consistent Theory of Three-Dimensional Convection in the Geomagnetic Tail," *J. Geophys. Res.* **88**, 6969 (1983).

[44] J. Birn and K. Schindler, "Computer Modeling of Magnetotail Convection," *J. Geophys. Res.* **90**, 3441 (1985).

[45] Lyon et al. (1981).

[46] J. Birn and E. W. Hones, Jr., "Three-Dimensional Computer Modeling of Dynamic Reconnection in the Geomagnetic Tail," *J. Geophys. Res.* **85**, 6802 (1981).

[47] H. Goldstein and K. Schindler, "Large Scale Collision-Free Instability of Two-Dimensional Plasma Sheets," *Phys. Rev. Lett.* **48**, 1466 (1982).

[48] M. Kiessling, R. P. Brinkmann, and K. Schindler, "Statistical Mechanics Approach to Stability of Current Carrying Plasmas," *Phys. Rev. Lett.* **56**, 143 (1986).

[49] W. Zwingmann, "Self-Consistent Magnetotail Theory: Equilibrium Structures Including Arbitrary Variation Along the Tail Axis," *J. Geophys. Res.* **88**, 9101 (1983).

[50] M. Hesse and K. Schindler, "Bifurcation of Current Sheets in Plasmas," *Phys. Fluids* (in press, 1986).

[51] K. Schindler and J. Birn, "On the Generation of Field-Aligned Plasma Flow at the Boundary of the Plasma Sheet," *J. Geophys. Res.* (submitted, 1985).

[52] D. N. Baker, S. J. Bame, R. D. Belian, W. C. Feldman, J. T. Gosling, P. R. Higbie, E. W. Hones, Jr., D. J. McComas, and R. D. Zwickl, "Correlated Dynamical Changes in the Near-Earth and Distant Magnetotail: ISEE-3," *J. Geophys. Res.* **89**, 3855 (1984).

[53] J. T. Gosling, D. N. Baker, S. J. Bame, E. W. Hones, Jr., D. J. McComas, R. D. Zwickl, J. A. Slavin, E. J. Smith, and B. T. Tsurutani, "Plasma Entry into the Distant Tail: ISEE-3," *Geophys. Res. Lett.* **11**, 1078 (1984).

[54] B. T. Tsurutani, D. E. Jones, R. P. Lepping, E. J. Smith, and D. G. Sibeck, "The Relationship Between IMF B_y and the Distant Tail Lobe and Plasma Sheet B_y," *Geophys. Res. Lett.* **11**, 1082 (1984).

[55] D. G. Sibeck, G. L. Siscoe, J. A. Slavin, E. J. Smith, B. T. Tsurutani, S. J. Bame, and R. P. Lepping, "The Distant Magnetotail Magnetopause and Boundary Layer," *J. Geophys. Res.* **90**, 9561 (1985).

[56] J. A. Slavin, E. J. Smith, D. G. Sibeck, D. N. Baker, R. D. Zwickl, and S.-I. Akasofu, "An ISEE-3 Study of Average and Substorm Conditions in the Distant Magnetotail," *J. Geophys. Res.* **90**, 10875 (1985).

[57] M. Scholer, D. Hovestadt, and B. Klecker, "Average Flow Between ~70 and ~220 R_e," *Geophys. Res. Lett.* **11**, 343 (1984).

[58] P. W. Daly, T. R. Sanderson, and K.-P. Wenzel, "Survey of Energetic (E > 35 keV) Ion Anisotropies in the Deep Geomagnetic Tail," *J. Geophys. Res.* **84**, 10733 (1984).

[59] R. D. Zwickl, D. N. Baker, S. J. Bame, W. C. Feldman, J. T. Gosling, E. W. Hones, Jr., D. J. McComas, B. T. Tsurutani, and J. A. Slavin, "Evolution of the Earth's Distant Magnetotail: ISEE-3 Electron Plasma Results," *J. Geophys. Res.* **89**, 11007 (1984).

[60] J. A. Slavin, P. W. Daly, E. J. Smith, T. R. Sanderson, K.-P. Wenzel, R. P. Lepping, and H. W. Kroehl, "Magnetic Configuration of the Distant Plasma Sheet: ISEE-3 Observations" (this volume, 1986).

[61] T. E. Eastman, E. W. Hones, Jr., S. J. Bame, and J. R. Asbridge, "The Magnetospheric Boundary Layer: Site of Plasma, Momentum, and Energy Transport from the Magnetosheath into the Magnetosphere," *Geophys. Res. Lett.* **3**, 685 (1976).

[62] W. J. Heikkila, "Magnetospheric Topology of Fields and Currents," in *Magnetospheric Currents*, T. A. Potemra, ed., American Geophysical Union, Washington, D.C., p. 208 (1984).

[63] P. H. Reiff, R. W. Spiro, and T. W. Hill, "Dependence of Polar Cap Potential Drop on Interplanetary Parameters," *J. Geophys. Res.* **86**, 7639 (1981).

[64] L. A. Frank, K. L. Ackerson, and R. P. Lepping, "On Hot Tenuous Plasmas, Fireballs, and Boundary Layers in the Earth's Magnetotail," *J. Geophys. Res.* **81**, 3855 (1976).

[65] M. Scholer, G. Gloeckler, B. Klecker, F. M. Ipavich, D. Hovestadt, and E. J. Smith, "Fast Moving Plasma Structures in the Distant Magnetotail," *J. Geophys. Res.* **89**, 6717 (1984).

[66] I. G. Richardson and S. W. H. Cowley, "Plasmoid-Associated Energetic Ion Bursts in the Deep Geomagnetic Tail: Properties of the Boundary Layer," *J. Geophys. Res.* **90**, 12133 (1985).

[67] J. A. Slavin, E. J. Smith, B. T. Tsurutani, D. G. Sibeck, H. G. Singer, D. N. Baker, J. T. Gosling, E. W. Hones, Jr., and F. L. Scarf, "Substorm Associated Traveling Compression Regions in the Distant Tail: ISEE-3 Geotail Observations," *Geophys. Res. Lett.* **11**, 657 (1984).

[68] K. W. Ogilvie and M. A. Copland, "Composition Measurements in the Magnetotail Between 60 and 240 R_e," *Geophys. Res. Lett.* **11**, 347 (1984).

DISCUSSION

W. I. Axford: It is important to remember that all our drawings of the magnetosphere tend to be 2-dimensional as well as time-stationary. The y-component of the interplanetary field, in particular, should have some interesting effects. First, it will tend to squash the tail at large distances from the Earth as pointed out by Dessler. Second, it tends to make asymmetric current and electric potential systems in the polar ionospheres, as observed. Third, it leads to nonconjugate reconnection in the sense that the ends of field lines are not at proper conjugate points as determined from the internal geomagnetic field (see Mendis and Axford, "Geomagnetic Conjugacy Variations," *Comments on Astrophysics and Space Science, 1970*). This effect has been observed in aurora where there is often pronounced nonconjugacy (east-west) depending on the degree of activity and especially on the poleward side of the auroral oval. This should be studied systematically with simultaneous N and S auroral imaging. Finally, the plasmoids that form as a result of tail reconnection will, in general, be flux ropes rather than regions of individually closed field lines. Lobe field lines that reconnect will tend to have a single twist resulting in Alfvén waves moving tailwards. This effect has been observed very often in comet tails.

T. W. Speiser: For the plasmoid near earth neutral line model: Lyons points out that the quiet arc (presumably mapped to the boundary layer) brightens at the substorm onset. This could fit in with the boundary layer model if one imagines that the rapid (explosive?) growth occurs first when the boundary layer grazes on or impinges on the neutral line. Axford has previously pointed out that the reconnection electric field should increase drastically when the boundary (and lobe plasma) is involved.

VIII. SUMMARY

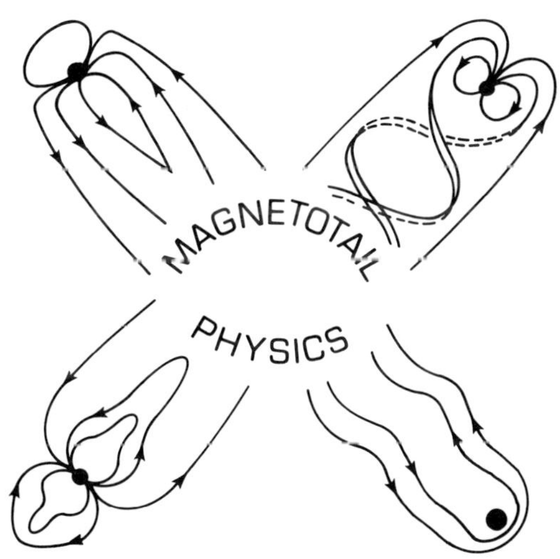

CONCLUDING REMARKS

S.-I. Akasofu*

As an introduction, a personal note describing magnetospheric physics at about the time of the discovery of the magnetotail is presented. Then, seven longstanding major problems associated with magnetotail physics are singled out and discussed. As a suggested observational effort in the future, the need for a balanced advance of macroscopic (global and remote sensing) and microscopic (local and in situ) observations is stressed.

HISTORICAL NOTE (PERSONAL) AND NEW FINDINGS

It was March 26, 1961 when the Explorer 10 satellite traversed a "boundary" at about a distance of 22 R_e, leaving the region where the earth's magnetic field has a dominant role and entering the region where the solar wind prevails. I remember clearly the scene in a room at the Goddard Space Flight Center, where Jim Heppner, Joe Cain, and Norman Ness were discussing whether this "boundary" should be identified as the Chapman-Ferraro boundary, namely the magnetopause; their paper was published in 1963.[1] At that time, I was busy computing the magnetic field of the ring-current belt, using a newly installed IBM 7090 computer there, confirming that the storm main phase is produced by a storm time Van Allen (ring current) belt.

In an attempt to put the discovery of the magnetotail in the proper perspective in the history of magnetospheric physics, the identification of the Chapman-Ferraro boundary by Explorer 10 was an important event. I remember that Chapman was very pleased with the news by saying that it is remarkable that a theory can be confirmed as late as 30 years after it was formulated. Chapman was also convinced that the Explorer 10 observation made it all the more clear that their theory must be the foundation upon which the formation mechanism of the ring-current belt and the cause of geomagnetic storms should be constructed.

However, it soon became apparent to both Chapman and me that what the Chapman-Ferraro theory basically indicates is that the solar wind simply blows around the earth, so that solar-wind energy cannot be transferred into the Chapman-Ferraro cavity (the magnetosphere); note that Chapman and Ferraro[2] assumed a non-magnetized solar wind. We found many events during which the earth encountered a strong solar-wind flow, but the ring current was too weak to be observable; those were geomagnetic "storms" accompanied by a large storm sudden commencement (SSC) and a long initial phase, but not by any appreciable main phase.[3,4] Thus, we came to the conclusion that a simple plasma flow around the earth alone cannot produce a geomagnetic storm and the ring-current belt. This conclusion seemed to outrage a number of our colleagues who felt that such a statement would negate all efforts made to that date; one of them even wrote to Chapman saying that his reputation was at stake.

On the other hand, in the same paper, we added a suggestion that there must be something in a fully ionized plasma flow in producing the ring-current belt by stating that "the variety of development of the storms seems to suggest some intrinsic differences between the solar streams far beyond what we would expect from a mere difference between their pressures. The nature of their intrinsic differences is at present unknown." A little later, I suggested that so long as our thinking is based on the Chapman-Ferraro theory, the ring current could be formed deep in the magnetosphere only if the solar wind contained a significant amount of neutral hydrogen atoms; being neutral, they have no difficulty in crossing the magnetopause and becoming ionized after charge exchange with protons in the magnetosphere. However, it is now quite obvious that our "unknown" factor is found out to be the IMF B_z component.

The year 1961 was a very important year in the history of magnetospheric physics. Dungey[5] added an important ingredient, the interplanetary magnetic field, to the Chapman-Ferraro theory. His theory may be considered as a bold "outline painting" of the magnetosphere for the reasons given shortly. Axford and Hines[6] proposed a convective flow of plasma within the magnetosphere. Both are still the basic guiding concepts in the present magnetospheric physics.

The pioneering in-situ measurements of the magnetotail reported by Ness and his colleagues[7-9] came when we were stuck with the "unknown" factor. It added an entirely new dimension to magnetospheric physics. An early excitement of the discovery of the magnetotail had led us to the hope that the tail processes would solve all major problems, including the formation of the ring-current belt and the acceleration of auroral particles. Thus, we thought that all the difficulties we had faced at that time would be removed by this discovery. I recall vividly a similar excitement associated with the discovery of the Van Allen belts in 1959; many people thought that the mystery of the aurora was solved by the discovery!

*Geophysical Institute, University of Alaska, Fairbanks, Fairbanks, Alaska 99775-0800.

During the last twenty years, various features in the magnetotail have been discovered and added to the original outline painting by Dungey. We have added the plasma sheet, the plasma mantle, the low-latitude boundary layer, etc, and the plasma sheet is divided into the central plasma sheet (CPS) and the boundary plasma sheet (BPS). Hones[10] has suggested that plasmoids are formed during substorms. The plasmoid formation has been extensively studied theoretically by Birn and Schindler[11,12] and most recently by Lee et al.[13] and Min and Okuda.[14] The distant magnetotail has been extensively surveyed by the ISEE 3 satellite.

Both Dungey's theory and the theory of Axford and Hines were immediately adapted to this newly discovered region of the magnetosphere. In particular, the magnetotail has been considered to be an ideal region for magnetic reconnection, since the magnetic fields in the northern and southern halves of the magnetotail are approximately antiparallel. Thus, reconnection theories of solar flares were soon adapted to the magnetotail situation. This step had an immediate implication that solar-wind energy must be stored in the magnetotail in the form of magnetic energy until a spontaneous reconnection process occurs explosively (see Fig. 1), converting the magnetic energy to substorm energies. More specifically, magnetic reconnection is thought to convert magnetic energy into the ring-current energy and auroral

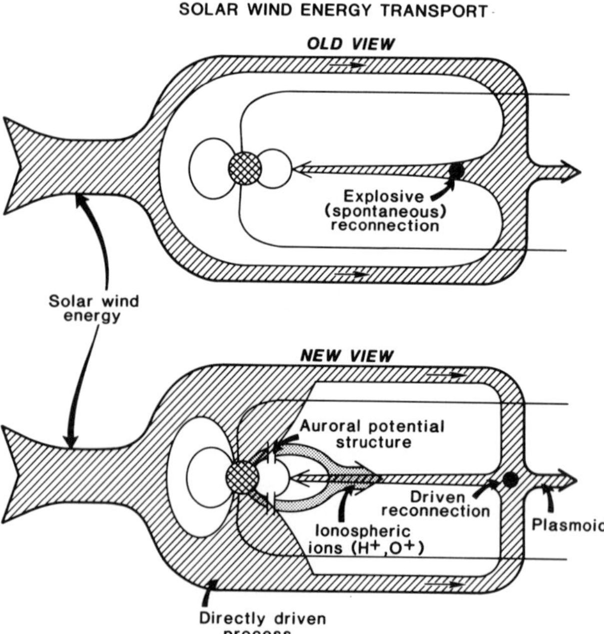

Figure 1—A schematic illustration indicating some important conceptual changes in the energy flow from the solar wind to the polar ionosphere. In the past, it had been tacitly assumed that substorm energy was supplied only by spontaneous and explosive reconnection in the magnetotail. In the present view, the energy to the polar ionosphere is supplied directly by a dynamo process in the near-earth magnetopause and also by a directly driven (or forced) reconnection in the magnetotail. The ionosphere is also an important supplier of plasma particles in the magnetotail.

energy. It was also thought that the earthward plasma flow (convection) generated by the explosive reconnection "injects" plasma particles deep into the trapping region, resulting in the ring-current belt. It was also thought that auroral particles are accelerated by reconnection processes.

Since then, there has been a strong trend among theorists to isolate the reconnection region from the rest of the magnetosphere and only to attempt to find conditions that can make reconnection explosive. Naturally, they paid little attention to other parts of the magnetosphere; for example, the solar-wind conditions, the IMF, and the ionosphere were often ignored in their theories.

However, it has become evident during the last several years that time variations of the southward component of the IMF are highly correlated with the auroral electrojet index AE and also that time variations of the power of the solar-wind magnetosphere dynamo correlate well with those of the total energy dissipation rate of the magnetosphere.[15] These findings have suggested that the solar wind has at least comparable control on processes within the magnetosphere with the internal (spontaneous and explosive) processes. This conclusion may be interpreted that the directly driven (or forced) reconnection is at least as important as, or more important than, the spontaneous one.[16] Further, a sudden brightening of auroral arcs at substorm onset could simply be due to a positive feedback effect of the magnetosphere-ionosphere coupling, rather than to explosive reconnection.[17]

One of the interesting reports during the conference was that the outward flow speed in the distant magnetotail correlates well with the AE index.[18] Since the IMF B_z component is known to be well correlated with the AE index, it is expected that the flow speed is also well correlated with the IMF B_z component. If this flow is supposed to result from reconnection, this flow-speed/AE relationship suggests also that the reconnection is directly driven (controlled) by the IMF B_z (see Fig. 1).

Perhaps one of the most unexpected features in the magnetotail in the last decade is the finding of ionospheric ions, such as H^+ and O^+. During this conference, Gloeckler et al.[19] reported that ⅓ to ⅔ of H^+ and O^+ in the magnetotail are of ionospheric origin. Thus, the ionosphere is as important as the solar wind in supplying the plasma into the magnetotail, so that the magnetotail cannot be isolated from the ionosphere. Further, this fact also urges us to make a major conceptual change. In the past, the energy and plasma were supposed to flow only from the magnetotail to the ionosphere. The ionosphere was supposed to be only a passive layer. The above finding is an important indication that this is not really the case. Note also that even a significant part of the ring-current particles is of ionospheric origin, as they are initially accelerated upward toward the plasma sheet and subsequently "injected" back toward the earth.[20] Another unexpected finding was that there is not much plasma flow[21] in the vicinity of the "neutral sheet." On the other hand, it appears that the low latitude boundary layer and the boundary plasma sheet are a source of many interesting processes in the magnetotail and the ionosphere.

After all, however, it is those excitements and disappointments that have advanced our understanding of the magnetotail. Through such processes, the relative importance of the magnetotail with respect to other parts of the magnetosphere has been accurately assessed. Thanks to all who have contributed to this field of science.

LONGSTANDING MAJOR PROBLEMS

In this section, I have selected several longstanding major problems and discuss them briefly.

Source Region of Region 1 Current

The importance of understanding this problem is that Region 1 current is considered to be the primary current generated by the solar-wind/magnetosphere dynamo. Therefore, Region 1 current is directly related to the power supply process, namely the solar-wind/magnetosphere interaction. Since most magnetotail processes are electromagnetic in nature, their power supply must be a dynamo. There is little doubt also that the kinetic energy of the solar wind is the ultimate source of energy for magnetospheric processes. Thus, it is a dynamo process that must be responsible for converting the solar-wind kinetic energy to the electromagnetic energy. Therefore, these two inferences lead us to the same conclusion that the solar-wind/magnetosphere coupling constitutes a dynamo. Unfortunately, this power-supply process has only qualitatively been described in terms of the so-called "flux transfer" from the dayside magnetosphere to the magnetotail. However, such a qualitative description can tell us little about physical processes associated with the dynamo and also the generation mechanism of Region 1 current. We must attempt to understand the solar-wind/magnetosphere coupling as a dynamo process.

Anomalous Resistivity

It has long been argued that the MHD description of magnetotail phenomena requires "resistivity," particularly in modeling reconnection processes. However, since the plasma is essentially collisionless, the resistivity must be an "anomalous" one. We have made little progress in identifying the actual plasma processes involved in the "anomalous resistivity." As mentioned earlier, there appears to be little electrostatic noise in the vicinity of the neutral sheet, so that it is not certain whether or not the anomalous resistivity, if any, would arise from wave-particle interactions.

During the conference, several MHD studies of the global aspects of magnetospheric dynamics were presented. A three-dimensional MHD simulation study appears to be making good progress (cf. Ref. 22). However, these studies must rely on a somewhat artificial anomalous resistivity. It is not certain that the agreement of the simu-

lation results with satellite observations justifies the use of an artificial anomalous resistivity or that it can be considered as a support for specific morphological models based on satellite observations. In the conclusion session, T. Sato reminded the audience specifically of this particular point.

Acceleration of MeV Particles

The sun is known to be a source of protons of a few BeV. The magnetosphere is a source of protons of a few MeV. The acceleration mechanism of these high-energy particles in the sun and the magnetosphere is one of the most important problems in cosmic electrodynamics. It has long been suggested that MeV particles are produced by reconnection processes, but the basic physics associated with the acceleration has not been elucidated. An early consideration was that an electric field can accelerate particles along the X-type neutral line. However, a quantitative examination of the trajectories of such particles in the vicinity of the X-line indicates that those particles would not remain long enough along the X-line to be accelerated to MeV particles.[23] They can be accelerated along the O-type neutral line, but may not be able to get out from the middle of the line. It has been suggested that the electric field might result from a rapid change of the magnetic field ($\partial B/\partial t$), but no specific process to make $\partial B/\partial t$ large has been specified. Since this particular acceleration is such an important subject, its understanding will be a major contribution to cosmic electrodynamics.

Two-Dimensional Mentality

A two-dimensional approximation of a three-dimensional feature may or may not be a good one. The degree of success of a two-dimensional approximation depends on individual cases. The magnetosphere has a three-dimensional structure, but magnetotail phenomena have often been described in a two-dimensional cross section (e.g., the noon-midnight meridian cross section). Unless proven otherwise, such an approximation may indeed be a useful one, but the attitude we should avoid is that some true three-dimensional features may be thrown out as anomalous or trivial, if they are bothersome to a two-dimensional model. For example, it has long been pointed out that the B_y (dawn-dusk) component of the magnetic field is often at least comparable to the B_x and B_z components,[24] but this fact has been ignored in constructing a model of the plasmoid structure.

Convection Pattern and Plasma Flows

Since the convection model envisaged by Axford and Hines was adapted in the magnetotail region, there has been little success in determining the global convection pattern by satellite observations. Further, field-aligned flows and the $E \times B$ flows are often not distinguished in many reports. The $E \times B$ flow is supposed to be fastest near the neutral sheet region, but the flow is often absent there.[21] It appears that fast flows are present in the BPS during substorms and that they are strongly field aligned.[25]

"Injection" of Ring-Current Particles

There have been a number of excellent satellite observations of protons and electrons inside the inner boundary of the plasma sheet near the equatorial plane. Among them, major changes of the flux and energy spectrum of the particles during substorms are called altogether the "injection" events. Great complexities of such individual events have been described in detail in a number of papers. It appears that at least two processes are at work in those events and that it is not easy to distinguish their effects at a few single points (satellites). A unified theory may be needed to take into account all important processes, including H^+ and O^+ ions that are accelerated upward in the auroral potential structure and subsequently injected.

Relationship Between Auroral Phenomena and Tail Phenomena

We are making progress in modeling the magnetospheric structure and finding the counterpart in the polar ionosphere. The modeling of the magnetotail was pioneered by Williams and Mead[26] and Ness and Williams.[9] It has generally been accepted that the so-called "polar cap" is the projection, along the magnetic field lines, of the high-latitude region and that the auroral oval as a whole is the projection of the plasma sheet. However, we have not confirmed the exact correspondence experimentally, particularly in dynamic situations.

There is a tacit assumption among many workers that an auroral arc is the projection of the X-line (along the geomagnetic field lines). For example, it has been stated by many workers that a sudden brightening of an auroral arc is related to the formation of X-line at a distance of 10 R_e or less. However, there has been no serious attempt during the last two decades to explain how the X-line can produce an auroral arc. It appears that the neutral line is still considered by many as a magic line, and thus the auroral potential structure is ignored by most magnetotail researchers. Since the magnetotail is not detached from the rest of the magnetosphere, magnetotail researchers should pay more attention to recent progress made in the area of the magnetosphere-ionosphere coupling.

We have not yet identified tail phenomena associated with the poleward expanding bulge of the aurora. The plasma sheet is known to become thin during the expansive phase, instead of becoming thick. It is hoped that magnetotail phenomena will soon be studied in conjunction with auroral images taken from the Dynamics Explorer satellite.

SUGGESTED OBSERVATIONAL EFFORTS IN THE FUTURE

ISTP

Our immediate need is a systematic, multipoint observation of the magnetosphere, including global auroral imaging. It is for this reason that we must make every effort to execute successfully the ISTP.

The satellite program for the ISTP has already been well planned, but the ground-based observations must be coordinated soon. Systematic ground-based observations, such as incoherent scatter radars and magnetometers, are an integral part of the global observational program of the magnetosphere. During the IMS, the seven meridian chains of magnetometers played a vital role, providing the basic frame of reference (the global distribution of the potential, ionospheric currents, field-aligned currents, the joule heat production rate, etc.) in studying satellite data. This was demonstrated during the CDAW-6 study. It is hoped that improved magnetometer meridian chains will be operated during the ISTP.

Need for Innovative Remote Sensing Methods

It will not be too early to plan innovative methods of studying the magnetosphere as post ISTP projects. During the last two decades, we have made an intensive effort to make in situ observations of the magnetosphere. Perhaps this is an important opportunity to consider remote sensing methods in studying the global structure of the magnetosphere, the global plasma flow pattern, etc. In a sense, we have initiated remote sensing methods during the IMS. The Dynamic Explorer satellite has imaged the aurora from a great distance. An optical device aboard the same satellite remote-sensed the upper atmospheric temperature. The IMS meridian chains of magnetometers were used as a remote sensing tool of electromagnetic conditions of the ionosphere.

In spite of the great usefulness of in situ observations, the imaging of the entire magnetosphere is vital in studying global aspects of the magnetosphere. Are there any atomic or molecular emissions that can observe the entire magnetosphere from a distant satellite? It will be extremely exciting if we can observe the magnetosphere like a comet. The moon can also provide an ideal observing site during the half-moon periods for this purpose. The emissions do not have to be visible ones. Similarly, is there also any ground-based radio method that can accomplish the same thing? One example of the powerfulness of the remote sensing in this connection would be the mapping of the galactic arms by the 21 cm radio emission from the neutral hydrogen storms. Is there any radio method, either ground-based or satellite-based, to map the plasma flow in the magnetosphere? It should be mentioned in this connection that the interplanetary scintillation (IPS) method has succeeded in mapping the interplanetary shock structure as it propagates from the sun to the earth.

Also, is there any method to map the direction and/or intensity of the magnetic field in the magnetosphere from a lunar observatory or a distant satellite? Is there any method to determine the equatorial crossing distance of the closed field lines? Can starlight be used for this purpose? Also, is there any way to distinguish open and closed field lines?

It is important to realize that there is a limit to the information one can obtain from in situ observations of plasmas and magnetic fields by satellites. First of all, a better understanding of both microscopic and macroscopic processes is needed for a better understanding of global aspects of the magnetosphere. Secondly, relying only on in situ observation, we may eventually need more than ten satellites monitoring magnetospheric processes simultaneously, but we may not be able to afford them. This is not to say that satellite efforts should be abandoned. My point is that the past in situ efforts should be supplemented by remote sensing methods. That is to say, *a balanced advance of macroscopic (global) and microscopic (local) observations is essential for the future progress of magnetospheric physics*. It is interesting to note that we have almost an opposite situation in comet studies.

In summary, magnetospheric physicists are encouraged to familiarize themselves with the great progress of high technology in recent years. Some technological advances may be able to overcome difficulties of several orders of magnitude in remote sensing measurements.

ACKNOWLEDGMENT—The work reported here was supported in part by a grant from the National Science Foundation, Atmosphere Sciences Section, ATM83-12515 and in part by a grant from the National Aeronautics and Space Administration, NSG-7447.

REFERENCES

[1] J. P. Heppner, N. F. Ness, C. S. Scearce, and T. L. Skillman, "Explorer 10 Magnetic Field Measurements," *J. Geophys. Res.* **68**, 1 (1963).

[2] S. Chapman and V. C. A. Ferraro, "A New Theory of Magnetic Storms. Part I: The Initial Phase," *Ten. Magn. Atmos. Elect.* **36**, 77 (1931).

[3] S. I. Akasofu and S. Chapman, "Magnetic Storms: The Simultaneous Development of the Main Phase (DR) and of Polar Magnetic Substorms (DP)," *J. Geophys. Res.* **68**, 3155 (1963).

[4] S.-I. Akasofu, S. Chapman, and D. Venkatesan, "The Main Phase of Great Magnetic Storms," *J. Geophys. Res.* **68**, 3345 (1963).

[5] J. W. Dungey, "Interplanetary Magnetic Field and the Auroral Zone," *Phys. Rev. Lett* **6**, 47 (1961).

[6] W. I. Axford and C. O. Hines, "A Unifying Theory of High-Latitude Geophysical Phenomena and Geomagnetic Storms," *Can. J. Phys.* **39**, 1433 (1961).

[7] N. F. Ness, C. S. Scearce, and J. B. Seek, "Initial Results of the IMP 1 Magnetic Field Experiments," *J. Geophys. Res.* **69**, 3531 (1964).

[8] N. F. Ness, "The Earth's Magnetic Tail," *J. Geophys. Res.* **70**, 2989 (1965).

[9] N. F. Ness and D. J. Williams, "Correlated Magnetic Tail and Radiation Belt Observations," *J. Geophys. Res.* **71**, 322 (1965).

[10] E. W. Hones, Jr., "Plasma Sheet Behavior During Substorms," in *Magnetic Reconnection in Space and Laboratory*, Geophysical Monograph 30, p. 178, American Geophysical Union, Washington, D.C. (1984).

[11] J. Birn and I. Schindler, "Computer Modeling of Magnetotail Convection," *J. Geophys. Res.* **88**, 6230 (1985).

[12] K. Schindler and J. Birn, "Self-Consistent Theory of Time-Dependent Convection in the Earth's Magnetotail," *J. Geophys. Res.* **87**, 2263 (1961).

[13] L. C. Lee, Z. F. Fu, and S.-I. Akasofu, "A Simulation Study of Driven Reconnection Processes and Magnetospheric Storms and Substorms," *J. Geophys. Res.* **90**, 10896 (1985).

[14] K. Min, H. Okuta, and T. Sato, "Numerical Studies on Magnetotail Formation and Driven Reconnection," *J. Geophys. Res.* **90**, 4035 (1985).

[15] D. N. Baker, E. W. Hones, Jr., J. B. Payne, and W. C. Feldman, "A High Time Resolution Study of Interplanetary Parameter Correlations with AE," *J. Geophys. Res.* **86**, 179 (1981).

[16] R. A. Wolf, S.-I. Akasofu, S. W. H. Cowley, R. L. McPherron, G. Rostoker, G. L. Siscoe, and B. U. O. Sonnerup, "Coupling Between the Solar Wind and the Earth's Magnetosphere: Summary Comments," (to be published, 1985).

[17] S.-I. Akasofu, "Some Critical Issues on Magnetospheric Substorms," *Planet. Space Sci.* (in press, 1986).

[18] J. A. Slavin, E. J. Smith, P. W. Daly, T. R. Sanderson, K.-P. Wenzel, R. P. Lepping, and H. W. Kroehl, "ISEE-3 Plasma Sheet Observations: Magnetic Flux Transfer and Global Magnetosphere Reconnection Rate" (this volume).

[19] G. Gloeckler, D. C. Hamilton, F. M. Ipavich, W. Studemann, B. Wilken, G. Kremser, and D. Hovestadt, "Composition of Thermal and Suprathermal Ions in the Near-Earth Plasma Sheet," *Proceedings on Chapman Conference on Magnetotail Physics*, p. 2 (1985).

[20] J. B. Cladis and W. E. Francis, "The Polar Ionosphere as a Source of the Storm Time Ring Current," *J. Geophys. Res.* **90**, 3465 (1985).

[21] C. Y. Huang, L. A. Frank, and T. E. Eastman, "Plasma Flows near the Neutral Sheet of the Magnetotail" (this volume).

[22] R. J. Walker, T. Ogino, and M. Ashour-Abdalla, "An MHD Simulation of Reconnection at the Magnetopause and in the Magnetotail" (this volume).

[23] J. S. Wagner, P. C. Gray, J. R. Kan, T. Tajima, and S.-I. Akasofu, "Particle Dynamics in Reconnection Field Configurations," *Planet. Space Sci.* **29**, 391 (1981).

[24] S.-I. Akasofu, A. T. Y. Lui, C.-I. Meng, and M. Haurwitz, "Need for a Three-Dimensional Analysis of Magnetic Fields in the Magnetotail During Substorms," *Geophys. Res. Lett.* **5**, 283 (1978).

[25] A. T. Y. Lui, L. A. Frank, K. L. Ackerson, C.-I. Meng, and S.-I. Akasofu, "Plasma Flows and Magnetic Field Vectors in the Plasma Sheet During Substorms," *J. Geophys. Res.* **83**, 3849 (1978).

[26] D. J. Williams and G. D. Mead, "Nightside Magnetotail Configuration as Observed from Trapped Electrons at 1100 am," *J. Geophys. Res.* **70**, 3017 (1965).